AF794553

APPLICATIONS OF MODERN HEURISTIC OPTIMIZATION METHODS IN POWER AND ENERGY SYSTEMS

IEEE Press
445 Hoes Lane
Piscataway, NJ 08854

IEEE Press Editorial Board
Ekram Hossain, *Editor in Chief*

Jón Atli Benediktsson	Xiaoou Li	Jeffrey Reed
Bimal Bose	Peter Lian	Diomidis Spinellis
David Alan Grier	Andreas Molisch	Sarah Spurgeon
Elya B. Joffe	Saeid Nahavandi	Ahmet Murat Tekalp

APPLICATIONS OF MODERN HEURISTIC OPTIMIZATION METHODS IN POWER AND ENERGY SYSTEMS

Edited by

KWANG Y. LEE
ZITA A. VALE

WILEY

IEEE PRESS

Copyright © 2020 by The Institute of Electrical and Electronics Engineers, Inc. All rights reserved.

Published by John Wiley & Sons, Inc., Hoboken, New Jersey.

Published simultaneously in Canada.

No part of this publication may be reproduced, stored in a retrieval system, or transmitted in any form or by any means, electronic, mechanical, photocopying, recording, scanning, or otherwise, except as permitted under Section 107 or 108 of the 1976 United States Copyright Act, without either the prior written permission of the Publisher, or authorization through payment of the appropriate per-copy fee to the Copyright Clearance Center, Inc., 222 Rosewood Drive, Danvers, MA 01923, (978) 750-8400, fax (978) 750-4470, or on the web at www.copyright.com. Requests to the Publisher for permission should be addressed to the Permissions Department, John Wiley & Sons, Inc., 111 River Street, Hoboken, NJ 07030, (201) 748-6011, fax (201) 748-6008, or online at http://www.wiley.com/go/permissions.

Limit of Liability/Disclaimer of Warranty: While the publisher and author have used their best efforts in preparing this book, they make no representations or warranties with respect to the accuracy or completeness of the contents of this book and specifically disclaim any implied warranties of merchantability or fitness for a particular purpose. No warranty may be created or extended by sales representatives or written sales materials. The advice and strategies contained herein may not be suitable for your situation. You should consult with a professional where appropriate. Neither the publisher nor author shall be liable for any loss of profit or any other commercial damages, including but not limited to special, incidental, consequential, or other damages.

For general information on our other products and services or for technical support, please contact our Customer Care Department within the United States at (800) 762-2974, outside the United States at (317) 572-3993 or fax (317) 572-4002.

Wiley also publishes its books in a variety of electronic formats. Some content that appears in print may not be available in electronic formats. For more information about Wiley products, visit our web site at www.wiley.com.

Library of Congress Cataloging-in-Publication data is applied for

hardback: 9781119602293

Cover Design: Wiley
Cover Images: Electricity pylons © Sam Robinson/Getty Images,
Offshore farm © T.W. van Urk/Shutterstock

Set in 10/12pt Times by SPi Global, Pondicherry, India

10 9 8 7 6 5 4 3 2 1

FSC
www.fsc.org
MIX
Paper from responsible sources
FSC® C013604

DISCLAIMER

The Editors are not endorsing evolution as a scientific fact, in that species evolve from one kind to another. The term "evolutionary" in the evolutionary computation (EC) simply means that the characteristics of an individual changes within the population of the same species, as observed in the nature.

CONTENTS

PREFACE

Heuristic search and optimization is a new and modern approach for solving complex problems that overcome many shortcomings of traditional optimization techniques. Heuristic optimization techniques are general-purpose methods that are very flexible and can be applied to many types of objective functions and constraints. Recently, these new heuristic tools have been combined among themselves, and new methods have emerged that combine elements of nature-based methods, or which have their foundation in stochastics and simulation methods. Developing solutions with these tools offers two major advantages: development time is much shorter than when using more traditional approaches, and the systems are very robust, being relatively insensitive to noisy and/or missing data/information known as uncertainty.

In competitive electricity market along with automation, heuristic optimization methods are very useful. As electric utilities are trying to provide smart solutions with economical, technical (secure, stable, and good power quality), and environmental goals, there are several challenging issues in the smart grid solutions such as, but not limited to, forecasting of load, price, ancillary services; penetration of new and renewable energy sources; bidding strategies of participants; power system planning and control; operating decisions under missing information; increased distributed generations and demand response in the electric market; tuning of controller parameters in varying operating conditions, etc. Risk management and financial management in the electric sector are concerned with finding an ideal trade-off between maximizing the expected returns and minimizing the risks associated with these investments.

The objective of this book is to review the state-of-the-art technologies in the modern heuristic optimization techniques and present case studies how these techniques have been applied in these complex power and energy systems problems. Empathies will be given to applications rather than theory and the organization of book will be on application basis rather than tools.

The book is composed of six chapters: Chapter *2* gives an overview of applications of evolutionary computation techniques in power and energy systems, including fundamentals of genetic algorithms, evolutionary programming and strategies, simulated annealing, particle swarm optimization, artificial bee colony search algorithm, and tabu search.

Chapter *3* gives an overview of the applications in power system planning and operation problems, such as unit commitment, economic dispatch, active power multi-objective optimal dispatch, hydrothermal coordination, maintenance scheduling, load flow, optimal power flow, transmission system expansion planning, and OPF test bed and performance evaluation of modern heuristic optimization techniques.

Chapter *4* gives an overview of the applications in power system and power plant control problems, such as voltage control, load frequency control with optimization and stability, control of FACTS devices, hybrid of analytical and heuristic techniques for FACTS devices, power system automation, power plant control, predictive control in large-scale power plant, and industrial power plant control.

Chapter *5* gives an overview of the applications in distribution systems, such as active distribution network planning, optimal selection of distribution system architecture, conservation voltage reduction planning, dynamic distribution network expansion planning with demand side management, capacitor placement in distribution systems, network reconfiguration, distribution system restoration, group-based PSO for system restoration, parameter identification of dynamic equivalents for distribution networks, and parameter estimation for distribution transformers.

Chapter *6* gives an overview of the applications in integration of renewable energy in smart grid, such as renewable energy sources, operation and control of smart grid, compliance of reactive power requirements in wind power plants, photovoltaic controller design, demand side management and demand response, solar power forecasting, load demand and solar generation forecast for PV integrated smart buildings, multi-objective planning of public electric vehicle charging stations, and dispatch modeling incorporating maneuver components, wind power, and electric vehicles.

Chapter *7* gives an overview of the applications of modern heuristic optimization techniques in electricity markets, such as bidding strategies, market analysis and clearing with market simulator, electricity market forecasting with artificial neural networks and support vector machines, fuzzy optimization (FO), and FO-based simultaneous bidding of V2G in ancillary service markets.

Kwang Y. Lee
Waco, TX, USA

Zita A. Vale
Porto, Portugal

CONTRIBUTORS

Ali T. Al-Awami, King Fahd University of Petroleum & Minerals, Dhahran, Saudi Arabia
David L. Alvarez, Universidad Nacional de Colombia, Bogotá, Colombia
Alexandre P. Alves da Silva, Vale S.A., Rio de Janeiro, Brazil
Kyungsung An, SK Telecom, Seoul, Korea
Eduardo N. Asada, University of São Paulo, São Carlos, Brazil
Wenlei Bai, ABB Enterprises Software Inc. Houston, TX, USA
Jens C. Boemer, Electric Power Research Institute, Seattle, USA
Luiz Eduardo Borges da Silva, Itajuba Federal University, Itajuba, MG, Brazil
Peter A. N. Bosman, Centrum Wiskunde and Informatica, Amsterdam, The Netherlands
Leonel Carvalho, INESC TEC, Porto, Portugal
Hsiao-Dong Chiang, Cornell University, Ithaca, NY, USA
Jaeseok Choi, Gyeongsang National University, Jinju, Korea
Jinda Cui, Lehigh University, Bethlehem, PA, USA
Anna Carolina R.H. da Silva, Eletrobras, Rio de Janeiro, Brazil
Ibrahim Eke, Kirikkale University, Kirikkale, Turkey
Ahmed Elsayed, Florida International University, Miami, FL, USA
István Erlich, University of Duisburg-Essen, Duisburg, Germany
Pedro Faria, Polytechnic of Porto, Porto, Portugal
Malihe Maghfoori Farsangi, Shahid Bahonar University of Kerman, Kerman, Iran
Pavlos S. Georgilakis, National Technical University of Athens (NTUA), Athens, Greece
Marinus O. W. Grond, Centrum Wiskunde and Informatica, Amsterdam, The Netherlands
Digvijay Gusain, TU Delft, Delft, The Netherlands
Nikos D. Hatziargyriou, National Technical University of Athens (NTUA), Athens, Greece
Kyeon Hur, Yeonsei University, Seoul, Korea
Nikolaos C. Koutsoukis, National Technical University of Athens (NTUA), Athens, Greece
Germano Lambert-Torres, Gnarus Institute, Itajuba, MG, Brazil
Han La Poutré, Centrum Wiskunde and Informatica, Amsterdam, The Netherlands
Kwang Y. Lee, Baylor University, Waco, TX, USA

Yeonchan Lee, Gyeongsang National University, Jinju, Korea
João Bosco A. London Jr, University of São Paulo, São Carlos, Brazil
Ngoc Hoang Luong, Centrum Wiskunde and Informatica, Amsterdam, The Netherlands
Liangyu Ma, North China Electric Power University, Baoding, China
Vladimiro Miranda, INESC TEC / University of Porto, Porto, Portugal
Sukumar Mishra, Indian Institute of Technology Delhi, New Delhi, India
Osama Mohammed, Florida International University, Miami, FL, USA
H. Morais, INESC-ID / University of Lisbon, Lisbon, Portugal
Hiroyuki Mori, Meiji University, Nakano-city, Tokyo, Japan
Mithulananthan Nadarajah, The University of Queensland, Brisbane, Queensland, Australia
Koichi Nara, Ibaraki University, Ibaraki, Japan
Mario Ndreko, TenneT TSO GmbH, Bayreuth, Germany
Hossein Nezamabadi-pour, Shahid Bahonar University of Kerman, Kerman, Iran
Narayana Prasad Padhy, Indian Institute of Technology Roorkee, Roorkee, Uttarakhand, India
Peter Palensky, TU Delft, Delft, The Netherlands
Jong-Bae Park, Konkuk University, Seoul, Korea
Tiago Pinto, Polytechnic of Porto, Porto, Portugal
Deepak Pullaguram, National Institute of Technology, Warangal, Telangana, India
Muhammad Qamar Raza, The University of Queensland, Brisbane, Queensland, Australia
Sergio Rivera, Universidad Nacional de Colombia, Bogotá, Colombia
Andrés Romero, Universidad Nacional de San Juan, San Juan, Argentina
Rubén Romero, São Paulo State University, Ilha Solteira, São Paulo, Brazil
Jose Rueda, TU Delft, Delft, The Netherlands
Camila Paes Salomon, Itajuba Federal University, Itajuba, MG, Brazil
Filipe O. Saraiva, Federal University of Pará, Belém, Brazil
Dushyant Sharma, Indian Institute of Technology Jodhpur, Rajasthan, India
Ruifeng Shi, North China Electric Power University, Beijing, China
Sishaj Pulikottil Simon, National Institute of Technology Tiruchirappalli, Tamilnadu, India
S. N. Singh, Madan Mohan Malaviya University of Technology, Gorakhpur, India
João Soares, Polytechnic of Porto, Porto, Portugal
Aldir S. Souza, State University of Piauí, Teresina, Piauí, Brazil
T. Sousa, Technical University of Denmark, Lyngby, Denmark
Wei Sun, University of Central Florida, Orlando, FL, USA
Masato Takahashi, Fuji Electric Co., Hino-city, Tokyo, Japan
Aimilia-Myrsini Theologi, Jedlix Smart Charging, Rotterdam, The Netherlands
Zita A. Vale, Polytechnic of Porto, Porto, Portugal
Kumar Venayagamoorthy, Clemson University, Clemson, SC, USA
E.M. Voumvoulakis, National Technical University of Athens (NTUA), Athens, Greece

Shuo Wang, China Electric Power Planning & Engineering Institute, Beijing, China
Xiao Wu, Southeast University, Nanjing, China
Shu Xia, North China Electric Power University, Beijing, China
Tianshi Xu, Tianjin University, Tianjin, PRC
Yong-Feng Zhang, University of Jinan, Jinan, China
Ming Zhou, North China Electric Power University, Beijing, China
Qun Zhou, University of Central Florida, Orlando, FL, USA

LIST OF FIGURES

LIST OF TABLES

ABOUT THE EDITORS

Kwang Y. Lee received the BS degree in Electrical Engineering in 1964 from Seoul National University, MS degree in Electrical Engineering in 1968 from North Dakota State University, and PhD degree in Systems Science in 1971 from Michigan State University. He is currently Professor and Chair of the Electrical and Computer Engineering Department at Baylor University, Waco, Texas. Before receiving his current calling, he worked as faculty at Michigan State University, Oregon State University, University of Houston, and Pennsylvania State University from 1972 to 2007.

His research interests include control, operation, and planning of power and energy systems; intelligent control and their applications to power and energy systems; and modeling, simulation, and control of micro-grids with renewable and distributed energy sources. He has published over 700 papers in archived journals and refereed conference proceedings and has authored chapters in several books. He has coauthored *Intelligent Network Integration of Distributed Renewable Generation* (Springer 2017) and *Probabilistic Power System Expansion Planning with Renewable Energy Resources and ESS* (Wiley 2020) and has coedited *Modern Heuristic Optimization Techniques: Theory and Applications to Power Systems* (Wiley 2008) and *Applications of Modern Heuristic Optimization Methods in Power and Energy Systems* (Wiley 2020).

He was elected as a Fellow of IEEE in January 2001 for his contributions to the development and implementation of intelligent system techniques for power plants and power systems control. Dr. Lee has served as Editor of IEEE Transactions on Energy Conversion, Associate Editor of IEEE Transactions on Neural Networks, and Associate Editor of IFAC Journal on Control Engineering Practice. He has served as the Chair of the IFAC Power & Energy Technical Committee (TC 6.3) and IEEE PES Working Group on Modern Heuristic Optimization Techniques.

Zita A. Vale received her Diploma in Electrical Engineering in 1986 and her PhD in 1993, both from the University of Porto. She is currently Professor of the Electrical Engineering Department at the Engineering School of the Polytechnic of Porto, Portugal.

She works in the area of Power and Energy Systems, with a particular interest in the application of Artificial Intelligence techniques. She has participated in more than 60 funded projects mainly related to the development and use of Knowledge-Based systems, Multi-Agent systems, Machine Learning, Data Analysis and Mining, Particle Swarm Intelligence, and Genetic Algorithms. The main application fields of these projects comprise Electricity Markets, Smart Grids, Renewable Energy, Storage, Electric Vehicles, and Demand Flexibility and Response. She has published over 800 works, including over 100 papers in scientific journals and over 500 papers in scientific conferences.

She is a Senior Member of IEEE and chaired several Task Forces, Working Groups, and Sub-Committee on Intelligent Systems of the IEEE Power & Energy Society.

CHAPTER 1

INTRODUCTION

Zita A. Vale[1], João Soares[1], and Kwang Y. Lee[2]
[1]Polytechnic of Porto, Porto, Portugal
[2]Baylor University, Waco, TX, USA

1.1 BACKGROUND

Artificial intelligence (AI) has been used to address a wide range of problems in power systems, using a large set of AI techniques [1, 2]. A wide set of power system problems can be seen as optimization problems [3–6] for which computational intelligence (CI) techniques[1] can be of great help as mathematical approaches have difficulties to handle nonlinearities in large-scale problems, namely when time efficiency is an important issue. Recent evolution in the power industry, namely the sector deregulation and the intensification of distributed generation, made power system optimization problems even more complex. On one hand, distributed generation and the increasing number of market players of diverse nature make geographical dispersion even more important than before. On the other hand, the goals of distributed generators, in particular, and of market players, in general, require a new insight into power system optimization problems as optimal solutions must seek from the point of view of each relevant player [7]. The evolution has been accompanied by introduction of smart grid technologies that enabled the increase of renewable energy sources (wind and solar photovoltaic) and electric vehicles (EVs) [8]. This evolution has been effective to reduction in the carbon emissions [9, 10] but at the cost of increasing the complexity in power systems [11–14].

The AI family techniques most proposed to address difficult problems in power systems are CI techniques, inspired by nature and animal behavior. The

[1] CI is a subset of AI, which is a family of problem-solving and problem-stating approaches that attempt to exhibit or mimic the intelligence observed in nature.

Applications of Modern Heuristic Optimization Methods in Power and Energy Systems,
First Edition. Edited by Kwang Y. Lee and Zita A. Vale.
© 2020 by The Institute of Electrical and Electronics Engineers, Inc.
Published 2020 by John Wiley & Sons, Inc.

CI is commonly used for different purposes, but three main categories of its application can be highlighted, namely optimization, learning/modeling, and control. Artificial neural networks (ANN) are widely used for learning and forecasting, fuzzy systems (FS) are generally applied to control problems and decision-making, and evolutionary computation (EC) is suited for optimization problems where modern heuristic optimization (MHO) falls. The EC includes a variety of algorithms, including: ant colony [15], chaotic bat algorithm [16], differential evolution (DE) [17], genetic algorithm (GA) [18], particle swarm optimization (PSO) and multi-objective particle swarm optimization (MOPSO) [19], and non-dominated sorting genetic algorithm (NSGA-II and NSGA-III) [20, 21]. Although ANN, FS, and EC can target different applications, many real-world problems require a combination of different methods from those categories in order to achieve the most effective solution. Figure 1.1.1 depicts the evolution of the number of publications in the CI field (EC, ANN, and FS) between 2002 and 2018 in the power systems and smart grid field according to SCOPUS database [22]. ANN and EC have registered the most relevant increase in the number of contributions. The FS contributions in this field are also increasing but not ahead of EC since early 2000s. The EC where MHO falls is behind ANN in number of contributions but ahead of FS contributions in the field of power systems and smart grid.

The EC optimization methods differ from most of the classical optimization methods as they are general-purpose methods to explore the solution space. This means that they are not specifically designed to adequately approach a specific problem and that they use limited knowledge about the specific problem to be addressed. They are based on concepts and behaviors that were usually not considered relevant for the classic optimization techniques.

The interest in these heuristic-based methods has been increasing due to their adequate characteristics to solve complex real-world problems. Power system

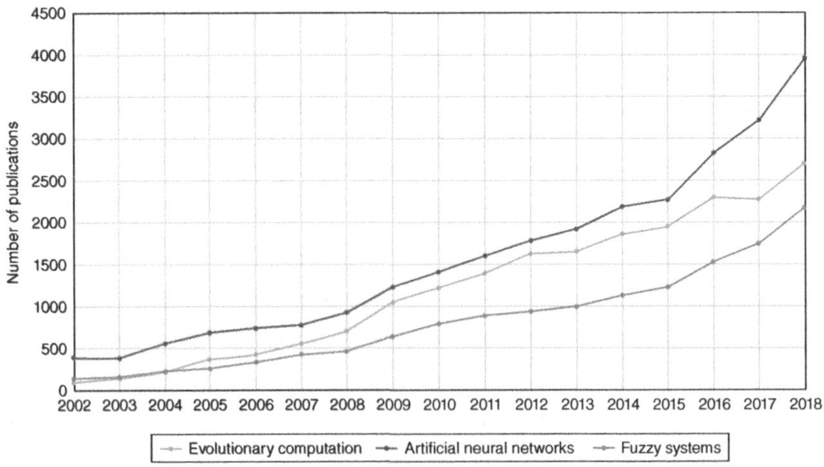

Figure 1.1.1 Publications of CI in power systems and smart grid (2002–2018).

problems' intrinsic complexity and large dimension make MHO a very relevant issue for the power system community.

Although there is not a widely accepted definition for MHO methods, one can say that these methods seek near-optimal solutions at a reduced computational effort without guaranteeing the optimality and feasibility [5]. Using this definition as a basis, it is simple to conclude that heuristic optimization methods do not guarantee that the optimal solution is found. Moreover, it is not usually possible to evaluate how close to optimality the obtained solution is. Despite heuristic optimization methods' characteristics can be pointed as limitations, in practice, MHO methods allow to obtain high-quality solutions for real-world problems that were otherwise practically insoluble or, at least, very difficult to solve. The most important advantages that can be summarized from MHO to solve real complex problems are:

- their ability to find good-quality solutions in much shorter computing times than traditional methods;
- their robustness, namely in what concerns their low sensitivity to noisy and/or missing data.

In some demanding applications, one or more MHO techniques are used together with traditional approaches (such as operational research and statistical methods) resulting in efficient hybrid applications.

Some power system problems, namely those concerning planning and scheduling, require the consideration of a large set of constraints and variables. In addition, the models often require the consideration of nonlinear functions to provide more accurate real-world solutions, which increase exponentially the complexity of the problem. In problems for which these constraints are rather complex, the use of constraint logic programming (CLP) appears as an adequate approach largely used in commercial applications. The CLP allows to easily model the problem and can be combined with other techniques including MHO to solve the problems more efficiently [23]. This chapter provides an introduction to EC and describes some of the most prominent MHO algorithms used in power systems field.

The rest of the book will cover overview of MHO applications in power and energy systems (Chapter 2), specific but a significant set of relevant problems in power systems topics tackled with MHO, namely including planning and operation (Chapter 3), power plant control (Chapter 4), distribution systems (Chapter 5), renewables and smart grid (Chapter 6), and electricity markets (Chapter 7).

1.2 EVOLUTIONARY COMPUTATION: A SUCCESSFUL BRANCH OF CI

EC is a vital part of CI (and a very successful branch of it) that includes algorithms for global optimization motivated by biological and evolutionary processes [24]. Evolutionary problem solving is based on the concepts subjacent to natural

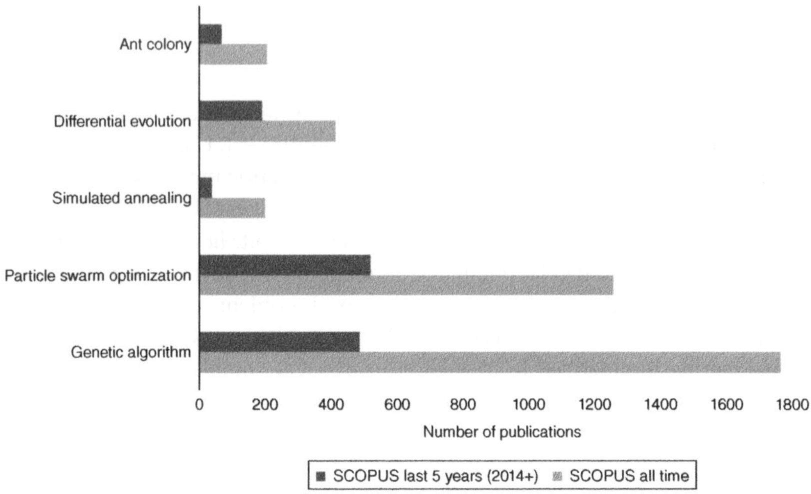

Figure 1.2.1 Number of publications by EC methods in power systems and smart grid.

evolution processes. The EC simulates these processes on a computer resulting in stochastic optimization techniques. It is not limited to the research of evolutionary algorithms (EAs) but also includes swarm intelligence (SI), bio- and nature-inspired algorithms, or natural computation.

The EA embraces a large set of algorithms that usually differ in the way the information is represented and naturally its implementation details. Among the several EAs available in the literature, the main research topics correspond to evolution strategies (ES) [25, 26], GAs [27, 28], and genetic/programming (GP) [29, 30].

Figure 1.2.1 presents the number of publications retrieved in the SCOPUS database in the field of power systems and smart grid organized by different EC methods. The EC algorithms with the most publications are ranked in this order (descendent) GA, PSO, DE, ant colony optimization (ACO), and simulated annealing (SA). GA holds the record of the most used EC algorithm in the power systems field. One reason of GA's large success is because it was one of the first MHO to be introduced. In addition, it has gained popularity due to successful applications and significant efforts made to improve GA ability to cope with single-objective and multi-objective problems [31]. However, if we consider the publications of the last five years, PSO overtakes GA in this rank. These data suggest that PSO has gained more attention lately than GA for the power system problems. The reason behind may lie in PSO's simplicity of implementation (less tuning parameters) when compared to GA, and while PSO has been introduced in 1995, GA has been introduced in 1985, which has contributed to its later success [32].

Table 1.2.1 depicts relevant research papers, namely the highly cited ones for each EC algorithm in the power system and smart grid field since 2014. The highest cited papers are from DE and PSO in this group. The DE presents the highest cited paper from the group, namely [36] with 126 citations, and the PSO, [42], with 115

TABLE 1.2.1 Research Documents by EC Algorithm 2014–Present

		Citations	
EC Algorithm	No. Documents	Reference	Citations
Ant colony	68	[33]	85
		[34]	20
		[35]	16
Differential evolution	189	[36]	126
		[37]	114
		[38]	58
Simulated annealing	37	[39]	32
		[40]	13
		[41]	10
Particle swarm optimization	517	[42]	115
		[43]	103
		[44]	100
Genetic algorithm	483	[45]	84
		[46]	76
		[47]	61

citations. The GA presents papers with fewer citations than PSO and DE. The most cited GA paper in this field is [45] with 84 citations, which also reveals a less active area of research compared with PSO in the last few years.

The EAs can be distinguished by their specific characteristics, which make them a bit special and different from other meta-heuristics. They are technically called population-based meta-heuristics, since they rely on a set of candidate individuals (population). Hence, it can be said that EAs rely on a population-based optimization process. A representation which is a mapping from the state space of possible solutions to a state space of encoded solutions is generally used. It begins by initializing a population of candidate solutions and proceeds evolving these solutions through an iterative process. The solutions in EA are iteratively updated during each iteration (commonly known as generations). The evaluation of each individual of the population is calculated by the fitness function at each iteration/generation. The solutions in the population with the lowest fitness evaluation are probabilistically removed. New solutions are then generated by a particular operation according to the EA logic and replaced by weaker individuals in the population. Selection is the means used to determine the composition of the next generation. The selection operator acts on individuals in the current population. In the selection process, some individuals are eliminated and the remaining survive and become parents of individuals in the next generation. Moreover, some individuals can be chosen multiple times for originating next generation's individuals. Selection methods can be deterministic or stochastic. Considering a specific population, deterministic selection will always eliminate the same individuals, whereas stochastic methods generate a probability mass function over the possible compositions of the next iteration. Deterministic selection is usually faster than stochastic selection, and also usually leads to faster convergence. However, it must be noted that this may not be an advantage as the obtained solution can be of lower quality than the one that could be obtained using stochastic selection.

Adopting the principle of natural selection as seen in nature, EA can evolve the population toward a near-optimal solution. In practice, EA rarely leads to the best solutions in the real-world problems because stopping criteria prevent the EA to run long enough. However, high-quality solutions can be obtained in short periods of time so that they can be useful for practical use.

Different from EA, SI are meta-heuristics motivated by the "intelligence" observed in the collective behavior of agents. Nature-/bio-inspired meta-heuristics are based on the natural/biological processes. SI mimics the interaction and behavior of agents with each other on a given environment. Natural systems including the ant colonies and bird flocking follow this idea and survive using the "collective intelligence." Examples of such algorithms include artificial bee colony (ABC) [48], ACO [49], cuckoo search (CS) [50], or PSO [51]. Nature-inspired algorithms mimic physical or even chemical phenomena. Examples of those algorithms are gravitational search algorithms (GSA), SA, harmony search (HS), and others. A well-conducted taxonomy and classification of these algorithms can be seen in [32].

A knowledge approach can be used together with EC algorithms. Knowledge concerning the best places to look for good solutions can be incorporated both in the initialization of possible solutions and in the design of variation operators.

The EC is very versatile and suitable for addressing the large majority of optimization problems. Varying the representation, the population size, the variation operators, the selection mechanism, the evaluation function, the initialization, and other aspects allows using a diversity of search procedures. The performance of a method in terms of computation time depends on the choices concerning these decisions and on their higher or lower adequacy to the problem being addressed.

1.2.1 Genetic Algorithm

GA uses an evolutionary programming (EP) approach based on genetic, biological processes, with the goal of finding the best solution of combinatorial problems. In fact, this type of algorithm can only guarantee an optimal local solution but has advantages as it requires less computational resources than traditional approaches [52].

The GAs are based on the concepts of natural selection and genetics. It is an optimization algorithm that differs from conventional optimization techniques by adopting population genetics to guide the optimization search.

Figure 1.2.2 illustrates a flowchart of the GA process. The GAs use a population of solutions to which selection and recombination strategies are applied

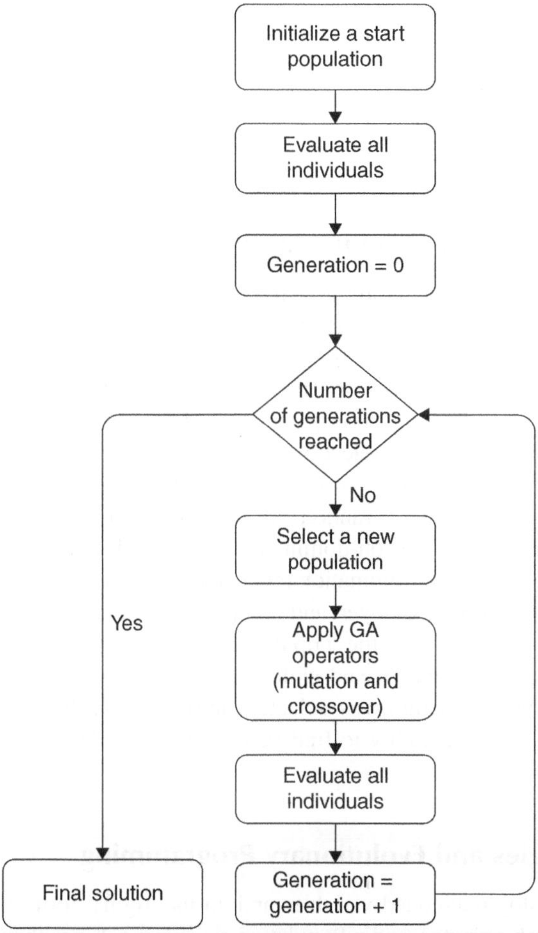

Figure 1.2.2 Genetic algorithm process flowchart. Source: adapted from [53].

in an iterative process, with the aim of obtaining improved solutions. They use genetic operators such as crossover and mutation to obtain the next generation, using the current one.

The GA requires that the user adequately tunes a set of parameters so that they are efficient to solve a specific problem. It is one of the heuristic optimization method most used for power system applications [4]. This can be explained because they are relatively simple to use and present several advantages over traditional optimization techniques:

1. The search is based on a set of solutions and not a single solution, reducing the possibility of convergence to local minima.
2. It uses only rough information of the objective function to guide the search. There is no requirement for derivatives or other auxiliary functions and it

does not impose restrictions such as differentiability and convexity on the objective function.

3. It uses a coding of parameters instead of the parameters themselves what enables evolution from one iteration to the next one with minimum computations.

1.2.2 Non-dominated Sorting Genetic Algorithm II

The NSGA-II proposed in 2002, by Deb et al. [20], has been applied to various problems and is considered one of the champions in multi-objective optimization [54]. It has addressed some problems of the earlier proposal, NSGA, namely by reducing the high computational complexity of non-dominated sorting, from $O(mN^3)$ to $O(mN^2)$, introducing elitism and less parameters. The crossover and mutation operators remain as in GA, but selection operator works differently [55]. The selection is done with the help of a crowd-comparison operator, based on ranking and crowding distance. Initially, a random parent population is created. The population space is sorted based on the non-domination. Then, each solution is assigned a fitness rank based on the non-domination level. The new generation is created using the tournament selection, crossover, and mutation. Elitism is enabled by comparing the current individuals of the population with the best non-dominated solution. In the next step, parent and children are merged to form a new set of individuals and next generation is selected among this collection [56]. Like MOPSO, NSGA-II has the ability to find Pareto-optimal solutions in a single run.

1.2.3 Evolution Strategies and Evolutionary Programming

Evolution strategies (ES) employ real-coded variables and, in its original form, it relied on Mutation as the search operator, and a Population size of one. Since then it has evolved to share many features with GA. The major similarity between these two types of algorithms is that they both maintain populations of potential solutions and use a selection mechanism for choosing the best individuals from the population. The main differences are: ES operate directly on floating point vectors while classical GAs operate on binary strings; GAs rely mainly on recombination to explore the search space, while ES uses mutation as the dominant operator; and ES is an abstraction of evolution at individual behavior level, stressing the behavioral link between an individual and its offspring, while GAs maintain the genetic link.

EP is a stochastic optimization strategy similar to GA, which places emphasis on the behavioral linkage between parents and their offspring, rather than seeking to emulate specific genetic operators as observed in nature. The EP is similar to ES, although the two approaches developed independently. Like both ES and GAs, EP is a useful method of optimization when other techniques such as gradient descent or direct analytical discovery is not possible. Combinatorial and real-valued function optimization in which the optimization surface or fitness landscape is "rugged", possessing many locally optimal solutions, is well suited for EP.

1.2.4 Simulated Annealing

SA is based on the annealing process seen in the thermodynamics and metallurgies. The analogy between the annealing of material to its lowest energetic state and an optimization problem gave place to a heuristic optimization method that is able to deal with arbitrary problems and cost functions, to refine the present solution, and that is simple to implement even for complex problems.

The SA was proposed in 1983 by Kirkpatrick et al. [57]. The SA algorithm has the advantage of requiring less computational resources when compared with population-based algorithms because it generates a single neighborhood solution [58, 59]. The SA has proved to be effective in the solution of a large number of optimization problems [39, 60, 61]. It searches the set of all possible solutions and allows moves toward solutions using a random scheme, reducing in this way the chance of convergence to a local optimum. The SA is usually classified as a trajectory-based method because it determines at each iteration a new single solution.

The method uses a control parameter T that takes the role of the temperature and uses a cooling scheme to determine how this parameter is decreased. Initial and final values of the temperature must be specified. The application of the method also requires the choice of the space of feasible solutions, the form of the cost function, and the neighborhood structure employed. All these should be selected carefully as they determine the effectiveness and the performance.

During the cooling of the metal, the temperature decreases in steps. In each step, the equilibrium presents many configurations of electrons, which correspond to the value of the control variables. After selecting the best configuration of the step, the objective function is evaluated. In the next temperature step, a new best configuration is obtained. If the value of the objective function (energy value) is better than the one of the preceding step, this configuration is accepted as the best. If the result of the current step is worse than the previous one, the probability of the current configuration is calculated using the Boltzmann distribution, which depends on the energy of the step and on a constant. If this probability is higher than a determined value randomly generated, the current configuration is accepted as the best, though the solution itself is not better. After finding a value of the objective function, all the constraints are verified and wrong control variables are corrected. The process proceeds until the minimum temperature is reached [59]. In a large combinatorial optimization problem, an appropriate perturbation mechanism, cost function, solution space, and cooling schedule are required in order to find an optimal solution with SA. The SA is effective in network reconfiguration problems for large-scale distribution systems, and its search capability becomes more significant as the system size increases. Moreover, the cost function with a smoothing strategy enables the SA to escape more easily from local minima and to reach rapidly to the vicinity of an optimal solution.

1.2.5 Particle Swarm Optimization

The PSO [62] belongs to the category of SI methods [63]. The PSO is a computational algorithm inspired on natural systems, namely on the behavior of groups. In each iteration, a set of solutions known as particles move in the search space according to rules that depend on three factors: inertia (particles are influenced to move in the direction they were in previous iteration), memory (particles are influenced to move in the direction of the best solution found so far in their trajectory), and cooperation (particles are influenced to move in the direction of the global best solution).

Later PSO approaches [19, 64] use slight modifications from the traditional one [62]. The strategic parameters, inertia, memory, and cooperation, introduced in [65], are used as part of the mutation process. The mutation of the strategic parameters is applied directly to the original swarm rather than the replicated swarm as in [65]. The replication and selection of the particles implemented in [65] can increase computation time.

The inertia of particle controls the exploration and exploitation abilities of PSO. When the velocity is too high, the particles move fast and could move beyond the aimed global solution. On the contrary, if the velocity is too low, the particles could be trapped into a local optimum. A possible approach to achieve faster convergence and avoid the problems described above is to adopt an inertia term that varies with the number of iterations and limit the maximum velocity of particles.

1.2.6 Quantum Particle Swarm Optimization

Quantum PSO (QPSO) is a population-based stochastic method with different characteristics of those found in PSO [66]. The introduced exponential distribution of positions makes QPSO global convergent. The traditional PSO relies on the convergence to the global best (best solution of the swarm) particle, independent of the position of other particles. On the other hand, in the QPSO with mean best position, each particle cannot converge to global best position without considering its colleagues [66, 67]. Figure 1.2.3 illustrates how differently the lagging phenomena of QPSO and PSO particles are. In the figure, the big circle represents the global best position while the little circles represent the other particles, and the little circles with vertical lines represent the lagged particles. The arrows around the little circles represent the possible directions of the particles; the big arrowhead points to the direction in which the particle moves with high probability.

In the PSO, each particle flies toward the global best position without waiting for other particles. Each particle converges to the global best position independently without waiting for other particles. The influence of the lagged particles on the other particles is very little since the only connection is the global best position. In QPSO the lagged particles have greater influence on the other particles through mean best position. In the QPSO method the lagged particles are not abandoned by the swarm. The lagged particles affect the mean best position and

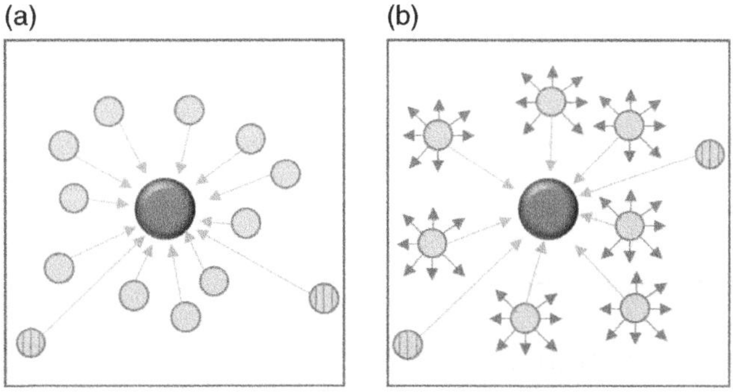

Figure 1.2.3 Lagged particles and waiting phenomena in (a) PSO and (b) QPSO. Source: adapted from [67].

therefore the lagged particles are shifted toward the rest of the swarm. The mean best position is intrinsically related with the movement equation of QPSO. The particles' distribution affects the convergence rate; however, QPSO can provide stronger global search ability than traditional PSO [66].

1.2.7 Multi-objective Particle Swarm Optimization

The PSO is recognized for its high speed of convergence, while being easily adapted to multi-objective problems. Its analogy with EAs makes it suitable for using a Pareto ranking scheme. The individual best solutions can be used to store non-dominated solutions (NDS), which is analog to the elitism mechanism found in EAs [68]. The basic idea of the Pareto front is to pick up the set of points that are Pareto efficient, i.e. the points that are not dominated by other. This set of points are the Pareto optimal set.

The MOPSO is an advanced optimization algorithm to solve multi-objective problems [68] to handle the difficult optimization problems. It is demonstrated to outperform other important multi-objective optimization evolutionary algorithms (MOEAs) such as NSGA-II, Pareto archived evolutionary strategy (PAES), and microGA in several benchmark functions [68, 69]. The MOPSO adopts an external repository similar to the adaptive grid of PAES and uses a mutation operator aiming to explore the remote region of the search space and the full range of each decision variable. To improve the algorithm, mutation of the strategic parameters is implemented based on evolutionary particle swarm optimization (EPSO) [65], instead of the fixed parameters as in original MOPSO. This modification aims to improve the cover rate of the Pareto front and introduce higher diversity in the search procedure. The flowchart of MOPSO is presented in Figure 1.2.4. Two types of mutation occur during the search loop, namely mutation of the velocity coefficients and mutation in the position of some particles (randomly selected). The algorithm stops after the defined number of iterations is reached; this setup is widely used in other multi-objective meta-heuristic-based algorithms.

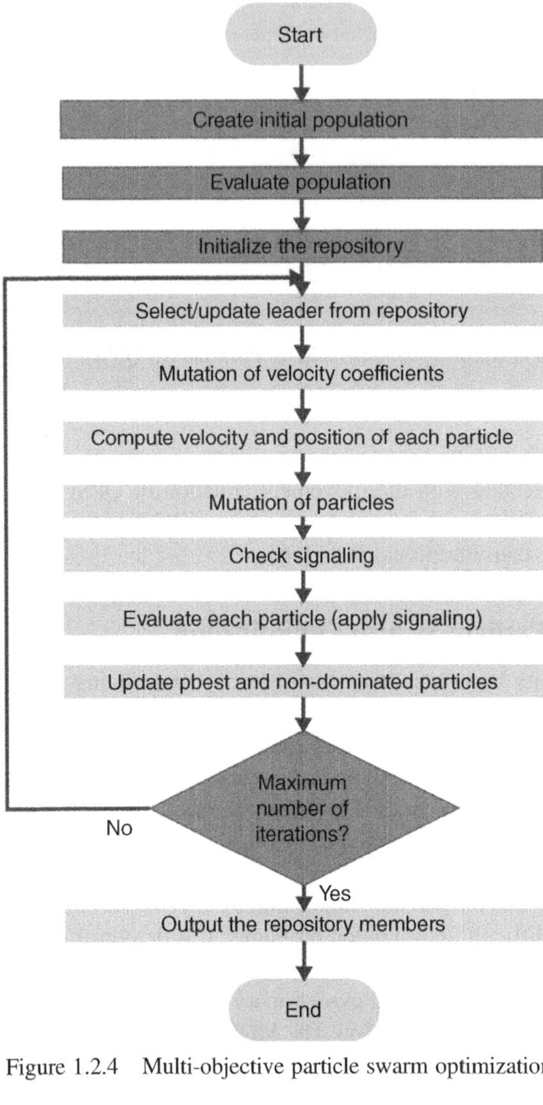

Figure 1.2.4 Multi-objective particle swarm optimization flowchart.

1.2.8 Particle Swarm Optimization Variants

The traditional PSO relies on externally fixed particles' velocity limits, inertia, memory, and cooperation weights without changing these values along the swarm search process (PSO iterations). In very complex problems this can compromise solution diversity because swarm movements are limited to the initially fixed velocities and weights.

To overcome this limitation, several enhanced versions of the classic PSO have been proposed [65, 70–72]. There are many other variants of PSO, some of which are related to more specific problems (e.g. as multi-objective optimization functions) [73, 74]. In [65] the authors introduced mutation of the strategic parameters (inertia, memory, and cooperation) and selection by stochastic tournament. The method is called evolutionary particle swarm optimization (EPSO) and proved to be proficient in several optimization problems. The authors also propose replicating the particles in order to increase the probability of finding more solutions that enhance the diversity of the search space. In [71] the authors propose a modification of the velocity equation to include particle's bad experience component besides the global best memory introduced earlier. The bad experience component helps to remember its previously visited worst position. The method is called nw particle swarm optimization (NPSO). The authors claim superiority over conventional PSO in terms of convergence and robustness properties. Execution time is slightly higher when compared with classic PSO due to the additional computation requirements to process bad experience component. There is no mutation process as in EPSO.

Discipline of quantum mechanics and analysis of the particle behavior in PSO motivated a new version named QPSO. The iterative equation of QPSO is somewhat different from PSO. The method uses a movement equation based on a quantum delta potential well model. With the previous best points and the help of the mean best position of the swarm it aims to enhance the global search ability of the particle. Without the velocity parameters of the traditional PSO, QPSO is easier to adjust, making it straightforward to implement for a given problem. Benchmark problems have demonstrated that QPSO can in some cases outperform the well-established variants of PSO [66, 75, 76].

1.2.9 Artificial Bee Colony

One of the recently proposed heuristic algorithms called ABC has drawn researchers' attention due to its simplicity and robustness. The ABC is based on intelligent behavior of honeybees. It was first developed by Karaboga in 2005 [77]. This method is a population-based optimization algorithm which has been demonstrated competitive to other methods because of the advantage of controlling fewer parameters and its robustness [78–82].

The ABC is as simple as PSO and DE algorithms, and uses only common control parameters such as colony size and maximum cycle number. There are three types of bees in the ABC system: employed bee, onlooker bee, and scout bee. The aim of all bees is to find the best food source (possible solution) with highest nectar (fitness value); in other words, artificial bees fly in a multidimensional search space to find the global optimal. Employed bees search for food sources based on their memory and the information gathered on food sources is shared with onlooker bees. Onlooker bees tend to choose good sources with higher nectar and further explore new food sources around the selected food sources. Scout bees abandon old food sources and randomly start a new one in order to avoid local minimum.

As the ABC has proven its robust, efficient, and simple characteristics, it has been widely implemented in solving a range of optimization problems in recent years such as job shop scheduling and machine timetabling problems [83]. The ABC was also implemented to tune the PI controllers' parameters in microgrid power electronics control [84]. In [85, 86] the authors applied the ABC or improved ABC in power system problems such as optimal power flow (OPF) or modified OPF which integrated wind power and storage devices.

1.2.10 Tabu Search

Tabu search (TS) is basically a gradient-descent search with memory [5]. The memory preserves a number of previously visited states along with a number of states that might be considered unwanted. This information is stored in a Tabu list. The definition of a state, the area around it, and the length of the Tabu list are critical design parameters. In addition to these Tabu parameters, two extra parameters are often used: aspiration and diversification. Aspiration is used when all the neighboring states of the current state are also included in the Tabu list. In that case, the Tabu obstacle is overridden by selecting a new state. Diversification adds randomness to this otherwise deterministic search. If the TS is not converging, the search is reset randomly.

TS consists of a meta-heuristic procedure used to manage heuristic algorithms that perform local search. Meta-heuristics consist of advanced strategies that allow the exploitation of the search space by providing means of avoiding being entrapped into local optimal solutions. As it happens with other combinatorial approaches, TS carries out a number of transitions in the search space aiming to find the optimal solutions or a range of near-optimal solutions. The name *Tabu* is related to the fact that in order to avoid revisiting certain areas of the search space that have already been explored, the algorithm turns these areas Tabu (or forbidden). It means that for a certain period of time (the Tabu tenure) the search will not consider the examination of alternatives containing features that characterize the solution points belonging to the area declared *Tabu*.

TS was developed from concepts originally used in AI. Unlike other combinatorial approaches such as GAs and SA, its origin is not related to biological or physical optimization processes [87]. TS was originally proposed by Fred Glover in the early 1980s and has ever since been applied with success to a number of complex problems in science and engineering. Applications to electric power network problems are already significant and growing. These include, for example, the long-term transmission network expansion problem and distribution planning problems such as the optimal capacitor placement in primary feeders [88–92].

Compared with SA and GAs, TS explores the solution space in a more aggressive way; i.e. it is greedier than those algorithms. TS algorithms are initialized with a configuration (or a set of configurations when the search is performed in parallel), which becomes the current configuration. At every iteration of the algorithm, a neighborhood structure is defined for the current configuration; a move is then made to the best configuration in this neighborhood, i.e. in a minimization problem, the algorithm switches to the configuration presenting the smallest cost. Normally only the most promising neighbors are evaluated, otherwise the problem could become intractable. Unlike gradient-type algorithms used for local search, the neighborhood in TS is updated dynamically. Another difference is that transitions to configurations with higher cost are allowed (this gives the method the ability to move out of local minimum points). An essential feature of TS algorithms is the direct exclusion of search alternatives temporarily classed as forbidden (Tabu). Consequently, the use of memory becomes crucial in these algorithms: one has to keep track of the Tabu's restrictions.

Other mechanisms of TS are the *intensification* and *diversification*: by the intensification mechanism the algorithm does a more comprehensive exploration of attractive regions which may lead to a local optimal point; by the diversification mechanism, on the other hand, the search is moved to previously unvisited regions, something that is important in order to avoid local minimum points. TS consists of a set of principles (or functions) applied in an integrated way to solve complex problem in an intelligent manner.

REFERENCES

1. Vale, Z.A. (2009). Intelligent power system. In: *Wiley Encyclopedia of Computer Science and Engineering*. Hoboken, NJ: Wiley.
2. Eremia, M., Liu, C.C., and Edris, A.A. (2016). *Advanced Solutions in Power Systems: HVDC, FACTS, and AI Techniques*. Wiley-IEEE Press.
3. Rau, N.S. (2003). *Optimization Principles: Practical Applications to the Operation and Markets of the Electric Power Industry*. Wiley-IEEE Press.
4. Zhu, J. (2015). *Optimization of Power System Operation*. Wiley-IEEE Press.
5. Lee, K.Y. and El-Sharkawi, M.A. (2008). *Modern Heuristic Optimization Techniques: Theory and Applications to Power Systems*, vol. 39. Wiley.
6. Rayward-Smith, V.J., Osman, C., Reeves, C.R., and Smith, G.D. (1996). *Modern Heuristic Search Methods*. Wiley.
7. Rezaee Jordehi, A. (2016). Allocation of distributed generation units in electric power systems: a review. *Renew. Sustain. Energy Rev.* 56: 893–905.
8. Oren, S. (2012). *Renewable Energy Integration and the Impact of Carbon Regulation on the Electric Grid*. IEEE.
9. European Commission (2016). *EU Energy, Transport and GHG Emissions Trends to 2050: Reference Scenario 2016*. Luxembourg: European Commission.
10. Foster, V. and Bedrosyan, D. (2014). *Understanding CO_2 Emissions from the Global Energy Sector*. Washington, DC: The World Bank.
11. Arias, N.B., Franco, J.F., Lavorato, M., and Romero, R. (2017). Metaheuristic optimization algorithms for the optimal coordination of plug-in electric vehicle charging in distribution systems with distributed generation. *Electr. Power Syst. Res.* 142: 351–361.
12. Colak, I., Sagiroglu, S., Fulli, G. et al. (2016). A survey on the critical issues in smart grid technologies. *Renew. Sustain. Energy Rev.* 54: 396–405.
13. Liang, H. and Zhuang, W. (2014). Stochastic modeling and optimization in a microgrid: a survey. *Energies* 7 (4): 2027–2050.

14. Abujarad, S.Y., Mustafa, M.W., and Jamian, J.J. (2017). Recent approaches of unit commitment in the presence of intermittent renewable energy resources: a review. *Renew. Sustain. Energy Rev.* 70: 215–223.
15. Dorigo, M. (2016). Editorial: ten years of swarm intelligence. *Swarm Intell.* 10 (4): 245–246.
16. Adarsh, B.R., Raghunathan, T., Jayabarathi, T., and Yang, X.-S. (2016). Economic dispatch using chaotic bat algorithm. *Energy* 96: 666–675.
17. Yao, E., Wang, H., Wang, L. et al. (2017). Multi-objective optimization and exergoeconomic analysis of a combined cooling, heating and power based compressed air energy storage system. *Energy Convers. Manag.* 138: 199–209.
18. Li, B., Roche, R., and Miraoui, A. (2017). Microgrid sizing with combined evolutionary algorithm and MILP unit commitment. *Appl. Energy.* 188: 547–562.
19. Soares, J., Fotouhi Ghazvini, M.A., Silva, M., and Vale, Z. (2016). Multi-dimensional signaling method for population-based metaheuristics: solving the large-scale scheduling problem in smart grids. *Swarm Evol. Comput.* 29: 13–32.
20. Deb, K., Pratap, A., Agarwal, S., and Meyarivan, T. (2002). A fast and elitist multiobjective genetic algorithm: NSGA-II. *IEEE Trans. Evol. Comput.* 6 (2): 182–197.
21. Yuan, X., Tian, H., Yuan, Y. et al. (2015). An extended NSGA-III for solution multi-objective hydro-thermal-wind scheduling considering wind power cost. *Energy Convers. Manag.* 96: 568–578.
22. Elsevier (2016). Scopus. *SCOPUS* [Online]. https://www.elsevier.com/solutions/scopus (accessed April 2019).
23. Gavanelli, M. and Rossi, F. (2010). Constraint logic programming. In: *A 25-Year Perspective on Logic Programming*, vol. 6125 (eds. A. Dovier and E. Pontelli), 64–86. Berlin: Springer.
24. Fogel, D.B. (2005). *Evolutionary Computation*. Hoboken: Wiley.
25. Rechenberg, I. (1978). Evolutionsstrategien. In: *Simulationsmethoden in der Medizin und Biologie* (eds. B. Schneider and U. Ranft), 83–114. Berlin: Springer.
26. Beyer, H.-G. and Schwefel, H.-P. (2002). Evolution strategies – a comprehensive introduction. *Nat. Comput.* 1 (1): 3–52.
27. Holland, J.H. (1962). Outline for a logical theory of adaptive systems. *J. ACM* 9 (3): 297–314.
28. Golberg, D.E. (2006). *Genetic Algorithms*. Pearson Education India.
29. Koza, J.R. (1992). *Genetic Programming: On the Programming of Computers by Means of Natural Selection*. MIT Press.
30. Fogel, M.J., Owens, L.J., and Walsh, A.J. (1966). *Artificial Intelligence Through Simulated Evolution*. Wiley.
31. Deb, K. and Jain, H. (2014). An evolutionary many-objective optimization algorithm using reference-point-based nondominated sorting approach, Part I: solving problems with box constraints. *IEEE Trans. Evol. Comput.* 18 (4): 577–601.
32. Fister, I. Jr., Yang, X.-S., Fister, I. et al. (2013). A brief review of nature-inspired algorithms for optimization. *Elektroteh. vestn.* 80 (3): 116–122.
33. Kefayat, M., Lashkar Ara, A., and Nabavi Niaki, S.A. (2015). A hybrid of ant colony optimization and artificial bee colony algorithm for probabilistic optimal placement and sizing of distributed energy resources. *Energy Convers. Manag.* 92: 149–161.
34. Castillo Salazar, C.A., Shih, M.Y., and Conde Enríquez, A. (2015). Adaptive directional overcurrent relay coordination using ant colony optimisation. *IET Gener. Transm. Distrib.* 9 (14): 2040–2049.

35. Yang, S., Wu, M., Yao, X., and Jiang, J. (2015). Load modeling and identification based on ant colony algorithms for EV charging stations. *IEEE Trans. Power Syst.* 30 (4): 1997–2003.
36. Mohanty, B., Panda, S., and Hota, P.K. (2014). Controller parameters tuning of differential evolution algorithm and its application to load frequency control of multi-source power system. *Int. J. Electr. Power Energy Syst.* 75: 205–214.
37. Wang, Y., Li, H.X., Huang, T., and Li, L. (2014). Differential evolution based on covariance matrix learning and bimodal distribution parameter setting. *Appl. Soft Comput. J.* 18: 232–247.
38. Ghasemi, M., Ghanbarian, M.M., Ghavidel, S. et al. (2014). Modified teaching learning algorithm and double differential evolution algorithm for optimal reactive power dispatch problem: a comparative study. *Inf. Sci. (NY).* 278: 231–249.
39. Sousa, T., Vale, Z., Carvalho, J.P. et al. (2014). A hybrid simulated annealing approach to handle energy resource management considering an intensive use of electric vehicles. *Energy* 67: 81–96.
40. Velik, R. and Nicolay, P. (2016). Energy management in storage-augmented, grid-connected prosumer buildings and neighborhoods using a modified simulated annealing optimization. *Comput. Oper. Res.* 66: 248–257.
41. Sousa, T., Soares, T., Morais, H. et al. (2016). Simulated annealing to handle energy and ancillary services joint management considering electric vehicles. *Electr. Power Syst. Res.* 136: 383–397.
42. Jiang, S., Ji, Z., and Shen, Y. (2014). A novel hybrid particle swarm optimization and gravitational search algorithm for solving economic emission load dispatch problems with various practical constraints. *Int. J. Electr. Power Energy Syst.* 55: 628–644.
43. Aman, M.M., Jasmon, G.B., Bakar, A.H.A., and Mokhlis, H. (2014). A new approach for optimum simultaneous multi-DG distributed generation Units placement and sizing based on maximization of system loadability using HPSO (hybrid particle swarm optimization) algorithm. *Energy* 66: 202–215.
44. Bahrami, S., Hooshmand, R.A., and Parastegari, M. (2014). Short term electric load forecasting by wavelet transform and grey model improved by PSO (particle swarm optimization) algorithm. *Energy* 72: 434–442.
45. Moeini-Aghtaie, M., Dehghanian, P., Fotuhi-Firuzabad, M., and Abbaspour, A. (2014). Multiagent genetic algorithm: an online probabilistic view on economic dispatch of energy hubs constrained by wind availability. *IEEE Trans. Sustain. Energy* 5: 699–708.
46. Ismail, M.S., Moghavvemi, M., and Mahlia, T.M.I. (2014). Genetic algorithm based optimization on modeling and design of hybrid renewable energy systems. *Energy Convers. Manag.* 85: 120–130.
47. Vrionis, T.D., Koutiva, X.I., and Vovos, N.A. (2014). A genetic algorithm-based low voltage ride-through control strategy for grid connected doubly fed induction wind generators. *IEEE Trans. Power Syst.* 29 (3): 1325–1334.
48. Karaboga, D. and Basturk, B. (2007). A powerful and efficient algorithm for numerical function optimization: artificial bee colony (ABC) algorithm. *J. Glob. Optim.* 39 (3): 459–471.
49. Dorigo, M. and Di Caro, G. (1999). Ant colony optimization: a new meta-heuristic. *Proceedings of the 1999 Congress on Evolutionary Computation-CEC99 (Cat. No. 99TH8406)* 2: 1470–1477.
50. Yang, X.-S. and Deb, S. (2009). Cuckoo search via levy flights. *2009 World Congress on Nature & Biologically Inspired Computing (NaBIC)*, Coimbatore, India (9–11 December 2009), pp. 210–214.

51. Eberhart, R. and Kennedy, J. (1995). A new optimizer using particle swarm theory. *MHS'95. Proceedings of the Sixth International Symposium on Micro Machine and Human Science*, Nagoya, Japan (4–6 October 1995), pp. 39–43.
52. Sivanandam, S.N. and Deepa, S.N. (2008). *Introduction to Genetic Algorithms*. Springer-Verlag.
53. Vale, Z.A., Ramos, C., Faria, P. et al. (2010). Comparison between deterministic and meta-heuristic methods applied to ancillary services dispatch. In: *Trends in Applied Intelligent Systems. IEA/AIE 2010, Pt I*, Lecture Notes in Computer Science, vol. 6096 (eds. N. García-Pedrajas, F. Herrera, C. Fyfe, et al.), 731–741.
54. Chaube, A., Benyoucef, L., and Tiwari, M.K. (2010). An adapted NSGA-2 algorithm based dynamic process plan generation for a reconfigurable manufacturing system. *J. Intell. Manuf.* 23 (4): 1141–1155.
55. Mandal, D., Pal, S.K., and Saha, P. (2007). Modeling of electrical discharge machining process using back propagation neural network and multi-objective optimization using non-dominating sorting genetic algorithm-II. *J. Mater. Process. Technol.* 186 (1–3): 154–162.
56. Fotouhi, M.A., Soares, J., Horta, N. et al. (2015). A multi-objective model for scheduling of short-term incentive-based demand response programs offered by electricity retailers. *Appl. Energy* 151: 102–118.
57. Kirkpatrick, S., Gelatt, C.D., and Vecchi, M.P. (1983). Optimization by simulated annealing. *Science.* 220 (4598): 671–680.
58. (a) Hannachi, A., Jolliffe, I.T., Stephenson, D.B., and Trendafilov, N. (2006). In search of simple structures in climate: simplifying EOFs. *Int. J. Climatol.* 26 (1): 728.
(b) Aarts, E. and Korst, J. (1989). *Simulated Annealing and Boltzmann Machines: A Stochastic Approach to Combinatorial Optimization and Neural Computing*. Wiley.
59. van Laarhoven, P.J.M. and Aarts, E.H.L. (1987). *Simulated Annealing: Theory and Applications*. Reidel.
60. Sousa, T., Morais, H., Castro, R., and Vale, Z. (2016). Evaluation of different initial solution algorithms to be used in the heuristics optimization to solve the energy resource scheduling in smart grids. *Appl. Soft Comput.* 48: 491–506.
61. Ciornei, I. and Kyriakides, E. (2013). Recent methodologies and approaches for the economic dispatch of generation in power systems. *Int. Trans. Electr. Energy Syst.* 23 (7): 1002–1027.
62. Kennedy, J. and Eberhart, R. (1995). Particle swarm optimization. In: *Proceedings of ICNN'95 – International Conference on Neural Networks, Perth, WA, Australia (1995)*, vol. 4, 1942–1948. IEEE.
63. Webb, B. (2002). Swarm intelligence: from natural to artificial systems. *Connect. Sci.* 14 (2): 163–164.
64. Soares, J., Silva, M., Sousa, T. et al. (2012). Distributed energy resource short-term scheduling using Signaled Particle Swarm Optimization. *Energy* 42 (1): 466–476.
65. Miranda, V., Keko, H., and Jaramillo, A. (2007). EPSO: evolutionary particle swarms. In: *Advances in Evolutionary Computing for System Design*, Studies in Computational Intelligence, vol. 66 (eds. L.C. Jain, V. Palade and D. Srinivasan), 139–167. Berlin: Springer.
66. Sun, J., Lai, C.H., and Wu, X.J. (2011). *Particle Swarm Optimisation: Classical and Quantum Perspectives*. Taylor & Francis.
67. Sun, J., Fang, W., Palade, V. et al. (2011). Quantum-behaved particle swarm optimization with Gaussian distributed local attractor point. *Appl. Math. Comput.* 218 (7): 3763–3775.

68. Coello, C.A.C., Pulido, G.T., and Lechuga, M.S. (2004). Handling multiple objectives with particle swarm optimization. *Evol. Comput. IEEE Trans.* 8 (3): 256–279.
69. Tripathi, P.K., Bandyopadhyay, S., and Pal, S.K. (2007). Multi-Objective Particle Swarm Optimization with time variant inertia and acceleration coefficients. *Inf. Sci. (NY)* 177 (22): 5033–5049.
70. del Valle, Y., Venayagamoorthy, G.K., Mohagheghi, S. et al. (2008). Particle swarm optimization: basic concepts, variants and applications in power systems. *IEEE Trans. Evol. Comput.* 12 (2): 171–195.
71. Selvakumar, A.I. and Thanushkodi, K. (2007). A new particle swarm optimization solution to nonconvex economic dispatch problems. *IEEE Trans. Power Syst.* 22 (1): 42–51.
72. Lee, T.Y. (2007). Optimal spinning reserve for a wind-thermal power system using EIPSO. *IEEE Trans. Power Syst.* 22 (4): 1612–1621.
73. Aghajani, G.R., Shayanfar, H.A., and Shayeghi, H. (2015). Presenting a multi-objective generation scheduling model for pricing demand response rate in micro-grid energy management. *Energy Convers. Manag.* 106: 308–321.
74. Ali, H. and Khan, F.A. (2013). Attributed multi-objective comprehensive learning particle swarm optimization for optimal security of networks. *Appl. Soft Comput.* 13 (9): 3903–3921.
75. Coelho, L.D. and Mariani, V.C. (2008). Particle swarm approach based on quantum mechanics and harmonic oscillator potential well for economic load dispatch with valve-point effects. *Energy Convers. Manag.* 49 (11): 3080–3085.
76. Sun, J., Fang, W., Wang, D.J., and Xu, W.B. (2009). Solving the economic dispatch problem with a modified quantum-behaved particle swarm optimization method. *Energy Convers. Manag.* 50 (12): 2967–2975.
77. Karaboga, D. (2005). An Idea Based on Honey Bee Swarm for Numerical Optimization. Erciyes University, Kayseri, Turkey. *Tech. Rep. TR06.*
78. Karaboga, D. and Akay, B. (2009). A comparative study of artificial bee colony algorithm. *Appl. Math. Comput.* 214: 108–132.
79. Tasgetiren, M.F., Pan, Q.-K., Suganthan, P.N., and Chen, A.H.L. (2011). A discrete artificial bee colony algorithm for the total flowtime minimization in permutation flow shops. *Inf. Sci.* 181: 3459–3475.
80. Pan, Q.K., Tasgetiren, M.F., Suganthan, P.N., and Chua, T.J. (2011). A discrete artificial bee colony algorithm for the lot-streaming flow shop scheduling problem. *Inf. Sci.* 181: 2455–2468.
81. Cain, M.B., O'Neill, R.P., and Castillo, A. (2012). History of Optimal Power Flow and Formulations. *Federal Energy Regulatory Commission Tech. Rep.*
82. Lee, K.Y. and El-Sharkawi, M.A. (eds.) (2008). *Modern Heuristic Optimization Techniques with Applications to Power Systems*. New York: Wiley-IEEE Press.
83. Fong, C.W., Asmuni, H., and McCollum, B. (2015). A hybrid swarm-based approach to university timetabling. *IEEE Trans. Evol. Comput.* 19 (6): 870–884.
84. Bai, W., Abedi, M.R., and Lee, K.Y. (2016). Distributed generation system control strategies with PV and fuel cell in microgrid operation. *Control Eng. Pract.* 53: 184–193.
85. Adaryani, M.R. and Karami, A. (2013). Artificial bee colony algorithm for solving multi-objective optimal power flow problem. *Elect. Power Energy Syst.* 53: 219–230.
86. Bai, W. and Lee, K.Y. (2016). Modified optimal power flow on storage devices and wind power integrated system. In: *2016 IEEE Power Engineering Society General Meeting*, 1–5. IEEE.
87. Glover, F. and Laguna, M. (1997). Tabu search. In: *Handbook of Combinatorial Optimization* (eds. D.Z. Du and P.M. Pardalos), 2093–2229. Kluwer Academic Publishers.

88. Wen, F.S. and Chang, C.S. (1997). Transmission network optimal planning using the Tabu search method. *Electr. Power Syst. Res.* 42 (2): 153–163.
89. Gallego, R.A., Monticelli, A., and Romero, R. (1998). Comparative studies of non-convex optimization methods for transmission network expansion planning. *IEEE Trans. Power Syst.* 13 (3): 822–828.
90. Wang, Y.C., Yang, H.T., and Huang, C.L. (1996). Solving the capacitor placement problem in a radial distribution system using Tabu search. *IEEE Trans. Power Syst.* 11 (4): 1868–1873.
91. Gallego, R.A., Monticelli, A., and Romero, R. (2001). Optimal capacitor placement in radial distribution networks. *IEEE Trans. Power Syst.* 16 (4): 630–637.
92. da Silva, E.L., Ortiz, J.M.D., De Oliveira, G.C., and Binato, S. (2001). Transmission network expansion planning under a Tabu search approach. *IEEE Trans. Power Syst.* 16 (1): 62–68.

CHAPTER 2

OVERVIEW OF APPLICATIONS IN POWER AND ENERGY SYSTEMS

Zita A. Vale and João Soares

Polytechnic of Porto, Porto, Portugal

2.1 APPLICATIONS TO POWER SYSTEMS

Power systems are the backbone of modern society. Without their complex infrastructure, our society cannot outlive [1]. Reforms in the energy sector brought new horizons, thanks to the efforts of many researchers worldwide. For example, considering the immense proliferation of renewable energy sources, mainly wind and solar generation and increasing penetration of electric vehicles (EVs) [2]. In fact, these resources have been contributing to reduction in the carbon footprint and increased sustainability [3–5]. However, power systems have become more complex and difficult to understand with those new technologies [6–9]. These will translate into new constraints, more precisely into specific mathematical equations in optimization problems, which will make them harder to solve [10]. In this context, modern heuristic optimization (MHO) has emerged to tackle some of the most prominent problems in this field. This section starts with a general overview of the research produced in the last few years (regarding quantity, relevant cites, and authors) and later moves on to specific overview of applications in unit commitment (UC), economic dispatch (ED), forecasting, among other problems.

In the last few years, research effort has been made to tackle aforementioned complexity and overcome the challenges brought by the smart grid era in modern power systems. The scientific research generated in the last few years is quite extensive. We investigate the numbers of research since 2010 in the SCOPUS

Applications of Modern Heuristic Optimization Methods in Power and Energy Systems, First Edition. Edited by Kwang Y. Lee and Zita A. Vale.
© 2020 by The Institute of Electrical and Electronics Engineers, Inc.
Published 2020 by John Wiley & Sons, Inc.

TABLE 2.1.1 Research Documents Produced in 2010–2018 According to SCOPUS

Search Term	Documents 2010–2018[a]	Top 3 Relevant Articles 2010–2018[a]		Top 5 Authors 2010–2018
		Reference	Citations	Author (No. Articles)
"Smart grid" OR "smart distribution network"	29 121	[12] [13] [14]	1 902 1 504 1 430	Javaid, N. (108); Vale, Z. (93); Mouftahm, H.T. (88); Morais, H. (76); Li, H. (65)
"Renewable generation" OR "renewable energy sources"	22 940	[15] [16] [17]	4 459 1 475 1 051	Guerrero, J.M. (73); Blaabjerg, F. (50); Senjuy, T. (50); Catalão, J.P.S. (43); Duic, N. (49)
"Smart grid" AND "electric vehicle"	2 857	[18] [19] [20]	1 245 781 715	Soares, J. (28); Mouftah, H.T. (27); Morais, H. (26); Vale, Z. (26); Masoum, M.A.S. (25)
"Demand response" OR "demand-side management"	12 223	[14] [21] [19]	1 430 1 177 781	Javaid, N. (101); Faria, P. (95); Vale, Z. (82); Lehtonen, M. (52); Catalão, J.P.S. (57)
"Smart grid" AND "optimization"	4 878	[14] [22] [21]	1 430 1 186 1 177	Javaid, N. (84); Khan, Z.A. (30); Soares, J. (28); Giannakis, G.B. (27); Morais, H. (27)
"Energy resource management" OR "energy management system"	6 937	[13] [14] [19]	1 504 1 430 781	Vale, Z. (81); Javaid, N. (51); Morais, H. (46); Guerrero, J.M. (39); Soares, J. (35)

[a]Excluding documents are already available for 2019. 2018 maybe incomplete.

database [11] (i.e. in February 2019) regarding different research topics in the power system. Table 2.1.1 shows the summary of research articles produced since 2010 (excluding 2019) filtered by different search terms. The search terms have been combined with similar variants within the power system field to observe results as accurate as possible. We looked for each search terms in the abstract, title, and paper's keywords. In addition, we investigate the most cited articles in each search criteria as well the most noticeable research authors by the number of articles produced. A total of 29 121 documents (2010–2018) have been found in the SCOPUS database by using the search term "smart grid" OR "smart distribution network" in the abstract, title, and keywords; and 22 940 if the search terms are "renewable generation" OR "renewable energy sources." The most cited article we found within this search is [15]. The interest in grid storage applications is noticeable and nowadays considered like the holy grail of modern power systems, since it can solve many of the current problems. In particular, [15] reviews battery systems for grid applications including redox-flow, sodium–sulfur, and lithium batteries. The interest of storage for grid applications is related with diversified factors, including managing peak demands, grid reliability, postponing grid investments, and the integration of renewable energy sources.

The search for the term "smart grid" AND "electric vehicle" returns 2857 documents. In spite of their complementary, electric vehicle in the smart grid context is yet a growing research field not as mature as the research observed in other topics, such as demand response (DR), which return 12 223 documents (for the term "demand response" OR "demand-side management"). Regarding smart grid optimization ("smart grid" AND "optimization"), the results return 4878 documents. Energy resource management returns 6937 documents. Articles [14, 19] appear often in those search retrievals. The work in [14] anticipated an autonomous and distributed demand-side management among system users. Game theory is used and an energy consumption scheduling is formulated. The envisaged advantages are simplicity of implementation and better privacy since users do not need to reveal all data to energy aggregators. Reference [19] reviews the application of DR and benefits to smart grid. The paper discusses real industrial case studies and research projects.

In the following part of this section, we provide an overview of MHO applications to particular problems of power system, namely: UC, ED, forecasting, and other particular problems like the maintenance scheduling and power flow.

2.1.1 Unit Commitment

The UC is a very important decision-making process established long before deregulation and markets in power systems. It can be described as a 0/1 process (being 0 OFF and 1 ON) of the generation units scheduling for a defined period of time. It can be stated that the difference between ED and UC is that ED is usually performed after the UC is made, to allocated load demand to assigned generators. UC and ED can also appear jointly in the same model in modern energy management systems [23]. The objective of the UC is to maximize social welfare in the case of demand bidding or the overall committed cost of the resources to meet load and reserve requirements. The UC must respect the resources' characteristics such as minimum up and down time [24].

The large-scale combinatorial nature of the UC problem reduces the possibility of using rigorous mathematical models to solve the problem. Therefore, many methods use some sort of approximation. Variable time resolution and variable modeling complexity are used in order to reduce computational requirements. Hydropower is an important source of renewable energy and one of the most important energy storage technology in power systems. Hydrothermal coordination is important in power systems with large hydropower generation. All the dependencies in the hydrothermal coordination lead to a heavy optimization problem that is suitable to be solved by MHO.

Maintenance scheduling is a preventive outage schedule in power systems with the goal to find the best timing and duration for generation units' maintenance in order to minimize fuel costs and maximize the availability of the system while satisfying several complex constraints.

Congestion management is a critical task for the system operator. Congestion management models are thus required in market-based solutions to maintain high-level reliability. Typically, there are two types of congestion management problems, based on technical cost-free methods and non-cost based on economics. Due to the fact that congestion management problems are of nonlinear nature and aimed to solve large-scale power systems of many electric buses, they are adequate for MHO.

State estimation of an electric power system can be defined as the vector of the voltage magnitudes and angles for every network buses. The available techniques in the literature correspond to heuristic and mathematical programming methods [23–28]. The stochastic model in [25] is presented to solve the UC problem with uncertain information about wind generation. An adaptive particle swarm optimization (PSO) algorithm with a based scenario generation and reduction algorithm to model the uncertainties is described. This approach demonstrates that is possible to achieve lower risk in the day-ahead UC with a MHO rather than a deterministic model alone. Recent approaches adopt uncertainty factors in the UC and multi-objective goals [26, 27, 29]. For instance, in [27], a UC and ED model within a unique framework is proposed with up to 24-hour-ahead horizon using MHO. Lately work show that MHO, namely biased random key genetic algorithm (GA), has been applied with success to multi-objective UC, addressing both cost and environmental factors (pollutant minimization) [29].

2.1.2 Economic Dispatch

ED is a problem of paramount importance in power system's operation as it can play a crucial role in the economics of the system. The aim of the ED is to schedule the production of the generation resources and minimize the operation cost while satisfying several constraints [30–34]. Typically, the ED is a nonlinear problem due to non-convex and nonlinear cost functions, ramp limits, discontinuous prohibited operation zones, and power losses.

Traditional approaches to solve ED are not able to provide an optimal solution, because they are local search techniques such as interior point method, dynamic programming, gradient method, and lambda-iteration method [33–36].

Heuristic methods such as GA, PSO, differential evolution (DE), evolutionary programming (EP), Tabu search (TS), and mean–variance optimization (MVO) have been applied to solve the ED successfully [30, 31, 37–43].

PSO approach has been used for solving the ED with generators ramp rate limits, prohibited operating zone, and non-smooth cost functions [38]. The results are compared with GA and the feasibility and superiority of the method is demonstrated with three different test power systems. Quantum-inspired version of PSO has been applied to solve the ED problem using the harmonic oscillator potential well [30]. The results are promising when compared with traditional versions of PSO and other older optimization algorithms in the literature. Mutation-enabled version of the quantum PSO (QPSO) shows to even improve earlier proposals applied to ED and tested in three different power systems [31]. Mutated QPSO version for ED is said to be stable and efficient than any other tested optimization algorithm so far. Reference [39] proposes an improved GA to solve the ED with valve-point loadings and change fuels. The proposed algorithm is compared with conventional GA and promises to be more effective. In [40], a TS method is applied to solve the ED. The method uses a flexible memory system to avoid getting stuck in local optima and enables to accelerate convergence and optimization speed. The MVO approach has also shown interesting results in order to solve the ED non-convex problem. ED MVO uses Kuhn–Tucker condition and swap process to improve the global minimum searching capability [41]. The method is applied to the large-scale power system of Korea. The results are compared with state-of-the-art methods. In [32], a multi-objective bacterial foraging is applied to solve the ED with two conflicting objectives and the IEEE 30-bus system is used to test the developed algorithm. A comprehensive survey regarding the methodologies and approaches for the ED problem can be found in [44] for both regulated and deregulated markets. Multi-area ED has also been proposed. Reference [43] covers a comprehensive study regarding use of MHO (DE, simulated annealing [SA], GA, and EP) to this problem. The study points out that DE achieves the lowest cost and SA requires the least amount of processing time for the tested MHO. The lowest execution time is possible in SA due to the fact that it is not a population-based MHO but rather a single individual meta-heuristic.

2.1.3 Forecasting in Power Systems

Modern power and energy system integrates a diverse number of energy resources, mostly from renewable sources. In fact, the electrical sector faces new challenges extended to the different involved participants. In last few decades, all over the world the electricity demand has been increasing and several concerns regarding supply security and reliability have been a priority [45].

Typically, in most electricity markets the distribution and supply services have been unbundled, and the electricity retailers are operating within a competitive environment, and with freedom in formulating the tariff offers. Each tariff offer can be formulated with reference to a specific customer class defined for several attributes, such as the voltage level, electric energy consumption, etc. [46].

In forecasting of power systems, we can highlight the main subfields: load, price, wind, and solar forecasting. In addition, we can distinguish between short-, mid-, and long-term forecasting [47–49]. Yet, the literature suggests that research on short-term forecasting for power systems is more intensive. In the last few decades, the research has been devoted to point predictions. The recent advent of smart grid and renewables has brought new sources of uncertainty in future supply, demand, and price. Researchers and academics all over the world have realized the importance of probabilistic forecasting (price, load, etc.) for the new reality. Thus, those methods are now more important for operation and planning than ever [50].

Techniques usually applied to forecasting are related to machine learning and data-mining techniques. MHO has been applied in this field as a hybrid method of parameter tuning [51]. PSO has been used to estimate long-term load pattern as an optimization problem. Results suggest that PSO performs better than the least-square method for the long-term load pattern estimation [52]. Artificial immune system has been used as a learning regression algorithm for short-term load forecasting (STLF) with local feature selection [53]. Results indicate that it is a strong competitor for other popular methods, including regression and neuronal networks. Hybridizations have also shown promising results. Modifications to fruit algorithm in combination with neuronal networks have been proposed to enhance the weight coefficients of those networks and improve results [54]. In addition, PSO has been used for parameterization of an extreme learning method to train the preprocessed dataset. Research concludes that the proposed hybrid method performs well even in microgrids with high load fluctuations. Despite limitations of local minimization, PSO has also been used jointly with support vector machines with the proposed load forecasting [55].

The implementation of the energy markets brought the need to deal with load forecasting as an advantageous tool to support commercial trades between buyers and suppliers. Several load forecasting methods are exploited and well documented. In [56] is proposed a methodology based on neuronal networks supported by detailed information related to consumers' typical behavior and climatic information, for forecasting the hourly electric load values for the following day. In [57], a modern approach that takes advantage of hourly information to create more accurate and defensible forecasts is proposed in order to support long-term load forecasting. A generic strategy for STLF based on the support vector regression machines can be found in [58].

Another challenge that electricity markets' managers and participants' face is the electricity market price forecast. In [59], a new approach to short-term electricity price forecasting is presented. The proposed method is derived by integrating the kernel principal component analysis (KPCA) method with the local informative vector machine (IVM), which can be derived by combining the IVM with the local regression method. The results show that the proposed method can improve the price forecasting accuracy and provides a much better prediction performance in comparison with other recently published approaches.

With the increasing of wind power as a renewable energy source, the energy sector needs to integrate this intermittent power source into the power grid. Thus, wind power forecasting at the day before and at the day of delivery may be required. In [60], the reader may find an overview of several wind power prediction models for very short term, short term, medium term, and long term. The studies available in the literature have been evaluated and criticized in consideration with their prediction accuracies and deficiencies. In this paper it is shown that adaptive neuro-fuzzy inference systems, neural networks, and multilayer perceptrons give better results in wind power predictions. In [61], a review of wind power and wind speed forecasting methods with different time horizons can be found. The higher penetration of renewable resources in the power grid, mainly by solar energy integration, requires reliable forecast information. In fact, this integration can offer a better quality of service if the solar irradiance variation can be predicted with great accuracy. In [62], an in-depth review of the current methods used to forecast solar irradiance in order to facilitate selection of the appropriate forecast method according to needs is presented.

2.1.4 Other Applications in Power Systems

Besides the UC and ED problems, MHO have been applied with success to other areas in power system operation, such as hydrothermal coordination, maintenance scheduling, congestion management, optimal power flow, state estimation, etc.

Hydropower is an important source of renewable energy and one of the most important energy storage technology in power systems. Hydrothermal coordination is important in power systems with large hydropower generation. All the dependencies in the hydrothermal coordination lead to a heavy optimization problem that is suitable to be solved by MHO. In [63–65] evolutionary-based techniques such as GAs are presented for the hydrothermal scheduling problem and the results demonstrate their success in industrial-grade power systems. Recently, works with newer MHO techniques have shown that algorithms like cuckoo search can reduce the number of tweaking parameters (as seen in GA) but deal effectively with non-convex problems' hydrothermal scheduling [66]. Those newer MHO methods include not only cuckoo search but also gravitational search algorithm, chaotic DE, among others [67–69].

Maintenance scheduling is a preventive outage schedule in a power system with the goal to find the best timing and duration for generation unit's maintenance in order to minimize fuel costs and maximize the availability of the system while satisfying several complex constraints. Although this problem falls into the realm of planning and does requires an immediate response, MHO techniques have also played a significant role to deal with non-convexities and nonlinearities [70]. In [71–73] several evolutionary-based techniques are applied successfully to the maintenance scheduling problem in traditional and restructured power systems. Recent survey highlights the importance of heuristics due to the NP-Hardness characteristic of the problem and the size of the real-world instances. Those heuristics include not only GA but also PSO, SA, clonal selection algorithm, teaching–learning algorithm, TS, ant colony, and constraint programing [70].

Congestion management is a critical task for the system operator. The restructure of the power industry led to more intensive usage of the distribution and transmission grid not predicted at the design stage [74]. The lack of coordination between different players of the network makes this situation worse.

Congestion management models are thus required in market-based solutions to maintain high-level reliability. Typically, there are two types of congestion management problems, based on technical cost-free methods and non-cost based on economics. Due to the fact that congestion management problems are of nonlinear nature and aimed to solve large-scale power systems of many electric buses, they are adequate for MHO. In [75–78] several different methods are proposed, including using EVs and flexible AC transmission system (FACTS) device to control congestion. Recently, DR has gained importance to address congestion management issues [79]. PSO, GA, DE, and other MHO have been widely used for optimization purposes in congestion management [74, 80, 81].

Optimal power flow is one most researched problems in power system fields [82]. It is an important part tool for many applications within power system studies. The optimal power flow is a large-scale constrained nonlinear optimization problem where the control variables are the generator real powers, transformer taps, reactive power of switchable VAR sources, and generator bus voltages while the problem-dependent variables include the load bus voltages, the generator reactive powers, and the line power flows [83]. Recently, optimal power flow in smart grid has gained attraction with several different approaches (including decentralized) being proposed [84]. The characteristics of optimal power flow make the problem suitable for MHO. In fact, several applications of heuristic optimization have been applied successfully to optimal power flow problem including PSO, GAs, gravitational search algorithm, gray wolf optimization, teaching–learning algorithm, and black-hole-based optimization [85–90].

State estimation of an electric power system can be defined as the vector of the voltage magnitudes and angles for every network buses. This problem can be classified as a real-time algorithm with high dimensionality that requires large amounts of processing power and storage [91, 92]. The latest challenges in this field include dealing with cyberattacks (false data and false attacks) and larger amounts of uncertainty in the state estimation especially focused on smart grid applications [92–94]. Phasor measurement unit (PMU) has been suggested as one the most promising equipment for improving state estimation accuracy despite its investment cost [95]. MHO has been less employed in this field; however, some applications can still be found. For instance, identification of bad errors in state estimator by using MHO, namely TS [96]. Other MHO techniques, including PSO and GA-based techniques, have been used to address the problem of state estimation in [97, 98].

2.2 SMART GRID APPLICATION COMPETITION SERIES

In this section of the book we present a smart grid application that has been developed by GECAD/Polytechnic of Porto. This smart grid application has been launched jointly with other universities as an worldwide algorithm competition held inside renowned international conferences and events. The first competition

TABLE 2.2.1 Summary of the GECAD Smart Grid Application Competition Editions

Edition	Name and Location	Case Study	Winning Criteria
PES GM 2017	Optimal scheduling of distributed energy resources (Chicago, IL, USA)	33-bus 1800 EVs 180-bus 6000 EVs (no uncertainty, network constrains)	Lowest mean fitness function in the two cases
WCCI 2018	Evolutionary computation in uncertain environments: a smart grid application (Rio de Janeiro, Brazil)	25-bus 34 EVs (uncertainty 100 scenarios)	Lowest rank based on lowest mean fitness plus standard deviation over each scenario
CEC & GECCO 2019	Evolutionary computation in uncertain environments: a smart grid application (Prague, Czech Republic and Wellington, New Zealand)	25-bus 34 EVs (uncertainty 500 scenarios)	Lowest rank based on lowest mean fitness plus standard deviation over each scenario

of this smart grid application was launched in Power and Energy Society General Meeting (PES GM 2017).[1] The second edition was launched in World Congress on Computational Intelligence (WCCI 2018)[2] and the current third edition was launched at two major venues, namely in Congress on Evolutionary Computation (CEC 2019) and Genetic and Evolutionary Computation Conference (GECCO 2019).[3] Table 2.2.1 presents a summary of the features of each edition. The first edition featured two large-scale case studies with a high penetration of energy resources. The second edition in WCCI 2018 incorporated uncertainty represented by 100 scenarios. In CEC/GECCO edition a more challenging case study was introduced. The uncertainty is then represented by 500 scenarios with a higher level of uncertainty on the photovoltaic (PV) production. A maximum of 50 000 evaluations per run (31 runs) was allowed in each edition.

2.2.1 Problem Description

The problem considers an energy aggregator with aims of procuring energy needs from distributed resources and the electricity market. The aggregator looks for the minimization of operational costs while making revenues from selling energy in available electricity markets. Moreover, it may use its own assets, e.g. energy storage systems (ESS), to supply the load demand. In PES GM 2017 edition,

[1] PES GM 2017 edition: http://sites.ieee.org/psace-mho/2017-smart-grid-operation-problems-competition-panel.

[2] WCCI 2018 edition: http://www.gecad.isep.ipp.pt/WCCI2018-SG-COMPETITION.

[3] CEC/GECCO edition: http://www.gecad.isep.ipp.pt/ERM2019-Competition.

the network constraints are included to validate the performance of the network [99]. In addition, a V2G feature that allows the use of energy in the battery of EVs is also possible. The energy aggregator establishes bilateral energy contracts with those who seek electricity supply, e.g. residential and industry customers. In this case, it is assumed that the aggregator does not make profits from the supply of energy to fixed loads and EVs charging. The main idea is that the optimization software can perform the energy resource scheduling of the dedicated resources in the day-ahead context for the 24 hours of the following day.

Since the aggregator performs the scheduling of resources for the day-ahead (i.e. the next 24 hours), it relays in the forecast of weather conditions (to predict renewable generation), load demand, EV trips, and market prices. However, the assumption of "perfect" or "highly accurate" forecast might bring catastrophic consequences into the operation of the grid when the realizations do not follow the expected predictions.

Due to this situation, it is desired that the aggregator determines solutions that are robust to the uncertainty inherent in some parameters and the environment. Four aspects of uncertainty that affect the performance of a solution are considered in WCCI 2018, CEC/GECCO 2019 edition,[4] namely: (i) weather conditions, (ii) load forecast, (iii) planned EVs' trips, and (iv) market prices.

Therefore, the aggregator should find solutions that provide not only an optimal (or near-optimal) value of operational costs but also those solutions must have the characteristic of being as less sensible as possible to the variations of the uncertain parameters. In [100], uncertainty in evolutionary computation is classified into four categories, namely noise, robustness, fitness approximation, and time-varying fitness functions. This competition lays in the category of robustness, in which the design variables (or environmental parameters in this particular case) are subject to perturbations or changes after the optimal solution has been determined (i.e. the realizations of uncertain parameters).

To incorporate the uncertainty of parameters, Monte Carlo simulation (MCS) is used to generate a large number of possible scenarios using probability distribution functions of the forecast errors (obtained from historical data). A high number of scenarios increases the accuracy of the model but comes with a computation cost associated with a large number of variations in the parameters. Due to this, a reduction technique [101] is used to maintain a reasonably small number of scenarios while keeping the main statistical characteristics of the initial scenarios set (Figure 2.2.1).

2.2.2 Best Algorithms and Ranks

In this section the best algorithms and ranks for each competition edition are presented. In PES GM 2017 edition, five participants submitted their results while in WCCI 2018 edition, nine participants. Table 2.2.2 shows the best three

[4] PES GM 2017 edition does not consider uncertainty.

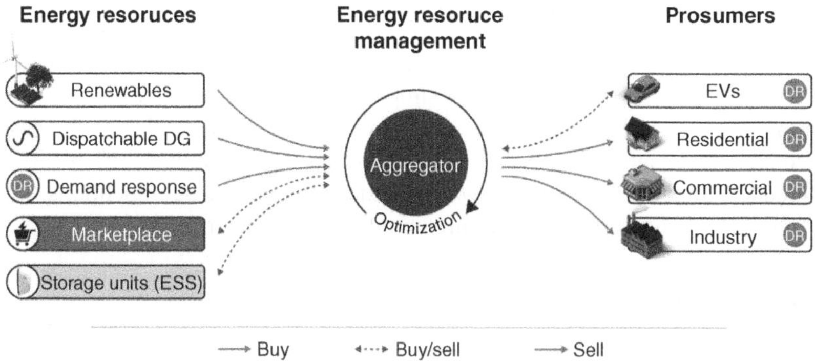

Figure 2.2.1 Overview of the aggregator energy management problem.

TABLE 2.2.2 Main Results of PES GM 2017 and WCCI 2018 Edition

Edition	Name and location	Best algorithms	Score
PES GM 2017	Optimal scheduling of distributed energy resources (Chicago, IL, USA)	1st Variable neighborhood search (VNS) algorithm	−8649.99
		2nd Modified chaotic biogeography-based optimisation (CBBO) with random sinusoidal migration	−8040.46
		3rd Cross entropy method and evolutionary particle swarm optimization (CEEPSO)	−7735.34
WCCI 2018	Evolutionary computation in uncertain environments: a smart grid application (Rio de Janeiro, Brazil)	1st VNS-DEEPSO (VNS combined with DEPSO)	18.21
		2nd Enhanced velocity differential evolutionary PSO (DEEPSO)	19.57
		3rd Chaotic evolutionary particle swarm optimization/particle swarm optimization with global best perturbation	24.89

algorithms and respective score for the each past edition of the competition. CEC/GECCO 2019 results will be disclosed after the events. Variable neighborhood search (VNS) and PSO variants are frequent top-ranked algorithms in this smart grid application competition. A combination of VNS with PSO variant called differential evolutionary PSO (which is a combination of DE and PSO) was proposed at WCCI 2018 edition for the problem with uncertainty. The combination of VNS with DEEPSO ranked first among the nine participants in this edition.

2.2.3 Further Information and How to Download

The complete information of the smart grid application can be checked individually in the webpage of each edition. For simplicity purposes we also created a webpage that unifies all the competition series into one single page. Hence, the best-ranked algorithms and problems of each edition mentioned in this section can be downloaded directly from http://www.gecad.isep.ipp.pt/ERM-competitions. We aim to update this webpage with future competitions' outcomes.

REFERENCES

1. Soares, J., Pinto, T., Lezama, F., and Morais, H. (2018). Survey on complex optimization and simulation for the new power systems paradigm. *Complexity* 2018: 1–32.
2. Biresselioglu, M.E., Demirbag Kaplan, M., and Yilmaz, B.K. (2018). Electric mobility in Europe: a comprehensive review of motivators and barriers in decision making processes. *Transp. Res. A Policy Pract.* 109: 1–13.
3. European Commission (2016). *EU Energy, Transport and GHG Emissions Trends to 2050: Reference Scenario 2016*. Luxembourg: European Commission.
4. Foster, V. and Bedrosyan, D. (2014). *Understanding CO_2 Emissions from the Global Energy Sector*. Washington, DC: The World Bank.
5. Dincer, I. and Acar, C. (2018). Potential energy solutions for better sustainability. In: *Exergetic, Energetic and Environmental Dimensions* (ed. I. Dincer, C.O. Colpan and O. Kizilkan), 3–37. Elsevier.
6. Arias, N.B., Franco, J.F., Lavorato, M., and Romero, R. (2017). Metaheuristic optimization algorithms for the optimal coordination of plug-in electric vehicle charging in distribution systems with distributed generation. *Electr. Power Syst. Res.* 142: 351–361.
7. Colak, I., Sagiroglu, S., Fulli, G. et al. (2016). A survey on the critical issues in smart grid technologies. *Renew. Sustain. Energy Rev.* 54: 396–405.
8. Liang, H. and Zhuang, W. (2014). Stochastic modeling and optimization in a microgrid: a survey. *Energies* 7 (4): 2027–2050.
9. Abujarad, S.Y., Mustafa, M.W., and Jamian, J.J. (2017). Recent approaches of unit commitment in the presence of intermittent renewable energy resources: a review. *Renew. Sustain. Energy Rev.* 70: 215–223.
10. De Lara, M., Carpentier, P., Chancelier, J., and Leclère, V. (2014). *Optimization Methods for the Smart Grid*. World Energy Council.
11. Elsevier (2016). Scopus. *SCOPUS* [Online]. https://www.elsevier.com/solutions/scopus (accessed April 2019).
12. Guerrero, J.M., Vasquez, J.C., Matas, J. et al. (2011). Hierarchical control of droop-controlled AC and DC microgrids – a general approach toward standardization. *IEEE Trans. Ind. Electron.* 58 (1): 158–172.
13. Farhangi, H. (2010). The path of the smart grid. *IEEE Power Energy Mag.* 8 (1): 18–28.
14. Mohsenian-Rad, A.H., Wong, V.W.S., Jatskevich, J. et al. (2010). Autonomous demand-side management based on game-theoretic energy consumption scheduling for the future smart grid. *IEEE Trans. Smart Grid* 1 (3): 320–331.
15. Dunn, B., Kamath, H., and Tarascon, J.M. (2011). Electrical energy storage for the grid: a battery of choices. *Science* 334 (6058): 928–935.
16. Larcher, D. and Tarascon, J.M. (2015). Towards greener and more sustainable batteries for electrical energy storage. *Nat. Chem.* 7: 19–29.

17. Gutfleisch, O., Willard, M.A., Brück, E. et al. (2011). Magnetic materials and devices for the 21st century: stronger, lighter, and more energy efficient. *Adv. Mater.* 23: 821–842.
18. Kamaya, N., Homma, K., Yamakawa, Y. et al. (2011). A lithium superionic conductor. *Nat. Mater.* 10 (9): 682–686.
19. Siano, P. (2014). Demand response and smart grids – a survey. *Renew. Sustain. Energy Rev.* 30: 461–478.
20. Skyllas-Kazacos, M., Chakrabarti, M.H., Hajimolana, S.A. et al. (2011). Progress in flow battery research and development. *J. Electrochem. Soc.* 158 (8): R55.
21. Palensky, P. and Dietrich, D. (2011). Demand side management: demand response, intelligent energy systems, and smart loads. *IEEE Trans. Ind. Informat.* 7 (3): 381–388.
22. Güngör, V.C., Sahin, D., Kocak, T. et al. (2011). Smart grid technologies: communication technologies and standards. *IEEE Trans. Ind. Informat.* 7 (4): 529–539.
23. Nemati, M., Braun, M., and Tenbohlen, S. (2018). Optimization of unit commitment and economic dispatch in microgrids based on genetic algorithm and mixed integer linear programming. *Appl. Energy* 210: 944–963.
24. Padhy, N.P. (2004). Unit commitment – a bibliographical survey. *IEEE Trans. Power Syst.* 19 (2): 1196–1205.
25. Pappala, V.S., Erlich, I., Rohrig, K., and Dobschinski, J. (2009). A stochastic model for the optimal operation of a wind-thermal power system. *IEEE Trans. Power Syst.* 24 (2): 940–950.
26. Ruiz, P.A., Philbrick, C.R., Zak, E. et al. (2009). Uncertainty management in the unit commitment problem. *IEEE Trans. Power Syst.* 24 (2): 642–651.
27. Pappala, V.S. and Erlich, I. (2010). A variable-dimension optimization approach to unit commitment problem. *IEEE Trans. Power Syst.* 25 (3): 1696–1704.
28. Rahman, D.F., Viana, A., and Pedroso, J.P. (2014). Metaheuristic search based methods for unit commitment. *Int. J. Electr. Power Energy Syst.* 59: 14–22.
29. Roque, L., Fontes, D., and Fontes, F. (2017). A metaheuristic approach to the multi-objective unit commitment problem combining economic and environmental criteria. *Energies* 10 (12): 2029.
30. Coelho, L.D. and Mariani, V.C. (2008). Particle swarm approach based on quantum mechanics and harmonic oscillator potential well for economic load dispatch with valve-point effects. *Energy Convers. Manag.* 49 (11): 3080–3085.
31. Sun, J., Fang, W., Wang, D.J., and Xu, W.B. (2009). Solving the economic dispatch problem with a modified quantum-behaved particle swarm optimization method. *Energy Convers. Manag.* 50 (12): 2967–2975.
32. Pandi, V.R., Panigrahi, B.K., Hong, W.C., and Sharma, R. (2014). A multiobjective bacterial foraging algorithm to solve the environmental economic dispatch problem. *Energy Sources Part B Econ. Plan. Policy* 9 (3): 236–247.
33. Wood, A. and Wollenberg, B. (2013). *Power Generation Operation and Control*, 3e. Wiley.
34. Report, I.C. (1971). Present practices in the economic operation of power systems. *IEEE Trans. Power Appar. Syst.* PAS-90 (4): 1768–1775.
35. Liang, Z.X. and Glover, J.D. (1992). A zoom feature for a dynamic-programming solution to economic-dispatch including transmission losses. *IEEE Trans. Power Syst.* 7 (2): 544–550.
36. Vargas, L. and Quintana, V.H. (1994). Optimal reactive dispatch through interior-point methods – discussion. *IEEE Trans. Power Syst.* 9 (1): 144–145.

37. Sun, J., Lai, C.H., and Wu, X.J. (2011). *Particle Swarm Optimisation: Classical and Quantum Perspectives*. Taylor & Francis.
38. Zwe-Lee, G. (2003). Particle swarm optimization to solving the economic dispatch considering the generator constraints. *IEEE Trans. Power Syst.* 18 (3): 1187–1195.
39. Chao-Lung, C. (2005). Improved genetic algorithm for power economic dispatch of units with valve-point effects and multiple fuels. *IEEE Trans. Power Syst.* 20 (4): 1690–1699.
40. Whei-Min, L., Fu-Sheng, C., and Ming-Tong, T. (2002). An improved tabu search for economic dispatch with multiple minima. *IEEE Trans. Power Syst.* 17 (1): 108–112.
41. Kim, M.J., Song, H.Y., Park, J.B. et al. (2013). An improved mean-variance optimization for nonconvex economic dispatch problems. *J. Electr. Eng. Technol.* 8 (1): 80–89.
42. Song, H.Y., Park, Y.G., Roh, J.H. et al. (2012). Immune-PSO for economic dispatch with valve point effect. In: *Management, Manufacturing and Materials Engineering, Parts 1 and 2*, vol. 452–453 (ed. P. Liu), 1054–1058. Stafa-Zurich: Trans Tech Publications Ltd.
43. Pattanaik, J.K., Basu, M., and Dash, D.P. (2017). Review on application and comparison of metaheuristic techniques to multi-area economic dispatch problem. *Prot. Control Mod. Power Syst.* 2 (1): 17.
44. Ciornei, I. and Kyriakides, E. (2013). Recent methodologies and approaches for the economic dispatch of generation in power systems. *Int. Trans. Electr. Energy Syst.* 23 (7): 1002–1027.
45. Catalão, J.P.S. (2017). *Electric Power Systems: Advanced Forecasting Techniques and Optimal Generation Scheduling*. CRC Press.
46. Chicco, G. (2012). Overview and performance assessment of the clustering methods for electrical load pattern grouping. *Energy* 42 (1): 68–80.
47. Boroojeni, K.G., Amini, M.H., Bahrami, S. et al. (2017). A novel multi-time-scale modeling for electric power demand forecasting: from short-term to medium-term horizon. *Electr. Power Syst. Res.* 142: 58–73.
48. Kaboli, S.H.A., Selvaraj, J., and Rahim, N.A. (2016). Long-term electric energy consumption forecasting via artificial cooperative search algorithm. *Energy* 115: 857–871.
49. Azad, H.B., Mekhilef, S., and Ganapathy, V.G. (2014). Long-term wind speed forecasting and general pattern recognition using neural networks. *IEEE Trans. Sustain. Energy* 5 (2): 546–553.
50. Nowotarski, J. and Weron, R. (2017). Recent advances in electricity price forecasting: a review of probabilistic forecasting. *Renew. Sustain. Energy Rev.* 81 (10): 1548–1568.
51. Fallah, S., Deo, R., Shojafar, M. et al. (2018). Computational intelligence approaches for energy load forecasting in smart energy management grids: state of the art, future challenges, and research directions. *Energies* 11 (3): 596.
52. AlRashidi, M.R. and EL-Naggar, K.M. (2010). Long term electric load forecasting based on particle swarm optimization. *Appl. Energy* 87 (1): 320–326.
53. Dudek, G. (2017). Artificial immune system with local feature selection for short-term load forecasting. *IEEE Trans. Evol. Comput.* 21 (1): 116–130.
54. Xiao, L., Shao, W., Liang, T., and Wang, C. (2016). A combined model based on multiple seasonal patterns and modified firefly algorithm for electrical load forecasting. *Appl. Energy* 167: 135–153.
55. Hong, W.-C. (2009). Chaotic particle swarm optimization algorithm in a support vector regression electric load forecasting model. *Energy Convers. Manag.* 50 (1): 105–117.
56. Sousa, J.C., Jorge, H.M., and Neves, L.P. (2011). Forecasting the next day load profile using load profiling information and meteorological variables. In: *Proceedings of the 2011 3rd International Youth Conference on Energetics (IYCE)*, 1–6. IEEE.

57. Hong, T., Wilson, J., and Xie, J. (2014). Long term probabilistic load forecasting and normalization with hourly information. *IEEE Trans. Smart Grid* 5 (1): 456–462.
58. Ceperic, E., Ceperic, V., and Baric, A. (2013). A strategy for short-term load forecasting by support vector regression machines. *IEEE Trans. Power Syst.* 28 (4): 4356–4364.
59. Elattar, E.E. (2013). Day-ahead price forecasting of electricity markets based on local informative vector machine. *IET Gener. Transm. Distrib.* 7 (10): 1063–1071.
60. Colak, I., Sagiroglu, S., and Yesilbudak, M. (2012). Data mining and wind power prediction: a literature review. *Renew. Energy* 46: 241–247.
61. Soman, S.S., Zareipour, H., Malik, O., and Mandal, P. (2010). A review of wind power and wind speed forecasting methods with different time horizons. In: *North American Power Symposium 2010 (NAPS 2010)*. IEEE.
62. Diagne, M., David, M., Lauret, P. et al. (2013). Review of solar irradiance forecasting methods and a proposition for small-scale insular grids. *Renew. Sustain. Energy Rev.* 27: 65–76.
63. Gil, E., Bustos, J., and Rudnick, H. (2003). Short-term hydrothermal generation scheduling model using a genetic algorithm. *IEEE Trans. Power Syst.* 18 (4): 1256–1264.
64. Zoumas, C.E., Bakirtzis, A.G., Theocharis, J.B., and Petridis, V. (2004). A genetic algorithm solution approach to the hydrothermal coordination problem. *IEEE Trans. Power Syst.* 19 (3): 1356–1364.
65. Sinha, N., Chakrabarti, R., and Chattopadhyay, P.K. (2003). Fast evolutionary programming techniques for short-term hydrothermal scheduling. *Electr. Power Syst. Res.* 66 (2): 97–103.
66. Nguyen, T.T., Vo, D.N., and Truong, A.V. (2014). Cuckoo search algorithm for short-term hydrothermal scheduling. *Appl. Energy* 132: 276–287.
67. Basu, M. (2014). Improved differential evolution for short-term hydrothermal scheduling. *Int. J. Electr. Power Energy Syst.* 58: 91–100.
68. Gouthamkumar, N., Sharma, V., and Naresh, R. (2015). Disruption based gravitational search algorithm for short term hydrothermal scheduling. *Expert Syst. Appl.* 42 (20): 7000–7011.
69. Zhang, J., Lin, S., and Qiu, W. (2015). A modified chaotic differential evolution algorithm for short-term optimal hydrothermal scheduling. *Int. J. Electr. Power Energy Syst.* 65: 159–168.
70. Froger, A., Gendreau, M., Mendoza, J.E. et al. (2016). Maintenance scheduling in the electricity industry: a literature review. *Eur. J. Oper. Res.* 251 (3): 695–706.
71. Burke, E.K. and Smith, A.J. (2000). Hybrid evolutionary techniques for the maintenance scheduling problem. *IEEE Trans. Power Syst.* 15 (1): 122–128.
72. Elyas, S.H., Sarookolaee, S.B., and Foroud, A.A. (2011). A novel approach for maintenance scheduling of generation units in restructured power system. In: *Innovative Smart Grid Technologies – India (ISGT India), 2011 IEEE PES*, 374–379. IEEE.
73. Yare, Y., Venayagamoorthy, G.K., and Aliyu, U.O. (2008). Optimal generator maintenance scheduling using a modified discrete PSO. *IET Gener. Transm. Distrib.* 2 (6): 834–846.
74. Peesapati, R., Yadav, V.K., and Kumar, N. (2018). Flower pollination algorithm based multi-objective congestion management considering optimal capacities of distributed generations. *Energy* 147: 980–994.
75. Li, R.Y., Wu, Q.W., and Oren, S.S. (2014). Distribution locational marginal pricing for optimal electric vehicle charging management. *IEEE Trans. Power Syst.* 29 (1): 203–211.
76. Visalakshi, S. and Baskar, S. (2011). Multiobjective decentralized congestion management using modified NSGA-II. *Arab. J. Sci. Eng.* 36 (5): 827–840.

77. Ferreira, M.J., Vale, Z.A., and Cardoso, J. (2007). A congestion management and transmission price simulator for competitive electricity markets. In: *2007 IEEE Power Engineering Society General Meeting*, vol. 1–10, 4066–4073. New York: IEEE.
78. Reddy, K.R.S., Padhy, N.P., and Patel, R.N. (2006). Congestion management in deregulated power system using FACTS devices. In: *Power India Conference, 2006 IEEE*, 8. IEEE.
79. Huang, S., Wu, Q., Shahidehpour, M., and Liu, Z. (2018). Dynamic power tariff for congestion management in distribution networks. *IEEE Trans. Smart Grid* 10 (2): 2148–2157.
80. Esfahani, M.M., Cintuglu, M.H., and Mohammed, O.A. (2017). Optimal real-time congestion management in power markets based on particle swarm optimization. In: *2017 IEEE Power & Energy Society General Meeting*, 1–5. IEEE.
81. Suganthi, S.T., Devaraj, D., Ramar, K., and Hosimin Thilagar, S. (2018). An improved differential evolution algorithm for congestion management in the presence of wind turbine generators. *Renew. Sustain. Energy Rev.* 81: 635–642.
82. Capitanescu, F. (2016). Critical review of recent advances and further developments needed in AC optimal power flow. *Electr. Power Syst. Res.* 136: 57–68.
83. Abido, M.A. (2002). Optimal power flow using particle swarm optimization. *Int. J. Electr. Power Energy Syst.* 24 (7): 563–571.
84. Abdi, H., Beigvand, S.D., and La Scala, M. (2017). A review of optimal power flow studies applied to smart grids and microgrids. *Renew. Sustain. Energy Rev.* 71: 742–766.
85. Chang, Y.C., Lee, T.Y., Chen, C.L., and Jan, R.M. (2014). Optimal power flow of a wind-thermal generation system. *Int. J. Electr. Power Energy Syst.* 55: 312–320.
86. Reddy, S.S., Bijwe, P.R., and Abhyankar, A.R. (2014). Faster evolutionary algorithm based optimal power flow using incremental variables. *Int. J. Electr. Power Energy Syst.* 54: 198–210.
87. Shaw, B., Mukherjee, V., and Ghoshal, S.P. (2014). Solution of reactive power dispatch of power systems by an opposition-based gravitational search algorithm. *Int. J. Electr. Power Energy Syst.* 55: 29–40.
88. Bouchekara, H.R.E.H. (2014). Optimal power flow using black-hole-based optimization approach. *Appl. Soft Comput. J.* 24: 879–888.
89. Bouchekara, H.R.E.H., Abido, M.A., and Boucherma, M. (2014). Optimal power flow using teaching-learning-based optimization technique. *Electr. Power Syst. Res.* 114: 49–59.
90. Kahourzade, S., Mahmoudi, A., and Bin Mokhlis, H. (2015). A comparative study of multi-objective optimal power flow based on particle swarm, evolutionary programming, and genetic algorithm. *Electr. Eng.* 97 (1): 1–12.
91. Expósito, A.G., Gomez-Exposito, A., Conejo, A.J., and Canizares, C. (2016). *Electric Energy Systems: Analysis and Operation*. CRC Press.
92. Weng, Y., Negi, R., Faloutsos, C., and Ilic, M.D. (2017). Robust data-driven state estimation for smart grid. *IEEE Trans. Smart Grid* 8 (4): 1956–1967.
93. Liu, X. and Li, Z. (2017). False data attacks against AC state estimation with incomplete network information. *IEEE Trans. Smart Grid* 8 (5): 2239–2248.
94. Zhao, J., Zhang, G., and Jabr, R.A. (2017). Robust detection of cyber attacks on state estimators using phasor measurements. *IEEE Trans. Power Syst.* 32 (3): 2468–2470.
95. Zhao, J., Zhang, G., Das, K. et al. (2016). Power system real-time monitoring by using PMU-based robust state estimation method. *IEEE Trans. Smart Grid* 7 (1): 300–309.

96. Asada, E.N., Garcia, A.V., and Romero, R. (2005). Identifying multiple interacting bad data in power system state estimation. In: *Power Engineering Society General Meeting, 2005*, vol. 1, 571–577. IEEE.
97. Naka, S., Genji, T., Yura, T., and Fukuyama, Y. (2003). A hybrid particle swarm optimization for distribution state estimation. *IEEE Trans. Power Syst.* 18 (1): 60–68.
98. Tungadio, D.H., Jordaan, J.A., and Siti, M.W. (2016). Power system state estimation solution using modified models of PSO algorithm: comparative study. *Meas. J. Int. Meas. Confed.* 92: 508–523.
99. Lezama, F., Soares, J., Vale, Z. et al. (2018). 2017 IEEE competition on modern heuristic optimizers for smart grid operation: testbeds and results. *Swarm Evol. Comput.* 44: 420–427.
100. Jin, Y. and Branke, J. (2005). Evolutionary optimization in uncertain environments – a survey. *IEEE Trans. Evol. Comput.* 9 (3): 303–317.
101. Soares, J., Canizes, B., Fotouhi Gazvhini, M.A. et al. (2017). Two-stage stochastic model using Benders' decomposition for large-scale energy resources management in smart grids. *IEEE Trans. Ind. Appl.* 53: 5905–5914.

CHAPTER 3

POWER SYSTEM PLANNING AND OPERATION

Sishaj Pulikottil Simon

National Institute of Technology Tiruchirappalli, Tamilnadu, India

3.1 INTRODUCTION

Power flow analyses are one of the fundamental studies in power systems. They are used in transmission system planning, solving the unit commitment and economic dispatch (ED) problems, computing system congestion, steady-state operating points, while also being extensively used in contingency planning. A simple power flow calculation provides a set of operational set-points, given a set of initial conditions. System operation can be optimized by using various optimization algorithms. This chapter provides implementation of various optimization algorithms to various power system problems that utilize power flow calculations. Section 3.1 focuses on transmission system expansion planning. It puts emphasis on improvements on the classical models starting from the classical constructive heuristic algorithms (CHA) to re-evaluate the currently used meta-heuristics in terms of performance and efficacy. Section 3.2 looks into advancements in unit commitment problem (UCP) formulations and models used in these formulations. Section 3.3 presents an economic dispatch problem. Solutions are found using genetic algorithm (GA) and particle swarm optimization (PSO) and compared. Section 3.4 provides an insight into application of differential evolution (DE) in power systems. The problem is formulated as active power optimal power dispatch. Section 3.5 provides a study case of hydrothermal coordination based on meta-heuristics. Section 3.6 illustrates the generator maintenance scheduling (GMS) problem and solves it using GA. Section 3.7 uses a hybrid PSO to solve load flow problem while Section 3.8 uses artificial bee colony (ABC) optimization for solving the optimal power flow (OPF) problem. Section 3.9 highlights an OPF

Applications of Modern Heuristic Optimization Methods in Power and Energy Systems, First Edition. Edited by Kwang Y. Lee and Zita A. Vale.
© 2020 by The Institute of Electrical and Electronics Engineers, Inc.
Published 2020 by John Wiley & Sons, Inc.

test bed and evaluates the performance of a meta-heuristic optimization (MHO) algorithm, mean–variance mapping optimization (MVMO), on a test case.

3.2 UNIT COMMITMENT

Narayana Prasad Padhy[1] *and Sishaj Pulikottil Simon*[2]

[1]*Indian Institute of Technology Roorkee, Roorkee, Uttarakhand, India*
[2]*National Institute of Technology, Tiruchirappalli, Tamilnadu, India*

3.2.1 Introduction

The consumption of electricity is dynamic in nature [1–4]. For example, electricity is consumed at a lower rate between midnight and early morning than during the day. Therefore, the power consumed by a load center varies from time to time. Also, power consumed by an industrial area, for example, on a Monday will be higher than the power consumed by the same area on Saturday and Sunday. Also, the power consumed by a residential area will be higher, for example, on a Sunday compared to power consumed by the same area on Monday. Therefore, the load pattern on a power-generating station varies from day to day and follows a weekly cycle [5]. If power-generating stations are to supply the maximum demand continuously, then they had to be "ON" all time. However, keeping more generating stations "ON" than the required number will not be economical for the power system. This makes the generator to operate at its minimum generating limit and will be inefficient. Therefore, the system operator needs to determine which units should be turned ON "Committed" and for how long. Determination of the schedule (ON/OFF status and amount of power generated) of generating units within a power system results in great saving for electric utilities [6, 7]. Therefore, the unit commitment problem is formulated in order to minimize the total operating cost, satisfying the system, the unit, and the operational constraints [8].

3.2.2 Problem Formulation

The unit commitment problem can be formulated to minimize the operational cost subject to constraints [9] such as power balance constraint, generator operating limit constraint, ramp rate constraints, spinning reserve (SR) constraints, minimum up time constraint, must run unit constraint, and must out unit constraint [7, 10]. The problem can be formulated as follows:

Minimize:

$$\mathrm{OC} = \sum_{i=1}^{N} \sum_{t=1}^{T} \mathrm{FC}_{it}(P_{it}) + \mathrm{MC}_{it}(P_{it}) + \mathrm{ST}_{it} + \mathrm{SD}_{it} \tag{3.2.1}$$

where

T is the number of schedule intervals,

N is the number of generators,

$FC_{it}(P_{it})$ is the fuel cost,

$MC_{it}(P_{it})$ is the maintenance cost,

ST_{it} is the startup cost,

SD_{it} is the shutdown cost of generator i at period t.

The fuel cost $FC_{it}(P_{it})$ is expressed as follows [11, 12]:

$$f_i(P_{it}) = a_i + b_iP_{it} + c_iP_{it}^2 \tag{3.2.2}$$

where a_i, b_i, and c_i are the cost coefficients.

The maintenance cost $MC_{it}(P_{it})$ is expressed as follows [1, 10]:

$$MC_{it}(P_{it}) = BM_{it} + IM_{it}*P_{it} \tag{3.2.3}$$

where

BM_{it} is the base maintenance cost,

IM_{it} is the incremental maintenance cost.

The startup cost ST_{it} is expressed as follows [1, 12]:

$$ST_{it} = TS_{it} + \left(1 - e^{\left(\frac{Dow_{it}}{BC_{it}}\right)}\right)BS_{it} + MS_{it} \tag{3.2.4}$$

where

TS_{it}: turbine startup cost,

BS_{it}: boiler startup cost,

MS_{it}: startup maintenance cost,

Dow_{it}: number of hours down,

BC_{it}: boiler cool down coefficient.

The shutdown cost is expressed as follows:

$$SD_{it} = KP_{it} \tag{3.2.5}$$

where K is the incremental shutdown cost.

Subject to:

A. Power balance constraint: The sum of power generated by all committed generating units should be equal to the sum of demand (D_t) and losses (PL_t) at any instance (t). This is known as the power balance constraint and expressed as in Eq. (3.2.6):

$$\sum_{i=1}^{N} U_{it}P_{it} = D_t + PL_t \tag{3.2.6}$$

$$\text{where } U_{it} = \begin{cases} 1 & \text{if unit } i \text{ is "on" at instance "}t\text{"} \\ 0 & \text{if unit } i \text{ is "off" at instance "}t\text{"} \end{cases}$$

B. Generator operating limit constraint: Equipments in a generating unit "i" is designed to generate a maximum real power ($P\text{max}_i$). Also, there is a minimum amount of power ($P\text{min}_i$) that should be generated by a generating unit, if it is turned on. Therefore, a generating unit can only be operated within a range, for which the generating unit is designed. This generator operating limit constraint is expressed as in Eq. (3.2.7).

$$P\min_i \le P_{it} \le P\text{max}_i \text{ for } i = 1, 2, \ldots, N \tag{3.2.7}$$

C. Ramp rate constraints: In a generating unit the pressure and temperature within equipments, such as boiler and turbine, should not vary drastically [13]. These mechanical constraints are translated into electrical constraints, known as the ramp rate constraints. Ramp rate constraints limit the change in power generated by the generating unit within permissible values which will ensure safety of the equipments.

$$P\min_{it} = \max\left\{P\min_i, P_{i(t-1)} - \text{Ud}_i\right\} \tag{3.2.8}$$

$$P\max_{it} = \min\left\{P\max_i, P_{i(t-1)} - \text{Up}_i\right\} \tag{3.2.9}$$

where Up_i and Ud_i are ramp up and ramp down limits, respectively.

D. Spinning reserve: SR requirements are necessary in the operation of a power system if load interruption is to be minimal [14]. This necessity is due partly to certain outages of equipment. SR requirements may be specified in terms of excess megawatt capacity or some form of reliability measures.

E. Crew constraints: Certain plants may have limited crew size which prohibits the simultaneous starting up and/or shutting down of two or more units at the same plant. Such constraints would be specified by the times required to bring a unit online and to shut down the unit.

F. Minimum up time: A unit must be on for certain number of hours before it can be shut down.

G. Minimum down time: A unit must be off for certain number of hours before it can be brought online.

H. Must run units: These units include prescheduled units which must be online, due to operating reliability and/or economic considerations.

I. Must out units: Units which are on forced outages and maintenance are unavailable for commitment.

3.2.3 Advancement in UCP Formulations and Models

The fast-changing scenarios of the power industry require solution methodologies used to solve the UCP to be more efficient. Factors such as competition, customer satisfaction, power quality, increase in energy needs, increase in renewable energy generation, reliability, security, and environmental constraints necessitate continuous improvement of the methodologies. The current generation UCP models are those that incorporate the above requirements, which are listed below.

Profit-based Models

In a deregulated environment, the electricity is a commodity and therefore, the power system utilities try to maximize the profit. Here, the objective is maximizing profit rather than minimizing fuel cost [1, 15, 16]. The objective function in profit-based models is expressed below:

$$\text{maximize profit} = \sum_{i=1}^{N}\sum_{t=1}^{T} P_{it} f_{pt} U_{it} - \text{OC} \tag{3.2.10}$$

where f_{pt} is the forecasted price for period t.

The constraints for the conventional UCP model are applicable for the profit-based models as well. However, the power balance constraint is modified as follows:

$$\sum_{i=1}^{N} U_{it} P_{it} \geq D_t + \text{PL}_t \tag{3.2.11}$$

Network Constrained Models

The power transfer limit of overhead transmission lines (OTLs) is an important constraint for power systems' planning and operation [17]. This constraint plays an essential role in the secure and economic management of power systems. Moreover, the system constraints such as reactive power injection limits, voltage magnitude limits, and voltage angle limits are also incorporated into an UCP. Here, the conventional UCP is extended by incorporating system constraints in parallel with the conventional constraints of the UCP.

1. Inequality constraint on reactive power generation Qg_{it} at each PV bus:

$$\text{Qg min}_i \leq \text{Qg}_{it} \leq \text{Qg max}_i \tag{3.2.12}$$

   where Qg min_i and Qg max_i are, respectively, minimum and maximum value of reactive power at PV bus i.

2. Inequality constraint on voltage magnitude V_{it} of each PQ bus [18]:

$$\text{VI min}_i \leq \text{VI}_{it} \leq \text{VI max}_i \tag{3.2.13}$$

   where VI min_i and VI max_i are, respectively, minimum and maximum voltage magnitudes allowed at bus i.

3. Inequality constraint on phase angle ϕ_{it} at each bus:

$$\phi\text{min}_i \leq \phi_{it} \leq \phi\text{max}_i \quad \text{for } i = 1, 2, \ldots, N \tag{3.2.14}$$

   where ϕmin_i and ϕmax_i are, respectively, minimum and maximum voltage angles allowed at bus i.

4. MVA flow limit on transmission line [18]:

$$\mathrm{MVA}_{ij} \leq \mathrm{MVA\,max}_{ij} \text{ for } i = 1, 2, \ldots, N \tag{3.2.15}$$

where

MVA_{ij} is the power transferred between transmission line connecting bus "i" and "j,"

$\mathrm{MVA\,max}_{ij}$ is the maximum rating of the transmission line.

Multi-objective Models

Emission Constraint Model Increased awareness of limiting environmental pollution caused by thermal power plants because of O_2, SO_x, and NO_x emissions is a crucial issue faced in the UCP [19]. Owing to these emissions, destruction is caused to nature and its life forms [20]. Strategies to reduce the atmospheric emissions are the installation of pollutant cleaning equipment, switching to low emission fuels, and replacement of the aged fuel-burners and generator units.

The first three options require installation of new equipment and/or modification of existing equipment, which involves considerable capital outlay and, hence, can be considered as long-term options. The emission dispatching option is an attractive short-term alternative in which the emission, in addition to the fuel cost objective, is to be minimized. Thus, the UCP is handled as a multi-objective optimization problem with two objectives, namely the operation cost minimization and the emission cost minimization [21]. In recent years, this option has received much attention [22, 23]. The objective function to minimize the emission cost [24] is

$$E(P_{it}) = \sum_{t=1}^{T}\sum_{i=1}^{N} 10^{-2}\left(\alpha_i + \beta_i P_{it} + \gamma_i {P_{it}}^2\right) + \varepsilon_i e^{(\lambda_i P_{it})} \tag{3.2.16}$$

where α_i, β_i, γ_i, ε_i, and λ_i are emission coefficients of the ith generator emission characteristics.

Security Constraint Model In recent years, reliability of power systems is a major focus of the system operators [25, 26]. The increase in reliability level will increase the reserve cost (RSC) and thereby a proportional increase in the total system cost is observed. Here, fixing the limits of reliability level is again conflicting in nature in order to obtain a trade-off solution between cost and reliability level. Therefore, the UCP is handled as multi-objective optimization problem with two objectives, namely the operation cost minimization and RSC minimization, without violating the SR constraint. The objective function to maximize reliability is given in Eq. (3.2.17) [27]:

$$\sum_{t=1}^{T}\sum_{i=1}^{N}[R_{it}\sigma_{it}] + \sum_{d=1}^{\mathrm{ND}}[\mathrm{DR}_d \sigma_{\mathrm{drr}}] \tag{3.2.17}$$

where

R_{it} is the SR provided by the generating unit,

σ_{it} is the rate of SR,

DR_d is the amount of power obtained by load curtailment through load curtailment at bus d,

σ_{drr} is the rate of power obtained through demand response (DR).

The SR constraint is formulated in Eq. (3.2.18). Here, SR_t is the spinning reserve requirement at instance t.

$$\sum_{i=1}^{N}(P\max_i I_{it}) \geq D_t + PL_t + SR_t \qquad (3.2.18)$$

SR requirement in Eqs. (3.2.17 and 3.2.18) is assessed according to the desired level of reliability. Therefore, SR requirement should satisfy either one of the reliability constraints, namely the loss of load probability (LOLP) and expected energy not served (EENS). These indices at any load center "*d*" at any instance "*t*" should be less than the prespecified value as defined in the following equations:

$$\text{LOLP} \leq \text{LOLP}_{td}^{\text{spec}} \qquad (3.2.19)$$

$$\text{EENS} \leq \text{EENS}_{td}^{\text{spec}} \qquad (3.2.20)$$

UCP Model Considering Security and Emission

UCP is solved considering three objective functions, namely operation cost minimization, emission cost minimization, and RSC minimization subject to all constraints including the reliability constraints.

Combined Power Generation Models

The ever-increasing electric power demand has forced the modern power system utilities to extract more amount of available renewable and nonrenewable sources of energy. The sources of mechanical power, commonly known as the prime mover, may be hydraulic turbines at waterfalls, steam turbines whose energy comes from the burning of coal and nuclear fuel, gas turbines, or occasionally internal combustion engines burning oil [27, 28]. With today's emphasis on environmental consideration and conservation of fossil fuels, many alternate energy sources are being used to some extent such as solar power, geothermal power, wind power, tidal power, and biomass. Therefore, commitment scheduling has to be done with combined power generation models such as hydrothermal [29, 30], thermal-nuclear [31], thermal-wind [32, 33], etc. Even though there are advances in technologies, abundance of free resources, falling capital cost, and commercial viability of using alternate energy sources such as wind, hydro, tidal, solar, etc., are intermittent and unpredictable and may cause serious threat to power system security. Hence, considering the changes in the generation mix and market operation of power systems, commitment scheduling of generating units has to be done with higher priority on system security and minimizing the operation cost. Objective function considering thermal-wind coordination is expressed in Eq. (3.2.21):

$$\text{minimize}\left\{\sum_{i=1}^{N}\sum_{t=1}^{T}\text{FC}_{it}(P_{it}) + \text{MC}_{it}(P_{it}) + \text{ST}_{it} + \text{SD}_{it} + \sum_{j=1}^{NW} W_j P_{Wj}\right\} \qquad (3.2.21)$$

where

NW is the number of wind turbines,

P_{Wj} is power generated by wind power-generating unit "j,"

W_j is the feed in tariff for wind-power generating unit "j."

UCP in Smart-Grid and Deregulated Environment

As the amount of renewable power generation increases around the world, impacts on power system reliability and cost have increased. Among the renewable resources, wind and solar energy converters have been recognized as the most promising means of electric power generation in future. Several countries across the world have understood the importance and started the integration of renewable resources with conventional power utilities. In addition, in deregulated environment there is provision for incentives to customers designed to induce lower electricity use at times of high market prices or when the system reliability is jeopardized. DR refers to actions taken by the system operator to respond to a shortage of supply for a short duration of time in the future. SR obtained through DR together with the reserve generated by the online committed units helps in satisfying the system reliability in an efficient manner. In recent days, UCP is modeled to integrate the various distributed energy sources, energy storage equipment, and DR [26].

Hybrid Models

Hybrid models employ more than one technique to solve the problem in hand. In this hybrid system, one technique can complement the other to enhance its performance and accuracy. They can be classified into sequential, auxiliary, and embedded hybrid systems [34]. Sequential hybrid systems make use of techniques in a pipeline-like fashion. Thus, one technique's output becomes another's input, and so on. In [35], a hybrid dynamic programming (DP)-artificial neural network (ANN) is used to solve UCP in a two-step process. First, an ANN is used to generate a preschedule according to the input load profile. Then, a dynamic search is performed at those stages where the commitment states are not certain. Auxiliary hybrid system calls the other as "subroutine" to process or manipulate information needed by it. In references [36, 37], PSO is combined with the language relaxation (LR) method to optimally set the Lagrange multipliers (i.e. LR employs a PSO to optimize its structural parameters) in solving the UCP. In embedded hybrid systems, the participating technologies are integrated in such a manner that they appear entwined.

3.2.4 Solution Methodologies, State-of-the-Art, History, and Evolution

Various approaches ranging from simple rule of thumb to highly complicated methods have been developed to solve the optimal UCP. The development of UCP and its solution methodologies dates back to 1960s [1]. Later, with the

TABLE 3.2.1 Evolution of UCP Solution Methodology

Year	Method	Year	Method
1963	Integer programming [38]	1990	Artificial neural networks [46]
1966	Exhaustive enumeration [39]	1990	Simulated annealing [47]
1966	Dynamic programming [40]	1991	Heuristic expert system-based approaches [48]
1973	Lagrangian relaxation [41]	1996	Genetic algorithm [49]
1979	Linear programming [42]	1996	Evolutionary programming [50]
1980	Priority list method [43]	1997	Fuzzy systems [51]
1982	Branch and bound method [44]	2002	Ant colony search algorithm [52]
1987	Expert systems [45]	2014	Fire fly algorithm [25, 26]
		Hybrid models	
1985	Dynamic programming with successive approximation [53]	1999	Integrated genetic algorithm, tabu search, and simulated annealing [55]
1991	Fuzzy-based dynamic programming [54]	2001	ANN-dynamic programming [56]

improvement and increased penetration of optimization, soft computing, and computation technology, the problem formulation and solution methodologies have faced drastic improvement in the past few decades. The improvement in the UCP formulation and solution methodologies are still valid research problems to obtain more optimal solutions in lesser time than the predecessors. There are so many solution methodologies adopted to solve the UCP. The evolution of UCP solution methodologies is given in Table 3.2.1, and certain notable methodologies are discussed as follows.

Exhaustive Enumeration Method

In the exhaustive enumeration method, all possible combinations of generating unit status are enumerated and the combination which yields the least operation cost is selected as the solution of the UCP. The method provides an accurate solution for small power systems. However, the method is not suitable for large power systems due to large computational burden.

Priority List and Dynamic Programming Methods

In this method, the full load average production cost for each unit is calculated. A priority list is prepared in which the unit having the least full load average production cost is given the first priority and the priority reduces as the full load average production cost increases. The solution obtained from the priority list arranged in order of the full load average cost will result in optimum dispatch and commitment only if zero no load costs is assumed, input–output characteristics of all the generating units are linear between zero output and full load output, and startup costs are fixed. However, in DP, it systematically evaluates a large number of possible decisions in a multistep problem. A subset of possible decisions is associated with each sequential problem step and a single one must be selected, i.e. a single decision must be made in each problem step and a cost is associated with each possible decision. The objective is to make a decision in each problem step, which minimizes the total cost for all the decisions made. Though DP approaches are flexible enough to modify and reduce the dimensionality when compared to the method of complete enumeration, it still suffers from the "curse of dimensionality" as the size of the problem increases.

Nonconventional Heuristic-based Approaches

Since the UCP is a hard combinatorial in nature, more computational requirement is needed to solve it. The minimum total generation cost for the UCP in a reasonable time is very crucial. The cost-efficient scheduling has to be found at least for a time horizon of 24 hours (1 day). If the N number of generators is present in a system, then the number of combinations available for a schedule interval is $2^N - 1$. The number of combinations available to schedule for T schedule intervals are $(2^N - 1)^T$. For instance, consider a system with two generating units and the schedule is to be obtained for a time horizon of three hours with a schedule interval of one hour. The total number of solution (paths) to be enumerated in order to obtain a feasible solution is 27 ($[2^2 - 1]^3$). The paths to be enumerated are shown in Figure 3.2.1. Therefore, if 10 generating units are available and the number of schedule interval is 24, then the number of available paths is 1.725 898 332 285 254 $\times 10^{72}$.

Therefore, finding the minimum cost path out of all above combinations by satisfying all the required constraints is a challenging task as the solution space is very large. The exact solution of the UCP can be obtained by complete enumeration, or multistage dynamic or integer programming (IP) methods. However, they require enormous computational time, which increases exponentially with the number of generating units. Also, the memory requirement is very large,

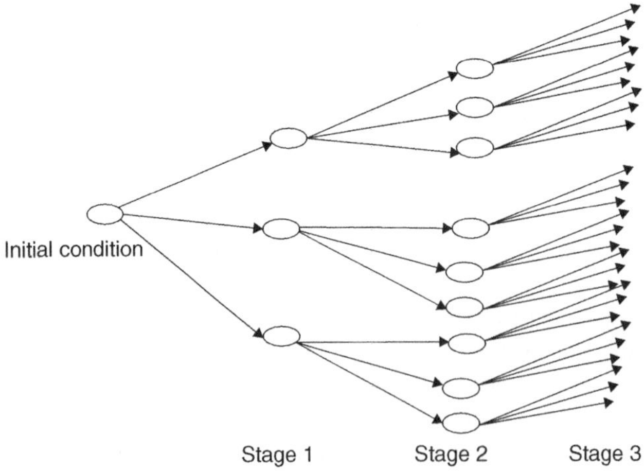

Figure 3.2.1 UCP solution paths.

which the computational equipment cannot handle. Therefore, heuristic algorithms such as single solution-based hill climbing algorithm and population-based methods including GA and swarm intelligence methods are also used in solving the UCP. These heuristic methods cannot guarantee the global optimum solution. However, a near-optimal solution is always guaranteed. Though, the cost function of thermal power-generating unit is conventionally assumed to be linear, it is nonlinear, non-convex, and non-smooth due to multiple steam admission valves in the turbine. The nonlinear cost function of a generating unit is given in Eq. (3.2.22):

$$f_i(P_{i,t}) = a_i + b_iP_{i,t} + c_iP_{i,t}^2 + |e_i \sin(f_i(P_{\min,i} - P_{i,t}))| \tag{3.2.22}$$

where a_i, b_i, c_i, e_i, and f_i are the cost coefficients of the generator i. Heuristic algorithms such as the GA and soft computing techniques such as fuzzy logic and ANN are applied widely to solve nonlinear problems with large solution space. A generic solution approach is followed while solving the UCP using the heuristic soft computing techniques is outlined as a flowchart in Figure 3.2.2. Here, initially a set of probable solutions are generated randomly. Each solution represents the on/off status of generating units in all schedule intervals. The number of initially

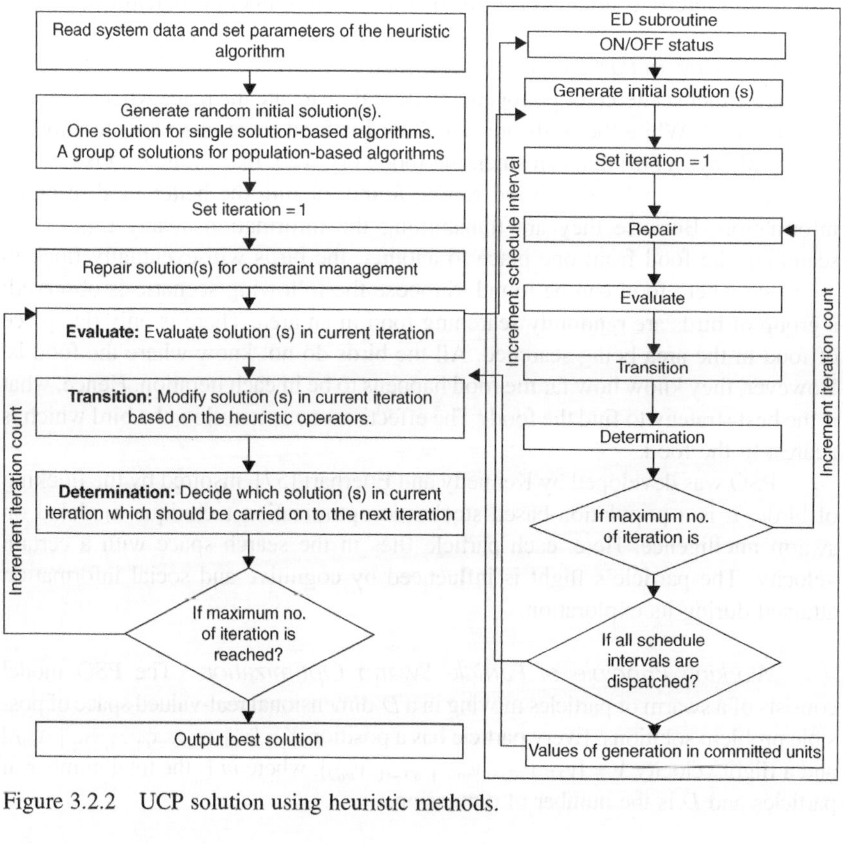

Figure 3.2.2 UCP solution using heuristic methods.

generated probable solutions is based on the heuristic algorithm employed. If a single solution-based algorithm like hill climbing algorithm or the simulated annealing (SA) is used, then one initial solution is generated randomly.

If a population-based algorithm is used, then a set of probable solutions are generated randomly. Randomly generated solution(s) may not satisfy the constraints of the problem. Therefore, the solutions are tailored to satisfy the constraints. The objective functions (say, the operational cost) of each probable solution are obtained and their fitness values are calculated. The solutions in the current iterations are modified by the heuristic operator used in the heuristic algorithm (for example, mutation and crossover for GA and update of velocity and position for swarm-based algorithm). This step is generically termed as transition. Thereafter, the determination is carried out. This step determines if the modification carried out by the transition operator may be accepted or not. Here, usually, the modifications for certain probable solutions are retained and certain probable solutions are ignored. This determination is generally based on improvement in the solution resulted by the transition. The evaluation, transition, and determination are carried out till a preset maximum number of iteration is reached.

UCP Solution Using Particle Swarm Optimization

This section presents an approach to solve unit commitment problem using PSO.

Overview of Particle Swarm Optimization While searching for food, the birds are either scattered or go together before they locate the place where they can find the food. While the birds are searching for food from one place to another, there is always a bird that can smell the food very well, that is, the bird is perceptible of the place where the food can be found, having the better food resource information. Because they are transmitting the information at any time while searching the food from one place to another, the birds will eventually flock to the place where food can be found. Suppose the following scenario is observed: a group of birds are randomly searching food in an area. There is only one piece of food in the area being searched. All the birds do not know where the food is. However, they know how far the food happens to be in each iteration. Hence, what is the best strategy to find the food? The effective one is to follow the bird which is nearest to the food.

PSO was developed by Kennedy and Eberhart [57], inspired by the flocking of birds. It is a population-based stochastic optimization technique modeled on swarm intelligence. Here, each particle flies in the search space with a certain velocity. The particle's flight is influenced by cognitive and social information attained during its exploration.

Working Principle of Particle Swarm Optimization The PSO model consists of a swarm of particles moving in a D-dimensional real-valued space of possible problem solutions. Every particle has a position $X = [x_{11}\ x_{12} \ldots x_{m-1,\,D-1}\ x_{mD}]$ and a flight velocity $V = [v_{11}\ v_{12} \ldots v_{m-1,\,D-1}\ v_{mD}]$, where m is the total number of particles and D is the number of dimension.

At each iteration t, the velocity is updated and the particle is moved to a new position. This new position is simply calculated as the sum of the previous position and the new velocity:

$$X_{pq} = X_{pq} + V_{pq} \tag{3.2.23}$$

The update of the velocity from the previous velocity to the new velocity is determined by

$$V_{pq} = WV_{pq} + C_1R_1\left(XPb_{pq} - X_{pq}\right) + C_2R_2\left(XGb - X_{pq}\right) \tag{3.2.24}$$

where

- XGb is the best position known for all particles,
- XPb is the best position for the given particle,
- W is known as the inertia weight,
- R_1 and R_2 are random numbers.

The step-by-step procedure to solve a general optimization problem using PSO is given below.

1. Set the number of particles (M), cognitive scaling factor (C_1), social scaling factor (C_2), initial inertia (WI), final inertia (WF), maximum number of iterations (MaxIt), and iteration count $k = 1$.
2. Initialize: Randomly generate M number of initial particles. The position of each of the particles is represented by $X = [x_{11}\ x_{12} \ldots x_{m-1,\,D-1}\ x_{mD}]$. Initialize the velocity of each of the particles as zero, $V = [0\ 0\ 0\ \ldots\ 0_M]$.
3. Evaluate: Find the fitness value of each particle.
4. Update the fitness of local best particle and the global best particle using Eqs. (3.2.25) and (3.2.26), respectively.

$$\text{Pbfit}_p^k = \min\left\{\text{Pbfit}_p^{k-1}, \text{fit}_p^k\right\} \tag{3.2.25}$$

$$\text{Gbfit} = \min\left\{\text{Pbfit}_i^k, \ldots, \text{Pbfit}_p^k, \ldots, \text{Pbfit}_{Np}^k\right\} \tag{3.2.26}$$

5. Update the local best position of particles (Pbl_p) and the global best particle (Gb).
6. Transition: For each of the particle, update velocity and position using Eqs. (3.2.27) and (3.2.28), respectively.

$$V_{pq} = WV_{pq} + C_1R_1\left(XPb_{pq} - X_{pq}\right) + C_2R_2\left(XGb_{pq} - X_{pq}\right) \tag{3.2.27}$$

$$X_{pq} = X_{pq} + V_{pq} \tag{3.2.28}$$

7. If the iteration count k = MaxIt (maximum number of iterations), then terminate the search. Else, go to Step 3.

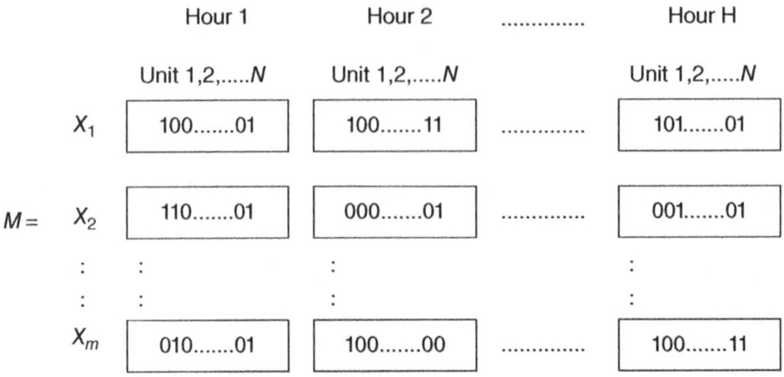

Figure 3.2.3 Initial generation of population.

Implementation of PSO to Solve UCP In UCP, binary numbers 1 and 0 are used to indicate the unit status ON/OFF. Therefore, the relevant variables are interpreted in terms of changes of probabilities, so that the X values will take either 0 or 1 [63]. The step-by-step procedure in solving UCP using PSO is given below.

1. Set the number of particles (N), cognitive scaling factor (C_1), social scaling factor (C_2), initial inertia (WI), final inertia (WF), maximum number of iterations (MaxIt), and iteration count (k).
2. Initialize: Randomly generate particles of M initial particle position represented by a binary string. Initialize randomly an initial position $M = [X_1, X_2, X_3, \ldots, X_m]$ of m solutions in the multidimensional solution space where m represents the total number of particles. Each solution of X is represented by the D-dimensional vector. Here, D is equal to $N*H$. The particles of M initial solution with D-dimensional vector are given in Figure 3.2.3.
3. Evaluate: Find the fitness value of each particle. Calculate the generation cost added together with its shutdown, startup cost, and ramping costs. Generally, the fitness of the particle is inversely proportional to this cost.
4. Update the fitness and position of local best particle and the global best particle.
5. Calculate the velocity of each particle (V_{pq}) for current iteration.
6. Transition: The velocity V_{pq} will be compared with a probability threshold. If V_{pq} is higher than the threshold, the individual (X_{pq}) is more likely to choose 1, and lower values favor selection of 0 choice. The probability threshold is always chosen between 0 and 1.
7. If the iteration count k = MaxIt (maximum number of iterations), then terminate the search. Else, go to Step 3.

ED Subproblem For each of the probable solution, the objective function (operating cost) is evaluated by solving ED problem at each schedule interval. This evaluation is based on the ON/OFF status of the probable solution given by the master problem. The same steps in the master problem (generation, evaluation, transition, and determination) are carried out in the ED subproblem to obtain optimum dispatch for each schedule interval. However, the solution representation, solution space, and set of constraints are different for the master and the slave programs.

Constraint Management and Repair Strategy When an initial solution is randomly generated or if modified due to transition, there is no guarantee that all constraints will be satisfied. Therefore, all constraints should be checked at each iteration in the algorithm. Few constraint management strategies are discussed below.

1. **For unit commitment master problem:** Whenever the probable solution representing the commitment status for each time interval is generated randomly or modified due to transition, violation of minimum up/down time constraints and SR constraints has to be checked as follows. The randomly generated commitment status for each time interval is checked for the violation of minimum up/down time constraints and reliability constraints.

   Step 1: If the reliability level is met, then go to Step 4. Otherwise, go to next step.

   Step 2: Add DR reserve in addition to the reserve generated by the online committed unit to satisfy the reliability level. If the reliability level is not met, then set DR = 0 and then go to next step. Otherwise, go to Step 4.

   Step 3: The less expensive units which are in the OFF state are identified and turned ON. Then, go to Step 1.

   Step 4: If the reliability constraint is satisfied, then the minimum up and down time constraints are checked for each unit over the scheduling horizon in each interval. If there is any violation in the minimum up or down time constraint, then the repair mechanism is used to overcome the violation. For instance, let us assume that the Ton and Toff for a hypothetical unit is 4 and 5. For a scheduling interval of 12 hours, if the actual off time for unit 1 is 3 hours (5th–7th hour), then it violates the Toff constraint. In this case, the unit status before 5th hour or after 7th hour can be made 0. By doing this change, if it violates the Ton constraint, then the status of the units are made 1 during the violated down time period.

2. **For economic dispatch subproblem:** Power balance mismatch (PBM) is the difference between the sum of power generated by all generators and the sum of load demand and transmission loss. If the power balance constraint is violated (PBM > Tol) while solving ED subproblem, then power generated by each generating unit is modified to make the PBM less than the specified tolerance. Here, the PBM is equally shared between all the

generating units. For example, let $N = 3$, probable solution $X = [47\ 50\ 60]$, and the power demand $D = 150$ MW and losses corresponding to the probable solution PL = 1 MW. Then, the PBM will be equal to 6 MW. Therefore, the position of the probable solution is varied as = [45 48 58]. If generator operating limit constraints are violated, then the generator which violates the limit will be brought back to its corresponding limit. For example, if the minimum limit on the first generator is 46 MW, then the position of probable solution is modified as [46 48 58]. Modification of generating unit output carried on satisfying one constraint leads to violation of the other constraint. In order to satisfy all the constraints simultaneously, the above corrections are repeated for a preset number of iterations. Here, the generating unit which undergoes correction for operating limit constraints is restrained from further PBM sharing. Once the repair strategy is completed, the probable solutions which violate constraints will be penalized during evaluation.

Industrial Practices and Software Packages

If a utility has the responsibility for satisfying the electricity demand, then its most important short-term constraint is that the total generation must equal the forecast demand in each time interval. Therefore, security-constrained unit commitment (SCUP) problem is solved by the utility to obtain the schedule. On the other hand, prime objective of a generating company (GENCO) in a deregulated electricity market may be profit maximization. It may not need to match its demands exactly, but it will have to decide at what prices to offer its generating plant to the market at different times of the day.

The role of the independent system operator (ISO) in a deregulated electricity market is to coordinate various GENCOs and distribution companies (DISCOs) connected to the market. One of the prime roles of the ISO is to prepare day-ahead generation schedules. The steps involved in obtaining the schedule are given below.

1. GENCOs obtain the forecasted demand or the forecasted price from the ISO.
2. GENCO schedules its generators by solving profit based unit commitment (PBUCP) problem to obtain maximum profit and submit bids to the ISO.
3. The ISO solves security constrained unit commitment problem (SCUCP) considering generation requirements, reserve requirements, and transmission constraints along with the generation bids received from various GENCOs and load bids received from DISCOs. Based on the SCUCP solution, the ISO sets market clearing price (MCP) and approves/rejects bids. The MCP along with the approved load and generation dispatch are sent to the GENCOs.
4. The GENCO revises its bids by rescheduling its generators by solving the UCP considering demand and profit.

Steps 3 and 4 are repeated till a specified time on the scheduling day till the ISO and GENCOs come to a consensus [64]. Any deviation in the schedule in

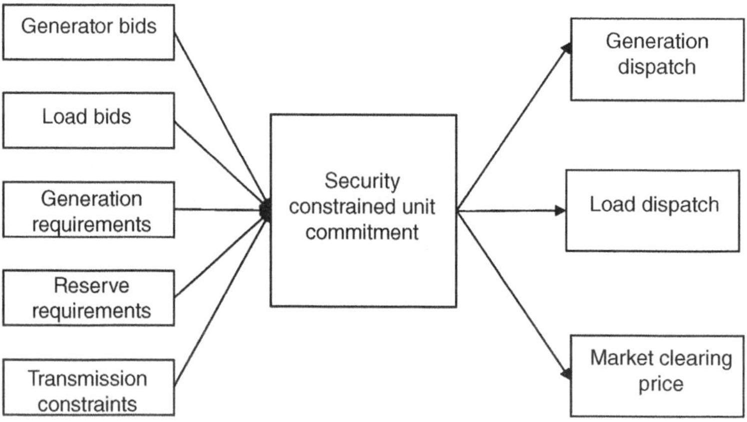

Figure 3.2.4 ISO activities.

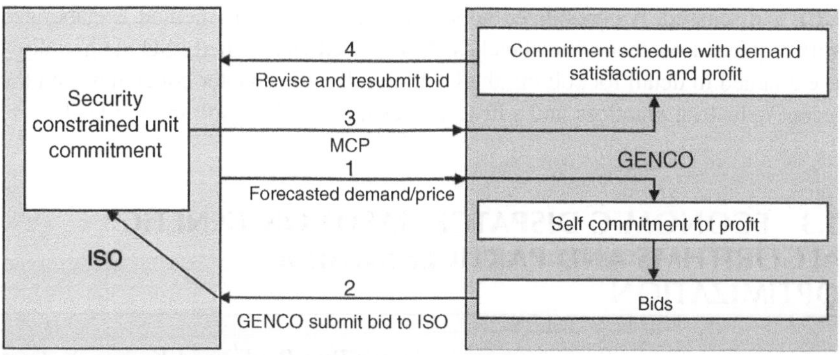

Figure 3.2.5 Co-ordination between ISO and GENCO.

terms of power generation and consumption is corrected by appropriate online control mechanisms such as automatic generation control (AGC), load shedding, DR, and online UCP in a real-time energy market. These online correcting mechanisms are costlier compared to day-ahead scheduling. Prime roles of ISO and the coordinating activities of ISO and GENCOs are shown in Figures 3.2.4 and 3.2.5.

Different commercially available softwares are used to obtain UCP schedule by power system utilities. Few of them are listed below.

Powerop™: Powerop is based on a proprietary solution algorithm developed by Power Optimisation, which is a multistage version of the mixed integer linear programming (MILP) method. Powerop are used by power system operators to schedule and dispatch the generating units in their power systems, to produce generation schedules that take into account all the important constraints on the power system and on the generating units. One of the notable clients using Powerop is the British Electricity Trading and Transmission Arrangements Limited in the British electricity market.

Siemens Spectrum Power™ SCUC: Siemens Spectrum Power SCUC solves a security-constrained unit commitment problem to obtain generation schedules. It is also used to examine the physical constraints on the grid and send electricity across the most efficient pathways.

PLEXOS™: PLEXOS is a simulation software that can be used for scheduling in electricity, gas, and oil industries to obtain schedules. The software has the option to solve problems such as OPF problem, hydrothermal coordination, gas electricity coordination, and maintenance optimization along with conventional UCP.

3.2.5 Conclusions

This chapter discusses the UCP in traditional, deregulated, and smart-grid environment. The objectives and constraints are presented suitably in the above scenarios. The various solution methodologies, state-of-the-art, history, and evolution of the UCP are presented. Both conventional and nonconventional way of solving the UCP is discussed. A generalized population-based heuristic method is elaborated with required flowcharts and sketches. Also, a well-established PSO methodology is explained in detail for solving the UCP. Finally, the chapter concludes with the recent industrial practices and software packages used for solving the UCP.

3.3 ECONOMIC DISPATCH BASED ON GENETIC ALGORITHMS AND PARTICLE SWARM OPTIMIZATION

Jong-Bae Park[1] *and Kwang Y. Lee*[2]

[1]*Konkuk University, Seoul, Korea*

[2]*Baylor University, Waco, TX, USA*

3.3.1 Introduction

Most power system operation and planning problems have complex nonlinear characteristics and heavy constraints to tackle. To formulate the problems well and find their solution(s), numerous marked mathematical methods have been suggested and successfully applied during the past few decades. However, there still exist some unsolved issues such as finding the global optimum solution. Therefore, alternative solution techniques to the problem including artificial intelligence applications to optimization problems have been attracting much interest. In recent years, GAs and PSO have been considered for a wide range of application in complex power system analysis. The theoretical foundation of GAs was established by the development of Holland [58] and De Jong [59]. GAs can provide an alternative solution to adaptive search techniques and they have proven to be robust and effective over a broad range of problems [58–62].

Classical GAs and their refinements have been successfully applied to various power system optimization problems such as ED [63–65], power system planning [66–69], and others [70, 74, 75]. However, longer computation time to provide good solution(s) is needed in traditional GA-based frameworks than in conventional analytical approaches. Also, they have a limitation that they cannot always converge to the global optimal solution in some complex problems [60, 71]. To overcome these problems inherent in traditional GAs to some extent, improvements over the conventional GAs are suggested in many papers to effectively solve an ED problem [63, 78, 81–84].

Recently, Kennedy and Eberhart suggested a PSO based on the analogy of swarm of bird and school of fish [57]. The PSO mimics the behavior of individuals in a swarm to maximize the survival of the species. In PSO, each individual makes his decision using his own experience together with other individuals' experiences [86]. The algorithm, which is based on a metaphor of social interaction, searches a space by adjusting the trajectories of moving points in a multidimensional space. The individual particles are drawn stochastically toward the position of present velocity of each individual, their own previous best performance, and the best previous performance of their neighbors [87]. The main advantages of the PSO algorithm are summarized as: simple concept, easier implementation, robustness to control parameters, and computational efficiency when compared with mathematical algorithm and other heuristic optimization techniques.

Recently, PSO has been successfully applied to various fields of power system optimization such as power system stabilizer design [88], reactive power and voltage control [86], and dynamic security border identification [89]. The original PSO mechanism is directly applicable to the problems with continuous domain and without any constraints. Therefore, it is necessary to revise the original PSO to reflect the equality/inequality constraints of the variables in the process of modifying each individual's search. After a modified PSO (MPSO) was published to solve non-smooth ED problems, many researcher efforts have been made to improve these problems. Almost all of the improved PSO replace a part of MPSO algorithms or add helpful idea from other theories. In this study, an alternative approach is described to the non-smooth ED problem using a MPSO [95] and an improved PSO from MPSO, which focuses on the treatment of the equality and inequality constraints when modifying each individual's search and compared with penalty attractive and repulsive PSO (PARPSO) [98] and GA.

This chapter is composed of the following sections: In Section 3.3.2, we describe general explanation of fundamental GAs and PSO. In Section 3.3.3, the ED problem formulation is described with its objective function and constraints. GA implementations to the ED problem are discussed in terms of, encoding method, constraints treatment, improved genetic operators, and advanced fitness function evaluation in Section 3.3.4 and PSO implementations to ED problem are described focusing how to treat the constraints and to improve the convergence speed in Section 3.3.5. Finally, in Section 3.3.6, GAs and PSO are tested on a sample power system and their results are compared with each other.

3.3.2 Fundamentals of Genetic Algorithms and Particle Swarm Optimization

Fundamentals of Simple Genetic Algorithm (SGA)

GAs, first introduced by Holland in 1975 [58] and systematically improved by De Jong [59], have become one of the important tools for practical and robust optimization methods during the past decades [60, 61, 72]. GAs are search or optimization methods based on the conjecture of biological evolutionary hypothesis. They are probabilistic algorithms that start with a population of randomly generated candidates and evolve toward better solutions by applying genetic operations (i.e. selection, crossover, and mutation), modeled on the genetic processes occurring in nature.

The SGA of Holland has become the substantial fundamentals to the advanced and improved GAs that followed for practical applications. In this section, we will briefly discuss on the SGA mechanisms focused on encoding mechanism, fitness function, selection, crossover and mutation, and generation cycle [72].

The encoding mechanism in SGA is based on the binary strings where the variables are treated as discrete integer or fixed-point real numbers [60, 72, 73]. The fitness value of each string in SGA is originally based on the value of objective function. Since the ranges of fitness value in optimization problems can be varied widely, the normalized values are normally used. The selection method in SGA, which mimics the nature's survival-of-the-fittest mechanism, is based on the roulette wheel selection scheme as the proportionate selection scheme. The primary genetic operator in GA is the crossover operator, which combines the features of two parent structures to form two similar offspring. In SGA, first two strings are selected randomly but proportional to their fitness values in the mating pool under the probability of crossover, and then an arbitrary crossover position along the two strings is selected, beyond which the crossover takes place. This kind of crossover is called as the one-point crossover which is used in SGA mechanism.

Mutation is an occasional alteration of a value at a particular position in the string under the probability of mutation. In SGA, the mutation is used as the secondary operator to restore lost genetic information.

Two genetic operators of GAs mentioned above, crossover and mutation, are used to produce better offspring. Also, other components such as the population size, the value of control parameters, the formation of fitness function, etc., are required. Figure 3.3.1 illustrates the basic structure of simple genetic algorithm (SGA) of Holland [58, 72].

Fundamentals of Particle Swarm Optimization

Kennedy and Eberhart [85] developed a PSO algorithm based on the behavior of individuals (i.e. particles or agents) of a swarm. Its roots are in zoologist's modeling of the movement of individuals (e.g. fishes, birds, or insects) within a group.

```
Simple Genetic Algorithm ()
[
Create Initial Population at Random;
Evaluate Each Chromosome Using Fitness Function;
Set Generation as 1;
While Generation <= Maximum_Generation
  [
  Select Chromosome for Next Generation;
  Perform Crossover and Mutation;
  Evaluate Each Chromosome;
  Generation = Generation + 1;
  ]
]
```

Figure 3.3.1 Structure of simple genetic algorithm.

It has been noticed that members within a group seem to share information among them, a fact that leads to increased efficiency of the group [90]. The PSO algorithm searches in parallel using a group of individuals similar to other AI-based heuristic optimization techniques [91]. An individual in a swarm approaches to the optimum or a quasi-optimum through its present velocity, previous experience, and the experience of its neighbors.

In a physical n-dimensional search space, the position and velocity of individual i are represented as the vectors $X_i = (x_{i1}, \ldots, x_{in})$ and $V_i = (v_{i1}, \ldots, v_{in})$, respectively, in the PSO algorithm. Let $\text{Pbest}_i = \left(x_{i1}^{\text{Pbest}}, \ldots, x_{in}^{\text{Pbest}}\right)$ and $\text{Gbest}_i = \left(x_{i1}^{\text{Gbest}}, \ldots, x_{in}^{\text{Gbest}}\right)$, respectively, be the best position of individual i and its neighbors' best position so far. Using the information, the updated velocity of individual i is modified under the following equation in the PSO algorithm:

$$V_i^{k+1} = \omega V_i^k + c_1 \text{rand}_1 \left(\text{Pbest}_i^k - X_i^k\right) + c_2 \text{rand}_2 \left(\text{Gbest}^k - X_i^k\right) \tag{3.3.1}$$

where

V_i^k and X_i^k are velocity and position of individual i at iteration k,

Pbest_i^k and Gbest^k best position of individual i and best position of the group until iteration k, respectively,

ω, c_1 and c_2 symbolize parameter and factors of weight,

rand_1 and rand_2 random numbers between 0 and 1, respectively.

In Eq. (3.3.1), the weight parameter ω is given as a function as follows:

$$\omega = \omega_{\max} - \frac{\omega_{\max} - \omega_{\min}}{\text{Iter}_{\max}} \times \text{Iter} \tag{3.3.2}$$

where

$\omega_{\max}$ and $\omega_{\min}$ express initial and final weights respectively,

$\text{Iter}_{\max}$ symbolized the maximum iteration number.

Finally, Iter represents the current iteration number.

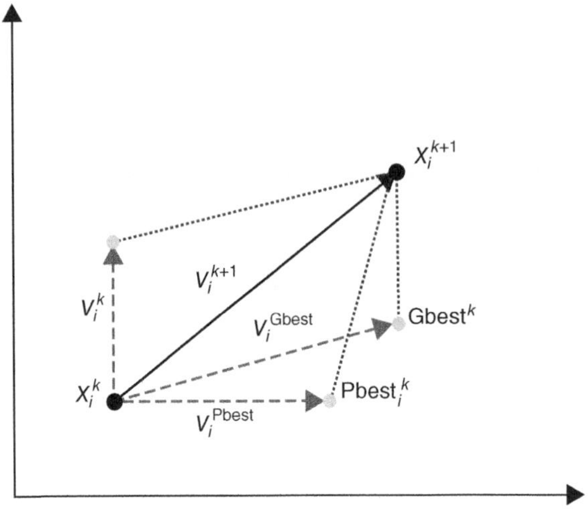

Figure 3.3.2 Search mechanism of PSO.

Each individual moves from the current position to the next one by the modified velocity in (3.3.1) using the following equation:

$$X_i^{k+1} = X_i^k + V_i^{k+1} \tag{3.3.3}$$

The search mechanism of the PSO using the modified velocity and position of individual i based on (3.3.1) and (3.3.2) is illustrated in Figure 3.3.2.

3.3.3 Economic Dispatch Problem

The objective of conventional ED problems is to find the optimal combination of power generations that minimizes total generation cost while satisfying an equality constraint and several inequality constraints [76, 77].

The most simplified type of the objective function in the ED problem can be expressed as the summation of all generating units' operating cost in the shape of smooth function:

$$\text{Minimize} \sum_{i \in I} F_i(P_i) \tag{3.3.4}$$

$$F_i(P_i) = \alpha_i + \beta_i P_i + \gamma_i P_i^2$$

where α_i, β_i, and γ_i represent cost coefficients of generating unit-i, P_i the electrical output of generating unit-i, finally I indicates the set for all generating units.

While minimizing the total generation cost, the following constraints should be satisfied. For energy balance, the following equality constraint should be satisfied. That is, the total generated power should be equal to the total load demand plus the total transmission network losses.

$$\sum_{i \in I} P_i = D + P_{\text{loss}} \tag{3.3.5}$$

where

D implies the total system demand,

P_{loss} means the total network losses.

However, the transmission losses are not considered in this study.

Also, generation power from each unit should be between its maximum and minimum limits. That is, the following inequality constraints for all generating units should be satisfied.

$$P_{i,\min} \le P_i \le P_{i,\max} \quad \forall i \in I \tag{3.3.6}$$

where

$P_{i,\min}$ corresponds to the minimum output of unit-i,

$P_{i,\max}$ corresponds to the maximum output of unit-i.

To consider a more realistic and accurate representation of the objective function, non-smooth cost functions with a few shapes have been applied to ED problem [65, 79, 80]. That is, the objective function of an ED problem has discontinuous and non-differentiable points according to valve loading, change of fuels and prohibited zones. Therefore, it is more realistic to treat the cost function as a set of piecewise quadratic functions as illustrated in Figure 3.3.3 when considering the multi-fuel problems, which is defined as follows [79]:

$$F_i(P_i) = \begin{cases} \alpha_{i1} + \beta_{i1}P_i + \gamma_{i1}P_i^2 & \text{if } P_i^{\min} \le P_i \le P_{i1} \\ \alpha_{i2} + \beta_{i2}P_i + \gamma_{i2}P_i^2 & \text{if } P_{i1} \le P_i \le P_{i2} \\ \vdots & \vdots \\ \alpha_{im} + \beta_{im}P_i + \gamma_{im}P_i^2 & \text{if } P_{im-1} \le P_i \le P_i^{\max} \end{cases} \tag{3.3.7}$$

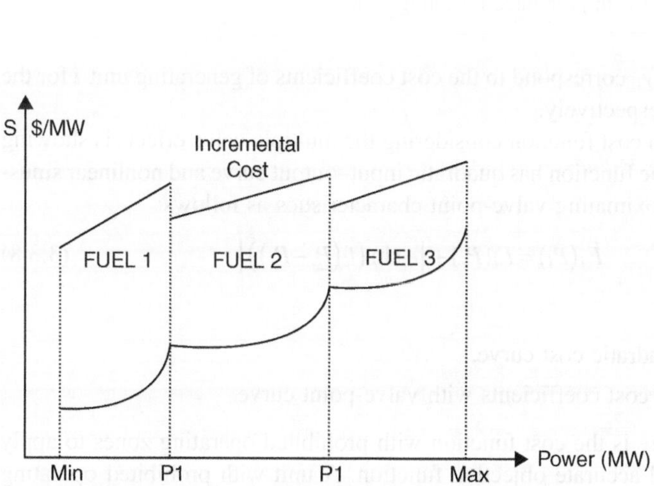

Figure 3.3.3 Piecewise quadratic cost function of a generator.

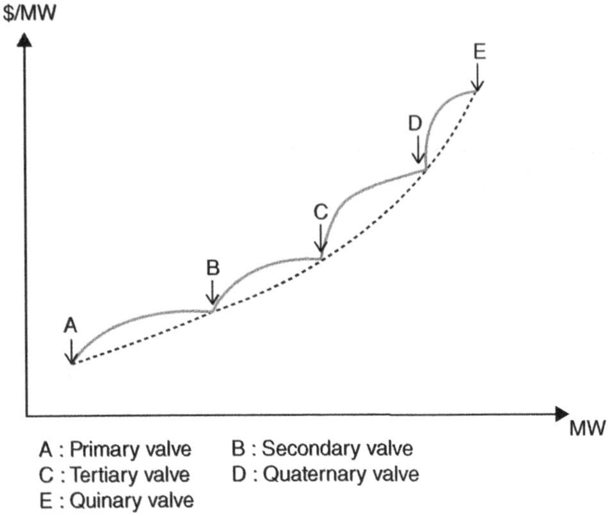

Figure 3.3.4 Cost function with five valves.

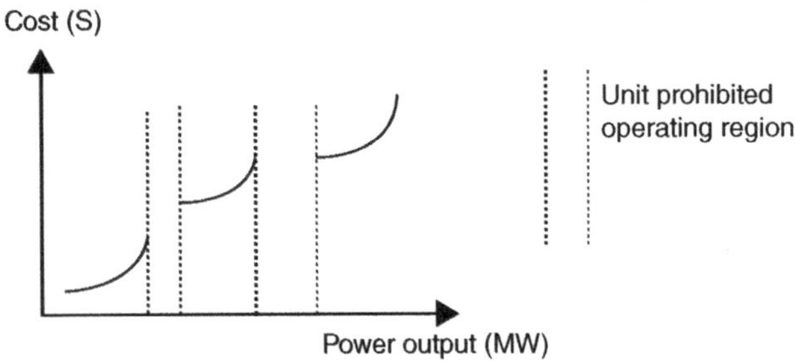

Figure 3.3.5 A unit with prohibited operating zone.

where α_{ij}, β_{ij}, and γ_{ij} correspond to the cost coefficients of generating unit-i for the jth power level, respectively.

Non-smooth cost function considering the multiple valve effects is showing in Figure 3.3.4. The function has quadratic input–output curve and nonlinear sinusoid function approximating valve-point characteristics as follows:

$$\hat{F}_i(P_i) = F_i(P_i) + \left|e_i \sin\left(f_i\left(\underline{P}_i - P_i\right)\right)\right| \tag{3.3.8}$$

where

$F_i(P_i)$ is quadratic cost curve,

e_i and f_i are cost coefficients with valve-point curve.

Finally, there is the cost function with prohibited operating zones to apply more realistic and accurate objective function. A unit with prohibited operating zones has a discontinuous input–output power generation characteristic as illustrated in Figure 3.3.5. In the real world, the whole operating range of a unit is not always available for load allocation due to physical limitations such as faults in the machine itself or other associated auxiliaries. Classical quadratic cost curves were used and, for units with prohibited operating zones, additional operating range constraints were imposed.

3.3.4 GA Implementation to ED

Until now, many studies have been carried out on the application of GA to the ED problem. In this section, a brief comparison of the studies [63, 78, 81–84] is made in terms of the encoding method, constraints handling, genetic operation, fitness function, and so on.

Encoding Method

Two types of encoding schemes, that is, series encoding and embedded encoding have been explored in [63] and [81]. Both utilize the binary number {0,1}. Illustrations of these encoding methods are shown in Figure 3.3.6. The first encoding simply stacks each unit's output value in MW structure in cascade with every other string. The second method uses the same scheme for representation and decoding as the first, except that the assigned gene structures are embedded within each other through the string.

In [91], Bakirtzis et al. have developed a GA version that produces better results. This is based on the observation that when N units with valve points are dispatched, at least $N-1$ units operate at a valve point in the optimum solution.

That is, at most one reference unit operates on a free point, between valve points, in order to satisfy the equality power balance constraint. Therefore, $N-1$ unit is described by a very smaller bit, say four bits, allowing only the valve

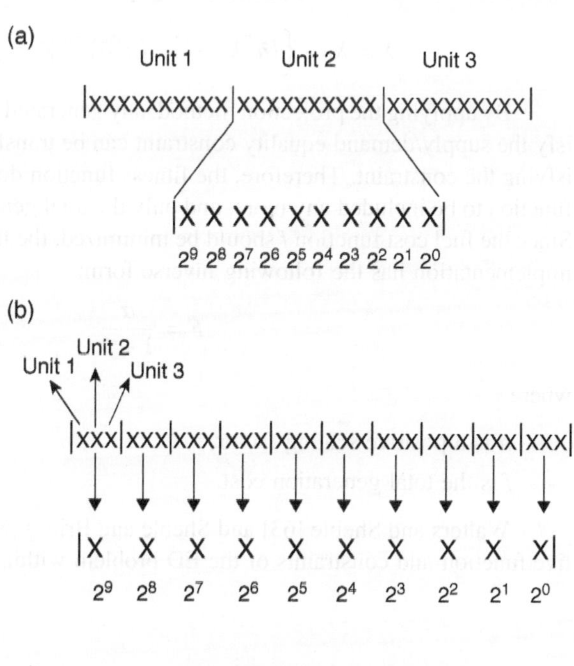

Figure 3.3.6 Encoding schemes. (a) Series encoding and (b) embedded encoding.

point value of each unit to be encoded in it. In other words, every unit except reference is allowed to produce a small number of discrete power output values and these discrete values are chosen from the set of the unit's valve points.

In [83, 84], for the implementation of the dispatch problem, the unit power output was used as the main decision variable, and therefore the loading range of each unit was represented by a binary number. This representation implicitly takes care of the unit minimum and maximum loading limits since the binary representation is made to cover the ratings between the limits. A binary series encoding scheme was used in [83, 84] since it has been reported that the series encoding can provide a better ED solution than embedded one does [81].

Constraints Handling

To implement GA in the ED problem, electrical outputs covering all generating units are selected as encoding variables. By applying a linear mapping strategy in the random creation of a population, generating unit's inequality constraint (3.3.6) can easily be satisfied. However, the strings that satisfy the inequality constraints do not always produce offsprings satisfying the inequality constraints due to genetic operations. In these cases, offsprings not satisfying the inequality constraints are discarded, and genetic operations are repeated until they produce offsprings satisfying the inequality constraints.

However, it is not easy to treat the equality constraint (3.3.5). Therefore, the penalty function approach has been widely applied to include the equality constraint in the fitness function.

In [87], Park et al. proposed a projection method to create chromosomes always satisfying the equality constraint. Literally, if X_0 is not satisfied by constraint $\overline{a}^T X = b$, i.e. $\overline{a}^T X_0 \neq b$, then by applying the following transformation of X_0 into X, the constraint is satisfied exactly. That is,

$$X = X_0 - \left\{ \left(\overline{a}^T X_0 - b\right) \cdot \left(\overline{a}^T \overline{a}\right)^{-1} \right\} \overline{a} = \text{proj}(X_0) \tag{3.3.9}$$

By applying the projection method, any generated string X_0 that does not satisfy the supply/demand equality constraint can be transformed to the string X satisfying the constraint. Therefore, the fitness function does not require the penalty function to be included any more, and only the total generation cost is considered. Since the fuel cost function f should be minimized, the fitness function (F) for GA implementation has the following inverse form:

$$F = \frac{\alpha}{1 + f} \tag{3.3.10}$$

where

α is a positive scaling factor,

f is the total generation cost.

Walters and Sheble [63] and Sheble and Brittig [81] implemented an objective function and constraints of the ED problem within the fitness function. Unit limits are automatically satisfied by the normalization process in the chosen encoding method, which only allows solutions in the operating range of the units. Equality power balance constraint was incorporated in the augmented objective function as a form of penalty term for the conservation of energy constraint, and was handled by the fitness function evaluation through the weighting factor.

Bakirtzis et al. [91] handled the equality constraint based on the observation that when N units with valve points are dispatched, at least $N-1$ units operate at a valve point in the optimum solution. That is, at most one reference unit operates on a free point, between valve points, in order to satisfy the equality power balance constraint. Inequality constraints regarding units' operating limits were handled such that strings are encoded to automatically satisfy the inequality constraint.

Using the Lagrange function method where constraints are implemented by imposing penalties in individuals that violate them, Song and Chou [92] included the constraints of ED problem in the fitness function formulation. Both equality and inequality constraints are taken into consideration during the optimization process. That is, if the output of a generator unit is over its upper limit or under lower limit, a penalty factor is added into the power balance equation, followed by redistributing the portion of PBM error into each unit if any power balance error occurs. The revised power balance equation and the redistributed power output of each unit are as follows:

$$\text{lerr} = \left| \sum_{i \in I} P_i - D - P_{\text{loss}} \right| + \sum_{i \in I} \rho_i P_i \tag{3.3.11}$$

$$P_{i\text{new}} = \frac{P_i}{\sum_{i \in I} P_i} (D + P_{\text{loss}}) \tag{3.3.12}$$

where

lerr is the revised power balance equation,

ρ_i is a power balance penalty factor for unit-i.

The system constraints to be satisfied are those of matching the load demand and reserve requirements with power generation. Orero and Irving [93] took care of the unit minimum and maximum loading limits in the problem encoding, and they used a penalty function approach to treat the constraints of the unit prohibited operating zones and the power balance equation.

Genetic Operations

In [63], a one-point crossover and mutation operation was used. They used both series and embedded encoding schemes for the ED problem solution and compared the two encoding methods in terms of generation cost and performance. In [81], several different techniques that enhance program efficiency and accuracy such as mutation and crossover prediction, elitism, and interval approximation, were proposed. When crossover is needed, a binary string which is the same length

```
Create a binary string. c. randomly that is the same length
as the population string

i=1
DO WHILE (i ≤ String Length)
  Choose two parents, p1 and p2, randomly

  For the offspring c1,
  IF [c[i] = 1] THEN
    c1[i] = p1[i]
  ELSE IF [c[i] = 0] THEN
    c1[i] = p2[i]

  For the offspring c2,
  IF [c[i] = 1] THEN
    c2[i] = p2[i]
  ELSE IF [c[i] = 0] THEN
    c2[i] = p1[i]
  i = i+1
END DO
```

Figure 3.3.7 Pseudocode for the new crossover technique.

as the population string is used to cue the two parent strings as to whether the child will get its bit value from the first parent or the second one. For example, if random string bit value in any string position is a “1,” the first parent gives its bit value to the child, however if it is a “0” the second parent donates its bit value in the corresponding string position. In order to create the second child, a complement method is followed. That is, if the random string bit value is a “0” the first parent gives its bit value to the child, however if it is a “1” the second parent inherits its bit value in the corresponding string position. A pseudo code for this new crossover technique is illustrated in Figure 3.3.7.

In addition, elitism was used to improve the performance capability of conventional GA such that early solutions are saved by ensuring the survival of the most fit strings in each population.

To accelerate the local gradient descend, so-called phenotype mutation that is based on perturbation of real power vector (not binary vector) was introduced in [82]. This new operator was designed only for the best genotype of every generation, so that it was not expected to load the algorithm with many more fitness evaluations. And, to handle the combinatorial problem in the population realizing the existence of valve point, unit swap and power slide operator were newly introduced.

An advanced version of GA was also proposed in [82], where $N-1$ unit can be described by a small number of bits, say four bits, allowing only the valve point value of each unit to be encoded. Therefore, every unit except the reference is allowed to produce a small number of discrete power output values and these discrete values are chosen from the set of the unit's valve points. This is based on the observation that when N units with valve points are dispatched, at least $N-1$ units operate at a valve point in the optimum solution, and, at most one reference unit operates on a free point, between valve points, in order to satisfy the equality power balance constraint.

An advanced version of GAs was proposed in [83] and it showed more accurate and less time-consuming property. This algorithm combines a strategy involving local search algorithm and GAs. Moreover, several techniques, such as elite policy, adaptive mutation prediction, nonlinear fitness mapping and different crossover techniques, were explored to enhance program efficiency and accuracy.

To save early solution by ensuring the survival of the most fit strings in the current population to be carried over to the next generation, elitism technique was adopted. Comparing the results of the most recent population to the elite population, elitism combines the two populations and determines the best results from both populations. Various crossover techniques including one-point, two-point, and uniform crossover were explored and compared. As a result, two-point crossover was reported to yield a better result in ED applications. By using the geometric mean probability theory to work out the probability of mutation, adaptive mutation prediction was explored and much of the computation effort can be saved without sacrificing accuracy.

In [84], a deterministic crowding GA was presented to maintain population diversity as the algorithm proceeds. In the crowding GA, the new population is created by letting the children population replace the parents that are most similar to them. This crowding GA is inspired by a corresponding ecological phenomenon, where similar members of a natural population compete for the same resources. A pseudo code for the deterministic crowding GA is shown in Figure 3.3.8.

```
DO WHILE (GEN. ≤ MAX.GENERATION)
  Reshuffle population
  i=1
  DO WHILE (i ≤ population size)
    Choose two parents, p1 and p2, randomly, without replacement
    Perform crossover and mutation to produce offspring c1 and c2
    Evaluate fitness, f, of parents and offspring

    IF Distance (p1,c1)+Distance(p2,c2)≤
                                  Distance (p1,c2)+Distance(p2,c1)
    THEN
      IF (f(c1)>f(p1)) replace p1 with c1
      IF (f(c2)>f(p2)) replace p2 with c2
    ELSE
      IF (f(c2)>f(p1)) replace p1 with c2
      IF (f(c1)>f(p2)) replace p2 with c1
    i=i+2
  END DO
END DO
```

Figure 3.3.8 Pseudocode for the deterministic crowding GA.

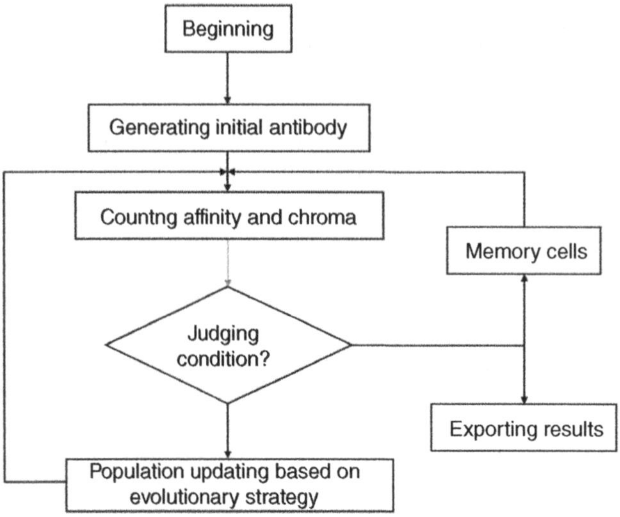

Figure 3.3.9 Artificial immune genetic algorithm flowchart.

Artificial immune genetic algorithm (AIGA) was proposed in [50]. In the AIGA, some characteristics of biology immune system, such as learning, memory, diversity and recognizes are brought into GA. Algorithm flowchart is shown in Figure 3.3.9.

In [101], GA based on the Lagrange method was presented. This algorithm exploits the augmented Lagrange formulation of the ELD and performs two iterative loops. In the inner loop a GA with classical structure is performed to minimize the augmented Lagrange function. An advantage of this algorithm is that the augmented Lagrange function can be scaled to avoid the ill condition which leads to some difficulties in searching for the problem solution. The flowchart of GA based on the Lagrange method is showed in Figure 3.3.10.

Fitness Function

Walters and Sheble defined the fitness function as the following form [63]:

$$\mathrm{Fit} = \mathrm{sf}_1[(1-\%\mathrm{obj})^{\mathrm{sp}_1}] + \mathrm{sf}_2[(1-\%\mathrm{lerr})^{\mathrm{sp}_2}] \tag{3.3.13}$$

$$\mathrm{obj} = \sum_{i\in I} \alpha_i + \beta_i P_i + \gamma_i P_i^2 + \mid e_i \sin(f_i(\underline{P}_i - P_i)) \mid \tag{3.3.14}$$

$$\mathrm{lerr} = \left| \sum_{i\in I} P_i - D - P_{\mathrm{loss}} \right| \tag{3.3.15}$$

where

sf is a scaling factor for emphasis within the function itself,

sp is a scaling power factor for emphasis over the entire population,

%obj is a percentage value of the objective function,

%lerr is percentage value of the string's constraint equation error.

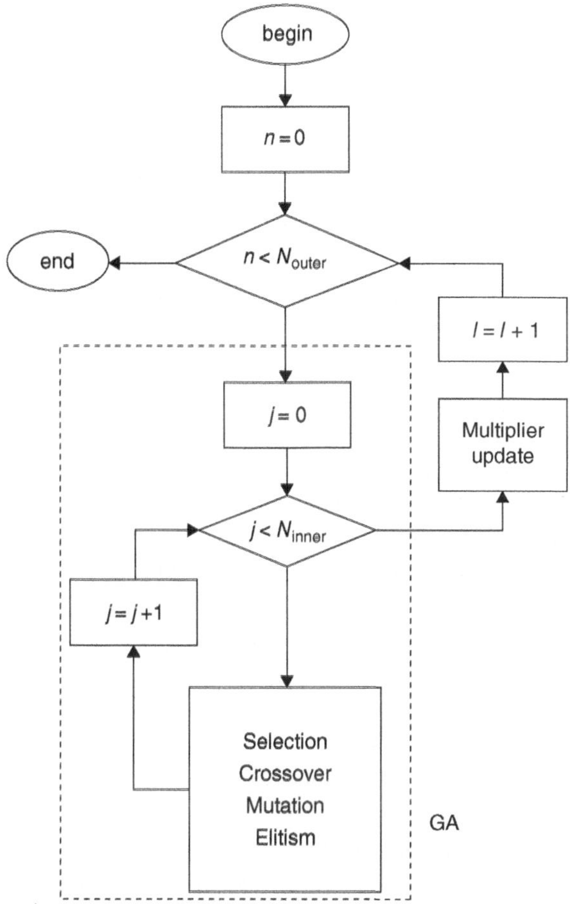

Figure 3.3.10 Flowchart of the genetic algorithm based on the Lagrange method.

Sheble and Brittig [90] also explored the same fitness function of Walters and Sheble selected as in (3.3.12), except that objective function of the ED problem was defined as a smooth function of quadratic cost curve.

In [92], Song and Chou designed the fitness function based on the percentage rating. The lowest cost operating solution is the point at which the minimum incremental cost rate (λ) is equal and simultaneously satisfies the specified demand. An error term was introduced and a measure of this error was calculated.

$$\lambda_{\text{err}} = \sum_{i \in I} \left| \lambda_{\text{avg}} - \lambda_i \right| \tag{3.3.16}$$

where

λ_{err} is an incremental cost error term,

λ_i is an incremental cost for unit-i,

λ_{avg} is an average incremental cost of individual string.

By combining the two errors resulting from (3.3.15) and (3.3.16), the fitness function becomes:

$$\text{Fit} = \text{sf}_1[(1 - \%\lambda\text{err})^{\text{sp}_1}] + \text{sf}_2[(1 - \%\text{lerr})^{\text{sp}_2}] \qquad (3.3.17)$$

where

$\%\lambda$err is the percentage of incremental cost error,

%lerr is the percentage of power balance error,

sf is a scaling factor for emphasis within function,

sp is a scaling power factor for emphasis over the entire population.

Orero and Irving [93] explored the use of a GA for the solution of an ED problem where some of the units have prohibited operating zones. The unit minimum and maximum loading limits were embedded in the problem encoding process. Other constraints to be considered are the power balance equation and the unit prohibited operating zones. These two constraints are handled using a penalty function approach. Penalty terms are incorporated in the fitness function, and are set to reduce the fitness of the string according to the magnitude of the violation. The problem objective function is formulated as follows:

$$\min \sum_{i \in I} F_i(P_i) + \Psi\left[\sum_{i \in I} P_i - (D + P_{\text{loss}})\right] + \Phi\left[\sum_{k=1}^{n_i} P_i(\text{zone}_k)\right] \qquad (3.3.18)$$

where

Ψ is the penalty function for not satisfying the load demand,

Φ represents the penalty function for a unit loading fallen within a prohibited operating zone,

n_i is the number of prohibited zones for unit-i.

Multistage Method and Directional Crossover

GAs simulate the analogous biological mechanism. Each chromosome has its adequate number of bits and each bit has the corresponding genetic information. In continuous variable optimization problems, numerous bits are needed to obtain precise optimal solutions. As a kind of a local search method, a multistage (MS) method was proposed by Park et al. [87], which divides solution space into N segments and finds the best solution point from generation to generation in the current stage. Some advantages are observed when this method is applied. First, it requires smaller bits than the traditional encoding schemes. Second, as the stage is increased, solution space is subdivided further, and more accurate solution can be obtained.

In addition, a directional crossover (DC) was also presented [78] to overcome the limitation of traditional GA mechanism that it cannot guarantee the convergence to the optimal point and sometimes causes the loss of critical patterns [60, 71]. The directional operator helps to produce more competitive offsprings when combined with conventional operators.

3.3.5 PSO Implementation to ED

Recently, PSO have been successfully applied to various fields of power system optimization such as power system stabilizer design [88], reactive power and voltage control [86], and dynamic security border identification [89]. Yoshida et al. [95] suggested a MPSO to control reactive power and voltage considering voltage security assessment. Since the problem was a mixed-integer nonlinear optimization problem with inequality constraints, they applied the classical penalty method to reflect the constraint-violating variables. Abido [88] developed a revised PSO for determining the optimal values of parameters for power system stabilizers.

The practical ED problems with valve-point and multi-fuel effects are represented as a non-smooth optimization problem with equality and inequality constraints, and this makes the problem of finding the global optimum difficult. To solve this problem, many salient methods have been proposed such as a mathematical approach [79], DP [76], evolutionary programming (EP) [65, 92, 93], Tabu search (TS) [94], neural network approaches [57, 80], and GA [96].

In this section, an alternative approach is proposed to the non-smooth ED problem using a MPSO [95] and improved PSO from MPSO [97, 98], which focuses on the treatment of the equality and inequality constraints when modifying each individual's search. Additionally, to accelerate the convergence speed, a dynamic search-space reduction strategy is devised based on the distance between the best position of the group and the inequality boundaries.

Constraints Handling

In this section, a new approach to implement the PSO algorithm will be described in solving the ED problems. Especially, a suggestion will be given on how to deal with the equality and inequality constraints of the ED problems when modifying each individual's search point in the PSO algorithm. The process of the MPSO algorithm can be summarized as follows:

Step 1: Initialization of a group at random while satisfying constraints

Step 2: Velocity and position updates while satisfying constraints

Step 3: Update of Pbest and Gbest

Step 4: Activation of space reduction strategy

Step 5: Go to Step 2 until satisfying stopping criteria.

In [104], Park et al. proposed the simple heuristic method for handling the inequality/equality constraints keeping the randomness of each individual. The position of each individual is modified by (3.3.3) in PSO. The resulting position of an individual is not always guaranteed to satisfy the inequality constraints due to over/under velocity. If any element of an individual violates its inequality

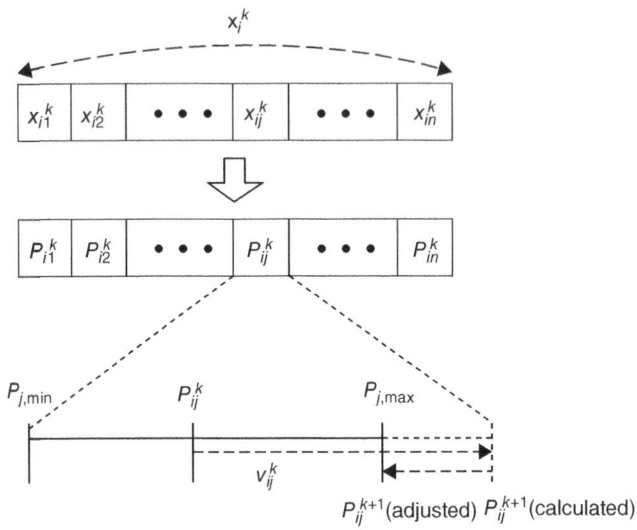

Figure 3.3.11 Adjustment strategy for an individual's position within boundary.

constraint due to over/under speed then the position of the individual is fixed to its maximum/minimum operating point as illustrated in Figure 3.3.11 and this can be formulated as following equation:

$$P_{ij}^{k+1} = \begin{cases} P_{ij}^{k+1} + v_{ij}^{k+1} & \text{if} \quad P_{ij,\min} \le P_{ij}^{k+1} + v_{ij}^{k+1} \le P_{ij,\max} \\ P_{ij,\min} & \text{if} \quad P_{ij}^{k+1} + v_{ij}^{k+1} < P_{ij,\min} \\ P_{ij,\max} & \text{if} \quad P_{ij}^{k+1} + v_{ij}^{k+1} > P_{ij,\max} \end{cases} \tag{3.3.19}$$

Although the aforementioned method always produces the position of each individual satisfying the inequality constraints (3.3.6), the problem of equality constraint still remains to be resolved. Therefore, it is necessary to develop a new strategy such that the summation of all elements in an individual (i.e. $\sum_{j=1}^{n} P_{ij}^k$) is equal to the total system demand. To resolve the equality constraint problem without intervening the dynamic process inherent in the PSO algorithm, we propose the following heuristic procedures:

Step 1: Set $j = 1$. Let the present iteration be k.

Step 2: Select an element (i.e. generator) of individual i at random and store in an index array $A(n)$.

Step 3: Modify the value of element j using (3.3.6), (3.3.7) and (3.3.10).

Step 4: If $j = n - 1$ then go to Step 5, otherwise $j = j + 1$ and go to Step 2.

Step 5: The value of the last element of individual i is determined by subtracting $\sum_{j=1}^{n-1} P_{ij}^k$ from D. If the value is not within its boundary then adjust the value using (3.3.10) and go to Step 6, otherwise go to Step 8.

Step 6: Set l = 1.

Step 7: Readjust the value of element l in the index array $A(n)$ to the value satisfying equality condition (i.e. $D - \sum_{j=1, j \neq l}^{n} P_{ij}^{k}$). If the value is within its boundary then go to Step 8; otherwise, change the value of element l using (3.3.11). Set $l = 1 + 1$, and go to Step 7. If $l = n + 1$, go to Step 1.

Step 8: Stop the modification procedure.

In [105], Park et al. proposed the Chaotic Sequence PSO (CSPSO). In order to overcome the existing drawbacks of PSO to some extents, proposed an improved PSO (IPSO) framework combining the chaotic sequences and the crossover potation. Although this paper proposed PSO-based approach for the non-convex ED problems with heavy constraints, in order to help understanding, this study only applied the chaotic sequences and light constraints. The chaotic sequences combined with the linearly decreasing inertia weights are suggested as new dynamic inertia weights in PSO. The employment of chaotic sequences in PSO can improve the global searching capability by preventing premature convergence through increased diversity of the population.

One of the dynamic systems evidencing chaotic behavior is the iterator called the logistic map [107], shows equation is described as following equation:

$$\gamma_k = \mu \cdot \gamma_{k-1} \cdot (1 - \gamma_{k-1}) \tag{3.3.20}$$

where

γ is a control parameter and has a real value between (3.3.7),

γ_k is the chaotic parameter at iteration k.

Despite the apparent simplicity of the equation, the solution exhibits a rich variety of behaviors. The value of μ determines whether γ is stabilized at a constant size, oscillates between a limited sequence of sizes, or behaves chaotically in an unpredictable pattern. The system (3.3.20) is deterministic, and displays chaotic behaviors when $\mu = 4.0$ and $\gamma_0 \notin \{0, 0.25, 0.50, 0.75, 1.0\}$.

Eberhart et al. made a significant improvement in the performance of the PSO with a linearly varying inertia weights over the iterations (i.e. IWA in the form of (3.3.2)), which is widely used in PSO.

$$\mathrm{cw}^k = w^k \cdot \gamma_k \tag{3.3.21}$$

where

cw^k is a chaotic weight at iteration k,

w^k is the weight factor from the IWA,

γ_k is the chaotic parameter whereas the weight in the original IWA decreases monotonously from $w_{\max}$ to $w_{\min}$, the proposed chaotic weight decreases and oscillates simultaneously as shown in Figure 3.3.12. This algorithm

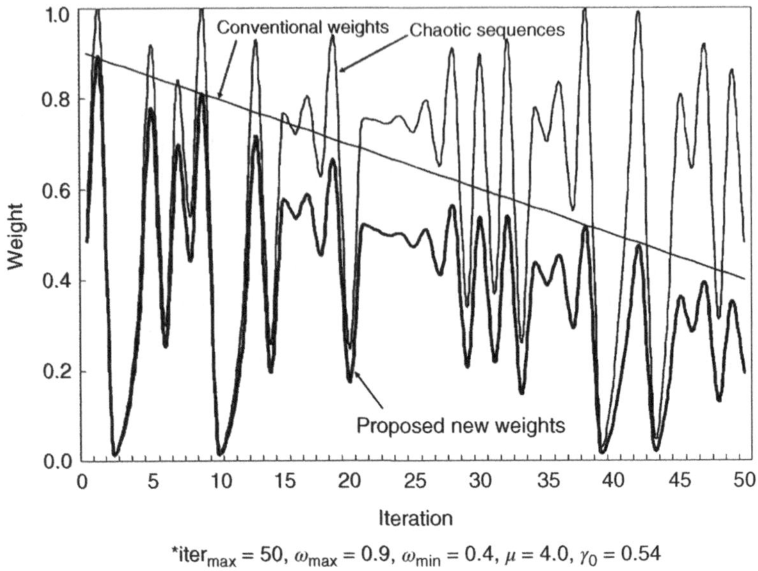

Figure 3.3.12 Comparison of inertia weights for IWA and CIWA.

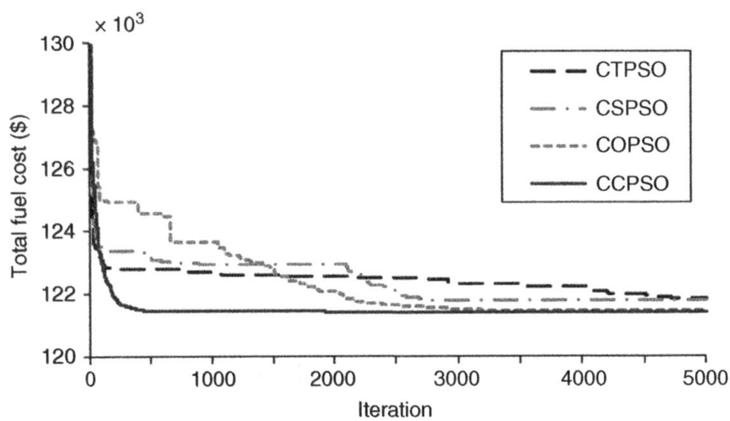

Figure 3.3.13 Convergence characteristics of the CSPSO.

can apply to a part of MPSO's weight parameter to improve result that is illustrated in Figure 3.3.13:

In [97], Park et al. proposed four strategies for comparison of convergence characteristics.

CTPSO: The conventional PSO with the proposed constraint treatment strategy.

CSPSO: PSO with chaotic sequences.

COPSO: PSO with crossover operation.

CCPSO: PSO with both chaotic sequences and crossover operation.

Since COPSO, CCPSO are proposed for heavy constraints (i.e. ramping up/down constraint), this study does not consider heavy constraints.

In [106], Baek et al. described the non-smooth ED problem using PARPSO. In this paper, we have introduced a new penalty factor to improve the attractive and repulsive PSO (ARPSO). When a particle determines a next move, it considers three kinds of experiences, such as the particle's best position, the global best position, and the global worst position. Giving penalty to a particle close to the global worst position makes the particle to have repulsive behavior from the worst particle. The mechanism of proposed PSO with Gworst (global worst) is illustrated in Figure 3.3.14.

Diversity determines each particle's exploring and exploiting behavior. Distance from each particle to the global best particle is changed every step so diversity is also changed every step. Diversity definition is as follows:

$$\text{diversity}^k(S) = \frac{1}{|S|}\sum_{i=1}^{|S|} \text{div}_i^k \tag{3.3.22}$$

$$\text{div}_i^k = \begin{cases} 0, \text{if} \left|\text{Pbest}_i^k - P_i\right| < \left|\text{Gbest}^k - P_i\right| \\ 1, \text{otherwise} \end{cases} \tag{3.3.23}$$

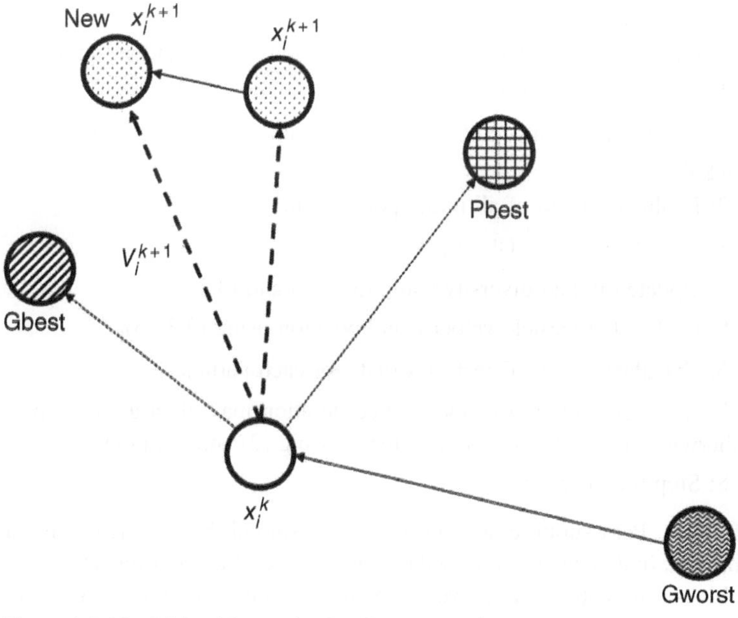

Figure 3.3.14 PSO with penalty for Gworst search mechanism.

Diversity is a variable to control each particle's behavior. Equation (3.3.22) and (3.3.23) explain how to determine diversity value, where individual i, S is the swarm, and $|S|$ is swarm size as the number of individual i. If most individuals are close to Pbest_i^k, diversity will have a small value. Diversity is an average value of div and it has 0–1 value.

$$0 < \text{diversity}(S) < 1 \tag{3.3.24}$$

To determine the diversity effect to particle's next direction, based on (3.3.22) and (3.3.23), the velocity of individual i of the PARPSO algorism is rewritten as

$$V_i^{k+1} = w_1 V_i^k + \text{dir}^{k+1} \times \left(c_1 r_1 \left(\text{Pbest}_i^k - x_i^k\right) + c_2 r_2 \left(\text{Gbest}^k - x_i^k\right)\right) - w_2 c_3 r_3 \left(\text{Gworst}^k - x_i^k\right) \tag{3.3.25}$$

$$\text{dir}^{k+1} = \begin{cases} -1, \text{if}\left(\text{dir}^k > 0\right) \;\&\; \left(\text{diversity}^k < d_{\text{low}}\right) \\ 1, \text{if}\left(\text{dir}^k < 0\right) \;\&\; \left(\text{diversity}^k > d_{\text{high}}\right) \\ \text{dir}^k, \text{ otherwise} \end{cases} \tag{3.3.26}$$

$$w_2 = \frac{\min\left(\left|\text{Gworst}^k - x_i^k\right|\right)}{\left|\text{Gworst}^k - x_i^k\right|} \text{ for all } i \tag{3.3.27}$$

where d_{low} and d_{high} influence the efficiency of MARPSO in Niu et al. [108] The value of parameters are set as $d_{\text{low}} = 0.2$, $d_{\text{high}} = 0.8$, and dir is a weight parameter for decision in the next direction. w_1 is constant and step inertia, which is set to decrease in every iteration. The penalty factor w_2 increases when the particle and Gworst^k are getting closer, moving the particle far away from Gworst^k. Additional parameter c_3 is an acceleration constant and r_3 is a random number.

The process of the PARPSO algorism can be summarized as follows:

Step 1: Initialize random population of individuals as proposed, velocity, and weight.

Step 2: Evaluate fitness of all initial population.

Step 3: Calculate Pbest_i^k, Gbest^k, Gworst^k.

Step 4: Update div and diversity using (3.3.22) and (3.3.23).

Step 5: Update the particle velocity and position using (3.3.25).

Step 6: Calculate Pbest_i^k, Gbest^k, Gworst^k for each particle.

Step 7: If current iteration is not reached iteration max, then go to Step 4; otherwise, update the weight parameter using (2) and iteration +1.

Step 8: Stop the procedure.

In [110], A. Parassuram et al. solved the non-smooth ED problem using a hybrid technique. In this paper, they used two approaches, the algorithm synergistically combines PSO with a very powerful member of the evolutionary algorithm (EA), to propose a new technique.

The procedure of DE is almost the same as that of the GA whose main process has mutation, crossover, and selection. The main difference between DE and GA lies in the mutation process. In GA, mutation is caused by the small changes of the genes, whereas in DE, the arithmetic combination of the selected individuals carries out mutation. DE maintains a population of constant size that consists of PN real-valued vector X_i^G, $i = 1, 2, \ldots,$ PN, where i indicates the index of the individual and G is the generation. The evolution process of the DE algorithm is as follows:

$$X_{j,i}^0 = \text{rand}_{j,i}[0, 1] \cdot \left(X_j^{(U)} \cdot X_j^{(L)}\right) + X_j^{(L)} \tag{3.3.28}$$

To construct a starting point for the optimization process, the population with PN individuals should be initialized. Usually, the population is initialized by randomly generated individuals within the boundary constraints, where $i = 1, 2, \ldots,$ PN, $j = 1, 2, \ldots,$ VD, VD is the variable dimension, $X_j^{(L)}$ and $X_j^{(U)}$ are the lower and upper boundary of the j component, respectively, and $\text{rand}_{j,i}[0, 1]$ denotes a uniformly distributed random value in the range [0, 1].

For each target vector, or parent vector X_i^G, a mutant vector is generated according to:

$$V_i^{G+1} = X_{n1}^G + F \cdot \left(X_{n2}^G \cdot X_{n3}^G\right) \tag{3.3.29}$$

where random indices $n1$, $n2$, and $n3$ are integers, mutually different, and chosen to be different from the running index i. In the initial DE scheme, the parameter i is a real and constant factor during the entire optimization process, and the variable range is $F \in [0, 2]$.

The trial vector u_i^{G+1} is generated using the parent and mutated vectors as follows:

$$u_{j,i}^{G+1} = \begin{cases} V_{j,i}^{G+1}, \text{if } \text{rand}_{j,i}(0, 1) \leq \text{CR or } j = k \\ x_{j,i}^G, \text{otherwise} \end{cases} \tag{3.3.30}$$

where $k \in \{1, 2, \ldots, \text{VD}\}$ is the randomly selected index chosen once for each i, and CR is the parameter that is a real-valued crossover factor in the range [0, 1] and controls the probability that a trial vector component comes from the randomly chosen, mutated vector $V_{j,i}^{G+1}$, instead of the current vector $X_{i,j}^G$. If CR is 1, then the trial vector u_i^{G+1} is the replica of the mutated vector V_i^{G+1}.

To decide the population for the next generation, the trial vector u_i^{G+1} and the target vector x_i^G are compared, and the individual of the next generation x_i^{G+1} is decided according to the following rule for minimization problems:

$$x_i^{G+1} = \begin{cases} u_i^{G+1}, \text{if } f\left(u_i^{G+1}\right) \leq f\left(x_i^G\right) \\ x_i^G, \text{otherwise} \end{cases} \tag{3.3.31}$$

The feature of DE selection scheme is that a trial vector is compared with only one individual, not all the individuals in the current population. Due to the greedy selection scheme, all the individuals of the next generation are as good as or better than their counterparts in the current generation.

The procedure for hybrid PSO is as follows:

Step 1: Read the given data.

Step 2: Generate the initial random population.

Step 3: Evaluate the fuel cost using fitness function for all individuals.

Step 4: Check whether a stopping criterion occurred or not. If occurred, compare the fuel cost of each individual and find optimal solution. Then, go to Step 8.

Step 5: The population was splitted according to hybridization coefficient (HC) value if HC is within 0 and 1. Apply the splitted population for PSO and DE algorithm.

Step 6: IF HC = 0, apply PSO algorithm for pure PSO and if HC = 1, apply DE algorithm for pure DE.

Step 7: Evaluate the new resulting population and go to Step 3.

Step 8: Stop.

Dynamic Space Reduction Strategy

To accelerate the convergence speed to the solutions in PSO algorithms, Park et al. proposed the dynamic space reduction strategy. This strategy is activated in the case when the performance is not increased during a prespecified iteration period. In this case, the search space is dynamically adjusted (i.e. reduced) based on the "distance" between the Gbest and the minimum and maximum output of generator *j*. To determine the adjusted minimum/maximum output of generator *j* at iteration *k*, the distance is multiplied by the predetermined step-size Δ and subtracted (added) from the maximum (minimum) output at iteration *k* as described in (3.3.32) and the mechanism of dynamic space reduction is illustrated in Figure 3.3.15.

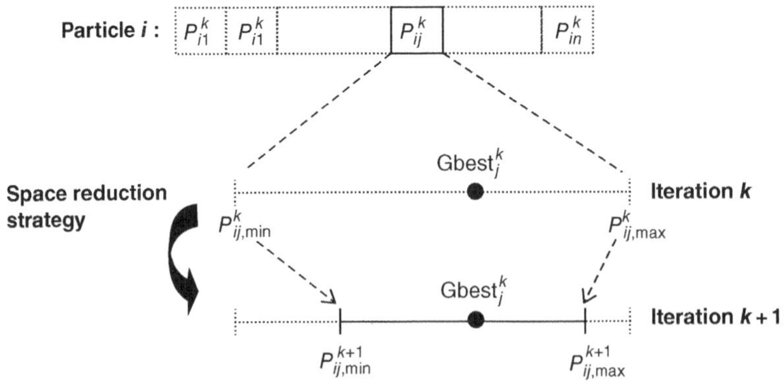

Figure 3.3.15 Schematic of dynamic space reduction strategy.

$$
\begin{aligned}
P_{j\max}^{k+1} &= P_{j\max}^{k} - \left(P_{j\max}^{k} - \mathrm{Gbest}_j^k\right) \times \Delta \\
P_{j\min}^{k+1} &= P_{j\min}^{k} + \left(P_{j\min}^{k} - \mathrm{Gbest}_j^k\right) \times \Delta
\end{aligned}
\tag{3.3.32}
$$

3.3.6 Numerical Example

GA Implementation to ED with Smooth Cost Function

In this section, we will briefly present the results from the GAs using MS method and DC and compare those from the simple GAs [60] and the conventional lambda-iteration method [76]. The objective of this section is to show that well-devised improvement strategies can provide better solutions over the conventional SGA. Therefore, various improvement strategies introduced in this chapter may produce better solutions than the SGA and may provide the global solution.

For numerical tests, we have considered the ED problem with six generating units whose input data are given in Table 3.3.1.

The solutions obtained from the GAs using multistage and DC are compared with those of SGA and the conventional lambda-iteration method. Numerical tests are performed without considering the transmission loss for the total demand of 1680 MW. Three cases assumed are given in Table 3.3.2: (Case I) Goldberg's SGA, (Case II) applying the DC strategy to the SGA structure, and (Case III) applying both the MS method and the DC strategy to the SGA. Based on empirical

TABLE 3.3.1 Cost Coefficient of the Test System

	α_i	β_i	γ_i	$P_{i,\max}$ (MW)	$P_{i,\min}$ (MW)
U1	561.0	7.92	0.001 56	600.0	150.0
U2	310.0	7.85	0.001 94	400.0	100.0
U3	78.0	7.97	0.004 82	200.0	50.0
U4	102.0	5.27	0.002 69	500.0	100.0
U5	51.0	9.90	0.001 72	350.0	40.0
U6	178.0	8.26	0.006 93	280.0	100.0

TABLE 3.3.2 Basic Characteristics and Parameters in Each Case

Case	I (SGA)	II (SGA + DC)	III (SGA + DC + MS)
Crossover probability	0.9	0.9	0.9
Mutation probability	0.05	0.05	0.05
Encoding	6-bit (decimal)	6-bit (decimal)	2-bit (decimal)
Average convergence iteration number	400	400	210
DC probability	—	0.15	0.15

TABLE 3.3.3 Objective Function Values in Each Case

Population Size	Case I	Case II	Case III
40	14 924.9	14 916.8	14 901.0
60	14 922.2	14 906.4	14 902.1
80	14 906.2	14 900.9	14 911.2
100	14 920.3	14 902.1	14 953.2
120	14 909.6	14 905.3	14 897.7
140	14 921.7	14 906.4	14 908.8
160	14 907.3	14 917.2	14 901.6
180	14 926.5	14 900.4	14 899.0
200	14 924.4	14 899.4	14 902.1
220	14 967.7	14 897.9	14 896.9
240	14 925.7	14 898.1	14 911.6
260	14 921.2	14 899.5	14 902.9
280	14 911.3	14 908.2	14 903.1
300	14 907.0	14 904.9	14 899.2
320	14 909.2	14 900.6	14 896.6
340	14 904.4	14 902.8	14 900.0
360	14 924.7	14 898.2	14 897.4
380	14 901.9	14 905.1	14 901.9
400	14 936.9	14 896.5	14 904.8

studies, various GA parameters are set as follows and the resulting average iteration number in each case is given in Table 3.3.2.

The projection method to cope with the equality constraint [78] is applied for all the cases. In order to see the performance results, the population size is gradually increased, and the results are summarized in Table 3.3.3. From Table 3.3.3, we can see that the results of Cases II and III provide a better solution than the SGA does, except one place with a population size of 380.

Regarding the convergence, Case III with the average of 210 iterations shows the best performance, while Cases I and II show similar performance with the average of 400 iterations. Therefore, one can improve the performance of GAs by applying the various strategies or the combination of the strategies introduced in this chapter.

PARPSO Implementation to ED with Smooth/Non-Smooth Cost Function

In this section, improved PARPSO [98] has been applied to ED problems where the objective functions can be either smooth or non-smooth. The results obtained from the PARPSO are compared with those of other methods: the numerical lambda-iteration method (NM) [76], the hierarchical numerical method (HM)

TABLE 3.3.4 Cost Coefficient of Test System with Quadratics Cost Functions

Unit	a_i	b_i	γ_i	$P_{i,\min}$	$P_{i,\max}$
1	561.0	7.92	0.001 562	150.0	600.0
2	310.0	7.85	0.001 94	100.0	400.0
3	78.0	7.97	0.004 82	50.0	200.0

TABLE 3.3.5 Comparison of Simulation Results of Each Method

Unit	NM	MHNN	IEP (pop = 10)	MPSO (par = 10)	PARPSO (par = 10)
1	393.170	393.800	393.170	393.170	393.170
2	334.604	333.100	334.603	334.604	334.610
3	122.226	122.300	122.227	122.226	122.220
TP	850.000	849.200	850.000 00	850.000 00	850.000 00
TC	8 194.356 12	8 187.000 00	8 194.356 14	8 194.356 12	8 194.356 12

TABLE 3.3.6 Cost Coefficient of Test System with Valve Point Effects

Unit	a_i	b_i	γ_i	e_i	f_i	$P_{i,\min}$	$P_{i,\max}$
1	561.0	7.92	0.001 562	300.0	0.031 5	100.0	600.0
2	310.0	7.85	0.001 94	200.0	0.042	100.0	400.0
3	78.0	7.97	0.004 82	150.0	0.063	50.0	200.0

[79], the GA [95], the TS [94], the EP [92, 93], the modified Hopfield neural network (MHNN) [80], the adaptive Hopfield neural network (AHNN) [57], and MPSO [95].

The PARPSO is applied to an ED problem with three generators with the quadratic cost functions. Table 3.3.4 shows the input data of the system and the system demand considered is 850 MW. Table 3.3.5 shows the comparison of the results from PARPSO, MPSO [95], NM [76], improved evolutionary programming (IEP) [65], and MHNN [80], where the values of parameters are set as $\omega_{\max} = 0.8$, $\omega_{\min} = 0.4$, $c_1 = c_2 = 2$, $c_3 = 0.4$, $\Delta = 0.31$, and $d_{\text{low}} = 0.2$, $d_{\text{high}} = 0.8$.

As seen in Table 3.3.5, the PARPSO has provided the global solution, the same result of the lambda-iteration method and MPSO, exactly satisfying the equality and inequality constraints. In addition, the PARPSO is applied to two ED problems with three generators, where valve-point effects are considered. The input data for the three-generator system are given in Table 3.3.6. Note that the values of parameters are set as $\omega_{\max} = 0.8$, $\omega_{\min} = 0.4$, $c_1 = c_2 = 2$, $c_3 = 0.4$,

TABLE 3.3.7 Comparison Simulation Results of Each Method

Unit	GA	IEP (pop = 20)	EP	MPSO (par = 20)	PARPSO (par = 20)
1	300.00	300.23	300.26	300.27	349.47
2	400.00	400.00	400.00	400.00	400
3	150.00	149.77	149.74	149.73	100.53
TP	850.00	850.00	850.00	850.00	850.00
TC	8237.60	8234.09	8234.07	8234.07	8220.93

$\Delta = 0.31$, and $d_{low} = 0.2$, $d_{high} = 0.8$ as applied to the test system with quadratic cost functions.

Here, the total demand for the three-generator is set as 850 MW. It was reported in [94] that the global optimum solution without PAROSO found for the three-generator system is \$8234.07. PARPSO found the considered global optimum solution for this system with \$8220.93. The obtained results for the three-generator system are compared with those from GA [96], IEP [65], EP [92], and MPSO [95] as shown in Table 3.3.7.

It shows that the PARPSO has succeeded in finding the best global solution presented in [94], always satisfying the equality and inequality constraints.

Finally, the PARPSO has also been applied to the ED problem with 10 generators where the multiple-fuel effects are considered. In this case, the objective function is represented as the piecewise quadratic cost function. The input data and related constraints of the test system are given in [57, 65, 79]. In this case, the total system demand is varied from 2400 to 2700 MW with 100 MW increments.

For these problems, the same parameter determination strategy is adopted as the case of valve-point loading problems. The resulting values of parameters are $\omega_{max} = 0.8$, $\omega_{min} = 0.4$, $c_1 = c_2 = 2$, $c_3 = 0.4$, $\Delta = 0.31$ and $d_{low} = 0.2$, $d_{high} = 0.8$. The best results from the PARPSO are compared with those of HM [79], IEP [65], MHNN [80], AHNN [57], and MPSO [95] and given in Tables 3.3.8–3.3.11. In this case, the global solution is not known, or it may be impossible to find the global solution with the numerical approach for piecewise quadratic cost functions.

As seen in Tables 3.3.8–3.3.11, the PARPSO [98] has always provided better solutions than HM [79] (except for 2600 MW case), IEP [65], MHNN [80], and MPSO [95]. Furthermore, it has provided solutions satisfying the equality and inequality constraints while HM [79] and MHNN [80] do not satisfy the equality constraint.

When compared with AHNN and MPSO, the PARPSO has provided better solution for all demands (2400–2700 MW). Note that the fuel types and dispatch levels from the PARPSO are quite different from those of other approaches. Although the PARPSO has almost the same algorithm as the MPSO, the difference between each particle's behavior characteristic and total cost of PARPSO is 80–100 (\$) less than other approaches.

TABLE 3.3.8 Comparison Simulation Results of Each Method (Demand = 2400 MW)

		HM		MHNN		AHNN		IEP (pop = 30)		MPSO (par = 30)		PARPSO (par = 30)	
S	U	F	GEN	F	GEN	F	GEN	F	GEN	F	GEN	F	Gen
1	1	1	193.2	1	192.7	1	189.1	1	190.9	1	189.7	1	170.9
	2	1	204.1	1	203.8	1	202.0	1	202.3	1	202.3	1	184.8
	3	1	259.1	1	259.1	1	254.0	1	253.9	1	253.9	1	245.7
	4	3	234.3	2	195.1	3	233.0	3	233.9	3	233.0	3	231.0
2	5	1	249.0	1	248.7	1	241.7	1	243.8	1	241.8	1	237.5
	6	1	195.5	3	234.2	1	233.0	3	235.0	3	233.0	3	235.1
	7	1	260.1	1	260.3	1	254.1	1	253.2	1	253.3	1	222.1
3	8	3	234.3	3	234.2	3	232.9	3	232.8	3	233.0	3	232.9
	9	1	325.3	1	324.7	1	320.0	1	317.2	1	320.4	3	440
	10	1	246.3	1	246.8	1	240.3	1	237.0	1	239.4	1	200
TP			2401.2		2399.8		2400.0		2400.0		2400.0		2400.0
TC			488.500		487.87		481.700		481.779		481.723		422.08

TABLE 3.3.9 Comparison Simulation Results of Each Method (Demand = 2500 MW)

		HM		MHNN		AHNN		IEP (pop = 30)		MPSO (par = 30)		PARPSO (par = 30)	
S	U	F	GEN	F	GEN	F	GEN	F	GEN	F	GEN	F	Gen
1	1	2	206.6	2	206.1	2	206.0	2	203.1	2	206.5	1	178.0
	2	1	206.5	1	206.3	1	206.3	1	207.2	1	206.5	1	210.8
	3	1	265.9	1	265.7	1	265.7	1	266.9	1	265.7	1	248.0
	4	3	236.0	3	235.7	3	235.9	3	234.6	3	236.0	3	213.9
2	5	1	258.2	1	258.2	1	257.9	1	259.9	1	258.0	1	223.1
	6	3	236.0	3	235.9	3	235.9	1	236.8	3	236.0	3	236.7
	7	1	269.0	1	269.1	1	269.6	1	270.8	1	268.9	1	251.2
3	8	3	236.0	3	235.9	3	235.9	3	234.4	3	235.9	3	237.4
	9	1	331.6	1	331.2	1	331.4	1	331.4	1	331.5	3	440
	10	1	255.2	1	255.7	1	255.4	1	254.9	1	255.1	1	260.9
TP			2501.1		2499.8		2500.0		2500.0		2500.0		2500.0
TC			526.700		526.13		526.230		526.304		526.239		463.8

TABLE 3.3.10 Comparison Simulation Results of Each Method (Demand = 2600 MW)

		HM		MHNN		AHNN		IEP (pop = 30)		MPSO (par = 30)		PARPSO (par = 30)	
S	U	F	GEN	F	GEN	F	GEN	F	GEN	F	GEN	F	Gen
1	1	2	216.4	2	215.3	2	215.8	2	213.0	2	216.5	2	250
	2	1	210.9	1	210.6	1	210.7	1	211.3	1	210.9	1	205.3
	3	1	278.5	1	278.9	1	279.1	1	283.1	1	278.5	1	249.9
	4	3	239.1	3	238.9	3	239.1	3	239.2	3	239.1	3	232.6
2	5	1	275.4	1	275.7	1	276.3	1	279.3	1	275.5	1	246.9
	6	3	239.1	3	239.1	3	239.1	1	239.5	3	239.1	3	245.4
	7	1	285.6	1	286.2	1	286.0	1	283.1	1	285.7	1	270.1
3	8	3	239.1	3	239.1	3	239.1	3	239.2	3	239.1	3	241.4
	9	1	343.3	1	343.5	1	342.8	1	340.5	1	343.5	3	440
	10	1	271.9	1	272.6	1	271.9	1	271.9	1	272.0	1	218.4
TP			2600.0		2599.8		2600.0		2600.0		2600.0		2600.0
TC			574.030		574.26		574.370		574.473		574.381		508.560

TABLE 3.3.11 Comparison Simulation Results of Each Method (Demand = 2700 MW)

		HM		MHNN		AHNN		IEP (pop = 30)		MPSO (par = 30)		PARPSO (par = 30)	
S	U	F	GEN	F	GEN	F	GEN	F	GEN	F	GEN	F	Gen
1	1	2	218.4	2	224.5	2	225.7	2	219.5	2	218.3	2	228.9
	2	1	211.8	1	215.0	1	215.2	1	211.4	1	211.7	1	230
	3	1	281.0	3	291.8	1	291.8	1	279.7	1	280.7	1	238.7
	4	3	239.7	3	242.2	3	242.3	3	240.3	3	239.6	3	265
2	5	1	279.0	1	293.3	1	293.7	1	276.5	1	278.5	1	253.5
	6	3	239.7	3	242.2	3	242.3	1	239.9	3	239.6	3	236.7
	7	1	289.0	1	303.1	1	302.8	1	289.0	1	288.6	1	305.3
3	8	3	239.7	3	242.2	3	242.3	3	241.3	3	239.6	3	265
	9	3	429.2	1	355.7	1	355.1	3	425.1	3	428.5	3	440
	10	1	275.2	1	289.5	1	288.8	1	277.2	1	274.9	1	236.9
TP			2702.2		2699.7		2700.0		2700.0		2700.0		2700.0
TC			625.180		626.12		626.240		623.851		623.809		554.326

3.3.7 Conclusions

This chapter presents the fundamentals of GAs and PSO and their applications to ED problems in power systems. GAs have become one of the important tools for practical and robust optimization methods during the past decade in power system problems including ED problems. However, to deal with the limitations of the algorithm, various improvements have been explored in terms of GA implementation to ED such as encoding scheme, genetic operation variation, fitness function evaluation, and constraints handling.

Moreover, this study presents a new approach to non-smooth ED problems based on the PSO algorithm. The strategies for handling constraints are incorporated into the PSO framework in order to provide the solutions satisfying the inequality constraints. The equality constraint in the ED problem is resolved by reducing the degree of freedom by one at random. The strategies for handling constraints are devised while preserving the dynamic process of the PSO algorithm. Additionally, the dynamic search-space reduction strategy is applied to accelerate the convergence speed.

Numerical examples are tested on the sample ED problem by using GAs, and the results demonstrate that GAs have shown considerable success in providing good solutions in the ED problem. In addition, we can obtain that the MPSO has provided the global solution satisfying the constraints with a very high probability for the ED problems with smooth cost functions. For the ED problems with non-smooth cost functions due to the valve-point effects, the MPSO has also provided the global solution with a high probability for the three-generator system. In the case of non-smooth function problem due to multi-fuel effects, the MPSO has shown superiority to the conventional numerical method, the conventional Hopfield neural network, and the EP approach, while providing very similar results with the MHNN.

3.4 DIFFERENTIAL EVOLUTION IN ACTIVE POWER MULTI-OBJECTIVE OPTIMAL DISPATCH

Ming Zhou and Shu Xia

North China Electric Power University, Beijing, China

3.4.1 Introduction

With the increasing integration of renewable energy into power systems as well as the operational requirements of power systems, the influence of renewable energy on both the environment and power grid security has been attracting considerable attention. The optimal active power dispatch model of electric power systems has gradually been transformed from a single-objective optimization model to a multiple-objective optimization model with pollutant discharge level [111, 112], operational risk [113], accommodation capability [114], and other indicators included. Traditionally, the weights are set up manually for solving a multi-objective optimization problem in order to transform it into single-objective problems. However, in practice, it is difficult to determine these weights. In recent years, GAs, PSO, DE, and other intelligent optimization algorithms have been adopted to solve multi-objective optimization problems. The development of multi-objective optimization algorithms has provided another approach for solving multi-objective optimization problems. For such algorithms, it is not necessary to set up the weights; instead, a set of evenly distributed Pareto optimal solutions is obtained. Decision-makers can select one or more optimal solutions by adopting multiple-attribute decision-making according to their requirements. Among those solutions, two typical algorithms include the improved strength Pareto evolutionary algorithm (SPEA2) and the improved non-dominated sorting genetic algorithm (NSGA-II).

This chapter explores the application of the DE algorithm in the optimal active power dispatch of electric power systems. Owing to its good performance, the DE algorithm has been adopted by many researchers to solve optimal dispatch problems of electric power systems [115, 116]. Reference [117] has introduced the Pareto non-inferior grade and crowding distance sorting in NSGA-II on the basis of DE, with the multi-objective DE algorithm being proposed as well. The search performance of the algorithm is better than that of NSGA-II for many testing problems. However, in the differential evolution for multi-objective optimization (DEMO) algorithm, the randomly selected testing points may lead to "prematurity." Moreover, the crossover and mutation operations have high data dependency. The performance can be further improved on the basis of the fact that crowding distance sorting may cause uneven distribution of the Pareto optimal solution set.

Based on the research background presented above, the chaotic search strategy, adjustment strategy of the control parameter, and dynamic crowding distance sorting are introduced in this chapter. Moreover, the improved differential evolution for multi-objective optimization (IDEMO) algorithm is proposed and its application to multi-objective modeling of active power optimization for wind power integrated power systems is analyzed. The IEEE 118-node system is taken as an example to compare the IDEMO with three other multi-objective algorithms from the aspects of external solution, C index, and S index in order to verify that the proposed algorithm can obtain an evenly distributed Pareto optimal solution set and provide good candidate projects for optimal active power dispatch of power systems. Furthermore, Technique for Order Performance by Similarity to Ideal Solution (TOPSIS) is used for sorting and providing a final plan for optimal active power dispatch.

3.4.2 Differential Evolution for Multi-Objective Optimization

Basic Concepts

DE is based on the principle of "survival of the fittest" in biological populations. It is an intelligent optimization algorithm that adopts real-number coding for random search in continuous spaces. DE is an intelligent population-based optimization algorithm with the advantages of easy usage, strong robustness, and good global search capability. Therefore, it has been widely employed in practice. The crossover operation of the DE algorithm is conducive to maintaining the diversity of the Pareto optimal solution set, and its mechanism of "greedy" selection is conducive to convergence speed enhancement.

Conflicts occur between the different objectives of multi-objective optimization problems, and it is impossible to ensure that all objective values are optimized at the same time. Therefore, only one Pareto optimal solution set can be obtained. A certain set Y_f refers to the feasible set of optimal active power dispatch. Several basic concepts for the Pareto optimum are discussed below.

Definition 3.1 (Pareto domination) Suppose that y_1 and y_2 are both feasible solutions. When all the objective values corresponding to y_1 are not inferior to those of y_2, and one of the objective values of y_1 is superior to those of y_2, i.e.

$$\{\forall i \in \{1,2,\dots,M\}, f_i(y_1) \le f_i(y_2)\} \wedge \{\exists i \in \{1,2,\dots,M\}, f_i(y_1) < f_i(y_2)\} \tag{3.4.1}$$

It can be concluded that y_1 dominates y_2 with notation $y_1 \prec y_2$. In the above equation, M is the number of objective functions and $f_i(\cdot)$ represents the ith objective function.

Definition 3.2 (Pareto optimal solution) If $y_e \in Y_f$ and there is no $y \in Y_f$ that satisfies $y \prec y_e$, then y_e is the Pareto optimal solution or the non-dominated solution.

Definition 3.3 (Pareto optimal solution set) If all the elements of the set are Pareto optimal solutions, then P_E is the Pareto optimal solution set or the non-dominated solution set.

Definition 3.4 (Pareto frontier) When the set is composed of target components corresponding to the Pareto optimal solution set, it is called P_F or the Pareto frontier, i.e.

$$P_F = \{F(y_e) = [f_1(y_e), f_2(y_e), \dots, f_M(y_e)] | y_e \in P_E\} \tag{3.4.2}$$

Procedure of DEMO Algorithm

The procedure of the DEMO algorithm includes initialization, mutation, crossover, and selection operations. Among these operations, the selection operation adopts a setup of non-inferior level, crowding distance calculation, and so on in NSGA-II in order to make it applicable for solving multi-objective problems and obtaining a Pareto optimal solution set.

Suppose that the population size in the DEMO algorithm is N_p. The specific operations of initialization, mutation, and crossover are described below.

Initialization The DEMO operation is suitable for a population consisting of multiple individuals. Therefore, initialization of the population is carried out.

The initial population $S = \{x_1, x_2, x_d, \ldots, x_D\}$, $x_d \in R^J$; for the dth individual, $x_d = (x_{d,1}, x_{d,2}, \ldots, x_{d,J})$, where J is a variable dimension. In general, each component of the individual vector x_d is generated as follows:

$$x_{d,j} = x_{d,j}^{\min} + r_{\text{rand}} \cdot \left(x_{d,j}^{\max} - x_{d,j}^{\min}\right) \tag{3.4.3}$$

where

$x_{d,j}$: jth component of x_d,

$x_{d,j}^{\max}$: upper limits of the jth component,

$x_{d,j}^{\min}$: lower limits of the jth component,

r_{rand}: random number between 0 and 1.

Mutation The DEMO mutation operation is the basis for the algorithm, which is used to generate new individuals. The mutation operation of the dth individual of the kth generation is given by

$$v_d^k = x_{d1}^k + F \cdot \left(x_{d2}^k - x_{d3}^k\right) \tag{3.4.4}$$

where

v_d^k: intermediate individuals obtained in the mutation,

F: mutation factor,

$x_{d1}^k, x_{d2}^k, x_{d3}^k$: three different individuals randomly selected in the contemporary parent population with $d1$, $d2$, and $d3$ being distinct from d.

Crossover The crossover operation aims to improve the dissimilarity of new individuals as offspring. In the crossover operation, the jth component of the dth individual in the offspring depends on the intermediate individual $v_d^k = \left(v_{d,1}^k, v_{d,2}^k, \ldots, v_{d,J}^k\right)$ and the current individual x_d^k. The specific operation is given by

$$u_{d,j}^k = \begin{cases} v_{d,j}^k, & \phi_j \le C_R \\ x_{d,j}^k, & \phi_j > C_R \end{cases} \tag{3.4.5}$$

where

u_d^k: tentative offspring, and $u_d^k = \left(u_{d,1}^k, u_{d,2}^k, \ldots, u_{d,J}^k\right)$,

ϕ_j: randomly controlled parameter, and $\phi_j \in [0, 1]$,

C_R: crossover factor.

Selection In the selection operation, the experimental individuals u_d^k should compete with the parent individual xk. The operation criteria are as follows:

1. If one individual can dominate another, the dominated one will enter the new population and the non-dominated one will be eliminated.

2. If the two individuals do not dominate each other, they will enter the new population at the same time.

After the above operation, a population with a scale between N_p and $2N_p$ will be obtained. The quality level and crowding distance are calculated as well.

The setup process of the quality level is as follows. First, in the new population, the optimal solution is obtained, and its quality level is 1. Then, in the remaining population, the optimal solution is obtained, and its quality level is 2. Similarly, the quality level of all the individuals is obtained.

The crowding distance of the boundary solution of individuals with the same quality level can be set as a large number. For the points that are not on the boundaries, the crowding distance is expressed as

$$\sigma_d = \sum_{m=1}^{M} \sigma_{d,m} \tag{3.4.6}$$

$$\sigma_{d,m} = \frac{f_m\left(x_d^{\text{next}}\right) - f_m\left(x_d^{\text{front}}\right)}{f_m^{\max} - f_m^{\min}} \tag{3.4.7}$$

where

M: number of objectives,

σ_d: crowding distance of the dth individual,

$\sigma_{d,m}$: crowding distance of the dth individual in the mth objective component,

$f_m\left(x_d^{\text{next}}\right)$: next adjacent values of $f_m(x_d)$,

$f_m\left(x_d^{\text{front}}\right)$: previous adjacent values of $f_m(x_d)$,

$f_m^{\max}$: upper limits of the mth objective component,

$f_m^{\min}$: lower limits of the mth objective component.

At the same quality level, all the individuals together with the dth individual will be sorted in ascending order according to the mth component value of the objective.

The new population is sorted according to the quality level of the individual and the crowding distance. The criteria are (i) the individual with lower non-inferior level will be in the front, and (ii) if the non-inferior level is the same, the individual with small crowding distance will be in the front. Further, after sorting, N_p individuals are selected from the population as the next generation of the parent population.

Improvement Measurement of DEMO

Chaos Initialization Strategy For population initialization of DEMO, a random method is adopted. Thus, the groups generated will be distributed unevenly with the utilization ability of the algorithm for the initial group being reduced. Reference [118] exploits the randomness and ergodicity of chaotic signals to introduce an initial group of the chaotic model in one-dimensional logistic mapping, and it is given by

$$r_{g+1} = \mu r_g \left(1 - r_g\right) \tag{3.4.8}$$

where

r_g: value after the gth chaos evolution,

μ: control parameter, whose value generally is 4.

However, the evenness of the chaotic mapping is poor and the density of the mapping points on the edges is relatively higher, while it is relatively lower in the middle of the interval. Therefore, mixed mapping of the linear power function with better evenness is adopted in this work [115]. It is expressed as

$$r_{g+1} = \begin{cases} r_g^p, r_g \in (0, a) \\ \mu r_g \left(1 - r_g\right), r_g \in [a, b] \\ r_g^q, r_g \in (b, 1) \end{cases} \tag{3.4.9}$$

and $0 < a < b < 1, 0 < p < 1, q > 1$. In general, a is 0.2, b is 0.8, p is 0.5, and q is 15.

The specific operation process for the initialization strategy of the chaotic model is as follows:

First, $r_{1,j}$ (j = 1,2, …, J) of r_g is randomly generated as shown in (3.4.9), where J is the number of components of an individual. Then, each component $r_{d+1,\,j}$ of vector $r_{d+1} = (r_{d+1,1}, \ldots, r_{d+1,j}, \ldots, r_{d+1,J})$ is generated, i.e.

$$r_{d+1,j} = \begin{cases} r_{d,j}^p, r_{d,j} \in (0, a) \\ \mu r_{d,j} \left(1 - r_{d,j}\right), r_{d,j} \in [a, b], \quad d = 1, 2, \ldots, N_{\mathrm{p}} - 1 \\ r_{d,j}^q, r_{d,j} \in (b, 1) \end{cases} \tag{3.4.10}$$

Finally, $r_{d,i}$ is used to modify (3.4.3) with the jth component of the dth individual in the population being obtained:

$$x_{d,j} = x_{d,j}^{\min} + r_{d,j} \cdot \left(x_{d,j}^{\max} - x_{d,j}^{\min}\right) \tag{3.4.11}$$

Adjustment Strategy of Control Parameter In DEMO, the setup of the mutation factor F and crossover factor C_{R} has a significant impact on the convergence performance and search efficiency of the algorithm. In DE, there are various parameter control strategies. Reference [115] uses fitness variance as the adaptive adjusting parameter. Reference [119] adjusts the parameter by comparing the value of objective functions of the experimental individual and the parent individual. Reference [120] adopts a random method and evolutionary algebra as the basis for adjustment, which is an effective approach. The method considered in Reference [121] is sufficiently feasible to be applied in DEMO. However, this method makes the value of F fall within [0.5, 1]. For the problem of optimal active power dispatch, this value is not reasonable; the value scope of F is determined by actual problems. Therefore, this chapter proposes the following adjustment method of the control parameter based on Reference [120].

$$F_d^k = F_{\min} + r_{\mathrm{rand}} \cdot (F_{\max} - F_{\min}) \tag{3.4.12}$$

$$C_{\mathrm{R}d}^k = (C_{\mathrm{Rmax}} - C_{\mathrm{Rmin}}) \cdot \frac{(K_{\max} - k)}{K_{\max}} \tag{3.4.13}$$

where

F_d^k: mutation factor of the dth individual in the kth generation,
$C_{\mathrm{R}d}^k$: cross factor of the dth individual in the kth generation,
$F_{\max}$: upper limits of the mutation factor,
$F_{\min}$: lower limits of the mutation factor,
C_{Rmax}: upper limits of the cross factor,
C_{Rmin}: lower limits of the cross factor,
$K_{\max}$: maximum number of iterations.

Sorting Strategy of Dynamic Crowding Distance In the same Pareto non-inferior grade, DEMO takes the crowding distance of individuals for sorting, i.e. the crowding distance of all the individuals is calculated, and then, the optimal solutions are selected according to the number required. This operation is fast but the results are relatively rough. Moreover, it may lead to uneven distribution of the solutions. As shown in Figure 3.4.1a, f_1 and f_2 represent the values of two different objective functions. For points A and B, all the solutions are not selected, which causes part of the optimum solutions to be lost. Therefore, this chapter adopts the crowding distance sorting strategy. Its operation process is as follows. First, some special non-inferior level l is obtained, i.e. the total number of individuals lower than this non-inferior level is smaller than N_{p}. The total number of individuals with this non-inferior level and the total number of individuals lower than this non-inferior level are greater than N_{p}. Then, the crowding distance of all the individuals in this non-inferior level is calculated, with the individual whose crowding distance is the smallest being excluded. Next, the crowding distance of all the remaining individuals is recalculated, with the individual whose crowding distance is the smallest being excluded. Therefore, the calculation continues until the number of remaining individuals is equal to the set number. According to the examples in Figure 3.4.1a, the results of the sorting strategy of dynamic crowding distance are shown in Figure 3.4.1b. The solutions selected are distributed evenly. Overall, although the computing speed of the sorting strategy of dynamic crowding distance is slightly lower than that of the sorting strategy of crowding distance, a population in which the individuals are more evenly distributed is obtained.

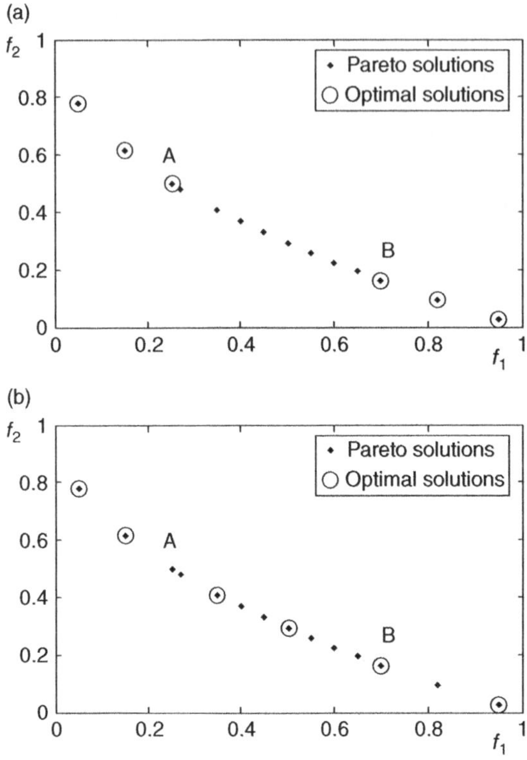

Figure 3.4.1 Solution distribution by different selection strategies. (a) Crowding distance strategy. (b) Dynamic crowding distance strategy.

TOPSIS

After a Pareto optimal solution set is obtained by solving the multi-objective optimization problem, a decision-making problem for multiple attributes should be selected. Currently, the most commonly used method is the TOPSIS, which approaches the ideal plan [122]. Its aim is to make the selected plan in the shortest distance to the ideal plan and the longest distance to the negative ideal plan. Suppose that there are m candidate plans for n attributes ($n = J$, $m = N_{\mathrm{p}}$). The TOPSIS calculation process is as follows:

1. *Normalized decision-making matrix*: If the decision-making matrix for the decision-making problem of multiple attributes is $P = \{p_{i,\ j}\}$ and the decision-making matrix after standardizing is $Z = \{z_{i,\ j}\}$, then

$$z_{i,j} = \frac{p_{i,j}}{\sqrt{\sum_{c=1}^{m} p_{c,j}^2}}, \quad i = 1, \ldots, m; j = 1, \ldots, n \tag{3.4.14}$$

2. *Establish the weighted normalized matrix*: If it is $H = \{h_{i,\ j}\}$, then

$$h_{i,j} = w_j z_{i,j}, \quad i = 1, \ldots, m; j = 1, \ldots, n \tag{3.4.15}$$

where w: weight vector, and $w = (w_1, \ldots, w_n)$, which is generally provided by the decision-maker.

3. *Determine the ideal solution h^+ and the negative ideal solution h^-*: Given by

$$h_j^+ = \begin{cases} \max_{i=1}^{m} h_{i,j}, j \text{ is the benefit-type attribute} \\ \min_{i=1}^{m} h_{i,j}, j \text{ is the cost-type attribute} \end{cases}, \quad j = 1, \ldots, n \tag{3.4.16}$$

$$h_j^- = \begin{cases} \max_{i=1}^{m} h_{i,j}, j \text{ is the cost-type attribute} \\ \min_{i=1}^{m} h_{i,j}, j \text{ is the benefit-type attribute} \end{cases}, \quad j = 1, \ldots, n \tag{3.4.17}$$

where

h_j^+ : jth attribute values of h^+,

h_j^- : jth attribute values of h^-.

4. *Determine the distance of each candidate plan to the ideal plan and negative ideal plan*: Given by

$$d_i^+ = \sqrt{\sum_{j=1}^{n} \left(h_{i,j} - h_j^+\right)^2}, \quad i = 1, \ldots, m \tag{3.4.18}$$

$$d_i^- = \sqrt{\sum_{j=1}^{n} \left(h_{i,j} - h_j^-\right)^2}, \quad i = 1, \ldots, m \tag{3.4.19}$$

5. *Comprehensive evaluation index of each candidate plan*: Given by

$$C_i = \frac{d_i^-}{d_i^+ + d_i^-}, \quad i = 1, \ldots, m \tag{3.4.20}$$

6. *Sort the candidate plans*: The candidate plans are sorted in descending order according to the value of C_i. The largest solution of C_i is used as the trade-off solution.

In the sorting process of the TOPSIS method, the weight of each attribute should be provided manually, which is manifested in (3.4.15). In many practical problems, it is difficult to determine the weights. Therefore, a weight selection method based on entropy can be adopted. In information theory, entropy is used to measure the degree of uncertainty. The greater the uncertainty of the variables, the larger is the entropy and the greater is the information required for clarification. Therefore, combined with the concept of information entropy, the contrast information of the data set is used to determine the weights of the attributes. After the matrix operation of normalized decision-making, i.e. (3.4.14), the selection process of weights based on entropy is as follows:

The entropy value e_j of the jth attribute in the candidate plan is calculated as

$$e_j = -\frac{1}{\ln m}\sum_{i=1}^{m} z_{i,j} \ln z_{i,j}, \quad j = 1, \ldots, n \tag{3.4.21}$$

The differential coefficient of the jth attribute is determined as

$$g_j = 1 - e_j, \quad j = 1, \ldots, n \tag{3.4.22}$$

The weight coefficient w_j of the jth attribute is obtained according to the differential coefficient:

$$w_j = \frac{g_j}{\sum_{j=1}^{n} g_j} \tag{3.4.23}$$

For the jth attribute, the smaller the entropy, the greater is the difference and the weight of the index is correspondingly larger. By contrast, the smaller the entropy, the greater is the difference and the weight of the index is correspondingly smaller. Thus, from the perspective of a certain attribute, when the results of all the plans are the same, this attribute will play a small role in decision-making.

Calculation Process of IDEMO Algorithm

The IDEMO algorithm and TOPSIS algorithm based on entropy are used in the multi-objective optimal active power dispatch model. The flowchart is shown in Figure 3.4.2.

The specific calculation steps are as follows

Step 1: Set up each parameter of the algorithm.

Step 2: According to (3.4.11), the population is initialized with the chaotic search strategy.

Step 3: According to (3.4.4) and (3.4.5), the population is mutated and the new population is obtained after the crossover operation. In this process, (3.4.12) and (3.4.13) are adopted to adjust the control parameter.

Step 4: The selection operation for the population is carried out by adopting merits level and dynamic crowding distance sorting.

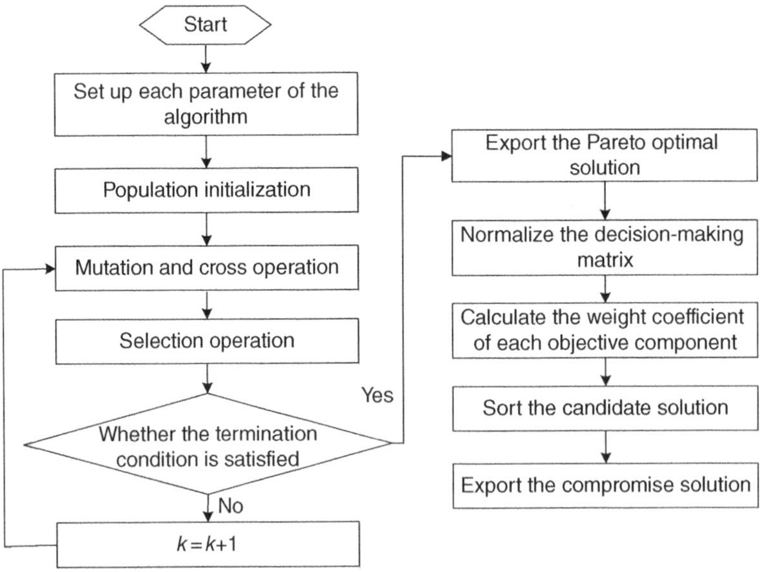

Figure 3.4.2 The calculation flowchart of IDEMO algorithm.

Step 5: Determine if the optimization condition is reached. If it is, the Pareto optimal solution set of the final generation is exported; otherwise, go back to Step 3 for calculation.

Step 6: The Pareto optimal solution set is used as the decision-making matrix and processed for normalization according to (3.4.14).

Step 7: Based on the theory of entropy, the weight coefficient of each objective component is calculated according to (3.4.21–3.4.23).

Step 8: The comprehensive evaluation index of each candidate plan is calculated by adopting the TOPSIS algorithm and combining it with (3.4.15) to (3.4.20). Moreover, sorting of the decision-making for the Pareto optimal solution set is performed according to the comprehensive evaluation index with the trade-off plan being exported as well.

3.4.3 Multi-Objective Model of Active Power Optimization for Wind Power Integrated Systems

Multi-Objective Function

The multi-objective model of active power optimization for a wind power integrated system can be expressed as follows, with the objective of reducing the fuel cost, pollutant emissions, and reserve risk for wind power and load uncertainty:

$$\min \quad y = [I_{\mathrm{C}}, I_{\mathrm{E}}, I_{\mathrm{R}}] \tag{3.4.24}$$

where

I_{C}: fuel cost,

I_{E}: pollutant emissions,

I_{R}: reserve risk index that is used to manifest the degree of load loss and the degree of wind power curtailment under different reserve capacities,

$$I_{\mathrm{C}} = \sum_{i=1}^{N_{\mathrm{I}}} \left[a_i + b_i P_{i,t} + c_i P_{i,t}^2 + \left| d_i \sin \left\{ e_i \left(P_i^{\min} - P_{i,t} \right) \right\} \right| \right] \tag{3.4.25}$$

$$I_{\mathrm{E}} = \sum_{i=1}^{N_{\mathrm{I}}} \left[\alpha_i + \beta_i P_{i,t} + \gamma_i P_{i,t}^2 + \eta_i \exp\left(\delta_i P_{i,t} \right) \right] \tag{3.4.26}$$

$$I_{\mathrm{R}} = \sum_{s} P_{\mathrm{rob}}^{s} \omega_1 \frac{V_{\mathrm{D}}^{s}}{P_{\mathrm{D}}} + \sum_{s} P_{\mathrm{rob}}^{s} \omega_2 \frac{V_{\mathrm{W}}^{s}}{P_{\mathrm{WR}}} \tag{3.4.27}$$

where

N_{I}: number of conventional generators,

$P_{i,t}$: capacity of unit i during time t,

$P_i^{\min}$: minimum power output of unit i,

a_i, b_i, c_i, d_i, e_i: fuel cost coefficients of unit i,

α_i, β_i, γ_i, η_i, δ_i: emission coefficients of pollutant gas of unit i,

s: possible operation scenarios that may occur because of wind power and load forecasting errors,

ω_1, ω_2: weight coefficients,

P^s_{rob}: probability for the sth scenario,

V^s_{D}: amount of load shedding in the sth scenario,

V^s_{W}: amount of curtailed wind in the sth scenario,

P_{D}: expected data of the load,

P_{WR}: total installed capacity of the wind power plant (WPP).

All the information can be obtained according to the detection data of pollution gas in the power plant by adopting the least squares method.

The greater the reserve amount for the dispatch plan, the smaller is the amount of load shedding and amount of curtailed wind and the lower is the risk indicator of reserve shortage. Otherwise, the smaller the reserve amount for the dispatch plan, the higher is the risk indicator of reserve shortage. Further, I_{R} can be obtained using a specific calculation method [123].

Constraints

1. Minimum and maximum power output of thermal units

$$P_i^{\min} \le P_{i,t} \le P_i^{\max} \tag{3.4.28}$$

$$-T_{10} r_i^{\text{d}} \le P_{i,t} - P_{i,t-1} \le T_{10} r_i^{\text{u}} \tag{3.4.29}$$

where

$P_i^{\max}$: maximum power output of unit i,

T_{10}: duration of 10 minutes,

r_i^{u}: unit i's up ramping rate (MW/min),

r_i^{d}: unit i's down ramping rate (MW/min).

2. Load balance constraint

$$\sum_{i=1}^{N_{\text{I}}} P_{i,t} + \sum_{w=1}^{N_{\text{W}}} P_{w,t} - P_{\text{D}t} - P_{\text{L}} = 0 \tag{3.4.30}$$

Here, power losses P_{L} are calculated by the B coefficient method,

$$P_{\text{L}} = \sum_{n=1}^{N_{\text{G}}} \sum_{m=1}^{N_{\text{G}}} P_{\text{G}n,t} B_{n,m} P_{\text{G}m,t} + \sum_{n=1}^{N_{\text{G}}} P_{\text{G}n,t} B_{n0} + B_{00} \tag{3.4.31}$$

where

N_W: number of wind plants,

$P_{w,t}$: wind plant w's scheduled output at time t,

P_{Dt}: load forecast at time t,

N_G: number of all generators, including thermal units and wind turbine generators, and $N_G = N_I + N_W$,

$P_{Gn,t}$, $P_{Gm,t}$: unit n's output and unit m's output, respectively, including thermal unit and wind unit,

$B_{n,m}$, B_{n0}, B_{00}: coefficients in the B coefficient method.

Reserve constraints

In ED, the units with adjusting capability should reserve an amount of capacity to deal with the uncertainties from load forecast and wind power forecast,

$$\sum_{i=1}^{N_I} X_i S_i^u \geq P_R^u \tag{3.4.32}$$

$$\sum_{i=1}^{N_I} X_i S_i^d \geq P_R^d \tag{3.4.33}$$

where

X_i: thermal unit i's real time adjust status, 1 means participating in adjust, 0 means not participating,

S_i^u: unit i's up adjusting capacity,

S_i^d: unit i's down adjusting capacity,

P_R^u: required up reserve which is used to cover the forecast error,

P_R^d: required down reserve which is used to cover the forecast error.

Moreover, S_i^u, S_i^d should meet the unit i's adjusting capability and output limit, that is,

$$\begin{cases} S_i^u \leq S_{i\max}^u \\ P_{i,t} + S_i^u \leq P_i^{\max} \end{cases} \tag{3.4.34}$$

$$\begin{cases} S_i^d \leq S_{i\max}^d \\ P_{i,t} - S_i^d \geq P_i^{\min} \end{cases} \tag{3.4.35}$$

where

$S_{i\max}^u$: unit i's up adjusting capability limit,

$S_{i\max}^d$: unit i's down adjusting capability limit.

Meanwhile, SR should cover one online unit's unexpected outage (i.e. the "$N-1$" criteria), that is,

$$\sum_{\substack{i=1 \\ i\neq j}}^{N_\mathrm{I}} P_{\mathrm{S}i} \geq P_\mathrm{R}^\mathrm{u} + P_{j,t}, \quad j = 1, 2, \ldots, N_\mathrm{I} \tag{3.4.36}$$

$$P_{\mathrm{S}i} = \min\left(T_{10} r_i^\mathrm{u}, P_i^\mathrm{max} - P_{i,t}\right) \tag{3.4.37}$$

where $P_{\mathrm{S}i}$ is SR provided by unit i.

3.4.4 Case Studies

Testing System

Based on the IEEE 118-bus system [131], the categories of the generators are corrected. The system includes 118 buses, 184 lines, 14 conventional generators, and 2 WPPs. WPP 1 is connected to bus 54 and WPP 2 is connected to bus 103. The two WPPs include 100 wind power generators. The units C_1, C_3, C_5, C_{10}, and C_{12} are involved in real-time adjustment. For further information about the parameters of the units, wind power generator, loading, and so on, the readers are referred to Reference [123].

Performance Analysis of IDEMO

In order to compare the performance of the objective optimization algorithms, SPEA2, NSGA-II, DEMO, and IDEMO are used to solve the multiple-objective power dispatch problem. The parameters of IDEMO are set as follows: population scale, $N_\mathrm{P} = 100$; maximum number of iterations, $K_\mathrm{max} = 1500$; upper and lower limits of variability factor, $C_\mathrm{Rmax} = 0.8$ and $C_\mathrm{Rmin} = 0.3$. The population scale and the maximum number of iterations of the other algorithms are the same as those of IDEMO. After 1500 generations, the four algorithms obtain a clear Pareto frontier and the spatial shape constructed by the Pareto frontier is also very similar. As shown in Figure 3.4.3, the four algorithms are all suitable for solving the multi-objective

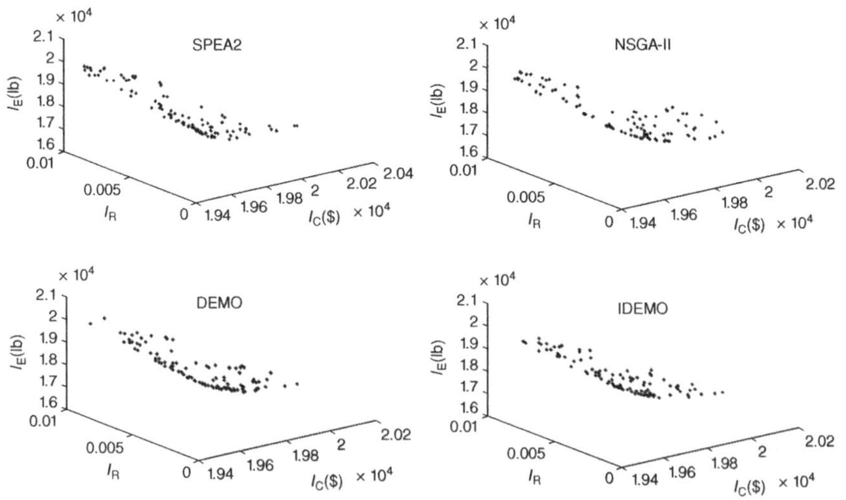

Figure 3.4.3 Pareto fronts of different algorithms.

optimal dispatch problem. The Pareto optimal solution set obtained can show the relationship between each objective in a better way and provide the decision-maker with more scope for choice. From the Pareto frontier, it is difficult to compare the performance of each algorithm. The external solution, *C* index, and *S* index are used to specifically analyze the advantages of IDEMO as discussed below.

Comparison and Analyses of External Solution In the Pareto optimal solution set, the optimal solution of one objective component is called the external solution. In general, the number of objective components is the same as the number of external solutions. It is defined as

$$x_m^k = \{x_i \mid \forall x_j \in P_{\mathrm{E}}^k : f_m(x_i) \leq f_m(x_j)\} \tag{3.4.38}$$

where

m: objective component,

k: evolutional generations,

P_{E}^k: Pareto optimal solution set of the kth generation.

The mth objective component corresponding to x_m^k is the optimum one. The global search performance, robustness, and computational speed of the algorithm can be evaluated by comparing the final value and its evolutionary process of the external solution for the objective components.

In order to analyze the external solution, the algorithm was independently run 30 times. For each objective function, each algorithm obtained 30 components corresponding to the external solution in the kth generation, i.e. the fuel cost $x_{\mathrm{C}}^{(k)}$, pollutant discharge level $x_{\mathrm{E}}^{(k)}$, and risk indicator of reserve shortage. The objective components $x_{\mathrm{C}}^{(1500)}$ $x_{\mathrm{E}}^{(1500)}$ $x_{\mathrm{R}}^{(1500)}$ corresponding to 30 external solutions obtained in the 1500th generation are obtained statistically (see Table 3.4.1). From Table 3.4.1, one can see that:

The optimal solution for the fuel cost and the best value of the optimal solution for the pollutant emissions obtained by IDEMO are \$19 436.43 and 16 910.82 lb, respectively, which are slightly better than the results of DEMO (\$19 437.21 and 16 916.62 lb). Moreover, they are significantly better than the results of SPEA2 (\$19 461.62 and 16 964.06 lb) and NSGA-II (\$19 450.76 and 16 920.17 lb), which implies that the global convergence performance of IDEMO is better.

The deviations between the best value and the worst value of the optimal solution of fuel cost and pollutant emissions based on 30 runs of IDEMO are 0.016 and 0.013%, respectively, which are significantly smaller than the deviations of the other three algorithms, implying that IDEMO is the most robust algorithm.

The average value of the optimal objective component of the fuel cost, which is obtained by calculating the external solution after 30 runs of each algorithm, is acquired statistically, and the evolution curve from the 100th generation to the 1500th generation is plotted as shown in Figure 3.4.4. As the results of previous generations of various algorithms cannot meet the relevant constraints, the results

TABLE 3.4.1 External Solutions at the Final Generation by Different Algorithms

	$I_C(x_C^{[1500]})$			$I_E(x_E^{[1500]})$			$I_R(x_R^{[1500]})$		
Algorithm	Best Value ($)	Worst Value ($)	Deviation (%)	Best Value (lb)	Worst Value (lb)	Deviation (%)	Best Value	Worst Value	Deviation (%)
SPEA2	19 461.62	19 479.37	0.091	16 964.06	17 020.25	0.331	0.003 33	0.003 33	0
NSGA-II	19 450.76	19 470.81	0.103	16 920.17	16 991.33	0.421	0.003 33	0.003 33	0
DEMO	19 437.21	19 442.26	0.026	16 916.62	16 951.25	0.205	0.003 33	0.003 33	0
IDEMO	19 436.43	19 439.48	0.013	16 910.82	16 937.79	0.160	0.003 33	0.003 33	0

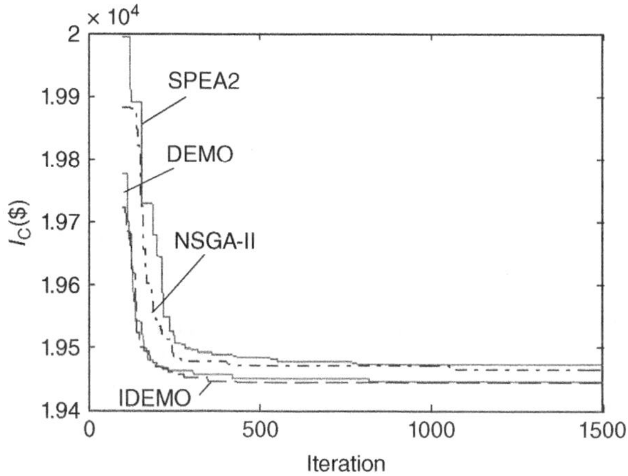

Figure 3.4.4 Convergence curves of outer solutions for cost.

of the previous 100 generations are not displayed. As seen from Figure 3.4.4, SPEA2 and DEMO show convergence in around 800 generations, while NSGA-II needs more than 1000 generations. The IDEMO algorithm obtains the approximate optimal solution at about 500 generations. The calculation speed is significantly faster than that of the other three algorithms.

Comparative Analysis of C Index The C index is used to compare the relative coverage rate of the two solutions in multi-objective optimization, and it can also be used to determine the quality of the solution set. It is defined as follows [125]:

$$C(Q_1, Q_2) = \frac{|\{g \in Q_2; \exists e \in Q_1 : e \prec g\}|}{|Q_2|} \tag{3.4.39}$$

where

Q_1, Q_2: two solution sets,

e, g: elements in Q_1 and Q_2, respectively,

C(Q_1, Q_2): ratio of solutions in Q_2 being dominated by Q_1, for example, $C(Q_1, Q_2) = 50\%$ implies that 50% of the solutions in Q_2 are dominated by those in Q_1.

Table 3.4.2 lists the average values of the final generation C indicator for 30 runs. It can be seen from the table that 78 and 49% of the solutions of the IDEMO algorithm can dominate the solutions of SPEA2 and NSGA-II, proving that the algorithm is of high quality.

Comparative Analysis of S Index The S index is used to measure the distribution of points on the solution set. The smaller the S index, the more even is the distribution of the solution set. It is defined as [126]:

$$S = \sqrt{\frac{1}{N_p - 1}\sum_{i=1}^{N_p}(\bar{d} - d_i)^2} \tag{3.4.40}$$

TABLE 3.4.2 Average of Metric C at the Final Generation

Algorithm	SPEA2	NSGA-II	DEMO	IDEMO
SPEA2	—	6	0	0
NSGA-II	39	—	2	0
DEMO	71	42	—	3
IDEMO	78	49	11	—

TABLE 3.4.3 Metric S at the Final Generation by Different Algorithms

Algorithm	Best Value (p.u.)	Worst Value (p.u.)	Deviation (p.u.)
SPEA2	0.0504	0.0697	0.0193
NSGA-II	0.0462	0.0589	0.0127
DEMO	0.0404	0.0469	0.0065
IDEMO	0.0323	0.0378	0.0055

$$d_i = \min\left\{\sum_{m=1}^{M} |f_m(x_i) - f_m(x_j)|\right\}, j = 1, 2, \ldots, N_p \tag{3.4.41}$$

where

N_p: population size,

d_i: neighbor vector distance of the ith individual,

$\bar{d}$: average value of d_i,

M: number of objective components,

$f_m(x)$: mth objective component value of individual x.

In the above equation, the influence of the dimension of each objective component is not considered. Therefore, (3.4.41) is modified as follows:

$$d_i = \min\left\{\sum_{m=1}^{M} \frac{|f_m(x_i) - f_m(x_j)|}{f_{m\max} - f_{m\min}}\right\}, j = 1, 2, \ldots, N_p \tag{3.4.42}$$

where

$f_{m\max}$: maximum values of the M objective components,

$f_{m\min}$: minimum values of the M objective components.

The statistics for the S indicators of the final generation of the Pareto frontier obtained after 30 runs are listed in Table 3.4.3. As shown in the table, the final generation S indicator obtained by IDEMO is between 0.0323 and 0.0378, which is significantly smaller than that of the other three optimization algorithms. This proves that in the Pareto frontier obtained by IDEMO, the solutions are distributed more evenly.

The above analyses show that the robustness, computation speed, and solution accuracy of the IDEMO algorithm are obviously superior to those of SPEA2 and NSGA-II and slightly superior to those of the DEMO algorithm. The uniformity of the solution of the IDEMO algorithm is obviously better than that of the other three algorithms. Therefore, the IDEMO algorithm can provide a better candidate for the multi-objective model of active power optimization dispatch for power systems.

3.4.5 Analyses of Dispatch Plan

In order to study the influence of different dispatch plans on the power grid, the following plans are designed:

Plan 1: In the Pareto frontier, the external solution corresponding to the minimum fuel cost.

Plan 2: In the Pareto frontier, the external solution corresponding to the minimum pollutant emission.

Plan 3: In the Pareto frontier, the external solution corresponding to the optimal risk indicator of reserve shortage.

Plan 4: In the Pareto frontier, the optimal solution obtained by using the TOPSIS decision-making method based on entropy.

The fuel costs, pollutant emissions, and risk indicator of reserve shortage of the four plans are summarized in Table 3.4.4. By comparing Plan 1 with Plan 2, it can be seen that if only the cost of fuel is taken into consideration, i.e. IC is \$19 436.43, the pollutant emissions of Plan 1 are higher than those of Plan 2 by 3500.91 lb. This is detrimental to environmental protection, and the reserve risks corresponding to Plan 1 and Plan 2 are too large, indicating that the fuel costs, pollutant emissions, and risk indicator of reserve shortage are counterbalanced by one another. For Plan 3, when the risk indicator of reserve shortage is the smallest, i.e. when all the units are able to provide the maximum reserve, there will be multiple dispatch plans. Here, one of them is randomly selected and listed in Table 3.4.4, and the optimal risk indicator obtained is 0.003 33. Plan 4 uses TOPSIS based on entropy, which is proposed to obtain a trade-off solution from an objective point of view. The weight coefficients of the fuel costs, pollutant emissions, and risk indicator of reserve shortage are wc = 0.01, we = 0.03, and wr = 0.96, respectively. The weight of the risk indicator of reserve shortage is found to be the largest, i.e. 96%,

TABLE 3.4.4 Optimization Results in Different Scheduling Modes

Dispatch Plan	Fuel Cost (\$)	Pollutant Emission (lb)	Risk Indicator of Reserve Shortage
Plan 1	19 436.43	20 411.73	0.007 67
Plan 2	20 138.99	16 910.82	0.008 83
Plan 3	19 676.85	17 813.84	0.003 33
Plan 4	19 783.64	17 423.03	0.003 33

in the final decision-making process, while the weights of fuel cost and pollutant emissions account for 1 and 3%, indicating that the risk indicator of reserve shortage has the greatest impact on decision-making. According to the trade-off solution, the corresponding fuel cost is $19 783.64, the pollutant emission is 17 423.03 lb, and the risk indicator of reserve shortage is 0.003 33; thus, the risk indicator of reserve shortage is the smallest. Therefore, it can be seen that TOPSIS based on entropy can coordinate the various objectives in a better way so as to realize the function of decision-making.

3.4.6 Conclusions

The IDEMO algorithm and the TOPSIS algorithm based on entropy were adopted to realize multi-objective optimal active power dispatch for wind power integrated systems with the fuel cost, pollutant emissions, and risk indicator of reserve shortage being taken into consideration. With regard to IDEMO, the chaotic search strategy, adjustment strategy of the control parameter, and dynamic crowding distance sorting were introduced, which provide a better candidate plan for multi-objective optimal active power dispatch. On the other hand, TOPSIS can coordinate each objective in a better way so as to realize the decision-making function of multi-objective optimal active power dispatch.

ACKNOWLEDGEMENT

This worked is supported in part by the National Natural Science Foundation of China under Grants 51577061. The authors like to acknowledge the contribution of Junyi Zhai.

3.5 HYDROTHERMAL COORDINATION

Alexandre P. Alves da Silva[1] and Anna Carolina R.H. da Silva[2]

[1]Vale S.A., Rio de Janeiro, Brazil
[2]Eletrobras, Rio de Janeiro, Brazil

3.5.1 Introduction

This chapter focuses on the monthly based long-term hydrothermal coordination task, which sets targets for the weekly based medium-term scheduling horizon. Subsequently, short-term operation planning, in an hourly basis, is defined upon medium-term targets. The present work considers a deterministic approach for inflows, with individual modeling of hydroelectric plants. Stochastic DP has also been employed to solve long-term hydrothermal coordination, because of its ability to handle stochastic inflows and nonlinear relations. However, for the case of multiple reservoirs, the exponential computational burden increase imposes the aggregation of reservoirs into a single reservoir system. On the other hand, one might claim that the aggregation procedure inaccuracy is offset by the uncertainties of the individual representation approach. Nevertheless, better river flow forecasting tools have been favoring the individual representation of hydroelectric plants' reservoirs.

Considering the characteristics of the hydrothermal problem formulation adopted in this work, it seems appropriate to investigate the application of meta-heuristics [127, 128] for tackling large-scale instances. The operational planning objective is defined as the minimization of the thermal units' costs, subject to load balance and hydraulic/thermal constraints. This problem is non-convex and exposed to the curse-of-dimensionality. The solution via meta-heuristics allows accurate representation of all functional relationships (spatial and temporal) among decision variables.

Meta-heuristics, in general, and GAs, in particular, have been facing two big challenges in order to become a very competitive alternative for dealing with large-scale non-convex optimization problems. The first challenge involves the setting of the search control parameters. General schemes for adapting the search control parameters have shown limited impact on the solution convergence process for large-scale problems. Because of that, customized tuning is almost always a useful resource. Additionally, GAs are usually challenged in terms of convergence robustness and efficiency when tackling high-dimensional search spaces [129]. In fact, hydrothermal coordination can be an optimization problem with such challenging characteristics. The contribution of this work is based on the use of domain knowledge to decompose the problem for improving the quality of the solution. The basic framework for addressing hydrothermal coordination in this work is illustrated in Figure 3.5.1.

3.5.2 Hydrothermal Coordination Formulation

Electric power is produced based on different energy sources. In countries where hydro plants are responsible for a great percentage of the total amount of generation, the hydrothermal scheduling problem becomes very important. In Brazil, in order to explore water resources more efficiently, the hydrothermal coordination problem has been solved in a centralized way. In this context, the basic challenge of operation planning is to minimize generation costs without jeopardizing water supply. The problem formulation is quite complex, which usually leads to solutions that do not take accurate modeling into account [130]. In the following sections the problem formulation is described. Transmission constraints have not been taken into account.

Thermal Plants Modeling

The operational costs for thermal plants are approximated by a monotonically increasing function with respect to the generated power. For long-term studies, a linear approximation is acceptable, i.e.

$$C_j(g_{j,t}) = a.g_{j,t} + b \tag{3.5.1}$$

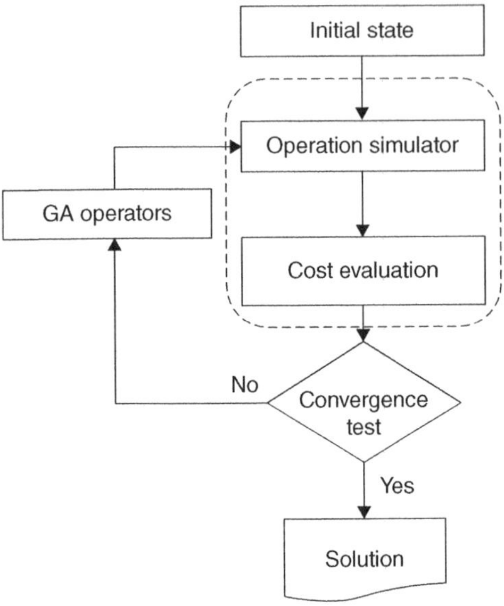

Figure 3.5.1 Simulation-based optimization for solving hydrothermal coordination.

where the parameter "a" is the incremental cost of the generation unit j and t is the index for the time interval. The thermal units are scheduled according to a priority list criterion based on production costs and the individual generators' limits.

Hydro Plants Modeling

In a monthly discretized basis, the decision variables are the average discharge flow rates from each reservoir, which allow the calculation of the thermal generation total cost for the whole planning horizon, i.e.

$$\min \sum_{i=1}^{T} \lambda_t \sum_{j=i}^{J} C_j(g_{j,t}) \tag{3.5.2}$$

Subject to:

$$D_t = G_t + P_t$$

$$G_t = \sum_{j=1}^{J} g_{j,t}$$

$$g_j^{\min} \le g_{j,t} \le g_j^{\max}$$

$$P_t = \sum_{i=1}^{I} p_{i,t}$$

$$x_{i,t} = x_{i,t-1} + \left(y_{i,t} + \sum_{k \in \Omega_i} u_{k,t} - u_{i,t} \right) \frac{\Delta t_t}{10^6}$$

$$h_{i,t} = \phi\left(x_{i,t}^{\text{avg}}\right) - \theta(u_{i,t}) - \text{loss}_{i,t}$$

$$x_{i,t}^{\text{avg}} = \frac{x_{i,t-1} + x_{i,t}}{2}$$

$$p_{i,t} = k_i h_{i,t} q_{i,t}$$

$$u_{i,t} = q_{i,t} + v_{i,t}$$

$$x_{i,t}^{\min} \leq x_{i,t} \leq x_{i,t}^{\max}$$

$$u_{i,t}^{\min} \leq u_{i,t} \leq u_{i,t}^{\max}$$

$$q_{i,t}^{\min} \leq q_{i,t} \leq q_{i,t}^{\max}(h_{i,t})$$

$$v_{i,t} \geq 0$$

$$\lambda_t = \frac{1}{(1+r)^t}$$

where, for each thermal plant j and for each hydro plant i,

$C_j(g_{j,t})$: plant j cost function;
G_t: total thermal generation during interval t;
P_t: total hydro generation during interval t;
D_t: load level during interval t;
T: number of time intervals;
I: number of hydroelectric plants;
J: number of thermal plants;
λ_t: discount factor for interval t;
r: discount rate for interval t;
k_i: productivity of hydro plant i $[MW/((m^3/s).\ m)]$;
$g_{j,t}$: generation of thermal plant j during interval t;
$p_{i,t}$: hydro unit i generation during interval t;
$x_{i,t}$: storage in the reservoir during interval t;
$q_{i,t}$: hydro turbine flow (unit i discharge) during interval t;
$v_{i,t}$: spillage during interval t;
$h_{i,t}$: average water head for unit i during interval t;
$u_{i,t}$: discharge flow rate for unit i during interval t;
$y_{i,t}$: inflow rate for unit i during interval t;
$\phi_i(x)$: forebay volume (water head fourth-order polynomial relationship);
$\theta_i(u)$: afterbay discharge (head polynomial);
$\text{loss}_{i,t}$: hydraulic loss during interval t;

$g_j^{\min}$: minimum generation for thermal unit j;

$g_j^{\max}$: maximum generation for thermal unit j;

$x_{i,t}^{\text{avg}}$: average volume for unit i during interval t;

$x_{i,t}^{\max}$: maximum storage at the end of interval t;

$x_{i,t}^{\min}$: minimum storage at the end of interval t;

Δt_t: length of interval t (one month); and

Ω_i: set of upstream hydro plants.

In order to apply the framework presented in Figure 3.5.1, the previous optimization problem formulation has to be transformed. This work adopts the additive penalization technique to transform the original constrained optimization problem into an unconstrained one, i.e.

$$\min \sum_{t=1}^{T}\left\{\lambda_t \sum_{j=1}^{J} C_j\left(g_{j,t}\right) + \sum_{i=1}^{I}\left[\alpha.w_1\left(x_{i,t}-x_{i,t}^{\max,\min}\right)^2 + \beta.w_2\left(u_{i,t}-u_{i,t}^{\max,\min}\right)^2\right]\right\} \tag{3.5.3}$$

where the w's are the penalty factors and

$$\begin{cases} \alpha = 1 \text{ if } x_{i,t}\langle x_{i,t}^{\min} \text{ or } x_{i,t}\rangle x_{i,t}^{\max} \\ \alpha = 0, \quad \text{otherwise} \end{cases}$$

and

$$\begin{cases} \beta = 1 \text{ if } u_{i,t}\langle u_{i,t}^{\min} \text{ or } u_{i,t}\rangle u_{i,t}^{\max} \\ \beta = 0, \quad \text{otherwise} \end{cases}$$

3.5.3 Problem Decomposition

In order to tackle hydrothermal coordination problems with high dimensionalities, some kind of problem decomposition should be applied to improve the solution quality. Therefore, modern heuristic optimization techniques have to be adjusted for dealing with problem decomposition. One popular technique for addressing the decomposition of the search space is coevolutionary cooperation [131] combined with GAs. Coevolutionary cooperation aims at simplifying the original problem by partitioning into subproblems. Each subproblem should aggregate the most interdependent decision variables. After independently solving each subproblem using GAs, the solutions interact and another round of subproblems is solved until convergence is reached. Convergence rate depends on the degree of interaction among variables in different subproblems.

The obvious advantage of coevolutionary cooperation is the possibility of trading the large original dimension by several lower dimensional search spaces [132]. However, the fast convergence of this technique relies on the hypothesis of a weak interaction among different subproblems. Fortunately, large-scale hydrothermal coordination problems usually allow decompositions based on both temporal and geographical aspects. The discharges (decision variables) of the hydro units located along the same river cascade are strongly affected by each other across the planning horizon. By the same token, all variables related to the same hydro plant (discharge, power generation, and water head) are highly dependent for the same time interval and across the planning horizon.

3.5.4 Case Studies

The test system has been selected for demonstrating the effectiveness of the decomposition approach. The system presents 14 hydro plants spread along parallel rivers cascades, according to Table 3.5.1.

The test system is part of the Brazilian electric power network and it is comprised of 14 reservoirs along different rivers cascades and of 6 thermal units. The search parameters have been customized for each simulation. As previously mentioned, the decision variables are the average monthly discharges for each hydro plant. The inflows for each reservoir are considered deterministic and the same assumption also applies to the loads.

The solution baseline is determined using the complete search space, i.e. without decomposing the original problem formulation. After that, decomposition schemes are tried in order to improve the baseline solution. After the adjustment of the search control parameters, each decomposition scheme is optimized 15 times for estimating the solution robustness.

TABLE 3.5.1 Hydro Plants' Data

Hydro Plant	Capacity (MW)	Max Storage (hm^3)	Downstream
Serra da Mesa	1 200	43 250	Tucuruí
Tucuruí E	4 000	32 013	—
Segredo	1 260	388	Salto Santiago
Salto Santiago	1 332	4 113	—
Jurumirim	9 776	3 165	Chavantes
Chavantes	414	3 041	Capivara
Capivara	608	5 725	—
Emborcação	1 192	13 056	Itumbiara
Itumbiara	2 280	12 454	São Simão
São Simão	1 710	5 540	Ilha Solteira
Furnas	1 312	17 217	Marimbondo
Marimbondo	1 488	5 260	Água Vermelha
Água Vermelha	1 398	5 169	Ilha Solteira
Ilha Solteira	3 444	5 516	—

The following decomposition schemes have been applied:

I. Fourteen subproblems with 12 variables each, with the discharges for the same hydro plant along the whole operation planning horizon, i.e. 12 months.

II. Four subproblems, with two of them using 24 variables, one subproblem with 36 variables, and one subproblem with 84 variables. These subproblems are related to the hydro units along the cascades of rivers Tocantins, Iguaçú, Paranapanema, and Paraná, respectively.

III. Delta grouping, a general decomposition scheme for non-separable function optimization as described in [133].

The GAs' simulations have been set as follows, which have been tuned after several tests:

a. For the complete search space, real coding and tournament selection have been adopted. Crossover has been implemented using weighted averages from the parents' corresponding variables. The population size is equal to 400, the best individual from each population is always saved for the next one, and mutation is based on Gaussian perturbation.

b. The subproblems have been tackled using real coding, proportional selection (roulette wheel), uniform crossover, populations with 100 individuals, an elitist strategy of saving the best individual, and Gaussian mutation.

For both cases (a and b), the percentage of the population generated via crossover is 80%. The mutation rate has been set to 1%. The initial population for each subproblem is randomly generated. However, the solution provided by the GA using the complete search space is used to set the other decision variables up. These complementary variables remain constant during the optimization process for constraints and cost evaluation purposes. After solving the subproblems for the first time, the same procedure is applied again until final convergence is reached (i.e. with the solutions for the $N-1$ other subproblems being used for setting up the complementary variables for each subproblem). Final convergence usually requires just a couple of interactions.

The solutions for the complete search space have provided a minimum cost of R$1.566 billion, an average of R$1.581 billion, and a standard deviation of R $14.663 million. The best solution has been used as a benchmark for the decomposition alternatives. Table 3.5.2 shows how each decomposition scheme compares with the best solution for the complete search space. It is worth mentioning that, in practice, the planning horizon is extended by one extra year. This avoids excessive use of water resources in the absence of future obligations.

The results above show the advantage of the problem partition by hydro plants. With this decomposition scheme, each hydro plant has its schedule determined for the whole planning horizon without changing the schedules for the other hydro plants, which are obtained from the previous round of simulations. The

TABLE 3.5.2 Simulations for the Decomposition Schemes

Trials	Grouping I	Grouping II	Delta
1	1.412B	1.469B	1.566B
2	1.408B	1.472B	1.566B
3	1.412B	1.502B	1.560B
4	1.435B	1.486B	1.563B
5	1.415B	1.475B	1.566B
6	1.422B	1.493B	1.560B
7	1.420B	1.479B	1.557B
8	1.417B	1.513B	1.558B
9	1.409B	1.495B	1.566B
10	1.412B	1.482B	1.560B
11	1.415B	1.476B	1.557B
12	1.422B	1.478B	1.558B
13	1.420B	1.502B	1.566B
14	1.410B	1.487B	1.566B
15	1.433B	1.491B	1.566B
Min	1.408B	1.469B	1.557B
Mean	1.417B	1.487B	1.562B
Std. dev.	8.128*E*+06	1.261*E*+07	3.845*E*+06
Gain (%)	11.2	6.6	0.6

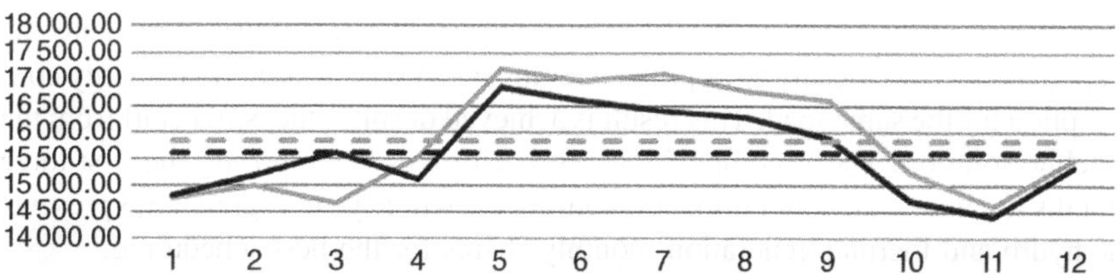

Figure 3.5.2 Hydro generation (MW) – Solution Baseline (gray) versus Grouping I (black).

optimization process continues until the schedules converge, which is equivalent to the successive approximation procedure applied by DP in reference [134]. The solution standard deviation for Grouping I also shows the robustness of this dimensionality reduction procedure. Moreover, it is true that the other decomposition schemes also provide improvement when compared with the solution for the original problem, although the non-domain-specific Delta grouping is by far the less efficient. The savings with fuel provided by the best solution from Grouping I is 158 million Reais (R$1.566B–1.408B).

Figure 3.5.2 displays the total hydro generation during the scheduling horizon. The black curve shows the total output from all hydro plants based on the

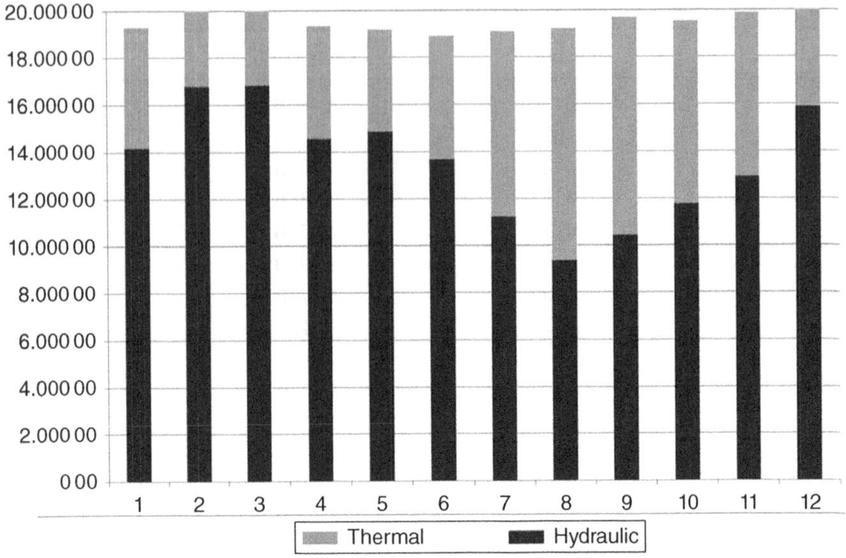

Figure 3.5.3 Monthly total hydro generation for the planning horizon (MW).

solution provided by Grouping I. The gray curve shows the equivalent information for the solution without problem partition.

The dotted lines represent the corresponding average power outputs for the operation planning horizon. The difference between the dotted lines represents the thermal energy savings by Grouping I solution, i.e. 5.04% less thermal energy consumption for the same load. This result is achieved despites the better start from the baseline solution. The decomposition approach saves water during the first three months for an overall better utilization during the whole year. Figure 3.5.3 presents the hydro and thermal generation monthly shares for the best scheduling.

3.5.5 Conclusions

This chapter introduced robust and efficient decomposition procedures for handling large-scale hydrothermal coordination problems via GAs. The coevolutionary cooperation approach has provided a useful framework for embedding domain knowledge in the problem decomposition criterion. Grouping I partition has been successfully tested against a literature benchmark. For a problem with such significant financial implications, the proposed solution technique has demonstrated great potential for providing big savings in operation costs. The application of the proposed technique to other energy systems is straightforward. Future work will focus on better mechanisms for local search, such as the ones provided by memetic algorithms [135]. This would allow further exploitation of all available domain-specific knowledge.

ACKNOWLEDGEMENT

This worked is supported in part by Eletrobras and GE Global Research. The authors would like to acknowledge the contribution of A.F. Amendola and K.C.A. Roberto.

3.6 META-HEURISTIC METHOD FOR GMS BASED ON GENETIC ALGORITHM

Jaeseok Choi and Yeonchan Lee

Gyeongsang National University, Jinju, Korea

3.6.1 History

The GMS problem is defined as when and what generator is decided for overhaul, as it is, shut down out of power system. Therefore, GMS is an extremely important operation planning decision for power system reliability and affects also generation expansion planning in impact.

The GMS has firstly been formulated as maximization optimal problem of minimum supply reserve rate at a time period by W.R. Christiaanse and A.H. Palmer in 1972 [136]. In same year, L.L. Garver proposed risk leveled GMS modeling considering forced outage rate of generators.

In 1975, Zurn and Quintana suggested successfully GMS using state space method. The GMS was firstly formulated as multi-objective style, which considers not only reliability maximization but also cost minimization. They solved it by dynamic programming successive approximation (DPSA). The proposed model is available for a large-scale system, which cannot be solved by DP, IP, and branch and bound method (B&B) that are not applicable for 20 units and 12 interval time scale systems because of huge computation times. In 1983, Yamayee et al. proposed a GMS model using cumulant method for probabilistic production cost simulation, which requires a large computation time [266].

In the 1990 period, GMS considering multi-objective, GMS using annealing method and decomposition for actual size system, GMS considering transmission lines constraints, and GMS considering load flow were studied [138–145].

The GMS problem is essentially formulated as non-convex function and finding the global optimal solution is very difficult. More recently, the GMS R&D pursues three view sides as shown in Figure 3.6.1.

First, in view point of methodology, GMS are focused on research and studies for practical solution and methodology using Benders decomposition method, meta-heuristic method, etc., even if quasi-optimal solution is obtained.

Second, in view point of objective, multi-objective-style GMS, which is considering reliability maximization, cost minimization, and benefit maximization in

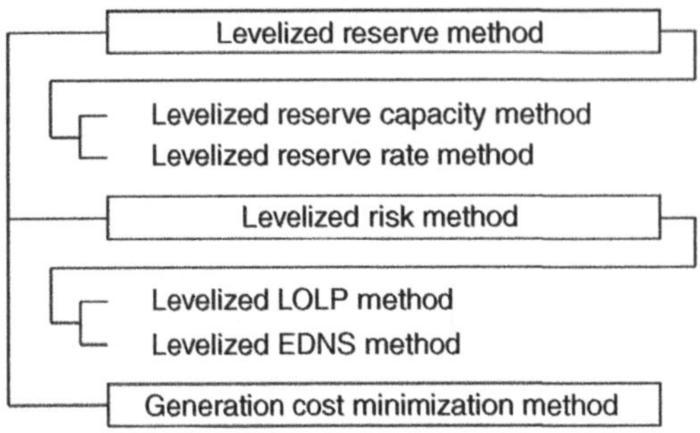

Figure 3.6.1 Methods for analyzing the generator maintenance scheduling problem.

electricity markets, and air pollution minimization, etc., installed user-friendly GMS system which is an interesting topic for GENCO.

Third, in view point of uncertainty, objective uncertainty as like as forced outage rate of generators but also ambiguity uncertainty is considered.

3.6.2 Meta-heuristic Search Method

A GMS problem including one generator, "Principle of Optimality" of Bellman-based forward dynamic programming (FDP) is formulated as in Eq. (3.6.1) [146, 147].

$$F^*\{X(k), k\} = \underset{U(k-1)}{\text{minimize}}[F\{X(k-1), U(k-1), k-1\} + F^*\{X(k-1), k-1\}] \tag{3.6.1}$$

where

X: state vector,

k: time period.

Therefore, the GMS is summarized to search optimal state vector, which are $U^*(1)$, $U^*(2)$, …, $U^*(k–1)$ satisfied with various constraints.

First step: Also, #i generator (unit) starts overhaul at K-stage (time period), the overhaul of the unit should be continued until $K+\text{MD}i$ stage. Therefore, overhaul of #i unit can be described in time space state vertex as shown in Figure 3.6.2. Also, the #i unit is able to be shifted in order to search better objective function value under assumption that other generators' overhaul are fixed. This method is called "time axis shift (TAS) method" in this study.

Next step: When the best overhaul stage (time) of the #i unit generator is decided using the TAS method, next unit, as it is, #$i+1$ unit will be searched as the same method under assumption that overhaul of #i unit and other units are fixed.

Although this method as like Figure 3.6.3, which is called "time axis shift meta-heuristic optimal (TASMHO) method," in this study, cannot certify always to obtain global optimal solution, it is very practical and available for large-scale

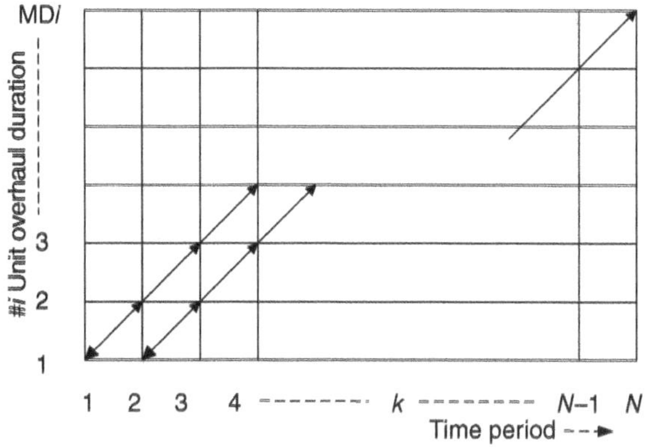

Figure 3.6.2 Configuration of the solutions searched in vertex by conventional DP.

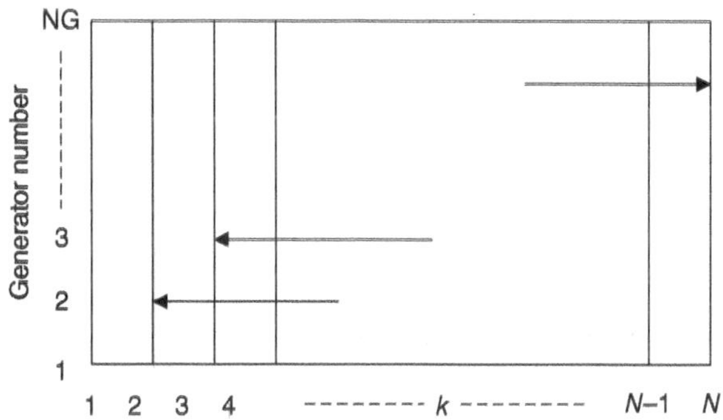

Figure 3.6.3 Configuration of the solutions searched by time axis shift method.

GMS system problem. Furthermore, GA combined with the time shifted meta-heuristic method can certify global optimal solution.

Example for TAS

Two generators (#1 unit and #2 unit), which have permissible overhaul period, as shown in Figure 3.6.4, and initial GMS of two generators, as shown in Figure 3.6.5, are tested using TAS method proposed in this study. Assuming that overhaul duration for unit GMS is two time periods and two units are installed on different plants, objective function is probabilistic production cost minimization in this example. The cost information evaluated for GMS cases is given in Table 3.6.1: "Overhaul as like as Figure 3.6.5." In Table 3.6.1 it means production cost of each time period under assumption in which #1 unit and #2 unit have overhaul planning as shown in Figure 3.6.5. "Only #1 unit on overhaul" in Table 3.6.1 means production cost of each time period under assumption in which #1 unit has overhaul planning during all periods, as it is, assumption case as same as #1 unit does not exist in the generation system. This table is very useful for iteration. Because it is just only one necessary for making this table before iteration using the TAS method.

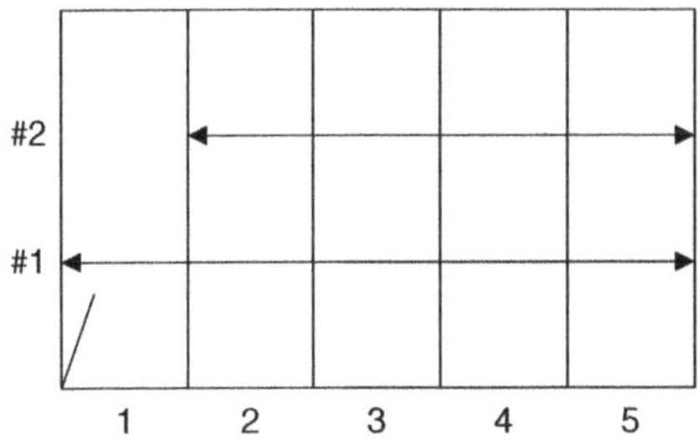

Figure 3.6.4 Possible time range of generator maintenance of the example problem.

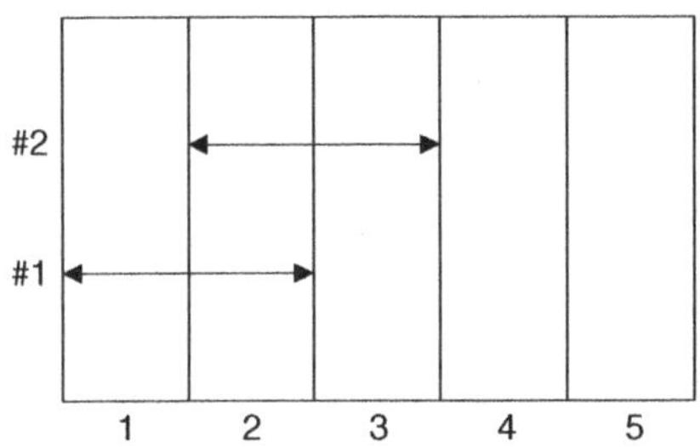

Figure 3.6.5 Initial time range of generator maintenance of the example problem.

TABLE 3.6.1 Production Cost Calculated in Each Case (US$ 10^5)

Time Period Number	1	2	3	4	5	ΣF_k
Overhaul as like as Figure 3.6.5	60	60	40	20	30	210
Nothing units on overhaul	**50**	**40**	30	20	30	170
Only #1 unit on overhaul	60	45	**40**	25	45	215
Only #2 unit on overhaul	65	50	40	30	**35**	220
Both (#1 and #2) units on overhaul	70	60	50	**35**	50	265

Analysis Using the TAS Method Just only iteration = 1 is tested in convenience.

Iteration = 1

#1 Unit overhaul decision: #1 Unit overhaul has four available cases as follows. Then, costs at each period come from appropriately selected cost set in Table 3.6.1. No more simulations, therefore, are necessary. For example, each time period cost of first case in which #1 unit overhaul has "3 and 4 period" comes from appropriately selected cost set in Table 3.6.2. This set is presented as underlined marked cost in Table 3.6.1.

Therefore, #1 unit overhaul is tentatively fixed in 3 and 4 time periods.

#2 Unit overhaul decision under condition that #1 unit overhaul is fixed in 3 and 4 time periods: #2 unit overhaul has three available cases as follows. Then,

TABLE 3.6.2 Optimal at 3 and 4 Period (US$ 10^5)

	Overhaul Cases						
Time Period	1	2	3	4	5	ΣF_k	Remark
1 and 2 time period	60	60	40	20	30	210	
2 and 3 time period	50	60	50	20	30	210	
3 and 4 time period	50	50	50	25	30	205	Optimal[a]
4 and 5 time period	50	50	40	25	45	210	

[a]It is tentative optimal at iteration = 1.

TABLE 3.6.3 Optimal at 4 and 5 Period (US$ 10^5)

	Overhaul Cases						
Time Period	1	2	3	4	5	ΣF_k	Remarks
2 and 3 time period	50	50	50	25	30	205	
3 and 4 time period	50	40	50	35	30	205	
4 and 5 time period	50	40	40	35	35	200	Optimal

costs at each period come from appropriately selected cost set in Table 3.6.1. For example, each time period cost of first case in which #2 unit overhaul has "4 and 5 period" comes from appropriately selected cost set in Table 3.6.3. This set is presented as bold marked cost in Table 3.6.1, and also, no more simulations are needed.

Therefore, #2 unit overhaul is fixed in 4 and 5 time periods tentatively for iteration = 1. After iteration = 1, total production cost is decreased as $210 \times 10^5 - 200 \times 10^5 = \text{US}\$10^6$.

Next iteration is processed in the same manner.

3.6.3 Flexible GMS

Introduction

Recently, flexible and adaptive solutions rather than robust and fixed optimal solutions are used for operation and expansion planning of power systems under deregulation environment [136]. It is the reason that the flexible and adaptive solutions are more effective than the latter under massive uncertainties and power system market with competition. It is certain that the methods based on fuzzy theory are more flexible than the optimal method with conventional crisp conditions [148, 149].

The GMS problem is an important planning problem that affects both economy and reliability for operation and planning of generating systems. Flexible and reasonable maintenance scheduling is able not only to raise supply and reserve rate but also to postpone the period of construction of generators [136, 150, 151]. Also, construction cost of generators, production cost, and maintenance scheduling cost can be reduced. Additionally, recent methods for considering the capacity of transmission line have been proposed in [152].

In the present study, we propose a new method for generator maintenance flexible scheduling using fuzzy search method that has been developed by the authors. It is expected that more flexible solution can be obtained because fuzzy set theory that can reflect the subjective decision of decision-maker is used in this study [153–156]. The main point of this study is to develop the algorithm of flexible GMS considering probability production cost, probabilistic reliability (expected demand not served [EDNS], LOLP), air pollutions of SO_2 and NO_x of thermal power generators, and supply reserve rate, which is deterministic reliability.

The practicability and effectiveness of the proposed approach is demonstrated by simulation studies of a real-size power system model.

Concept of Flexible Solution

Collapse of communism society pushed the world society system to capitalism naturally. Main characteristics of the capitalism are competition, deregulation, and price balance between supply and demand based on market. So, external conditions for power system scheduling problem demand growth, primary energy circumstances, and reliability of demand are becoming more and more unmanageable than ever before because of deregulation environment. Figure 3.6.6 means that the flexible GMS is one which, "although not necessarily gives the optimum solution for the basic forecasted conditions, yet can keep the reasonable scheduling solution from being significantly worsened by any assumed changes in the surrounding situations" [148, 149].

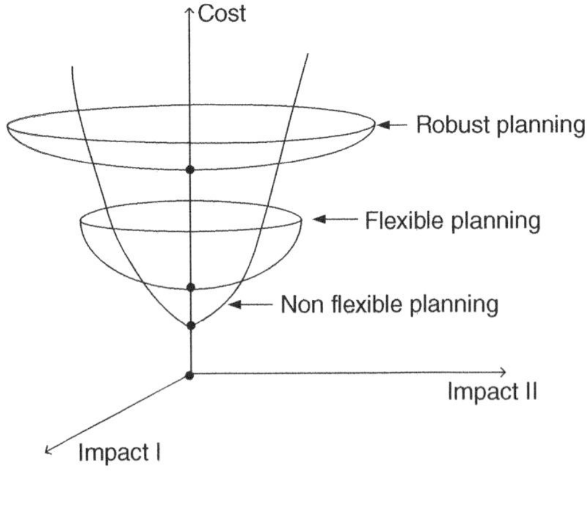

Figure 3.6.6 Concept of flexibility.

Fuzzy Search Method

The fuzzy decision set D resulting from fuzzy sets of p fuzzy constraints, C_1, C_2, …, C_p and fuzzy sets of q fuzzy goals, G_1,G_2, …, G_q is an intersection defined as follows [157–161]:

$$D = \left(\bigcap_{i=1}^{p} C_i \right) \cap \left(\bigcap_{j=1}^{q} G_i \right) \tag{3.6.2}$$

The membership function μ_D resulting from the membership function of fuzzy sets of goals and constraints is defined as follows:

$$\mu_D(x) = \min \left[\min_{i=1-p} \mu_c, \min_{j=1-q} \mu_c \right] \tag{3.6.3}$$

where min is an abbreviation of minimum.

If the fuzzy mathematical programming problem consists of finding the maximum point of the membership function of the fuzzy decision set D, the optimal solution can be obtained as

$$\mu_D(X^*) = \max \mu_D(X) \tag{3.6.4}$$

where

X^* is the optimal decision solution,

max is an abbreviation of maximum.

The vector form in Eq. (3.6.4) can be rewritten as

$$\mu_D\left(X_1^*, X_2^*, \ldots, X_N^*\right) = \max_{X_1 \to X_2} \mu_D(X_1, X_2, \ldots, X_N) \tag{3.6.5}$$

In order to solve this problem by using the fuzzy search method, the principle of optimality can be applied after Eq. (3.6.5) and can be reformulated as

$$\begin{aligned}\mu_D\left(X_1^*, X_2^*, \ldots, X_N^*\right) &= \max_{X_2,\ldots,X_N} \left\{ \max_{X_1} [\min(\mu_D(X_1), \mu_{F2}(X_2), \ldots, \mu_{FN-1}(X_{N-1}), \mu_{FN}(X_N)] \right\} \\ &= \max_{X_2,\ldots,X_N} \left[\min\left(\mu_D\left(X_1^*\right), \mu_{F2}(X_2), \ldots, \mu_{FN-1}(X_{N-1}), \mu_{FN}(X_N)\right] \end{aligned} \tag{3.6.6}$$

where X: decision variable, $F = G + C$ (algebraic sum of fuzzy sets). We can rewrite Eq. (3.6.6) as

$$\mu_D\left(X_1^*, X_2^*, \ldots, X_N^*\right) = \max_{Xn,\ldots,X_N} \left\{ \min \left[\mu_D\left(X_1^*, X_2^*, \ldots, X_{N-1}^*\right), \mu_{Fn}(X_n), \ldots, \mu_{FN}(X_N) \right] \right\} \tag{3.6.7}$$

We can also rewrite Eq. (3.6.7) as

$$\mu_D\left(X_1^*, X_2^*, \ldots, X_N^*\right) = \max_{Xn} \left\{ \min \left[\mu_D\left(X_1^*, X_2^*, \ldots, X_{N-1}^*\right), \mu_{Fn}(X_n) \right] \right\} \tag{3.6.8a}$$

$$\mu_D(S_n) = \max_{Xn} \left\{ \min \left[\mu_D(S_{n-1}), \mu_{Fn}(X_n) \right] \right\} \tag{3.6.8b}$$

where

$S_n = f(S_{n-1}, X_{n-1})$ $n = 1, 2, \ldots, N$,

S: state variable,

f: state transition function.

The Formulation Based on Fuzzy Search Method

Objective Functions

1. Minimization of probabilistic production cost F as [162]:

$$\text{Minimize } Z_1 = F\{E_{in}, \Phi_{in}(U_{in})\} = \sum_{n=1}^{NT}\sum_{i=1}^{NG}\{A_i E_{in} + B_i T\Phi_{in}(U_{in})\} \quad (3.6.9)$$

where

A_i is one-dimensional coefficient of fuel cost function (\$/MWh),

B_i is constant of fuel cost function (\$).

$$E_{in} = (1-q)T\int_{u_{i-1}}^{u_i} \Phi_{in-1}(X)dX \text{ (MWh)}$$

E_{in}: probabilistic generation energy of #i unit at #n stage

T: total period for study (hours)

i: number of the economic order of generators

$u_i = C_1 + C_2 + \cdots + C_i$ (MW)

C_i: capacity of #i unit

$u_0 = 0$

Φ_{in}: effective load duration curve

qi: forced outage rate of #i unit

Given the aspiration level of decision-maker for the probabilistic production cost, Eq. (3.6.9) can be represented as fuzzy goal function form as

$$Z_1 \lesssim Z_{01} \quad (3.6.10)$$

where Z_{01}: aspiration level of decision-maker for the production cost.

2. Minimization of maximum LOLP_n

If LOLP_n presents the loss of load probability of power system at #n stage, the objective as to minimize the LOLP of time stage that has maximum LOLP is defined as

$$\text{Minimize } Z_2 = \max(\text{LOLP}_n) = \max\{\Phi_{\text{NG}n}(U_{\text{NG}n})\} \text{ (p.u.)} \quad (3.6.11)$$

Also, Eq. (3.6.11) can be represented as fuzzy goal function form as

$$Z_2 \lesssim Z_{02} \quad (3.6.12)$$

where Z_{02}: aspiration level of decision-maker for LOLP.

3. Minimization of maximum $EDNS_n$
Also, as like as the LOLP, when $EDNS_n$ presents the expected demand not served by power system at #n stage, the objective as to minimize the EDNS of the stage that has maximum EDNS is defined as

$$\text{Minimize } Z_3 = \max(\text{EDNS}_n) = \max T \int_{U_{\text{NG}n}}^{\infty} \Phi_{\text{NG}n}(X)dX \text{ (MWh)} \qquad (3.6.13)$$

Also, Eq. (3.6.13) can be represented as fuzzy goal function form as

$$Z_3 \lesssim Z_{03} \qquad (3.6.14)$$

where Z_{03}: aspiration level of decision-maker for EDNS.

4. Minimization of air pollution
If AP_n means the total air pollution of SO_2 and NO_x that occur from thermal power generators operating at #n stage, the objective to minimize the air pollution volume of time stage that has maximum air pollution volume is defined as [163]:

$$\text{Minimize } Z_4 = \max(\text{AP}_n) = \max \sum_{i=1}^{\text{NG}} \{\text{TMW}_i(\text{SO}_{2in} + \text{NO}_{Xin})E_{in}\} \text{ (ppm)} \qquad (3.6.15)$$

where

TMW_i: necessary fuel consumption rate for a unit generating energy of the #i unit (ppm/ton)

SO_{2in}: discharge density of SO_2 of #i unit (ppm/ton)

NO_{xin}: discharge density of NO_x of #i unit (ppm/ton)

Also, Eq. (3.6.13) can be represented as fuzzy goal function form as

$$Z_4 \lesssim Z_{04} \qquad (3.6.16)$$

where Z_{04}: aspiration level of decision-maker for air pollution criterion.

5. Maximization of supply reserve rate
If RES_n presents the supply reserve rate of power system at #n stage, the objective to maximize the supply reserve rate of time stage that has minimum supply reserve rate is defined as

$$\text{Minimize } Z_5 = \min(\text{RES}_n) = \min\left\{\frac{\text{TCAP} - \text{MT}_n - L_{pn}}{L_{pn}} \times 100\right\} (\%) \qquad (3.6.17)$$

where

TCAP: total facility powers

MT_n: maintenance powers of #n stage

L_{pn}: peak load of #n stage

Also, Eq. (3.6.17) can be represented as fuzzy goal function form as

$$Z_5 \lesssim Z_{05} \qquad (3.6.18)$$

where Z_{05}: aspiration level of decision-maker for supply reserve rate.

Constraints

1. Boundary conditions

$$X(1) = \underline{0}$$
$$X(T+1) = \mathrm{col}[\mathrm{MD}_1, \mathrm{MD}_2, \mathrm{MD}_3, \ldots, \mathrm{MD}_{\mathrm{NG}}]^T \tag{3.6.19}$$

where

0: zero vector

MD_i: time period asked for maintenance of #i unit.

2. Constraints for maintenance of possible time period

$$U_i(t) = \begin{Bmatrix} 0 & t < \mathrm{MS}_i \text{ or } t > \mathrm{MF}_i + \mathrm{MD}_i \\ 1 & \mathrm{MS}_i \le t \le \mathrm{MF}_i + \mathrm{MD}_i \end{Bmatrix} \tag{3.6.20}$$

where

MS_i: starting time for maintenance of first possible maintenance time period of #i unit

MF_i: starting time for maintenance of last possible maintenance time period of #i unit

3. Maintenance crew constraint

$$\sum_{i \in P_k} U_i(t) \le 1 \tag{3.6.21}$$

where P_k: set of generators at #k generating plant.

4. Constraint of maintenance equipments

$$\sum_{i=1}^{\mathrm{NG}} U_i(t) \cdot M_{kli} \le \mathrm{MA}_k(t) \tag{3.6.22}$$

where

k: the number of the kinds of maintenance equipment ($k = 1, 2, \ldots, K$)

l: number of maintenance scheduled time of #i unit

$\mathrm{MA}_k(t)$: amount of #k maintenance equipment available within #t stage

M_{kli}: amount of #k maintenance equipment within #l maintenance time period of #i unit

Establishment of Membership Functions

1. Membership function of fuzzy set for the production cost is defined as

$$\mu_c\{X(t-1), \mu(t)\} = \begin{Bmatrix} 1 & : \Delta C(\cdot) \le 0 \\ e^{-W_c \Delta C\{X(t-1), \mu(t)\}} & : \Delta C(\cdot) > 0 \end{Bmatrix} \tag{3.6.23}$$

where

$\mu_c(\cdot)$: membership function of fuzzy set for the production cost,

$\Delta C(\cdot) = \{F(X(t)) - C_{asp}(t)\}/C_{asp}(t)$,

$C_{asp}(t)$: aspiration level for production cost at #t stage,

W_c: weighting factor of the membership function for production cost.

2. Membership functions of fuzzy set for the reliability (LOLP, EDNS) are defined as

$$\mu_R\{X(t-1), \mu(t)\} = \begin{Bmatrix} 1 & : \Delta R(\cdot) \le 0 \\ e^{-W_R \Delta R\{X(t-1), \mu(t)\}} & : \Delta R(\cdot) > 0 \end{Bmatrix} \tag{3.6.24}$$

where

$\mu_R(\cdot)$: membership function of fuzzy sets for reliability,

$\Delta R(\cdot) = \{\mathrm{RES}(X(t)) - \mathrm{REQ}(t)\}/\mathrm{REQ}(t)$,

$\mathrm{REQ}(t)$: aspiration level for reliability at #t stage,

W_R: weighting factor of the membership function for reliability.

3. Membership function of fuzzy set for the air pollution criterion is defined as

$$\mu_A\{X(t-1), \mu(t)\} = \begin{Bmatrix} 1 & : \Delta A(\cdot) \le 0 \\ e^{-W_A \Delta A\{X(t-1), \mu(t)\}} & : \Delta A(\cdot) > 0 \end{Bmatrix} \tag{3.6.25}$$

where

$\mu_A(\cdot)$: membership function of fuzzy sets for air pollution fuzzy set,

$\Delta A(\cdot) = \{\mathrm{AP}(X(t)) - \mathrm{ASP}(t)\}/\mathrm{ASP}(t)$,

$\mathrm{ASP}(t)$: aspiration level for air pollution at #t stage,

W_A: weighting factor of the membership function for air pollution.

4. Membership function of fuzzy set for the supply reserve rate is defined as

$$\mu_S\{X(t-1), \mu(t)\} = \begin{Bmatrix} 1 & : \Delta S(\cdot) \le 0 \\ e^{-W_S \Delta S\{X(t-1), \mu(t)\}} & : \Delta S(\cdot) > 0 \end{Bmatrix} \tag{3.6.26}$$

where

$\mu_S(\cdot)$: membership function of fuzzy sets for supply reserve rate,

$\Delta S(\cdot) = \{\mathrm{RSP}(X(t)) - \mathrm{RSP}(t)\}/\mathrm{RSP}(t)$,

$\mathrm{RSP}(t)$: aspiration level for supply reserve rate at #t stage,

W_S: weighting factor of the membership function for supply reserve rate.

Solution Procedure by the Fuzzy Search Method

Fuzzy decision set D applied to Eq.(3.6.25) can be formulated as

$$D = C \cap R_1 \cap R_2 \cap A \cap S \tag{3.6.27}$$

where

C: fuzzy set for probabilistic production cost,

R_1: fuzzy set for reliability LOLP,

R_2: fuzzy set for reliability EDNS,

A: fuzzy set for air pollution,

S: fuzzy set for supply reserve rate.

Therefore, using Eq. (3.6.27), we can obtain:

$$\mu_D(X(t)) = \max_{\mu\min(t) \leq \mu(t) \leq \mu\max(t)}[\min\{\mu_C(\cdot), \mu_{R_1}(\cdot), \mu_{R_2}(\cdot), \mu_A(\cdot), \mu_S(\cdot), \mu_D(X(t-1))\}] \tag{3.6.28}$$

where

$X(t) = X(t-1) + u(t)$,

$\mu_D(X(0)) = 1.0$,

$\mu_D(\cdot)$: membership function of fuzzy set for decision function.

Case Study

Input Data The proposed method was applied to the KEPCO system in 1997 and probabilistic production cost was calculated by the cumulant method. Figure 3.6.7 represents year load curve, which has weekly load peaks on case study year in the KEPCO system.

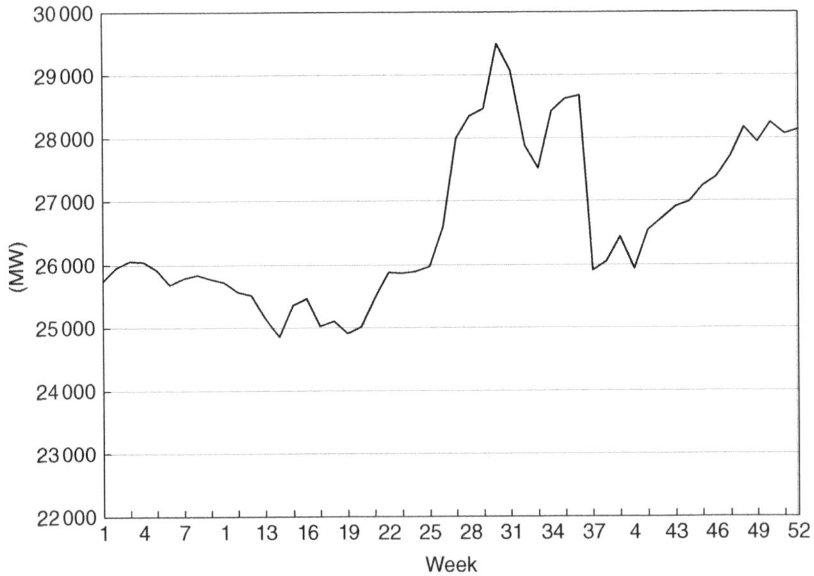

Figure 3.6.7 Year load curve (weekly load peaks).

TABLE 3.6.4 Aspiration Level and Weighting Factor

	Aspiration Level	Weighting Factor
Z_{01}	37 650 (10^8\$)	10.0
Z_{02}	0.50 (pu/week)	5.0
Z_{03}	150 000 (MWh/week)	2.0
Z_{04}	180 000 (10^8 ppm/week)	5.0
Z_{05}	7.5 (%)	2.0

TABLE 3.6.5 Input Data of Generators

NO	ANAME	TYP	PMN	PMX	CFT	A(i)	B(i)	C(i)	FOR	MS	MF	MD	MI	YEAR	MH	FCR	SO2	NOX
1	WSG 1	N/P	679	679	0.939	0.00005	1.0000	15.233	0.04408	1	52	8	1			0	0	0
2	WSG 2	N/P	700	700	0.939	0.00005	1.0000	15.233	0.04408	1	52	0	1	1997	5	0	0	0
3	GRI 1	N/P	580	580	1.849	0.00005	1.0000	15.233	0.04207	1	26	8	1			0	0	0
4	GRI 2	N/P	650	650	1.849	0.00005	1.0000	15.233	0.04674	1	26	8	1			0	0	0
5	GRI 3	N/P	950	950	1.849	0.00005	1.0000	15.233	0.07744	27	26	8	27			0	0	0
6	GRI 4	N/P	950	950	1.849	0.00005	1.0000	15.233	0.08893	27	26	8	27			0	0	0
7	YKG 1	N/P	950	950	1.849	0.00005	1.0000	15.233	0.08893	1	26	8	1			0	0	0
8	YKG 2	N/P	950	950	1.849	0.00005	1.0000	15.233	0.08893	27	26	8	27			0	0	0
9	YKG 3	N/P	1000	1000	1.849	0.00005	1.0000	15.233	0.08893	27	26	8	27			0	0	0
10	YKG 4	N/P	1000	1000	1.849	0.00005	1.0000	15.233	0.08893	27	26	8	27			0	0	0
11	UGN 1	N/P	950	950	1.849	0.00005	1.0000	15.233	0.08893	1	26	8	1			0	0	0
12	UGN 2	N/P	950	950	1.849	0.00005	1.0000	15.233	0.08893	27	26	8	27			0	0	0
13	ABRG 3	T/P	200	500	7.649	0.00058	1.6133	153.842	0.03234	1	26	6	1	1993	8	0.4030	700	450
14	ABRG 4	T/P	200	500	7.649	0.00058	1.6133	153.842	0.03347	27	26	6	27	1993	24	0.4030	700	450
15	BRG 1	T/P	200	475	7.649	0.00058	1.6133	153.842	0.03234	1	26	6	1			0.4030	700	450
16	BRG 2	T/P	200	475	7.649	0.00058	1.6133	153.842	0.03347	27	26	6	27			0.4030	700	450
17	ABRG 5	T/P	500	500	7.649	0.00058	1.6133	153.842	0.03347	27	26	6	27			0.4030	700	450
18	ABRG 6	T/P	500	500	7.649	0.00058	1.6133	153.842	0.03347	27	26	0	27	1997	32	0.4030	700	450
19	SCP 3	T/P	280	560	7.649	0.00059	1.5820	183.432	0.02050	27	26	8	27			0.4030	700	450
20	SCP 1	T/P	280	532	7.649	0.00059	1.5820	183.432	0.02050	27	26	8	27			0.4030	700	450
21	SCP 2	T/P	280	532	7.649	0.00059	1.5820	183.432	0.02050	27	26	8	27			0.4030	700	450
22	SCP 4	T/P	560	560	7.649	0.00059	1.5820	183.432	0.02050	27	26	8	27			0.4030	700	450
23	TAN 1	T/P	500	500	7.649	0.00059	1.5820	183.432	0.02050	27	26	8	27			0.4030	700	450
24	TAN 2	T/P	500	500	7.649	0.00059	1.5820	183.432	0.02050	27	26	0	27	1997	14	0.4030	700	450
25	TAN 3	T/P	500	500	7.649	0.00059	1.5820	183.432	0.02050	27	26	0	27	1997	31	0.4030	700	450
26	HDO 1	T/P	500	500	7.649	0.00059	1.5820	183.432	0.02050	27	26	0	27	1997	27	0.4030	700	450
27	HDO 2	T/P	500	500	7.649	0.00059	1.5820	183.432	0.02050	27	26	0	27	1997	45	0.4030	700	450
28	SCP 5	T/P	500	500	7.649	0.00059	1.5820	183.432	0.02050	27	26	0	27	1997	27	0.4030	700	450
				⋮				⋮								⋮		
120	HNH 7	S/T	100	100	12.221	0.00134	1.8737	53.723	0.03128	27	17	0	27	1997	27	0.0500	450	200
121	HNH 8	S/T	100	100	12.221	0.00134	1.8737	53.723	0.03128	27	17	0	27	1997	27	0.0500	450	200
122	HNH 9	S/T	100	100	12.221	0.00134	1.8737	53.723	0.03128	27	17	0	27	1997	27	0.0500	450	200
123	HNH10	S/T	100	100	12.221	0.00134	1.8737	53.723	0.03128	27	17	0	27	1997	29	0.0500	450	200
124	YWL 2	T/P	25	50	9.491	0.00485	2.7352	17.866	0.03749	1	52	4	1			0.4030	700	450
125	BPTG 6	G/T	88	88	11.879	0.00025	1.9636	66.170	0.00234	40	13	4	40			0.0500	450	200
126	BPTG 7	G/T	88	88	11.879	0.00025	1.9636	66.170	0.00234	40	13	4	40			0.0500	450	200
127	BPTG 8	G/T	88	88	11.879	0.00025	1.9636	66.170	0.00234	40	13	4	40			0.0500	450	200
128	BPTG 9	G/T	88	88	11.879	0.00025	1.9636	66.170	0.00234	40	13	4	40			0.0500	450	200
129	HLM 1	S/T	70	70	12.221	0.00134	1.8737	53.723	0.03128	27	17	4	27			0.0500	450	200
130	YWL 1	T/P	25	50	11.643	0.00485	2.7352	17.866	0.29191	1	52	4	1			0.4030	700	450
131	BUSN 7	G/T	0	50	7.649	0.00059	1.5820	183.432	0.02050	27	26	4	27			0.0500	450	200
132	HLM 2	S/T	35	35	12.221	0.00134	1.8737	53.723	0.03128	27	17	4	27			0.0500	450	200
133	JAE 1	T/P	30	30	7.649	0.00059	1.5820	183.432	0.02050	27	26	4	27			0.4030	700	450

Aspiration level and weighting factor for each objective function shown in Table 3.6.4 have been used for this case study.

Table 3.6.5 shows input data of generators for this case study.

Output Results It can be assumed that generators start to operate in the study year need not be maintained. Figure 3.6.8 shows convergence of the membership value of objective function according to iterations. Figure 3.6.8 shows that

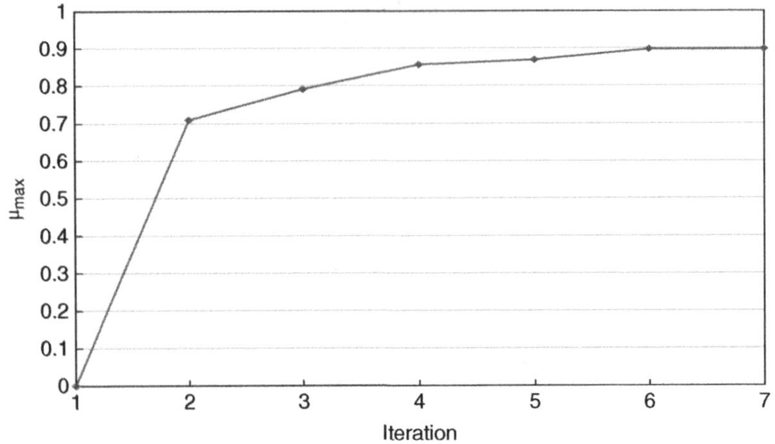

Figure 3.6.8 Convergence of the objective function (μ).

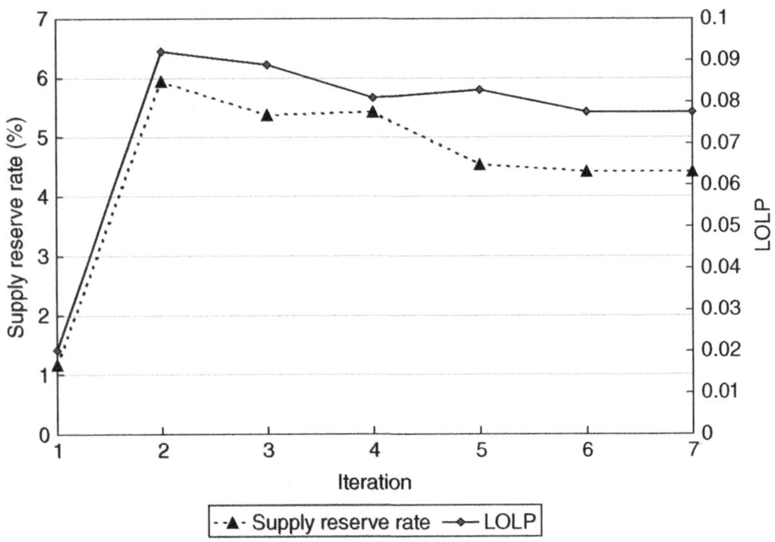

Figure 3.6.9 Standard deviation of supply reserve rate and LOLP.

the objective function which indicates the satisfaction degree of decision-maker converges to 0.9 after seven iterations.

The shapes of standard deviations of the supply reserve rate and LOLP with iteration are shown in Figure 3.6.9.

Also, the shapes of standard deviations of the EDNS with iteration are shown in Figure 3.6.10. Because these converged results can be obtained in this case study, we can carefully conclude that the proposed algorithm using fuzzy theory is a reasonable method for flexible maintenance scheduling.

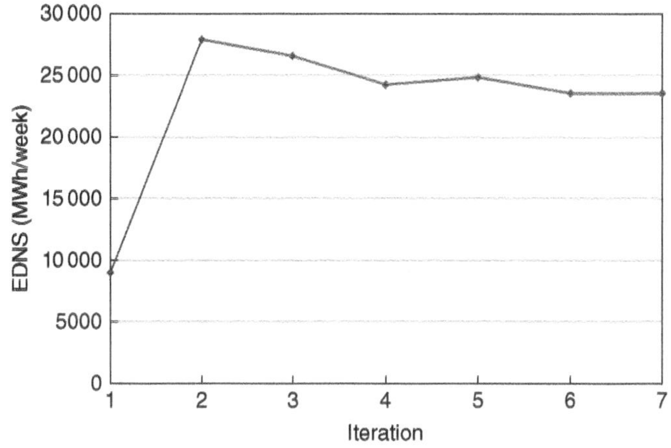

Figure 3.6.10 Standard deviation of EDNS according to iteration.

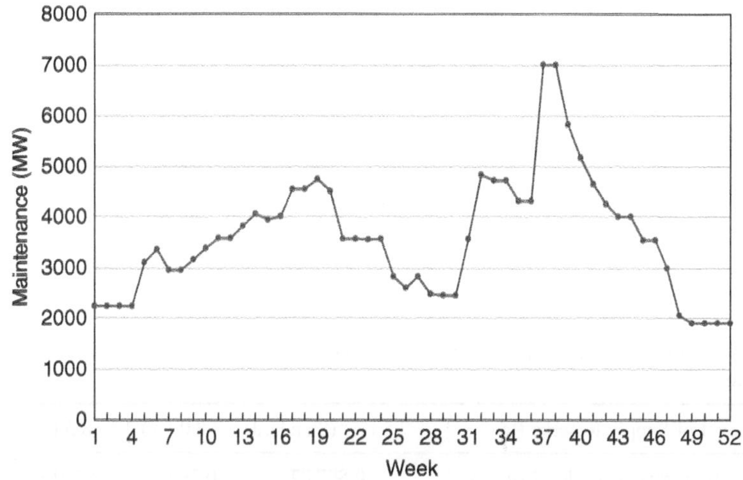

Figure 3.6.11 Maintenance powers at each week.

Maintenance powers at each week are shown in Figure 3.6.11. We can surely see that maintenance powers are allocated much more on seasons of spring and autumn than summer and winter with high load in Figure 3.6.11.

Table 3.6.6 shows results of GMS by this proposed method. In the table, dotted lines mean the weeks of mean possible maintenance scheduling and crosses mean the weeks of maintenance scheduling obtained by the proposed fuzzy search method.

Finally, comparisons of results by changing the aspiration level of supply reserve rate are shown in Table 3.6.7. They are interesting results that iteration numbers for the convergence are same as seven for all cases in Table 3.6.7. As

TABLE 3.6.6 Result

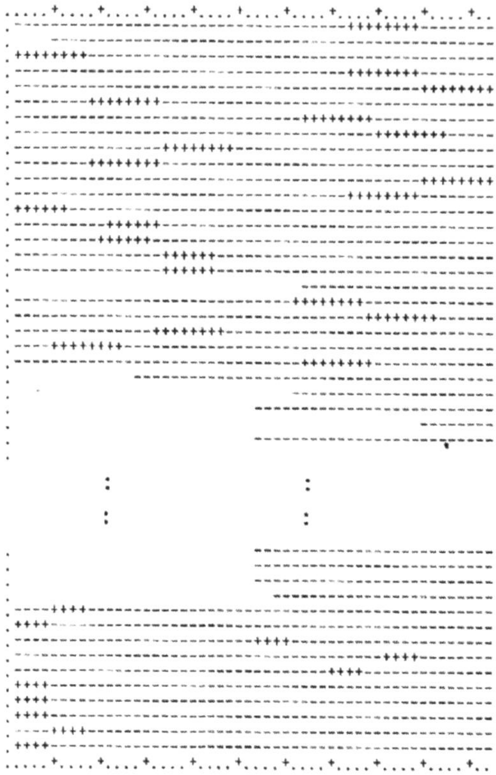

NUMG	NAME	IDD	CAP	NOPT
1	WSG	1	679.0	37
2	WSG	2	700.0	1
3	GRI	1	580.0	1
4	GRI	2	650.0	37
5	GRI	3	950.0	45
6	GRI	4	950.0	9
7	YKG	1	950.0	32
8	YKG	2	950.0	40
9	YKG	3	1000.0	17
10	YKG	4	1000.0	9
11	UGN	1	950.0	45
12	UGN	2	950.0	37
13	ABRG	3	500.0	1
14	ABRG	4	500.0	11
15	BRG	1	475.0	10
16	BRG	2	475.0	17
17	ABRG	5	500.0	17
18	ABRG	6	500.0	27
19	SCP	3	560.0	31
20	SCP	1	532.0	39
21	SCP	2	532.0	16
22	SCP	4	560.0	5
23	TAN	1	500.0	32
24	TAN	2	500.0	27
25	TAN	3	500.0	27
26	HDO	1	500.0	27
27	HDO	2	500.0	27
28	SCP	5	500.0	27
:	:			
120	HNH	7	100.0	27
121	HNH	8	100.0	27
122	HNH	9	100.0	27
123	HNH	10	100.0	27
124	YWL	2	50.0	5
125	BPTG	6	88.0	1
126	BPTG	7	88.0	27
127	BPTG	8	88.0	41
128	BPTG	9	88.0	35
129	HLM	1	70.0	1
130	YWL	1	50.0	1
131	BUSN	7	50.0	1
132	HLM	2	35.0	5
133	JAE	1	30.0	1
NUMG	NAME	IDD	CAP	NOPT

TABLE 3.6.7 Comparison of Results Obtained by Changing Aspiration Level of Supply Reserve Rate

Aspiration Level of Supply Reserve Rate	7.5 (%) (Standard)	8.0 (%)	8.5 (%)
Objective function satisfaction level (μ_{max})	**0.897 87**	0.799 46	0.720 14
Total production cost (10^9\$)	**3 378.11**	3 377.82	3 377.58
Maximum LOLP (pu)	**0.253 87**	0.253 87	0.253 87
EDNS sum (MWh)	**1 891 665**	1 907 536	1 919 940
Total air pollution (10^8ppm)	**88 715.8**	91 139.4	93 031.6
Mean of supply reserve rates (%)	**10.766 62**	10.764 15	10.754 05
Standard deviation of supply reserve rates (%)	**4.422 95**	4.439 21	4.492 24

the aspiration level of supply reserve rate increases from 7.5 to 8.5%, the value of objective function indicates the satisfaction level of decision-maker decreases from 0.9 to 0.7. A conclusion can be obtained that the reasonable aspiration level of supply reserve rate is 7.5% from this result. Another result can be obtained that variation curves of probabilistic production cost and LOLP are saturated because of their low initial aspiration levels. As aspiration levels of supply reserve rate increase, EDNS are much more and also total air pollution of SO_2 and NO_x increases because thermal power plants have to be operated more in order to guarantee the high supply reserve rate.

Lately, it can be concluded the standard deviation of supply reserve rate increases not much but small because of decreasing of the objective function.

3.6.4 User-Friendly GMS System

This section develops alternative user-friendly GMS considering not only probabilistic reliability maximization but also probabilistic production cost minimization. Furthermore, the proposed GMS system has various kinds of objective functions as like as CO_2 minimization. The probabilistic reliability objective includes loss of load expectation (LOLE), EENS, and energy index of reliability (EIR). Production cost is evaluated with consideration of uncertainties of generators. In actual system, case study describes effectiveness and user-friendly GMS simulation systems proposed in this paper. Additionally, the practicality and effectiveness of the proposed approach are demonstrated in the simulation for a real-size power system model in Korea South-East Power CO. (KOEN).

Introduction

In the world, GMS has well been decided using deterministic as well as probabilistic reliability objectives since long time years ago [164–167]. The reliability-based GMS has long history [136–154]. The reason is that reliability is a very important element for balancing electrical energy supply in power system operation. Generator needs shutdown duration for overhaul. It is equivalent with nothing of the overhaul generator. When power supply is unavailable for load balance, as it is, reliability is not satisfied with reliability criterion, and new generator construction will be required. It is too much of investment cost. Therefore, the reliability-based reasonable GMS is very important until now. In recent years, however, too high reserve rate and high reliability of power system has occurred as demand is saturated in developed countries such as South Korea or sluggish economics for too low oil price. In recent years, GMS meets an era of high reliability.

This paper proposes an alternative new user-friendly GMS simulation based on probabilistic production cost minimization instead of reliability under an era of high reliability condition. Additionally, this study tries CO_2 minimization as contribution to global environment. In addition, multi-objective GMS using fuzzy set theory is also applied [163]. The main objective functions considered in this paper are as follows, where k means interval time period number for GMS.

1. Maximization Minimum {Supply Reserve Ratek}
2. Minimization Maximum {LOLEk}
3. Minimization Total Probabilistic Production Cost
4. Multi-criteria Functions

As GMS is usually formulated as non-convex function, it is a very difficult problem. The user-friendly GMS proposed in this paper uses general GA based on MHO. Even if the initial value is very important for the meta-heuristic method, feasible solution in proposed GMS was always searched after a couple of iterations. It is expected that the proposed GMS is a very practical and effective tool. First, this paper is also trying conventional reliability objective. Second, probabilistic production cost simulation approach to GMS is tried. Third, CO_2 objective GMS is described. Finally, multi-objective function using fuzzy set theory is tried for proposed GMS in brief.

Objective Functions

Optimization of Supply Reservation Rate (SRR_n) It is possible to configure the optimal SRR_n (Z_1) as below that considers the reserved power level for stable supplying of electric power [165–168].

$$\text{Maximize } Z_1 = \text{Minimum}\,(\text{SRR}_n)\ (\%) \tag{3.6.29}$$

where SRR_n of n-th hours is defined as follows:

$$\text{SRR}_n = (\text{IC} - \text{MCAP}_n - \text{PD}_n) \times \frac{100}{\text{PD}_n}\ (\%) \tag{3.6.30}$$

where

IC: total installed capacity of generators (MW)

$MCAP_n$: maintenance capacity at nth week (MW)

PD_n: peak load at nth week (MW)

Optimization Production Cost When establishing the maintenance scheduling for generators, the total probabilistic power generation expenses for one year can be considered as an objective function. First, the probabilistic power generation amount can be as follows:

$$\text{SRR}_n = (\text{IC} - \text{MCAP}_n - \text{PD}_n) \times \frac{100}{\text{PD}_n}\ (\%) \tag{3.6.31}$$

$$E_{in} = (1-q)T\int_{u_{i-1}}^{u_i} \Phi_{in-1}(X)\mathrm{d}X\ (\text{MWh}) \tag{3.6.32}$$

where

i: number of the economic order of generators

$u_i = C_1 + C_2 + \cdots + C_i$ (MW)

C_i: capacity of #i unit

$u_0 = 0$

Φ_{in}: effective load duration curve

q_i: forced outage rate of #i unit

The probabilistic power generation expenses as an objective function(Z_2) using the probabilistic power generation amount can be calculated as follows:

$$\text{Minimize } Z_2 = F\{E_{in}, \Phi_{in}(U_{in})\} = \sum_{n=1}^{NT}\sum_{i=1}^{NG}\{A_i E_{in} + B_i T \Phi_{in}(U_{in})\} \text{ (₩)} \tag{3.6.33}$$

where

A_i = one-dimensional coefficient of fuel cost function (₩/MWh)

B_i = constant of fuel cost function (₩)

E_{in}: probabilistic generation energy of #i unit at #n stage

T: total period for study (hours)

Optimization Reliability (LOLE) When establishing the maintenance scheduling for generators, in order to consider the credibility, the objective function (Z_3) that minimizes the LOLE that is a probabilistic credibility index is as follows [163, 169]:

$$\text{Minimize } Z_2 = \max(\text{LOLE}_n) = \max\{\Phi_{\text{NG}n}(U_{\text{NG}n})\} \text{ (p.u.)} \tag{3.6.34}$$

Maximization of Satisfaction Level of Decision-Maker Using Fuzzy Set Theory [176–180]

Establishment of Membership Functions

1. Membership function of fuzzy set for the production cost is defined as:

$$\mu_c\{X(t-1), \mu(t)\} = \begin{Bmatrix} 1 & : \Delta C(\cdot) \le 0 \\ e^{-W_c \Delta C\{X(t-1), \mu(t)\}} & : \Delta C(\cdot) > 0 \end{Bmatrix} \text{ (p.u.)} \tag{3.6.35}$$

where

$\mu_c(\cdot)$: membership function of fuzzy set for production cost

$\Delta C(\cdot) = \{F(X(t))-C_{asp}(t)\}/C_{asp}(t)$

$C_{asp}(t)$: aspiration level for production cost at #t stage

W_c: weighting factor of the membership function for production cost

2. Membership functions of fuzzy set for the positive reliability (SRR) are defined as:

$$\mu_R\{X(t-1), \mu(t)\} = \begin{Bmatrix} 1 & : \Delta R(\cdot) \le 0 \\ e^{-W_{RSRR} \Delta R\{X(t-1), \mu(t)\}} & : \Delta R(\cdot) > 0 \end{Bmatrix} \tag{3.6.36}$$

where

$\mu_R(\cdot)$: membership function of fuzzy sets for reliability

$\Delta R(\cdot) = \{\text{SRR}(X(t)-\text{REQ}_{\text{SRR}}(t)\}/\text{REQ}_{\text{SRR}}(t)$

$\text{REQ}_{\text{SRR}}(t)$: aspiration level for reliability SRR at #t stage

W_{RSRR}: weighting factor of the membership function for reliability SRR

3. Membership functions of fuzzy set for the negative reliability (LOLE) are defined as:

$$\mu_R\{X(t-1),\mu(t)\} = \begin{Bmatrix} 1 & : \Delta R(\cdot) \le 0 \\ e^{-W_{RLOLE}\Delta R\{X(t-1),\mu(t)\}} & : \Delta R(\cdot) > 0 \end{Bmatrix} \quad (3.6.37)$$

where

$\Delta R(\cdot) = \{\text{LOLE}(X(t)\text{–REQ}_{\text{LOLE}}(t)\}/\text{REQ}_{\text{LOLE}}(t)$

$\text{REQ}_{\text{LOLE}}(t)$: aspiration level for reliability LOLE at #t stage

W_{RLOLE}: weighting factor of the membership function for reliability LOLE

4. Membership functions of fuzzy set for the positive reliability, EIR are defined as:

$$\mu_R\{X(t-1),\mu(t)\} = \begin{Bmatrix} 1 & : \Delta R(\cdot) \le 0 \\ e^{-W_{REIR}\Delta R\{X(t-1),\mu(t)\}} & : \Delta R(\cdot) > 0 \end{Bmatrix} \quad (3.6.38)$$

where

$\Delta R(\cdot) = \{\text{EIR}(X(t)\text{–REQ}_{\text{EIR}}(t)\}/\text{REQ}_{\text{EIR}}(t)$

$\text{REQ}_{\text{EIR}}(t)$: aspiration level for reliability EIR at #t stage

W_{REIR}: weighting factor of the membership function for reliability EIR.

Solution Procedure by the Fuzzy Search Method

1. Fuzzy decision set D can be formulated as:

$$D = C \cap R_1 \cap R_2 \cap R_3 \quad (3.6.39)$$

where

C: fuzzy set for production cost

R_1: fuzzy set for reliability SRR

R_2: fuzzy set for reliability LOLE

R_3: fuzzy set for reliability EIR

Therefore, decision membership function can be formulated:

$$\mu_D(X(t)) = \max_{\mu\min(t) \le \mu(t) \le \mu\max(t)} [\min\{\mu_C(\cdot),\mu_{R_1}(\cdot),\mu_{R_2}(\cdot),\mu_A(\cdot),\mu_S(\cdot),\mu_D(X(t-1))\}] \quad (3.6.40)$$

where

$X(t) = X(t-1) + u(t)$

$\mu_D(X(0)) = 1.0$

$\mu_D(\cdot)$: membership function of fuzzy set for decision function

Genetic Algorithm Method

As previously commented, the GMS is non-convex function problem. It is very difficult for searching optimal solution because local optimal points are sometimes obtained. Figure 3.6.12 describes the proposed GMS for searching global optima by using GA. Figure 3.6.13 is a flowchart.

Flow Chart

Figure 3.6.14 shows flowchart of the GMS system developed in this study briefly, where NG, MaxCap, CoEffs (a,b, and c), and FOR are total generator number, generator capacity (MW), fuel coefficients of quadratic function, and forced outage

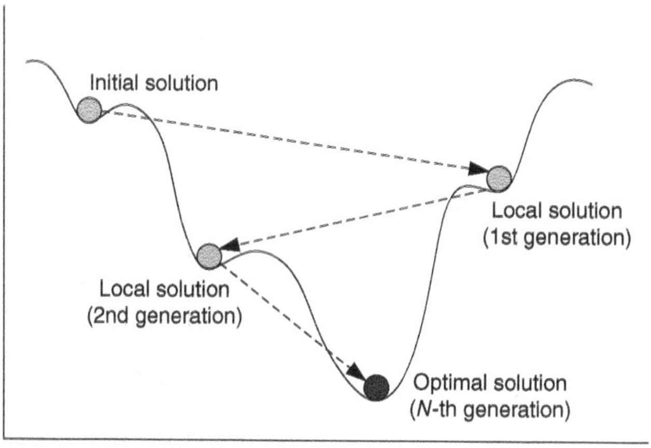

Figure 3.6.12 A search method using genetic algorithm.

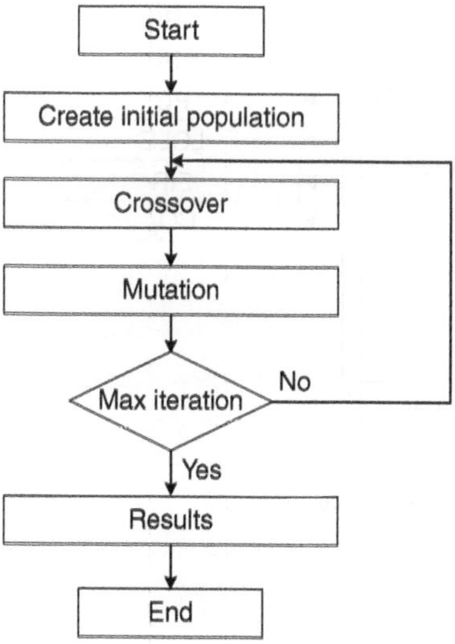

Figure 3.6.13 Flowchart of genetic algorithm.

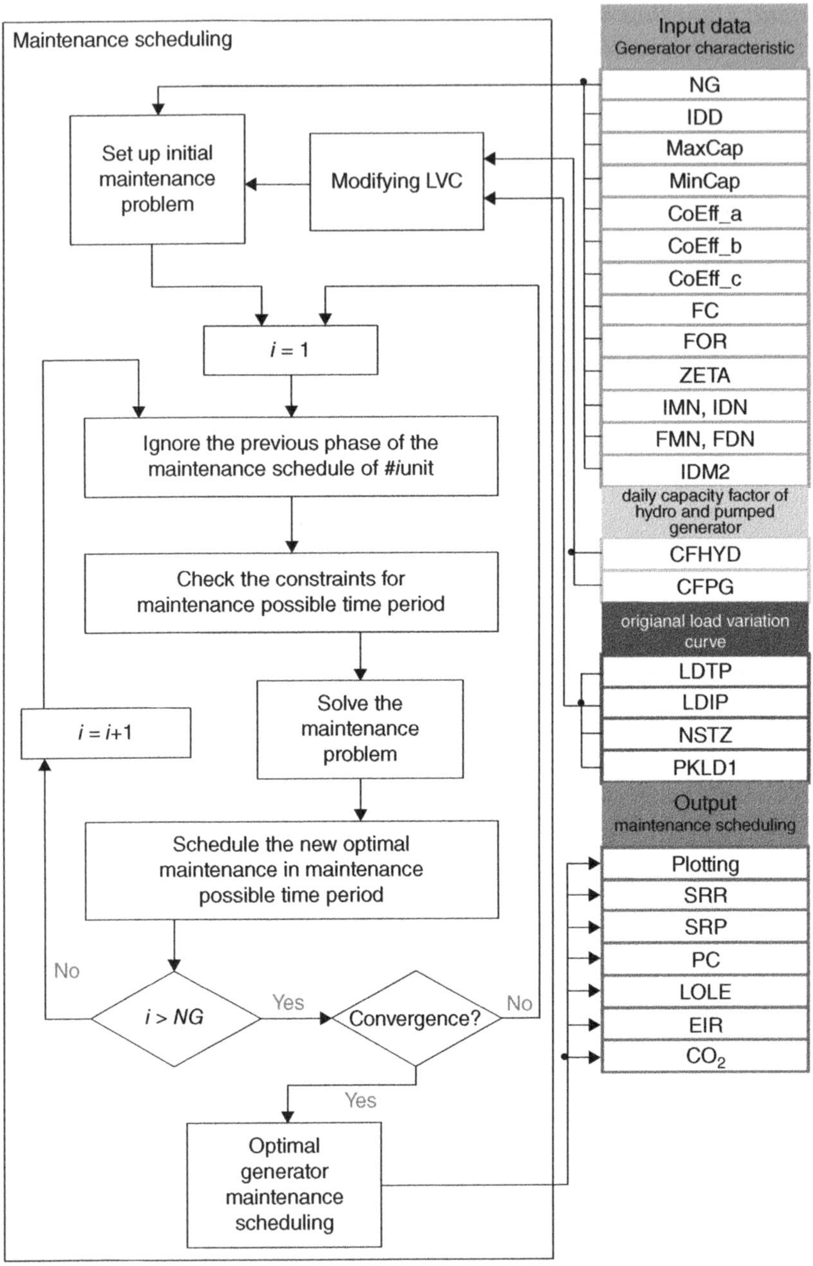

Figure 3.6.14 Generator maintenance scheduling system flowchart.

rate of generator as input data. Also, plotting, SRR, SRP, PC, LOLE, EIR, and CO_2 are output data for plotting, supply reserve rate, supply reserve power, probabilistic production cost, loss of load expectation, energy index of reliability, and CO_2 air pollution quantity. Left side box in Figure 3.6.14 describes GA-based meta-heuristic iteration method.

User-Friendly Visualization of GMS System

Figure 3.6.15 represents the starting screen of the GMS system. In each menu, the input for generator and various kinds of load variation curves are available. Main control parameters of the GMS are managed by project screen as shown in Figure 3.6.16. Constraints, objective function, target year, load, and other inputs can be set in the main project screen. Figure 3.6.17 shows an example of the running process that the GMS are searching the optimal point by GA based on meta-heuristic methodology.

Figure 3.6.18 shows user-friendly visualization results of power amount of each power generator, fuel cost, maintenance schedule, CO_2 emission quantity, etc., in the GMS. Black color in Figure 3.6.18 means overhaul period. Figure 3.6.19 is calendar style user-friendly visualization. Figure 3.6.20 represents circle-style diagram. Users can compare the results easily on the GMS tool

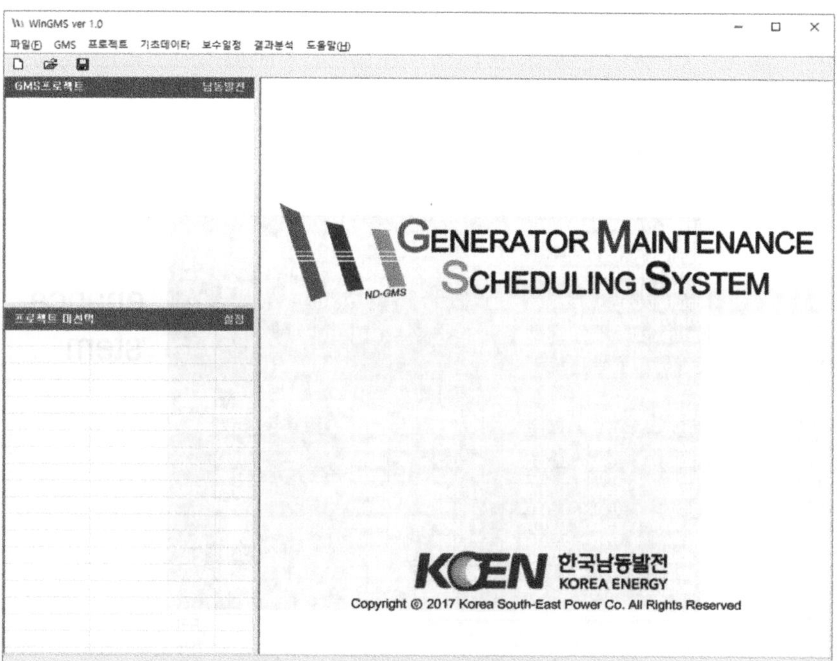

Figure 3.6.15 Starting screen of visualization of the GMS.

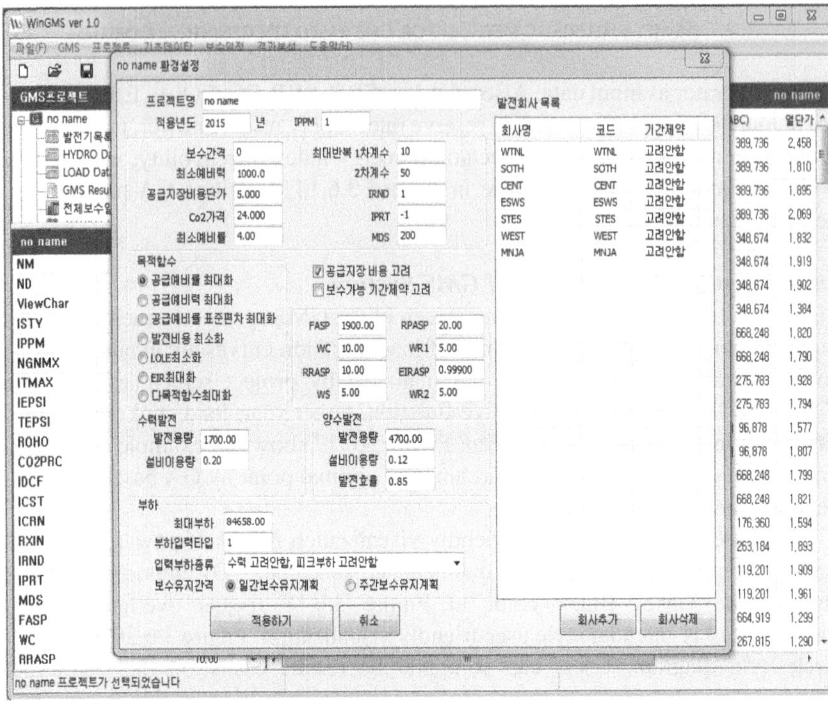

Figure 3.6.16 Preferences of the GMS.

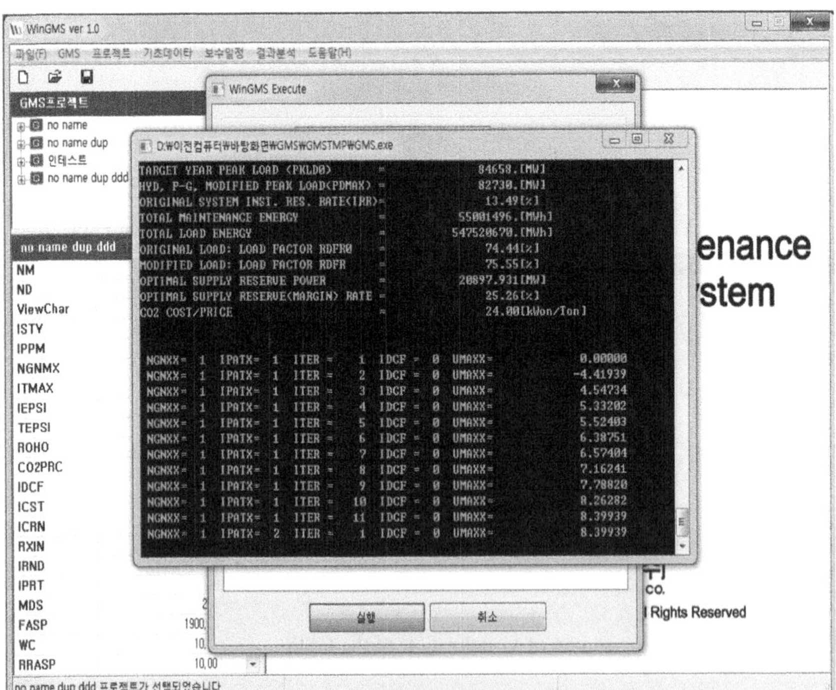

Figure 3.6.17 Running process of the GMS by GA.

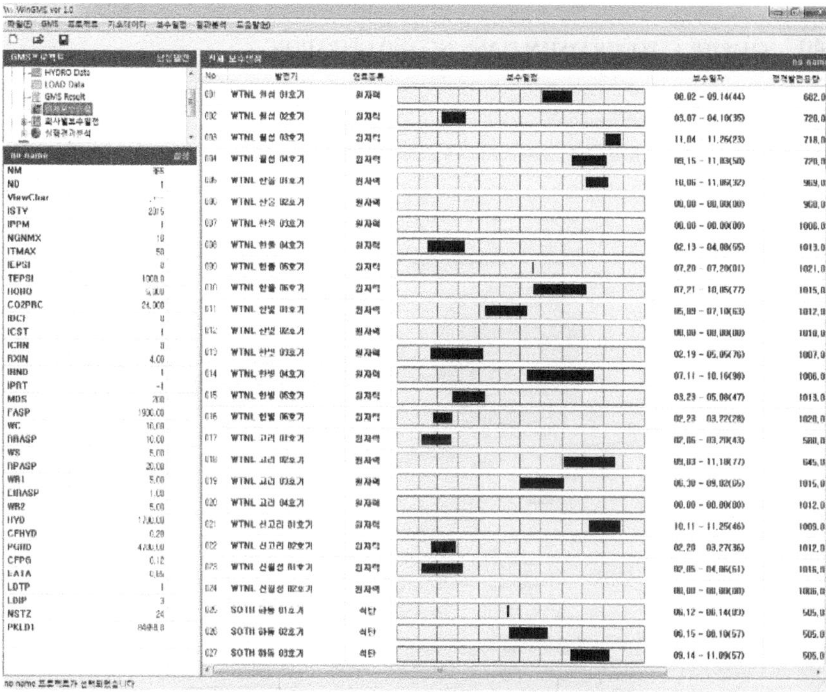

Figure 3.6.18 User-friendly visualization results of the GMS.

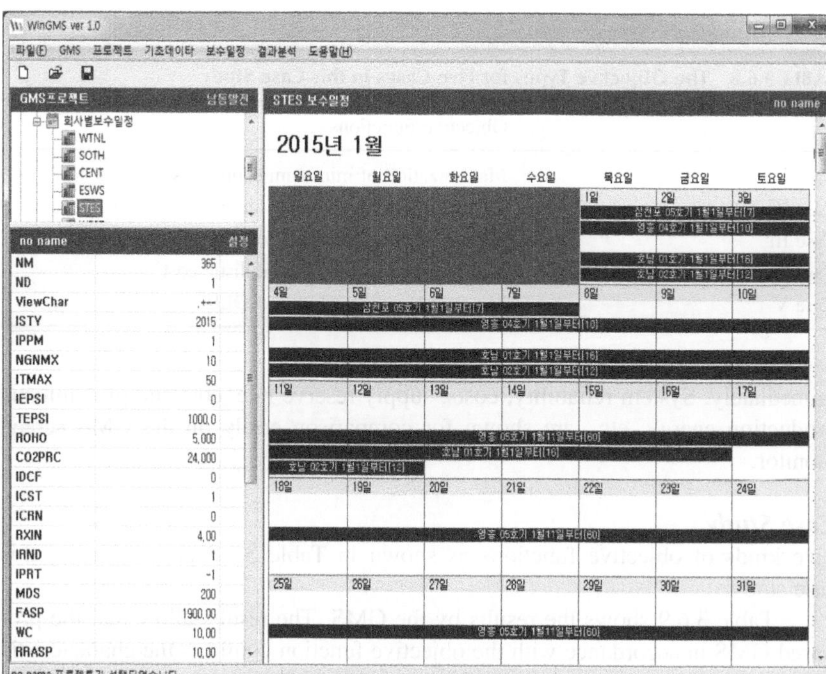

Figure 3.6.19 User-friendly visualization (calendar style) results of the GMS.

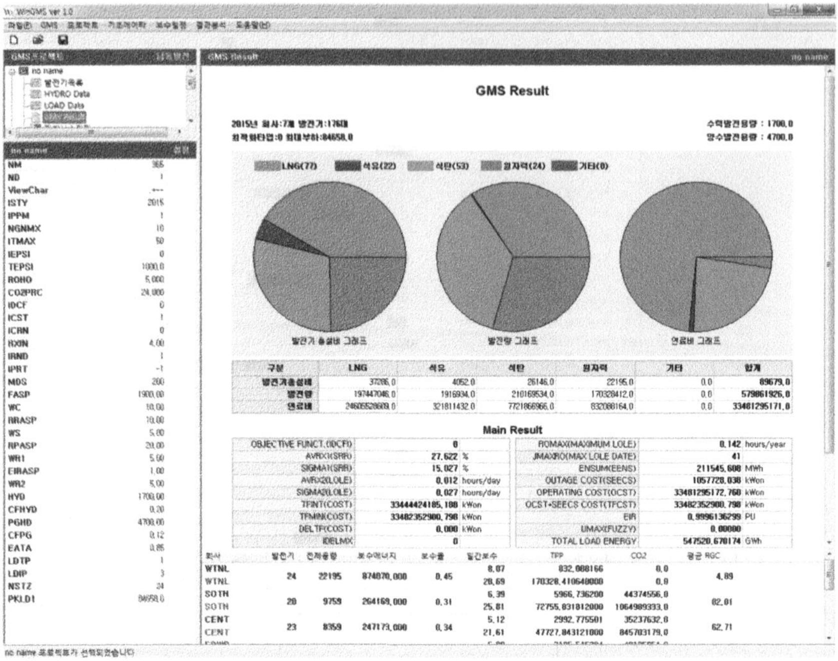

Figure 3.6.20 Total system result.

TABLE 3.6.8 The Objective Types for Five Cases in this Case Study

	Objective Functions
Case I	Maximization of minimum supply reserve rate
Case II	Maximization of minimum supply reserve power
Case III	Minimization of standard deviation in supply reserve rate
Case IV	Minimization of total probabilistic cost
Case V	Minimization of maximum LOLE

immediately. System reliability, costs, supply reserve rate, cost and probabilistic production energy, etc., are shown for comparison easily on the GMS screen monitor.

Case Study

Five kinds of objective functions as shown in Table 3.6.8 are studied in this case study.

Table 3.6.9 shows the results by the GMS. The results allow that the proposed GMS in accordance with the objective function confirms the characteristic variation of power system in accordance with objective functions.

TABLE 3.6.9 Results

	Case I	Case II	Case III	Case IV	Case V
SRR (%)	13.954	13.954	12.024	7.321	13.954
SRP (MW)	8 857.38	10 981.51	9 462.51	5 525.51	10 981.52
LOLE (h/yr)	0.444	0.353	0.154	2.828	0.064
EENS (MWh/yr)	27 364.57	22 817.11	13 146.96	138 635.5	11 000.07
EIR (PU)	0.999 947	0.999 956	0.999 974	0.999 734	0.999 979
TPC (10^6 won)	29 529 761	29 511 319	29 415 322	29 174 973	29 439 907
CO_2 (ton)	250 141 156	250 084 626	249 892 841	250 344 570	249 923 843

TABLE 3.6.10 Energy Production According to the Type of Fuel (MWh)

	Nuclear Power Generation	Coal Thermal Power Generation	Oil Thermal Power Generation	LNG Thermal Power Generation
Case I	170 328 412	210 154 296	397 391	169 990 219
Case II	170 328 412	210 154 223	311 452	170 080 787
Case III	170 328 412	210 169 530	141 155	170 245 441
Case IV	170 328 412	210 155 074	1 104 279	169 171 280
Case V	170 328 412	210 162 845	82 882	170 312 545

For example, SRR objective in Case IV has 7.3%, which is a dramatic decrease compared to the other cases. For another example, LOLE and EENS in Case V are 0.064 (hours/year) and 11 000.07 (MWh/year), respectively, which demonstrates excellent GMS results compared to the other cases.

As a result of analyzing power amount in accordance with the kind of fuel as shown in Table 3.6.10, nuclear power has consistent power amount regardless of the objective function. In the case of coal thermal power, there is a little variation in each case but it is not significant. Oil thermal power shows maximum power amount in Case IV and minimum amount in Case V.

The average probabilistic production energy percent mix for each generation kind is composed as shown in Figure 3.6.21. Nuclear, coal thermal, and LNG combined cycle generation energy are 31, 38, and 31%, respectively. It is less than 1% in case of oil thermal power.

3.6.5 Conclusion

This study proposes the user-friendly visualization system of the GMS based on not only probabilistic conventional reliability but also probabilistic production cost. Additionally, the user-friendly GMS using general GA combined with MHO is developed newly in this paper. Even if the initial value is very important

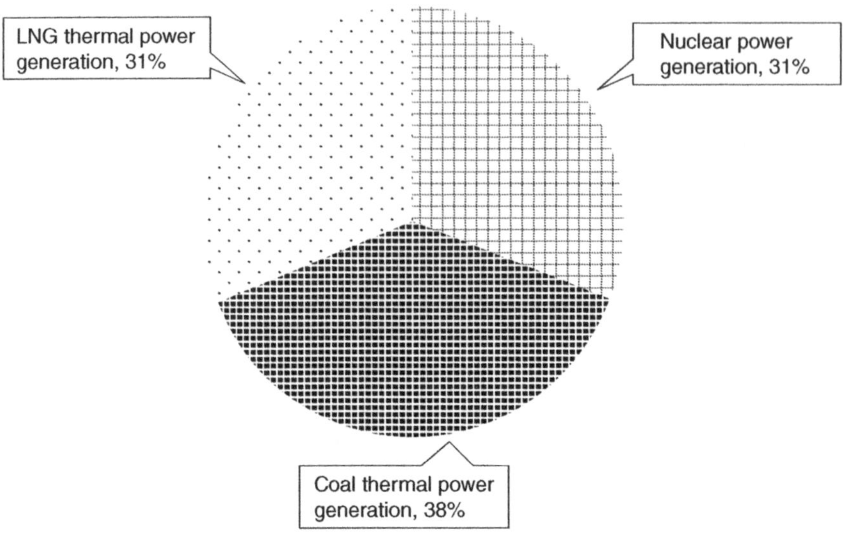

Figure 3.6.21 The share of power production average of all cases.

for the meta-heuristic method, feasible solution in proposed GMS was always searched after a couple of iterations. It is expected that the proposed GMS is a very practical and effective tool.

First, this paper proposes an alternative new user-friendly GMS based on probabilistic production cost minimization instead of reliability under an era of high reliability condition.

Second, objective functions of the GMS developed in this paper uses not only conventional reliability objective but also probabilistic production cost simulation approached GMS.

Third, assessment of CO_2 for objective function is used in the GMS. This study will be tried for CO_2 minimization as a contribution to global environment in near future.

Finally, multi-objective function using fuzzy set theory is formulated and the objective function is installed for the proposed GMS.

Therefore, it is able to know that the GMS can give different GMS depending on the objective functions. It is possible to predict what probabilistic production cost and reliability are when the new additional generators will be constructed. With these results, it is expected that the benefit for electric generation power co. (GENCOs) can be achieved by establishing the optimum GMS in case of establishment in GMS for power generators.

In the future, electric power industry will be composed of environmentally friendly generators such as renewable energy. However, these generators are not able to adjust the input and output power characteristic as conventional generator. Therefore, it may be necessary to establish the study of GMS for renewable energy generator and energy storage system as conventional generator.

ACKNOWLEDGEMENT

This work was supported by the Korean National Research Foundation (NRF) (No. #2012R1A2A2A01012803) and Korea South-East Power CO. (KOEN).

3.7 LOAD FLOW

Germano Lambert-Torres[1], Camila Paes Salomon[2], and Luiz Eduardo Borges da Silva[2]

[1]*Gnarus Institute, Itajuba, MG, Brazil*

[2]*Itajuba Federal University, Itajuba, MG, Brazil*

3.7.1 Introduction

Modern power system control centers have managed electrical power systems remotely and automatically, accomplishing functions such as AGC, state estimation, topology analysis, etc. Most of these functions require load flow studies, which are notably important for planning future expansion of power systems and determining the best operation of existing systems [170]. Load flow consists in providing the power system operation point in the steady state and is modeled by a set of nonlinear equations usually solved in control centers by using programs based on numerical computation [171, 172].

Among the numerical methods commonly applied to solve load flow, Newton–Raphson approach – and its variants – is highlighted, because of its good and quick convergence. Newton–Raphson approach is a tangent method relying on Jacobian matrix calculation; thus, it has some limitations. There are some difficulties because of the complex Jacobian matrix calculation and inversion and also the dependence on good initial estimated values to guarantee the convergence. Moreover, some changes in the power system characteristics, such as a higher resistance-to-reactance ratio (*R*/*X*), may complicate the load flow convergence [173, 174]. Finally, as previously mentioned, Newton–Raphson is a tangent method, so it is not able to solve non-convex problems. Thus, new approaches have been researched and applied to solve load flow.

Recently, some applications based on modern MHO methods have been used in control centers. Usually, because of the nature of these algorithms, they can overcome some limitations of traditional numerical methods' applications or provide solutions when the traditional methods fail. MHO is useful for problems with traditional methodologies to solve them but these methodologies have a high computational cost, problems with traditional methodologies to solve them but with some constraints in their application, and problems without any traditional methodologies to solve them. Moreover, they can run completely alone, in integrated systems and in fused systems [175].

PSO is a MHO technique that has been successfully applied to power system problems. PSO-based methodologies have provided good convergence properties, ease of implementation, and low computational time [176]. Moreover, PSO is not a tangent method, it does not depend on good initial estimative values to converge to a best solution and so it is able to solve even non-convex problems.

PSO-based approaches have been applied to normal and low-voltage multiple load flow solutions [177], OPF problem [176, 178–180], power system restoration [181], obtainment of the generator contributions to transmission system [182], voltage and reactive power control [183], reactive power control considering renewables using a fast-probabilistic power flow [184], among other problems in the context of electrical power systems. Several researches have also been implementing hybrid models, putting PSO together with other techniques. In [185], a hybrid PSO with mutation is applied to power loss minimization. Reference [186] presents a hybrid PSO methodology to solve the discrete OPF problem considering the valve loading effects. In [187], the PSO algorithm is presented for solving the optimal distribution system reconfiguration problem for power loss minimization. In [173] a chaotic PSO algorithm is presented with local search to the load flow calculation. Reference [188] proposed a hybrid algorithm based on combining fuzzy adaptive PSO and DE for non-convex ED.

This chapter approaches a hybrid PSO algorithm with mutation operation to solve the load flow problem. The chapter starts with an introduction about the load flow problem mathematical model. Then, an overview of PSO technique is presented. After this introductory part, a hybrid PSO methodology with mutation operation is presented in detail, with the possibility to readers reproduce the results. Numerical results are presented for a 6-bus and 14-bus electrical power systems.

3.7.2 Load Flow Analysis in Electrical Power Systems

Load flow is an electrical engineering known problem which determines the power system operation point in the steady state. The load flow – or power flow – problem consists in the obtainment of the buses voltages, and then in the calculation of the amount of power in the system generation buses as well as the power flow in the system branches. This problem is modeled by a set of nonlinear equations, which are commonly solved by numerical computational methods [171].

Electric power systems consist basically of buses and branches. The buses are the system nodes, defined by the same voltage level. The branches are the elements between the buses, which can be transmission lines or transformers, for instance. A power bus in the system has four variables, where two of them can be controlled and the other two are dependent on the system conditions. These variables are expressed by P, Q, $|V|$, and δ, where they mean active power in the bus, reactive power in the bus, magnitude of the bus voltage, and phase angle of the bus voltage, respectively. The values of P and Q are positive if the active power is injected in the bus – i.e. for generation buses, and negative if the active power is taken from the bus – i.e. for load buses.

The power system buses are classified according to the variables known a priori, in three main types:

1. *Type 1 or Type PQ*: P_i and Q_i are known a priori and $|V_i|$ and δ_i are calculated – usually, this type represents the load buses of the system. For the load buses, the values of P and Q are usually known and can be controlled. However, it is impossible to control the voltage values ($|V|$ and δ).
2. *Type 2 or Type PV*: P_i and $|V_i|$ are known a priori and Q_i and δ_i are calculated – usually, this type represents the generation buses. In a power plant, the values of P and $|V|$ can be controlled by the operator. If he/she increases the primary source of energy, the value of P increases together, and, vice-versa, it means, if he/she decreases the primary source of energy, the value of P decreases. The same occurs with the value of $|V|$ but in this case the operator changes the excitation system of the generator. However, for this type of bus, the operator cannot control the Q and δ values.
3. *Type 3 or Type Vδ (namely "Slack Bus")*: $|V_i|$ and δ_i are known a priori and P_i and Q_i are calculated – this type is a representation of the strongest generation bus of the system. The bus type 3 is generally only one in the power flow calculation. This bus gives a reference for the system (the values of $|V_i|$ and δ) and receives all balance power required from the system, it means, this generator needs to supply all power not given by the other generators to the system.

Let us consider the system presented in Figure 3.7.1, where there are three buses: two generation buses and one load bus. Let us define the bus 1 as slack bus, bus 2 as PV bus, and bus 3 as PQ bus. Moreover, there are three transmission lines, defined by the paths 1–2, 1–3, and 2–3.

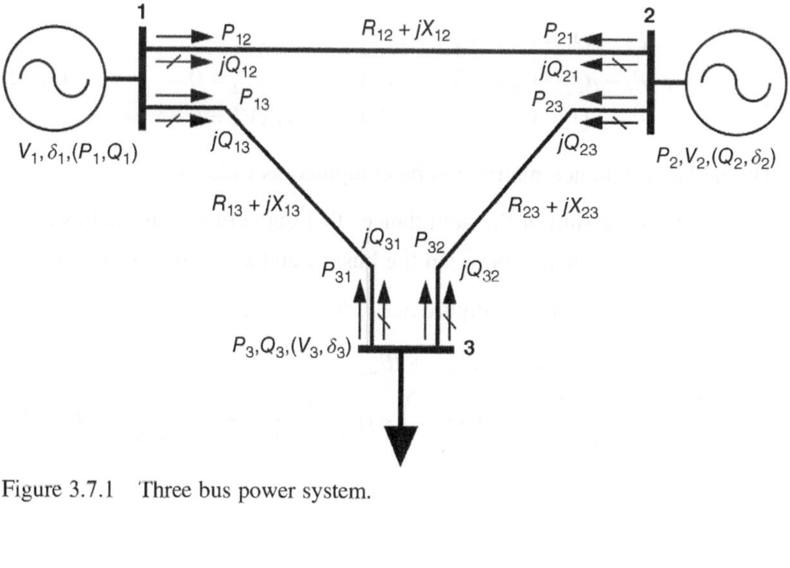

Figure 3.7.1 Three bus power system.

The load flow study consists in the determination of the possible power system operational states through the prior knowledge of some variables of the system buses. The goal is to compute the voltage of system buses in order to determine the power adjustments in the generation buses and the power flow in the system branches.

After the system steady state is obtained, it is possible to obtain the amount of power generation necessary to supply the power demand plus the power losses in the system branches. Besides, there are some constraints. The voltage levels must remain within the predefined boundaries and overloaded conditions and those in the stability limit must be prevented [189].

The power flow computation is made by iterative form and aims to find equilibrium of the known values (controlled values) and the calculated values. In the general form, the static power flow equations are given by (3.7.1) [172]:

$$P_i - jQ_i - y_{i1}V_1V_i^* - y_{i2}V_2V_i^* - \ldots - y_{in}V_nV_i^* = 0 \tag{3.7.1}$$

where

i: bus number ($i = 1, \ldots, n$),

P_i: active power generated or injected in the bus i,

Q_i: reactive power generated or injected in the bus i,

$|V_i|$: voltage magnitude of the bus i,

δ_i: voltage phase angle of the bus i,

V_i: voltage in the polar form, $V_i = |V_i|e^{j\delta i}$,

V_i^*: conjugate voltage, $V_i^* = |V_i|e^{-j\delta i}$,

y_{ik}: element of the nodal admittance matrix Y_{bus}, $k = 1, \ldots, n$.

For the system depicted in Figure 3.7.1, the static load flow equations are given by:

$$\begin{aligned} P_1 - jQ_1 - y_{11}V_1V_1^* - y_{12}V_2V_1^* - y_{13}V_3V_1^* = 0 \\ P_2 - jQ_2 - y_{21}V_1V_2^* - y_{22}V_2V_2^* - y_{23}V_3V_2^* = 0 \\ P_3 - jQ_3 - y_{31}V_1V_3^* - y_{32}V_2V_3^* - y_{33}V_3V_3^* = 0 \end{aligned} \tag{3.7.2}$$

The nodal admittance matrix can be computed as follows:

- if $i = k$, y_{ik} is the sum of the admittances that come out from the bus i;
- else y_{ik} is the admittance between the buses i and k, multiplied by -1.

For the system depicted in Figure 3.7.1, for instance,

$$y_{11} = G_{12} + jB_{12} + G_{13} + jB_{13} + G_{\text{sh},1} + jB_{\text{sh},1}$$

$$y_{11} = \left(\frac{R_{12}}{R_{12}^2 + X_{12}^2} + \frac{R_{13}}{R_{13}^2 + X_{13}^2} + G_{\text{sh},1}\right) - j\left(\frac{X_{12}}{R_{12}^2 + X_{12}^2} + \frac{X_{13}}{R_{13}^2 + X_{13}^2} - B_{\text{sh},1}\right) \tag{3.7.3}$$

$$y_{12} = (-1)(G_{12} + jB_{12}) = -\frac{R_{12}}{R_{12}^2 + X_{12}^2} + j\frac{X_{12}}{R_{12}^2 + X_{12}^2} \tag{3.7.4}$$

The power flow in the system branches is given by (3.7.5) [3]:

$$S_{ij} = P_{ij} + jQ_{ij} = V_i\left(V_i^* - V_j^*\right)Y_{ij}^* + V_iV_i^*Y_{\text{sh},i} \tag{3.7.5}$$

where

S_{ij} = complex apparent power between the buses i and j;

P_{ij} = real power between the buses i and j;

Q_{ij} = reactive power between the buses i and j;

V_i = bus i voltage;

V_j = bus j voltage;

$V_i^* = |V_i|e^{-j\delta i}$, i.e. the conjugate voltage;

$V_j^* = |V_j|e^{-j\delta j}$, i.e. the conjugate voltage;

Y_{ij} = admittance between the buses i and j;

$Y_{\text{sh},i}$ = shunt admittance of the bus i.

For the system depicted in Figure 3.7.1, for instance,

$$S_{12} = P_{12} + jQ_{12} = V_1\left(V_1^* - V_2^*\right)Y_{12}^* + V_1V_1^*Y_{\text{sh},1} \tag{3.7.6}$$

A complex and nonlinear equations system is represented by (3.7.1), whose solution is obtained through approximations using numerical methods. These methods adopt initial estimate values to the bus voltages and then apply the static power flow equations in successive iterations, looking for better approximations. The required accuracy determines the stop criterion. These methods usually take into consideration the minimization of the power mismatches of the system buses and the stop criterion can be related to a specified tolerance for the power mismatches. The real and reactive power mismatches are calculated as the sum of the injected power in the approached bus, as given by (3.7.7) and (3.7.8):

$$\Delta P_i = P_i - \sum_{k=1}^{n} P_{ik} \tag{3.7.7}$$

$$\Delta Q_i = Q_i - \sum_{k=1}^{n} Q_{ik} \tag{3.7.8}$$

where

ΔP_i: real power mismatch at bus i,

ΔQ_i: reactive power mismatch at bus i.

For the system depicted in Figure 3.7.1, for instance,

$$\begin{aligned} \Delta P_1 &= P_1 - P_{12} - P_{13} \\ \Delta Q_1 &= Q_1 - Q_{12} - Q_{13} \end{aligned} \tag{3.7.9}$$

3.7.3 Particle Swarm Optimization and Mutation Operation

Particle Swarm Optimization

Swarm intelligence is a kind of artificial intelligence based on the behavior of the animals living in groups and provided with some ability to interact among each other as well as the environment in which they are inserted. The individuals of the intelligent swarm are able to optimize a function or goal by collectively adapting to their environment. PSO consists of an optimization algorithm developed through the simulation of simplified social models as bird flocks flying randomly in search for food [176, 190, 191].

There are different variants of the PSO algorithm, but this chapter focuses on the global version of PSO (*Gbest* model). PSO is applied to optimize a collection of functions by using a population of individuals, i.e. a set of particles, where each one is a possible solution of the approached problem. These particles fly in the search space, each one having position and velocity parameters at each time instant. These parameters are provided with the same size as the problem dimension. Moreover, such particles have cognition about their own performances and also about their neighbors' performances. The best individual position of a particle is called *personal best*, and the best position of all particles is called *global best*. The particles are assessed through a specific *rule function* at each time instant. The rule function performs the interaction between the particles and the environment, and it is related to the problem modeling.

The particles' positions are updated based on their own personal exploration (personal best), best swarm overall experience (global best), and previous velocity vector according to (3.7.10) and (3.7.11).

$$v_{i,j}(t+1) = wv_{i,j}(t) + c_1 r_{1,j}\left[p_{i,j}(t) - x_{i,j}(t)\right] - c_2 r_{2,j}\left[g_j(t) - x_{i,j}(t)\right] \tag{3.7.10}$$

$$x_{i,j}(t+1) = x_{i,j}(t) + v_{i,j}(t) \tag{3.7.11}$$

where

i: the particle index,

j: dimension of the vector associated with the particle i,

t: iterations counter,

ni: total number of iterations,

$v_{i,j}(t)$: velocity of the dimension j of the particle i at iteration t,

$x_{i,j}(t)$: position of the dimension j of the particle i at iteration t,

r_1, r_2: random numbers between 0 and 1,

c_1, c_2: acceleration coefficients, both set to a value of 2.0,

$p(t)$: particle i personal best found at iteration t,

$g(t)$: global best found at iteration t,

w: velocity equation's inertia weight.

The inertia weight can be a constant or a function dependent on the iteration counter. The presented approach considers the inertia weight as given by (3.7.12) [185].

$$w = w_{\max} - \frac{(w_{\max} - w_{\min})}{ni} t \tag{3.7.12}$$

where

w: velocity equation's inertia weight;

$w_{\max}$: inertia weight maximum value, set to a value of 0.7;

$w_{\min}$: inertia weight minimum value, set to a value of 0.2.

The described process is iterative and it proceeds until all the particles converge to the achieved global best, which is adopted as the problem solution.

A general PSO algorithm can be described as follows [192]:

1. For each particle, the position and velocity vectors are randomly initialized.
2. Measure the personal best of each particle by using the rule function:

```
If f[xi,j(t)] f[pi,j(t)]
      Then pi,j(t) = xi,j(t)
End
```

3. Store the particle with the best fitness value (global best):

```
If f[pi,j(t)] f[gj(t)]
      Then gj(t) = pi,j(t)
End
```

4. Update velocity and position vectors according to (3.7.10), (3.7.11), and (3.7.12) for each particle.
5. Repeat Steps 2–4 until a stop criterion is satisfied.

Hybrid PSO with Mutation Operation

PSO is provided with several advantages over other optimization techniques like:

- it is a derivative-free algorithm;
- it is less sensitive to the nature of the objective function;
- it has less parameters to adjust, unlike other EAs;
- it is easy to implement and program;
- it can handle objective functions with stochastic nature;
- it is not largely dependent on a good initial estimative to converge;
- it has less computational time [191, 192].

However, PSO can suffer premature stagnation, which is a phenomenon when the particles stop to move in the search space even before they reach the optimal solution. A possible way to overcome this problem is to associate the PSO with features of other artificial intelligence techniques, composing a hybrid intelligent system. For instance, reference [185] proposed the use of GA mutation operation associated with PSO to deal with the stagnation problem.

GA is a search heuristic iterative procedure useful to optimization and search problems [61]. GA is inspired by natural evolution and uses a population of individuals, in which each one represents a candidate solution to the approached problem [193, 194]. The GA procedure is applied iteratively to decrease the search space and create new good solutions to the problem domain [195].

Mutation is a GA operation performed by altering one or more gene values in a chromosome, randomly and with an associated probability to occur. This operation is applied to better cover the problem search space, maintaining genetic diversity and also preventing the premature convergence [185].

3.7.4 Load Flow Computation via Particle Swarm Optimization with Mutation Operation

This section presents a methodology using a hybrid PSO algorithm with mutation operation for accomplishing load flow studies of electrical power systems. The intent of this chapter is to describe this MHO approach for load flow computation, which has the advantages of being an easier and more flexible implementation comparing with conventional computational routines. Moreover, it is potential to overcome some limitations found when executing load flow studies by using traditional numerical methods.

The proposed rule function is based on the minimization of the apparent power mismatches in the power system buses. The swarm particles' positions are modeled as the buses voltages, considering magnitudes and phase angles. Thus, they assume continuous values within the ranges specified in the power system input data. The particles' structure is depicted in (3.7.13).

$$x_i = [\delta_2, \delta_3, \ldots, \delta_k, \ldots, \delta_n, V_2, V_3, \ldots, V_k, \ldots, V_n] \tag{3.7.13}$$

where

i = particle index;

x_i = particle i position;

k = bus index;

n = total number of buses;

δ_k = voltage phase angle at bus k;

V_k = voltage magnitude at bus k.

The bus 1 is defined as the slack bus, so the particle position does not comprise it.

Rule Function

The rule function adopted in this approach is the apparent power mismatches of the system buses, which are calculated by:

$$\Delta S_i = \sqrt{(\Delta P_i)^2 + (\Delta Q_i)^2} \quad (3.7.14)$$

where

ΔS_i: apparent power mismatch at bus i;

ΔP_i: real power mismatch at bus i, given by (3.7.7);

ΔQ_i: reactive power mismatch at bus i, given by (3.7.8).

The goal of the optimization process is to minimize the arithmetic mean of the apparent power mismatches of the system buses, which is calculated by:

$$\Delta S_m = \frac{1}{N}\sum_{i=1}^{N} \Delta S_i \quad (3.7.15)$$

where

N: total number of buses of the power system;

ΔS_m: arithmetic mean of the apparent power mismatches of the system buses.

The rule function parameters, i.e. the arithmetic mean of the power mismatches of the system buses, are defined as *scores*. Each particle has a *personal score*, i.e. the score obtained by its personal best. The *global score* is the score associated with the global best. The *current score* is the score obtained by a particle at the current iteration of the process.

Algorithm Initialization

The first step of the proposed algorithm is the generation of the initial estimate values of the particle positions, velocities, personal best, global best, and fitness parameters.

In *Particle Positions*, the voltage phase angles, which are one kind of parameters of the particle positions, begin as random values within the specified boundary; and the voltage magnitudes, which are the other kind of parameters of the particle positions, depend on the bus type for the initial estimation. In the case of a PQ bus, the voltage magnitude begins as a random value within the specified boundary; in the case of a PV bus, the voltage magnitude is the rated value specified in the input data.

The *Initial Velocities* are null. The *Personal Best* starts as the associated particle position values; and the *Global Best* parameter starts as an arbitrary particle value. Finally, the *particles fitness* begins with large values in order to be minimized later.

PSO Procedure

After the initialization of all the parameters of the population, the PSO algorithm iterations begin. The procedure detailed as follows is performed for all iterations and for each particle.

1. The system buses voltages are assigned with the particles' values.
2. The reactive power of the PV buses is computed applying (3.7.1).
3. The active and reactive power of the $V\delta$ bus is computed applying (3.7.1).
4. The power flow in the system branches is calculated using (3.7.5).
5. The active and reactive power mismatches of each bus are calculated by using (3.7.7) and (3.7.8). The apparent power mismatches are calculated using (3.7.14).
6. The buses' apparent power mismatches arithmetic mean is calculated using (3.7.15). This is assigned as the particle current score.
7. The particle with the worst performance, i.e. the larger score, is obtained. This particle index is stored and it is used in the mutation operation.
8. The criterion of the personal best parameter update is applied as follows:

```
If score[x_i,j(t)] p_i,j(t)]
  Then p_i,j(t) = x_i,j(t)
End
```

   Once all particles have passed through the described routine, it proceeds through the steps described as follows:

9. The criterion of the global best parameter update is applied as follows:

```
If score[p_i,j(t)] _j(t)]
  Then g_j(t) = p_i,j(t)
End
```

10. The particles' positions and velocities are updated according to (3.7.10), (3.7.11). and (3.7.12).
11. The mutation operation is applied to the worst particle of the current iteration, i.e. the particle with the larger score [196]. This is performed by adding a random value to the particle position, according to (3.7.16).

$$mx(k) = x(k) + 0.1[(x_{max} - x_{min})r + x_{min}] \tag{3.7.16}$$

where

k: mutated particle index,

$x(k)$: particle position before the mutation operation,

$mx(k)$: particle position after the mutation operation,

r: random number between 0 and 1,

x_{max}: maximum value of the position, related to the specified boundary in the input data,

x_{min}: minimum value of the position, related to the specified boundary in the input data.

The iterations proceed until the required accuracy for the global score is reached or the iterations' counter reaches a maximum predefined value. The accuracy tolerance was adopted as 10^{-5} (p.u.). The execution runs a group of a fixed number of iterations. In the end of the iterations group, if the global score is larger than the tolerance, the algorithm runs another group of iterations. The process continues until a maximum number of iteration groups it can run. So, the final global best is adopted as the load flow solution. Figure 3.7.2 presents the flowchart of the proposed methodology.

In the proposed PSO methodology, each particle has a random initial estimative value and the PSO equations also use random values, so different acceptable solutions can be obtained for the same initial estimative. Thus, the proposed methodology can provide several acceptable results for the same load flow study, depending on the execution of the algorithm [196].

3.7.5 Numerical Results

This section presents numerical results for the proposed methodology of load flow computation. The proposed algorithm is evaluated in two case studies: a 6-bus [7] and the IEEE 14-bus power systems. A 1.66-GHz Intel® T1600 PC has been used for the software simulations. For each case study, five runs have been performed. The presented results are the best solution over these five runs.

6-Bus Power System Case Study

This case study is performed with the 6-bus power system presented in [7], as shown in Figure 3.7.3. This power system is provided with three generation buses and three load buses, totalizing six buses, and 11 transmission lines, which are the system branches.

The parameters used for this case study were: number of particles: 15; and maximum number of iterations: 4000 iterations, arranged into 40 groups of 100 iterations per group.

Figure 3.7.4 presents the decrease of the global score along the iterations. The global score starts with 0.559 868 (p.u.) at the first iteration and reaches 1.002 069*E*–05 (p.u.) at the end of the iterations. Figure 3.7.5 presents the obtained voltage profile of the system buses, considering voltage magnitudes and phase angles. Tables 3.7.1 and 3.7.2 present the load flow results for the 6-bus power system obtained through the application of the proposed PSO methodology.

14-Bus Power System Case Study

The second case study is performed with the IEEE 14-bus power system, as shown in Figure 3.7.6. This power system is provided with two generation buses, three synchronous condensers, and none load buses, totalizing 14 buses, and 20 system branches, composed of transmission lines and transformers.

The parameters used for this case study were: number of particles: 20; and maximum number of iterations: 6000 iterations, arranged into 40 groups of 150 iterations per group. Figure 3.7.7 presents the decrease of the global score along

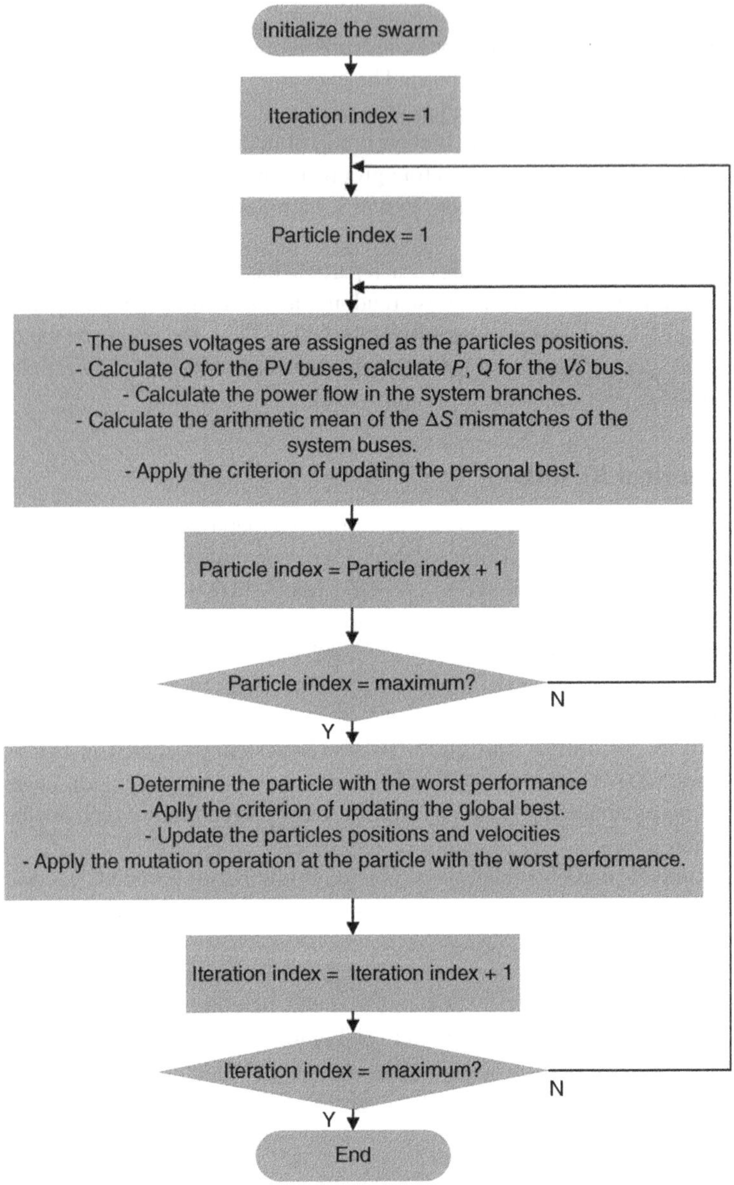

Figure 3.7.2 Flowchart of the proposed hybrid PSO algorithm.

the iterations. The global score starts with 1.028 145 (p.u.) at the first iteration and reaches 5.103 039*E*–05 (p.u.) at the end of the iterations. Figure 3.7.8 presents the obtained voltage profile of the system buses, considering voltage magnitudes and phase angles. Tables 3.7.3 and 3.7.4 present the load flow results for the 14-bus power system obtained through the application of the proposed PSO methodology.

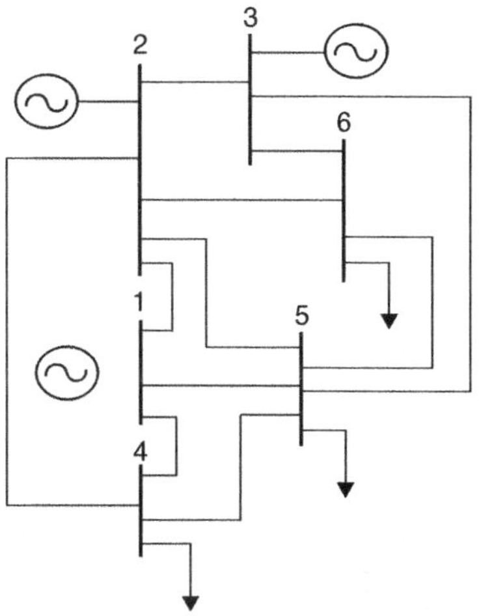

Figure 3.7.3 Six-bus power system [28].

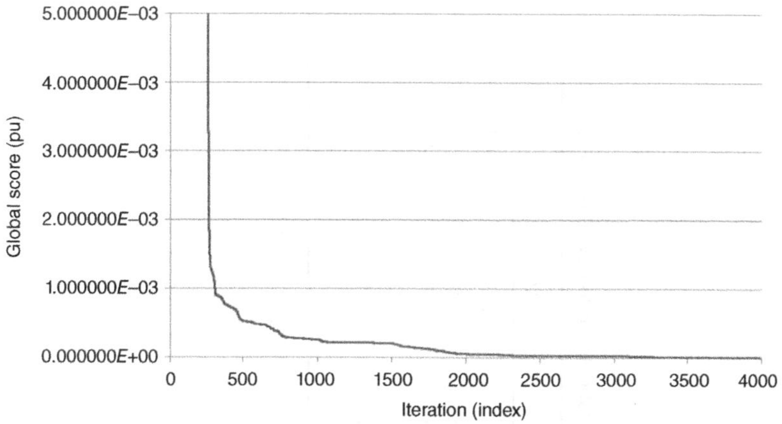

Figure 3.7.4 Decrease of global score for the 6-bus power system.

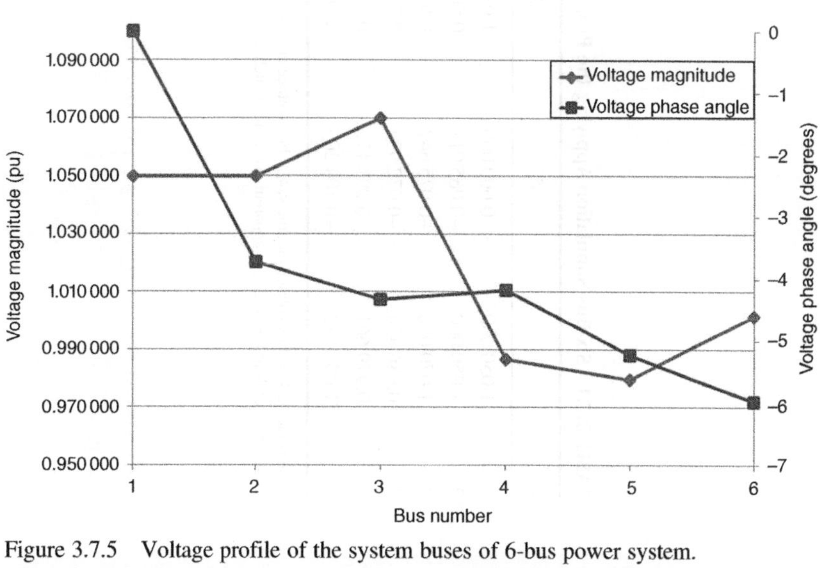

Figure 3.7.5 Voltage profile of the system buses of 6-bus power system.

TABLE 3.7.1 Six-Bus Simulation Applying the Proposed PSO Methodology: Buses' Parameters Results

k	V_k	δ_k	P_k	Q_k	ΔP_k	ΔQ_k	ΔS_k
1	1.050 000	0.000 000	1.084 407	0.231 198	−4.996 004E−16	3.226 586E−16	5.947 345E−16
2	1.050 000	−0.065 011	0.500 000	0.868 623	−3.136 684E−08	−6.182 554E−15	3.136 684E−08
3	1.070 000	−0.075 642	0.600 000	0.988 295	−1.413 960E−07	1.866 562E−15	1.413 960E−07
4	0.986 424	−0.072 934	−0.700 000	−0.700 000	−4.609 943E−06	1.059 944E−07	4.611 162E−06
5	0.979 661	−0.091 176	−0.700 000	−0.700 000	−2.938 276E−06	9.101 118E−08	2.939 685E−06
6	1.001 443	−0.104 203	−0.700 000	−0.700 000	−5.237 999E−05	−1.465 926E−06	5.240 050E−05

k = bus index, V_k = bus voltage module k, δ_k = bus voltage angle k, P_k = real power generated at bus k, Q_k = reactive power generated at bus k, ΔP_k = real power mismatch at bus k, ΔQ_k = reactive power mismatch at bus k, ΔS_k = apparent power mismatch at bus k.

TABLE 3.7.2 Six-Bus Simulation Applying the Proposed PSO Methodology: Load Flow in the System Branches

i	j	P_{ij}	Q_{ij}	P_{ji}	Q_{ji}
1	2	0.291 155	−0.133 932	−0.281 839	0.152 564
1	4	0.436 947	0.238 303	−0.425 713	−0.193 367
1	5	0.356 305	0.165 415	−0.345 107	−0.123 425
2	3	0.029 830	−0.089 712	−0.029 425	0.091 739
2	4	0.332 798	0.501 470	−0.316 370	−0.468 615
2	5	0.154 944	0.195 714	−0.149 292	−0.178 758
2	6	0.264 267	0.166 468	−0.258 074	−0.148 772
3	5	0.193 274	0.283 064	−0.180 961	−0.256 385
3	6	0.436 150	0.650 701	−0.425 431	−0.597 103
4	5	0.042 088	−0.003 962	−0.041 721	0.004 697
5	6	0.017 084	−0.076 548	−0.016 443	0.078 471

P_{ij} (ji) = real power in the branch composed by the buses i–j (j–i), Q_{ij} (ji) = reactive power in the branch composed by the buses i–j (j–i).

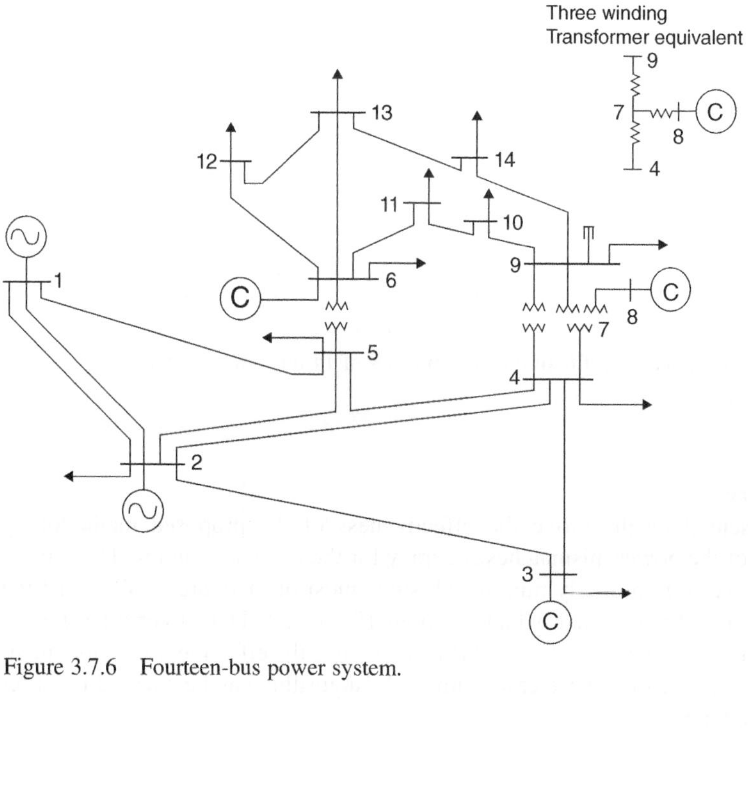

Figure 3.7.6 Fourteen-bus power system.

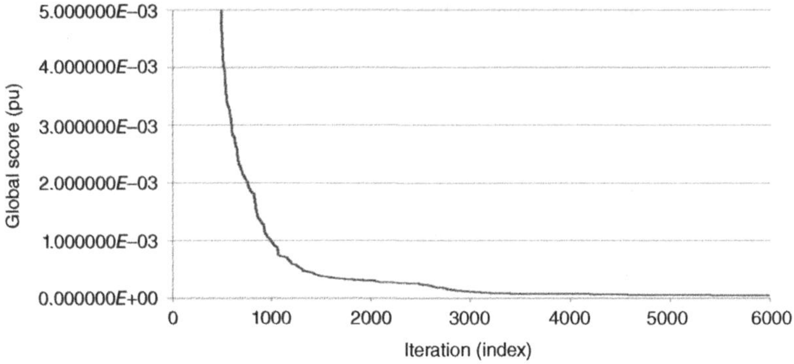

Figure 3.7.7 Decrease of global score for the 14-bus power system.

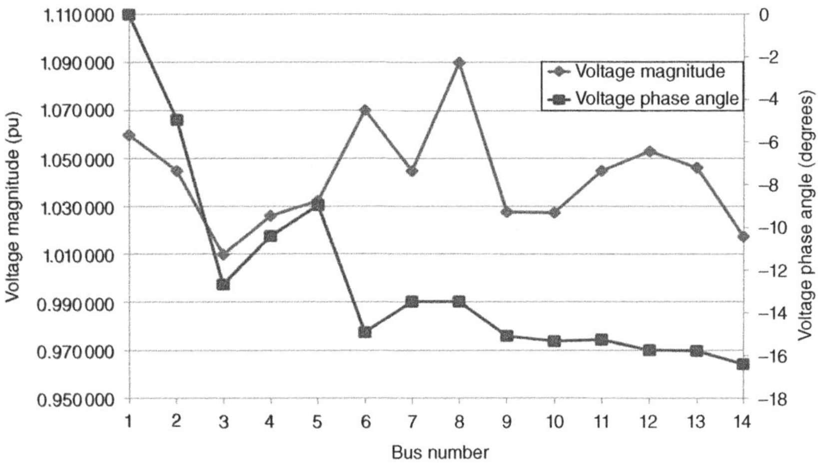

Figure 3.7.8 Voltage profile of the system buses of 14-bus power system.

Summary

The presented results prove the effectiveness of the proposed methodology because of the power mismatches accuracy for the two case studies. These parameters have been properly minimized, since most of them are smaller than the tolerance usually accepted, which is about 10^{-4} (p.u.). Thus, hybrid PSO algorithms can be applied to solve load flow and, once the effectiveness of the methodology is proven for these case studies, the algorithm can be adapted to larger power systems.

TABLE 3.7.3 Fourteen-Bus Simulation Applying the Proposed PSO Methodology: Buses' Parameters Results

k	V_k	δ_k	P_k	Q_k	ΔP_k	ΔQ_k	ΔS_k
1	1.060 000	0.000 000	2.326 004	–0.225 140	1.110 223*E*–16	3.608 225*E*–15	3.609 932*E*–15
2	1.045 000	–0.086 534	0.183 000	0.180 764	4.591 792*E*–07	2.241 263*E*–15	4.591 792*E*–07
3	1.010 000	–0.220 527	–0.942 000	0.010 519	2.357 969*E*–06	1.361 758*E*–15	2.357 969*E*–06
4	1.026 085	–0.180 974	–0.478 000	0.039 000	3.153 200*E*–07	–3.556 198*E*–07	4.752 811*E*–07
5	1.032 587	–0.156 204	–0.076 000	–0.016 000	2.245 127*E*–06	–2.420 918*E*–06	3.301 733*E*–06
6	1.070 000	–0.259 846	–0.112 000	0.416 225	3.963 632*E*–06	3.927 414*E*–15	3.963 632*E*–06
7	1.044 813	–0.234 835	0.000 000	0.000 000	7.036 756*E*–08	–1.701 521*E*–07	1.841 286*E*–07
8	1.090 000	–0.234 835	0.000 000	0.279 614	3.409 777*E*–07	–6.106 227*E*–16	3.409 777*E*–07
9	1.027 638	–0.263 116	–0.295 000	–0.166 000	–9.095 922*E*–06	–7.267 837*E*–06	1.164 291*E*–05
10	1.027 551	–0.267 460	–0.090 000	–0.058 000	–2.604 466*E*–07	–2.667 324*E*–07	3.727 983*E*–07
11	1.044 951	–0.265 658	–0.035 000	–0.018 000	1.867 445*E*–05	–4.468 844*E*–05	4.843 337*E*–05
12	1.052 981	–0.274 599	–0.061 000	–0.016 000	6.224 401*E*–04	–1.348 020*E*–04	6.368 699*E*–04
13	1.046 233	–0.274 857	–0.135 000	–0.058 000	1.123 377*E*–07	–1.168 538*E*–07	1.620 944*E*–07
14	1.017 438	–0.286 260	–0.149 000	–0.050 000	1.638 358*E*–06	–5.627 776*E*–06	5.861 406*E*–06

TABLE 3.7.4 Fourteen-Bus Simulation Applying the Proposed PSO Methodology: Load Flow in the System Branches

i	j	P_{ij}	Q_{ij}	P_{ji}	Q_{ji}
1	2	1.561 401	−0.172 642	−1.518 836	0.302 598
1	5	0.764 603	0.004 806	−0.736 489	0.111 248
2	3	0.726 497	0.060 098	−0.703 630	0.036 240
2	4	0.558 668	−0.044 915	−0.541 952	0.095 634
2	5	0.416 671	−0.046 816	−0.407 502	0.074 810
3	4	−0.238 372	0.003 148	0.242 105	0.006 380
4	5	−0.609 694	0.042 556	0.614 430	−0.027 616
4	7	0.275 986	−0.084 459	−0.275 986	0.101 005
4	9	0.155 555	0.003 526	−0.155 555	0.009 263
5	6	0.453 560	−0.129 764	−0.453 560	0.182 368
6	11	0.079 040	0.097 103	−0.077 740	−0.094 379
6	12	0.080 765	0.032 860	−0.079 949	−0.031 161
6	13	0.181 750	0.103 895	−0.179 217	−0.098 908
7	8	0.000 000	−0.268 023	0.000 000	0.279 614
7	9	0.275 985	0.167 018	−0.275 985	−0.156 531
9	10	0.047 932	−0.016 869	−0.047 855	0.017 076
9	14	0.088 617	−0.001 855	−0.087 671	0.003 867
10	11	−0.042 145	−0.075 076	0.042 721	0.076 424
12	13	0.018 327	0.015 296	−0.018 213	−0.015 193
13	14	0.062 431	0.056 101	−0.061 330	−0.053 861

3.7.6 Conclusions

This chapter presented a hybrid PSO methodology with mutation operation to solve load flow in electrical power systems. The PSO algorithm-based techniques are potential to solve problems whose traditional methodologies to solve them hold some constraints in their application and problems without any traditional methodologies to solve them. The main advantages of this presented methodology are the ease and flexibility of implementation, and its better convergence. PSO-based methods are not tangent methods, so they are able to solve even non-convex problems, unlike traditional numerical methods. The chapter provided an explanation of how to perform a PSO-based algorithm to solve load flow and the results presented for two power systems proved the effectiveness of the proposed methodology, since the power mismatches were small and lower than the usually accepted tolerance of 10^{-4} (p.u.). Finally, it is expected that this chapter has provided the reader with a comprehensive view of the use of hybrid PSO algorithms to solve load flow problems and supports them in developing new PSO-based methods for this application.

ACKNOWLEDGEMENT

The authors would like to thank CNPq, CAPES, FAPEMIG, and the Brazilian Electricity Regulatory Agency Research and Development (ANEEL R&D) for supporting this project.

3.8 ARTIFICIAL BEE COLONY ALGORITHM FOR SOLVING OPTIMAL POWER FLOW

Wenlei Bai[1], Kwang Y. Lee[2], and Ibrahim Eke[3]

[1]ABB Enterprises Software Inc. Houston, TX, USA
[2]Baylor University, Waco, TX, USA
[3]Kirikkale University, Kirikkale, Turkey

System operators and planners consider OPF as one of the most essential problems in power systems because of the detailed controls over power system which result in a significant cost reduction. The OPF was first proposed as early as 1960s by Carpentier and has grown into a powerful tool for system operation and planning [197–200]. The OPF problem, under the objective function of minimizing cost, takes into account the constraints of AC load flow at each node, transmission line capacity, voltage limits, etc., and controls real, reactive power, voltage, transformer taps, shunt compensators which lead to significant cost reduction. In other words, the aim of OPF is to optimize an objective function representing the total cost of generation, power losses, voltage stability, and/or other relevant information, while satisfying the system constraints [201].

Classical methods such as linear programming [202], quadratic programming [203], and interior-point method (IPM) [204, 205] rely on theoretical assumptions of convexity and they are very sensitive to the starting points. Therefore, these methods can be easily trapped into local optimum or diverge and do not offer much freedom in modifying objective functions. OPF becomes a highly nonlinear, non-convex, and large dimensional optimization problem after incorporating non-smooth, non-convex, nonlinear, and non-differential objective functions and constraints which is difficult, if not impossible, to solve by classical methods. Therefore, researchers have focused on developing more efficient and robust methods to handle OPF problems without simplifying the system. For this purpose, meta-heuristic methods such as GA [206], EP [207], TS [208], and PSO [105] have been placed attention on solving OPF. Reference [105] proposed an improved PSO to tackle the problem considering the valve-point effect on the regular quadratic fuel cost function. One of the recently proposed heuristic algorithms called ABC has drawn researchers' attention due to its simplicity and robustness. The ABC is based on intelligent behavior of honeybees. It was first developed by Karaboga in 2005 [209]. This method is a population-based optimization algorithm which has been demonstrated competitive to other methods because of the advantage of controlling fewer parameters and its robustness [210–212]. In this chapter, we apply ABC in OPF problems, starting from the application of minimizing basic quadratic cost to more complex objective functions. The procedures were demonstrated on IEEE 30-bus test system and comparison with other methods was also provided.

3.8.1 Optimization in Power System Operation

Power system optimization has evolved with developments in computing and optimization theory in decades. As early as in the first half of the twentieth century, the OPF was "solved" by empirical methods, rules of thumb, and primitive tools such as analog network analyzers [213]. With the computational aid gradually, the OPF problem was first formulated by Carpentier in 1962 and has proven to be a very difficult problem to solve.

There are generally three types when it comes to power system operation optimization problems in the literature: power flow (load flow), ED, and OPF. Here, we want to give readers a brief description of the differences/commons among these problems. Table 3.8.1 lists the major characteristics of power system operation problems. The power flow focuses on the generation, load and transmission network equations, and solves a nonlinear mathematical solution. However, the solution might not be optimal or physically feasible under certain constraints. For instance, the power flow equations do not consider generator reactive power and transmission line limits. Historically, ED has been the major tool for system operating and planning; however, the control variables for ED are only real power and the electrical network is solely represented by single equality constraint, the power balance equation [214]. ED fails to consider the power flow constraints in network. The OPF solves the optimal solution under a specified objective function subject to the power flow constraints, generator limits, transmission lines' thermal capacity, switching equipment limits, etc. From Table 3.8.1, there are various subproblems stemming from the three general types in order to meet specific situations, such as alternating current OPF (ACOPF), direct current OPF (DCOPF), security constrained ED (SCED), security constrained OPF (SCOPF), etc.

Note that OPF can be a fundamental tool for power system operation, and based on such tool various modified versions can be developed for specific purposes. For instance, nowadays with high penetration of renewable energy and storage devices, system operators are developing dispatch methods to meet the needs and the essence of such methods are still OPF problem.

3.8.2 The Optimal Power Flow Problem

Mathematically, solving an OPF problem is equivalent to finding a set of optimal decision/control vectors that minimizes an objective function under several constraints. The OPF problem to be considered is formulated as follows [221]:

TABLE 3.8.1 Major Types of Power System Problems

General Problem Type	Problem Name	Voltage Angle Constraint?	Bus Voltage Constraint?	Transmission Constraint?	Assumptions	Generator Costs?	Contingency Constraints?
PF	PF	N	Y	N	—	N	N
ED	ED	N	N	N	No transmission constraints	Y	N
OPF	ACOPF	Y	Y	Y	—	Y	N
OPF	DCOPF	N	N	Y	*V* is constant	Y	N
OPF	SCED	Y	N	Y	*V* is constant	Y	Y
OPF	SCOPF	Y	Depends	Y	Depends	Y	Y

$$\begin{aligned} \min \quad & f(x,u) \\ \text{s.t.} \quad & g(x,u) = 0 \\ & h(x,u) \leq 0 \end{aligned} \tag{3.8.1}$$

where

u: decision/control vectors,

x: state vectors,

f: objective functions,

g: equality functions,

h: inequality functions,

The vector u includes generator real power P_G except at slack bus, generator bus voltage V_G, transformer tap T, and shunt compensator Q_C at selected buses. The vector x includes real power P_{G1} at slack bus, voltage V_L at load bus, reactive power Q_G at generator bus, and transmission line loading S_L. There are four objective functions f chosen in the study: quadratic cost function, quadratic cost function with valve effect, power loss, and voltage stability. They are listed as follows, respectively:

$$f_1 = a_i P_{Gi}^2 + b_i P_{Gi} + c_i \tag{3.8.2}$$

$$f_2 = a_i P_{Gi}^2 + b_i P_{Gi} + c_i + |d_i \sin(e_i(P_{Gi,\min} - P_{Gi}))| \tag{3.8.3}$$

$$f_3 = \sum_{k=1}^{N_l} \frac{r_k}{r_k^2 + x_k^2} \left[V_i^2 + V_j^2 - 2V_i V_j \cos(\delta_i - \delta_j) \right] \quad \forall i, \forall j \tag{3.8.4}$$

$$f_4 = a_i P_{Gi}^2 + b_i P_{Gi} + c_i + \omega \sum_{i=1}^{N_{pq}} |V_i - 1| \tag{3.8.5}$$

The equality constraints g from (3.8.1) are the AC power flow balance equations at each bus representing that the power flowing into that specific bus is equal to the power flowing out, and the equations are defined as

$$\begin{aligned} P_i &= V_i \sum_{j=1}^{N} V_j Y_{ij} \cos(\delta_i - \delta_j - \theta_{ij}) \\ Q_i &= V_i \sum_{j=1}^{N} V_j Y_{ij} \sin(\delta_i - \delta_j - \theta_{ij}) \quad \forall i, \forall j \end{aligned} \tag{3.8.6}$$

Inequality constraints h are listed as generator limits, tap position of transformers, shunt capacitor constraints, security constraints, and load bus voltage and transmission line flows.

Generator limits:

$$\begin{aligned} & P_{Gi,\min} \leq P_{Gi} \leq P_{Gi,\max} \\ & Q_{Gi,\min} \leq Q_{Gi} \leq Q_{Gi,\max} \\ & V_{Gi,\min} \leq V_{Gi} \leq V_{Gi,\max} \quad i \in N_G \end{aligned} \tag{3.8.7}$$

Tap positions of transformers:

$$\mathrm{TP}_{i,\min} \leq \mathrm{TP}_i \leq \mathrm{TP}_{i,\max} \quad i \in N_\mathrm{T} \tag{3.8.8}$$

Shunt capacitors constraints:

$$Q_{\mathrm{c},\min} \leq Q_{\mathrm{c}i} \leq Q_{\mathrm{c}i,\max} \quad i \in N_\mathrm{c} \tag{3.8.9}$$

Security constraints on the limits of load bus voltage and transmission line flows:

$$\begin{aligned} &V_{\mathrm{L}i,\min} \leq V_{\mathrm{L}i} \leq V_{\mathrm{L}i,\max} \quad i \in N_\mathrm{pq} \\ &S_{Li} \leq S_{Li,\max} \quad i \in N_l \end{aligned} \tag{3.8.10}$$

where

a_i, b_i, c_i, d_i, e_i: fuel cost coefficients of the i-th unit,

$P_{\mathrm{G}i}$: real power of the i-th unit,

V_i: voltage magnitude at bus i,

r_k, x_k: the resistance and reactance of the transmission line k that links bus i and j,

V_i, V_j: voltages at bus i and j,

δ_i, δ_j : angles at bus i and j,

ω: the weighting factor,

N_pq: the number of PQ buses,

N_l: the total number of transmission lines,

N_G: the number of generators,

N_T: the number of tap-changing transformers,

Y_{ij}, θ_{ij}: the Y-bus admittance matrix elements,

$P_{\mathrm{G}i,\min}$, $P_{\mathrm{G}i,\max}$: the minimum/maximum real power limits of generating unit i,

$Q_{\mathrm{G}i,\min}$, $Q_{\mathrm{G}i,\max}$: the minimum/maximum reactive power limits of generating unit i,

$V_{\mathrm{G}i,\min}$, $V_{\mathrm{G}i,\max}$: the minimum/maximum voltage limits of generating unit i,

$TP_{i,\min}$, $TP_{i,\max}$: the limits of transformers,

$Q_{\mathrm{c}i,\min}$, $Q_{\mathrm{c}i,\max}$: the limits of shunt capacitors,

$V_{\mathrm{L}i,\min}$, $V_{\mathrm{L}i,\max}$: the limits of load bus voltage,

$S_{\mathrm{L}i,\max}$: the maximum line flow of transmission line i.

It is worth mentioning that the control variables (real power generation of PV buses, voltage at all generator buses, transformer tap settings, and shunt compensators) are randomly initialized within the feasible domain, while a penalty function is introduced in order to ensure that the dependent/state variables are in the feasible domain as well. In other words, penalty function is utilized to handle the inequality constraints. The penalty cost function is defined as

$$p(x_i) = \begin{Bmatrix} (x_i - x_{i,\max})^2 & \text{if} \quad x_i > x_{i,\max} \\ (x_{i,\min} - x_i)^2 & \text{if} \quad x_i < x_{i,\min} \\ 0 & \text{if} \quad x_{i,\min} \le x_i \le x_{i,\max} \end{Bmatrix} \tag{3.8.11}$$

where

$p(x_i)$ is the penalty function of dependent variable x_i at bus i.

The penalty cost increases with a quadratic form when dependent variables are exceeding the limits and the cost is zero if the constraints are not violated. For example, if one of the PQ bus voltage exceeds the limit, certain amount of penalty will be added, which leads to the increase of total cost and eventually this solution will be abandoned.

Thus, the augmented objective function by adding the penalty function of the slack bus, reactive power generation, PQ bus voltage, and transmission line capacity is described as:

$$F = f + C_p p(P_{G1}) + C_q \sum_{i=1}^{N_G} p(Q_{Gi}) + C_v \sum_{i=1}^{N_{pq}} p(V_{Li}) + C_s \sum_{i=1}^{N_l} p(S_{Li}) \tag{3.8.12}$$

where f is the original fuel cost function (f_1, f_2, or f_3 in this paper), C_p, C_q, C_v, and C_s, respectively, denote penalty factors of real power generation of slack bus, reactive power output of the generator buses, and PQ bus voltage and transmission line capacity.where

f: the original fuel cost function (f_1, f_2, f_3, or f_4 in this study),

C_p: penalty factors of real power generation of slack bus,

C_q: penalty factors of reactive power output of the generator buses,

C_v: penalty factors of *PQ* bus voltage,

C_s: penalty factors of transmission line capacity.

3.8.3 Artificial Bee Colony

Basically, ABC is a population-based search procedure inspired from the intelligent behavior of honeybees [209]. It is as simple as PSO and DE algorithms, and uses only common control parameters such as colony size and maximum cycle number. There are three types of bees in the ABC system: employed bee, onlooker bee, and scout bee. The aim of all bees is to find the best food source (possible solution) with highest nectar (fitness value); in other words, artificial bees fly in a multidimensional search space to find the global optimal. Employed bees search for food sources based on their memory and the information gathered on food sources is shared with onlooker bees. Onlooker bees tend to choose good sources with higher nectar and further explore new food sources around the selected food sources. Scout bees abandon old food sources and randomly start a new one in order to avoid local minimum.

As the ABC has proven its robust, efficient, and simple characteristics, it has been widely implemented in solving a range of optimization problems in recent years such as job shop scheduling and machine timetabling problems [215]. The ABC was also implemented to tune the PI controllers' parameters in microgrid power electronics control [216]. In references [198, 217], authors applied the ABC or improved ABC in power system problems such as OPF, modified OPF which integrated wind power and storage devices. Continuing this section, the detailed procedures of the ABC algorithm is presented.

Employed Bee Phase

Before entering into employed bee phase, initial possible solutions are generated from the searching space. After the initialization, the search process will be carried out in repeated cycles by those three types of bees. At initialization, each vector solution $X_i = \{X_{i,1}, X_{i,2}, \ldots, X_{i,D}\}$ is generated randomly within the limits of the control variables as follows:

$$X_{i,j} = X_{i,j\_\min} + \text{rand}(0,1) \times \left(X_{i,j\_\max} - X_{i,j\_\min}\right) \tag{3.8.13}$$

where

SN: the number of employed bees and onlooker bees, i is from 1 to SN,

D: the number of control variables, j is a random number from 1 to D,

$X_{i,j\_\min}$: the lower bounds for dimension j,

$X_{i,j\_\max}$: the upper bounds for dimension j,

rand(0,1): a uniformly distributed random number in (0,1).

In the employed bee phase, employed bees update the current solution based on neighborhood information and then evaluate the nectar (fitness) of the new food source. The update equation is defined as:

$$V_{i,j} = X_{i,j} + \Phi_{i,j} \times \left(X_{i,j} - X_{k,j}\right) \tag{3.8.14}$$

where

k: is an integer different from i, uniformly chosen from the range (1, SN),

$\Phi_{i,j}$: a random number from (−1, 1).

If the new source is better (higher nectar) than the old one, employed bee will memorize the new source and disregard the old. Otherwise, the old one will remain. Such scheme is simply known as greedy selection.

Onlooker Bee Phase

The onlooker bee phase starts when the food source information was shared from employed bees. In nature, onlooker bees tend to select the food source with higher nectar. The nectar information has been shared by employed bees. To mimic such phenomenon, the roulette wheel selection Scheme [215] is used in the onlooker bee phase, which ensures good food source will have a higher probability to be selected, and then onlooker bee will update those solutions. The roulette wheel selection scheme is defined in the following:

$$P_i = \frac{\text{fit}_i}{\sum_{j=1}^{\text{SN}} \text{fit}_j} \tag{3.8.15}$$

where

fit_i: the fitness value associated with solution i,

P_i: the probability associated with solution i.

The onlooker bee updates the selected solution by using Eq. (3.8.14) as the employed bees do and memorize the solution by greedy selection. This process will continue until every onlooker bee finishes its search.

Scout Bee Phase

After a predefined number of searching cycles, food sources become exhausted (inactive solution) if their quality could not be improved anymore, then the employed bee will become a scout bee to start a random direction to search for new food source. This process is to avoid local optima. In the original ABC, only one scout is allowed to occur in each cycle [209]. After finding a new food source, the scout bee will turn itself back to employed bee. Note that the random search by scout is also performed by function (3.8.13) same as in the initialization stage.

3.8.4 ABC for the OPF Problem

Decision and State Variables

As mentioned in previous sections, the OPF problem is to find the optimal decision variables so that the objective function can be optimized. The control/decision vector u consists of:

$$u = [P_{\text{G}}; V_{\text{G}}; T; Q_{\text{C}}] \tag{3.8.16}$$

where

P_{G}: real power output at PV buses,

V_{G}: bus voltage at PV and slack buses,

T: transformer tap settings,

Q_{C}: shunt compensators settings.

The state vector x consists of

$$x = [P_{\text{G1}}; V_{\text{L}}; Q_{\text{G}}; S_{\text{L}}] \tag{3.8.17}$$

where

P_{G1}: real power output at slack bus,

V_L: PQ bus voltage,

Q_G: generator reactive power output,

S_L: transmission line loadings.

It is necessary to clarify for those readers who are new to power system that there are three types of buses in the power system: slack bus, PV bus, and PQ bus. Slack bus is to balance the real and reactive power in the system while performing load flow calculations, it is also known as reference bus. PV bus is the node where real power P and voltage magnitude V are specified, it is also known as generator bus. PQ bus is the node where real and reactive power is specified, known as load bus.

Fitness Function

There are four types of objective functions ($f_1\,f_2\,f_3$ and f_4) in this study as defined in previous sections. The fitness value of one solution can be evaluated using the following equation [209]:

$$\text{fit}_i = \left\{ \begin{array}{ll} \dfrac{\alpha}{1+f} & \text{if } f \geq 0 \\ 1 + \text{abs}(f) & \text{if } f < 0 \end{array} \right\} \tag{3.8.18}$$

where

α: constant,

f: objective function of (3.8.12).

It can be easily seen that for the cost function f greater than zero (OPF problems), to minimize cost function is to maximize the fitness value.

Framework of ABC

The framework of the ABC algorithm is summarized as follows:

Step 1: Initialization:

1.1) Randomly generate SN points in the search space as feasible solution X_i by (3.8.13).

1.2) Run AC load flow and evaluate the fitness function (3.8.17).

Step 2: For all employed bees ($i = 1, \ldots, SN$):

2.1) Generate a candidate solution V_i by (3.8.14).

2.2) Run Load Flow and evaluate the fitness function, and calculate the probability p associated with its fitness by (3.8.15).

2.3) Choose a solution (from X_i and V_i) with better fitness value.

Step 3: For all onlooker bees (will only be executed under certain probability p):

3.1) Generate a new candidate solution V_i by (3.8.14).

3.2) Run AC Load flow and evaluate the fitness function.

3.3) Choose a solution (from X_i and V_i) with better fitness function.

Step 4: For all scout bees (they will be executed only after the maximum trial m). Note that the maximum trial m is a predefined number that if a certain food cannot be improved by an employed bee after m times, the employed bee will become a scout bee.

4.1) Replace X_i with a new random solution X_i by (3.8.13).

After initialization, the algorithm repeats the search processes of employed bees, onlooker bees, and scout bees by a predefined cycle.

3.8.5 Case Studies

All simulations were performed on a computer with 3.4 GHz Intel core i7 Processor and 8 GB RAM. Power flow was calculated by the Newton–Raphson method in MATPOWER package [218].

Test System Description

All four test cases were performed on the IEEE 30-bus system. The configuration of the system is shown in Figure 3.8.1. Reference [219] gives the data of the IEEE 30-bus test system, and control variables' limits can be found from [199]. There are total 24 control variables which consist of five real power output control at PV bus and voltage magnitudes control of all six generator buses, nine shunt compensators control for injecting reactive power, and four transformer tap controls. The six generators are shown in Figure 3.8.1. Buses 10, 12, 15, 17, 20, 21, 23, 24, and 29 are equipped with shunt compensators. In addition, lines 4–12, 6–9, 6–10, and 28–27 are equipped with tap-changing transformers as shown in Figure 3.8.1. The system

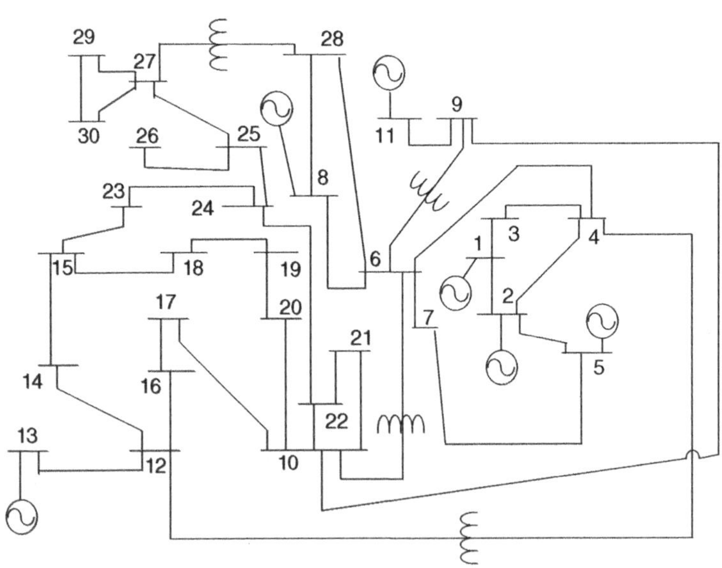

Figure 3.8.1 IEEE 30-bus system.

is at 100 MVA base with active power demand of 2.834 p.u. and reactive power demand of 1.262 p.u. The quadratic cost fuel cost coefficients were taken from [220].

The ABC Parameters for OPF

There are several parameters to be predetermined in the ABC algorithm (Table 3.8.2). For instance, the number of colony size is 200, in other words, the size of employed bee and onlooker bee is 100, respectively. The number of food sources is 100. The limited trials where a food source could not be improved and then the food source is abandoned by its employed bee is 100. The number of cycles for foraging (stop criteria) is 400 iterations and there are total 24 parameters to be optimized.

Case 1: Quadratic Cost Function

Case 1 is the standard OPF problem with quadratic cost function. The objective of this case is to minimize total generator fuel cost (3.8.2). Simulation was run 30 times in order to conduct statistical analysis. The minimum total cost from ABC is 799.904 $/h, with the maximum 801.518 $/h, the average 800.944 $/h, and 0.162 standard deviation. Results from other methods such as gravitational search algorithm (GSA), linearly decreasing inertia weight particle swarm optimization (LDI-PSO), enhanced genetic algorithm (EGA), modified differential evolution (MDE), and modified shuffle-frog leaping algorithm (MSFA) [206, 220–223] were compared with the results from ABC. The comparison including execution time is given in Table 3.8.3. Figure 3.8.2 shows the convergence properties of the ABC algorithm.

TABLE 3.8.2 Parameters for ABC Algorithm

Parameters	Values
Colony size	200
Food number	100
Limited trials	100
Maximum cycles	400
Parameters to be optimized	24

TABLE 3.8.3 Comparison of Case 1 Fuel Cost

Method	Min. ($/h)	Avg. ($/h)	Max. ($/h)	Std. Dev. (σ)	t (s)
ABC	799.904	800.944	801.518	0.162	39.8
GSA [220]	805.175	812.194	827.459	N/A	10.8
LDI-PSO [221]	800.734	801.557	803.869	N/A	N/A
EGA [206]	802.060	N/A	802.140	N/A	N/A
MDE [222]	802.376	802.382	802.404	N/A	23.3
MSFLA [223]	802.287	802.414	802.509	N/A	N/A

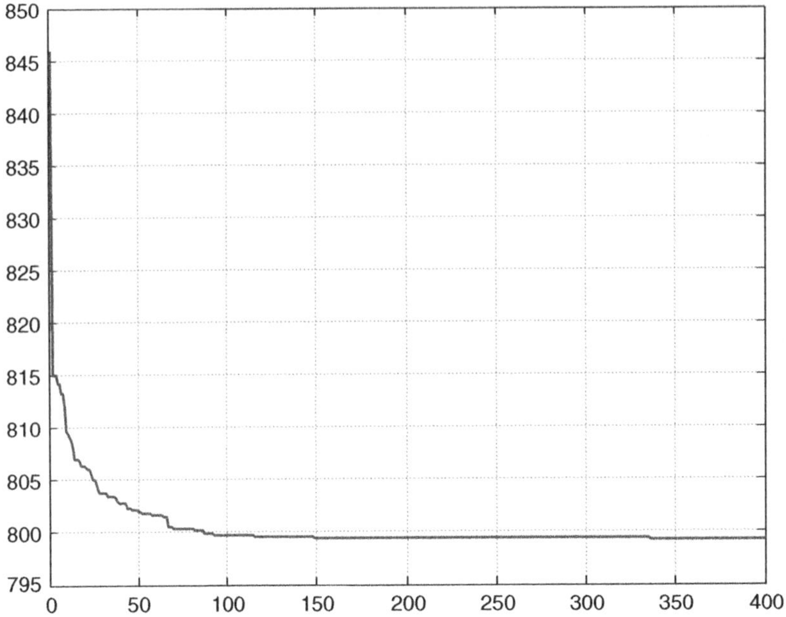

Figure 3.8.2 Convergence characteristics of ABC method in case 1.

Note that the computational time can be affected by several variables, such as the performance of computer, the complexity of algorithms, efficiency of code, etc. Although the ABC algorithm used in the study did not outperform some of other algorithms, we argue that the computational performance can be improved by using advanced computers or parallel computing methods. Therefore, computational time comparison is not the focal point in the study.

Case 2: Quadratic Cost Function with Valve Effect

In a real power plant, steam is controlled by valves to enter the turbine through separate nozzle groups. The best efficiency is achieved when each nozzle group operates at full output [221]. Therefore, in order to achieve highest possible efficiency for given output, valves are opened in sequence and this results in a rippled cost curve as in Figure 3.8.3. The objective function is given by (3.8.3). Table 3.8.4 shows the comparison with other methods.

From Table 3.8.4, the minimum total cost from ABC is 923.436 $/h, with the maximum 924.894 $/h, the average 924.124 $/h, and 0.562 standard deviation, and the convergence property is shown in Figure 3.8.4.

Case 3: Minimization of Power Loss

Real power loss is due to the power flowing through transmission lines which consist of resistance and reactance. It is apparent that minimizing real power loss is one of the major concerns for system operation. The objective function is given by (3.8.4). The convergence property is shown in Figure 3.8.5.

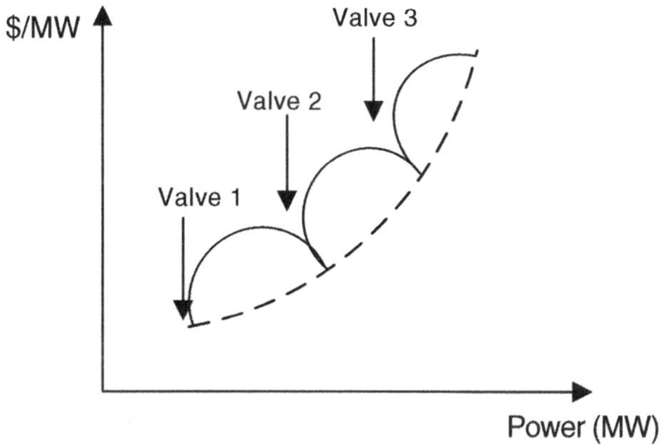

Figure 3.8.3 Effect of valve-point loading on a quadratic cost function.

TABLE 3.8.4 Comparison of Case 2 Fuel Cost

Method	Min. ($/h)	Avg. ($/h)	Max. ($/h)	Std. Dev. (σ)	t (s)
ABC	923.436	924.124	924.894	0.562	39.8
GSA [221]	929.724	930.925	932.049	N/A	9.83
MDE [222]	930.793	942.501	954.073	N/A	N/A

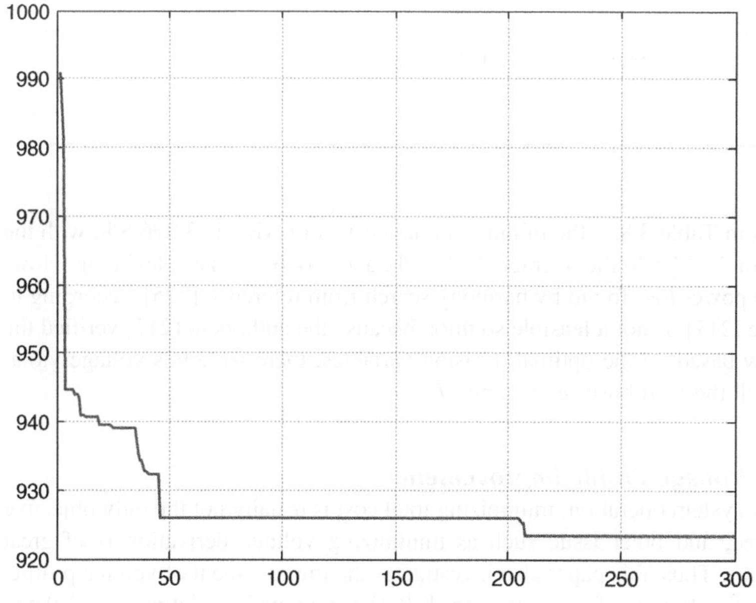

Figure 3.8.4 Convergence characteristics of ABC method in case 2.

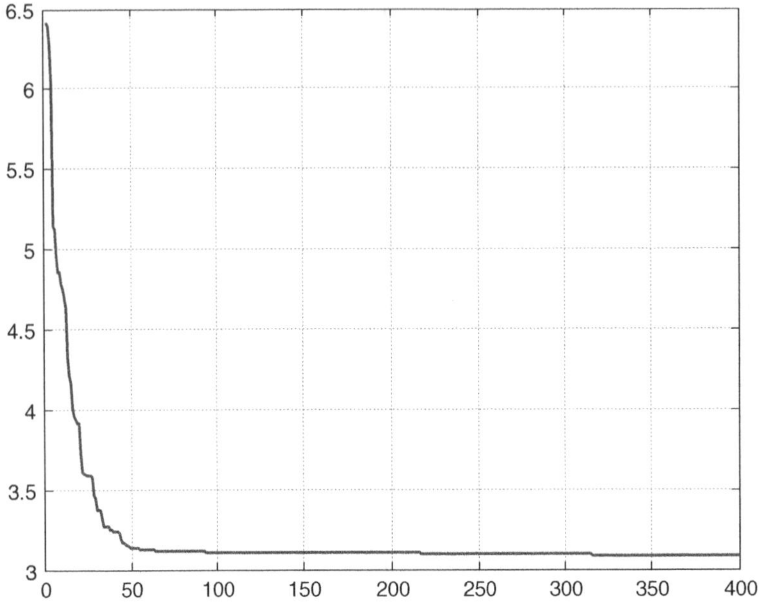

Figure 3.8.5 Convergence characteristics of ABC method in case 3.

TABLE 3.8.5 Comparison of Case 3 Power Loss

Method	Min. ($/h)	Avg. ($/h)	Max. ($/h)	Std. Dev. (σ)	t (s)
ABC	3.096	3.112	3.177	0.036	70.8
HS [225]	N/A	2.967	N/A	N/A	N/A
EGA [206]	N/A	3.201	N/A	N/A	N/A

From Table 3.8.5, the minimum total cost from ABC is 3.096 $/h, with the maximum 3.177 $/h, the average 3.112 $/h, and 0.036 standard deviation. However, the power loss found by harmony search from reference [225], according to reference [217], is not a feasible solution because the authors in [217] verified the load flow based on the optimal decision variables; there were bus voltage violations at all the load buses except bus 7.

Case 4: Voltage Profile Improvement

In power system operation, minimizing total cost is usually not the only objective considered, and other issue such as minimizing voltage derivation is of great importance. Thus, this paper also investigates the improvement of voltage profile. The objective function for minimizing all PQ bus voltage V deviating from 1.0 p.u. is described in (3.8.5). Figure 3.8.6 shows the convergence property.

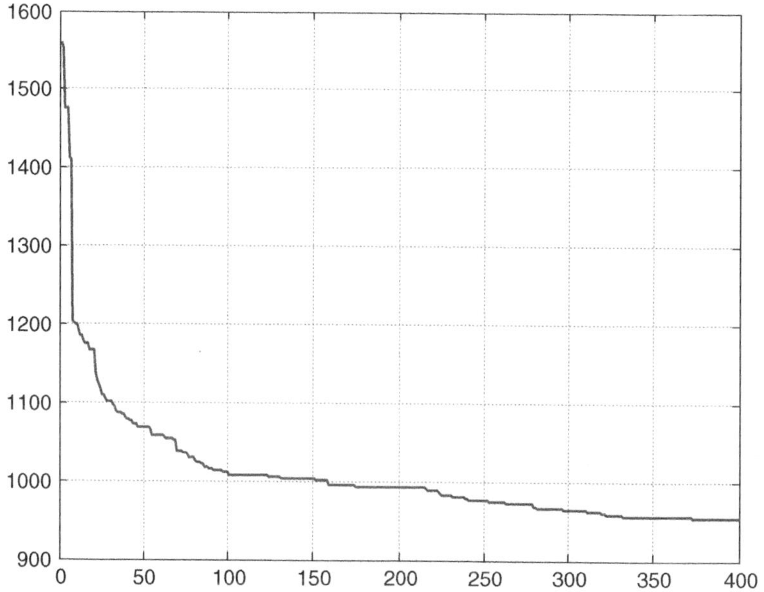

Figure 3.8.6 Convergence characteristics of ABC method in case 4.

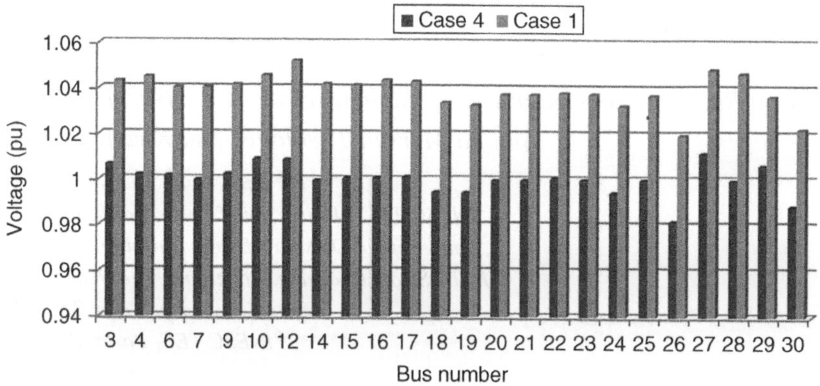

Figure 3.8.7 Voltage profiles for case 1 and case 4.

After 30 simulations, the minimum total cost found by ABC is 955.227 \$/h, with the maximum 958.147 \$/h, the average 956.824 \$/h, and 0.512 standard deviation. Since this is not a standard comparison case, no reference from other methods with the same system parameters can be found. Figure 3.8.7 compares the PQ bus voltage profiles with the case of minimizing basic quadratic cost function (case 1).

From Figure 3.8.7, it is obvious that by considering voltage improvement, the voltages of PQ buses stay close to 1 p.u.; on the other hand, the voltage deviates from 1 p.u. but within feasible limits without considering voltage improvement.

3.8.6 Conclusions

This chapter introduced the ABC algorithm for handling nonlinear, non-convex OPF problems. Due to its simplicity and less control parameters, the ABC has been applied in various problems, such as shop scheduling and machine timetabling problems. There are three phases in the ABC system: the employed bee, onlooker bee, and scout bee. They corporate to find the best food source which represents optimal solution. In this study, the ABC has been proven its efficiency and robustness over four test cases of OPF problems. The comparison of the results has shown that ABC outperforms other heuristic methods in terms of finding better solutions with fewer costs. In addition, with the help of high-speed computers or parallel computing algorithm, the computational burden can be further reduced, which enables the ABC algorithm promise as a useful tool for OPF problems.

ACKNOWLEDGEMENT

The Scientific and Technical Research Council of Turkey (TUBiTAK) is acknowledged for its financial support to the project stated in this paper.

3.9 OPF TEST BED AND PERFORMANCE EVALUATION OF MODERN HEURISTIC OPTIMIZATION

Jose Rueda[1], Leonel Carvalho[6], Vladimiro Miranda[2], Istvan Erlich[3], Aimilia-Myrsini Theologi[4], and Kwang Y. Lee[5]

[1]TU Delft, Delft, The Netherlands
[2]INESC TEC / University of Porto, Porto, Portugal
[3]University of Duisburg-Essen, Duisburg, Germany
[4]Jedlix Smart Charging, Rotterdam, The Netherlands
[5]Baylor University, Waco, TX, USA
[6]INESC TEC, Porto, Portugal

3.9.1 Introduction

Nowadays, the attention regarding the solution of power system optimization problems is focused on heuristic optimization algorithms. The nonlinear, non-convex, multimodal, discontinuous, and high-dimensional search space of modern heuristic techniques enables coping with inherent computational complexities. Besides the experience obtained so far, further research effort on evolutionary mechanisms is required in order to improve search exploration and exploitation.

The development of power system optimization test beds, where the general applicability and effectiveness of emerging heuristic optimization tools can be evaluated and compared with one another, is envisaged by the objectives of Working Group on Modern Heuristic Optimization under IEEE PES Analytic Methods in Power Systems (AMPS) Committee. In the first instance, a special panel was organized at the 2014 IEEE PES General Meeting, which concerned with a competition regarding the application of heuristic tools in solving OPF problems.

The aim of the competition was intended to compare the searching ability of different heuristic optimization algorithms based on statistical tests performed on results submitted by interested participants. For the automatic performance evaluation of OPF's objective function and constraints, as well as automatic collection and storage of the results, an encrypted file based on functionalities of MATLAB and MATPOWER has been prepared. Consequently, the OPF problems are considered as black box tasks, where different test cases should be investigated based on four selected test networks. Although all optimization test cases are resolvable based on preliminary tests, in some very complex tasks, the discovery of feasible solutions poses a major challenge. Thus, the implementation of particular heuristic optimization algorithm constitutes the focus point for all the participants. For constraint handling or treatment of discrete/binary optimization variables, associated with transformer and compensation devices, any special scheme is allowed to be included in the optimization algorithm. Like participants of the 2014 OPF competition, researchers who are interested in testing new optimization methods and constraint handling strategies are kindly encouraged to follow the implementation guidelines and submission procedure indicated in [226].

3.9.2 Problem Definition

The formulations of optimal reactive power dispatch (ORPD) and optimal active–reactive power dispatch (OARPD) problems presented in [227–229] are implemented in such a way that different OPF test cases can be performed for four selected test systems with different sizes and structural complexities.

The objective function and constraints of all OPF tasks, as well as for automatic collection and storage of results in formatted ASCII-files, have been calculated by the encrypted file *test_bed_OPF.p*. This file uses the functions for modeling and load flow calculation available in MATPOWER [230]. The code along with instructions on how to use it, as well as an implementation example with basic PSO algorithm, is contained in the zipped folder *test_bed_OPF_V13.zip* and is considered as black box. The folder can be downloaded from [226]. Each participant should work exclusively on the particular optimization algorithm to be used, and the adoption of any type of constraint handling technique is allowed. The routine for constraint handling calculates internally the fitness f' by using the following equation:

$$f' = f + \rho \sum_{i=1}^{n} \max\,[0, g_i]^2 \tag{3.9.1}$$

where ρ stands for the penalty factor, which is set to a high value, i.e. $1E+7$, for both ORPD and OARPD problems. The fitness calculation performed by the aforementioned encrypted file is exclusively intended for ascertaining fulfillment of constraints in the competition. The values of f' are automatically recorded at a predefined rate of 100 function evaluations and is used to ensure unbiased evaluation of algorithms' performance.

The folder *test_bed_OPF_V13.zip* also contains an exemplary file named as *rounding.m*, which constitutes an external function that can be employed as given or modified by the user to code discrete/binary optimization variables for rounding the real numbers (i.e. optimization variables). It is worth clarifying that the function syntax, i.e. *x_out* = *rounding*(*x_in*), should be kept, because it is called internally in *test_bed_OPF.p* before every function evaluation. Note that *x_in* denotes one individual component of the sequence of discrete/binary variables from the vector of optimization variables to be generated using the evolution scheme of your optimization algorithm (e.g. offspring generation). The file *main_commented.m* provides instructions to determine the indices (elements) of the vector of optimization variables defined as discrete/binary variables. Note that *test_bed_OPF.p* is configured to automatically round the values corresponding to the discrete/binary coded variables to the nearest integer, so this rounding approach will be internally used regardless of whether your algorithm uses a rounding strategy or not. If a rounded variable violates its boundary, it will be automatically fixed in *test_bed_OPF.p* to the corresponding limit. The user can also refer to *main_commented.m* file to gather how to obtain all power system and optimization related information, e.g. location of controllable transformer and compensation devices, problem dimensionality, bounds of optimization variables, and steps of discrete variables.

3.9.3 OPF Test Systems

In order to evaluate ORPD and OARPD problems, slightly modified versions of IEEE 57-, 118-, and 300-bus test systems are used. The data of each system have been structured in MATPOWER data format based on details given in [231] for system buses and branches. Branch thermal limits were defined based on reference values given in [232]. A test offshore WPP, which is based on the typical layout given in [229], is also used for the evaluation of ORPD problem. The characteristics of the different test systems are comparatively summarized in Table 3.9.1. The following sections provide the description of the optimization test cases to be performed for each system.

Optimization Variables

ORPD Problem The minimization of the total power transmission losses with the fulfillment of constraints for non-contingency ($N-0$) and selected $N-1$ conditions constitutes the target of ORPD test case. The constraints are associated with nodal balance of power, nodal voltages, allowable branch power flows, and generator power capability. In Table 3.9.2, the parameters, with which the different types of optimization variables are associated, are presented.

OARPD Problem The minimization of the total fuel cost with the fulfillment of constraints for non-contingency ($N-0$) and selected $N-1$ conditions

TABLE 3.9.1 Composition of the Benchmark Systems Used in the OPF Test Bed

Item/System		IEEE 57-Bus System	IEEE 118-Bus System	IEEE 300-Bus System	Offshore WPP
Generators		7	54	69	18
Loads		42	99	201	1
Lines/cables		63	177	304	20
Transformers	Stepwise	15	9	62	2
	Fixed tap	2	0	45	18
Shunt compensation	Switchable (on/off)	3	14	14	0
	Stepwise	0	0	0	1
	Continuous	0	0	0	1

TABLE 3.9.2 Optimization Variables Associated with Controllable Elements Listed in Table 3.9.1 for the ORPD Problems

Type of Variables	Associated Parameters
Continuous	Generator bus voltage set-points
Discrete	Stepwise adjustable onload transformer tap positions
Binary	Switchable shunt compensation devices

TABLE 3.9.3 Optimization Variables Associated with Controllable Elements Listed in Table 3.9.1 for OARPD Problems

Type of Variables	Associated Parameters
Continuous	Generator active power outputs and bus voltage set-points
Discrete	Stepwise adjustable onload transformers' tap positions
Binary	Switchable shunt compensation devices

constitutes the target of OARPD test case. The constraints are associated with nodal balance of power, nodal voltages, allowable branch power flows, generator power capability, and maximum active power output of slack generator. In Table 3.9.3, the parameters, with which the different types of optimization variables are associated, are presented.

IEEE 57-Bus System

In this case, the characteristics of each optimization task are presented in Table 3.9.4.

Only one scenario, i.e. peak load, is considered.

TABLE 3.9.4 Test Cases for the IEEE 57-Bus System

		ORPD Test Case	OARPD Test Case
Minimization of		Total active power transmission losses	Total fuel cost
Scenarios		1	1
Considered contingencies at branches		8 and 50	8 and 50
Number of function evaluations		50 000	50 000
Constraints for	Non-contingency conditions	178	178
	Each $N-1$ condition	177	177
Optimization variables	Continuous	7	13
	Discrete	15	15
	Binary	3	3

TABLE 3.9.5 Test Cases for the IEEE 118-Bus System

		ORPD Test Case	OARPD Test Case
Minimization of		Total active power transmission losses	Total fuel cost
Scenarios		1	1
Considered contingencies at branches		21, 50, 16, and 48	21, 50, 16, and 48
Number of function evaluations		100 000	100 000
Constraints for	Non-contingency conditions	492	492
	Each $N-1$ condition	491	491
Optimization variables	Continuous	54	107
	Discrete	9	9
	Binary	14	14

IEEE 118-Bus System

In this case, the characteristics of each optimization task are presented in Table 3.9.5:

Only one scenario, i.e. peak load, is considered.

IEEE 300-Bus System

In this case, the characteristics of each optimization task are presented in Table 3.9.6.

Only one scenario, i.e. peak load, is considered.

TABLE 3.9.6 Test Cases for the IEEE 300-Bus System

		ORPD Test Case	OARPD Test Case
Minimization of		Total active power transmission losses	Total fuel cost
Scenarios		1	1
Considered contingencies at branches		187, 176, and 213	187, 176, and 213
Number of function evaluations		300 000	300 000
Constraints for	Non-contingency conditions	651	651
	Each $N-1$ condition	950	950
Optimization variables	Continuous	69	137
	Discrete	62	62
	Binary	14	14

Offshore Wind Power Plant

The offshore WPP used as test system is presented in Figure 3.9.1 and corresponds to a commonly used topology for offshore WPPs that are located near the shore. The plant is connected to the main grid using an AC cable with transformers at both ends. The line reactor X_{sh2} on the offshore side is always connected to the cable, due to the large charging currents of the AC cables, while reactor X_{sh1} can be adjusted in a continuous manner to provide Var control on the onshore end of the cables. C_1 stands for the stepwise regulated capacitor and is intended for auxiliary reactive power support. The real part of the dummy load L_1 corresponds with WPP's currently generated active power, while the imaginary component with the reactive power requirement at the point of common coupling (PCC), which is denoted by q_{ref}. The total losses of the wind energy system and the fulfillment of the reactive power requirement at PCC are quantified by the difference between injection from the WPP and the fictitious consumption associated with L_1.

A control strategy is defined to continuously fulfill q_{ref} through optimum management of the available WPP's var sources during normal (steady state or quasi steady state) conditions, assuming the availability of data acquisition system to provide measurements related to the actual status of all wind generators, transformers, and compensation devices within the WPP. The reactive power requirements corresponding to the actual operating condition are defined as stepwise changes of q_{ref} throughout 24 hours in order to highlight the relevance of online optimal reactive power control problem. The stepwise changes of q_{ref} and the independent variability of WPP output power are shown in Figure 3.9.1 and eventually, ORPD problem has to be solved for every considered scenario, defined by each point of these curves. Since 15-minute intervals are considered for 24-hour time horizon, 96 different scenarios have been determined, of which some entailing

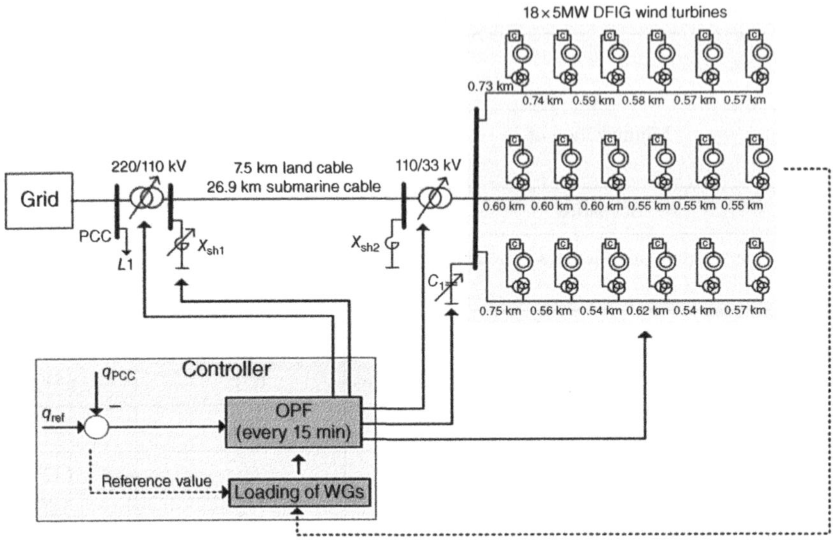

Figure 3.9.1 Offshore WPP with optimization-based management of reactive power sources.

TABLE 3.9.7 Test Cases for the Offshore Wind Power Plant

		ORPD Test Case
Minimization of		Total active power transmission losses
Constraints		123
Optimization variables	Continuous	19
	Discrete	3
Scenarios		96
Number of function evaluations		300 000

complex optimization tasks. The file main.m includes all the possible scenarios to be used for individual testing.

In this case, the target of ORPD is the minimization of the total power transmission losses with the fulfillment of constraints for non-contingency and selected $N-1$ conditions. Excluding the aforementioned ORPD test case for the IEEE test systems, here the constraints are associated with nodal balance of power, nodal voltages, allowable branch power flows, and difference between q_{ref} and q_{PCC} (Table 3.9.7).

The problem has 96 scenarios, which correspond to a forecasted wind speed series for 24 hours, with resolution of 15 minutes. In Table 3.9.8 the parameters of the offshore WPP, with which the different types of optimization variables, are presented.

TABLE 3.9.8 Optimization Variables for Offshore WPP's ORPD Test Case

Type of Variables	Associated Parameters
Continuous	Wind generator reactive power set-points and adjustment of $X_{\mathrm{sh}1}$
Discrete	Stepwise adjustable onload transformers' tap positions and stepwise adjustment of C_1
Binary	Switchable shunt compensation devices

3.9.4 Differential Evolutionary Particle Swarm Optimization: DEEPSO

DEEPSO, which is the acronym for differential evolutionary particle swarm optimization, is a variant of evolutionary particle swarm optimization (EPSO), an original meta-heuristic proposed by Vladimiro Miranda in 2002 [233]. In effect, EPSO can be seen as the PSO meta-heuristic with self-adaptation or the evolutionary strategies (ES) with an enhanced mutation/recombination operator. In EPSO, the progress toward the optimum is governed by Darwin's natural selection paradigm, i.e. by the joint application of reproduction (mutation/recombination) and selection, and by the social interactions within groups of animals in the pursuit of a common goal, which are associated with factors such as inertia, memory, and cooperation and the quantity of information exchanged. Accordingly, EPSO takes advantage of the natural selection and the self-adaption power typical of ES and the unique space exploration capabilities of PSO to speedup the rate of progression toward the optimum needing only few strategic parameters to be set. In addition, the standard EPSO formulation includes a probabilistic communication structure between individuals, which is named as Stochastic Star Communication Topology [234], to restrict the information exchanged regarding the location of the global best position found so far aiming to allow pure cognitive movements in some dimensions of the search space.

The improvement of DEEPSO [235] over EPSO consists on using a new recombination mechanism based on the information acquired during the optimization process. The additional source information consists of using a limited-size memory that stores past global best positions to generate new solutions based on the macro-gradient concept of DE. Note that EPSO already relies on past information during the search procedure, namely on the information stored in the memory of the individuals: DEEPSO proposes the replacement of this individual memory by a collective one to enable an improved perception of the optimization landscape. Accordingly, the (individual) memory term of EPSO's Movement Rule is replaced by the (collective) perception which has proven to result in superior performances for a wide range of linear, nonlinear, and mixed-integer optimization problems [235].

The set of procedures in DEEPSO are very similar to those in EPSO. The difference between the two meta-heuristics lies in the equation used to create new individuals: The Movement Rule. In EPSO, a new position, X_t, is created from its ancestor, X_{t-1}, from its best ancestor, X_b, from the best position ever found, X_{gb}, and from its current velocity, V_t, according to:

$$V_t = w_i^* V_{(t-1)} + w_m^* \left(X_b - X_{(t-1)}\right) + w_c^* \left(X_{gb}{}^* - X_{(t-1)}\right) \tag{3.9.2}$$

$$X_t = X_{t-1} + V_t \tag{3.9.3}$$

where t denotes the current generation, the superscript $*$ indicates that the corresponding parameter is subjected to a mutation process, and C corresponds to a diagonal matrix of Bernoulli variables with success probability P (probability P must be set by the user). Note that matrix C is sampled in every generation t.

The mutation of a generic weight w in DEEPSO follows the simple additive rule:

$$w^* = w + \tau N(0, 1) \tag{3.9.4}$$

where τ is the mutation rate (parameter τ must be set by the user), and $N(0, 1)$ is a number sampled from the standard Gaussian distribution. It is important to stress that weights must not become negative nor greater than 1 and are subjected to selection: the individuals that will be part of the next generation keep their weights. The same mutation process is applied for the global best position found so far to create a new attractor in its vicinity aiming to further increase EPSO/DEEPSO exploitative capabilities, especially, when social movements are dominant.

As previously mentioned, the Movement Rule of DEEPSO (see Figure 3.9.2) can be derived by replacing the memory term in (3.9.2) by the perception term as

$$V_t = w_i^* V_{t-1} + w_m^* (X_r - X_{t-1}) + w_c^* \left(X_{\mathrm{gb}}^* - X_{t-1}\right) \tag{3.9.5}$$

where X_r is a position different from X_{t-1}. The new position X_r can be obtained according to one of the following strategies:

1. Randomly sampled from the positions in the current population: Sg.
2. Randomly sampled from the positions in the memory of past bests: Pb.
3. Randomly sampled from the positions in the current population and from the positions in the memory of past bests: SgPb.
4. Randomly sampled as a uniform recombination from the positions in the current population: Sg – rnd.
5. Randomly sampled as a uniform recombination from the positions in the memory of past bests: Pb – rnd.
6. Randomly sampled as a uniform recombination from the positions of the current generation and from the positions in the memory of past bests: SgPb – rnd.

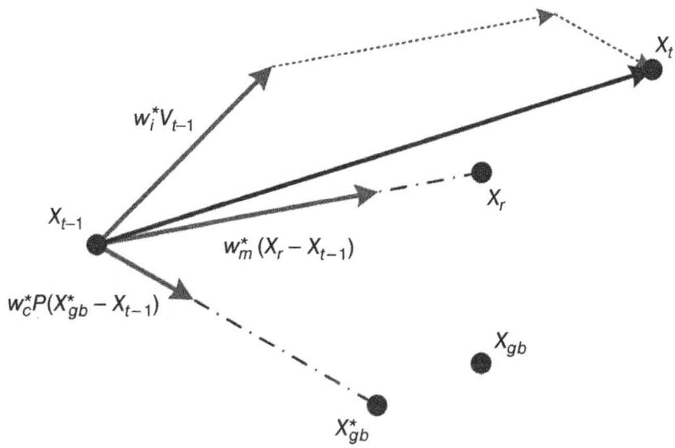

Figure 3.9.2 Simple 2D illustration of the Movement Rule in DEEPSO.

To compute V_t, which is according to (3.9.4), the fitness of the positions X_r and X_{t-1} must be taken into account: if X_r is better than X_{t-1}, then the individual must be attracted to X_r (in such case, the velocity is computed according to (3.9.4)); conversely, if X_r is worse than X_{t-1}, then the relative position of X_r and X_{t-1} in (3.9.4) must be swapped in order to repel the individual from X_r.

When uniform recombination is used, i.e. in the case of strategies 4, 5, and 6, new positions are created with unknown fitness. Bearing this in mind, two approaches can be used to determine the relative position of X_r and X_{t-1} in (3.9.4). In the first approach, which is denoted by the superscript ^{0}, the relative position of X_r and X_{t-1} is determined for every dimension $i = 1, 2, \ldots, N$ of the search space according to the fitness of the individual randomly selected to donate the ith dimension of X_r and the fitness of X_{t-1}. In such cases, DEEPSO follows strategy $\text{Sg} - \text{rnd}^0$, $\text{Pb} - \text{rnd}^0$, and $\text{SgPb} - \text{rnd}^0$. While not requiring an extra fitness evaluation, which can be seen as an advantage in terms of computational effort, this approach might lead to miss assessments of whether the individual should be attracted to or repelled from X_r when all dimensions of the search space are taken into account. An alternative approach, which is denoted by the superscript $^+$, consists on evaluating the new position X_r created to determine the relative position of X_r and X_{t-1} taking into account all the dimensions in the search space. For the problems addressed in this competition, it was found that strategy $\text{SgPb} - \text{rnd}^+$ yields the best performance.

A pseudocode for DEEPSO is as follows:

Pseudocode 1: DEEPSO

Select the perception strategy (Sg, Pb, $SgPb$, $Sg - rnd^0$, $Pb - rnd^0$, and $SgPb - rnd^0$, $Sg - rnd^+$, $Pb - rnd^+$, and $SgPb - rnd^+$), the maximum number of positions in the memory of past bests, the communication probability P, and the mutation rate τ

```
Initialize a population of k individuals, evaluate all
individuals, update the best individual X_gb, and update
memory of past bests
t = 1
While the convergence criteria, e.g. a maximum allowed
number of generations or a maximum number of generations
with no improvement, is not satisfied
  For all k individuals the population
    Compute the perception term X_r for the current individual
    according to the perception strategy selected
    Copy the current individual
    Mutate the strategic parameters w_i, w_m, w_c, and w_gb of
    the copy individual according to (3.9.4)
    Move the current individual and the copy according to
    (3.9.3) and (3.9.5)
    Evaluate the current individual and the copy
    Select the individual with the best fitness to be part
    of the next generation
  End For
  Update the best individual X_gb and the memory of past bests
t := t + 1
End While
```

DEEPSO is a meta-heuristic for real search spaces. Discrete variables can be obtained from real ones by using a simple rounding procedure. In the scope of this competition, an alternative mutation scheme was adopted to deal with discrete variables. This scheme, which is described in Pseudocode 2, is applied during the optimization process in tandem with that of DEEPSO. To select the mutation scheme to use, a Bernoulli distribution with probability *pDiscMut* was defined: if the Bernoulli variable takes the value 1 with probability *pDiscMut*, then Pseudocode 2 is used to generate a new population; when the value is 0 with probability 1–*pDiscMut*, then the mutation/recombination scheme of DEEPSO described in Pseudocode 1 is applied. Typically, *pDiscMut* should not assume a very large value (e.g. 0.25) and must be fine-tuned by the user.

Pseudocode 2: Simple Integer Variable Mutation Scheme

```
For all k individuals the population
  Copy the current individual
  Randomly select one of the discrete dimensions of the
copy individual
 If U(0, 1) > 0.5
  X_t^COPY(selected dimension) := X_t^COPY(selected dimension) + 1
 Else
  X_t^COPY(selected dimension) := X_t^COPY(selected dimension) - 1
 End If
```

```
    Evaluate the copy individual
    Select the individual (current or copy) with the best
    fitness to be part of the next generation
End For
```

Finally, the parameters of the fitness function were adjusted to ensure that penalties of different types of constraints violations have the same order of magnitude. As a matter of fact, there are different constraints violations with different orders of magnitude, namely the active power limits in REF bus (MW), the voltage magnitude limits in PQ buses (in p.u.), the reactive power limits in PV and REF bus (MVar), and the apparent power flow limit in transmission circuits (MVA). According to the constraint handling philosophy described, the fitness of the individual X_t is

$$f(X_t) = o(X_t) + a\left(\lambda_{\Delta P}|\Delta P| + \lambda_{\Delta V}\sum|\Delta V| + \lambda_{\Delta Q}\sum|\Delta Q| + \lambda_{\Delta S}\sum|\Delta S|\right) \tag{3.9.6}$$

where $o(X_t)$ is the objective to be attained (e.g. minimization of active power losses in the ORPD case or the minimization of the total production cost in the OARPD case), α and λ are scaling factors, $|\Delta P|$ is the absolute value of the active power violation in REF bus, $|\Delta V|$ is the absolute value of the voltage violation in the PQ buses, $|\Delta Q|$ is the absolute value of the reactive power violation in PV and REF buses, and $|\Delta S|$ is absolute value of the apparent power flow violation in the transmission circuits. The scaling factors λ are adjusted in every generation as well as the historical means of the corresponding penalty type (i.e. the historical mean of $|\Delta P|$, of $\sum|\Delta P|$, of $\sum|\Delta Q|$, and of $\sum|\Delta S|$). The adjustment of the scaling factors λ is according to the following process: the factor λ of the penalty with the largest historical mean is assigned the value 1 being the remainder scaling factors assigned a factor of 10 to ensure that all four historical means have the same order of magnitude. The scaling factor α, which must be set by the user, is the parameter responsible for driving the population away from unfeasible solutions.

3.9.5 Enhanced Version of Mean–Variance Mapping Optimization Algorithm: MVMO-PHM

MVMO belongs to the population-based optimization algorithms and can be applied in multi-objective mixed-integer nonlinear problems, such as optimal tuning of controllers, due to its conceptual simplicity, easy adaptability, and reduced human intervention. The enhanced performance of MVMO in comparison with other EAs appears in terms of convergence speed. This is mainly attributed to the so-called mapping function evolutionary operator, which adaptively switches the search priority between exploration to exploitation according to recorded statistics of the evolved best solutions so far in a continuously updated "solution archive" (i.e. adaptive memory).

Flowchart

In Figure 3.9.3, the overall algorithmic process of MVMO-PHM is presented. The procedure starts with the initialization of the parameter settings and the random generation of N_p candidate solutions between the [min, max] bounds of the optimization variables. The real min/max boundaries have to be normalized within [0, 1], due to the confined interval of MVMO-PHM search space. The optimization variables are only de-normalized for fitness evaluation or launching local search. Since, MVMO-PHM always generates a solution vector within the range [0, 1], the fulfillment of bound constraints is fully ensured. Local search is commenced after a given number of fitness evaluations $i_{\text{loc\_start}}$ for any new solutions belonging to the group of good particles and is based on IPM or sequential quadratic programming (SQP). A particle is selected for local search on a given probability γ_{LS}, that is,

$$\text{rand} < \gamma_{\text{LS}} \tag{3.9.7}$$

subject to

$$a_{\text{LS\_min}} < a < a_{\text{LS\_max}} = \frac{i}{i_{\max}} \tag{3.9.8}$$

where i stands for fitness evaluation number and rand denotes a uniform randomly generated number within the range [0, 1]. $\alpha_{\text{LS\_min}}$, $\alpha_{\text{LS\_max}}$ represents the range scheduled for local search.

Normalized Evolutionary Mechanism

Solution Archive: A Dynamic Knowledge Database The solution archive serves as the knowledge base for guiding the algorithm's searching direction and is updated whenever an improvement of the fitness function is achieved. The filling of the archive with the n-best child solutions obtained so far obeys a descending order of fitness over the iterations. The size of the archive remains constant during the entire process.

As illustrated in Figure 3.9.4, the solution archive records statistical measures like the mean $\overline{x_i}$, shape s_i, and d-factor d_i associated with each optimization variable, and whenever it is updated, these measures are recalculated. However, for the calculation of mean, a weighting of the old and new values is used according to the following equation:

$$\overline{x}_{\text{update}} = 0.1 \cdot \overline{x}_{\text{old}} + 0.9 \cdot \overline{x}_{\text{new}} \tag{3.9.9}$$

At the beginning, $\overline{x_{\text{ini}}}$ is set to 0.5, which value corresponds exactly in the middle of the search space. Random or user-defined definitions are also possible like other additional options.

Similarly, the tuning factors, which are represented by the initial values and the updating procedure of s_i, and d-factor d_i, allow some adaptation to the function to be optimized. However, in this study the updating of these values based on the archive only without weighting old and current values and the settings $s_i = 0$, $d_i = 1$

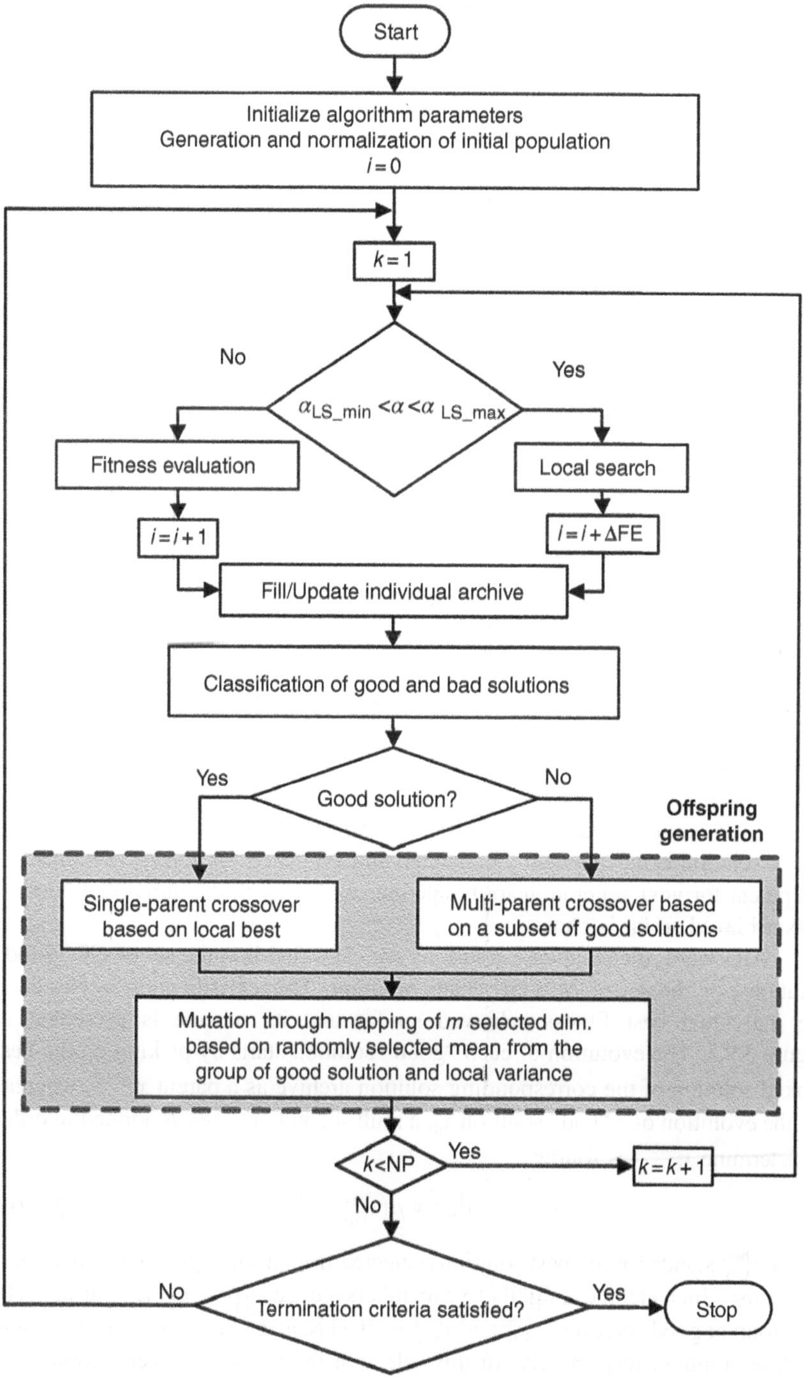

Figure 3.9.3 Algorithmic procedure of MVMO-PHM. The fitness evaluation and car date solution counters are denoted by i and k, whereas N_p, ΔFE, and rand stand for numb candidate solutions, number of fitness evaluations, and uniform random numbers b [0, 1], respectively.

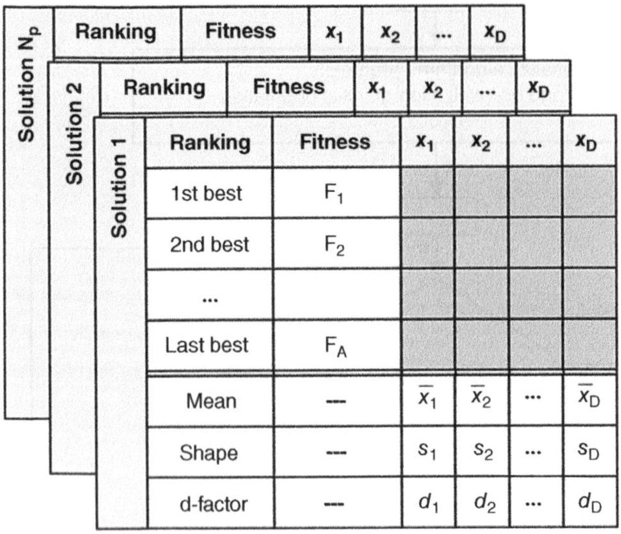

Figure 3.9.4 Layout of the set of solution archive.

has been used. All these parameters influence the geometric shape of the mapping function, which adaptively changes the search priority between exploration and exploitation.

Classification and Determination of Parent Solution Initially, each candidate solution is independently generated and evaluated for at least two runs. As the parent for next-generation solutions, the solution with the individual best fitness achieved so far is chosen.

Afterward, the candidate solutions are classified into the set of GP "good" solutions, or the set of $N_p - \text{GP}$ "bad" solution. This classification is based on the individual best fitness achieved so far and its scheme is presented in Figure 3.9.5. The evolution of each "good" solution starts by picking up the best ranked solution of the corresponding solution archive as a parent x_k^{parent}, whereas for the evolution of a "bad" solution x_k, a multi-parent criterion is applied in order to determine x_k^{parent} as follows:

$$x_k^{\text{parent}} = x_{\text{RG}}^{\text{best}} + \beta\left(x_{\text{GB}}^{\text{best}} - x_{\text{LG}}^{\text{best}}\right) \tag{3.9.10}$$

where $x_{\text{GB}}^{\text{best}}$ stands for the best solutions selected from a small group of global best solutions, which includes initially five members reduced to one in the final stage. In the group of good solutions, $x_{\text{LB}}^{\text{best}}$ and $x_{\text{RB}}^{\text{best}}$ are the last and a randomly selected intermediate solution, respectively. In this selection the border between "good" and "bad" particles exhibits a small variation and is not fixed in advance.

The mean values associated with x_k are required for subsequent mutation via the mapping function and its vector is determined randomly from the set of good solutions. The shape variables s_i and d_i are synthesized based on the corresponding

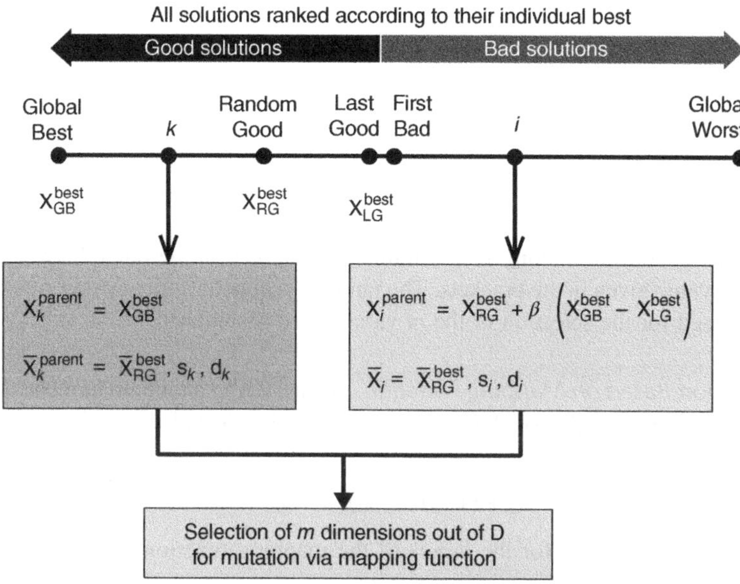

Figure 3.9.5 Procedure for parent selection in MVMO-PHM.

solution archive for both "good" and "bad" solutions. Finally, the factor β is computed as follows:

$$b = 1.1 + (\text{rand} + 0.5) \cdot 2.0 \tag{3.9.11}$$

$$\beta = \delta \cdot 3.0 \cdot b \cdot \left(1.0 + 2.5 \cdot a^2\right) \cdot \text{rand} - \left(1 - a^2\right) \cdot 0.8 \tag{3.9.12}$$

where δ is a tuning parameter, which can vary in the range of [0.1 – 20]. The factor β is redrawn and (3.9.10) is recalculated for any element of x_k^{parent} going outside the range [0, 1].

The relative number GP of solutions belonging to the group of good solutions is determined throughout the search process, as follows:

$$\text{GP} = \text{round}\left(N_\text{p} \cdot g_\text{p}^*\right) \tag{3.9.13}$$

$$g_\text{p}^* = g_{\text{p\_ini}}^* - a^2\left(g_{\text{p\_final}}^* - g_{\text{p\_ini}}^*\right) \tag{3.9.14}$$

This value is varied in a small range of 15% around g_p^* by using simplified normal distribution function. In the initial stage of the search process, where each solution is evaluated independently, Eq. (3.9.10) is not calculated. Note that GP is linearly narrowed down following the decrease from $g_{\text{p\_ini}}^*$ to $g_{\text{p\_final}}^*$.

Selection of Variables for Mutation Operation By performing mutation operation on m selected dimensions of $x_\text{p}^{\text{parent}}$, a child vector (array) $x^{\text{new}} = [x_1, x_2, x_3, \ldots, x_D]$ is created. D stands for the number of problem dimensions. For performing this operation, the mapping function based on the actual values of the parameters $\overline{x_i}$, s_i, and d_i associated with each solution is applied. Finally, throughout the search process the value of m is determined as follows:

$$m = \text{round}(m_{\text{final}} + \text{irand}(m^* - m_{\text{final}})) \tag{3.9.15}$$

$$m^* = \text{round}\left(m_{\text{ini}} + a^2(m_{\text{ini}} - m_{\text{final}})\right) \tag{3.9.16}$$

where irand() denotes a random integer number, generated in the range between zero and the value given in the brackets. The random-sequential strategy described in [236] is used for the selection of the m variables to be mutated.

Mutation Based on Mapping Function Considering a random number x_i^* sampled in the interval [0, 1], the new value of each selected dimension x_i is computed by using (3.9.17).

$$x_i = h_x + (1 - h_1 + h_0) \cdot x_i^* - h_0 \tag{3.9.17}$$

where h_x, h_1, and h_0 stand for the inputs of the mapping function based on the following substitutions:

$$h_x = h\left(x = x_i^*\right) \tag{3.9.18}$$

$$h_1 = h(x = 1) \tag{3.9.19}$$

$$h_0 = h(x = 0) \tag{3.9.20}$$

Note that the input and output of the mapping function are always within the range [0, 1]. The definition of the mapping h-function is given by (3.9.21).

$$h(\overline{x_i}, s_1, s_2, x) = \overline{x_i} \cdot (1 - e^{-x \cdot s_{i1}}) + (1 - \overline{x_i}) \cdot e^{-(1-x) \cdot s_{i2}} \tag{3.9.21}$$

It is worth highlighting that a new mapping function given by (3.9.22) and (3.9.23) will be used in future developments of MVMO. The new mapping function is defined slightly differently in the two halved of the random variable x_i^*, in order that function crosses the mean value $\overline{x}$ always at $x_i^* = 0.5$. Consequently, in both directions, the probability of the optimization variable decreasing or increasing is the same.

$$\begin{aligned}
&\text{for } x_i^* < 0.5 \\
&s_1^* = \frac{s_1}{(1 - \overline{x})} \\
&h_m = \overline{x} \cdot \left(1 - e^{-0.5 \cdot s_1^*}\right) \\
&h_f = \overline{x} \cdot \left(1 - e^{-x_i^* \cdot s_1^*}\right) \\
&h_c = (\overline{x} - h_m) \cdot 2 \cdot x_i^* \\
&x_i = h_f + h_c
\end{aligned} \tag{3.9.22}$$

$$
\begin{aligned}
&\text{for } x_i^* \geq 0.5\\
&s_2^* = \frac{s_2}{x}\\
&h_m = (1-\overline{x}) \cdot e^{-0.5 \cdot s_2^*}\\
&h_b = (1-\overline{x}) \cdot e^{-\left(1-x_i^*\right) \cdot s_2^*} + \overline{x}\\
&h_c = h_m \cdot 2 \cdot \left(1-x_i^*\right)\\
&x_i = h_b - h_c
\end{aligned}
\tag{3.9.23}
$$

Note that by using the function, described above, the mean value is always reached at 0.5 at the x-axis, which means that the probability of generating a new value of x_i that is greater or smaller than the mean value is equal. In addition, the new mapping is particularly indicated for local search.

In the first evaluation of every solution, the mean $\overline{x_i}$ is set to 0.5, while variance is set to 1.0, which corresponds with $s_i = 0$. After every update of the particle's solution archive for each selected optimization variable, as the optimization progresses, the parameters are recalculated. Both input and output of the mapping function cover the range [0, 1].

Additionally, f_s can be used to change the shape of the function and is increased as the optimization progresses from a small initial value (e.g. $f^*_{\text{s\_ini}} = 1$) up to a higher final value (e.g. $f^*_{\text{s\_final}} = 20$) by using the following equations:

$$f_{\text{s}}^* = f^*_{\text{s\_ini}} + a^2\left(f^*_{\text{s\_final}} - f^*_{\text{s\_ini}}\right) \tag{3.9.24}$$

$$f_{\text{s}} = f_{\text{s}}^*(1.0 + 4.0(\text{rand} - 0.2)) \tag{3.9.25}$$

The shape factors s_{i1} and s_{i2} of the variable x_i are assigned by using the procedure given in [237].

3.9.6 Evaluation Results

Codes and Parameters of DEEPSO and MVMO-PHM

The codes required for the execution of DEEPSO and MVMO-PHM algorithms in Matlab environment are available in [226]. The provided m-files indicate the adaptation of these algorithms to solve the optimization problems of the OPF test bed, and the user can also refer to the provided parameters for testing purposes. These parameters are also a reference for possible application of the algorithms in other optimization problems. Nevertheless, it should be kept in mind that the algorithms may require some parameter tuning.

Results

Table 3.3.9 presents the four best-performing algorithms that participated in the 2014 OPF Competition. The algorithms and their authors are listed below.

TABLE 3.9.9 Statistics of Results Obtained with Top 5 Algorithms

			Offshore WPP Scenario		IEEE 57 Test Case		IEEE 118 Test Case		IEEE 300 Test Case	
#	Abbr.	Stats.	**48**	**80**	**1**	**2**	**1**	**2**	**1**	**2**
1	**DEEPSO**	**Fbest**	1.449 3	2.637 3	25.014 9	41 688.675	117.875 8	135 492.35	391.436	721 377.54
		Worst	1.549 9	2.637 3	26.637 2	41 727.856	127.363 6	136 381.65	562.464 6	736 527.31
		Mean	1.464 7	2.637 3	25.572 5	41 697.581	119.279 9	135 890.72	414.623 9	723 041.1
		Median	1.458 1	2.637 3	25.497 9	41 696.456	118.997 6	135 848.17	400.175 7	722 361.28
		Std	0.021 088	7.48*E*–08	0.412 5	7.939 3	1.713 1	221.991	42.036	2 669.908 3
2	**ICDE**	**Fbest**	1.452 3	65 566.949	24.638 2	41 702.846	123.841 8	136 238.11	400.085 1	722 608.09
		Worst	1.563 8	754 941.31	26.913 2	41 805.352	135.695 5	173 785.34	69 818.834	754 493.47
		Mean	1.505 5	334 114.72	25.041 4	41 739.811 1	128.354 9	154 234.87	8 669.619 2	740 112.6
		Median	1.513 2	208 404.61	24.904 3	41 737.559	127.984 8	157 412.25	2 722.300 4	739 098.76
		Std	0.031 969	184 051.637 8	0.429 56	27.688 4	2.453 5	12 324.250 2	15 255.713 5	6 205.452 8
3	**MINLP**	**Fbest**	1.446 1	2.638 1	24.288 1	41 679.317	114.723 9	134 954.17	382.709 8	720 359.19
		Worst	1.446 1	2.638 1	24.288 1	41 679.317	114.723 9	134 954.17	382.709 8	720 359.19
		Mean	1.446 1	2.638 1	24.288 1	41 679.317	114.723 9	134 954.17	382.709 8	720 359.19
		Median	1.446 1	2.638 1	24.288 1	41 679.317	114.723 9	134 954.17	382.709 8	720 359.19
		Std	0	0	0	0	0	0	0	0
4	**CBGA**	**Fbest**	1.461 4	11 190 779	24.279 9	41 693.322	112.950 7	134 968.83	371.645 9	720 358.74
		Worst	1.584 9	18 475 064	25.779 6	41 729.981	120.383 9	137 704.02	382.132 2	5.306*E*+13
		Mean	1.484 9	16 314 553.1	25.034 9	41 710.516 9	115.377 4	135 108.498	374.470 9	1.828*E*+13
		Median	1.479	14 526 025	25.103 6	41 709.915	114.921	135 013.23	374.075 2	1.976*E*+13
		Std	0.023 87	2 192 622.375	0.381 04	11.058 3	1.718 2	483.535 4	2.255 4	1.589*E*+13
5	**MVMO***	**Fbest**	1.450 7	2.637 3	24.634 6	41 688.998	117.249 7	135 004.06	385.628 4	720 744.04
		Worst	1.529 8	2.637 3	25.129 6	41 732.323	118.874 5	135 137.07	415.539 1	731 004.15
		Mean	1.466 8	2.637 3	24.817 2	41 707.689	117.862	135 050.21	394.129 4	722 763.9
		Median	1.457	2.637 3	24.776 3	41 705.517	117.766 9	135 051.07	391.339 2	721 234.68
		Std	0.021 983	3.01*E*–08	0.125 04	10.558 1	0.423 79	27.501 4	8.221 9	2 706.895 6

[a]MVMO (i.e. its variant MVMO-PHM) was not considered as a participant algorithm, but it was used to produce reference results for the comparisons.

- Differential Evolutionary Particle Swarm Optimization (DEEPSO). Authors: Vladimiro Miranda, and Leonel Carvalho. Affiliation: INESC TEC, Portugal.
- Differential Evolution Algorithm with a Modified Archiving-based Adaptive Tradeoff Model for Optimal Power Flow (IC-DE). Authors: Ming Niu, Youwei Jia, Zhao Xu, and Kit Po Wong. Affiliation: Hong Kong Polytechnic University, Hong Kong, and University of Western Australia, Australia.
- Heuristic MINLP for Solving Optimal Power Flow Problems (MINLP). Author: Carleton Coffrin and Hassan Hijazi. Affiliation: National Information and Communication Technology Austria (NICTA)
- A Modified Chu-Beasley's Genetic Algorithm to Solve the Optimal Power Flow Problem (CBGA). Authors: Leonardo H. Macedo, Marcos J. Rider, John F. Franco, Rubén Romero, and Edgar Manuel Carreño. Affiliation: Sao Paulo State University, UNESP-Ilha Solteira, Brazil.

MVMO-PHM did not participate in the competition and has been used to produce reference results to compare the results submitted by the contestants. It is important to acknowledge that the authors of MVMO are also the organizers of the competition. This algorithm is developed by Jose Rueda (Delft University of Technology, Netherlands) and Istvan Erlich (University Duisburg-Essen).

It is worth highlighting that the codes and presentations describing the rationale of the above indicated algorithms, which were presented in the Panel organized for the 2014 IEEE PES General Meeting, are available for download in [226]. Please note that the software and the associated literature come with no warranty whatsoever. The use of the codes and the possible relation to any new publication must be duly acknowledged (e.g. citing the corresponding authors and their presentations).

Table 3.9.9 shows the top 4 algorithms after evaluation by the organizers of the competition. For each algorithm, the table provides the statistics (fbest–best fitness value, worst–worst fitness value, mean, median, and std. standard deviation) associated with the fitness value obtained in the last function evaluation among 31 runs, including the objective function value (denoted as o@fbest), and the sum of bound constraint violations (denoted as gvar@fbest) associated with fbest.

Note that, in general terms, all four contestant algorithms provided competitive and comparable results with respect to the results obtained by using MVMO. It is worth highlighting that the codes were verified and rerun in the cluster of computers of the University Duisburg-Essen. Although not shown in this chapter due to space constraints, it is worth highlighting that similar statistics (w.r.t. Table 3.9.9) were obtained.

Both DEEPSO and MINLP algorithms achieved the best mean fitness among all contestants. Nevertheless, the success of MINLP came at the expense of bound constraint violation (i.e. some variables outside the allowed min/max bound). Therefore, DEEPSO obtained the first place and was declared as the winner of the competition.

3.9.7 Conclusions

This chapter has introduced the definition and implementation of a test bed for ORPD and OARPD problems. This test bed has been used to launch the first competition on the application of modern heuristic optimization algorithm to power system problems. A comparative assessment carried out on submitted results by different contestants worldwide has shown the great potential of modern heuristic optimization algorithms to tackle the complexity of power system problems, i.e. high dimensionality, nonlinearity, discontinuity, multimodality, and non-convexity. Particularly, DEEPSO and MVMO constitute proven tools which can be generally used to solve OPF problems with different dimensionality and associated with power systems of different size.

Researchers working in heuristic optimization and are looking for testing their developments in power system field are kindly invited to try the 2014 OPF test bed. Guidelines and codes associated with the test bed can be downloaded from [226]. In case of beating the best results achieved so far (cf. Table 3.9.9), please submit the zipped folder, which is generated automatically from the encrypted file "*test_bed_OPF.p,*" to Prof. István Erlich (istvan.erlich@uni-due.de) and Dr. José L. Rueda (J.L.RuedaTorres@tudelft.nl) in order to perform comparisons with the results presented in the 2014 IEEE PES General Meeting and to update the table of best-performing algorithms. The implementation codes of the tested algorithm must also be submitted along with final results for full consideration in the evaluation. The submitted codes will be used for further tests, which are intended to crosscheck the submitted results. The submitted codes will be in the public domain through website of the Working Group on Modern Heuristic Optimization [238], and no intellectual property claims should be made.

The Working Group is currently working the development of new test beds concerning the optimization under uncertainty in smart-grid applications. Please visit [238] to follow the recent activities and contributions of the working group.

ACKNOWLEDGMENT

This work has received funding from the National Natural Science Foundation of China under Grant 51577061; the Eletrobras and GE Global Research; the Korean National Research Foundation (NRF) (No. #2012R1A2A2A01012803); Korea South-East Power CO. (KOEN); the Brazilian research agencies: CNPq, CAPES, and FAPEMIG; and the Scientific and Technical Research Council of Turkey (TUBiTAK).

The authors acknowledge the valuable contribution of Junyi Zhai (Section 3.4), A.F. Amendola and K.C.A. Roberto (Section 3.5), and Sebastian Wildenhues (Section 3.9).

3.10 TRANSMISSION SYSTEM EXPANSION PLANNING

Eduardo N. Asada[1], *Aldir S. Souza*[2], *and Rubén Romero*[3]

[1]*University of São Paulo, São Carlos, Brazil*
[2]*State University of Piauí, Teresina, Piauí, Brazil*
[3]*São Paulo State University, Ilha Solteira, São Paulo, Brazil*

3.10.1 Introduction

The best design of the electrical power network to satisfy technical and financial requirements has always been a challenge since the beginning of the power industry. There are three major actors who play the role in the system: generators, transmission system, and distribution utilities. Ideally, the best design of the system depends on the coordination among all of them, or at least between the generators and the transmission system. This concept led to many studies considering the composite planning of generation and transmission systems. One strategy is to separate the generation and transmission planning into two problems, which is commonly done in the current deregulated scenario of power systems. In this case, generation planning is carried out first and the result is sent to the transmission planning as the input data. In this context, the transmission planning has been modeled according to several needs, which mathematically results in complex nonlinear optimization problem with discrete variables. This type of problem is especially difficult to solve as the dimension of the system increases, in addition to complicating operation constraints which presents nonlinear relations.

Due to the difficulties in solving this problem with classical methods, heuristic algorithms, classical and hybrid algorithms, and meta-heuristics have been explored for decades. The current research scenario on this subject shows an advance of the two major areas regarding this topic: the mathematical modeling and the solution strategies. The progress of both areas relies on continuous advance on the computational power. For instance, nowadays, the classical approach to provide the best expansion plan which is based only on one temporal stage of planning with only the fixed or varying costs is not sufficient. More elaborate models, such as those that take into account investment risks or even a probabilistic oriented planning, are now under consideration. Therefore, comprehensive expansion models that take into account not only operation constraints but also different and variables scenarios that approximate closely to the projected operation scenario are thought. In this regard, the modern heuristic techniques are well suited to face these demanding scenarios with good solution possibilities.

This chapter addresses the long-term transmission system expansion planning model with emphasis on improvements on the classical model starting from the classical CHA in order to re-evaluate the currently used meta-heuristics in terms of performance and efficacy. Another important point is the current trend for the transmission expansion planning (TEP) problem and recent advances toward multi-objective optimization methods.

3.10.2 Transmission System Expansion Planning Models

The TEP can be seen as a system adequacy problem due to the constant growth of the power demand and the requirement of new energy sources and the necessity to connect both of them with infrastructure investment. The basic objective in expanding the system is to determine the best system topology that meets the desired operating conditions. Normally, the system updates consist of reinforcements by building new transmission lines, repowering system devices, embedding new technology to name a few. Due to the complexity involved in the whole planning, this problem has been divided into various smaller ones, yet complex, which are put together when the network is built or expanded. One of the problems consists in only determining the transmission lines and transformers that must be put into the system to satisfy the desired operating scenario aiming at some specific objective, such as the minimization of the expansion cost. If only static conditions, such as power flow constraints, are taken into account, this problem, although very simplified for a real deployment of transmission infrastructure, still results in complex mathematical problem. The reasons for this complexity are the nonlinear nature of the system operation, discrete decision variables, and enormous number of expansion possibilities.

In order to deal with this problem, in a not distant past, the only available tools for TEP have been a set of system analysis tools such as: power flow, short-circuit, and transient analysis [239] used by the planning engineer. Automated solutions would be provided by interactive computer programs that provide alternatives for system expansion. Basically, the program determines, based on the analysis of the system performance, which lines among many others would benefit the system operation mostly. For these algorithms, the new circuit additions are made based on indices which are formulated according to the expected behavior of the system. This type of interactive system planning represents the use of early CHA and has been quite used by the planners due to its simplicity. Many of the CHA have similar structure and differ only on the performance index, used as a guide for the component addition. It must be mentioned the context of the planning at that time, which has been conceived for a regulated power industry, where the major concern was to transfer the bulk generation to load centers under acceptable reliability conditions.

Moreover, the planner had access to information for planning due to the collaboration among utilities. The CHA are still in use in the power industry, albeit much modified to cope with the current requirements; three important references to these CHA are [240–242]. The deregulation and disaggregation of the power industry led to changes that created new requirements for the operation as well as for the planning area. In the new environment, there is no direct coordination between TEP and generation expansion planning and there is no certainty about generation and load due to the market-driven operation or great penetration of renewable energy sources. Therefore, vulnerability analysis, reliability analysis, and scenario analysis became important studies for the TEP problem. To start with the TEP problem, in the next section, two expansion models will be addressed.

3.10.3 Mathematical Modeling

According to the operation context and details needed for planning, the mathematical modeling may assume different formats and complexity. The complexity of the model changes depending on the objectives and the constraints. The constraints are due to the intrinsic nature of the electrical network which imposes stricter operations depending on the conditions of the electric power system operation. To discuss the new applications and new methods, this section will focus on the single-stage planning (static planning) with two models with different resolution complexity. Let us start with one of the most cited and solved model in the technical papers which is the DC expansion planning model [243, 244], formulated as follows:

$$\min v = \sum_{ij} c_{ij} n_{ij} + \alpha \sum_{l} r_i \tag{3.10.1}$$

Subject to

$$\begin{aligned} &Sf + g + r = d \\ &f_{ij} - \gamma_{ij}\left(n_{ij}^0 + n_{ij}\right)\left(\theta_i - \theta_j\right) = 0 \\ &|f_{ij}| \le \left(n_{ij}^0 + n_{ij}\right)\overline{f_{ij}} \\ &0 \le g \le \overline{g} \\ &0 \le r \le d \\ &0 \le n_{ij} \le \overline{n_{ij}} \\ &n_{ij} \text{ integer}, f_{ij} \text{ and } \theta_j \text{ unbounded} \end{aligned} \tag{3.10.1.1}$$

where v is the total expansion cost, c_{ij} is the cost of a line in path i–j, and n_{ij} is the number of lines added in path i–j and represents the decision variable of the optimization problem. $\overline{n_{ij}}$ is the maximum number of lines in path i–j, S is the transposed node-branch incidence matrix, f is the vector of power flows, f_{ij} is the total power flow in path i–j, $\overline{f}_{ij}$ is the maximum power of a line in path i–j, θ_j is the voltage angle in bus j, γ_{ij} is the susceptance of line in path i–j, d is the vector of demand in load buses, g is the vector of generation in generation buses, and r is the vector of artificial generation that acts as slack variables, to handle the infeasible solutions and α is the penalty factor. Feasible solutions must have the variable r set to zero (which also means no load shedding in the current solution). The first constraint of (3.10.1.1) is associated with the Kirchhoff's Current Law (power balance equation) and the second constraint with the Kirchhoff's Voltage Law (flow equation). The reminders of the constraints are related to limits of the transmission system components. In mathematical terms, the model (3.10.1) is a nonlinear mixed integer problem. If the integrality of variables n_{ij} is relaxed, the problem becomes a nonlinear problem.

Another interesting model is the hybrid linear model (HLM) [241], which is an approximation of the previous DC model. The HLM is obtained relaxing constraints $f_{ij} - \gamma_{ij}\left(n_{ij}^0 + n_{ij}\right)\left(\theta_i - \theta_j\right) = 0$ in the DC model to all new circuits added to the system. The constraints of (3.10.2) for the already existing circuits in the base topology are linear. The hybrid model is a mixed integer linear problem as follows:

$$\min v = \sum_{ij} c_{ij} n_{ij} + \alpha \sum_i r_i \tag{3.10.2}$$

Subject to

$$\begin{aligned}
& Sf + S^0 f^0 + g + r = d \\
& f_{ij}^0 - \gamma_{ij}\left(n_{ij}^0 + n_{ij}\right)\left(\theta_i - \theta_j\right) = 0 \quad \forall (i,j) \in \Omega_0 \\
& \left|f_{ij}\right| \le n_{ij}^0 \overline{f_{ij}} \in \Omega_0 \\
& \left|f_{ij}\right| \le n_{ij} \overline{f_{ij}} \in \Omega \\
& 0 \le g \le \overline{g} \\
& 0 \le r \le d \\
& 0 \le n_{ij} \le \overline{n_{ij}} \\
& n_{ij} \text{ integer}, f_{ij} \text{ and } \theta_j \text{ unbounded}
\end{aligned} \tag{3.10.2.1}$$

where S^0 is the branch-node incidence matrix formed by existent circuits in the base topology, f^0 is the vector with power flow of the circuits in the base topology with elements f_{ij}^0, S is the complete branch-node incidence matrix, and f is the vector with power flows in the new circuits to be added in the system. Ω_0 is the set of the circuits belonging to the initial (base) topology. From the previous model, the newly added circuits in the optimal solution are not obliged to follow the Kirchhoff's voltage law (KVL). An inconvenience with this approximation concerns the power flow flowing in a parallel circuit to the existing topology which may result in different flows in each circuit (the new circuit must follow only one Kirchhoff Law, and the circuit of the base topology both Kirchhoff Laws must be applied).

Despite the earlier fact, the interesting characteristic of this model is in the fact that in a heuristic search procedure, the selection of the lines is based on linear programming model; however, when the selected line is added, it becomes part of the topology for the next iteration. As a consequence, if the addition of the line is accepted, it results that the new topology complies the original DC model.

The transportation model applies further approximation to the DC model, which is to remove completely the second constraint of (3.10.2.1). In this case, the problem relies only on the Kirchhoff's Current Law, represented by the conservation of power on each bus. In this case, the optimization problem will provide the least cost expansion plan that provides the active power transportation to meet the demand. The solution of such simplified model does not satisfy the DC model, but its complexity still remains as a combinatorial nature and very hard to solve for large systems. The transportation model has been used in [240], which originated one of the constructive heuristics used for this problem.

When meta-heuristics are applied, a common approach is to let the method to manage the search for the best circuit additions that also satisfies the network constraints. It means the meta-heuristics only encode the solution proposal given by the number of lines added for expansion. Therefore, when the DC model is adopted, the variable n_{ij} as shown in the previous model is defined from the heuristic algorithm and the variables related to operation (f_{ij}, θ_j, g_k) are calculated after solving the resulting linear programming problem. Regarding the adequacy of the circuit modeling, a more precise model is the AC model which has been proposed in some works [245, 246]. The challenge in using the AC model is to deal with disconnected parts of the system and also to correctly consider the reactive power, which is normally modeled as "*a posteriori*" in the operation planning phase. A comprehensive list of transmission expansion models can be found in [243] and details related to the complexity of TEP models can be found in [245, 247, 248].

3.10.4 Challenges

Let us start with the basic least cost planning that defines the set of lines to be added into the system. In this problem, the idea is to define the minimum number of new lines or reinforcements and the location of them to satisfy some conditions (from economical to environmental). More elaborate models can be obtained when the planning is carried out in several stages, which defines the best moment for installation according to the economic scenario. In this case, the moment to install or to expand the system becomes a decision variable of the problem and the whole planning process must define the location, the amount and the stage to add the lines in (multiyear planning). Some other topics have been also considered as current trends in planning studies:

- Impact of new transmission technology
- Environmental issues
- Expansion under specific market environments
- Probabilistic planning
- Vulnerability analysis
- Application of efficient mathematical methods

It is possible to notice the increasing challenge to address adequately any of these topics in a planning formulation, especially knowing the nature of the basic problem, which is modeled as a nonlinear mixed integer problem. Due to the complexity of each subject, it is usual to carry these analyses separately after the selection of the most promising expansion alternatives is done. Various studies employ different techniques, from classical optimization methods (Benders decomposition [249], branch and bound [250], etc.) to meta-heuristics. The adoption of a solution method is linked with the corresponding mathematical model. Therefore, according to the complexity of the problem, one must adapt the solution methodology. This is especially true for heuristic-based methods. A measure of efficiency is needed to justify the application of one method or another, for instance, even though the meta-heuristic might be more powerful than simpler heuristics, the search strategies used in the method should be explored to the maximum, instead of using the basic search mechanisms present in the majority of the basic presentation of the basic presentation of the algorithm, otherwise even meta-heuristics can present low efficiency. An example regarding the evolution of the heuristics to the TEP problem will be shown in the next section.

3.10.5 Application of Meta-heuristics to TEP

Heuristics are optimization techniques developed to find feasible and good quality solutions to complex problems. This type of method consists of an algorithm with well-defined steps, which is usually developed for a specific problem to provide good solutions. They became extremely popular between the years 1960s and 1970s to solve complex mathematical problems, usually nonlinear, non-convex with discrete variables. One important feature, explained by the available computational power, was the simplicity and resolution speed. However, the majority, in terms of mathematical theory, lacks proof of optimality and convergence. Also, due to the simplicity they hardly find optimal solutions as the complexity of the problem increases. Here, the term complexity is used to explain the difficulty in finding the global optimal solution due to combinatorial explosion that the problem may face when the dimension increases or because of the mathematical modeling complexity due to discrete variables, nonlinear and non-differentiable objective functions, nonlinear constraints, and non-convex search space.

The heuristic algorithm can be extremely simple, and normally the procedures are conceived by observation of features which are present in the problem or based on some performance index. The classical heuristic algorithms applied to the TEP problem are [240–242]. These algorithms are classified as constructive heuristic which is an iterative process that builds the solution by adding a circuit (transmission line or transformer) to the current topology at a time. After the addition, the modified system is renamed as current configuration for the next iteration. The selection of the circuit is based on indices that provide measure of the efficacy in adding a specific circuit. The following characteristics are observed for the sensitivity index:

- It must quantify the benefit of adding a circuit.
- It is an index that presents local information, i.e. it identifies the best strategy (for the current configuration) for the current iteration, instead of a global information.
- As the local information used by the index usually does not match with information related to search of global optimum, it does not have the ability to find the global optimal solution.

The sensitivity index used to select the line for addition, according to the characteristics shown earlier, can be derived from the original optimization model. For instance, the decision variable n_{ij} is approximated as a continuous variable and the original mixed-integer problem is converted to a continuous linear or nonlinear problem depending on the model. Therefore, if a given topology is considered as a basis for the expansion, the solution of the corresponding problem results in fractional values of variables n_{ij}. For instance, the resolution of the problem may indicate some paths with the necessity of, say, $n_{ij} = 1.7$ transmission lines, or in another situation the result is zero for all n_{ij}, which means no circuit addition is needed.

A general algorithm of the CHA to the TEP problem is:

1. Select a system model (linear or nonlinear)
2. Assume a base topology as the current solution
3. Solve the corresponding optimization problem with variables n_{ij} as continuous
4. If all n_{ij} variables are zero, stop, otherwise continue
5. Calculate an index for all $n_{ij} \neq 0$, and order them according to a criterion
6. Select the path $i-j$ based on the ranking and add one line
7. Go to Step 2.

The proposal presented by Garver [2] applies the transportation model in Step 1 and the circuit ranking is done based on the active power flow in path $i-j$ calculated from $f_{ij} = n_{ij}\overline{f}_{ij}$, where $\overline{f}_{ij}$ is the maximum flow in the path $i-j$. This strategy means that the path that presents the largest flow demands a new transmission line.

The Villasana–Garver–Salon (VGS) algorithm [241] applies the HLM in Step 1 of the previous algorithm and the sensitivity index is the same of the Garver's algorithm. The difference in this case, which relies on the model, will result in a solution that is also feasible regarding the Kirchhoff's Voltage Law, which also means that the solution satisfies the DC model, although cannot be guaranteed as the global optimal solution of the DC model.

Minor enhancements on the previous algorithm can be applied by considering the costs of the circuits in the calculation of the performance index. Another change would be the addition of a circuit removal phase after the stopping of the constructive phase. As shown, the decision regarding the addition of the line is based on the approximate solution, which indicates the inclusion of fractional values of lines. However, the inclusion operation is carried out with integer values. This would result in the addition of excess lines in the circuit. Therefore, a simple procedure of removal and testing can eliminate the additional lines due to the approximation in the process.

Toward Efficient Constructive Heuristic Algorithm

The critical characteristic of conventional CHA is the premature convergence of the search to poor local optimal solutions when large systems are analyzed. One of the reasons for this behavior is related to the limited search strategy which cannot perceive other relations regarding cost and transmission capacity. For instance, in the beginning of the construction process, it may happen that lines with large transmission capacity but with high cost are selected for addition. In this case, the search is guided by costly lines and better options with similar transmission capacity are excluded from the solution. Another problem is observed at the final stages of the process, when most of the variables n_{ij} calculated from linear programming (LP) or nonlinear programming (NLP) model present small numeric values. In this case, the decision of adding a line might be based on small fractional values, which deviates the search from good solutions.

The first strategy to improve the algorithm is to consider the costs in the calculation of the performance index; however, a mechanism to allow testing different hypotheses during the search without forgetting the nature of a constructive algorithm should be thought. With this in mind, it is possible to develop an algorithm that stands between the classical branch and bound algorithm for linear or nonlinear mixed integer problem and advanced meta-heuristics. Based on the two facts mentioned earlier, when a costly transmission line is selected in path (p, q) (which happens normally in the beginning of the process) or when a line is selected based on small numerical value of variable n_{pq} (happens normally in the final stages of the search), the algorithm should test for alternative hypothesis. This idea leads to more flexible heuristic algorithm that admits testing other situations instead of only accepting the performance index for circuit selection. The structure of the algorithm can be similar to a divide-to-conquer strategy as seen in branch and bound algorithm, however, with simpler structure. In summary, the algorithm should analyze other expansion alternatives whenever the following situations occur:

1. The path (p, q) is chosen for circuit addition and the cost c_{pq} for expansion is above a predefined value.
2. The path (p, q) is chosen for circuit addition and the n_{pq} obtained from the LP or NLP resolution for current topology is below a predefined value.

When these situations occur, alternative analysis is triggered by the creation of a subproblem that sets n_{pq} to zero; in other words, the circuit indicated by the index is not accepted immediately and another better option should be tested. Figure 3.10.1 illustrates this situation.

v_1^k is the cost related to the circuits already selected by the search, v_2^k is the value of the objective function in node k, n_{pq} is the result from the LP or NLP model. At node k if the resulting n_{pq} is small, then the left branch which sets n_{pq} to zero is created. It means that although the performance index indicates the expansion in path (p, q) with a specified line, this decision is ignored, which generates the subproblem s. The node s can also be a result from the observation of the cost of the candidate circuit. The branch on the right side is a result of the acceptance of the performance index.

The basic CHA that applies the same structure of a branch and bound algorithm is shown next [251].

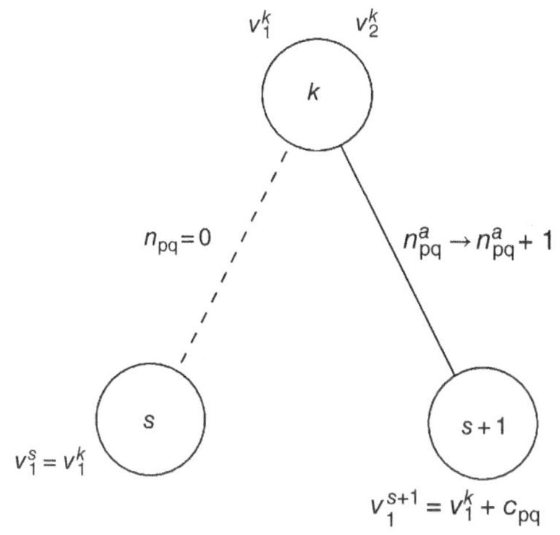

Figure 3.10.1 Simple divide-to-conquer strategy (branching).

1. Select the mathematical model of the CHA. Initialize the list of open problems and the initial problem to be solved. Set the limits n_{ij}^c and c_{ij}^c, which represent critical situation.
2. Solve the corresponding problem. If all resulting n_{ij} are integer, stop. Otherwise, store the p, q solution as inferior to the limit $v_{\text{inf}}.n_{\text{pq}} < n_{ij}^c$
3. Choose the variable for separation: Identify the circuit in a given path (p, q) indicated by the performance index. If and/or $c_{\text{pq}} > c_{ij}^c$, then generate a subproblem with the condition $n_{\text{pq}} = 0$. Generate a subproblem by adding a line in the path (p, q) indicated by the performance index and update v_1 (cost of already added circuits).
4. Select the next subproblem from the list of open problems. If the list is empty, then the incumbent solution is the solution. Otherwise, select the next problem according to the selected criterion (by LIFO – Last In–First Out or FIFO – Fist In–First Out). Solve the subproblem obtaining v_2. The inferior limit of the successor subproblem is updated as $v_{\text{inf}} = v_1 + v_2$.
5. Tree pruning test: After the selected subproblem is solved, three tests are applied to check whether the current node is able to provide better solutions, otherwise, it should be eliminated.
   a. Test 1: If $v_{\text{inf}} \geq v^*$, where v^* is the value of incumbent solution.
   b. Test 2: If the solution of corresponding LP (or NLP) is unfeasible.
   c. Test 3: If the solution of corresponding LP (or NLP) is equal to zero ($v_2 = 0$), then a feasible solution to the mathematical problem has been found. In this case, compare the objective function of the subproblem under analysis with the incumbent solution. If $v_1 < v^*$, then $v^* \leftarrow v_1$ and apply Test 1.

6. If the subproblem has not been pruned then, go to Step 3, otherwise go to Step 4.

Concerning the tree pruning phase, if node k has not been pruned, then the successor node s is generated whenever one of the critical conditions are detected after the selection of a circuit n_{pq}.The successor $s + 1$ is always generated and the search process goes through node $s + 1$. The subproblem s presents all constraints of node k plus the additional constraint $n_{\text{pq}} = 0$, which means that for all successors of this node, it is not possible to add more circuits in the path (p, q). The subproblem $s + 1$ has all constraints of node k plus the addition of a new circuit in the path (p, q). The variables v_1 and v_{inf} can be easily updated using the following relations:

$$v_1^s = v_1^k \tag{3.10.3}$$

$$v_{\text{inf}}^s = v^k + n_{\text{pq}} c_{\text{pq}} \tag{3.10.4}$$

$$v_1^{s+1} = v_1^k + c_{\text{pq}} \tag{3.10.5}$$

$$v_{\text{inf}}^{s+1} = v^k + \left(1 - n_{\text{pq}}\right) c_{\text{pq}} \tag{3.10.6}$$

Tests with the Enhanced CHA

Tests have been carried out with Brazilian Southern System, with 46 buses, 79 circuits, and 6880 MW of demand [252] and the reduced version of Brazilian Northern System with 87 buses, 183 circuits, and 20 600 MW of demand. The models tested inside the enhanced CHA were the transportation model from [240] and the HLM of [241].

The results for both systems showed very good results. For transportation model with generation redispatch with Garver's algorithm, the optimal solution for the case with generation redispatch is $v = 53\ \ 340\ \ 000$ (US$), which has been found by branch and bound algorithm by [250]. In this case, 260 LP problems have been solved. The simple application of Garver's original CHA resulted in a solution of $v = 63\ \ 163\ \ 000$ (US$) after solving 13 LP problems. The enhanced CHA in branch and bound structure obtained the optimal solution after solving 27 LP problems.

The observation of the sequence of line additions allows identifying the instant that led the search to the optimal solution. The Garver's algorithm selects the circuit addition in the following order: (20–21), (06–46), (05–06), (42–43), (20,21), (20–23), (05–06), which results in the following expansion:

$$n_{20-21} = 2, n_{42-43} = 1, n_{05-11} = 2, n_{46-11} = 1, n_{33-34} = 1$$

The constructive heuristic in branch and bound structure is set with parameters $n_{ij}^c = 0.35$ and $c_{ij}^c = 35\,000\,000$ (US$). The performance index indicates addition to the path (20–21), and a new subproblem is created and added to the list of open nodes. The subproblem sets path (06–46) to $n_{06-46} = 0$ due to the cost of the transmission line in this path. The performance index will lead to the option

(46–03); however, due to the cost of line and limit given by c_{ij}^c, an alternative solution with n_{46-03} set to zero must be analyzed. The next line additions are in path (20–21), (42–43), (05–11), (46–11). After this, a new branching occurs for circuit (19–21). Indicating the subproblems with $n_{19-21}=0$ and $n_{19-21}=n_{19-21}+1$ must be tested. The option with $n_{19-21}=0$ will lead to subproblems that will result in circuit addition in path (05–11) and path (33–34), thus obtaining the global optimal solution after the resolution of 27 subproblems, or LP problems.

The tests with HLM of [241], considering the Brazilian Southern system has also been successful. The solution of $v = 70\,289\,000$ (US$) has been found with parameters $n_{ij}^c = 0.35$ and $v^c = 35\,000\,000$, and 98 LP problems solved. The solution is the same global optimal solution found with Benders Decomposition.

When more complex system is used, for instance, the Brazilian North-northeastern system with 87 buses, the proposed heuristic in branch and bound structure, with Garver's algorithm (transportation model) with parameters $n_{ij}^c = 0.25$ and $c_{ij}^c = 60\,000\,000$ (US$) the best solution found with the heuristic in branch and bound structure was $v = 1\,245\,040\,000$ after solving 16 858 LP problems. It must be reminded that the optimal solution of transportation model for the Brazilian North-northeastern system is unknown.

Improving the CHA with Fuzzy Inference System

As observed in the previous case, the notion of expensive circuit or small value of decision variable n_{ij} that will trigger the creation of new subproblems in a branch and bound style tree depends on tentative and verification and on the system being solved. Therefore, an improvement on this method should allow the resolution of any systems in efficient way. An alternative is to use the fuzzy inference system to help the decision-making inside the algorithm that will lead to the best solution. The Mamdani-Type fuzzy inference system [253] is well known and presents the following components: fuzzification, fuzzy rules, inference system, and defuzzification. Basically, the main idea is to synthesize a set of linguistic control rules that represent approximate information of some quantity or variable for decision-making. The rules are of IF–THEN type, the Fuzzification process maps exact numerical values in grades of membership function in linguistic terms, and the inference system maps the fuzzy inputs to fuzzy output by following the rules. A proposal of the fuzzy system is based on three linguistic variables: value of n_{ij}, cost of transmission line $i-j$, and the variable branching that indicates whether or not to branch the problem and create new subproblems. The set of linguistic values for variable value of n_{ij} and cost of transmission line $i-j$ is composed by Tiny (TY), Small (SM), Medium (MD), Large (L), and Very Large (VL), which are represented by vectors of real numbers between 0 and 1. The variable branching has two fuzzy sets: branching (B) and not branching (NB). The universe of discourse of the variable value of n_{ij} is defined in the interval $[0, \max\{2, \max\{n_{ij}\}\}]$. For variable cost of line in path $i-j$ is in $[0, \max\{c_{ij}\}]$ and for variable branching is [0, 1]. The sets TY, MD, and VL can be modeled as trapezoidal membership function and the sets L and SM as triangular membership function. The sets B and NB are modeled as sigmoid function.

TABLE 3.10.1 Fuzzy Rules

		Value of n_{ij}				
Cost of transmission line $i-j$		**TY**	**SM**	**MD**	**L**	**VL**
	TY	B	NB	NB	NB	NB
	SM	B	NB	NB	NB	NB
	MD	B	B	NB	NB	NB
	L	B	B	B	NB	NB
	VL	B	B	B	B	B

The fuzzy rules must be defined to express the logical relation between the fuzzy variables of the problem. A proposal for this is to represent by 25 fuzzy rules which correspond to the combination of the variables shown earlier. Table 3.10.1 shows the relation: cost of transmission line $i-j$ with value of n_{ij} and the decision to apply or not the branching operation.

The fuzzy inference system combines the active rules for a specific data entry and generates a fuzzy output. The active rules are checked and then the aggregation is done to obtain a single fuzzy output. The min, max, or mean operator can be applied for this purpose. The next step is the defuzzification, which is the process of transforming the fuzzy output from the inference system to discrete values. There are different techniques, three of them are: Center of Area, Middle of Maximum, and First of Maximum. As an example, in [254] by changing the CHA in branch and bound structure [251], the decision to branch the problem with fuzzy inference system, the number of iterations to obtain a good solution changes. As shown in [254], the number of iterations resulted bigger, although the benefit in this case is the self-adjustment of the algorithm as the system changes. For this purpose, the use of the fuzzy inference system was successful because good solutions have been obtained automatically.

Meta-heuristics for Transmission Expansion Planning Problem

Meta-heuristics have appeared in the 1960s as a superior class of heuristics. They became popular due to the ability at finding good solutions to very complex problems and from the fact that many of them have been inspired from natural phenomena. The SA [244] has been one of the first meta-heuristics applied to the planning problem, followed by GA [263] and TS [264, 265]. A quick survey on the IEEE transactions on Power Systems and IET Generation, Transmission, and Distribution has shown that most of transmission planning papers with meta-heuristics employs GA or its variant (more than 30 papers). Other meta-heuristics such as SA, TS, GRASP, and PSO are used in about 10 papers. Traditional heuristics are also used and they are present in 10 papers, not to mention other optimization problems in power systems.

There are some points in the planning problem to be explored thoroughly with meta-heuristics. The reliability of the planning has been considered by the satisfaction of $N-1$ criterion for decades. However, this criterion is essentially deterministic and consequently, neither it considers the probabilistic nature of the outages, nor its impacts on the system. Probabilistic planning may contribute for a comprehensive planning process by adding the probabilistic reliability assessment of the expansion alternatives [255]. Therefore, the combination of the reliability assessment and economic analysis would result in the best expansion solution. Another important point is about planning a system that is capable to deal with congestion problem [256], which would require a versatile system. In this case, the planning might provide very expensive plans and a cost–benefit analysis should be needed. The use of new technologies such as FACTS devices introduces new characteristics in the operation of the system and may modify significantly the need for new transmission lines.

Another concern that has been raised recently is the vulnerability of the system regarding natural disasters or deliberate outages caused by human attacks [257, 258]. An interesting planning approach is the application of minimax regret concept, which is to adopt an expansion that would result in minimal worst-case regret due to the occurrence of the damaging event. An interesting characteristic is that it is independent of probabilities of the various outcomes, in opposition to the probabilistic planning, considering the regret is correctly calculated. Besides the transmission lines, this planning can also consider generators and other equipment failures.

It is possible to observe that some of the previous analysis may result conflicting and most of them are computational intensive. An approach is to consider two or more objectives simultaneously in a multi-objective framework [259, 260]. For this, the heuristics based on multi-objective optimization evolutionary algorithms (MOEA), are the ones with potential to be successfully employed for the TEP problem. Examples of MOEA algorithms are: NSGA-II (non-dominated sorting genetic algorithm) and SPEA-2. In order to deal with uncertainty and decision-making, these methods are merged with other techniques, such as fuzzy systems. In [260], three objective functions have been considered: the construction cost, congestion and load curtailment, and fuzzy decision-making are used to aid the selection of the best solution. It is possible to include in the objective function some reliability indices, such as the system average interruption duration index, or the system interruption frequency index.

Environmental impacts can also be considered in the planning problem in multi-objective framework [261]. It is also possible to adapt other algorithms such as PSO, TS, and SA for multi-objective approach. An important issue is how to model such objective function to make the solution corresponding to what is most beneficial to the system. Research on this topic is increasing fast as the multi-objective meta-heuristic evolves. At last, but not less important, the development of efficient algorithms is required for every heuristic method. Usually, it is hard to assess the quality of meta-heuristics, since the basic algorithm must be tailored according to the peculiarity of the problem. In general, meta-heuristics are composed of strategies which are, in general, heuristics. As observed in the earlier sections, some less complex heuristics which are modified to include more flexibility in the search can successfully obtain good solutions and occasionally similarly to computational intensive meta-heuristics. Therefore, it is important to observe some quality metrics that relates computational time, computation complexity, and the quality of the solution, in order to make the powerful algorithms such as the meta-heuristics efficient as they are expected to be. Another important point is the critical evaluation of the efficacy of such algorithms in statistical terms. The success on deploying these methods depends on statistical testing of the efficacy of such algorithms for various situations. Recent review on TEP problem can be found in [262].

3.10.6 Conclusions

This chapter presented the main challenges for the application of heuristics and meta-heuristics to the transmission system expansion planning problem. Some of the challenges highlighted in this chapter are related to the modeling of the problem, which despite its natural difficulty, has incorporated more challenges such as different operation scenarios, multiple objectives, economic and reliability analysis. In terms of the solution methods, some important points have been discussed. The industry demands reliable and efficient methods; therefore, a revision of the development of meta-heuristics in terms of computational efficiency must be presented. There is room for the improvement of classical meta-heuristics and the evolution of these algorithms is toward multi-objective optimization.

3.11 CONCLUSION

The section highlights the various power system problems based on power flow analyses such as transmission system planning, OPF, ED, unit commitment, etc. Such problems are formulated using models and solved using various powerful optimization techniques such as ABC, GA, PSO, and MVMO. In every case, use of these algorithms is shown to provide optimal results for the given problems. It is shown that power system problems may be single-objective or multi-objective and regardless, can be optimally solved using modern-day heuristic optimization algorithms.

REFERENCES

1. Padhy, N.P. (2004). Unit commitment – a bibliographical survey. *IEEE Transactions on Power Systems* 19 (2): 1196–1205.
2. Mantawy, H., Abdel-Magid, Y.L., and Seliin, S.Z. (1998). A simulated annealing algorithm for unit commitment. *IEEE Transactions on Power Systems* 13 (1): 197–204.
3. Motto, A.L. and Galiana, F.D. (2002). Equilibrium of auction markets with unit commitment: the need for augmented pricing. *IEEE Transactions on Power Systems* 17 (3): 798–805.
4. Merlin, A. and Sandrin, P. (1983). A new method for unit commitment at electricite de France. *IEEE Transactions on Power Systems* 102 (5): 1218–1225.

5. Saravanan, B., Das, S., Sikri, S., and Kothari, D.P. (2013). A solution to the unit commitment problem – a review. *IEEE Transactions on Power Systems* 7 (2): 117–124.
6. Shahidehpour, M., Yamin, H., and Li, Z. (2002). *Market Operations in Electric Power Systems: Forecasting, Scheduling, and Risk Management*. New York: Wiley.
7. Wood, A.J. and Wollenberg, B.F. (1996). *Power Generation, Operation and Control*, 2e. New York: Wiley.
8. Huang, Y., Pardalos, P.M., and Zheng, Q.P. (2017). *Electrical Power Unit Commitment: Deterministic and Two-Stage Stochastic Programming Models and Algorithms*. Springer.
9. Momoh, J.A. (2009). *Electric Power System Applications of Optimization*, 2e. CRC Press.
10. Padhy, N.P. (2000). Unit commitment using hybrid models: a comparative study for dynamic programming, expert system, fuzzy system and genetic algorithm. *International Journal of Electric Power and Energy System* 23 (8): 827–836.
11. Senjyu, T., Yamashiro, H., Shimabukuro, K. et al. (2003). Fast solution technique for large-scale unit commitment problem using genetic algorithm. *IEEE Transactions on Power Systems* 180 (6): 753–760.
12. Simon, S.P., Padhy, N.P., and Anand, R.S. (2005). Solution to unit commitment problem with spinning reserve and ramp rate constraints using ant Colony system. *Journal of Energy & Environment* 4: 21–35.
13. Wang, C. and Shahidehpour, S.M. (1993). Effects of ramp-rate limits on unit commitment and economic dispatch. *IEEE Transactions on Power Systems* 8 (3): 1341–1350.
14. Chandrasekaran, K. and Simon, S.P. (2013). Optimal deviation based firefly algorithm tuned fuzzy design for multi-objective UCP. *IEEE Transactions on Power Systems* 28 (1): 460–471.
15. Columbus, C.C., Chandrasekaran, K., and Simon, S.P. (2012). Nodal ant colony optimization for solving profit based unit commitment problem for GENCOs. *Applied Soft Computing* 12: 145–160.
16. Columbus, C.C. and Simon, S.P. (2013). Profit based unit commitment for GENCOs using parallel NACO in a distributed cluster. *Swarm and Evolutionary Computation* 10: 41–58.
17. Nick, M., Alizadeh-Mousavi, O., Cherkaoui, R., and Paolone, M. (2015). Security constrained unit commitment with dynamic thermal line rating. *IEEE Transactions on Power Systems* 37: 1–12.
18. Ma, H. and Shahidehpour, S.M. (1999). Unit commitment with transmission security and voltage constraints. *IEEE Transactions on Power Systems* 14 (2): 757–764.
19. Wang, S.J., Shahidehpour, S.M., Kirschen, D.S. et al. (1995). Short-term generation scheduling with transmission and environmental constraints using an augmented Lagrangian relaxation. *IEEE Transactions on Power Systems* 10 (3): 1294–1301.
20. Kuloor, S., Hope, G.S. and Malik, O.P. (1992). Environmentally constrained unit commitment. *IEE Proceedings C Generation, Transmission and Distribution* 139 (2): 122–128.
21. Gjengedal, T. (1996). Emission constrained unit-commitment (ECUC). *IEEE Transactions on Energy Conversion* 11 (1): 132–138.
22. Chandrasekaran, K. and Simon, S.P. (2012). Multi-objective unit commitment problem with reliability function using fuzzified binary real coded artificial bee colony algorithm. *IET Generation, Transmission & Distribution* 6 (10): 1060–1073.
23. Abido, M.A. (2006). Multiobjective evolutionary algorithms for electric power dispatch problem. *IEEE Transactions on Evolutionary Computation* 10 (3): 315–329.

24. Karthik, N., Parvathy, A.K., and Arul, R. (2019). Multi-objective economic emission dispatch using interior search algorithm. *International Transaction on Electrical Energy System* 29: e2683.
25. Chandrasekaran, K. and Simon, S.P. (2012). Network and reliability constrained unit commitment problem using binary real coded firefly algorithm. *International Journal of Electrical Power & Energy Systems* 43 (1): 921–932.
26. Chandrasekaran, K., Simon, S.P., and Padhy, N.P. (2014). SCUC problem for solar/thermal power system addressing smart grid issues using FF algorithm. *International Journal of Electrical Power & Energy Systems* 62: 450–460.
27. Lee, K.D., Day, J.T., Cooper, B.L., and Gibbons, E.W. (1983). A global optimization method for scheduling thermal generation, hydro generation and economy purchase. *IEEE Transactions on Power Systems* PAS-102: 1983–1986.
28. Nanda, J., Kothari, D.P., and Lingamurthy, K.S. (1988). Economic-emission load dispatch through goal programming techniques. *IEEE Transactions on Energy Conversion* 3: 26–32.
29. Rudolf, A. and Bayrleithner, R. (1999). A genetic algorithm for solving the unit commitment problem of a hydro-thermal power systems. *IEEE Transactions on Power Systems* 14: 1460–1468.
30. Iguchi, M. and Yamashiro, S. (2002). An efficient scheduling method for weekly hydro-thermal unit commitment. In: *IEEE Proceedings on Computers, Communications, Control and Power Engineering*, vol. 3, 1772–1777. TENCON.
31. Xu, Y.H., Wang, Z.Z., Sun, W. et al. (2012). Unit commitment model considering nuclear power plant load following. In: *The International Conference on Advanced Power System Automation and Protection*, 1828–1832. Beijing, China: IEEE.
32. Li, F. and Kuri, B. (2005). Generatrion scheduling in a system with wind power. In: *IEEE/PES Proceedings on Transmission and Distribution Conference and Exhibition: Asia and Pacific*, 1–6. IEEE.
33. Chen, C.-L. (2008). Optimal wind–thermal generating unit commitment. *IEEE Transactions on Energy Conversion* 23 (1): 273–280.
34. Rajasekaran, S. and Vijayalakshmi Pai, G.A. (2003). *Neural Networks, Fuzzy Logic, and Genetic Algorithms Synthesis and Applications*. New Delhi: Prentice-Hall of India Pvt. Ltd.
35. Ouyang, Z. and Shahidehpour, S.M. (1992). A hybrid artificial neural network-dynamic programming approach to unit commitment. *IEEE Transactions on Power Systems* 7: 236–242.
36. Balci, H.H. and Valenzuela, J.F. (2004). Scheduling electric power generators using particle swarm optimization combined with the Lagrangian relaxation method. *International Journal of Applied Mathematics and Computer Science* 14 (3): 411–421.
37. Sriyanyong, P. and Song, Y.H. (2005). Unit commitment using particle swarm optimization combined with Lagrange relaxation. *IEEE Proceedings of Power Engineering Society General Meeting* 3: 2752–2759.
38. Garver, L.L. (1963). Power generation scheduling by integer programming -development of theory. *AIEE Transactions* 2: 730–735.
39. Kerr, R.H., Scheidt, J.L., Fontana, A.J. Jr., and Wiley, J.K. (1966). Unit commitment. *IEEE Transactions on Power Apparatus and Systems* PAS-85 (5): 417–421.
40. Lowery, P.G. (1966). Generating unit commitment by dynamic programming. *IEEE Transactions on Power Apparatus and Systems* PAS-85 (5): 422–426.
41. Fisher, M.L. (1973). "Optimal solution of scheduling problems using Lagrange multipliers: part I," it operations research. *INFORMS* 21: 1114–1127.

42. Stott, B., Marinho, J.L., and Alsac, O. (1979). Review of linear programming applied to power system rescheduling. *IEEE PICA Conference Proceedings*, Cleveland, OH: 142–155.
43. Shoults, R.R., Chang, S.K., Helmick, S., and Grady, W.M. (1980). A practical approach to unit commitment, economic dispatch, and savings allocation for multiple-area Pool operation with import/export constraints. *IEEE Transactions on Power Systems* 3 (2): 625–633.
44. Lauer, G.S., Sandell, N.R. Jr., Bertsekas, D.P., and Posbergh, T.A. (1982). Solution of large-scale optimal unit commitment problems. *IEEE Transactions on Power Systems* PAS-101 (1): 79–86.
45. Mokhtari, S., Singh, J., and Wollenberg, B. (1987). A unit commitment expert system. In: *Proceedings of the PICA*, 400–405. IEEE.
46. Wang, C., Ouyang, Z., and Shahidehpour, S.M. (1990). Unit commitment by neural networks. *Proceedings of the 1990 American Power Conference* 52: 245–250.
47. Zhuang, F. and Galiana, F.D. (1990). Unit commitment by simulated annealing. *IEEE Transactions on Power Systems* PWRS-5 (1): 311–318.
48. Ouyang, Z. and Shahidehpour, S.M. (1991). Heuristic multi-area unit commitment with economic dispatch. *Proceedings of IEE, Part C* 138 (3): 242–252.
49. Sheble, G.B., Maifeld, T.T., Brittig, K. et al. (1996). Unit commitment by genetic algorithm with penalty methods and a comparison of Lagrangian search and genetic algorithm economic dispatch example. *International Journal of Electrical Power & Energy Systems* 18 (6): 339–346.
50. Yang, H.T., Yang, P.C., and Huang, C.L. (1996). Evolutionary programming based economic dispatch for units with non-smooth fuel cost functions. *IEEE Transactions on Power Systems* 11: 112–118.
51. Saneifard, S., Prasad, N.R., and Smolleck, H.A. (1997). A fuzzy logic approach to unit commitment. *IEEE Transactions on Power Systems* 12: 988–995.
52. Simon, S.P., Padhy, N.P., and Anand, R.S. An ant colony system approach for unit commitment problem. *International Journal of Electrical Power & Energy Systems* 28 (5): 315–323.
53. Van den Bosch, P.P.J. and Honderd, G. (1985). A solution of the unit commitment problem via decomposition and dynamic programming. *IEEE Transactions on Power Systems* PAS-104 (7): 1684–1690.
54. Su, C.C. and Hsu, Y.Y. (1991). Fuzzy dynamic programming: an application to unit commitment. *IEEE Transactions on Power Systems* PWRS-6 (3): 1231–1237.
55. Mantawy, A.H., Abdel-Magid, Y.L., and Selim, S.Z. (1999). Integrating genetic algoritms, Tabu search and simulated annealing for the unit commitment problem. *IEEE Transactions on Power Systems* 14: 829–836.
56. Hindsberger, X. and Ravn, H.F.M. (2001). Multiresolution modeling of hydro thermal systems. In: *Proceedings of 22nd IEEE Power Engineering Society International Conference Power Industry Computer Applications*, 5–10. IEEE.
57. Kennedy, J. and Eberhart, R. (1995). Particle swarm optimization. In: *Proceedings of the IEEE International Conference on Neural Networks*, 1942–1948. Piscataway, Perth, Australia.
58. Holland, J.H. (1975). *Adaptation in Natural and Artificial Systems*. Ann Arbor, MI: University of Michigan Press.
59. De Jong, K.A. (1980). Adaptive system design: a genetic approach. *IEEE Transactions on Systems, Man, and Cybernetics* SMC-10: 566–574.

60. Goldberg, D.E. (1989). *Genetic Algorithms in Search, Optimization, and Machine Learning*. Reading, MA: Addison-Wesley.
61. Davis, L. (ed.) (1991). *Handbook of Genetic Algorithms*. New York: Van Nostrand Reinhold.
62. Grefenstette, J.J. (1986). Optimization of control parameters for genetic algorithms. *IEEE Transactions on Systems, Man, and Cybernetics* SMC-16: 122–128.
63. Walters, D.C. and Sheble, G.B. (1993). Genetic algorithm solution of economic dispatch with valve point loading. *IEEE Transactions on Power Systems* 8 (3): 1325–1332.
64. Chen, P.H. and Chang, H.C. (1995). Large-scale economic dispatch by genetic algorithm. *IEEE Transactions on Power Systems* 10 (4): 1919–1926.
65. Park, Y.M., Won, J.R., and Park, J.B. (1998). A new approach to economic load dispatch based on improved evolutionary programming. *Engineering Intelligent Systems for Electrical Engineering and Communications* 6 (2): 103–110.
66. Iba, K. (1994). Reactive power optimization by genetic algorithm. *IEEE Transactions on Power Systems* 9 (2): 685–692.
67. Lee, K.Y., Bai, X., and Park, Y.M. (1995). Optimization method for reactive power planning by using a modified simple genetic algorithm. *IEEE Transactions on Power Systems* 10 (4): 1843–1850.
68. Park, Y.M., Park, J.B., and Won, J.R. (1996). A genetic algorithms approach for generation expansion planning optimization. In: *A Proceedings Volume of IFAC Symposium: Control of Power Plants and Power Systems 1995*, 257–262. the Netherlands: Pergamon.
69. Park, J.B., Park, Y.M., Won, J.R., and Lee, K.Y. (2000). An improved genetic algorithm for generation expansion planning. *IEEE Transactions on Power Systems* 15 (3): 916–922.
70. Lai, L.L. (1998). *Intelligent System Applications in Power Engineering: Evolutionary Programming and Neural Networks*. UK: Wiley.
71. Craig Potts, J., Giddens, T.D., and Yadav, S.B. (1994). The development and evaluation of an improved genetic algorithm based on migration and artificial selection. *IEEE Transactions on Systems, Man, and Cybernetics* 24 (1): 73–86.
72. Srinivas, M. and Patnaik, L.M. (1994). Genetic algorithms: a survey. *IEEE Computer* 27: 17–26.
73. Mitchell, M. (1996). *An Introduction to Genetic Algorithms*. The MIT Press.
74. Schaffer, J.D., Caruana, R.A., Eshelman, L.J., and Das, R. (1989). A study of control parameters affecting online performance of genetic algorithms for function optimization. In: *Proceedings of 3rd International Conference on Genetic Algorithms*, 51–60. Los Altos, CA: Morgan Kaufmann.
75. Yang, Q.W., Jiang, J.P., and Chen, G. (2000). How to select optimal control parameters for genetic algorithms. *Proceedings of the 2000 IEEE International Symposium on Industrial Electronics* 1: 37–41.
76. Renders, J.M. and Flasse, S.P. (1996). Hybrid methods using genetic algorithms for global optimization. *IEEE Transactions on Systems, Man, and Cybernetics – Part B: Cybernetics* 26 (2): 243–258.
77. Jiao, L. and Wang, L. (2000). A novel genetic algorithm based on immunity. *IEEE Transactions on Systems, Man, and Cybernetics – Part A* 30 (5): 552–561.
78. Parassuram, A., Deepa, S.N., and Karthick, M. (2011). A hybrid technique using particle swarm optimization and differential evolution to solve economic dispatch problem with valve-point effect. *2011 International Conference on Recent Advancements in Electrical, Electronics and Control Engineering* (December 2011).

79. Cao, Y.J. and Wu, Q.H. (1997). Convergence analysis of adaptive genetic algorithms. In: *Second International Conference on Genetic Algorithms in Engineering Systems: Innovations and Applications*, 85–89. IET.
80. Yamamura, M., Satoh, H., and Kobayashi, S. (1994). An analysis of crossover's effect in genetic algorithms. *IEEE World Congress on Computational Intelligence: Evolutionary Computation* 2: 613–618.
81. Wood, A.J. and Wollenberg, B.F. (1984). *Power Generation, Operation and Control.* New York, NY: Wiley.
82. Chowdhury, B.H. and Rahman, S. (1990). A review of recent advances in economic dispatch. *IEEE Transactions on Power Systems* 5 (4): 1248–1259.
83. Happ, H.H. (1977). Optimal power dispatch – a comprehensive survey. *IEEE Transactions on Power Apparatus and Systems* PAS-96 (3): 841–854.
84. El-Hawary, M.E. and Christensen, G.S. (1979). *Optimal Economic Operation of Electric Power Systems.* Academic Press.
85. Ramanathan, R. (1985). Fast economic dispatch based on the penalty factors from Newton's method. *IEEE Transactions on Power Systems* PAS-104 (7): 1624–1629.
86. Sheble, G.B. (1989). Real-time economic dispatch and reserve allocation using merit order loading and linear programming rules. *IEEE Transactions on Power Systems* 4 (4): 1414–1420.
87. Park, J.B., Won, J.R., Park, Y.M., and Lee, K.Y. (1999). Economic dispatch solutions using an improved genetic algorithm based on the multi-stage method and directional crossover. *Engineering Intelligent Systems for Electrical Engineering and Communications* 7 (4): 219–225.
88. Lin, C.E. and Viviani, G.L. (1984). Hierarchical economic dispatch for piecewise quadratic cost functions. *IEEE Transactions on Power Apparatus and Systems* PAS-103 (6): 1170–1175.
89. Park, J.H., Kim, Y.S., Eom, I.K., and Lee, K.Y. (1993). Economic load dispatch for piecewise quadratic cost function using Hopfield neural network. *IEEE Transactions on Power Systems* 8 (3): 1030–1038.
90. Sheble, G.B. and Brittig, K. (1995). Refined genetic algorithm – economic dispatch example. *IEEE Transactions on Power Systems* 10 (1): 117–124.
91. Bakirtzis, A., Petridis, V., and Kazarlis, S. (1994). Genetic algorithm solution to the economic dispatch problem. *IEE Proceedings - Generation, Transmission and Distribution* 141 (4): 377–382.
92. Song, Y.H. and Chou, C.S.V. (1997). Advanced engineered-conditioning genetic approach to power economic dispatch. *IEE Proceedings - Generation, Transmission and Distribution* 144 (3): 285–292.
93. Orero, S.O. and Irving, M.R. (1996). Economic dispatch of generators with prohibited operating zones: a genetic algorithm approach. *IEE Proceedings - Generation, Transmission and Distribution* 143 (6): 529–534.
94. Happ, H.H. (1974). Optimal power dispatch. *IEEE Transactions on Power Systems* PAS-93: 820–830.
95. Yoshida, H., Kawata, K., Fukuyama, Y. et al. (2000). A particle swarm optimization for reactive power and voltage control considering voltage security assessment. *IEEE Transactions on Power Systems* 15 (4): 1232–1239.
96. Clerc, M. and Kennedy, J. (2002). The particle swarm-explosion, stability, and convergence in a multidimensional complex space. *IEEE Transactions on Evolutionary Computation* 6 (1): 58–73.

97. Abido, M.A. (2002). Optimal design of power-system stabilizers using particle swarm optimization. *IEEE Transactions on Energy Conversion (ICNN'95)* 17 (3): 406–413.
98. Kassabalidis, I.N., El-Sharkawi, M.A., Marks, R.J. et al. (2002). Dynamic security border identification using enhanced particle swarm optimization. *IEEE Transactions on Power Systems* 17 (3): 723–729.
99. Ciuprina, G., Ioan, D., and Munteanu, I. (2002). Use of intelligent-particle swarm optimization in electromagnetics. *IEEE Transactions on Magnetics* 38 (2): 1037–1040.
100. Kennedy, J. and Eberhart, R.C. (2001). *Swarm Intelligence*. San Francisco, CA: Morgan Kaufmann.
101. Sinha, N., Chakrabarti, R., and Chattopadhyay, P.K. (2003). Evolutionary programming techniques for economic load dispatch. *IEEE Transactions on Evolutionary Computations* 7 (1): 83–94.
102. Lin, W.M., Cheng, F.S., and Tsay, M.T. (2002). An improved Tabu search for economic dispatch with multiple minima. *IEEE Transactions on Power Systems* 17 (1): 108–112.
103. Lee, K.Y., Sode-Yome, A., and Park, J.H. (1998). Adaptive Hopfield neural network for economic load dispatch. *IEEE Transactions on Power Systems* 13 (2): 519–526.
104. Park, J.B., Lee, K.S., Shin, J.R., and Lee, K.Y. (2005). A particle swarm optimization for economic dispatch with non-smooth cost functions. *IEEE Transactions on Power Systems* 20 (1): 34–42.
105. Park, J.-B., Jeong, Y.-W., Shin, J.-R., and Lee, K.Y. (2010). An improved particle swarm optimization for nonconvex economic dispatch problems. *IEEE Transactions on Power Systems* 25 (1): 156–166.
106. Baek, M.-K., Park, J.-B., and Lee, K.Y. (2016). *An Improved Attractive and Repulsive Particle Swarm Optimization for Nonconvex Economic Dispatch Problems*. IFAC Workshop on Control of Transmission and Distribution Smart Grid.
107. Caponetto, R., Foruna, L., Fazzino, S., and Xibilia, M.G. (2003). Chaotic sequences to improve to performance of evolutionary algorithms. *IEEE Transactions on Power Systems* 7 (5): 289–304.
108. Niu, G., Chen, B., and Zeng, J. (2010). Repulsive particle swarm optimization base on new diversity. In: *IEEE Chines Control and Decision Conference*, 815–825.
109. Hong-da, L., Zhong-li, M., Sheng, L., and Hai, L. (2006). A new solution to economic load dispatch using immune genetic algorithm. In: *2006 IEEE Conference on Cybernetics and Intelligent Systems*. IEEE Press.
110. Binetti, G., Naso, D., and Turchiano, B. (2015). Genetic algorithm based on the lagrange method for the non-convex economic dispatch problem. In: *2015 IEEE 20th Conference on Emerging Technologies & Factory Automation (ETFA)*. IEEE Press.
111. Cheng-Chien, K. (2010). Wind energy dispatch considering environmental and economic factors. *Renewable Energy* 35 (10): 2217–2227.
112. Mondal, S., Bhattacharya, A., and Halder, S. (2013). Multi-objective economic emission load dispatch solution using gravitational search algorithm and considering wind power penetration. *International Journal of Electrical Power and Energy Systems* 44 (1): 282–292.
113. Yao, Y. and Yu, J.L. (2011). Multi-objective hybrid optimal dispatch of power systems considering reserve risk due to wind power. *Automation of Electric Power Systems* 35 (22): 118–124.
114. Wang, K., Zhang, B.M., Zhou, Y., and Li, J.F. (2011). Real-time dispatch in wind power integrated system based on chaos quantum-behaved particle swarm optimization. *Automation of Electric Power Systems* 35 (22): 141–146.
115. Liu, Z.F. and Zhang, J.H. (2008). An improved differential evolution algorithm for solving the distribution problem of power economic load. *Proceedings of the CSEE* 28 (10): 100–105.

116. Chiou, J.P. (2009). A variable scaling hybrid differential evolution for solving large-scale power dispatch problems. *IET Generation, Transmission & Distribution* 3 (2): 154–163.
117. Robic, T. and Filipic, B. (2005). *DEMO: Differential Evolution for Multi-Objective Optimization, Lecture Notes in Computer Science*, 520–530. Berlin: Springer.
118. Niu, D.P., Wang, F.L., He, D.K., and Jia, M.R. (2009). Chaotic differential evolution for multiobjective optimization. *Control and Decision* 24 (3): 361–370.
119. Xia, S., Zhang, J.H., and Liu, Z.F. (2010). A binary adaptive differential evolution algorithm for unit commitment. *Power System Technology* 34 (4): 110–114.
120. Das, S., Abraham, A., and Konar, A. (2008). Automatic clustering using an improved differential evolution algorithm. *IEEE Transactions on Systems, Man, and Cybernetics Part A: Systems and Humans* 38 (1): 218–237.
121. Kukkonen, S. and Deb, K. (2006). Improved pruning of non-dominated solutions based on crowding distance for bi-objective optimization problems. In: *2006 IEEE Congress on Evolutionary Computation*, 1178–1186. Vancouver, Canada.
122. Hwang, C.L. and Yoon, K. (1981). *Multiple Attribute Decision Making: Methods and Applications: A State-of-the-Art Survey*. New York: Spinger-Verlag.
123. Xia, S. (2014). *Model and Method for Medium and Short-Term Power Dispatch Suitable for Power System of Wind Power Integration*. North China Electric Power University.
124. Vaisakh, K., Praveena, P., Rao, S.R.M., and Meah, K. (2012). Solving dynamic economic dispatch problem with security constraints using bacterial foraging PSO-DE algorithm. *International Journal of Electrical Power and Energy Systems* 39 (1): 56–67.
125. Li, Z.H. and Duan, X.Z. (2010). Comparison and analysis of multiobjective evolutionary algorithm for reactive power optimization. *Proceedings of the CSEE* 30 (10): 57–65.
126. Schott, R.J. (1995). *Fault Tolerant Design Using Single and Multicriteria Genetic Algorithm Optimization*. Cambridge: Massachusetts Institute of Technology.
127. Franz, A. and Zimmermann, J. (2016). A Two-Stage Heuristic Approach for Solving the Long-Term Unit Commitment Problem with Hydro-Thermal Coordination in Large-Scale Electricity Systems. In: *Proceedings of IEEE International Conference on Industrial Engineering and Engineering Management*, 70–74. Bali.
128. Zoumas, C.E., Bakirtzis, A.G., Theocharis, J.B., and Petridis, V. (2004). A genetic algorithm solution approach to the hydrothermal coordination problem. *IEEE Transactions on Power Systems* 19 (3): 1356–1364.
129. Merz, P. and Freisleben, B. (1998). On the effectiveness of evolutionary search in high-dimensional NK-landscapes. In: *Proceedings of IEEE Congress on Computational Intelligence*, 741–745. Anchorage.
130. Piekutowski, M.R., Litwinowicz, T., and Frowd, R.J. (1994). Optimal short-term scheduling for a large-scale cascaded hydro system. *IEEE Transactions on Power Systems* 9 (2): 805–811.
131. Potter, M. and De Jong, K.A. (2000). Cooperative coevolution: an architecture for evolving coadapted subcomponents. *Evolutionary Computation* 8 (1): 1–29.
132. Yang, Z. and Tang, K. (2008). Multilevel cooperative coevolution for large scale optimization. In: *Proceedings of IEEE Congress on Evolutionary Computation*, 1663–1670. Hong Kong.
133. Omidvar, M.N., Li, X., and Yao, X. (2010). Cooperative co-evolution with Delta grouping for large scale non-separable function optimization. In: *Proceedings of IEEE Congress on Evolutionary Computation*, 1–8. Barcelona.

134. Wood, A., Wollenberg, B., and Sheblé, G.B. (2014). *Power Generation, Operation and Control*, 3e, 225. Wiley.
135. Valenzuela, J. and Smith, A.E. (2002). A seeded memetic algorithm for large unit commitment problems. *Journal of Heuristics* 8 (2): 173–195.
136. Christiaanse, W.R. and Palmer, A.H. (1972). A technique for the automated scheduling of the maintenance of generating facilities. *IEEE Transactions on Power Apparatus and Systems* 1: 137–144.
137. Choi, J.S. and Song, K.Y. (1985). A study on generator maintenance scheduling. In: *Proceedings on Summer Meeting of KIEE*, 148–151.
138. Stremel, J.P., Jenkins, R.T., Babb, R.A. et al. (1980). Production costing using the cumulant method of representing the equivalent load curve. *IEEE Transactions on Power Systems* PAS-99 (5): 1947–1956.
139. Mukerji, R. and Parker, J.H. (1991). Power plant maintenance scheduling: optimizing economics and reliability. *IEEE Transactions on Power Systems* 6 (2): 476–483.
140. Satoh, T. and Nara, K. (1991). Maintenance scheduling by simulated annealing method. *IEEE Transactions on Power Systems* 6 (2): 850–857.
141. Chen, L. and Toyoda, J. (1991). Optimal generating unit maintenance scheduling for multi-area system with network constraints. *IEEE Transactions on Power Systems* 6 (3): 1168–1174.
142. Yellen, J., Al-Khamis, T.M., Vemuri, S., and Lemonidis, L. (1992). A decomposition approach to unit maintenance scheduling. *IEEE Transactions on Power Systems* 7 (2): 726–733.
143. Silva, E.L., Morozowski, M., Fonseca, L.G.S. et al. (1995). Transmission constrained maintenance scheduling of generating units: a stochastic programming approach. *IEEE Transactions on Power Systems* 10 (2): 695–701.
144. Marwali, M.K.C. and Shahidehpour, S.M. (1998). Integrated generation and transmission maintenance scheduling with network constraints. *IEEE Transactions on Power Systems* 13 (3): 1063–1068.
145. Do, D. (1999). A study on the generator maintenance scheduling based on fuzzy theory. PhD thesis, Department of Electrical Engineering Graduate School, Gyeongsang National University, Advised by Professor. Jaeseok Choi.
146. Ilic, M., Galiana, F., and Fink, L. (1998). *Power Systems Restructuring: Engineering and Economics*. Kluwer Academic Pub.
147. Wood, A.J. and Wollenberg, B.F. (1984). Transmission system effects. In: *Power Generation Operation & Control*, 239–289. Wiley.
148. Billinton, R. (1970). *Power System Reliability Evaluation*, 39–91. Gordon & Breach, Science Pub.
149. Song, G. (1984). *Analysis and Operation of Power System*, 11–43, 344–416. Dongil Press (Written in Korean).
150. Billinton, R. and Allan, R.N. (1988). *Reliability Assessment of Large Electric Power Systems*. Kluwer Academic Publishers.
151. Wang, X. and McDonald, J.R. (1994). *Modern Power System Planning*. McGraw-Hill Book Company.
152. Garver, L.L. (1972). Adjusting maintenance scheduling to Levelize risk. *IEEE Transactions on Power Systems* PAS-91: 2057–2062.
153. Zürn, H.H. and Quintana, V.H. (1975). Generator maintenance scheduling via successive approximation dynamic programming. *IEEE Transactions on Power Apparatus and Systems* 94 (2): 665–671.

154. Dopazo, J.F. and Merrill, H.M. (1975). Optimal generator maintenance scheduling using integer programming. *IEEE Transactions on Power Apparatus and Systems* 94 (2): 1537–1545.
155. Zürn, H.H. and Quintana, V.H. (1977). Several objective criteria for optimal generator preventive maintenance scheduling. *IEEE Transactions on Power Apparatus and Systems* 96 (3): 984–992.
156. Khatib, H. (1979). Maintenance scheduling of generating facilities. *IEEE Transactions on Power Systems* PAS-98: 1604–1607.
157. Takahadshi, K. (1988). *Power Systems Flexibility Principles and Means, Available Methods at the Planning Stage*. Brussels: CIGRE SC-37.
158. Arakawa, F. (1989). *Energy Security Assessment Quantitative Analysis for Flexible System Planning*. Florence: CIGRE SC-37.
159. Choi, J.S., Do, D.H., and Lee, T.I. (1995). Generator maintenance scheduling using fuzzy multi-criterion function. In: *Proceedings on Autumn Meeting of KFIS*, 131–138. Korea Fuzzy Logic and Intelligent Systems Society.
160. Choi, J.S. and Do, D.H. (1997). Development of a method for the generator maintenance scheduling using fuzzy integer programming. *Journal of KFIS* 7 (5): 77–85.
161. Zimmermann, H.J. (1986). *Fuzzy Set Theory and Its Applications*, 220–234. Boston: Kluwer-Nijhohh.
162. Sakawa, M. (1993). *Fuzzy Sets and Interactive Multi-Objective Optimization*. New York: Plenum Press.
163. Kim, H., Moon, S., Choi, J. et al. (1999). Generator maintenance scheduling considering air pollution based on the fuzzy theory. *IEEE International Fuzzy Systems Conference Proceedings* III: 1759–1764.
164. Choi, J.S. (1995). *A Study on the Generator Maintenance Scheduling Considering Load Uncertainty & Multi-Criterion Function*. EESRI, Final Report.
165. Endrenyi, J., Aboresheid, S., Allan, R.N. et al. (2001). The present status of maintenance strategies and the impact of maintenance on reliability. *IEEE Transactions on Power Systems* 16 (4): 638–646.A Report of the IEEE/PES Task Force on Impact of Maintenance Strategy on Reliability of the Reliability, Risk and Probability Applications Subcommittee
166. Lee, B.Y. and Shim, K.B. (1992). A comparative study on optimal generation maintenance scheduling with marginal maintenance cost and Levelized risk method. *KIEE* 41 (1): 9–17.
167. Khatib, H. (2003). *Economic Evaluation of Projects in the Electricity Supply Industry, IEEE Power & Energy Series 44*. MPG Books Limited.
168. Kim, H.S., Moon, S.P., Do, D.H. et al. (1998). Generator maintenance scheduling considering air pollution. In: *Proceedings on Autumn Meeting of Western Gyeongnam Area of KIEE*, 57–60. The Korean Institute of Electrical Engineers.
169. Park, J., Wu, L., Choi, J. et al. (2009). *Fuzzy Set Theory Based Flexible Generator Maintenance Scheduling*. ISME 2009.
170. Paucar, V.L. and Rider, M.J. (2001). On the use of artificial neural networks for enhanced convergence of the load flow problem in power systems. In: *Proceedings of Intelligent Systems Applications to Power Systems*, 153–158. ISAP 2001.
171. Stott, B. (1974). Review of load flow calculation methods. *Proceedings of the IEEE* 62: 916–929.
172. Elgerd, O.I. (1975). *Electric Energy Systems Theory: An Introduction*. New Delhi: McGraw-Hill Publishing Co. Ltd.

173. Acharjee, P. and Goswami, S.K. (2009). Chaotic particle swarm optimization based reliable algorithm to overcome the limitations of conventional power flow methods. *Proceedings of IEEE/PE Power Systems Conference and Exposition* PSCE '09: 1–7.
174. Alves da Silva, A.P., Ferreira, C., de Souza, A.C.Z., and Lambert-Torres, G. (1997). A new constructive ANN and its applications to electric load representation. *IEEE Transactions on Power Systems* 12 (4): 1569–1575.
175. Salomon, C.P., Coutinho, M.P., de Moraes, C.H.V. et al. (2012). Applications of genetic algorithm in power system control centers. In: *Genetic Algorithms in Applications* (ed. R. Popa). InTech.
176. Swarup, K.S. (2006). Swarm intelligence approach to the solution of optimal power flow. *Journal of the Indian Institute of Science* 86: 439–455.
177. Acharjee, P. and Goswami, S.K. (12–15 October 2008). An adaptive particle swarm optimization based method for normal and low voltage multiple load flow solutions. In: *Joint International Conference on Power System Technology and IEEE Power India Conference*, 1–6. POWERCON 2008.
178. Oumarou, I., Jiang, D., and Yijia, C. (14–16 August 2009). Particle swarm optimization applied to optimal power flow solution. In: *Proceedings of 2009 Fifth Conference on Natural Computation, ICNC '09*, vol. 3, 284–288.
179. Kumar, C. and Raju, P. (2012). Constrained optimal power flow using particle swarm optimization. *International Journal of Emerging Technology and Advanced Engineering* 2 (2): 235–241.
180. Onate Yumbla, P.E., Ramirez, J.M., and Coello Coello, C.A. (2008). Optimal power flow subject to security constraints solved with a particle swarm optimizer. *IEEE Transactions on Power Systems* 23 (1): 33–40.
181. Lambert-Torres, G., Martins, H.G., Coutinho, M.P. et al. (2008). Particle swarm optimization applied to restoration of electrical energy distribution systems. *Lecture Notes in Computer Science* 5370: 228–238.
182. Vlachogiannis, J.G. and Lee, K.Y. (2005). Determining generator contributions to transmission system using parallel vector evaluated particle swarm optimization. *IEEE Transactions on Power Systems* 20 (4): 1765–1774.
183. Yoshida, H., Kawata, K., Fukuyama, Y. et al. (2001). A particle swarm optimization for reactive power and voltage control in electric power systems considering voltage security assessment. *IEEE Transactions on Power Systems* 15: 1232–1239.
184. Hong, Y.Y., Lin, F.J., Lin, Y.C., and Hsu, F.Y. (2014). Chaotic PSO-based VAR control considering renewables using fast probabilistic power flow. *IEEE Transactions on Power Delivery* 29 (4): 1666–1674.
185. Esmin, A.A.A., Lambert-Torres, G., and Zambroni de Souza, A.C. (2005). A hybrid particle swarm optimization applied to loss power minimization. *IEEE Transactions on Power Systems* 20: 859–866.
186. AlRashidi, M.R. and El-Hawary, M.E. (2007). Hybrid particle swarm optimization approach for solving the discrete OPF problem considering the valve loading effects. *IEEE Transactions on Power Systems* 22 (4): 2030–2038.
187. Esmin, A.A.A. and Lambert-Torres, G. (2012). Application of particle swarm optimization to optimal power systems. *International Journal of Innovative Computing, Information and Control* 8 (3A): 1705–1716.
188. Niknam, T., Mojarrad, H.D., and Nayeripour, M. (2011). A new hybrid fuzzy adaptive particle swarm optimization for non-convex economic dispatch. *International Journal of Innovative Computing, Information and Control* 7 (1): 189–202.

189. Anderson, P.M. and Fouad, A.A. (2003). *Power System Control and Stability*. Hoboken, NJ: Wiley.
190. Kennedy, J. and Eberhart, R. (1997). The particle swarm: social adaptation of knowledge. In: *Proceedings of IEEE International Evolutionary Computation ICEC'97*, 303–308. IEEE Press.
191. del Valle, Y., Venayagamoorthy, G.K., Mohagheghi, S. et al. (2008). Particle swarm optimization: basic concepts, variants and applications in power systems. *IEEE Transactions on Evolutionary Computation* 12 (2): 171–195.
192. AlRashidi, M.R. and El-Hawary, M.E. (2009). A survey of particle swarm optimization applications in electric power systems. *IEEE Transactions on Evolutionary Computation* 13 (4): 913–918.
193. Beasley, D., Bull, D., and Martin, R. (1993). *An Overview of Genetic Algorithms: Part 1, Fundamentals*", Technical Report. Inter-University Committee on Computing.
194. Goldberg, D.E. and Holland, J.H. (1988). Genetic algorithms and machine learning: introduction to the special issue on genetic algorithms. *Machine Learning* 3: 95–99.
195. Lambert-Torres, G., Martins, H.G., Coutinho, M.P. et al. (2009). Comparison between PSO and GA in system restoration solution. In: *Proceedings of the 15th International Conference on Intelligent System Applications to Power Systems*. ISAP 2009.
196. Salomon, C.P., Coutinho, M.P., Lambert-Torres, G., and Ferreira, C. (2011). Hybrid particle swarm optimization with biased mutation applied to load flow computation in electrical power systems. *Lecture Notes in Computer Science* 6728: 595–605.
197. Carpentier, J. (1962). Contribution to the economic dispatch problem. *Bulletion of the Society Francaise Electronics* 3 (8): 431–447. (in French).
198. Bai, W. and Lee, K.Y. (2016). *Modified Optimal Power Flow on Storage Devices and Wind Power Integrated System*, 1–5. IEEE Power Engineering Society General Meeting.
199. Lee, K.Y., Park, Y.M., and Ortiz, J. (1985). A united approach to optimal real and reactive power dispatch. *IEEE Transactions on Power Apparatus Systems* 104: 1147–1153.
200. Gan, D., Thomas, R.J., and Zimerman, R.D. (2000). Stability-constrained optimal power flow. *IEEE Transactions on Power Systems* 15 (2): 535–540.
201. Bai, W., Eke, I., and Lee, K.Y. (2015). Improved artificial bee colony based on orthogonal learning for optimal power flow. In: *18th Intelligent Systems Application to Power Systems Conference*, 1–6. IEEE.
202. Palomin, R. and Quintana, V.H. (1986). Sparse reactive power scheduling by a penalty function – linear programming technique. *IEEE Transactions Power System* 1 (3): 31–39.
203. Burchett, R.C., Happ, H.H., and Vierath, D.R. (1984). Quadratically convergent optimal power flow. *IEEE Transactions on Power Apparatus Systems* 103 (11): 3267–3275.
204. Wei, H., Sasaki, H., Kubokawa, J., and Yokoyama, R. (2002). An interior point nonlinear programming for optimal power flow problems with a novel data structure. *IEEE Transactions on Power Systems* 13 (3): 870–877.
205. Yan, X. and Quintana, V.H. (1999). Improving an interior-point-based OPF by dynamic adjustments of step sizes and tolerances. *IEEE Transactions on Power Systems* 14 (2): 709–717.
206. Kumari, M.S. and Maheswarapu, S. (2010). Enhanced genetic algorithm based computation technique for multi-objective optimal power flow solution. *Electrical Power and Energy System* 32: 736–742.

207. Sood, Y.R. (2007). Evolutionary programming based optimal power flow and its validation for deregulated power system analysis. *International Journal of Electrical Power & Energy Systems* 29 (1): 65–75.
208. Abido, M.A. (2002). Optimal power flow using Tabu search algorithm. *Electric Power Component and System* 30: 469–483.
209. Karaboga, D. (2005). *An Idea Based on Honey Bee Swarm for Numerical Optimization*. Turkey, Tech. Rep.- TR06: Erciyes University Kayseri.
210. Karaboga, D. and Akay, B. (2009). A comparative study of artificial bee Colony algorithm. *Applied Mathematics and Computer* 214: 108–132.
211. Tasgetiren, M.F., Pan, Q.-K., Suganthan, P.N., and Chen, A.H.L. (2011). A discrete artificial bee colony algorithm for the total flowtime minimization in permutation flow shops. *Information Sciences* 181: 3459–3475.
212. Pan, Q.K., Tasgetiren, M.F., Suganthan, P.N., and Chua, T.J. (2011). A discrete artificial bee colony algorithm for the lot-streaming flow shop scheduling problem. *Information Sciences* 181: 2455–2468.
213. Cain, M.B., O'Neill, R.P., and Castillo, A. (2012). *History of Optimal Power Flow and Formulations*. Federal Energy Regulatory Commission Tech. Report.
214. Lee, K.Y. and El-Sharkawi, M.A. (eds.) (2008). *Modern Heuristic Optimization Techniques with Applications to Power Systems*. New York: IEEE Press, Wiley.
215. Fong, C.W., Asmuni, H., and McCollum, B. (2015). A hybrid swarm-based approach to university timetabling. *IEEE Transactions on Evolutionary Computer* 19 (6): 870–884.
216. Bai, W., Abedi, M.R., and Lee, K.Y. (2016). Distributed generation system control strategies with PV and fuel cell in microgrid operation. *Control Engineering Practice* 53: 184–193.
217. Adaryani, M.R. and Karami, A. (2013). Artificial bee colony algorithm for solving multi-objective optimal power flow problem. *Electrical Power and Energy System* 53: 219–230.
218. MATPOWER. http://www.pserc.cornell.edu/matpower (accessed 1 May 2017).
219. Shoults, R.R. and Sun, D.T. (1982). Optimal power flow based upon P-Q decomposition. *IEEE Transactions on Power Apparatus Systems* PAS-101: 397–405.
220. Duman, S., Güvenç, U., Sönmez, Y., and Yörükeren, N. (2012). Optimal power flow using gravitational search algorithm. *Energy Conversion and Management* 59: 86–95.
221. Khorsandi, A., Hosseinian, S.H., and Ghazanfari, A. (2013). Modified artificial bee colony algorithm based on fuzzy multi-objective technique for optimal power flow problem. *Electric Power System Research* 95: 206–213.
222. Sayah, S. and Zehar, K.H. (2008). Modified differential evolution algorithm for optimal power flow with non-smooth cost function. *Energy Conversion and Management* 49: 3036–3042.
223. Niknam, T., Narimani, M.R., Jabbari, M., and Malekpour, A.R. (2011). A modified shuffle frog leaping algorithm for multi-objective optimal power flow. *Energy* 36: 6420–6432.
224. Akay, B. and Karaboga, D. (2012). Artificial bee colony algorithm for large scale problems and engineering design optimization. *Journal of Intelligent Manufacturing* 23: 1001–1104.
225. Sivasubramani, S. and Swarup, K.S. (2011). Multi-objective search algorithm for optimal power flow problem. *Electrical Power and Energy System* 33: 745–752.
226. Erlich, I., Lee, K.Y., Rueda, J.L., and Wildenhues, S. (2014). Competition on Application of Modern Heuristic Optimization Algorithms for Solving Optimal Power Flow Problems – Problem Definitions and Implementation Guidelines (February 2014)

[Online]. http://sites.ieee.org/psace-mho/panels-and-competitions-2014-opf-problems (accessed 6 September 2019).
227. Saraswat, A. and Saini, A. (2011). Optimal reactive power dispatch by an improved real coded genetic algorithm with PCA mutation. In: *International Conference on Sustainable Energy and Intelligent Systems (SEISCON 2011), Chennai*, 310–315. IET https://doi.org/10.1049/cp.2011.0380.
228. Huang, C.-M., Chen, S.-J., Huang, Y.-C., and Yang, H.-T. (2012). Comparative study of evolutionary computation methods for active-reactive power dispatch. *IET Generation, Transmission & Distribution* 6 (7): 636–645.
229. Pham, H.V., Rueda, J.L., and Erlich, I. (2014). Online optimal control of reactive sources in wind power plants. *IEEE Transactions on Sustainable Energy* 5 (2): 608–616.
230. Zimmerman, R.D., Murillo-Sánchez, C.E., and Thomas, R.J. (2011). MATPOWER: steady-state operations, planning and analysis tools for power systems research and education. *IEEE Transactions on Power Systems* 26 (1): 12–19.
231. http://www.ee.washington.edu/research/pstca (accessed 15 August 2013).
232. Song, Y.H. and Wang, X.F. (2003). *Operation of Market-Oriented Power Systems*, 146. London: Springer Verlag.
233. Miranda, V. and Fonseca, N. (2002). EPSO – best-of-two-worlds meta-heuristic applied to power system problems. In: *Proceedings of the 2002 Congress on Evolutionary Computation*. Honolulu, HI, USA.
234. Miranda, V., Keko, H., and Duque, A. (2008). Stochastic star communication topology in evolutionary particle swarms (EPSO). *International Journal of Computer Intelligent Research* 4 (2): 105–116.
235. Miranda, V. and Alves, R. (2013). Differential evolutionary particle swarm optimization (DEEPSO): a successful hybrid. In: *Proceedings of the 2013 BRICS Congress on Computational Intelligence and 11th CBIC Brazilian Congress on Computational Intelligence*. Ipojuca, Pernanbuco, Brazil.
236. Cepeda, J.C., Rueda, J.L., Erlich, I. et al. (2015). Mean–variance mapping optimization algorithm for power system applications in DIgSILENT power factory. In: *Power Factory Applications for Power System Analysis, Power Systems*, 267–295. Switzerland: Springer International Publishing.
237. Rueda, J.L. and Erlich, I. (2015). Testing MVMO on learning-based real-parameter single objective benchmark optimization problems. In: *Proceedings of 2015 IEEE Congress on Evolutionary Computation*, 1–7. Sendai, Japan.
238. Sishaj Pulikottil Simon, National Institute of Technology Tiruchirappalli, Tamilnadu, India, http://sites.ieee.org/psace-mho (accessed 6 September 2019).
239. Pereira, M.V.F., Pinto, L.M.V.G., Oliveira, G.C., and Cunha, S.H.F. (1987). *Composite Generation-Transmission Expansion Planning*. Rio de Janeiro, Brazil, Tech. Rep. EPRI EL-5179 (2473-9): Centro de Pesquisas de Energia Elétrica (CEPEL).
240. Garver, L.L. (1970). Transmission network estimation using linear programming. *IEEE Transactions on Power Apparatus Systems* PAS-89: 1688–1697.
241. Villasana, R., Garver, L.L., and Salon, S.J. (1985). Transmission network planning using linear programming. *IEEE Transactions on Power Apparatus Systems* 104 (2): 349–356.
242. Monticelli, A., Santos, A., Pereira, M.V.F. et al. (1982). Interactive transmission network planning using least-effort criterion. *IEEE Transactions on Power Apparatus Systems* 101 (10): 3919–3925.
243. Latorre, G., Cruz, R., Areiza, J., and Villegas, A. (2003). A classification of publications and models on transmission expansion planning. *IEEE Transactions on Power Systems* 18 (2): 938–946.

244. Romero, R., Gallego, R.A., and Monticelli, A. (1996). Transmission system expansion planning by simulated annealing. *IEEE Transactions on Power Systems* 11 (1): 364–369.
245. Rider, M.J., Garcia, A.V., and Romero, R. (2007). Power system transmission network expansion planning using AC model. *IET Generation, Transmission & Distribution* 1 (5): 731–774.
246. Alhamrouni, I., Khairuddin, A., Ferdavani, A.K., and Salem, M. (2014). Transmission expansion planning using AC-based differential evolution algorithm. *IET Generation, Transmission & Distribution* 8 (10): 1637–1644.
247. Fang, R. and Hill, D.J. (2003). A new strategy for transmission expansion in competitive electricity markets. *IEEE Transactions on Power Systems* 18 (1): 374–380.
248. Choi, J., Tran, T., El-Keib, A. et al. (2005). A method for transmission system expansion planning considering probabilistic reliability criteria. *IEEE Transactions on Power Systems* 20 (3): 1606–1615.
249. Romero, R. and Monticelli, A. (1994). A hierarchical decomposition approach for transmission network expansion planning. *IEEE Transactions on Power Systems* 9 (3): 1385–1391.
250. Haffner, S., Monticelli, A., Garcia, A. et al. (2000). Branch and bound algorithm for transmission system expansion planning using a transportation model. *IEE Proceedings - Generation, Transmission and Distribution* 147 (3): 149–156.
251. Romero, R., Asada, E., Carreno, E., and Rocha, C. (2007). Constructive heuristic algorithm in branch-and-bound structure applied to transmission network expansion planning. *IET Generation Transmission & Distribution* 1 (2): 318–323.
252. Romero, R., Monticelli, A., Garcia, A., and Haffner, S. (2002). Test systems and mathematical models for transmission network expansion planning. *IEE Proceedings – Generation, Transmission and Distribution* 149 (1): 27–36.
253. Mamdani, E. and Assilian, S. (1975). An experiment in linguistic synthesis with a fuzzy logic controller. *International Journal of Man-Machine Studies* 7 (1): 1–15.
254. Silva, S.A. and Asada, E.N. (2010). Combined heuristic with fuzzy system to transmission system expansion planning. *Electric Power Systems Research* 81: 123–128.
255. Leite da Silva, A.M., Rezende, L.S., Manso, L.A.F., and Anders, G.J. (2010). Transmission expansion planning: a discussion on reliability and "N−1" security criteria. In: *2010 IEEE 11th International Conference on Probabilistic Methods Applied to Power Systems (PMAPS)*. IEEE.
256. Shrestha, G.B. and Fonseka, P.A.J. (2004). Congestion-driven transmission expansion in competitive power markets. *IEEE Transactions on Power Systems* 19 (3): 1658–1665.
257. Arroyo, J.M., Alguacil, N., and Carrión, M. (2010). A risk-based approach for transmission network expansion planning under deliberate outages. *IEEE Transactions on Power Systems* 25 (3): 1759–1766.
258. Carrión, M., Arroyo, J.M., and Alguacil, N. (2007). Vulnerability-constrained transmission expansion planning: a stochastic programming approach. *IEEE Transactions on Power Systems* 22 (4): 1436–1445.
259. Arabali, A., Ghofrani, M., Etezadi-Amoli, M. et al. (2014). A multi-objective transmission expansion planning framework in deregulated power systems with wind generation. *IEEE Transactions on Power Systems* 29 (6): 3003–3011.
260. Maghouli, P., Hosseini, S.H., Buygi, M.O., and Shahidehpour, M. (2009). A multi-objective framework for transmission expansion planning in deregulated environments. *IEEE Transactions on Power Systems* 24 (2): 1051–1061.

261. Correa, C.A., Bolanos, R., Garces, A., and Molina, A. (2013). *Multiobjective Environmental Transmission Network Expansion Planning*. Grenoble: 2013 IEEE PowerTech (POWERTECH).
262. Hemmati, R., Hooshmand, R.-A., and Khodabakhshian, A. (2013). A comprehensive review of generation and transmission expansion planning. *IET Generation Transmission & Distribution* 7 (9): 955–964.
263. Gallego, R.A., Monticelli, A., and Romero, R. (1988). Transmission system expansion planning by extended genetic algorithm. *IEE Proceedings on Generation Transmission and Distribution* 145 (3): 329–335.
264. Wen, F. and Chang, C.S. (1997). Transmission network optimal planning using the tabu search method. *Electric Power Systems Research* 42: 153–163.
265. Gallego, R.A., Monticelli, A., and Romero, R. (2000). Tabu search algorithm for network synthesis. *IEEE Transactions on Power Systems* 15 (2): 490–495.
266. Yamayee, Z., Sidenblad, K., and Yoshimura, M. (1983). A computationally efficient optimal maintenance scheduling method. *IEEE Transactions on Power Apparatus and Systems* 102: 330–338.

CHAPTER 4

POWER SYSTEM AND POWER PLANT CONTROL

4.1 INTRODUCTION

In a power system, optimization has significant role for satisfactory and reliable operation to make the system more and more economical, environment friendly, and sustainable. Optimization techniques are often used at every level of power system, i.e. in planning, operation, and control for power generation and transmission. This chapter aims to highlight applications of meta-heuristic methods to achieve optimal power system operation and control.

Voltage and frequency are the two most important parameters in an AC power system. To maintain these parameters at their desired values, proper control of the system (particularly the active and the reactive powers) is essential. It is well understood that in a power transmission system, the active and reactive powers are decoupled which allows proper control of the frequency and the voltage by controlling active and reactive powers in the system, respectively. This chapter starts with the active power and frequency control problem, and shows the application of meta-heuristic methods in this area. It is shown how the operation of various elements like onload tap changer (OLTC), automatic voltage regulator (AVR), shunt compensators, etc., can be optimized to minimize the system losses while maintaining the desired system parameters like voltages and line flows at appropriate levels.

In modern power systems, system stability, power flow limits, and many other parameters can be significantly improved using flexible AC transmission system (FACTS) devices. Optimal sizing, placement, and utilization of these devices can help in achieving the desired system operation while minimizing the costs of the system. The steady-state operation of the power system needs to be secure enough to handle various wanted scenarios. In case if the system

Applications of Modern Heuristic Optimization Methods in Power and Energy Systems,
First Edition. Edited by Kwang Y. Lee and Zita A. Vale.
© 2020 by The Institute of Electrical and Electronics Engineers, Inc.
Published 2020 by John Wiley & Sons, Inc.

security is under threat, certain corrective actions are needed. Such actions like load shedding should be only done as per requirement and excessive steps can lead to new issues or can hamper system reliability and operating costs. All these optimization issues in operation of a power system can be solved using some analytical methods or meta-heuristic methods. The optimal operation of a power system is limited not only to generation and transmission but in other stages like fuel handling systems, steam boiler systems, and utilization of electric power as well. The modeling, control, and operation of all these stages need to be accurate and optimum from the perspective of system reliability and sustainability and therefore, it can be said that almost each and every step associated with power generation, transmission, and distribution must be optimized to achieve certain objectives like minimization of costs, minimization of losses, minimization of emissions while maintaining the required system health. This chapter illustrates how various meta-heuristic methods are applied to solve these optimization problems and the benefits obtained. As almost all of these optimization problems have their own challenges, this chapter will present a comparison of performance of different methods while solving a given problem and improvements in these methods to make them suitable for various power system operation-related optimization problems.

4.2 LOAD FREQUENCY CONTROL – OPTIMIZATION AND STABILITY

Sukumar Mishra[1] and Dushyant Sharma[2]

[1]Indian Institute of Technology Delhi, New Delhi, India
[2]Indian Institute of Technology Jodhpur, Rajasthan, India

This chapter introduces the concept of load frequency control for controlling frequency and tie-line power flow for interconnected power systems following load perturbations. The chapter demonstrates the gain tuning for each area secondary (also called the supplementary) controller using conventional iterative optimization method and modern heuristic optimization (MHO) techniques. The chapter also compares the effect of optimization of various control parameters. The illustration is done for a two-area thermal power system. The chapter emphasizes on proper selection of controller gains by showing their impact on stability of the system.

4.2.1 Introduction

In interconnected power systems, power is exchanged among multiple areas over tie-lines by which they are connected. The desired operating conditions include nominal frequency and tie-line power flow. These quantities are controlled by controlling the active power generation in each control area.

Load frequency control (LFC) or the automatic generation control (AGC) plays an important role in maintaining desired system frequency and tie-line flow during normal operating conditions and also under small perturbations. The LFC analysis has been done in different kinds of systems like hydro, thermal, and gas power plants for different generation rate constraints, different types of governors, and different types of secondary (or the supplementary) controller. For LFC-relevant studies, the secondary controller gain tuning is an important aspect as the controller performance is affected by the proper selection of controller gains. Optimal controller gains provide best dynamic response for a system. Traditionally, iterative methods have been used for selection of parameters. These methods are very complex and time-consuming and may even take too many iterations to converge. Heuristic optimization methods, on the other hand, provide various advantages and are easy to implement. Literature shows various studies done on optimization of secondary controller gains using various artificial intelligent techniques like GA, bacteria foraging (BF), firefly algorithm, etc. [1–39].

This chapter describes various cost functions applicable for tuning of controllers for LFC applications and then shows which all parameters related to LFC can be optimized using modern heuristic methods and compares the LFC performance for variation in these parameters by taking a two-area thermal power system; each area having different capacities.

4.2.2 Load Frequency Control

Power mismatch caused by load change happens all the time in power systems. When power mismatch occurs, there are mainly four main phases of actions associated with LFC. These are:

- extraction of kinetic energy from the rotating masses,
- power change of loads as load natural characteristics,
- generator power change by governor primary or the droop control, and
- generator power change by the secondary controller.

When power imbalance initially occurs (say due to increase in load), the mismatch in power is completely compensated by the extraction of kinetic energy of the rotating turbine-generating system leading to decrease in frequency. The power consumed by frequency-dependent loads changes with change in frequency. Due to reduction in frequency, loads consume less power than they originally need. This action is called load natural characteristic.

Immediately following the inertial response, the primary frequency control will act to reduce the power imbalance. Primary frequency control is the fast controller through which the governor responds to change in frequency. The primary control action is sufficient to make power balance in the system but is not capable to restore system frequency. A slower system called LFC (AGC) is brought out for full restoration. LFC is a central control system which collects data of the power system and theoretically calculates real power mismatch of a certain area. These power change instructions are then dispatched to generator units to gain full balance which will bring system frequency back to nominal through integral controller. This action is normally called the secondary/supplementary controller. Generator will change its power at certain speed after receiving power change instruction from its control center [33, 40, 41].

4.2.3 Components of Active Power Control System

Other than the primary and the secondary controller associated with each generating system, turbines and speed governors form the most important part of the active power control system. The three main kinds of turbines that are mainly considered in frequency control studies are steam turbine, hydraulic turbine, and the gas turbine. In this section, we shall discuss in brief about these components of an active power control system. A detailed description of these is beyond the scope of this chapter.

Steam Turbine

A steam-based power plant may consist of three sections of turbines: high pressure (HP), intermediate pressure (IP), and low pressure (LP) turbines. These turbines are of two types: reheat and non-reheat turbines. In reheat type, the steam passes through a reheater after passing through the HP turbine before passing through the LP and IP. In non-reheat type, the steam is directly used by the other turbines without being reheated. A non-reheat turbine may be treated as a first-order transfer function given by (4.2.1) while a reheat-type turbine transfer function depends on the fraction of power developed by individual turbines [40]. Reference [31] gives a simplified transfer function for single-stage reheat turbine given by (4.2.2).

$$\frac{\Delta T_{\mathrm{m}}}{\Delta C_{\mathrm{v}}} = \frac{1}{1 + sT_{\mathrm{CH}}} \tag{4.2.1}$$

$$\frac{\Delta T_{\mathrm{m}}}{\Delta C_{\mathrm{v}}} = \frac{1 + sK_{\mathrm{r}}T_{\mathrm{r}}}{(1 + sT_{\mathrm{r}})(1 + sT_{\mathrm{t}})} \tag{4.2.2}$$

where ΔT_{m} is the change in mechanical torque, ΔC_{v} is the change in control valve position, T_{CH} is the time constant of the main inlet volumes and steam chest, T_{r} is the reheat time constant, T_{t} is the turbine time constant, and K_{r} is the steam turbine reheat constant.

Hydraulic Turbine

An ideal hydraulic turbine transfer function is given by (4.2.3) [40].

$$\frac{\Delta P_{\mathrm{m}}}{\Delta G} = \frac{(1 - sT_{\mathrm{W}})}{\left(1 + {}^{sT_{\mathrm{W}}}/_{2}\right)} \tag{4.2.3}$$

where ΔP_{m} is the change in mechanical power, ΔG is the change in gate opening, and T_{W} is the water starting time constant.

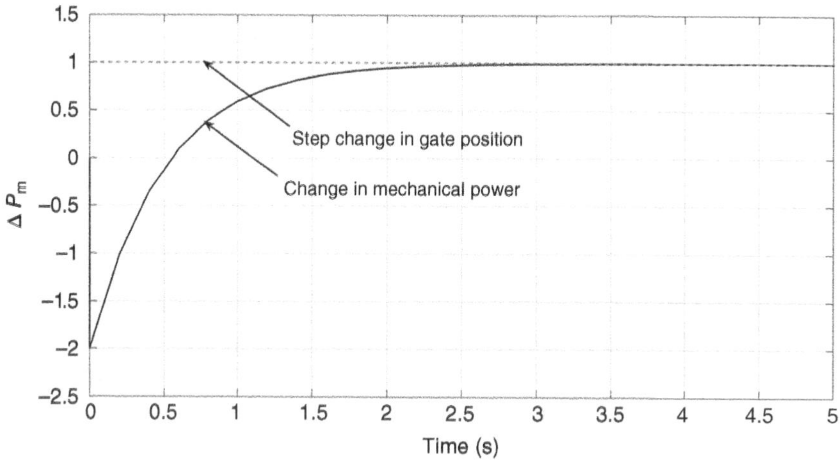

Figure 4.2.1 Turbine mechanical power variation following a step change in gate position.

For the transfer function given in (4.2.3), for a step change in gate position, the initial change in mechanical power is obtained by applying initial value theorem.

$$\Delta P_{\mathrm{m}}(0)=\lim_{s\to\infty}\left[\frac{s\left(\frac{1}{s}-T_{\mathrm{W}}\right)}{s\left(\frac{1}{s}+\frac{T_{\mathrm{W}}}{2}\right)}\right]=-2.0 \tag{4.2.4}$$

The time response for such a transfer function is given in Figure 4.2.1.

Thus, when subjected to a sudden change in gate position, rather than increasing, the turbine's output reduces by double the amount. This is due to water inertia because of which water flow does not change at the instant of change in gate position. A reduction in the pressure across the turbine is seen which leads to reduction in turbine power. This gives rise to the need of a transient droop which is explained later in this chapter.

Gas Turbine

The main constituents of a gas power plant are compressor, combustor, and gas turbine.

The compressor sucks air which is forced through the compression stage to increase the air pressure. This high-pressure air enters the combustor where it burns with the fuel and makes combustor gas at high temperature, pressure by which the gas turbine is driven.

A gas turbine control involves three controllers: speed, acceleration, and temperature. The exhaust high-temperature gas is used for a steam turbine in a combined cycle gas turbine (CCGT) power plant. In a CCGT plant, energy of the exhaust gas is collected by waste heat recovery boiler, where it produces steam by which the steam turbine is driven. The plant output power is the resultant of the power outputs of the two turbines [42, 43].

Governors

The speed governor is a transducer which controls the valve opening by means of servomotor as per the controller's instructions. Mechanical governors are widely used in most power plants. A mechanical governor can be treated as a first-order transfer function given by (4.2.5) [40].

$$\frac{\Delta C_{\mathrm{v}}}{\Delta V_{\mathrm{C}}} = \frac{1}{1 + sT_{\mathrm{G}}} \tag{4.2.5}$$

where ΔV_{C} is the controller instruction depicting the amount of change required in gate valve and T_{G} is the time constant of the mechanical governor.

In case of hydraulic turbine a temporary droop compensation having large resetting time is needed due to typical hydro turbine response to change in gate as discussed in the previous section. The transient droop is given by a transfer function:

$$G_{\mathrm{C}}(s) = \frac{(1 + sT_{\mathrm{R}})}{\left(1 + s\left(\frac{R_{\mathrm{T}}}{R_{\mathrm{P}}}\right)T_{\mathrm{R}}\right)} \tag{4.2.6}$$

where T_{R} is the reset time, R_{T} and R_{P} are the temporary and the permanent droop, respectively [40].

Electrohydraulic or electric governors for hydraulic turbines are now in operation which provides improved dynamic response. Reference [36] discusses development and operation of electric governors.

4.2.4 Designing LFC Structure for an Interconnected Power System

Obtaining the Plant Transfer Function

As discussed in Section 4.2.2, frequency deviation occurs in a power system due to mismatch in power generated by the turbines in the area and the power consumed by the loads. The change in frequency for any change (mismatch) of power can be related by a first-order transfer function obtained by linearizing the swing equation for a generator and taking the Laplace transform of the linearized equation [41].

$$\Delta P_{\mathrm{T}} - \Delta P_{\mathrm{D}} = \frac{2H}{f^0}\frac{\mathrm{d}}{\mathrm{d}t}(\Delta f) + D\Delta f \tag{4.2.7}$$

Laplace transformation of the above equation gives

$$\Delta P_{\mathrm{T}}(s) - \Delta P_{\mathrm{D}}(s) = \frac{2H}{f^0}s\Delta f(s) + D\Delta f(s) = D\left\{\frac{2H}{f^0 D}s + 1\right\}\Delta f(s) = K_{\mathrm{P}}\{T_{\mathrm{P}}s + 1\}\Delta f(s) \tag{4.2.8}$$

$$\Delta f(s) = \frac{K_{\mathrm{P}}}{1 + sT_{\mathrm{P}}}\{\Delta P_{\mathrm{T}}(s) - \Delta P_{\mathrm{D}}(s)\} \tag{4.2.9}$$

where $\Delta P_T(s)$ is the change in turbine power, $\Delta P_D(s)$ is the change in load, H is per unit inertia constant, f^0 is the initial (nominal) frequency, and D represents the frequency dependency of loads, $T_P = \frac{2H}{f^0 D}$ and $K_P = \frac{1}{D}$.

Primary Control Parameter (Speed Regulation)

The speed regulation parameter or the droop (generally denoted by R) determines the steady-state speed power characteristics of a generating unit. It is expressed in percent as a ratio of percent change in frequency to percent change in output power and is expressed as

$$R(\%) = \left(\frac{\omega_{NL} - \omega_{FL}}{\omega_0}\right) \times 100$$

where ω_{NL}, ω_{FL}, and ω_0 are steady-state speed at no load, steady-state speed at full load, and rated speed, respectively.

The primary controller regulates the valve to control the steam/water supply to the turbine to ensure that any power mismatch in the system is removed. However, the primary controller is not able to bring the frequency back to its nominal value.

Secondary Controller

The most commonly used secondary control scheme is the classical integral controller which varies the power set-point and ensures that the frequency is restored back to nominal. Many other classical controllers like PI, PID, ID, IDD, etc., were used in various literatures. In this chapter, we will concentrate on the most commonly used controller which is the integral controller only.

The combined effect of primary and secondary controller can be observed in Figure 4.2.2.

For a given R, the generator frequency and power are related to a straight line depending on the speed changer setting. The primary controller removes the power mismatch as per the characteristic set by speed changer but in this action the frequency gets deviated from the rated value. The secondary controller shifts the speed changer characteristics in such a way that rated frequency is obtained at the required power. The slope of the characteristics of the speed power response of governor changes with change in R.

Modeling the Change in Tie-line Power Flow

For the two-area system shown in Figure 4.2.3 having tie-line impedance of X_T and the bus voltages are V_1^0 and V_2^0 for area 1 and area 2, respectively, with angles δ_1^0 and δ_2^0.

The initial tie-line power flow from area 1 to area 2 (P_{12}^0) is given by

$$P_{12}^0 = \frac{|V_1^0||V_2^0|}{X_T} \sin\left(\delta_1^0 - \delta_2^0\right) \tag{4.2.10}$$

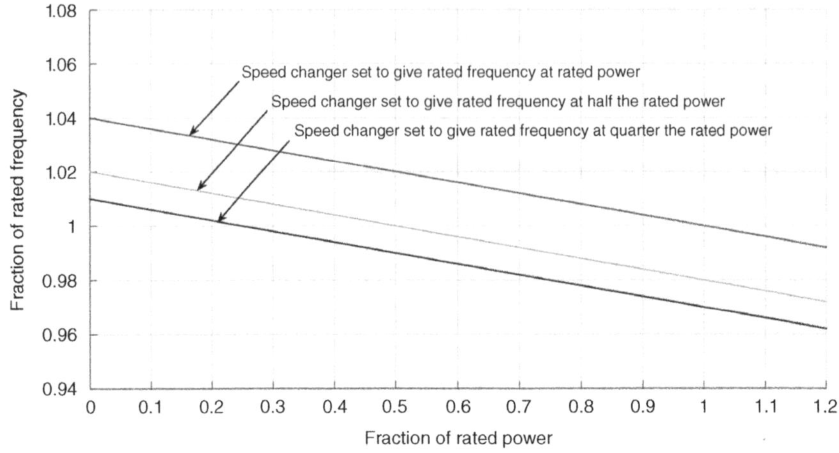

Figure 4.2.2 Speed power response of governor.

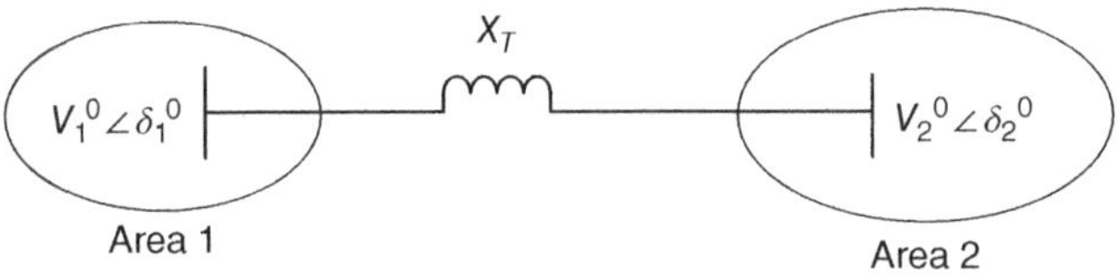

Figure 4.2.3 Two-area system.

By linearizing at the initial OPs V_1^0, V_2^0, δ_1^0, and δ_2^0, the change in power flow can be related to the change in electrical angles by the synchronizing coefficient (T^0) by [41]:

$$\Delta P_{12} = \frac{|V_1^0||V_2^0|}{X_T} \cos\left(\delta_1^0 - \delta_2^0\right)(\Delta\delta_1 - \Delta\delta_2) = T^0(\Delta\delta_1 - \Delta\delta_2) \qquad (4.2.11)$$

The term $T^0 = \frac{|V_1^0||V_2^0|}{X_T} \cos\left(\delta_1^0 - \delta_2^0\right)$ is known as the synchronizing coefficient, and (4.2.11) can also be written as

$$\Delta P_{12} = \frac{2\pi T^0}{s}\{\Delta f_1(s) - \Delta f_2(s)\} \qquad (4.2.12)$$

Frequency Bias in Secondary Control

The control signal that goes to the secondary controller in each area is known as the area control error (ACE). The ACE contains information on the frequency as well as the tie power deviation. The ACE for a two-area interconnected power system is given by

$$\text{ACE}_1 = \Delta P_{12} + B_1\Delta f_1 \qquad (4.2.13)$$

$$\text{ACE}_2 = \Delta P_{21} + B_2\Delta f_2 \qquad (4.2.14)$$

where ACE_i, ΔP_{ij}, B_i, and Δf_i represent the ACE of area i, change in tie-line flow from area i to area j, frequency bias factor in area i, and frequency deviation in area i.

It is desirable that supplementary controller action in each area should take place only for the changes occurred in the same area. To ensure this, the bias factor B_i is introduced in the secondary control loop., which is set to the composite frequency response characteristics (β_i) of the area [40].

$$B_i = \beta_i = D_i + \frac{1}{R_i} \tag{4.2.15}$$

Generation Rate Constraints

Due to mechanical and thermal stresses, the rate at which the generation (turbine power output) can be changed is limited. For thermal units it is of the order of 2–10% of maximum rating per minute while it is quite high for hydro units [31].

Modeling a Two-area Interconnected Thermal–thermal Power System

As discussed in previous sections, the LFC model of an interconnected power system comprises of governor, turbine, plant, tie power, rate limiter, and primary and secondary controller modeling. Considering a power system having two thermal areas with reheat turbine and having mechanical governors in both the areas, the system can be modeled as per (4.2.1)–(4.2.15) and is shown in Figure 4.2.4. The system parameters are given in Table 4.2.1.

The system shown in Figure 4.2.4 is studied for LFC performance by giving a 1% step load perturbation (SLP) in area 1 and 2% SLP in area 2.

Requirement of Heuristic Optimization Methods

As discussed in previous sections, the LFC model consists of a secondary integral controller. The LFC system performance depends on proper selection of the secondary controller gains. The controller is generally tuned to minimize the frequency and tie power deviations for any load change. To achieve this, the gains are selected such that an objective function (J) giving information on frequency and tie power deviations is minimized. Most commonly used objective function is the integral square error (ISE) given in (4.2.16). Other objective functions include integral time square error (ITSE) (4.2.17), weights in ISE (4.2.18) [44], etc. In further sections, we will use heuristic optimization techniques in LFC applications for optimizing the controller gains as per the different objective functions discussed in (4.2.16)–(4.2.18).

$$J_1 = \int_0^T \left(\Delta f_1^2 + \Delta f_2^2 + \Delta P_{12}^2\right) dt \tag{4.2.16}$$

$$J_2 = \int_0^T \left\{\left(\Delta f_1^2 + \Delta f_2^2 + \Delta P_{12}^2\right) \times t\right\} dt \tag{4.2.17}$$

$$J_3 = \int_0^T \left(\Delta f_1^2 + \Delta f_2^2 + W_{p12}\Delta P_{12}^2\right) dt \tag{4.2.18}$$

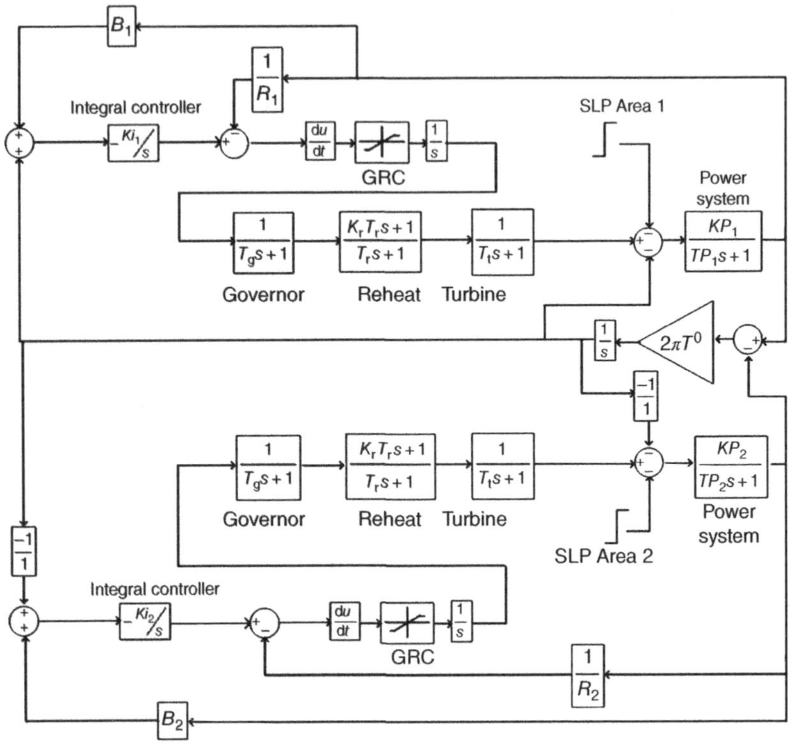

Figure 4.2.4 Two-area thermal power system.

TABLE 4.2.1 System Parameters of a Two-area Thermal–thermal Power System

Description	Symbol	Value
Nominal speed regulation parameter	R_1, R_2	2.4 Hz/pu MW
Hydraulic governor time constant	T_g	0.08 s
Thermal unit reheat steam turbine parameters	T_r	10 s
	T_t	0.3 s
	K_r	0.5
Thermal power system gain	KP_1, KP_2	120
Thermal power system time constant	TP_1, TP_2	20

where T is the simulation period, t is time, and W_{p12} is the weight given to tie power deviation.

Optimization of controller parameters is traditionally done by an iterative method as described below:

1. Define lower and upper limit for each parameter.
2. Define step size and tolerance.

3. Repeat Steps 3.1–3.4 for each parameter.
   3.1 Select a parameter.
   3.2 Keeping all other parameters constant, vary the selected parameter in steps from its minimum to maximum.
   3.3 For each set of values, obtain the objective function.
   3.4 Set the parameters to the values which give minimum function value.
4. Repeat Step 3 till variation in each parameter is less than the tolerance.

This method can lead to an optimal set of values. However, this comes with various limitations. The numerical complexity increases as the number of parameters increase. This may even lead to divergence and a converged optimal solution may not be obtained. These limitations demand simultaneous and random search of optimal set of controller parameters. Thus, MHO methods are of great use for parameter optimization for LFC-relevant studies.

Frequency of LFC Operation

For satisfactory operation of LFC, the ACE is sent to the secondary controller every 2–4 seconds. In this chapter, the results presented are obtained at 4 seconds ACE operation.

4.2.5 Parameter Optimization and System Performance

Optimization of Secondary Controller Gains and System Performance

The most important aspect for satisfactory operation of LFC is the optimization of secondary controller gains. In this section, we will observe the impact of the different objective functions presented earlier on the system performance. The gains are optimized using PSO while the other system parameters are kept constant as per the values specified in Table 4.2.1.

PSO is a stochastic swarm intelligence algorithm for optimization problem which is based on the inspiration of social behavior of bird flocks and is detailed in [45]. In PSO, the potential solutions, called particles, fly through the problem space by following the current optimum particles. The position of a particle is influenced by its own best position P_{best} allowing the local search in the neighborhood as well as the global best solution found by all the particles in the entire population G_{best}. The particle position x_i in the kth iteration is adjusted using (4.2.19).

$$x_i^{k+1} = x_i^k + v_i^{k+1} \tag{4.2.19}$$

where v_i is the velocity component with which the particle is moving. The velocity component is calculated by using

$$v_i^{k+1} = wv_i^k + c_1 r_1 \left(P_{\text{best}i} - x_i^k\right) + c_2 r_2 \left(G_{\text{best}} - x_i^k\right) \tag{4.2.20}$$

where c_1 and c_2 are the acceleration coefficients which provide the acceleration in the direction of G_{best} and P_{best}, w is the inertia weight, r_1, r_2 are uniform random

TABLE 4.2.2 Optimized Controller Gains with Different Cost Functions

Cost Function Controller Gain	ISE	ITSE	Weighted ISE
Ki_1	0.069	0.101	0.071
Ki_2	0.765	0.089	0.157

numbers between (0, 1), P_{besti} is the particle i respective best position, and G_{best} is the global best position among all particles in the entire population.

The optimization here is equivalent to finding a set of optimal controller gains over a range that minimizes an objective function. The possible objective functions are given in (4.2.16)–(4.2.18). The optimization problem is to minimize J_1, J_2 or J_3. Each objective function will give different sets of controller gains which are tabulated in Table 4.2.2. A set of optimum controller gains is obtained in the range 0–1. The tie-line power weight W_{p12} is kept at 50 in this case.

Using PSO, the secondary controller gains are optimized by the following iterative method:

1. Initialize all the particles such that all set of gains or the parameters to be optimized within their limits.
2. For each particle, obtain the cost function (called the fitness value) using any one of the objective functions J_1, J_2, or J_3, from the simulation of the system for the duration T.
3. For first iteration, assign the so-obtained fitness value and the gains as the best ones while for other iterations, check for each particle if the current fitness value is better than the P_{besti} or not.
4. If current fitness value is better, then assign current fitness value to $P_{besti,}$ otherwise keep the previous one.
5. Compare best particle's P_{besti} to G_{best} and replace G_{best}, if $P_{besti} > G_{best}$.
6. Update velocity and position of each particle using (4.2.20) and (4.2.19) respectively.
7. Check for convergence by using suitable criteria which can either be the maximum number of iterations or the condition when all the particles reach the same point.
8. If convergence is achieved, select the gains corresponding to G_{best} as the optimum gains; or else go to Step 2.

The system response and LFC results with optimized gains are shown in Figure 4.2.5 and are discussed in this section.

Following a SLP of 1% in area 1 and 2% in area 2, the frequency response is nearly same for all the three types of cost functions. However, tie power response is improved when weighted cost function is used for optimization. The weight given to tie power response helps in improving the tie power response in the interconnected power system.

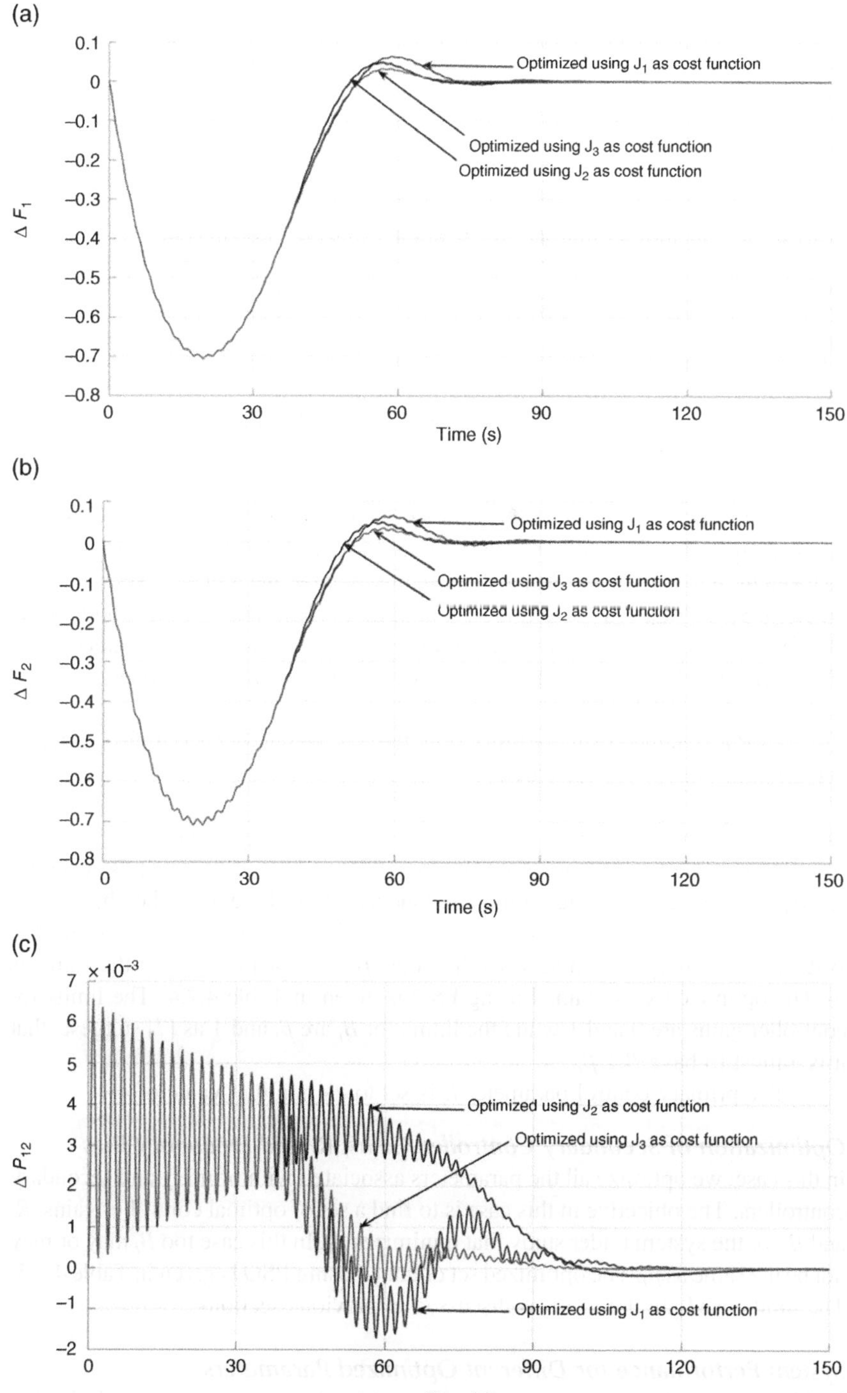

Figure 4.2.5 System performance with different cost functions.

TABLE 4.2.3 Optimized Controller Gains and *R* Using ISE Cost Function

Parameter	Ki_1	Ki_2	R_1	R_2
Optimized value	0.101	0.088	2.997	3.542

TABLE 4.2.4 Optimized Controller Gains and B_i Using ISE Cost Function

Parameter	Ki_1	Ki_2	B_1	B_2
Optimized value	0.037	0.464	0.794	0.790

Optimization of Secondary Controller Gains and Speed Regulation Parameter

Other than the gains, few other parameters can be obtained using optimization. One such optimization is to find the values of R, which is generally set to 4% in most cases, along with the secondary controller gains. Any of the objective functions can be taken in this case. In all further study in this chapter, we will use ISE as the cost function. So the objective is to find a set of optimal controller gains and optimal set of R for the system under study that minimizes J_1. The optimized set obtained using PSO is given in Table 4.2.3. The upper and lower limits for the controller gains and R is set to 0–1 and 4–10%.

As R is obtained using optimization, the B_i given in (4.2.15) is also varied as per the optimized R for each area.

Optimization of Secondary Controller Gains and Frequency Bias

Another possible optimization could be to find the values of B_i, which is generally set to β_i in most cases, along with the secondary controller gains. The objective in this case is to find a set of optimal controller gains and optimal set of B_i for the system under study that minimizes J_1. Thus, B_i may or may not be the same as β_i. The optimized set obtained using PSO is given in Table 4.2.4. The limits for controller gains are 0 and 1 while the limits for B_i are β_i and 1 as [7] suggests that it is a must to have $B_i \geq \beta_i$.

The primary control parameter R is set to its nominal value of 4%.

Optimization of Secondary Controller Gains, R*, and Frequency Bias*

In this case, we optimize all the parameters associated with primary and secondary controllers. The objective in this case is to find a set of optimal controller gains, R, and B_i for the system under study that minimizes J_1. In this case too B_i may or may not be the same as β_i. The optimized set obtained using PSO is given in Table 4.2.5. The limits used are the same as discussed in previous sections.

System Performance for Different Optimized Parameters

In this case the LFC performance is compared for the various sets of optimization presented in Sections "Optimization of Secondary Controller Gains and System Performance" to "System Performance for Different Optimized Parameter." The results are shown in Figure 4.2.6.

TABLE 4.2.5 Optimized Controller Gains, *R*, and B_i Using ISE Cost Function

Parameter	Ki_1	Ki_2	R_1	R_2	B_1	B_2
Optimized value	0.037	0.066	2.711	2.779	0.983	0.502

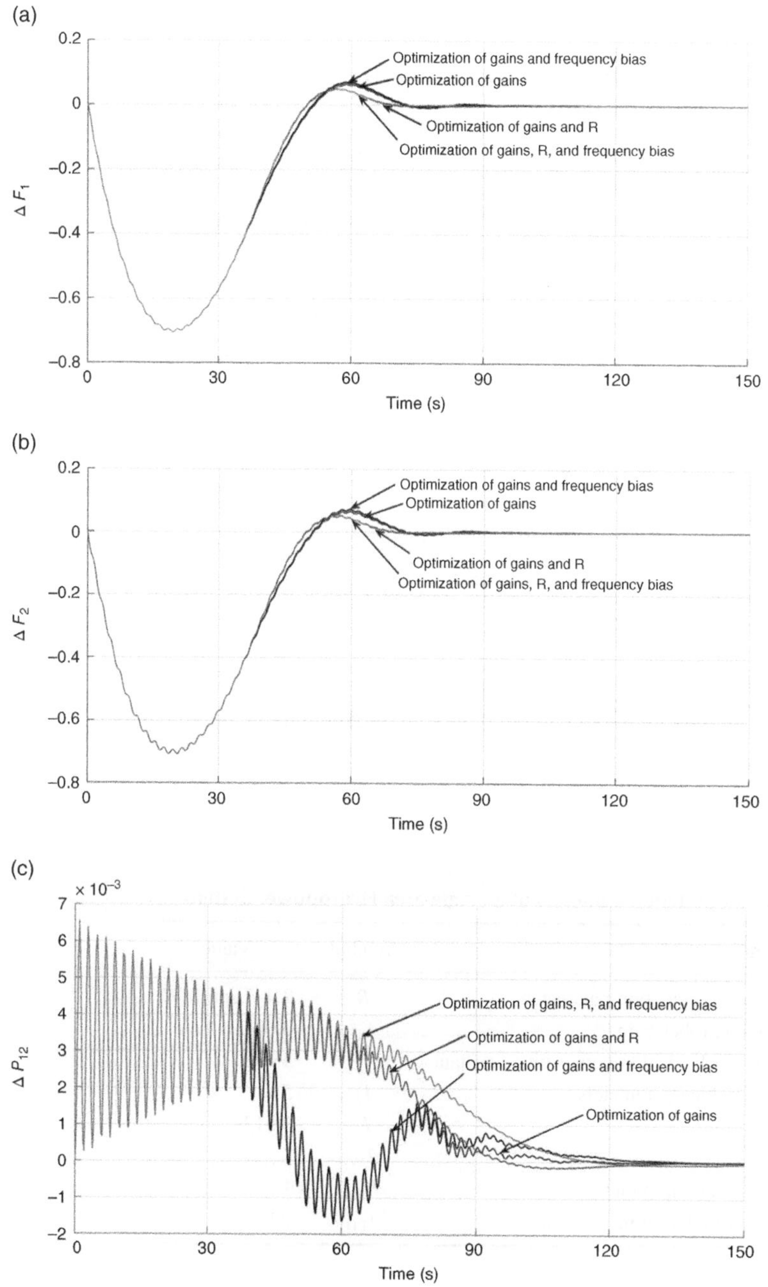

Figure 4.2.6 System performance with different parameter optimization.

The different optimization gives almost same dynamic responses; the best is obtained when all the parameters are simultaneously optimized. In this case the system observes minimum tie power oscillations. Further advantage obtained with simultaneous optimization of all parameters is that there is no necessity of keeping strictly lower R (4%).

4.2.6 System Stability in the Presence of Communication Delay

Earlier, dedicated communication networks (with minimum time delay) were used for information transfer for LFC, but with more use of open and distributed channel communication the time delay in AGC systems has increased up to 15 seconds [46–48]. The impact of time delay in the LFC execution is studied for a single-area hydro system shown in Figure 4.2.7 whose parameters are given in Table 4.2.6. It is observed that time delay affects the frequency regulation of the system as shown in Figure 4.2.8. The results are obtained for a step load change at 50 seconds with secondary controller gain fixed at 0.04.

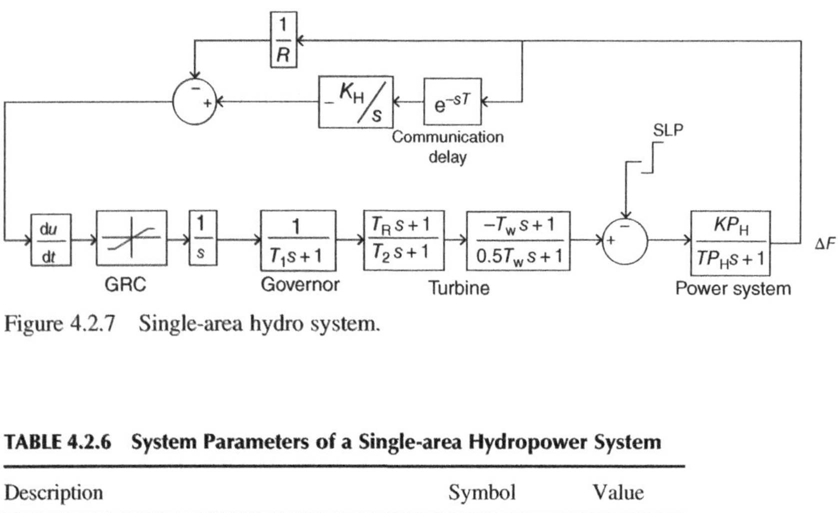

Figure 4.2.7 Single-area hydro system.

TABLE 4.2.6 System Parameters of a Single-area Hydropower System

Description	Symbol	Value
Nominal speed regulation parameter	R	2.4 Hz/pu MW
Secondary controller gain	K_H	0.04
Hydro unit mechanical governor time constant	T_1	48.75 s
Hydro unit turbine parameters	T_R	5 s
	T_2	0.513 s
	T_W	1 s
Hydro power system gain	K_{PH}	120
Hydro power system time constant	T_{PH}	20

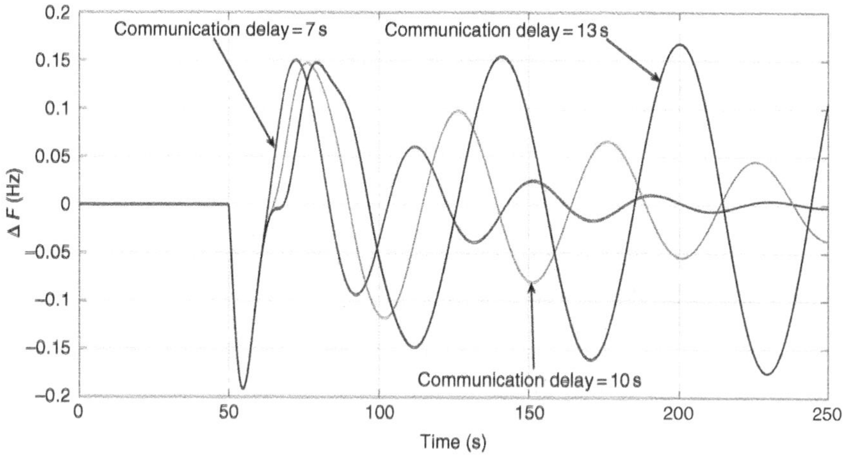

Figure 4.2.8 Impact of communication delay on frequency.

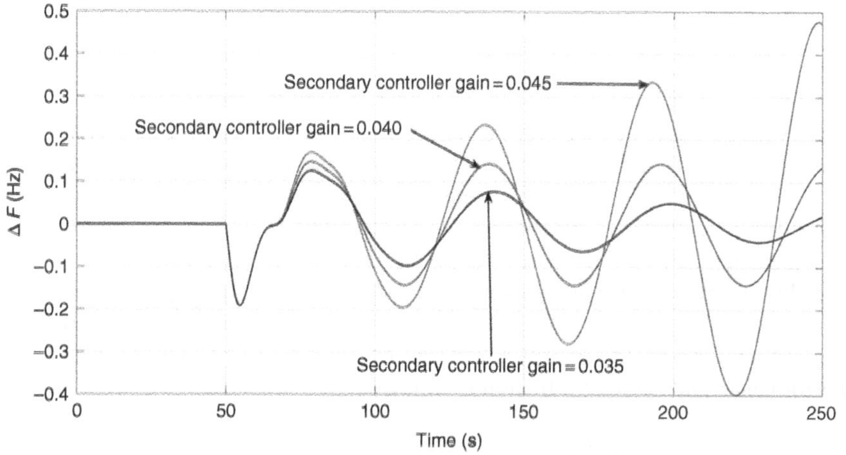

Figure 4.2.9 Impact of secondary controller gain in the presence of communication delay.

It is important to note from the figure that as the time delay increases, the system settling time also increases and beyond a particular time delay (called the critical time delay/delay margin) the system becomes unstable. Various methods of finding the delay margin can be found in [49–57] and their discussion is beyond the scope of this book.

An important result obtained for system with communication delay is that the stability of the system depends quite heavily on the secondary controller gain. The same is shown in Figure 4.2.9 and Table 4.2.7. In Figure 4.2.9, the impact of controller gain at a communication delay of 12.435 seconds on system frequency is observed. It is seen that while the system is marginally stable when gain is 0.040, it

TABLE 4.2.7 Variation of Communication Delay as a Function of Secondary Controller Gain

Secondary controller gain	0.035	0.040	0.045
Delay margin (s)	15.323	12.435	10.003

is stable when gain is 0.035 and unstable when gain is 0.045. In order to observe the effect of controller gain on delay margin, the variation of delay margin with controller gain is tabulated in Table 4.2.7.

Thus, proper selection of secondary controller through optimization plays a vital role in satisfactory operation of the system. Therefore, in this modern internet age, where the application of open channel communication is increasing, it is required that optimization of different parameters be carried out with added constraints keeping in view the stability of the system in the presence of large communication delays.

4.2.7 Conclusions

This chapter provides the basics of LFC and shows how MHO tools can be applied to solve LFC problems. Primary and secondary controller parameter tuning can be done with iterative methods but the iterative method provides various drawbacks which are overcome by the use of heuristic optimization tools like the PSO.

ISE, ITSE, weighted ISE, or any other cost function can be chosen as the objective. The objective function is chosen such that it contains the information on frequency and tie-line power deviations. The use of weighted ISE (in which weights are given in tie-line power deviations) improves the tie power response. It is also seen that not only gains but many other parameters associated with primary and secondary control action can be optimized to obtain the best LFC performance and achieving flexibility in the control structure.

4.3 CONTROL OF FACTS DEVICES

S. N. Singh

Madan Mohan Malaviya University of Technology, Gorakhpur, India

4.3.1 Introduction

The limitations of a power transmission network arising from environmental, right-of-way, and cost problems are fundamental to both bundled and unbundled power systems. Patterns of generation that result in heavy flows tend to incur greater losses, and to threaten stability and security, ultimately make certain generation patterns economically undesirable. Hence, there is an interest in better utilization of available power system capacities by installing new devices such as FACTS. The concepts of FACTS were first defined by Hingorani, N. G. in 1988 [58]. The FACTS controllers can control the interrelated parameters that govern the operation of transmission systems including series impedance, shunt impedance, current, voltage, phase angle, and the damping of oscillations at various frequencies below the rated frequency. By providing added flexibility, FACTS controllers enable a line to carry power closer to its thermal rating. Mechanical switching needs to be supplemented by rapid-response power electronics. It must be emphasized that FACTS is an enabling technology, and not a one-on-one substitute for mechanical switches.

FACTS devices [59–63] provide new control facilities, both in steady-state power flow control and dynamic stability control. The possibility of controlling power flow in an electric power system without generation rescheduling or topological changes can improve the performance considerably. Using controllable components such as controllable series capacitors, phase shifters, and unified power flow controllers (UPFCs) [64–68], line flows can be changed in such a way that thermal limits are not violated, losses minimized, stability margin increased, contractual requirement fulfilled, etc., without violating specified power dispatch [69, 70]. The increased interest in these devices is essentially due to two reasons. Firstly, the recent development in high-power electronics has made these devices cost effective and secondly, increased loading of power systems, combined with deregulation of power industry, motivates the use of power flow control as a very cost-effective means of dispatching specified power transactions. Several emerging issues in competitive power market [71], namely, as congestion management, enhancement of security and available transfer capability (ATC) of the system, transmission pricing, etc., have been restricting the free and fair trade of electricity in the open power market. FACTS devices can play a major role in these issues. Moreover, it is important to ascertain the location for placement of these devices because of their considerable costs.

The electric utilities for transmitting and distributing power are entering into a period of change. Competition has been introduced in power systems around the world based on the premise that it will increase the efficiency of this industrial sector and reduce the cost of electrical energy of consumers. Power systems, around the world, have been forced to operate in almost their full capacities due to the environmental and/or economical constraints to build new generation centers and transmission lines. The amount of electric power that can be transmitted between two locations through a transmission network is limited by security constraints. Power flows should not be allowed to increase to a level where a random event could cause the network to collapse because of angular instability, voltages instability, or cascaded outages. When such a limit is reached, the system is said to be congested. Managing congestion to minimize the restrictions of the transmission network in the market of electrical energy has thus become the central activity of power system operators.

FACTS are the name given to the application of power electronics devices to the control of flows and other quantities in the power system. Semiconductor technology enabled the manufacture of powerful thyristors and later of new elements such as the gate turnoff (GTO) thyristors and insulated-gate bipolar transistors (IGBT) [71]. Development based on the semiconductor devices first established high-voltage DC (HVDC) transmission technology as an alternative to long-distance AC transmission. HVDC technology, in turn, has provided the basis for the development of FACTS equipment, which can solve problem in AC transmission.

4.3.2 Role of FACTS

The FACTS is akin to high-voltage DC and related thyristor developments, designed to overcome the limitations of the present mechanically controlled AC power transmission systems. By using reliable and high-speed power electronic controllers, the technology offers six opportunities for increased efficiency of utilities.

1. Greater control of power so that it flows on the prescribed transmission routes.
2. Secure loading of transmission lines to levels nearer their thermal limits.
3. Greater ability to transfer between controlled areas.
4. Prevention of cascading outages.
5. Damping of power system oscillation.
6. Renewable energy sources interconnection to the existing grid.

The driving force for new and more cost-effective FACTS equipment is the development of semiconductor devices. The most powerful are thyristors, which can have a blocking capability of more than 10 kV and carry current up to 4 kA. However, the GTO devices offer additional advantage for interrupting the current. These devices permit the use of forced commutated converters, which are advantageous in building FACTS equipment with more advanced characteristics. The IGBT devices are used for converters in the lower rating ranges, mainly to be used in medium and low-voltage network. The advantage of these devices is that they allow switching frequencies in the higher range. The idea of FACTS is explained in Figure 4.3.1, which shows schematic diagram of an AC interconnection between two systems. The active power transmitted between the systems (lossless) is defined by the given equation where U_1 and U_2 are the voltages at both ends of the transmission. X is the equivalent impedance of the transmission, and $\delta_1 - \delta_2$ is the phase angle difference between both systems. From the equation, it is evident that the transmitted power is influenced by three parameters: voltage, impedance, and voltage angle difference. FACTS devices can influence one or more of these parameters as shown in figure, and thereby control the power flow.

Advantages of FACTS controllers in an AC system are shown in Table 4.3.1 and have been realized or are still under development for application. They can be used for load flow control, voltage control, and stability improvement in transmission system as well as for additional special applications. The advantage of

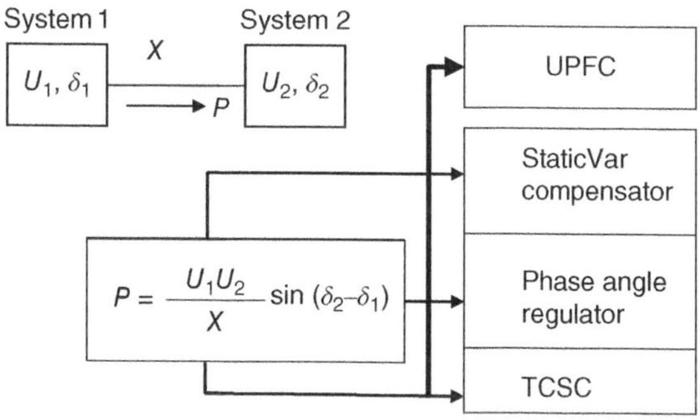

Figure 4.3.1 Representation of different controllers.

TABLE 4.3.1 Comparison of Advantages of FACTS

Device Name	Load Flow Control	Voltage Control	Transient Stability	Oscillation Damping
SVC	*	***	*	**
TCSC	**	*	***	**
SSSC	***	**	***	**
UPFC	***	***	***	***

* small; ** medium; *** strong

FACTS is that combining a variety of different equipment can create different new members of the FACTS family.

After years of rapid development, many types of FACTS devices have been put forward. Some of them are already brought into operation or being constructed. However, much work is still needed to fully make the use of the new characteristics of FACTS devices.

4.3.3 Static Modeling of FACTS devices

The FACTS technology is new, except for the static VAR compensators (SVCs). However, given the proven commercially available state-of-the-art, FACTS technology can provide services related to stability, voltage control, and network loading control. It can also enhance available transmission capacity for given transmission facilities or upgraded and new transmission facilities. Potentially for the future, if economically viable short-time storage becomes available, it can also serve the functions of rapid frequency control and may be even system restarts, but this is somewhat down the road and we must rely on generation to cover for the frequency control and system restarts.

Figure 4.3.2 shows a simple transmission line represented by its lumped π-equivalent parameters connected between bus-i and bus-j. Let complex voltages

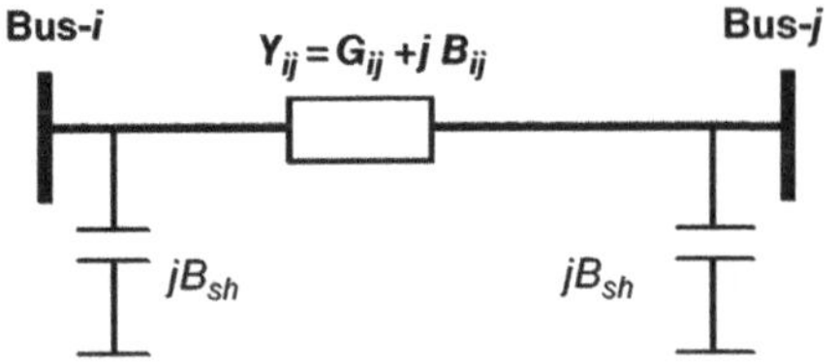

Figure 4.3.2 Model of transmission line.

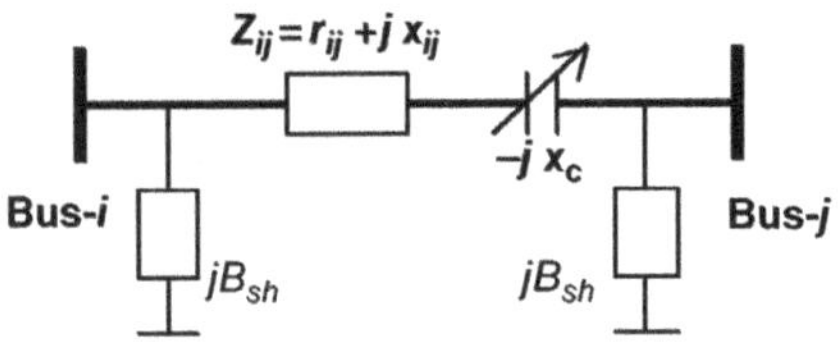

Figure 4.3.3 Model of TCSC.

at bus-i and bus-j be $V_i\angle\delta_i$ and $V_j\angle\delta_j$, respectively. The real and reactive power flows from bus-i to bus-j (P_{ij} and Q_{ij}) can be written as

$$P_{ij} = V_i^2 G_{ij} - V_i V_j \left[G_{ij}\cos\left(\delta_{ij}\right) + B_{ij}\sin\left(\delta_{ij}\right)\right] \tag{4.3.1}$$

$$Q_{ij} = -V_i^2\left(B_{ij} + B_{sh}\right) - V_i V_j \left[G_{ij}\sin\left(\delta_{ij}\right) - B_{ij}\cos\left(\delta_{ij}\right)\right] \tag{4.3.2}$$

where $\delta_{ij} = \delta_i - \delta_j$. Similarly, the real and reactive power flow from bus-j to bus-i (P_{ji} and Q_{ji}) is

$$P_{ji} = V_j^2 G_{ij} - V_i V_j \left[G_{ij}\cos\left(\delta_{ij}\right) - B_{ij}\sin\left(\delta_{ij}\right)\right] \tag{4.3.3}$$

$$Q_{ji} = -V_j^2\left(B_{ij} + B_{sh}\right) + V_i V_j \left[G_{ij}\sin\left(\delta_{ij}\right) + B_{ij}\cos\left(\delta_{ij}\right)\right] \tag{4.3.4}$$

Thyristor-Controlled Series Compensator

The model of a transmission line with a thyristor-controlled series compensator (TCSC) connected between bus-i and bus-j is shown in Figure 4.3.3 [72, 73]. During the steady state the TCSC can be considered as a static reactance $-jx_c$. The real and reactive power flows from bus-i to bus-j (P_{ij}^c and Q_{ij}^c), and from bus-j to bus-i (P_{ij}^c and Q_{ij}^c) of a line having series impedance z_{ij} ($=r_{ij} + j\,x_{ij}$) and a series reactance ($-jx_c$) are:

$$P_{ij}^c = V_i^2 G'_{ij} - V_i V_j \left(G'_{ij}\cos\delta_{ij} + B'_{ij}\sin\delta_{ij}\right) \tag{4.3.5}$$

$$Q_{ij}^c = -V_i^2\left(B'_{ij} + B_{\mathrm{sh}}\right) - V_i V_j \left[G'_{ij}\sin\left(\delta_{ij}\right) - B'_{ij}\cos\left(\delta_{ij}\right)\right] \tag{4.3.6}$$

$$P_{ji}^c = V_j^2 G'_{ij} - V_i V_j \left(G'_{ij}\cos\delta_{ij} - B'_{ij}\sin\delta_{ij}\right) \tag{4.3.7}$$

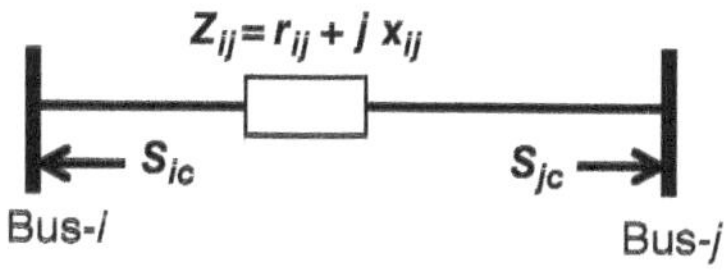

Figure 4.3.4 Injection model of TCSC.

$$Q_{ji}^c = -V_j^2\left(B'_{ij} + B_{sh}\right) + V_iV_j\left[G'_{ij}\sin\left(\delta_{ij}\right) + B'_{ij}\cos\left(\delta_{ij}\right)\right] \tag{4.3.8}$$

The active power loss (P_{Lk}^c) in line-k connected between bus-i and bus-j is

$$P_{Lk}^c = V_i^2 G'_{ij} + V_j^2 G'_{ij} - 2V_iV_jG'_{ij}\cos\delta_{ij} \tag{4.3.9}$$

where

$$G'_{ij} = \frac{r_{ij}}{r_{ij}^2 + \left(x_{ij} - x_c\right)^2} \quad \text{and} \quad B'_{ij} = \frac{-\left(x_{ij} - x_c\right)}{r_{ij}^2 + \left(x_{ij} - x_c\right)^2}$$

The change in the line flow due to series capacitance can be represented as a line without series capacitance with power injected at the receiving and sending ends of the line as shown in Figure 4.3.4. The real power injections at bus-i (P_{ic}) and bus-j (P_{jc}) can be expressed, using Eqs. (4.3.1)–(4.3.8), as

$$P_{ic} = V_i^2\Delta G_{ij} - V_iV_j\left[\Delta G_{ij}\cos\delta_{ij} + \Delta B_{ij}\sin\delta_{ij}\right] \tag{4.3.10}$$

$$P_{jc} = V_j^2\Delta G_{ij} - V_iV_j\left[\Delta G_{ij}\cos\delta_{ij} - \Delta B_{ij}\sin\delta_{ij}\right] \tag{4.3.11}$$

Similarly, the reactive power injections at bus-i (Q_{ic}) and bus-j (Q_{jc}) can be expressed as

$$Q_{ic} = -V_i^2\Delta B_{ij} - V_iV_j\left[\Delta G_{ij}\sin\delta_{ij} - \Delta B_{ij}\cos\delta_{ij}\right] \tag{4.3.12}$$

$$Q_{jc} = -V_j^2\Delta B_{ij} + V_iV_j\left[\Delta G_{ij}\sin\delta_{ij} + \Delta B_{ij}\cos\delta_{ij}\right] \tag{4.3.13}$$

where

$$\Delta G_{ij} = \frac{x_c r_{ij}\left(x_c - 2x_{ij}\right)}{\left(r_{ij}^2 + x_{ij}^2\right)\left(r_{ij}^2 + \left(x_{ij} - x_c\right)^2\right)} \quad \text{and} \quad \Delta B_{ij} = \frac{-x_c\left(r_{ij}^2 - x_{ij}^2 + x_c x_{ij}\right)}{\left(r_{ij}^2 + x_{ij}^2\right)\left(r_{ij}^2 + \left(x_{ij} - x_c\right)^2\right)}$$

Thyristor Controlled Phase Angle Regulator

The static model of a thyristor-controlled phase angle regulator (TCPAR) and transmission line between bus-i and bus-j is shown in Figure 4.3.5. The real and reactive power flow equations with TCPAR are

$$P_{ij}^s = V_i^2T^2G_{ij} - V_iV_jT\left[G_{ij}\cos\left(\delta_{ij} + \phi\right) + B_{ij}\sin\left(\delta_{ij} + \phi\right)\right] \tag{4.3.14}$$

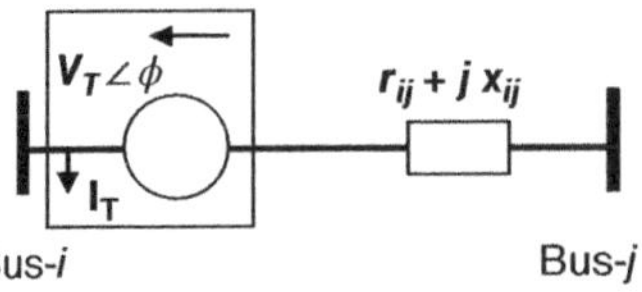

Figure 4.3.5 Equivalent circuit of TCPAR.

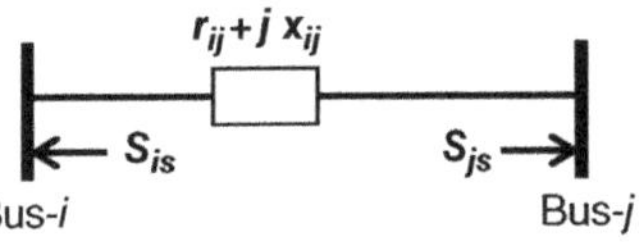

Figure 4.3.6 Injection model of TCPAR.

$$Q_{ij}^s = -V_i^2T^2B_{ij} - V_iV_jT\left[G_{ij}\sin\left(\delta_{ij}+\phi\right) - B_{ij}\cos\left(\delta_{ij}+\phi\right)\right] \tag{4.3.15}$$

$$P_{ji}^s = V_j^2G_{ij} - V_iV_jT\left[G_{ij}\cos\left(\delta_{ij}+\phi\right) - B_{ij}\sin\left(\delta_{ij}+\phi\right)\right] \tag{4.3.16}$$

$$Q_{ji}^s = -V_j^2B_{ij} + V_iV_jT\left[G_{ij}\sin\left(\delta_{ij}+\phi\right) + B_{ij}\cos\left(\delta_{ij}+\phi\right)\right] \tag{4.3.17}$$

where $T = \sec\phi$. The real power loss (P_{Lk}^s) in the line having TCPAR can be expressed as

$$P_{Lk}^s = V_i^2T^2G_{ij} + V_j^2G_{ij} - 2\,V_iV_jTG_{ij}\cos\left(\delta_{ij}+\phi\right) \tag{4.3.18}$$

From basic circuit theory, the injected equivalent circuit of Figure 4.3.6 can be obtained. The injected active power at bus-i (P_{is}) and bus-j (P_{js}), and reactive powers (Q_{is} and Q_{js}) of a line having a phase shifter are

$$P_{is} = -V_i^2K^2G_{ij} - V_iV_jK\left[G_{ij}\sin\delta_{ij} - B_{ij}\cos\delta_{ij}\right] \tag{4.3.19}$$

$$P_{js} = -V_iV_jK\left[G_{ij}\sin\delta_{ij} + B_{ij}\cos\delta_{ij}\right] \tag{4.3.20}$$

$$Q_{is} = V_i^2K^2B_{ij} + V_iV_jK\left[G_{ij}\cos\delta_{ij} + B_{ij}\sin\delta_{ij}\right] \tag{4.3.21}$$

$$Q_{js} = -V_iV_jK\left[G_{ij}\cos\delta_{ij} - B_{ij}\sin\delta_{ij}\right] \tag{4.3.22}$$

where $K = \tan\phi$.

Static Representation of UPFC

The equivalent circuit of UPFC placed in line-k connected between bus-i and bus-j is shown in Figure 4.3.7 and control vector diagram in Figure 4.3.8. UPFC [65–67] has three controllable parameters, namely the magnitude and the angle of inserted voltage (V_T, ϕ_T) and the magnitude of the current (I_q). All the bold face quantities represent the complex variables.

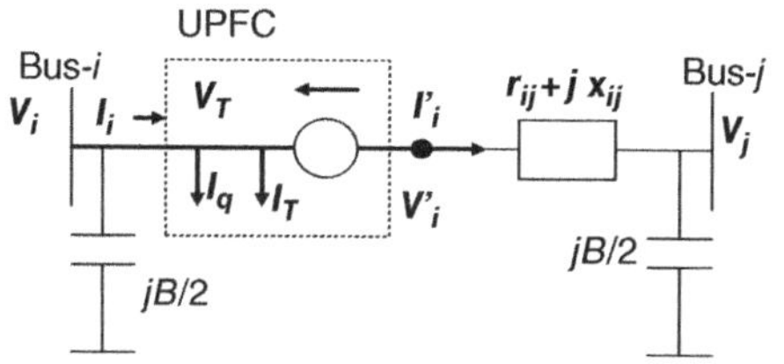

Figure 4.3.7 Equivalent circuit of UPFC.

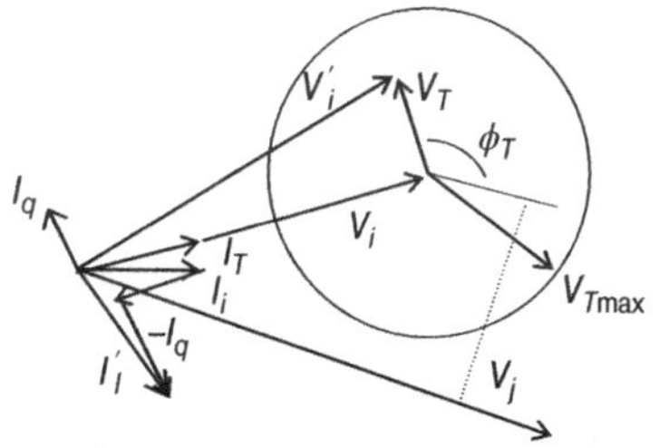

Figure 4.3.8 Vector diagram of UPFC.

Based on the principle of UPFC and the vector diagram, the basic mathematical relations can be given as

$$V_i' = V_i + V_T,\quad \mathrm{Arg}(I_q) = \mathrm{Arg}(V_i) \pm \frac{\pi}{2},\quad \mathrm{Arg}(I_T) = \mathrm{Arg}(V_i),\quad I_T = \frac{\mathrm{Re}\left[\boldsymbol{V}_T \boldsymbol{I}_i'^{*}\right]}{\boldsymbol{V}_i} \tag{4.3.23}$$

The power flow equations from bus-i to bus-j and from bus-j to bus-i can be written as

$$\boldsymbol{S}_{ij} = P_{ij} + jQ_{ij} = \boldsymbol{V}_i \boldsymbol{I}_{ij}^{*} = \boldsymbol{V}_i \left(\frac{j\boldsymbol{V}_i B}{2} + \boldsymbol{I}_T + \boldsymbol{I}_q + \boldsymbol{I}_i' \right)^{*} \tag{4.3.24}$$

$$\boldsymbol{S}_{ji} = P_{ji} + jQ_{ji} = \boldsymbol{V}_j \boldsymbol{I}_{ji}^{*} = \boldsymbol{V}_j \left(\frac{j\boldsymbol{V}_j B}{2} - \boldsymbol{I}_i' \right)^{*} \tag{4.3.25}$$

Active and reactive power flows in the line having UPFC can be written, with above Eqs. (4.3.23)–(4.3.25), as

$$\begin{aligned} P_{ij} = {} & \left(V_i^2 + V_T^2\right) g_{ij} + 2V_i V_T g_{ij} \cos\left(\phi_T - \delta_j\right) - V_j V_T \left[g_{ij} \cos\left(\phi_T - \delta_j\right) \right. \\ & \left. + b_{ij} \sin\left(\phi_T - \delta_j\right)\right] - V_i V_j \left(g_{ij} \cos\delta_{ij} + b_{ij} \sin\delta_{ij} \right) \end{aligned} \tag{4.3.26}$$

$$P_{ji} = V_j^2 g_{ij} - V_j V_T \left[g_{ij} \cos\left(\phi_T - \delta_j\right) - b_{ij} \sin\left(\phi_T - \delta_j\right)\right] - V_i V_j \left(g_{ij} \cos\delta_{ij} - b_{ij} \sin\delta_{ij} \right) \tag{4.3.27}$$

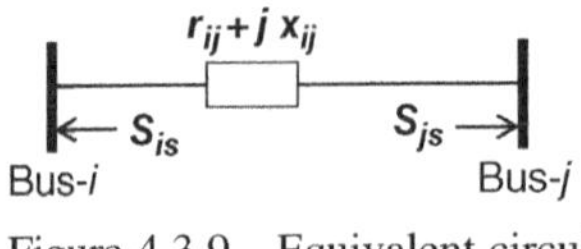

Figure 4.3.9 Equivalent circuit injection model of UPFC.

$$Q_{ij} = -V_iI_q - V_i^2\left(b_{ij} + \frac{B}{2}\right) - V_iV_T\left[g_{ij}\sin(\phi_T - \delta_i) + b_{ij}\cos(\phi_T - \delta_i)\right] - V_iV_j\left(g_{ij}\sin\delta_{ij} - b_{ij}\cos\delta_{ij}\right) \tag{4.3.28}$$

$$Q_{ji} = -V_j^2\left(b_{ij} + \frac{B}{2}\right) - V_jV_T\left(g_{ij}\sin(\phi_T - \delta_j) - b_{ij}\cos(\phi_T - \delta_j)\right) + V_iV_j\left(g_{ij}\sin\delta_{ij} + b_{ij}\cos\delta_{ij}\right) \tag{4.3.29}$$

From basic circuit theory, the injected equivalent circuit of Figure 4.3.9 can be obtained. The injected active power at bus-i (P_{iu}) and bus-j (P_{ju}), and reactive powers (Q_{iu} and Q_{ju}) of a line having a UPFC are

$$P_{iu} = -V_T^2g_{ij} - 2V_iV_Tg_{ij}\cos(\phi_T - \delta_i) + V_jV_T\left[g_{ij}\cos(\phi_T - \delta_j) + b_{ij}\sin(\phi_T - \delta_j)\right] \tag{4.3.30}$$

$$P_{ju} = V_jV_T\left[g_{ij}\cos(\phi_T - \delta_j) - b_{ij}\sin(\phi_T - \delta_j)\right] \tag{4.3.31}$$

$$Q_{iu} = V_iI_q + V_iV_T\left[g_{ij}\sin(\phi_T - \delta_i) + b_{ij}\cos(\phi_T - \delta_i)\right] \tag{4.3.32}$$

$$Q_{ju} = -V_jV_T\left[g_{ij}\sin(\phi_T - \delta_j) - b_{ij}\cos(\phi_T - \delta_j)\right] \tag{4.3.33}$$

Static Representation of Generalised Unified Power Flow Controller

The equivalent circuit of generalised unified power flow controller (GUPFC) placed in line-l having impedance $r_{ij} + jx_{ij}$ ($=1/(g_{ij} + jb_{ij})$) connected between bus-i and bus-j and in line-m having impedance $r_{ik}+jx_{ik}$ ($=1/(g_{ik} + jb_{ik})$) connected between bus-i and bus-k is shown in Figure 4.3.10. Let there be p (>2) numbers of lines connected at bus-i. GUPFC has five controllable parameters, namely the magnitude and the angle of inserted voltage (V_{s1}, ϕ_{s1}) in line-l, the magnitude and the angle of inserted voltage (V_{s2}, ϕ_{s2}) in line-m, and the magnitude of the current (I_q). The current in shunt converter can be delineated into two components viz. the current (I_T) in phase with the voltage at bus-i and current (I_q) in quadrature with the voltage at exciting substation.

Based on the principle of GUPFC operation and the circuit diagram, the basic mathematical relations can be written as

$$\boldsymbol{I}_{ij} = \left(\boldsymbol{V}_i + \boldsymbol{V}_{s1} - \boldsymbol{V}_j\right)\boldsymbol{y}_{ij} \tag{4.3.34}$$

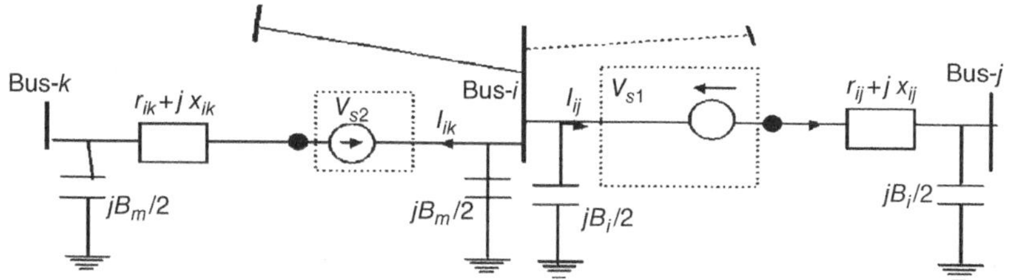

Figure 4.3.10 Equivalent circuit of GUPFC.

$$I_{ik} = (V_i + V_{s2} - V_k)y_{ik} \tag{4.3.35}$$

$$\operatorname{Arg}(I_q) = \operatorname{Arg}(V_i) \pm \frac{\pi}{2}, \quad \operatorname{Arg}(I_T) = \operatorname{Arg}(V_i) \tag{4.3.36}$$

$$I_T^* = -\frac{\operatorname{Re}\left[V_{s1}I_{ij}^* + V_{s2}I_{ik}^*\right]}{V_i} \tag{4.3.37}$$

The power injection at bus-i can be written as

$$S_i = P_i + jQ_i = V_iI_{ij}^* + V_iI_{ik}^* + V_i(I_T + jI_q)^* + \sum_{\substack{i=1 \\ \neq j,k}}^{p} V_iI_{ip}^* + V_iI_{sh}^* \tag{4.3.38}$$

where I_{sh} is the shunt current due to line charging.

The effect of GUPFC can be represented as injected power with the network without GUPFC as shown in Figure 4.3.11. The injected complex powers S_{ig} ($=P_{ig} + jQ_{ig}$) at bus-i, S_{jg} ($=P_{jg} + jQ_{jg}$) at bus-j, and S_{kg} ($=P_{kg} + jQ_{kg}$) at bus-k can be written as

$$S_{ig} = S_i^0 - S_i = -\left\{V_iV_{s1}^*y_{ij}^* + V_iV_{s2}^*y_{ik}^* + V_i(I_T + jI_q)^*\right\} \tag{4.3.39}$$

$$S_{jg} = S_j^0 - S_j = V_jV_{s1}^*y_{ij}^* \tag{4.3.40}$$

$$S_{kg} = S_k^0 - S_k = V_kV_{s2}^*y_{ik}^* \tag{4.3.41}$$

where S^0 is the complex power injection when there was no GUPFC.

From Eq. (4.3.39), the real and reactive power injections at bus-i can be derived as

$$P_{ig} = -\operatorname{Re}\left\{V_iV_{s1}^*y_{ij}^* + V_iV_{s2}^*y_{ik}^*\right\} - V_iI_T^* \tag{4.3.42}$$

$$Q_{ig} = -\operatorname{Im}\left\{V_iV_{s1}^*y_{ij}^* + V_iV_{s2}^*y_{ik}^*\right\} + V_iI_q \tag{4.3.43}$$

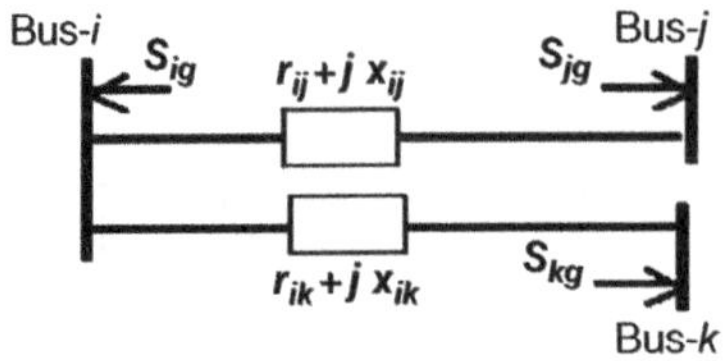

Figure 4.3.11 Injection model of GUPFC.

From Eq. (4.3.37), we have

$$V_i I_T^* = \text{Re}\left\{ \begin{array}{l} V_{s1}\angle\phi_{s1} \times \left(V_i\angle\delta_i + V_{s1}\angle\phi_{s1} - V_j\angle\delta_j\right)^* \times y_{ij}^* + \\ V_{s2}\angle\phi_{s2} \times \left(V_i\angle\delta_i + V_{s2}\angle\phi_{s2} - V_k\angle\delta_k\right)^* \times y_{ik}^* \end{array} \right\}$$

$$= \text{Re}\left\{ \begin{array}{l} \left(V_{s1}V_i\angle(\phi_{s1}-\delta_i) + V_{s1}^2 - V_{s1}V_j\angle(\phi_{s1}-\delta_j)\right) \times \left(g_{ij} - jb_{ij}\right) + \\ \left(V_{s2}V_i\angle(\phi_{s2}-\delta_i) + V_{s2}^2 - V_{s2}V_k\angle(\phi_{s2}-\delta_k)\right) \times \left(g_{ik} - jb_{ik}\right) \end{array} \right\}$$

Thus,

$$\begin{aligned} V_i I_T = {} & V_{s1}^2 g_{ij} + V_{s2}^2 g_{ik} + V_{s1}V_i\left[g_{ij}\cos(\phi_{s1}-\delta_i) + b_{ij}\sin(\phi_{s1}-\delta_i)\right] \\ & - V_{s1}V_j\left[g_{ij}\cos(\phi_{s1}-\delta_j) + b_{ij}\sin(\phi_{s1}-\delta_j)\right] + V_{s2}V_i\left[g_{ik}\cos(\phi_{s2}-\delta_i) + b_{ik}\sin(\phi_{s2}-\delta_i)\right] \\ & - V_{s2}V_k\left[g_{ik}\cos(\phi_{s2}-\delta_k) + b_{ik}\sin(\phi_{s2}-\delta_k)\right] \end{aligned} \tag{4.3.44}$$

The real and imaginary values of $V_i V_{s1}^* y_{ij}^*$ can be written as

$$\text{Re}\left(V_i V_{s1}^* y_{ij}^*\right) = V_i V_{s1}\left(g_{ij}\cos(\delta_i - \phi_{s1}) + b_{ij}\sin(\delta_i - \phi_{s1})\right) \tag{4.3.45}$$

$$\text{Im}\left(V_i V_{s1}^* y_{ij}^*\right) = V_i V_{s1}\left(g_{ij}\sin(\delta_i - \phi_{s1}) - b_{ij}\cos(\delta_i - \phi_{s1})\right) \tag{4.3.46}$$

The injected active and reactive powers at bus-i will be

$$\begin{aligned} P_{ig} = {} & -V_{s1}^2 g_{ij} - V_{s2}^2 g_{ik} - 2V_{s1}V_i g_{ij}\cos(\phi_{s1}-\delta_i) - 2V_{s2}V_i g_{ik}\cos(\phi_{s2}-\delta_i) \\ & + V_{s1}V_j\left[g_{ij}\cos(\phi_{s1}-\delta_j) + b_{ij}\sin(\phi_{s1}-\delta_j)\right] \\ & + V_{s2}V_k\left[g_{ik}\cos(\phi_{s2}-\delta_k) + b_{ik}\sin(\phi_{s2}-\delta_k)\right] \end{aligned} \tag{4.3.47}$$

$$\begin{aligned} Q_{ig} = {} & V_i I_q + V_i V_{s1}\left[g_{ij}\sin(\phi_{s1}-\delta_i) + b_{ij}\cos(\phi_{s1}-\delta_i)\right] \\ & + V_i V_{s2}\left[g_{ik}\sin(\phi_{s2}-\delta_i) + b_{ik}\cos(\phi_{s2}-\delta_i)\right] \end{aligned} \tag{4.3.48}$$

Similarly, the real and reactive powers injections at bus-j and bus-k can be derived as

$$P_{jg} = V_j V_{s1}\left(g_{ij}\cos(\phi_{s1}-\delta_j) - b_{ij}\sin(\phi_{s1}-\delta_j)\right) \tag{4.3.49}$$

$$P_{kg} = V_k V_{s2}\left(g_{ik}\cos(\phi_{s2}-\delta_k) - b_{ik}\sin(\phi_{s2}-\delta_k)\right) \tag{4.3.50}$$

$$Q_{jg} = -V_j V_{s1}\left(g_{ij}\sin\left(\phi_{s1}-\delta_j\right) + b_{ij}\cos\left(\phi_{s1}-\delta_j\right)\right) \tag{4.3.51}$$

$$Q_{kg} = -V_k V_{s2}(g_{ik}\sin(\phi_{s2}-\delta_k) + b_{ik}\cos(\phi_{s2}-\delta_k)) \tag{4.3.52}$$

Equations (4.3.53)–(4.3.56) are derived for three converters (one shunt and two series) and can be generalized for multi-converter UPFC, where one shunt converter is connected at bus-i and n series converters are connected between the lines connected at bus-i, as

$$P_{ig} = -\left(\sum_n V_{sn}^2 g_{in} + 2V_{sn}V_i g_{in}\cos(\phi_{sn}-\delta_i) - V_{sn}V_n[g_{in}\cos(\phi_{sn}-\delta_n) + b_{in}\sin(\phi_{sn}-\delta_n)]\right) \tag{4.3.53}$$

$$Q_{ig} = V_i I_q + \sum_n V_i V_{sn}[g_{in}\sin(\phi_{sn}-\delta_i) + b_{in}\cos(\phi_{sn}-\delta_i)] \tag{4.3.54}$$

$$P_{ng} = V_n V_{sn}(g_{in}\cos(\phi_{sn}-\delta_n) - b_{in}\sin(\phi_{sn}-\delta_n)) \tag{4.3.55}$$

$$Q_{ng} = -V_n V_{sn}(g_{in}\sin(\phi_{sn}-\delta_n) + b_{in}\cos(\phi_{sn}-\delta_n)) \tag{4.3.56}$$

where $n = j, k, \ldots$

4.3.4 Power Flow Control using FACTS

The use of FACTS devices enables the stepless control of power flow over selected lines or areas of an interconnected power system. This becomes more and more important as large power is transported over long distance.

Power Flow Equations with FACTS

Power flow analysis is of great importance in planning and designing the future expansion of power system as well as in determining the best operation of existing system. The power flow problem consists of a given transmission network where all lines are represented by Π-equivalent circuit and transformers by an ideal transformer in series with an impedance. The principal information obtained from a power flow analysis is the magnitude and phase angle of the voltage at each bus and thus, the real and reactive power flowing in each line. Instead, the power flow problem is mathematically linear, a nonlinear relationship between voltage and currents at each bus are calculated. Accordingly, analysis tools with strong convergence characteristics are essential; therefore, there are a number of papers that use a powerful method to arrive at the reliable solution. The Newton–Raphson method has a very good convergence (quadratic) rate; thus, this method is particularly suited for applications involving large systems requiring very accurate solutions.

The load flow control by FACTS devices [74, 75] is promising as it incorporates all three attributes for controlling the transmitted power flow. With the adoption of FACTS in power systems, the traditional power flow analysis faces new challenges in modeling and solution techniques. It is also necessary for investigating the control ability of the FACTS during steady- and dynamic-state analysis. FACTS devices are incorporated into an existing Newton–Raphson load flow algorithm so that the entire network including FACTS is derived. As far as load flow computation is concerned, it carries an intrinsic drawback with the calculation that requires a few more prespecified system states such as power flow of the FACTS-embedded transmission line and its specified bus voltage to be regulated. However, since no one has a priori knowledge of the operational conditions of the FACTS, the prespecified power flow and voltage values might lead to situation that no meaningful result can be obtained [76, 77].

The load flow equations with FACTS devices can then be obtained. It is assumed that FACTS is embedded in a transmission line connected between node-l and node-m. Therefore, for the FACTS-embedded transmission line, the load flow equations can be expressed as

$$P_{Gi} - P_{Li} = \sum_{j=1}^{n} V_i V_j \left(G_{ij} \cos \delta_{ij} + B_{ij} \sin \delta_{ij} \right) \tag{4.3.57}$$

$$Q_{Gi} - Q_{Li} = \sum_{j=1}^{n} V_i V_j \left(G_{ij} \cos \delta_{ij} - B_{ij} \sin \delta_{ij} \right) \tag{4.3.58}$$

where $i = 1,2, \ldots, n$, but $i \neq l$ and m.

$$P_{Gl} - P_{Ll} = \sum_{j=1}^{n} V_l V_j \left(G_{lj} \cos \delta_{lj} + B_{lj} \sin \delta_{lj} \right) + \widetilde{P}_{ls} \tag{4.3.59}$$

$$Q_{Gl} - Q_{Ll} = \sum_{j=1}^{n} V_l V_j \left(G_{lj} \sin \delta_{lj} - B_{lj} \cos \delta_{lj} \right) + \widetilde{Q}_{ls} \tag{4.3.60}$$

$$P_{Gm} - P_{Lm} = \sum_{j=1}^{n} V_m V_j \left(G_{mj} \cos \delta_{mj} + B_{mj} \sin \delta_{mj} \right) + \widetilde{P}_{ms} \tag{4.3.61}$$

$$Q_{Gm} - Q_{Lm} = \sum_{j=1}^{n} V_m V_j \left(G_{mj} \cos \delta_{mj} - B_{mj} \sin \delta_{mj} \right) + \widetilde{Q}_{ms} \tag{4.3.62}$$

where n is the total number of buses. P_{Gi}, Q_{Gi}, P_{Li} and Q_{Li} ($\forall i$) are the respective real and reactive power of generator and load of node-i and the values of $\widetilde{P}_{ls}$, $\widetilde{Q}_{ls}$, $\widetilde{P}_{ms}$, $\widetilde{Q}_{ms}$ for TCSC will be

$$\widetilde{P}_{ls} = V_l^2 \Delta G_{lm} - V_l V_m \left[\Delta G_{lm} \cos \delta_{lm} + \Delta B_{lm} \sin \delta_{lm} \right] \tag{4.3.63}$$

$$\widetilde{P}_{ms} = V_m^2 \Delta G_{lm} - V_l V_m \left[\Delta G_{lm} \cos \delta_{lm} - \Delta B_{lm} \sin \delta_{lm} \right] \tag{4.3.64}$$

$$\widetilde{Q}_{ls} = -V_m^2 \Delta B_{lm} - V_l V_m \left[\Delta G_{lm} \sin \delta_{lm} - \Delta B_{lm} \cos \delta_{lm} \right] \tag{4.3.65}$$

$$\widetilde{Q}_{ms} = -V_m^2 \Delta B_{lm} + V_l V_m [\Delta G_{lm} \sin \delta_{lm} + \Delta B_{lm} \cos \delta_{lm}] \tag{4.3.66}$$

For TCPAR, it will be

$$\widetilde{P}_{ls} = -V_l^2 K^2 G_{lm} - V_l V_m K [G_{lm} \sin \delta_{lm} - B_{lm} \cos \delta_{lm}] \tag{4.3.67}$$

$$\widetilde{P}_{ms} = -V_l V_m K [G_{lm} \sin \delta_{lm} + B_{lm} \cos \delta_{lm}] \tag{4.3.68}$$

$$\widetilde{Q}_{ls} = V_l^2 K^2 B_{lm} + V_l V_m K [G_{lm} \cos \delta_{lm} + B_{lm} \sin \delta_{lm}] \tag{4.3.69}$$

$$\widetilde{Q}_{ms} = -V_l V_m K [G_{lm} \cos \delta_{lm} - B_{lm} \sin \delta_{lm}] \tag{4.3.70}$$

For UPFC, it will be

$$\widetilde{P}_{ls} = -V_T^2 g_{lm} - 2V_l V_T g_{lm} \cos(\phi_T - \delta_m) + V_m V_T [g_{lm} \cos(\phi_T - \delta_m) + b_{lm} \sin(\phi_T - \delta_m)] \tag{4.3.71}$$

$$\widetilde{P}_{ms} = V_m V_T [g_{lm} \cos(\phi_T - \delta_m) - b_{lm} \sin(\phi_T - \delta_m)] \tag{4.3.72}$$

$$\widetilde{Q}_{ls} = V_l I_q + V_l V_T \left[g_{lm} \sin(\phi_T - \delta_l) + \left(b_{lm} + \frac{B}{2} \right) \cdot \cos(\phi_T - \delta_l) \right] \tag{4.3.73}$$

$$\widetilde{Q}_{ms} = -V_m V_T [g_{lm} \sin(\phi_T - \delta_m) + b_{lm} \cos(\phi_T - \delta_m)] \tag{4.3.74}$$

Thus, the relationship are obtained for small variations in V and δ, by forming the total differentials:

$$\begin{bmatrix} \Delta P \\ \Delta Q \end{bmatrix} = J_1 \begin{bmatrix} \Delta \delta \\ \Delta V \end{bmatrix} + J_2 \begin{bmatrix} \Delta \delta \\ \Delta V \end{bmatrix} \tag{4.3.75}$$

where J_1 is the normal N–R power flow Jacobian matrix and J_2 is the partial derivative matrix of injected power with respect to the variables. When bus-l and bus-m are PQ buses, the matrix J_2 may have 10 nonzero elements and if bus-l is a PV bus corresponding elements of row and column will not exist. When more than one FACTS devices are installed in the network, their effects are added to matrix J_2. In this situation the nonzero elements may be more than 10. Now we can see that the power flow can be solved by the N–R method in the normal way, except the small differences in J matrix and power mismatch equations.

The elements of J_2 for UPFC are given below:

$$\frac{\partial P_{ls}}{\partial \delta_l} = -2V_l V_T g_{lm} \sin(\phi_T - \delta_l) \tag{4.3.76}$$

$$\frac{\partial P_{ls}}{\partial \delta_m} = V_m V_T [g_{lm} \sin(\phi_T - \delta_m) - b_{lm} \cos(\phi_T - \delta_m)] \tag{4.3.77}$$

$$\frac{\partial P_{ms}}{\partial \delta_m} = V_m V_T [g_{lm} \sin(\phi_T - \delta_m) + b_{lm} \cos(\phi_T - \delta_m)] \tag{4.3.78}$$

$$\frac{\partial P_{ls}}{\partial V_l} = -2V_T g_{lm} \cos(\phi_T - \delta_l) \tag{4.3.79}$$

$$\frac{\partial P_{ls}}{\partial V_m} = V_T[g_{lm} \cos(\phi_T - \delta_m) + b_{lm} \sin(\phi_T - \delta_m)] \tag{4.3.80}$$

$$\frac{\partial P_{ms}}{\partial V_m} = V_T[g_{lm} \cos(\phi_T - \delta_m) - b_{lm} \sin(\phi_T - \delta_m)] \tag{4.3.81}$$

$$\frac{\partial Q_{ls}}{\partial \delta_l} = -V_l V_T[-g_{lm} \cos(\phi_T - \delta_l) + b_{lm} \sin(\phi_T - \delta_l)] \tag{4.3.82}$$

$$\frac{\partial Q_{ms}}{\partial \delta_m} = -V_m V_T[-g_{lm} \cos(\phi_T - \delta_m) + b_{lm} \sin(\phi_T - \delta_m)] \tag{4.3.83}$$

$$\frac{\partial Q_{ls}}{\partial V_l} = I_q + V_T[g_{lm} \sin(\phi_T - \delta_l) + b_{lm} \cos(\phi_T - \delta_l)] \tag{4.3.84}$$

$$\frac{\partial Q_{ms}}{\partial V_m} = -V_T[g_{lm} \sin(\phi_T - \delta_m) + b_{lm} \cos(\phi_T - \delta_m)] \tag{4.3.85}$$

$$\frac{\partial P_{ms}}{\partial \delta_l} = 0; \quad \frac{\partial P_{ls}}{\partial V_l} = 0 \tag{4.3.86}$$

$$\frac{\partial Q_{ls}}{\partial V_m} = 0; \quad \frac{\partial Q_{ms}}{\partial \delta_l} = 0; \quad \frac{\partial Q_{ls}}{\partial \delta_m} = 0; \quad \frac{\partial Q_{ms}}{\partial V_l} = 0 \tag{4.3.87}$$

With the help of Eqs. (4.3.76)–(4.3.87), the power flow Jacobian matrix can be modified and power flow equations can be solved by the conventional N–R method.

Direct Control of Power Flow using UPFC

The problem of direct load flow control by using UPFC is relatively well addressed. There are two questions to be answered. First, is how to set the control parameters of UPFC to realize control objective and another one is how to calculate the associated load flow.

Let UPFC be connected between bus-l and bus-m, and real and reactive powers at bus-m in line l–m are controlled to P_c and Q_c. To achieve this objective, the following modification can be done in (4.3.59)–(4.3.62).

$$P_{Gl} - P_{Ll} = \sum_{\substack{j=1 \\ j \neq m}}^{n} V_l V_j \left(G_{lj} \cos\delta_{lj} + B_{lj} \sin\delta\right) + P_{lm} \tag{4.3.88}$$

$$Q_{Gl} - Q_{Ll} = \sum_{\substack{j=1 \\ j \neq m}}^{n} V_l V_j \left(G_{lj} \sin\delta_{lj} - B_{lj} \cos\delta_{lj}\right) + Q_{lm} \tag{4.3.89}$$

$$P_{Gm} - P_{Lm} = \sum_{\substack{j=1 \\ j \neq l}}^{n} V_m V_j \left(G_{mj} \cos \delta_{mj} + B_{mj} \sin \delta_{mj} \right) + P_c \tag{4.3.90}$$

$$Q_{Gm} - Q_{Lm} = \sum_{\substack{j=1 \\ j \neq l}}^{n} V_m V_j \left(G_{mj} \cos \delta_{mj} - B_{mj} \sin \delta_{mj} \right) + Q_c \tag{4.3.91}$$

Since P_c and Q_c are set as constant for the given control requirement, the real and reactive powers at bus-l in line l–m can be calculated as follows [78]:

$$P_{lm} = \left(\frac{P_c^2 + Q_c^2}{V_m^2} + \left(\frac{B}{2} \right)^2 V_m^2 + \left(\frac{B}{2} \right) Q_c \right) r_{lm} - P_c \tag{4.3.92}$$

$$Q_{lm} = -I_q V_l - \left[\left(E_1 V_m^{-1} + F_1 V_m \right) \sin \delta_{lm} - \left(E_2 V_m^{-1} + F_2 V_m \right) \cos \delta_{lm} \right] V_l \tag{4.3.93}$$

where

$$E_1 = C_x P_c + C_y Q_c, \quad E_2 = C_y P_c - C_x Q_c, \quad F_1 = (B/2) C_y$$

$$F_2 = -\left(\frac{B}{2} \right) (1 + C_x), \quad C_x = \left(1 - \frac{x_{lm} B}{2} \right) \quad \text{and} \quad C_y = \frac{r_{lm} B}{2}$$

The algorithm for finding the solution does not place any restrictions, i.e. bus-l and bus-m can be PQ-bus, PV bus, or even slack bus. If bus-l is PQ-bus, I_q should be given and if bus-l is PV bus, Eqs. (4.3.89) and (4.3.93) can determine I_q. Thus, the load flow equations can be solved by decoupled N–R load flow program. After load flow converges, the control settings of the UPFC can be computed and can be given as follows:

$$V_T = \frac{\sqrt{\left(P_e^2 + Q_e^2 \right) \left(r_{lm}^2 + x_{lm}^2 \right)}}{V_m} \tag{4.3.94}$$

$$\phi_T = \text{Arc tan} \left[-\frac{(Q_c - Q_f)}{(P_c - P_f)} \right] - \text{Arc tan} \left(\frac{x_{lm}}{r_{lm}} \right) + \delta_m \tag{4.3.95}$$

where

$$P_f + jQ_f = \left(\frac{V_m - V_l}{r_{lm} + jx_{lm}} + j\frac{BV_m}{2} \right)^* V_m$$

4.3.5 Optimal Power Flow Using Suitability FACTS devices

Due to considerable costs of FACTS devices, it is important to optimally locate in the power system. There are several methods for finding optimal locations of FACTS devices in both vertically integrated and unbundled power systems [79–85]. In [83], the optimal locations of FACTS devices are obtained by solving the economic dispatch (ED) problem plus the cost of these devices making the assumption that all lines, initially, have these devices. In the presence of bilateral/multilateral contracts, it would be difficult to use this objective. Moreover, these papers do not suggest a simple and reliable method for determining the optimal location in the deregulated environment.

Congestion in a transmission system, whether vertically organized or unbundled, cannot be permitted except for very short duration, for fear of cascade outages with uncontrolled loss of load. Some corrective measures such as outage of congested branches (lines or transformers), using FACTS devices, operation of transformer taps, re-dispatch of generation, and curtailment of pool loads and/or bilateral contracts can relieve congestion. If there is no congestion, the placement of FACTS devices, from the static point of view, can be decided on the basis of reducing losses but this approach is inadequate when congestion occurs. A method based on the real power performance index (PI) can be considered for security and stability reasons.

Suitable methods to determine the optimal locations of TCSC, TCPAR, and UPFC based on the sensitivity of three objectives: loss on a transmission line in which a device is installed, the total system real power loss, and the real power flow PI, are also suggested. Previous studies have utilized dynamic considerations such as improving the stability and damping the oscillations for the placement of the FACTS devices. The static considerations based on the following objectives are:

- reduction in the real power loss of a particular line-k (P_{Lk}),
- reduction in the total system real power loss (P_{LT}), and
- reduction in the real power flow performance index (PI).

Line Loss Sensitivity Indices (Method-1)

Define the sensitivity a_k^c of transmission loss (P_{Lk}) on a series-compensated line-k with respective series capacitive reactance (x_{ck}), and the sensitivity a_k^s with respect to phase angle shift ϕ_k, on a phase regulated line, as follows:

$$a_k^c = \left.\frac{\partial P_{Lk}}{\partial x_{ck}}\right|_{x_{ck}=0} = \text{line loss sensitivity with respect to TCSC placed in line-}k(k=1,\ldots,N_l) \quad (4.3.96)$$

$$a_k^s = \left.\frac{\partial P_{Lk}^s}{\partial \phi_k}\right|_{\phi_k=0} = \text{line loss sensitivity with respect to TCPAR placed in line-}k(k=1,\ldots,N_l) \quad (4.3.97)$$

Hence, from Eqs. (4.3.9) and (4.3.18), at a base load flow, we obtain:

$$a_k^c = \left.\frac{\partial P_{Lk}^c}{\partial x_{ck}}\right|_{x_c=0} = -2\cdot\left[V_i^2 + V_j^2 - 2\cdot V_i\cdot V_j\cos\delta_{ij}\right]\cdot G_{ij}\cdot B_{ij} \quad (4.3.98)$$

$$a_k^s = \left.\frac{\partial P_{Lk}^s}{\partial \phi_k}\right|_{\phi_k = 0} = \left[2 \cdot V_i \cdot V_j \sin \delta_{ij}\right] \cdot G_{ij} \tag{4.3.99}$$

Total System Loss Sensitivity Indices (Method-2)

The exact loss formula of a system having N buses is

$$P'_{LT} = \sum_{j=1}^{N}\sum_{k=1}^{N}\left[\alpha_{jk}\left(P_jP_k + Q_jQ_k\right) + \beta_{jk}\left(Q_jP_k - P_jQ_k\right)\right] \tag{4.3.100}$$

where P_j and Q_j, respectively, are the real and reactive power injected at bus-j and α, β are the loss coefficients defined by

$$\alpha_{jk} = \frac{r_{jk}}{V_jV_k}\cos\left(\delta_j - \delta_k\right) \quad \text{and} \quad \beta_{jk} = \frac{r_{jk}}{V_jV_k}\sin\left(\delta_j - \delta_k\right) \tag{4.3.101}$$

where r_{jk} is the real part of the j–kth element of $[Z_{\text{bus}}]$ matrix. This total loss if FACTS device, one at a time, is used, can be written as follows (the symbols on the right-hand side are defined in Eqs. (4.3.10), (4.3.11), (4.3.19), and (4.3.20):

$$P_{LT} = \begin{cases} P'_{LT} - \left(P_{ic} + P_{jc}\right) & \text{for TCSC} \\ P'_{LT} - \left(P_{is} + P_{js}\right) & \text{for TCPAR} \\ P'_{LT} - \left(P_{iu} + P_{ju}\right) & \text{for UPFC} \\ P'_{LT} - \left(P_{iu} + P_{ju} + P_{ku}\right) & \text{for } G\text{UPFC} \end{cases} \tag{4.3.102}$$

The total system real power loss sensitivity factors with respect to the parameters of TCSC and TCPAR can be defined as

$$b_k^c = \left.\frac{\partial P_{LT}}{\partial x_{ck}}\right|_{x_{ck}=0} = \text{total loss sensitivity with respect to TCSC placed in line-}k(k = 1,\ldots,N_l) \tag{4.3.103}$$

$$b_k^s = \left.\frac{\partial P_{LT}}{\partial \phi_k}\right|_{\phi_k=0} = \text{total loss sensitivity with respect to TCPAR placed in line-}k(k = 1,\ldots,N_l) \tag{4.3.104}$$

These factors are computed using Eq. (4.3.104) at a base load flow solution. Consider a line-k connected between bus-i and bus-j. The total system loss sensitivity w.r.t. TCSC and TCPAR can be derived as

$$b_k^c = \left.\frac{\partial P_{LT}}{\partial P_i}\frac{\partial P_i}{\partial x_{ck}}\right|_{x_{ck}=0} + \left.\frac{\partial P_{LT}}{\partial P_j}\frac{\partial P_j}{\partial x_{ck}}\right|_{x_{ck}=0} + \left.\frac{\partial P_{LT}}{\partial Q_i}\frac{\partial Q_i}{\partial x_{ck}}\right|_{x_{ck}=0} + \left.\frac{\partial P_{LT}}{\partial Q_j}\frac{\partial Q_j}{\partial x_{ck}}\right|_{x_{ck}=0} - \left.\left(\frac{\partial P_{ic}}{\partial x_{ck}} + \frac{\partial P_{jc}}{\partial x_{ck}}\right)\right|_{x_{ck}=0} \tag{4.3.105}$$

$$b_k^s = \left.\frac{\partial P_{LT}}{\partial P_i}\frac{\partial P_i}{\partial \phi_k}\right|_{\phi_k=0} + \left.\frac{\partial P_{LT}}{\partial P_j}\frac{\partial P_j}{\partial \phi_k}\right|_{\phi_k=0} + \left.\frac{\partial P_{LT}}{\partial Q_i}\frac{\partial Q_i}{\partial \phi_k}\right|_{\phi_k=0} + \left.\frac{\partial P_{LT}}{\partial Q_j}\frac{\partial Q_j}{\partial \phi_k}\right|_{\phi_k=0} - \left.\left(\frac{\partial P_{is}}{\partial \phi_k} + \frac{\partial P_{js}}{\partial \phi_k}\right)\right|_{\phi_k=0} \tag{4.3.106}$$

where

$$\frac{\partial P_{LT}}{\partial P_i} = 2\sum_{m=1}^{N}(\alpha_{im}P_m - \beta_{im}Q_m) \quad \text{and} \quad \frac{\partial P_{LT}}{\partial Q_i} = 2\sum_{m=1}^{N}(\alpha_{im}Q_m + \beta_{im}P_m)$$

The terms $\left.\frac{\partial P_i}{\partial x_{ck}}\right|_{x_{ck}=0}$, $\left.\frac{\partial P_j}{\partial x_{ck}}\right|_{x_{ck}=0}$, $\left.\frac{\partial P_i}{\partial \phi_k}\right|_{\phi_k=0}$, and $\left.\frac{\partial P_j}{\partial \phi_k}\right|_{\phi_k=0}$ can be obtained using Eqs. (4.3.10), (4.3.11), (4.3.19), and (4.3.20), respectively and are as

$$\left.\frac{\partial P_i}{\partial x_{ck}}\right|_{x_{ck}=0} = \left.\frac{\partial P_{ic}}{\partial x_{ck}}\right|_{x_{ck}=0} = \left(V_i^2 - V_iV_j\cos\delta_{ij}\right)\left.\frac{\partial \Delta G_{ij}}{\partial x_{ck}}\right|_{x_{ck}=0} - V_iV_j\sin\delta_{ij}\left.\frac{\partial \Delta B_{ij}}{\partial x_{ck}}\right|_{x_{ck}=0} \tag{4.3.107}$$

$$\left.\frac{\partial P_j}{\partial x_{ck}}\right|_{x_{ck}=0} = \left.\frac{\partial P_{jc}}{\partial x_{ck}}\right|_{x_{ck}=0} = \left(V_j^2 - V_iV_j\cos\delta_{ij}\right)\left.\frac{\partial \Delta G_{ij}}{\partial x_{ck}}\right|_{x_{ck}=0} + V_iV_j\sin\delta_{ij}\left.\frac{\partial \Delta B_{ij}}{\partial x_{ck}}\right|_{x_{ck}=0} \tag{4.3.108}$$

$$\left.\frac{\partial P_i}{\partial \phi_k}\right|_{\phi_k=0} = \left.\frac{\partial P_{is}}{\partial \phi_k}\right|_{\phi_k=0} = V_iV_j\left(G_{ij}\sin\delta_{ij} - B_{ij}\cos\delta_{ij}\right) \tag{4.3.109}$$

$$\left.\frac{\partial P_j}{\partial \phi_k}\right|_{\phi_k=0} = \left.\frac{\partial P_{js}}{\partial \phi_k}\right|_{\phi_k=0} = -V_iV_j\left(G_{ij}\sin\delta_{ij} + B_{ij}\cos\delta_{ij}\right) \tag{4.3.110}$$

where

$$\left.\frac{\partial \Delta G_{ij}}{\partial x_{ck}}\right|_{x_{ck}=0} = 2G_{ij}B_{ij} \quad \text{and} \quad \left.\frac{\partial \Delta B_{ij}}{\partial x_{ck}}\right|_{x_{ck}=0} = B_{ij}^2 - G_{ij}^2$$

Using (4.3.12), (4.3.13), (4.3.21), and (4.3.22), the derivative of the reactive power injections with respect to FACTS parameters can be derived as

$$\left.\frac{\partial Q_i}{\partial x_{ck}}\right|_{x_{ck}=0} = \left.\frac{\partial Q_{ic}}{\partial x_{ck}}\right|_{x_{ck}=0} = -\left(V_i^2 - V_iV_j\cos\delta_{ij}\right)\left.\frac{\partial \Delta B_{ij}}{\partial x_{ck}}\right|_{x_{ck}=0} - V_iV_j\sin\delta_{ij}\left.\frac{\partial \Delta G_{ij}}{\partial x_{ck}}\right|_{x_{ck}=0} \tag{4.3.111}$$

$$\left.\frac{\partial Q_j}{\partial x_{ck}}\right|_{x_{ck}=0} = \left.\frac{\partial Q_{jc}}{\partial x_{ck}}\right|_{x_{ck}=0} = -\left(V_j^2 - V_i V_j \cos\delta_{ij}\right)\left.\frac{\partial \Delta B_{ij}}{\partial x_{ck}}\right|_{x_{ck}=0} + V_i V_j \sin\delta_{ij}\left.\frac{\partial \Delta G_{ij}}{\partial x_{ck}}\right|_{x_{ck}=0} \tag{4.3.112}$$

$$\left.\frac{\partial Q_i}{\partial \phi_k}\right|_{\phi_k=0} = \left.\frac{\partial Q_{is}}{\partial \phi_k}\right|_{\phi_k=0} = V_i V_j\left(G_{ij}\cos\delta_{ij} + B_{ij}\sin\delta_{ij}\right) \tag{4.3.113}$$

$$\left.\frac{\partial Q_j}{\partial \phi_k}\right|_{\phi_k=0} = \left.\frac{\partial Q_{js}}{\partial \phi_k}\right|_{\phi_k=0} = -V_i V_j\left(G_{ij}\cos\delta_{ij} - B_{ij}\sin\delta_{ij}\right) \tag{4.3.114}$$

The sensitivity factors b_k^c and b_k^s can now be found by substituting Eqs. (4.3.111)–(4.3.114) in Eqs. (4.3.105) and (4.3.106), respectively.

The total system real power loss sensitivity factors with respect to the control parameters of UPFC placed in line-k can be defined as

$$b_1^k = \left.\frac{\partial P_{LT}}{\partial V_T}\right|_{V_T=0} = \text{total loss sensitivity with respect to} V_T$$

$$b_2^k = \left.\frac{\partial P_{LT}}{V_T \partial \phi_T}\right|_{\phi_T=0} = \text{total loss sensitivity with respect to } \phi_T$$

$$b_3^k = \left.\frac{\partial P_{LT}}{\partial I_q}\right|_{I_q=0} = \text{total loss sensitivity with respect to } Iq$$

These factors are computed using (4.3.102) at a base load flow solution. Consider a line-k connected between bus-i and bus-j. The total system loss sensitivity with respect to control parameters of UPFC can be derived as

$$\begin{aligned} b_1^k = {} & \left.\frac{\partial P_{LT}}{\partial P_i}\frac{\partial P_i}{\partial V_T}\right|_{V_T=0} + \left.\frac{\partial P_{LT}}{\partial P_j}\frac{\partial P_j}{\partial V_T}\right|_{V_T=0} + \left.\frac{\partial P_{LT}}{\partial Q_i}\frac{\partial Q_i}{\partial V_T}\right|_{V_T=0} + \left.\frac{\partial P_{LT}}{\partial Q_j}\frac{\partial Q_j}{\partial V_T}\right|_{V_T=0} \\ & - \left.\left(\frac{\partial P_{is}}{\partial V_T} + \frac{\partial P_{js}}{\partial V_T}\right)\right|_{V_T=0} \end{aligned} \tag{4.3.115}$$

$$\begin{aligned} b_2^k = {} & \left.\frac{\partial P_{LT}}{\partial P_i}\frac{\partial P_i}{V_T\partial \phi_T}\right|_{\phi_T=0} + \left.\frac{\partial P_{LT}}{\partial P_j}\frac{\partial P_j}{V_T\partial \phi_T}\right|_{\phi_T=0} + \left.\frac{\partial P_{LT}}{\partial Q_i}\frac{\partial Q_i}{V_T\partial \phi_T}\right|_{\phi_T=0} \\ & + \left.\frac{\partial P_{LT}}{\partial Q_j}\frac{\partial Q_j}{V_T\partial \phi_T}\right|_{\phi_T=0} - \frac{1}{V_T}\left.\left(\frac{\partial P_{is}}{\partial \phi_T} + \frac{\partial P_{js}}{\partial \phi_T}\right)\right|_{\phi_T=0} \end{aligned} \tag{4.3.116}$$

$$\begin{aligned} b_3^k = {} & \left.\frac{\partial P_{LT}}{\partial P_i}\frac{\partial P_i}{\partial I_q}\right|_{I_q=0} + \left.\frac{\partial P_{LT}}{\partial P_j}\frac{\partial P_j}{\partial I_q}\right|_{I_q=0} + \left.\frac{\partial P_{LT}}{\partial Q_i}\frac{\partial Q_i}{\partial I_q}\right|_{I_q=0} + \left.\frac{\partial P_{LT}}{\partial Q_j}\frac{\partial Q_j}{\partial I_q}\right|_{I_q=0} \\ & - \left.\left(\frac{\partial P_{is}}{\partial I_q} + \frac{\partial P_{js}}{\partial I_q}\right)\right|_{I_q=0} \end{aligned} \tag{4.3.117}$$

The terms $\left.\frac{\partial P_i}{\partial V_T}\right|_{V_T=0}$, $\left.\frac{\partial P_j}{\partial V_T}\right|_{V_T=0}$, $\left.\frac{\partial P_i}{V_T \partial \phi_T}\right|_{\phi_T=0}$, $\left.\frac{\partial P_j}{V_T \partial \phi_T}\right|_{\phi_T=0}$, $\left.\frac{\partial P_i}{\partial I_q}\right|_{I_q=0}$, and $\left.\frac{\partial P_j}{\partial I_q}\right|_{I_q=0}$ can be obtained using Eqs. (4.3.30) and (4.3.31), respectively, and are given below.

$$\left.\frac{\partial P_i}{V_T \partial \phi_T}\right|_{\phi_T=0} = \left.\frac{\partial P_{is}}{V_T \partial \phi_T}\right|_{\phi_T=0} = -2V_i g_{ij} \sin(\delta_i) + V_j\left(g_{ij} \sin \delta_j + b_{ij} \cos \delta_j\right) \tag{4.3.118}$$

$$\left.\frac{\partial P_i}{\partial I_q}\right|_{I_q=0} = \left.\frac{\partial P_{is}}{\partial I_q}\right|_{I_q=0} = 0 \tag{4.3.119}$$

$$\left.\frac{\partial P_j}{\partial V_T}\right|_{V_T=0} = \left.\frac{\partial P_{js}}{\partial V_T}\right|_{V_T=0} = V_j\left(g_{ij} \cos \delta_j + b_{ij} \sin \delta_j\right) \tag{4.3.120}$$

$$\left.\frac{\partial P_j}{V_T \partial \phi_T}\right|_{\phi_T=0} = \left.\frac{\partial P_{js}}{V_T \partial \phi_T}\right|_{\phi=0} = V_j\left(g_{ij} \sin \delta_j + b_{ij} \cos \delta_j\right) \tag{4.3.121}$$

$$\left.\frac{\partial P_j}{\partial I_q}\right|_{I_q=0} = \left.\frac{\partial P_{js}}{\partial I_q}\right|_{I_q=0} = 0 \tag{4.3.122}$$

Using (4.3.32) and (4.3.33), the derivative of the reactive power injections with respect to UPFC control parameters can be derived as

$$\left.\frac{\partial Q_i}{\partial V_T}\right|_{V_T=0} = \left.\frac{\partial Q_{is}}{\partial V_T}\right|_{V_T=0} = V_i\left[-g_{ij} \sin \delta_i + b_{ij} \cos \delta_i\right] \tag{4.3.123}$$

$$\left.\frac{\partial Q_i}{V_T \partial \phi_T}\right|_{\phi_T=0} = \left.\frac{\partial Q_{is}}{V_T \partial \phi_T}\right|_{\phi_T=0} = V_i\left[g_{ij} \cos \delta_i + b_{ij} \sin \delta_i\right] \tag{4.3.124}$$

$$\left.\frac{\partial Q_i}{\partial I_q}\right|_{I_q=0} = \left.\frac{\partial Q_{is}}{\partial I_q}\right|_{I_q=0} = V_i \tag{4.3.125}$$

$$\left.\frac{\partial Q_j}{\partial V_T}\right|_{V_T=0} = \left.\frac{\partial Q_{js}}{\partial V_T}\right|_{V_T=0} = -V_j\left(-g_{ij} \sin \delta_j + b_{ij} \cos \delta_j\right) \tag{4.3.126}$$

$$\left.\frac{\partial Q_j}{V_T \partial \phi_T}\right|_{\phi_T=0} = \left.\frac{\partial Q_{js}}{V_T \partial \phi_T}\right|_{\phi_T=0} = -V_j\left(g_{ij} \cos \delta_j + b_{ij} \sin \delta_j\right) \tag{4.3.127}$$

$$\left.\frac{\partial Q_j}{\partial I_q}\right|_{I_q=0} = \left.\frac{\partial Q_{js}}{\partial I_q}\right|_{I_q=0} = 0 \tag{4.3.128}$$

The sensitivities corresponding to the GUPFC are similar to the UPFC.

Real Power Flow PI Sensitivity Indices (Method-3)

The severity of the system loading under normal and contingency cases can be described by a real power line flow PI, as given below.

$$PI = \sum_{m=1}^{N_l} \frac{w_m}{2n} \left(\frac{P_{lm}}{P_{lm}^{\max}} \right)^{2n} \tag{4.3.129}$$

where P_{lm} is the real power flow and $P_{lm}^{\max}$ is the rated capacity of line-m, n is the exponent, and w_m is a real non-negative weighting coefficient which may be used to reflect the importance of lines.

PI will be small when all the lines are within their limits and reach a high value when there are overloads. Thus, it provides a good measure of severity of the line overloads for a given state of the power system. Most of the work on contingency selection algorithms utilize the second-order performance indices, which, in general, suffer from masking effects. The lack of discrimination, in which the PI for a case with many small violations may be comparable in value to the index for a case with one huge violation, is known as *Masking effect*. By most of the operational standards, the system with one huge violation is much more severe than that with many small violations. Masking effect to some extent can be avoided by using higher order performance indices, that is $n > 1$. Here, the value of the exponent has been taken as 2 and $w_i = 1.0$.

The real power flow PI sensitivity factors with respect to the parameters of TCSC and TCPAR can be defined as

$$c_k^c = \left. \frac{\partial \text{PI}}{\partial x_{ck}} \right|_{x_{ck}=0} = \text{PI sensitivity with respect to TCSC placed in line-}k$$

$$c_k^s = \left. \frac{\partial \text{PI}}{\partial \phi_k} \right|_{\phi_k=0} = \text{PI sensitivity with respect to TCPAR placed in line-}k$$

The real power flow PI sensitivity factors with respect to the control parameters of UPFC can be defined as

$$c_1^k = \left. \frac{\partial \text{PI}}{\partial V_T} \right|_{V_T=0} = \text{PI sensitivity with respect to } V_T$$

$$c_2^k = \left. \frac{\partial \text{PI}}{V_T \partial \phi_T} \right|_{\phi_T=0} = \text{PI sensitivity with respect to } \phi_T$$

$$c_3^k = \left. \frac{\partial \text{PI}}{\partial I_q} \right|_{I_q=0} = \text{PI sensitivity with respect to } Iq$$

The real power flow PI sensitivity factors with respect to the control parameters of GUPFC can be defined as

$$c_1^k = \left.\frac{\partial \text{PI}}{\partial V_{s1}}\right|_{V_{s1}=0} = \text{PI sensitivity with respect to } V_{s1}$$

$$c_2^k = \left.\frac{\partial \text{PI}}{V_{s1}\partial \phi_{s1}}\right|_{\phi_{s1}=0} = \text{PI sensitivity with respect to } \phi_{s1}$$

$$c_3^k = \left.\frac{\partial \text{PI}}{\partial I_q}\right|_{I_q=0} = \text{PI sensitivity with respect to } Iq$$

c_1^k and c_2^k with respect to V_{s2} and ϕ_{s2}, respectively, will be the same as c_1^k and c_2^k with respect to V_{sI} and ϕ_{sI}.

Using Eq. (4.3.129), the sensitivity of PI with respect to FACTS device parameter X_k (x_{ck} for TCSC; ϕ_k for TCPAR; V_T, ϕ_T, and I_q for UPFC; V_{sI}, ϕ_{sI}, and I_q for GUPFC) connected between bus-i and bus-j can be written as

$$PI = \sum_{m=1}^{N_l} \frac{w_m}{2n}\left(\frac{P_{lm}}{P_{lm}^{\max}}\right)^{2n} \tag{4.3.130}$$

The real power flow in line-m (P_{lm}) can be represented in terms of real power injections using DC power flow equations, where s is slack bus, as

$$P_{lm} = \begin{cases} \sum\limits_{\substack{n=1\\ n\neq s}}^{N} S_{mn}P_n & \text{for } m \neq k \\ \sum\limits_{\substack{n=1\\ n\neq s}}^{N} S_{mn}P_n + P_j & \text{for } m = k \end{cases} \tag{4.3.131}$$

Using (4.3.26) and (4.3.131), the following relationship can be derived:

$$\frac{\partial P_{lm}}{\partial X_k} = \begin{cases} \left(S_{mi}\dfrac{\partial P_i}{\partial X_k} + S_{mj}\dfrac{\partial P_j}{\partial X_k}\right) & \text{for } m \neq k \\ \left(S_{mi}\dfrac{\partial P_i}{\partial X_k} + S_{mj}\dfrac{\partial P_j}{\partial X_k}\right) + \dfrac{\partial P_j}{\partial X_k} & \text{for } m = k \end{cases} \tag{4.3.132}$$

The terms $\left.\frac{\partial P_i}{\partial x_{ck}}\right|_{x_{ck}=0}$, $\left.\frac{\partial P_j}{\partial x_{ck}}\right|_{x_{ck}=0}$, $\left.\frac{\partial P_i}{\partial \phi_k}\right|_{\phi_k=0}$, and $\left.\frac{\partial P_j}{\partial \phi_k}\right|_{\phi_k=0}$ can be derived similar to Eqs. (4.3.107)–(4.3.110), respectively, for TCSC and TCPAR and thus, the sensitivity factors c_k^c and c_k^s. The terms $\left.\frac{\partial P_i}{\partial V_T}\right|_{V_T=0}$, $\left.\frac{\partial P_j}{\partial V_T}\right|_{V_T=0}$, $\left.\frac{\partial P_i}{V_T\partial \phi_T}\right|_{\phi_T=0}$, $\left.\frac{\partial P_j}{V_T\partial \phi_T}\right|_{\phi_T=0}$, $\left.\frac{\partial P_i}{\partial I_q}\right|_{I_q=0}$, and $\left.\frac{\partial P_j}{\partial I_q}\right|_{I_q=0}$ can be derived similar to Eqs. (4.3.107)–(4.3.108) for UPFC and thus, the sensitivity factors.

Observe that line-k, from bus-i to bus-j, is the line containing the series converter of GUPFC, P_{jg}, therefore, is the addition flow, at bus-j, in the line containing the GUPFC, due to the presence of the device. The derivatives of real and reactive powers with respect to control parameters of GUPFC are

$$\left.\frac{\partial P_{ig}}{\partial V_{s1}}\right|_{V_{s1}=0} = -2V_i g_{ij}\cos(\phi_{s1}-\delta_i) + V_j\left(g_{ij}\cos(\phi_{s1}-\delta_j) + b_{ij}\sin(\phi_{s1}-\delta_j)\right) \tag{4.3.133}$$

$$\left.\frac{\partial P_{ig}}{V_{s1}\partial\phi_{s1}}\right|_{\phi s1=0} = -2V_i g_{ij}\sin\delta_i + V_j\left(g_{ij}\sin\delta_j + b_{ij}\cos\delta_j\right) \tag{4.3.134}$$

$$\left.\frac{\partial P_{ig}}{\partial I_q}\right|_{I_q=0} = 0 \tag{4.3.135}$$

$$\left.\frac{\partial P_{jg}}{\partial V_{s1}}\right|_{V_{s1}=0} = V_j\left(g_{ij}\cos\delta_j + b_{ij}\sin\delta_j\right) \tag{4.3.136}$$

$$\left.\frac{\partial P_{jg}}{V_{s1}\partial\phi_{s1}}\right|_{\phi_{s1}=0} = V_j\left(g_{ij}\sin\delta_j - b_{ij}\cos\delta_j\right) \tag{4.3.137}$$

$$\left.\frac{\partial P_{jg}}{\partial I_q}\right|_{I_q=0} = 0 \tag{4.3.138}$$

The derivatives of real power injection with respect to phase angle of GUPFC are considered around zero although the phase angle in GUPFC can be controlled from 0^0 to 360^0. The angle difference for both ends of line is generally very small and it is limited to 30^0 due to stability reasons. In a practical power system, control of an angle of UPFC or GUPFC is generally not high. Therefore, the derivatives with respect to phase angle around zero are correct. However, it can be calculated around any angle, as derivation is very simple.

Criteria for Optimal Location

The FACTS device should be placed on the most sensitive lines. With the sensitive indices computed for each type of the FACTS devices, the following criterion can be used for their optimal placement:

a. The TCSC should be placed in a line (k) having the most negative sensitivity index.

b. The TCPAR should be placed in a line (k) having the largest absolute value of the sensitivity factor.

c. UPFC should be placed based on the sensitivity related to all three parameters: the most negative sensitivity index with respect to V_T; having the largest absolute value of the sensitivity factors w.r.t to phase angle and I_q.

d. GUPFC should be placed in the lines connected at a bus having more than two elements (lines, transformers, and/or generators). The choice of lines should be based on the sensitivities suitable for UPFC locations.

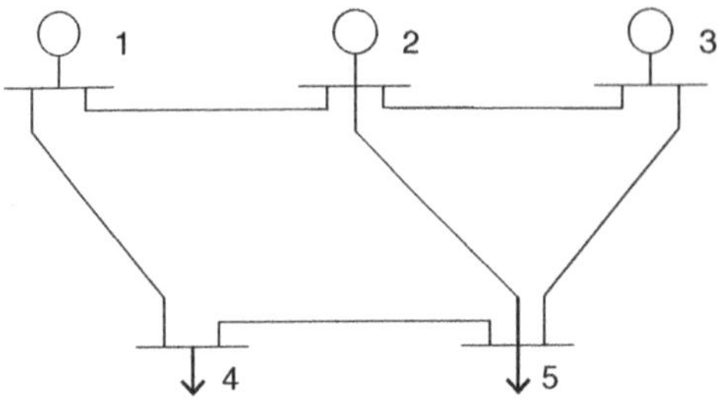

Figure 4.3.12 5-Bus system.

Additional criteria can also be used while deciding the optimal placement of FACTS devices that the TCSC, TCPAR, UPFC, and GUPFC should not be placed with generating transformer, even though the sensitivity is the highest.

Case Studies

To establish the effectiveness of the proposed methods for TCSC, TCPAR, and UPFC, it has been tested on a 5-bus system. The 5-bus system consists of three generator buses and two load buses shown in Figure 4.3.12. The two circuits 1–2 and 3–5 are of impedance $0.0258 + j0.866$ p.u. each while other four lines have an impedance of $0.0129 + j0.0483$ p.u. each, all to a 100 MVA base. The line flow limit is set to 100 MW. Bus-1 has been taken as the reference bus.

Location of TCSC and TCPAR: Sensitivities were calculated for each FACTS device (TCSC and TCPAR) placed in every line one at a time for the same operating conditions. The sensitivities of real power loss of line in which the FACTS device is installed (method-1), total system real power loss (method-2), and real power PI (method-3) with respect to TCSC and TCPAR are presented in Table 4.3.2. The highest negative sensitivities in case of TCSC and the highest absolute value of sensitivities in case of TCPAR are presented in bold type.

From Table 4.3.2 (column 3), it can be seen that the placement of TCSC in lines, as expected, will increase the loss of those lines as sensitivity factors a_k^c of the lines are positive. The increase of losses is more in those lines that are highly loaded as compared to lightly loaded lines. The sensitivity factors a_k^s that are computed for TCPAR are also positive. Again, obviously the increase of the real power flows in line will increase the loss of that line. The power flow and losses can be reduced with help of TCPAR using negative phase shift. The reduction of power flow with placement of TCPAR in line-2 will be the largest as the sensitivity factor corresponding to this line is the highest.

The magnitudes of sensitivity factors b_k^c and b_k^s are small, that is, reduction in total system loss will be less which can be seen from Table 4.3.2 (method-2). The magnitude of sensitivity of total system real power loss with respect to TCSC placed in line-2 is the highest followed by placing TCSC in line-4 which is not the same for TCPAR. This indicates that placement of TCSC in line-2 will reduce

TABLE 4.3.2 Sensitivities of 5-Bus System

Line-k		Method-1		Method-2		Method-3	
No		TCSC	TCPAR	TCSC	TCPAR	TCSC	TCPAR
b_k^c	$i-j$	a_k^c	a_k^s	b_k^c	b_k^s	c_k^c	c_k^s
1	2–1	0.2082	0.3385	0.0406	−0.0236	−1.096	−1.791
2	2–5	0.5856	**0.5209**	**−0.0577**	−0.0757	4.163	**3.843**
3	3–5	0.3027	0.3823	0.0427	0.0470	**−2.312**	−3.065
4	5–4	0.5192	0.4992	−0.054	−0.0880	1.577	1.527
5	1–4	0.4903	0.4713	−0.053	−0.0695	−1.493	−1.475
6	3–2	0.0374	0.1010	0.0229	**0.0970**	0.703	2.747

the total system real power loss more than the placement in other lines. The magnitude of sensitivity factor (b_k^s) of line-6 with respect to TCPAR is the highest which is a positive value. This indicates that placement of TCPAR in line-6 with negative phase shift will reduce the total system real power loss more than placement in other lines. This shows that method-2 is more appropriate for the placement of FACTS devices when there is no congestion. As method-1 and method-2 do not consider the loading of the lines, they are not suitable for congestion management. In the event of congestion, it is more important, for secure operation of the system, to alleviate the overloads instead of reducing the losses in the system.

The sensitivities of the real power flow PI with respect to TCSC and TCPAR control parameters has been computed and are shown in Table 4.3.2 (method-3). From the load flow, it was found that real power flows in lines 2 and 4 were 1.10 and 1.03 p.u., respectively, which are more than their line-loading limits. It can be observed from Table 4.3.2 that placement of TCSC in line-3 is suitable for reducing the PI. Placement of TCSC in line-5 will reduce the PI value but it will be less effective than placing a TCSC in line-3 as can be seen from its sensitivity factors. The placement of a TCSC in line-3 will increase the total loss of the system. However, the TCSC placed in line-5 will reduce the loss along with reduction of the PI.

Table 4.3.2 (column 8, c_k^s) shows that placement of TCPAR in line-2 is more sensitive than the placement in other lines. This sensitivity is positive which indicates that phase angle shift of the TCPAR should be negative. Placing of TCPAR in line-2 will reduce the loading of lines 2 and 4 (heavily loaded lines) but it will increase the loading of lines 1 and 3 that are underloaded. Table 4.3.2 also shows that the placement of TCPAR in line-3 in the next choice as the magnitude of sensitivity factors is the second highest.

To check the effectiveness of the proposed method-3, the line-loading limit of line-4 has been increased to 1.50 p.u. and the sensitivity factors calculated for TCSC and TCPAR are given in Table 4.3.3. The sensitivity of line-3 is more negative compared to other lines. The placement of TCSC in line-5 gives positive sensitivity, that is, the PI value will increase if a TCSC is placed in that line. This is opposite to the previous case when the line loading of all the lines were the same

TABLE 4.3.3 PI Sensitivities of 5-Bus System for Different Loading Limits

	Lines					
Factors	2–1	2–5	3–5	5–4	1–4	3–2
c_k^c	0.491	2.276	**−2.729**	−0.754	0.964	0.894
c_k^s	0.804	2.103	**−3.627**	−0.732	0.953	3.461

(see Table 4.3.3, column 7, c_k^c). The placement of TCSC in line-4 will now reduce the PI but will not when TCSC is placed in line-5. The absolute value of sensitivity c_k^s corresponding to line-6 is the second highest which is slightly less, in magnitude, than line-3. The placement of TCPAR in line-6 will be the next choice but with negative phase shift. The placement of TCPAR in line-2 that was the first choice for the line-loading limit of 1.0 p.u., is now third choice after lines 3 and 6.

Location of UPFC: Sensitivities were calculated for each control parameters of UPFC placed in every line one at a time for the same operating conditions. The sensitivities of total system real power loss (method-1) and real power flow PI (method-2) with respect to UPFC control parameters are presented in Table 4.3.4. The highest negative sensitivities b_1^k and b_2^k, and the highest absolute value of sensitivity b_3^k are presented in bold type.

The magnitudes of sensitivity factors b_1^k are small, that is, reduction in total system loss will be less which can be seen from Table 4.3.4 (method-1). For voltage magnitude control, line-4 is suitable as its sensitivity is more negative than other lines. The magnitude of sensitivity of total system real power loss with respect to phase angle (b_2^k) of UPFC placed in line-2 is the highest followed by line-4. This indicates that placement of UPFC in line-2 will reduce the total system real power loss more than the placement in other lines which is a positive value. This indicates that placement of UPFC in line-2 with negative phase shift will reduce the total system real power loss. The sensitivity factor b_3^k is almost same for each line, which is due to uniform voltage profile of the system. The sensitivity for lines 3 and 4 are the highest negative. As method-1 does not consider the loading of the lines, it is not suitable for congestion management. In the event of congestion, it is more important, for secure operation of the system, to alleviate the overloads instead of reducing the losses in the system. This shows that method-1 is only appropriate for the placement of this device when there is no congestion.

The sensitivities of the real power flow PI with respect to UPFC control parameters has been computed and are shown in Table 4.3.4 (method-2). From the load flow, it was found that real power flows in lines 2 and 4 were 1.15 and 1.04 p.u., respectively, which are more than their line-loading limits. It can be observed from Table 4.3.4 that the sensitivity of PI with respect to V_T for line-2 is the highest but it is positive, which indicates that increase in V_T will increase the PI, thus congestion of the system. Since the value of V_T cannot be negative, it is not suitable for PI

TABLE 4.3.4 Sensitivities of 5-Bus System

Line-k		Method-1			Method-2	
Nodes	i–j	b_1^k	b_2^k	b_3^k	c_1^k	c_2^k
1	2–1	0.0016	0.2947	−0.6824	−1.409	−5.229
2	2–5	0.0498	**0.5114**	−0.6824	2.186	**9.108**
3	3–5	0.1073	0.3183	**−0.6890**	−1.317	−4.684
4	5–4	**−0.1526**	0.4987	−0.6670	1.087	4.993
5	1–4	−0.1220	0.4223	−0.6693	**−1.525**	−4.847
6	3–2	−0.1100	−0.0167	**−0.6890**	−0.252	4.603

TABLE 4.3.5 PI Sensitivities of 5-Bus System for Different Loading Limits

	Lines					
Factors	2–1	2–5	3–5	5–4	1–4	3–2
c_1^k	0.268	1.604	2.083	−0.854	0.114	**−2.049**
c_2^k	−1.281	**6.493**	−5.438	1.609	0.798	5.664

reduction. However, c_1^k for line-5 is the most negative and thus suitable for PI reduction with control of V_T.

Table 4.3.4 (column 7, c_2^k) shows that placement of UPFC in line-2 is more sensitive than the placement in other lines. This sensitivity is positive which indicates that phase angle shift of the UPFC should be negative. Placing of UPFC in line-2 will reduce the loading of lines 2 and 4 (heavily loaded lines) but it will increase the loading of lines 1 and 3 that are underloaded. Table 4.3.4 also shows that the placement of UPFC in line-1 with phase angle control is the next choice as the magnitude of sensitivity factors is the second highest. The sensitivity factor c_3^k is always zero because it cannot control the real power flow of the line as it is in 90° phase with input voltage. Placement of UPFC in lines 2–5 will also reduce the total system real power loss.

To check the effectiveness of the proposed method-2, the line-loading limit of line-4 has been increased to 1.50 p.u. and the sensitivity factors calculated for UPFC control parameters are given in Table 4.3.5. The magnitude of sensitivity of PI with respect to phase angle of UPFC for line-2 is still higher than other lines but the value is less than that obtained for uniform line loading of 1.0 p.u. The absolute value of sensitivity c_2^k corresponding to line-6 is the second highest which is slightly higher, in magnitude, than line-3. The placement of UPFC in line-6 will be the next choice but also with negative phase shift. The placement of UPFC in line-1 that was the second choice for the line-loading limit of 1.0 p.u., is now the fifth choice. For voltage control V_T, the placement of UPFC in line-3 gives positive sensitivity, that is, the PI value will increase with increase of voltage V_T. Line-6 is suitable for this whereas line-5 was the choice for the case when the line loadings of all the lines were the same (see Table 4.3.4).

Optimal Power Dispatch with FACTS Devices

Transmission dispatch in an unbundled environment will be a mix of pool and bilateral/multilateral transactions. The optimal dispatch will be the delivery of all bilateral and multilateral transactions in full and to supply of all pool demand at least cost without any security violations. This case can be termed the normal condition. Mathematically, the normal dispatch problem can be written as

$$\min_{P_{pi}} \sum_{i \in I_G} C_i(P_{pi}) - \sum_{j \in I_D} B_j(D_{pj}) \tag{4.3.139}$$

where

I_G = a set of pool generator buses
I_D = a set of pool load buses
P_{pi} = active power of pool generator-i
C_i = bid price of pool generator-i
D_{pj} = active power of pool load-j
B_j = bid price of pool load-j

Operating Constraints

a. ***Equality constraint:*** Power flow equations corresponding to both real and reactive power balance equations are the equality constraints which can be written, for all the buses except buses i and j in which FACTS controller is connected, as

$$P_l = P_{gl} - P_{dl} = \sum_{m=1}^{N_b} V_l V_m [G_{lm} \cos(\delta_l - \delta_m) + B_{lm} \sin(\delta_l - \delta_m)] \tag{4.3.140}$$

$$Q_l = Q_{gl} - Q_{dl} = \sum_{m=1}^{N_b} V_l V_m [G_{lm} \sin(\delta_l - \delta_m) - B_{lm} \cos(\delta_l - \delta_m)] \tag{4.3.141}$$

$$l = 1, 2, \ldots, N_b; \quad l \neq i, j$$

For buses i and j, the equality constraints can be written as

$$P_i = P_{gi} - P_{di} = \sum_{m=1}^{N_b} V_i V_m [G_{im} \cos(\delta_i - \delta_m) + B_{im} \sin(\delta_i - \delta_m)] - P_{is} \tag{4.3.142}$$

$$Q_i = Q_{gi} - Q_{di} = \sum_{m=1}^{N_b} V_i V_m [G_{im} \sin(\delta_i - \delta_m) - B_{im} \cos(\delta_i - \delta_m)] - Q_{is} \tag{4.3.143}$$

$$P_j = P_{gj} - P_{dj} = \sum_{m=1}^{N_b} V_j V_m \left[G_{jm} \cos\left(\delta_j - \delta_m\right) + B_{jm} \sin\left(\delta_j - \delta_m\right)\right] - P_{js} \quad (4.3.144)$$

$$Q_j = Q_{gj} - Q_{dj} = \sum_{m=1}^{N_b} V_j V_m \left[G_{jm} \sin\left(\delta_j - \delta_m\right) - B_{jm} \cos\left(\delta_j - \delta_m\right)\right] - Q_{js} \quad (4.3.145)$$

where

P_i = real power injection at bus-i

Q_i = reactive power injection at bus-i

P_{gi} = real power generation at bus-i

Q_{gi} = reactive power generation at bus-i

P_{di} = real power load at bus-i

Q_{di} = reactive power load at bus-i

V_i = voltage magnitude at bus-i

δ_i = load angle at bus-i

$Y_{ij} = G_{ij} + B_{ij} = i - j$ th element of Y-bus matrix

N_q = number of reactive power sources in the system

b. ***Inequality Constraints:***

i. *Power generation limit*: This includes the upper and lower real power limit of generators.

$$P_{gi}^{\min} \le P_{gi} \le P_{gi}^{\max} \quad i = 1, 2, 3, \ldots, N_g \quad (4.3.146)$$

where $P_{gi}^{\min}$ and $P_{gi}^{\max}$ are the minimum and maximum limits of real power generation at bus-i, respectively.

ii. *Reactive power generator limit*: Let $Q_{gi}^{\max}$ and $Q_{gi}^{\min}$ be the maximum and minimum reactive power generation limits of reactive source generator-i, respectively, mathematically it can be written as:

$$Q_{gi}^{\min} \le Q_{gi} \le Q_{gi}^{\max} \quad i = 1, 2, 3, \ldots, N_q \quad (4.3.147)$$

iii. *Voltage limit:* This includes the upper ($V_i^{\max}$) and lower ($V_i^{\min}$) limits on the bus voltage magnitude.

$$V_i^{\min} \le V_i \le V_i^{\max} \quad i = 1, 2, 3, \ldots, N_b \quad (4.3.148)$$

iv. *Phase angle limits:* The phase angle at each bus should be between lower ($\delta_i^{\min}$) and upper ($\delta_i^{\max}$) limits.

$$\delta_i^{\min} \le \delta_i \le \delta_i^{\max} \quad i = 1, 2, 3, \ldots, N_b \quad (4.3.149)$$

These limits may vary depending upon the problem under consideration. Imposing phase angle limits at load buses is another way of limiting the power flow in the transmission lines and as for generator buses; this is done for stability reasons.

v. *Line flow limits*: These constraints represent the maximum power flow in a transmission line and are usually based on thermal and dynamic stability considerations. Let P_{li}^{max} be the maximum active power flow in line-*i*, respectively. The line flow limit can be written as

$$P_{li}^{\max} \geq P_{li} \quad i = 1, 2, 3 \ldots N_l \tag{4.3.150}$$

vi. *FACTS control parameter limits:* For the TCSC, the capacitive reactance should be within limits and for TCPAR, the phase angle must be in limit.

$$0 \leq x_{ck} \leq x_{ck}^{\max} \tag{4.3.151}$$

$$0 \leq \phi_T \leq 2\pi \tag{4.3.152}$$

The voltage magnitude (V_T) and phase angle (ϕ_T) of series voltage of UPFC must lie within limit. Mathematically, it can be written as

$$0 \leq V_T \leq V_{T\max} \tag{4.3.153}$$

$$0 \leq \phi_T \leq 2\pi \tag{4.3.154}$$

Reactive power component of shunt current (I_q) should also be less than this rating.

$$I_q^{\min} \leq I_q \leq I_q^{\max} \tag{4.3.155}$$

The series voltage magnitudes (V_{s1} and V_{s2}) and phase angle (ϕ_{s1} and ϕ_{s2}) of series voltage of GUPFC must lie within limit. Mathematically, it can be written as

$$0 \leq V_{s1} \leq V_{s1\max} \tag{4.3.156}$$

$$0 \leq V_{s2} \leq V_{s2\max} \tag{4.3.157}$$

$$0 \leq \phi_{s1} \leq 2\pi \tag{4.3.158}$$

$$0 \leq \phi_{s2} \leq 2\pi \tag{4.3.159}$$

Mathematically, Eqs. (4.3.135)–(4.3.157) can be written as

$$\text{Min } F(x, u, p) \tag{4.3.160}$$

subject to

$$h(x, u, p) = 0 \text{ (equality constraints)} \tag{4.3.161}$$

$$g(x, u, p) > 0 \text{ (inequality constraints)} \tag{4.3.162}$$

where

x = state vector, i.e. V, δ, FACTS parameters

u = control parameters viz. P_{gi}, Q_{gi}

p = fixed parameters viz. P_{di}, Q_{di}

Congestion Management by Optimizing FACTS Device Location

Existence of network constraints dictates the finite amount of power that can be transferred between two points on the electric grid. In practice, it may not be possible to deliver all bilateral and multilateral contracts in full and to supply all pool demand at least cost due to violation of operating constraints such as voltage limits and line overloads (congestion). Congestion management [86, 87] essentially has to do with rationing of transmission access. Rationing has to follow a user-pay philosophy where willingness to pay so as not to be constrained is an indicator of the importance that the parties to a transaction place on unfettered dispatch. The objective function of the dispatch problem in a system with bilateral and multilateral dispatches only, that is without pool loads, would be:

$$\min f(u,x) = [(u-u^o)\cdot A]\cdot w\cdot\left[(u-u^o)^T\cdot A\right] \tag{4.3.163}$$

where

$\boldsymbol{w}$ = a diagonal matrix whose elements are willingness-to-pay price premiums to avoid transmission curtailment

$\boldsymbol{u}$ = a set of control variables consisting of active power injected or extracted at generator buses and load buses, respectively

$\boldsymbol{u}^o$ = the desired or target value of u (bilateral and multilateral contracts)

$\boldsymbol{x}$ = the set of dependent variables

$\boldsymbol{A}$ = a constant matrix reflecting curtailment strategies used by market participants

The willingness-to-pay parameters w are essentially customer driven and bear a relationship to the value of lost load concept. Users with more sensitive loads can bid higher value of w and hence reduced curtailment.

When pool and bilateral transaction coexist, the following mathematical model describes an optimal "curtailment" problem where the dispatch of all pool demand and all bilateral transaction in full would have resulted in the violation of operating constraints.

$$\begin{aligned}\min \ & \sum_{i\in I_G} C_i(P_{pi}) - \sum_{j\in I_D} B_j(D_{pj}) + \sum_{j\in I_D} w_{Dj}\left(D_{pj}-D_{pj}^o\right)^2 \\ & + \sum_{k\in I_T} w_{Dk}(P_{tk}-P_{tk}^o)^2 - \sum_{k\in I_T} C_{tk}(P_{tk})\end{aligned} \tag{4.3.164}$$

subject to

$$\widetilde{\boldsymbol{L}}(\boldsymbol{P}_p,\boldsymbol{D}_p,\boldsymbol{P}_t,\boldsymbol{Q},\boldsymbol{V},\boldsymbol{\theta},X_k) = 0 \tag{4.3.165}$$

$$\widetilde{\boldsymbol{G}}(\boldsymbol{P}_p,\boldsymbol{D}_p,\boldsymbol{P}_t,\boldsymbol{Q},\boldsymbol{V},\boldsymbol{\theta},X_k) \le 0 \tag{4.3.166}$$

where

I_G = a set of pool generator buses
I_D = a set of pool load buses
P_{pi} = active power of pool generator-i
C_i = bid price of pool generator-i
D_{pj} = active power of pool load-j
B_j = bid price of pool load-j
$\boldsymbol{P_p}$ = vector of pool power injections
$\boldsymbol{D_p}$ = vector of pool power extractions
$\boldsymbol{P_t}$ = vector of bilateral contracts
$\boldsymbol{Q}$ = vector of reactive powers
$\boldsymbol{V},\boldsymbol{\theta}$ = vector of voltage magnitudes and angles, respectively
D^0_{pj} = desired value of pool demand at bus-j
P^0_{tk} = desired value of bilateral contract-k
C_{tk} = charge for delivering P_{tk}
$\boldsymbol{I_T}$ = set of bilateral/multilateral transactions $\boldsymbol{P_t}$
X_k = FACTS device control parameter

The objective function (4.3.164) measures the cost of production less social benefit as in the conventional optimal dispatch problem, and is then augmented by two terms to model economic dysfunctionality of congestion-induced shortages and a final term to model transmission charges for contract-type transactions.

Equality constraint (4.3.165) is basically a set of contracted transaction relationships in addition to conventional power balance equations while expression (4.3.166) is a set of inequality constraints indicating the magnitude of pool demands (mathematically upper limits) in addition to the usual system operating constraints such as bus voltage levels, line overloads, and generation limits. The weights, w_{Dj} and w_{tk}, are introduced to accommodate the interests of both pool and bilateral participants during congestion by levying extra charges to reduce curtailments. These weights, therefore, are called willingness-to-pay factors. Expressions (4.3.165) and (4.3.166) constitute an augmented constraint set for the deregulated power system.

For optimal location of FACTS devices for congestion management, Eqs. (4.3.165) and (4.3.166) are solved with FACTS control parameter as another control variable placing one device at a time in most sensitive individual lines. The final placement of these devices in lines is decided on the basis of minimum optimal objective function.

The Minimized Production Cost along with the FACTS Devices Cost

Due to high cost of FACTS devices, it is necessary to use cost–benefit analysis to analyze whether a new FACTS device is cost effective among several candidate locations when actually installed. Therefore, the objective function changed to the minimized production cost along with the FACTS devices cost. The TCSC cost in line-k can be given by

$$C_{\text{TCSC}}(k) = c{\cdot}x_c(k){\cdot}P_l^2{\cdot}\text{Base\_power} \tag{4.3.167}$$

where c is the unit investment cost of FACTS ($/MVAR), $x_c(k)$is the series capacitive reactance, and P_l is the power flow in line-k. It is assumed that the cost of TCPAR is same for all the lines. Therefore, for TCPAR the objective function will be the same.

The objective function for placement of FACTS will be:

$$\min_{P_{pi}} \sum_{i \in I_G} C_i(P_{pi}) - \sum_{j \in I_D} B_j(D_{pj}) + C^t{}_{\text{FACTS}} \tag{4.3.168}$$

where

$$C^t_{\text{FACTS}} = \frac{\alpha{\cdot}C_{\text{FACTS}}(k)}{8760} \quad \text{and} \quad \alpha = \frac{r(1+r)^n}{(1+r)^n - 1} \tag{4.3.169}$$

where α = the capital recovery factor (CRF); r = the interest rate; n = the capital recovery plan.

Enhancement of Total Transmission Capability by FACTS Devices

In deregulated power systems, total transfer capability (TTC) analysis is presently a critical issue either in the operating or planning because of increased area interchanges among utilities. A recent ruling by Federal Energy Regulatory Commission (FERC) requires electric utilities in the United States to provide transmission services for wholesale customers and under this ruling, utilities would be required to post information on ATC of their transmission networks. Such information will help power marketers, sellers, and buyers in reserving transmission services. Utilities, therefore, would have to determine the adequacy of their ATCs to ensure that the system reliability is maintained. According to North American Electric Reliability Corporation (NERC) documentation, ATC is defined to be composed of the following terms, viz. TTC, existing transmission commitment (ETC), transmission reliability margin (TRM), and capacity benefit margin (CBM).

A 1996 report by NERC establishes a framework for determining ATCs of the interconnected networks for a commercially viable wholesale electricity market. The ATC principles stated include the following:

- ATC calculations must recognize time-varying power flow conditions and simultaneous transfers and parallel path flows throughout the transmission network.
- ATC calculations must recognize the dependency of ATC on the point of power injection, the direction of power transfers, and the point power extraction.
- ATC calculation must produce commercially viable results and computed ATCs must give reasonably accurate and dependable indication of transfer capabilities available to the electric power market.

The ATC is composed of the following terms. The TTC is the total transfer capability between two areas and the ETC is the sum of the existing transmission commitment between those areas. The transmission reliability margin (TRM) is the amount of transmission capability necessary to ensure that the interconnected system is secure under a reasonable range of uncertainty. The cost benefit margin (CBM) is the amount of the transmission transfer capability reserved by load-serving entities to ensure access to generation from interconnected systems to meet generation reliability requirements. ATC is defined as:

$$\mathrm{ATC} = \mathrm{TTC} - \mathrm{TRM} - \mathrm{ETC} - \mathrm{CBM} \tag{4.3.170}$$

TTC can be calculated by solving optimal power flow (OPF) algorithms. To compute, for a given system state, the TTC from one location (bus-i) to another location (bus-j), the following procedure is used:

i. Obtain base case by running the OPF for the given system load and generation and obtain the flows on the selected transmission path between the two buses.
ii. Connect a generator of high capacity at bus-i. Connect a generator and a sink (load) at bus-j of the same capacity as of generator at bus-i.
iii. Set the higher cost of generator at bus-j than at bus-i. By fixing the other generation at the base level, run OPF to obtain the generation at bus-i.
iv. The output of the generator at bus-i will be the TTC.

This approach can be used to calculate the TTC with and without FACTS devices.

Enhancement of Power Systems' Security by FACTS Devices

Stressed power system, either due to increased loading or due to severe contingencies, often lead to situation where the system no longer remains in the secure operating region. Under these situations, it is the primary objective of the operator to apply control action to bring the power system into the secure region. Any delay or unavailability of suitable control, the system may become unstable. The security of power system can be defined as its ability to withstand a set of severe but credible contingencies and to survive transition to an acceptable new steady-state condition. In the present-day power system, there will be an increase in number of situations where power flow equations have either no real solution (unsolvable case) or solution with violating operating limits such as voltage limit (insecure case), particularly, in contingency analysis and planning applications [88–90].

Since insecure cases often represent the most severe threats to secure system operation, it is important that the FACTS devices should enhance the system security along with the other control devices. Few papers have dealt with the power system insecurity and possible control to bring the system back to secure state. The possibility of controlling power flow in an electric power system without

TABLE 4.3.6 Sensitivities (c_k^c) of 5-Bus System

Line			Outage of Lines					
No	*i–j*	Base Case	1–2	2–5	3–5	5–4	1–4	3–2
1	1–2	−0.481	—	0.089	−3.540	0.000	0.001	−0.359
2	2–5	6.868	6.929	—	16.65	4.947	27.83	1.952
3	3–5	**−2.569**	**−4.520**	0.030	—	**−3.226**	**−13.59**	0.033
4	5–4	0.033	0.022	0.317	−1.585	—	0.972	0.079
5	1–4	−1.702	0.027	1.494	**−9.680**	0.001	—	**−1.414**
6	3–2	0.722	0.914	**−0.926**	−0.370	0.842	0.173	—

generation rescheduling or topological changes can improve the performance considerably. The increased interest in these devices is essentially due to increased loading of power systems, combined with deregulation of power industry, and motivates the use of power flow control as a very cost-effective means of dispatching specified power transactions. It is important to ascertain the location for placement of these devices because of their considerable costs.

The proposed method has been tested on a 5-bus system as shown in Figure 4.3.12. Sensitivities were calculated for TCSC placed in every line one at a time for the same operating conditions. The sensitivities of real power PI with respect to TCSC for base case and different line outages are presented in Table 4.3.6. The highest negative sensitivities are presented in bold type.

From the load flow, it was found that real power flows in line-2 for base case was 1.267 p.u., which is more than its line-loading limit. It can be observed from Table 4.3.6 that placement of TCSC in line-3 is suitable for reducing the overloading of line 2–5. Placement of TCSC in line-5 will reduce the PI value but it will be less effective than placing a TCSC in line-3 as can be seen from its sensitivity factors.

From Table 4.3.7, it can be seen that line 2–5 is getting overloaded in most of the case except line 3–2 outage. In case line 2–5 outage, line 3–5 gets overloaded. The line overloading is presented in bold face. Table 4.3.6 indicates that the placement of TCSC in line 3–5 is suitable for four cases followed by line 1–4 which is suitable for two cases in which line 3–2 outage is not causing any overloading of the system. Placement of TCSC in line 3–2 is the next choice, which is suitable for only one case that is line 2–5 outage.

Sensitivity factor (c_k^s) for different outages are presented in Table 4.3.8. Table 4.3.8 shows that placement of TCPAR in line-2 is more sensitive for base case than the placement in other lines. This sensitivity is positive which indicates that phase angle shift of the TCPAR should be negative. Placing of TCPAR in line-2 will reduce the loading of line-2 (heavily loaded line) but it will increase the loading of lines 4 and 6 that are underloaded. Table 4.3.8 also shows that the placement of TCPAR in line-3 or line-6 is the next choice as the magnitude of sensitivity

TABLE 4.3.7 Line Flows (in p.u.) of 5-Bus System

Line		Outage of Lines					
i–j	Base Case	1–2	2–5	3–5	5–4	1–4	3–2
1–2	0.193	—	0.715	0.328	0.165	0.687	0.159
2–5	**1.267**	**1.420**	—	**1.840**	**1.288**	**1.906**	**1.132**
3–5	0.722	0.778	**1.446**	—	0.729	0.938	0.890
5–4	0.028	0.159	0.584	0.182	—	0.800	0.004
1–4	0.833	0.646	**1.399**	0.989	0.800	—	0.800
3–2	0.170	0.122	0.576	0.895	0.163	0.050	—

TABLE 4.3.8 Sensitivities (c_k^s) of 5-Bus System

Line		Outage of Lines					
i–j	Base Case	1–2	2–5	3-5	5–4	1–4	3–2
1–2	2.150	—	−0.132	10.39	0.00	0.000	1.855
2–5	5.457	5.000	—	8.975	3.880	14.55	1.716
3–5	−3.526	−5.848	0.823	—	−4.382	−14.19	0.032
5–4	2.051	0.084	−0.530	9.605	—	1.027	1.777
1–4	−2.057	0.039	0.999	−9.570	0.00	—	−1.772
3–2	3.662	5.924	0.387	−0.460	4.427	15.48	—

factors is the second highest. The sensitivity for line-3 and line-6 is almost same, as these lines are series with each other for phase angle control. Outage of line 1–2, the line 3–5 is suitable but the sensitivity of line 2–5 is also very close to this line.

For the outage of line 2–5, the line 1–4 becomes more sensitive to other lines. It can be seen in Table 4.3.8. To check the effectiveness of the proposed method-3, the line 2–5 is more suitable than other lines for all the outage cases, except line 2–5 in which it has to be placed. The sensitivity of line-2 is positive which shows that the phase angle should be in such a way that it should oppose the flow of power in line 2–5.

To check the effectiveness of the proposed method, the line loading has been increased by changing the generation schedule. 60 MW has decreased the generation at bus-2 while keeping the load constant at buses 4 and 5. The change in generation will be taken care by slack bus generator. The line flows have been changed and now line 1–4 is also critical which can be seen from Table 4.3.9. The sensitivity factors calculated for TCSC and TCPAR are given in Table 4.3.9. The placement of TCSC in line-5 gives positive sensitivity, that is, the PI value will increase if a TCSC is placed in that line. This is opposite to the previous base case. The

TABLE 4.3.9 PI Sensitivities of 5-Bus System for Different Loading Limits

	Lines					
Factors/flows	2–1	2–5	3–5	5–4	1–4	3–2
Flows	0.157	1.081	0.660	0.273	1.081	0.234
c_k^c	−0.044	2.219	−1.678	0.145	0.550	0.674
c_k^s	−0.544	2.025	−2.499	0.622	0.522	2.671

placement of TCSC in line 3–5 will now reduce the PI but will not when TCSC is placed in line-5. The sensitivity of line 2–1 is also negative and the magnitude is very small; hence, it is not suitable for the TCSC placement. Only line 3–5 is suitable for TCSC placement.

The absolute value of sensitivity c_k^s corresponding to line-6 is the highest which is slightly less, in magnitude, than line-3. The placement of TCPAR in line-6 will be the choice but with negative phase shift. The placement of TCPAR in line-2 that was the first choice for the previous base case is now the third choice after lines 3 and 6. Any reduction in line 2–5 flow will increase the flow of line 1–4 also and thus will cause congestion in the system. Similarly, placement of TCPAR in line 1–4 for reduction in the line flow will increase the flow of line 2–5 as well. The suitable choice will be line 3 or 6. Increasing flow in line 3–5 or decrease flow in line 3–2 will reduce the power flow in lines 2–5 and 1–4, simultaneously.

4.3.6 Use of Particle Swarm Optimization

The PSO, as described in Section 4.3.3, has been used to solve FACTS placement (Eq. 4.3.168). The placement of FACTS placement using PSO has been examined on modified IEEE 14-bus system consisting of four real power generator buses and one voltage-regulating bus. Generators at buses 1, 2, 3, and 4 bid into the pool. Generator at bus-1 has been taken as reference bus. Voltage magnitudes at load buses are kept within the range of 0.95–1.10. The prices bid by generators are given in Table 4.3.10 where P is in MW and \$ is a monetary unit which may be scaled by any arbitrary constant without affecting the results.

The generation schedule obtained from optimal dispatch using PSO without considering the line flow limit and FACTS devices was 86.0, 108.0, 103.7, and 200.4 MW for generators 1, 2, 3, and 4, respectively. With this generation schedule, it was found that the real power flows in lines were within the rating limit, except line-2 which was 96.8 MW. The rating of each line is given in Table 4.3.11. Sensitivities were calculated for FACTS devices (TCSC and TCPAR) placed in every line one at a time for this operating condition. The sensitivities of real power PI with respect to TCSC and TCPAR are presented in Table 4.3.11.

TABLE 4.3.10 Generator Data

Generator	Bid Price ($/h)	Specified Voltage
1	$0.005\ P^2 + 3.71\ P$	1.08
2	$0.005\ P^2 + 3.52\ P$	1.08
3	$0.005\ P^2 + 3.89\ P$	1.08
4	$0.005\ P^2 + 2.45\ P$	1.08

TABLE 4.3.11 Sensitivity Factors and Rating of Lines

Line -*k*				TCSC	TCPAR
No.	*i*–*j*	Line Rating (MW)	Power Flow (MW)	(c_k^c)	(c_k^s)
1	8–3	150.0	23.5	–0.585	–2.683
2	9–6	150.0	49.9	2.721	5.535
3	9–7	150.0	27.9	–0.950	–3.372
4	4–8	150.0	100.3	–3.978	–4.066
5	2–8	150.0	96.8	1.435	1.502
6	1–9	150.0	90.3	0.441	0.482
7	8–9	150.0	0.6	0.015	–0.265
8	4–2	90.0	96.8	2.502	3.460
9	2–1	90.0	6.9	0.053	0.943
10	6–5	50.0	0.0	0.000	0.000
11	2–9	150.0	95.0	1.419	1.485
12	6–7	50.0	49.9	2.722	5.645
13	7–10	50.0	14.4	0.442	2.658
14	3–11	150.0	59.8	–1.280	–2.139
15	3–12	50.0	19.7	–0.302	–1.511
16	3–13	50.0	45.2	0.877	1.865
17	7–14	50.0	5.6	0.015	–0.077
18	11–10	50.0	4.7	–1.129	–2.407
19	12–13	50.0	3.3	–0.037	–1.520
20	13–14	50.0	20.2	0.066	0.277

Sensitivity factor of TCSC for line-4 is the most negative than the other lines and hence the most suitable for the TCSC placement. Branches 2 and 12 are the most sensitive for TCPAR placement but have not been considered due to connecting the tertiary winding of the transformer. The next choices for placement of TCPAR would be line-4 and line-8. Table 4.3.11 also shows that the placement of TCPAR in line-8 is the other choice as the magnitude of sensitivity factors is the next highest. This sensitivity is positive which indicates that phase angle shift of the TCPAR should be negative.

TABLE 4.3.12 Optimal Generation Schedule

			TCPAR in	
Generators	Base Case	TCSC in Line-4	Line-4	Line-8
G1	89.9 + j 27.8	87.0 + j 28.8	86.1 + j 23.6	86.0 + j 27.3
G2	107.5 + j 88.6	110.2 + j 88.5	108.1 + j 72.7	107.8 + j 126.1
G3	110.7 + j 80.2	104.8 + j 80.9	102.9 + j 74.7	103.6 + j 79.6
G4	190.2 − j 1.4	196.6 − j 1.1	201.2 + j 2.6	201.1 − j 24.3
Total Cost ($/h)	1948.68	1948.18	1943.1	1945.1

The optimal dispatch, described in Eq. (4.3.168), with TCSC in line-4, TCPAR in line-4 and in line-8, taken one at a time, was obtained and presented in Table 4.3.12. Table 4.3.12 also shows the optimal dispatch without FACTS devices (base case). The optimal value of capacitive reactance of TCSC placed in line-4 was 0.0132 p.u. This indicates that 5.93% of compensation is optimal. However, the optimal value of phase angle of TCPAR placement in line-4 and line-8 were 5.77° and –3.95°, respectively. The placement of TCPAR in line-4 gives less pool generation price compared to other cases. The variation of optimal pool real power generation with each FACTS device is very small.

FACTS device cost is not considered in the case of TCPAR which will be the same for all the cases for fixed range of phase shifters.

4.3.7 Conclusions

FACTS devices have the ability to control the power flow in the network by which the static and dynamic performance of the system can be improved significantly. These devices can play a major role in both regulated and deregulated power systems. However, these devices were used mainly for improving the dynamic performances in vertically integrated power systems. Due to present trends in power system restructuring to provide competition in the electric supply industry, the role of these devices is not only limited to improve the dynamic performance but also static performances such as congestion management, enhancement of power transfer capability, improving the system security, etc.

The locations of FACTS devices are very important as these devices are costly and their proper location may improve the performance and the economic operation of the system. By including FACTS devices in the OPF problem, significant reduction in both total real power loss and total reactive power loss is obtained. The FACTS devices also help in reducing the generation cost in the deregulated energy market. Implementation of FACTS devices into the OPF is an effective way to simulate real and reactive power spot market. Both real and reactive power spot prices are expressed as a combination of different price components. Cost of the FACTS devices can also be allocated according to the reactive power loading condition in the system.

4.4 HYBRID OF ANALYTICAL AND HEURISTIC TECHNIQUES FOR FACTS DEVICES

Kwang Y. Lee[1], *Malihe Maghfoori Farsangi*[2], *and Hossein Nezamabadi-pour*[2]

[1]*Baylor University, Waco, TX, USA*
[2]*Shahid Bahonar University of Kerman, Kerman, Iran*

This chapter addresses applications of intelligent techniques on FACTS devices in terms of steady state and transient improvement of power systems. Taking advantages of the FACTS devices depends greatly on how these devices are placed in the power system, namely on their location and size. In a practical power system, allocation of the devices depends on a comprehensive analysis of steady-state stability, transient stability, small signal stability, and voltage stability. Moreover, other practical factors such as cost and installation conditions also need to be considered. Thus, allocation could be a multi-objective problem where finding a solution is not simple by analytical methods. Therefore, controlling of FACTS devices using heuristic methods has been paid attention to assure the security of the system in terms of voltage and angle stability. This chapter aims to show that, due to many attractive features of intelligent techniques, the heuristic methods are becoming popular for solving complex problems such as applications of FACTS devices in terms of steady state and transient improvement of transmission systems.

4.4.1 Introduction

In the last decades, efforts have been made to find the ways to assure the security of power system in terms of steady state and transient stability, and it became possible by finding an attractive technology such as FACTS. Applications of this technology started with the SVC since 1970, and were followed by the TCSC. Then, advances in power electronics devices allowed the use of the second generation of FACTS devices based on the self-commutated voltage-sourced converter (VSC) using GTO thyristor technology. It includes the STATic synchronous COMpensator (STATCOM), static synchronous series compensator (SSSC), VSC-based static phase shifter (SPS), UPFC, and interline power flow controller (IPFC).

It was shown by researchers around the world that FACTS devices are good choices to improve the voltage profile in power systems that operate near their steady-state stability limits and may result in voltage instability [91]. Furthermore, many studies have been also carried out on the use of FACTS devices in improving transient and angle stability [92–99].

Taking advantages of the FACTS devices depends greatly on how these devices are placed in the power system, namely on their location and size. In view of this, different works are reported around the world to find a way to place FACTS devices optimally in power systems.

In addition to their primary function of FACTS devices, the supplementary damping control action can be also added and how to utilize their control capabilities effectively as stabilizing aids is becoming very important. Many articles can be found that try to answer the question of which feedback signal could result in the FACTS devices having the maximum effect on the system damping [100].

Over the last decades there has been a growing interest in algorithms inspired from the observation of natural phenomenon. It has been shown by many studies that these algorithms are good replacement as tools to solve complex computational problems [101–115]. Various heuristic approaches have been adopted by researchers including GA, evolutionary strategies and programming, differential evolution, Tabu search (TS), simulated annealing (SA), ant colony system, PSO, immune system, etc.

In view of this, this chapter aims to show the applicability of different heuristics methods such as SA, TS, GA, PSO, and guaranteed convergence PSO (GCPSO) on FACTS devices to improve the performance of the power system.

Advantages of employing heuristic optimization techniques will be presented. Two examples will be described to demonstrate their applications on FACTS devices: (i) using of SA, TS, GA, and PSO in placing a single three SVCs in a power system and (ii) enhancement of TTC by placing of TCSC using PSO, GCPSO, and GA.

4.4.2 Heuristic Algorithms

A brief description is given for some heuristic algorithms.

Simulated Annealing

SA is a derivative-free optimization technique that simulates the physical annealing process in the field of combinatorial optimization [113, 114]. Annealing is the physical process of heating up a solid until it melts, followed by slowly cooling it down by decreasing the temperature of the environment in steps to obtain a perfect structure corresponding to a minimum energy state. The SA is a global search strategy, which tries to avoid local minima by accepting worse solutions with a probability. The probability to accept a solution is defined according to the Metropolis distribution:

$$p_i = \begin{cases} 1 & \text{if } f(y) < f(x) \\ \exp\left(\dfrac{f(x) - f(y)}{T}\right) & \text{otherwise} \end{cases} \tag{4.4.1}$$

where x and y are an initial and new solutions, respectively, and $f(\cdot)$ is the evaluation of the objective function at a solution. The SA starts from an initial solution x and then a solution y is generated. If y has been improved from x, it will be accepted; otherwise, y will be accepted as the current configuration with a probability proportional to the difference in the value of the objective function, $f(x) - f(y)$, on temperature T as shown in (4.4.1).

Tabu Search

TS algorithm is used to solve combinatorial optimization problem. This mechanism is somewhat similar to SA, but the main difference between the two search algorithms is that TS uses a flexible memory to store the information and data of the solutions in each iteration. This is to get to the lower objective function values with the help of the information stored, while special features are added to escape from being trapped in the local minima.

The TS searches the neighborhood of the current solution to find the next solution with more improvement in the value of objective function. In order to avoid returning to the local optimum, TS uses a list which is called Tabu List. The elements of the tabu list are called tabu moves. Tabu list stores the moves in a data structure such as finite length, and restricts the local search algorithm in reusing those moves for some iteration. The number of iterations a move is kept in the list is called tabu list size. Since the tabu list may forbid certain worthy or interesting moves found so far, these tabu moves will be accepted as the next move and will be released from the tabu list if they are judged to be worthy [113].

Genetic Algorithm

GA has desirable characteristics as an optimization tool and offers significant advantages over traditional methods. It is inherently robust and has been shown to efficiently search the large solution space containing discrete or discontinuous parameters and nonlinear constraints, without being trapped in local minima [115, 116].

The GA may be used to solve a combinatorial optimization problem. The GA searches for a solution inside a subspace of the total search space. Thus, it is able to give a good solution of a certain problem in a reasonable computation time. The optimal solution is sought from a population of solutions using random process. Applying to the current population, the following three operators create a new generation: reproduction, crossover, and mutation. The reproduction is a process dependent on an objective function to maximize or minimize, which depends on the problem. A typical simple GA is described in detail in [116].

Standard PSO

In PSO, each particle moves in the search space with a velocity according to its own previous best solution and its group's previous best solution [45]. The dimension of the search space can be any positive integer. Each particle updates its position and velocity with the following two equations:

$$X_i(t+1) = X_i(t) + V_i(t+1) \tag{4.4.2}$$

where $X_i(t)$ and $V_i(t)$ are vectors representing the position and velocity of the ith particle, respectively, and

$$\begin{aligned} V_{i,j}(t+1) = {} & wV_{i,j}(t) + c_1 r_{1,j}\left(pb_{i,j} - X_{i,j}(t)\right) \\ & + c_2 r_{2,j}\left(gb_j - X_{i,j}(t)\right) \end{aligned} \tag{4.4.3}$$

where $j \in 1, 2, \ldots, d$ represents the dimension of the particle; $0 \leq w < 1$ is an inertia weight determining how much of the particle's previous velocity is preserved; c_1 and c_2 are two positive acceleration constants; $r_{1,\ j}$, $r_{2,\ j}$ are two uniform random sequences sampled from $U(0, 1)$; pb_i is the personal best position found by the ith particle; and gb is the best position found by the entire swarm so far.

The PSO has been proven to be very effective for static and dynamic optimization problems. But in some cases, it converges prematurely without finding even a local optimum. Standard PSO may converge at the early stage; the best particle moves based only on the inertia term since $X_i = pb_i = gb$ at the time step when it became the best. Later, its position may improve where $X_i = pb_i = gb$ holds again. Also, its position will be worse where it will be drawn back to $pb_i = gb$ by the social component. Therefore, it is possible for the inertia weight to drive all velocities to zero before the swarms manage to reach a local extremum. When all the particles collapse with zero velocity on a given position in the search space, then the swarms have converged, but this does not mean that the algorithm has converged on a local extremum. It merely means that all the particles have converged on the best position discovered so far by the swarm. This phenomenon is referred to as stagnation [117]. Thus, it is possible for the standard PSO to converge prematurely without finding even a local extremum.

Guaranteed Convergence PSO

The GCPSO was introduced by Van den Bergh and Engelbrecht [117] to address the issue of premature convergence to solutions that are not guaranteed to be local extrema. The modifications to the standard PSO involve replacing the velocity update (4.4.3) of only the best particle with the following equation:

$$V_{i,j}(t+1) = wV_{i,j}(t) - X_{i,j}(t) + pb_{i,j} + \rho(t)r_j \tag{4.4.4}$$

where r_j is a sequence of uniform random numbers sampled from $U(-1,1)$ and $\rho(t)$ is a scaling factor determined using

$$\rho(0) = 1.0$$

$$\rho(t+1) = \begin{cases} 2\rho(t) & \text{if \# successes} > s_c \\ 0.5\rho(t) & \text{if \# failures} > f_c \\ \rho(t) & \text{otherwise} \end{cases} \tag{4.4.5}$$

where s_c and f_c are tunable threshold parameters.

Whenever the best particle improves its personal best position, the success count is incremented and the failure count is set to 0, and vice versa. The success and failure counters are both set to 0 whenever the best particle changes. These modifications cause the best particle to perform a directed random search in a non-zero volume around its best position in the search space.

4.4.3 SVC and Voltage Instability Improvement

Voltage stability is the ability of a power system to maintain acceptable voltages at all system buses under normal operation as well as following a disturbance. Voltage stability can be categorized as the large-disturbance voltage stability and the small-disturbance voltage stability. The large-disturbance voltage stability is the ability of the system to control the voltage after being subjected to a large disturbance such as system faults, and loss of load or generation. The small signal voltage stability is the ability of the system to control voltage after being subjected to a small perturbation, such as gradual changes in loads.

In this chapter, two techniques are used for analysis of voltage stability in a study system. The study system and brief explanations of two techniques are given below.

Study system

The system shown in Figure 4.4.1 consists of 16 machines and 68 buses for 5 interconnected areas. The first nine machines, G1 to G9, constitute the simple representation of Area 1. Next four machines G10 to G13, represent Area 2. The last three machines, G14 to G16, are the dynamic equivalents of the three large neighboring areas interconnected to Area 2. The sub transient reactance model for the generators, the first-order simplified model for the excitation systems, and the linear models for the loads and AC network are used. A power system stabilizer (PSS) is placed on machine 9 [100].

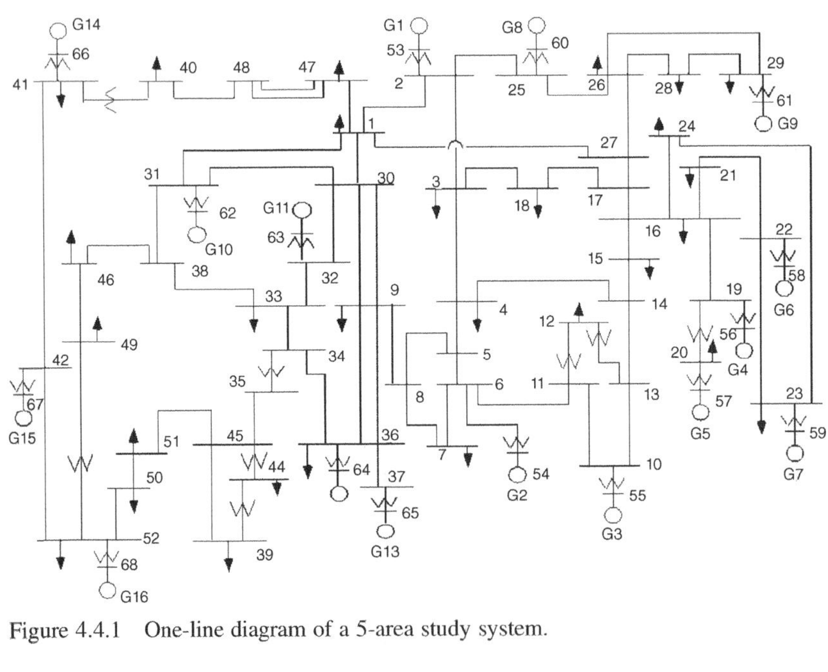

Figure 4.4.1 One-line diagram of a 5-area study system.

Placement Using Critical Modes of Voltage Instability (Modal Analysis)
Modal or eigenvalue analysis involves the computation of eigenvalues and eigenvectors of the system near the point of voltage collapse in order to identify different modes through which the system could come to the voltage collapse. There is no need to drive the system precisely to its "nose point," where the power flow does not converge to ensure that a maximum level of stress is reached.

In the study system, all loads are increased gradually near to the point of voltage collapse. Using modal analysis, it is found that the weakest area in this power system is in the study system near bus 40, as illustrated in Figure 4.4.2. Figure 4.4.3 shows the profile of the voltage when the system is heavily stressed and reaches the point of collapse. Based on the modal analysis, bus 40 is a good candidate to place the first SVC. Using the modal analysis only the weak point can be determined, and furthermore, it cannot give any information regarding the optimum size of the SVC.

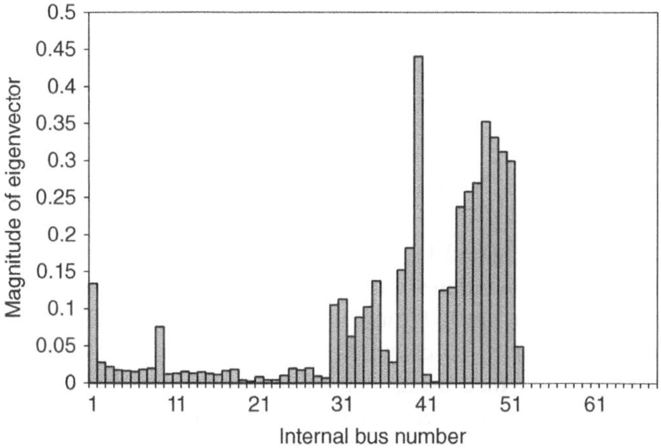

Figure 4.4.2 The critical eigenvector and the corresponding bus number.

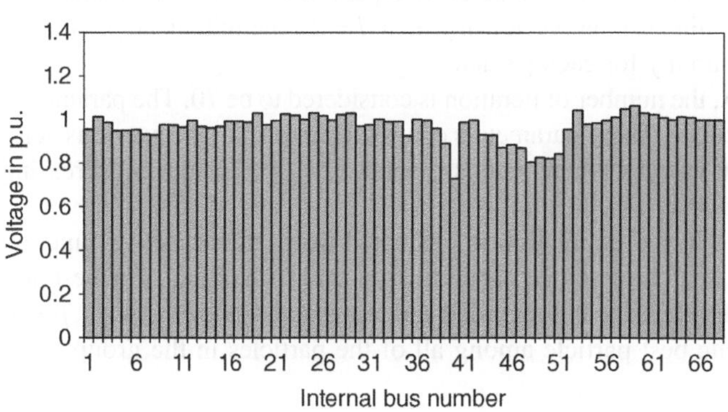

Figure 4.4.3 Bus voltage magnitude profile when system is heavily stressed.

However, the size is playing an important role for the SVC to be effective. In view of this, heuristic techniques are applied to place SVC in the system optimally.

Heuristic Methods

Placing of SVC using SA, TS, GA, and PSO starts from an initial load. All loads are increased gradually near to the point of voltage collapse, all at once.

The goal of the optimization is to find the best location of SVC where the optimization is made on two parameters: its location and size. The placing using the GA and PSO is given below in more detail, but only the obtained results by SA and TS are given. Therefore, using GA, a configuration is considered with two genes. The first gene is related to the location of SVC. The second gene is related to the size of the SVC. The number of chromosomes is set to be 50.

The chromosomes evolve through successive iterations, called generations. During each generation, the chromosomes are evaluated with some measure of fitness, which is calculated from the objective function.

In this chapter, SVC is placed based on its primary function, which is the voltage stability. For a level of loads, the following objective function is minimized:

$$\text{obj} = \sum \text{abs}|V_i - V_{\text{ref}\,i}|^6 \tag{4.4.6}$$

where V_i is the voltage magnitude and $V_{\text{ref}i}$ is the nominal voltage at bus i.

Moving to a new generation is done from the results obtained for the old generation. A biased roulette wheel is created from the obtained values of the objective function of the current population. To create the next generation, new chromosomes, called offsprings, are formed using a crossover operator and a mutation operator. In this paper, one point crossover is applied with the crossover probability $p_c = 0.9$ and the mutation probability is selected to be $p_m = 0.005$. Also, the number of iterations is considered to be 70, which is the stopping criterion.

In the PSO algorithm, n particles are generated randomly where n is selected to be 50. Since optimizations are made on two parameters, its location and size, each particle is a d-dimensional vector in which $d = 2$. The initialization is made on the position randomly for each particle.

As in the GA, the number of iteration is considered to be 70. The parameters in (4.4.3) must be tuned. These parameters control the impact of the previous velocities on the current velocity where, in this chapter, $c_1 = c_2 = 2$ and w is decreasing linearly from 0.9 to 0.1.

Each particle in the population is evaluated using the objective function defined by (4.4.6), searching for the particle associated with obj_{best}. The best previous position of the ith particle is recorded and represented as: $pb_i = (pb_{i,1}, pb_{i,2})$ and the index of the best particle among all of the particles in the group is for the gb.

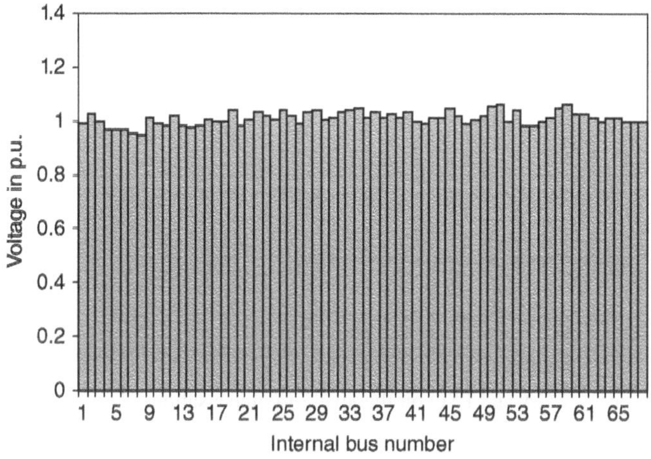

Figure 4.4.4 Voltage profile of the system after placing 145 Mvar at bus 40.

Using the *gb* and pb_i, particle velocity and position is updated according to (4.4.2) and (4.4.3). Also, in each iteration, the *gb* and pb_i are updated.

To locate SVC by heuristic techniques (SA, TS, GA, and PSO), suitable buses are selected based on 20 independent runs, under different random seeds. All algorithms except SA identify bus 40 as the bus vulnerable to the voltage collapse, which is the same result obtained by the modal analysis. Other buses are given by the algorithms such as 47, 48, 49, and 50 with different Mvar sizes. Also, TS, GA, and PSO found bus 40, but the algorithms give different Mvar sizes, where among them, 145 Mvar size is chosen.

Also, it is found that the convergence rates of GA and PSO are better than SA and TS. Figure 4.4.4 shows bus voltage magnitude profile of the stressed system after placing a 145 MVAR SVC at bus 40.

The results obtained by GA and PSO are averaged over independent runs. The average best-so-far and average level of compensation of each run are recorded and averaged over 20 independent runs. To have a better clarity, the convergence characteristics in finding the location and size of a SVC by GA and PSO is given in Figures 4.4.5 and 4.4.6. These figures show that PSO has a better feature to find the optimal solution.

After placing the first SVC at bus 40, the second SVC is going to be placed only by GA and PSO since they have a better convergence rate. Once again the loads are gradually increased. Using the modal analysis, as shown in Figure 4.4.7, the second worst bus is identified which is bus 49. Figure 4.4.8 shows the voltage profile when system is under stress. Again GA and PSO are used to find the optimal size of the SVC at bus 49, which yields 150 Mvar. Since Figure 4.4.7 shows that both buses 49 and 50 push the system to voltage instability almost at the same level of stress, the third SVC is placed at bus 50.

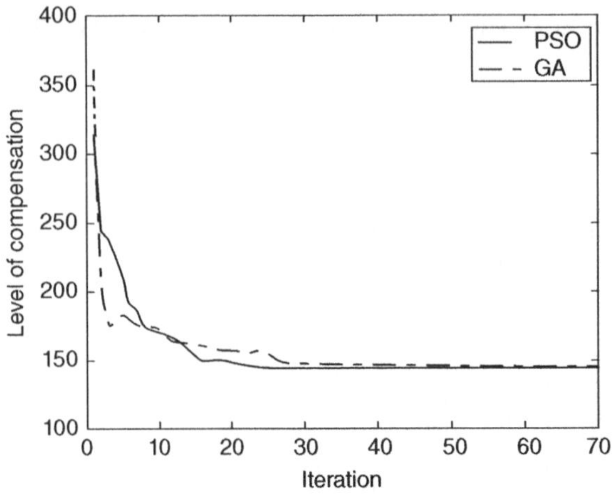

Figure 4.4.5 Convergence of GA and PSO on the level of compensation.

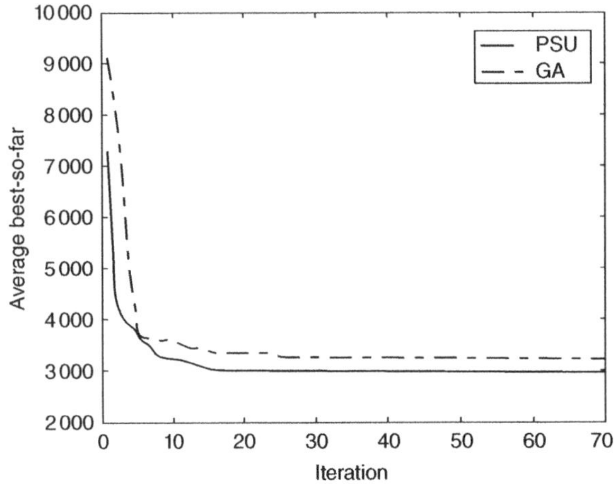

Figure 4.4.6 Convergence of GA and PSO on the average best-so-far.

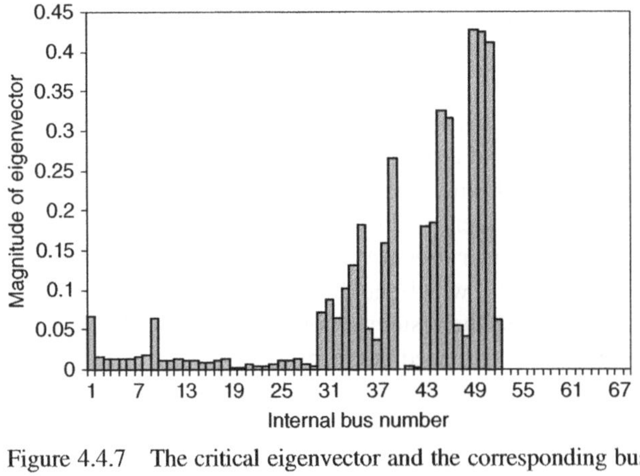

Figure 4.4.7 The critical eigenvector and the corresponding bus number in system.

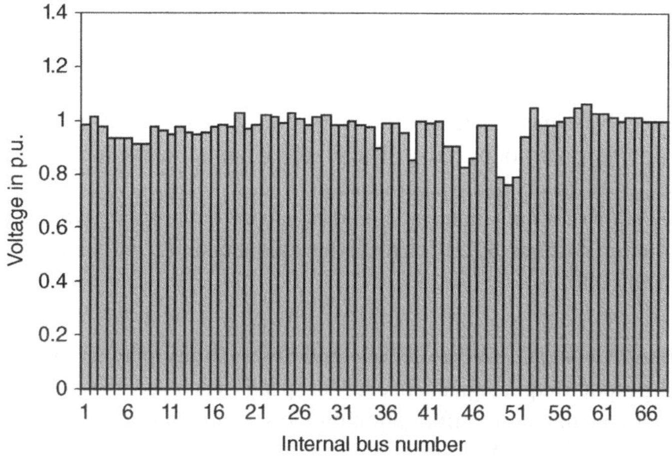

Figure 4.4.8 Bus voltage magnitude profile when system is heavily stressed in system.

By now three SVCs are placed in the study system based on the voltage stability. For the rest of the work, choosing the best input of the supplementary controllers of SVCs to damp the inter-area modes is investigated.

4.4.4 FACTS Devices and Angle Stability Improvement

Angle stability is the ability of synchronous machines of an interconnected power system to remain in synchronism following a disturbance. Instability may occur by increasing angular swings of some generators leading to their loss of synchronism with other generators following a disturbance. The rotor angle stability problem involves the study of the electromechanical oscillations.

Locating controllable power system devices, such as FACTS devices, is based on the issues unrelated to the damping of oscillations in the system. For instance, an SVC improves transmission system voltage, thereby enhancing the maximum power transfer limit; the TCSC control reduces the transfer impedance of a long transmission line, enhancing the maximum power transfer limit. Therefore, how to utilize their control capabilities effectively as stabilizing aids is becoming very important. Damping of the electromechanical oscillations by FACTS devices can be done in the following two steps:

1. Finding the most suitable supplementary input signals for the FACTS devices for damping improvement.
2. Designing a controller with a defined structure using heuristic methods.

Stabilizing signals for FACTS devices are selected based on the following stability indicators: right-half plane zeros (RHP-zeros), Hankel singular values (HSV), relative gain array (RGA), and minimum singular value (MSV) [100]. These indicators are summarized briefly as below.

Right-Half Plane Zeros The RHP-zeros limit the achievable performance of a feedback loop. Also, from the root locus analysis, it can be seen that the locations of zeros are not changed by feedback, but the pole locations are changed by feedback. As the feedback gain increases, the closed-loop poles move from the open-loop poles to the open-loop zeros. Therefore, if some zeros are in the RHP, the increased gain makes the system unstable.

Thus, the selection of inputs–outputs should be carried out in such a way that the plant has a minimum number of RHP-zeros, which are required not to lie within the closed-loop bandwidth.

Hankel Singular Values Controllability and observability play an important role in selecting input–output signals. In order to specify which combination of the input and output contains more information on the system internal states, one possible approach is to evaluate observability and controllability indices of the system, such as HSV that reflects the joint controllability and observability of the states.

HSV can be found by solving the following Lyapunov equations for the minimal realization of the state space system (A, B, C, D):

$$\begin{aligned} AP + PA^T + BB^T &= 0 \\ A^TQ + QA + C^TC &= 0 \end{aligned} \tag{4.4.7}$$

Here, P and Q are the solution to the above Lyapunov equations, and

$$\begin{aligned} &P = Q = \operatorname{diag}(\sigma_1, \sigma_2, \ldots, \sigma_n) \\ &\sigma_1 \geq \sigma_2 \geq \cdots \geq \sigma_n \geq 0 \end{aligned}$$

where the singular values σ_i (HSV) are ordered in descending order, and the first singular value is the largest and others are decreasing monotonically.

In the above realization the value of each singular value σ_i is associated with the state x_i, and the size of the singular value is a relative measure of the contribution that the corresponding state makes to the input–output behaviors of the system. Therefore, if $\sigma_i \geq \sigma_{i+1}$, then the state x_i affects the input–output behavior more than x_{i+1} does.

In choosing input and output signals, the HSV can be calculated for each combination of inputs and outputs, and the candidate with the largest HSV shows better controllability and observability properties. It means that this candidate can give more information about system internal states.

Relative Gain Array The RGA defined by $\Lambda = G(s) \times G^{-1}(s)^T$, where $G(s)$ is a multivariable plant with m inputs and m outputs, provides useful information for the pairing of inputs and outputs. Input and output variables should be paired so that the diagonal elements of the RGA are unity as close as possible. It is desired that Λ has small elements and for a diagonal dominance, $\Lambda - I$ to be small. These two objectives can be considered in the single objective, known as RGA-number defined as follows:

$$\text{RGA-number} = \|\Lambda - I\|_{\text{sum}} = \sum_{i=j} |1 - \lambda_{ij}| + \sum_{i \neq j} |\lambda_{ij}| \tag{4.4.8}$$

For the analysis of selection of input–output, lower RGA-number is more preferred in a control structure.

Minimum Singular Value Considering a multi-input multi-output (MIMO) system with transfer function G with n inputs and m outputs, G can be defined in terms of singular value decomposition as follows:

$$G = U \Sigma V^H \tag{4.4.9}$$

where $\Sigma = \begin{bmatrix} \Sigma_1 & 0 \\ 0 & 0 \end{bmatrix}$ is an $m \times n$ matrix and Σ_1 is defined as

$$\Sigma_1 = \begin{bmatrix} \sigma_1 & 0 & \cdots & 0 \\ 0 & \sigma_2 & \cdots & 0 \\ \vdots & \vdots & \ddots & \vdots \\ 0 & 0 & \cdots & \sigma_k \end{bmatrix} \tag{4.4.10}$$

in which the nonnegative singular values $\sigma_1 \geq \sigma_2 \geq \cdots \sigma_i \geq \cdots \sigma_k \geq 0$ are placed diagonally in a descending order with $k = \min\{m, n\}$ and σ_i^2 is an eigenvalue of $G^H G$, where G^H is the complex conjugate transpose of G.

The ratio between the maximum (σ_1) and the minimum (σ_k) singular values, σ_1/σ_k, indicates the degree of directionality of the system. If this ratio is big due to small value of σ_k, it shows that the system is ill-conditioned, which indicates a large sensitivity of the system to uncertainty. To avoid ill-conditioning, it is desired that the MSV σ_k be as large as possible when selecting the input–output signal.

4.4.5 Selection of Supplementary Input Signals for Damping Inter-area Oscillations

The input signal to the SVCs used for supplementary control should be responsive to the mode of oscillations to be damped. This can be carried out by using several different input–output controllability analyses such as HSV, RHP-zeros, RGA, and MSV. Once the SVC is placed in the system, the choices for the stabilizing signal could be in a wide range of local and global signals.

Making decision on the final supplementary input signal has been done based on pre-fault and post-fault conditions. The post-fault condition for the study system is when a three-phase fault occurred at bus 2 and line 1–2 will be disconnected.

As explained in the previous section, three SVCs are optimally placed based on modal analysis, GA, and PSO: the first one at bus 40, the second one at bus 49, and the third one at 50. The choice of stabilizing signal for these three SVCs could be as follows:

SVC at bus 40: $I_{40\text{–}41}$; $I_{40\text{–}48}$; $I_{48\text{–}47}$; $I_{41\text{–}42}$; $P_{40\text{–}41}$; $P_{40\text{–}48}$; $P_{48\text{–}47}$; $P_{41\text{–}42}$.

SVC at bus 49: $I_{49\text{–}46}$; $I_{49\text{–}52}$; $I_{46\text{–}38}$; $P_{49\text{–}46}$; $P_{49\text{–}52}$; $P_{46\text{–}58}$.

SVC at bus 50: $I_{50\text{–}51}$; $I_{50\text{–}52}$; $I_{51\text{–}45}$; $P_{50\text{–}51}$; $P_{50\text{–}52}$; $P_{51\text{–}45}$.

From the above choice of candidate signals, it is desired to select only one signal for each SVC. For this, there will be 288 possible combinations.

To determine the best stabilizing signal, the controllability and observability indices are used through the HSV. A comparison of modal observabilities should be done with care. The modal observability of the line current must not be compared with the modal observability of the line power. For this purpose, signals are categorized in different groups, where each group is made of the same type of signals; for example, magnitude of current with the magnitude of current, power with power, and then the HSV of the different signals in each group is compared. The candidate having larger HSV is more preferable than other candidates. In other words, the candidate having larger HSV contains more information about the system internal states than other candidates.

To check the HSV for the 288 sets of candidates, the sets are grouped into eight groups, each group with a combination of current or power signals for each of the three SVCs ($2^3 = 8$).

The 288 sets of candidates are reduced to 69 sets after checking the HSV and RHP-zeros. The results are summarized in Table 4.4.1. The table lists the 69 sets of candidates, which passed the screening of HSV and RHP-zeros, i.e. they have relatively larger HSV and do not produce RHP-zeros. Since 69 sets are still too many to choose from, additional stability tests are necessary to reduce the number of sets significantly. For these 69 candidates, the RGA-number is calculated over the frequency spectrum for both pre-fault and post-fault conditions. Candidate sets 3, 6, 9, 34, 37, and 40 are first selected since they have relatively smaller RGA-number. The RGA-number of these candidates is shown in Figures 4.4.9–4.4.14. Among the six sets, the set 40 is shown to be relatively larger compared to the rest of the sets, and thus the set 40 will be discarded. To demonstrate the difference between the RGA-number of the selected candidates and other candidates, a comparison between the sets 37, 22, and 31 is made, which is shown in Figures 4.4.11 and 4.4.14.

The 69 sets of candidates are now reduced to only five (5) sets based on the RGA-number: sets 3, 6, 9, 34, and 37. Five sets are still too many and one last test is performed, namely the calculation of MSV. The MSV is calculated for the five sets for both pre-fault and post-fault conditions and tabulated in the last two columns in Table 4.4.1. As mentioned in the previous section, it is desirable to have the MSV as large as possible. This condition excludes sets 3 and 34 from the final list because they have relatively smaller MSV. Thus, now only three sets are left for the finalist: sets 6, 9, and 37. Among the sets 6 and 9, the candidate set 6

TABLE 4.4.1 Selected Candidate Sets that Passed the HSV and RHP-zero Tests in Study System

Number of Candidate sets	Candidate-Controlled Outputs	Minimum Singular Value	
		Pre-fault	Post-fault
1	I_{40-41}; I_{49-46}; I_{50-51}	—	—
2	$I_{40\text{-}41}$; I_{49-52}; I_{50-51}	—	—
3	I_{40-41}; I_{46-38}; I_{50-51}	1.4310	1.5326
4	I_{40-48}; I_{49-46}; I_{50-51}	—	—
5	I_{40-48}: I_{49-52}; I_{50-51}	—	—
6	I_{40-48}; I_{46-38}; I_{50-51}	4.1529	3.9693
7	I_{48-47}; I_{49-46}; I_{50-51}	—	—
8	I_{48-47}; I_{49-52}; I_{50-51}	—	-
9	I_{48-47}; I_{46-38}; I_{50-51}	1.9890	2.0982
10	I_{41-42}; I_{49-46}; I_{50-51}	—	—
11	I_{41-42}; I_{49-52}; I_{50-51}	—	—
12	I_{41-42}; I_{46-38}; I_{50-51}	—	—
13	I_{48-47}; I_{49-46}; P_{50-51}	—	—
14	I_{40-41}; P_{49-46}; I_{50-51}	—	—
15	I_{40-41}; P_{49-52}; I_{50-51}	—	—
16	I_{40-41}; P_{46-38}; I_{50-51}	—	—
17	I_{40-41}; P_{46-38}; I_{50-52}	—	—
18	I_{40-41}; P_{46-38}; I_{51-45}	—	—
19	I_{40-48}; P_{49-46}; I_{50-51}	—	—
20	I_{40-48}; P_{49-52}; I_{50-51}	—	—
21	I_{40-48}; P_{46-38}; I_{50-51}	—	—
22	I_{48-47}; P_{49-46}; I_{50-51}	—	—
23	I_{48-47}; P_{49-52}; I_{50-51}	—	—
24	I_{48-47}; I_{46-38}; I_{50-51}	—	—
25	I_{48-47}; I_{46-38}; I_{51-451}	—	—
26	I_{41-42}; P_{49-46}; I_{50-51}	—	—
27	I_{41-42}; P_{49-52}; I_{50-51}	—	—
28	I_{41-42}; P_{46-38}; I_{50-51}	—	—
29	I_{40-41}; P_{46-38}; P_{50-51}	—	—
30	I_{40-41}; P_{46-38}; P_{50-52}	—	—
31	I_{40-41}; P_{46-38}; P_{51-45}	—	—
32	P_{40-41}; I_{49-46}; I_{50-51}	—	—
33	P_{40-41}; I_{49-52}; I_{50-51}	—	—
34	P_{40-41}; I_{46-38}; I_{50-51}	2.9999	1.1348
35	P_{40-48}; I_{49-46}; I_{50-51}	—	—
36	P_{40-48}: I_{49-52}; I_{50-51}	—	—
37	P_{40-48}; I_{46-38}; I_{50-51}	**2.4820**	**2.3811**
38	P_{48-47}; I_{49-46}; I_{50-51}	—	—

(Continued)

TABLE 4.4.1 (Continued)

		Minimum Singular Value	
Number of Candidate sets	Candidate-Controlled Outputs	Pre-fault	Post-fault
39	$P_{48–47}$; $I_{49–52}$; $I_{50–51}$	—	—
40	$P_{48–47}$; $I_{46–38}$; $I_{50–51}$	—	0.0801
41	$P_{41–42}$; $I_{49–46}$; $I_{50–51}$	—	—
42	$P_{41–42}$; $I_{49–52}$; $I_{50–51}$	—	—
43	$P_{40–41}$; $P_{49–46}$; $I_{50–51}$	—	—
44	$P_{40–41}$; $P_{49–52}$; $I_{50–51}$	—	—
45	$P_{40–41}$; $P_{46–38}$; $I_{50–51}$	—	—
46	$P_{40–41}$; $P_{46–38}$; $I_{50–52}$	—	—
47	$P_{40–41}$; $P_{46–38}$; $I_{51–45}$	—	—
48	$P_{40–48}$; $P_{49–46}$; $I_{50–51}$	—	—
49	$P_{40–48}$; $P_{49–52}$; $I_{50–51}$	—	-
50	$P_{40–48}$; $P_{46–38}$; $I_{50–51}$	—	—
51	$P_{40–48}$; $P_{46–38}$; $I_{50–52}$	—	—
52	$P_{40–48}$; $P_{46–38}$; $I_{50–45}$	—	—
53	$P_{48–47}$; $P_{49–46}$; $I_{50–51}$	—	—
54	$P_{48–47}$; $P_{49–52}$; $I_{50–51}$	—	—
55	$P_{48–47}$; $I_{46–38}$; $I_{50–51}$	—	—
56	$P_{48–47}$; $I_{46–38}$; $I_{50–52}$	—	—
57	$P_{48–47}$; $I_{46–38}$; $I_{51–451}$	—	-
58	$P_{41–42}$; $P_{49–46}$; $I_{50–51}$	—	—
59	$P_{41–42}$; $P_{49–52}$; $I_{50–51}$	—	—
60	$P_{41–42}$; $P_{46–38}$; $I_{50–51}$	—	—
61	$P_{40–41}$; $P_{46–38}$; $I_{50–51}$	—	—
62	$P_{40–41}$; $P_{46–38}$; $I_{50–52}$	—	—
63	$P_{40–41}$; $P_{46–38}$; $I_{51–45}$	—	—
64	$P_{40–48}$; $P_{46–38}$; $P_{50–51}$	—	—
65	$P_{40–48}$; $P_{46–38}$; $P_{50–52}$	—	—
66	$P_{40–48}$; $P_{46–38}$; $P_{51–45}$	—	—
67	$P_{48–47}$; $P_{46–38}$; $P_{50–51}$	—	—
68	$P_{48–47}$; $P_{46–38}$; $P_{50–52}$	—	—
69	$P_{48–47}$; $P_{46–38}$; $P_{51–45}$	—	—

has larger MSV. However, its RGA-number is larger compared to set 9. Therefore, set 6 is discarded from the three finalists.

Among the two finalists, the RGA-number of set 9 is relatively larger compared to set 37 for pre-fault and post-fault. Therefore, the candidate set 37 is the final choice since it has a better RGA-number while the MSV is relatively large. That is, three input signals, P40–48, I46–38, and I50–51, are selected as stabilizing input signals for the SVCs at buses 40, 49, and 50, respectively.

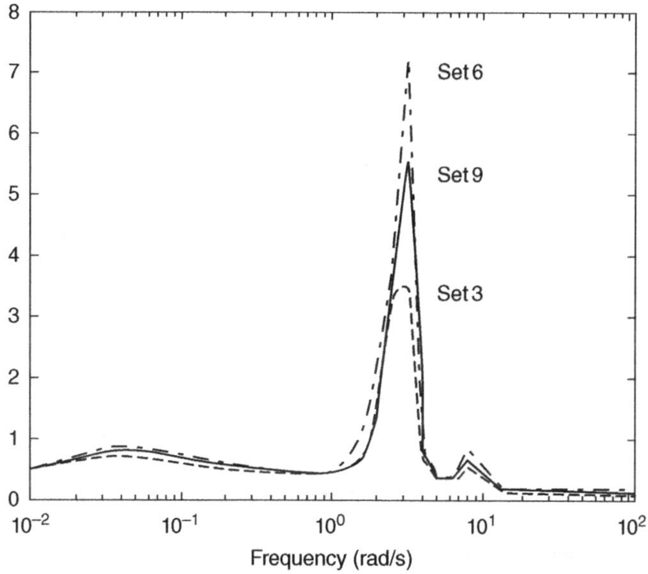

Figure 4.4.9 The RGA-number of candidate sets 3, 6, and 9 (pre-fault).

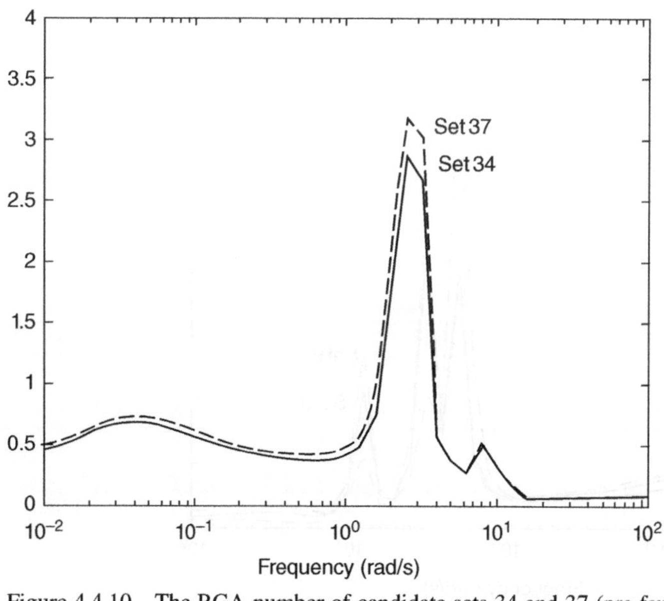

Figure 4.4.10 The RGA-number of candidate sets 34 and 37 (pre-fault).

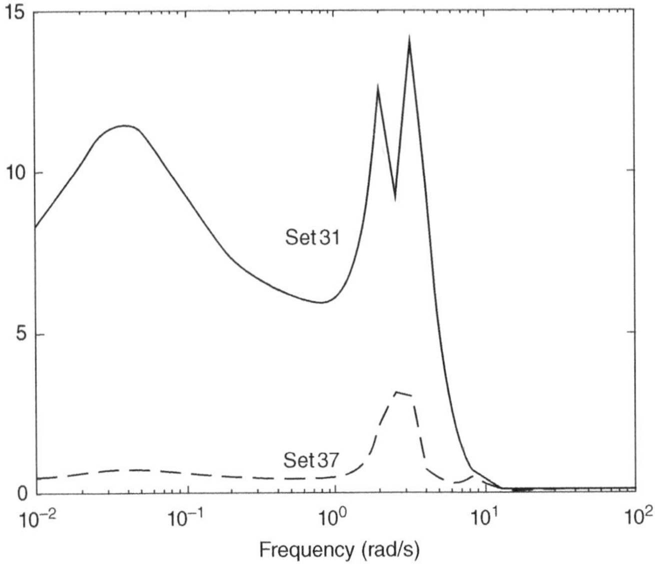

Figure 4.4.11 The RGA-number of candidate sets 31 and 37 (pre-fault).

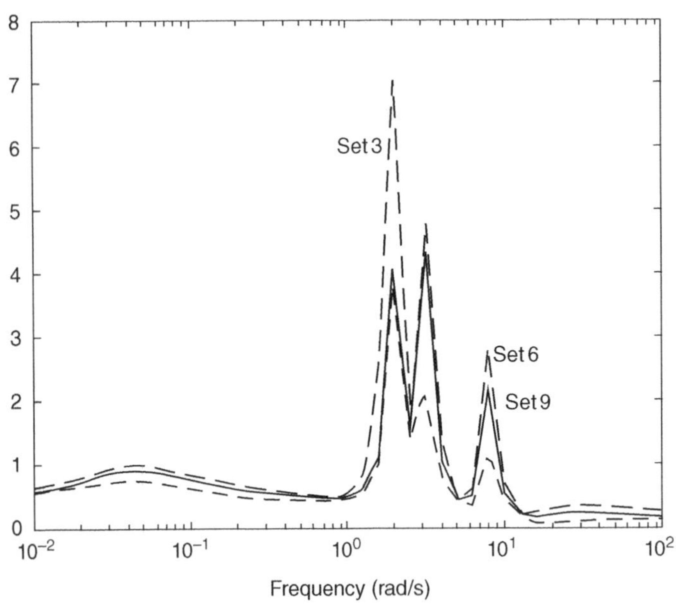

Figure 4.4.12 The RGA-number of candidate sets 3, 6, and 9 (post-fault).

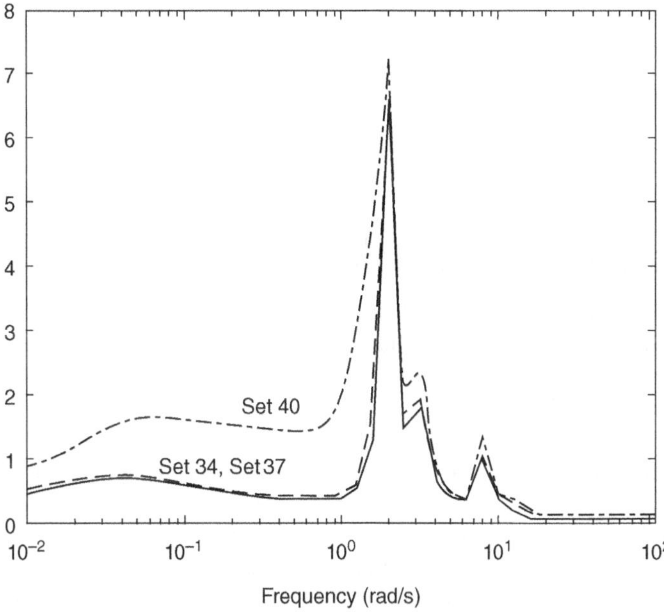

Figure 4.4.13 The RGA-number of candidate sets 34, 37, and 40 (post-fault).

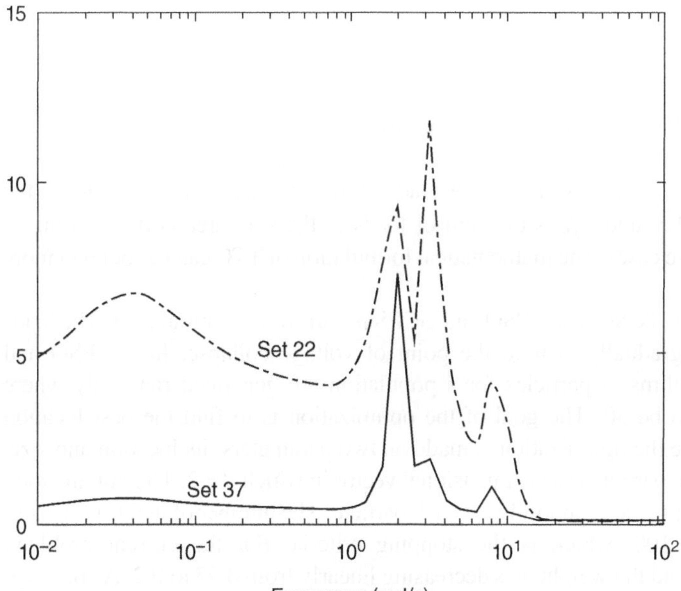

Figure 4.4.14 The RGA-number of candidate sets 22 and 37 (post-fault).

Once the most suitable input signals is determined for a FACTS device for damping improvement, a controller with a defined structure can be designed using heuristic methods.

Many articles can be found that addressed the application of FACTS devices in improving the angle stability. Due to page limitation, the reader is referred to reference [118].

4.4.6 TCSC and Improvement of Total Transfer Capability

One good example of the effects of FACTS devices on steady state is improvement of TTC. Both series and shunt devices could improve TTC, but the series devices are more effective. The TTC improvement using placing of TCSC by PSO, GCPSO, and GA is given below.

The goal is finding the location of TCSC in Figure 4.4.1 and its parameters in order to increase the TTC. In Figure 4.4.1, the power transfers from area 2 to other areas (areas 1, 3, and 5). Also, the power transfers from area 4 to 3 and 5 to 4.

For the TTC calculations, one area is considered as the source area and the other is considered as the sink area where TTC is a directional quantity from the source to the sink. The scenario that is used for TTC calculation is load/generation method (LG) so that the loads in the sink area are increased and the source area will compensate for this increase by increasing its generation.

The TTC level in each case (normal or contingency case) is calculated as follows:

$$\mathrm{TTC} = \sum_{i \in \mathrm{Sink}} P_{Di}(\lambda_{\max}) - \sum_{i \in \mathrm{Sink}} P^0_{Di} \tag{4.4.11}$$

where λ is scalar parameter representing the increase in bus load or generation, $\lambda = 0$ corresponds to no transfer (base case) and $\lambda = \lambda_{\max}$ corresponds to the maximal transfer, $P_{Di}(\lambda_{\max})$ is the sum of loads in the sink area corresponding to the maximal transfer, and P^0_{Di} is the sum of loads in the sink area corresponding to no transfer (base case). The mathematical formulation of TTC can be found in more detail in [119].

Placing of TCSC using PSO and GCPSO start from an initial load. The loads are increased gradually near to the point of voltage collapse. In the PSO and GCPSO algorithms, n particles for a population are generated randomly where n is selected to be 30. The goal of the optimization is to find the best location of TCSC where the optimization is made on two parameters: its location and size. Therefore, each particle is a d-dimensional vector in which $d = 2$. The initialization is made on the position randomly for each particle. The number of iterations is considered to be 100, which is the stopping criteria. For the current problem, $c_1 = c_2 = 2.05$ and the weight w is decreasing linearly from 0.95 to 0.2. Also, $s_c = f_c$ are equal to 5.

Both versions of PSO find line 52–42 for TCSC placement, and its value is a capacitance of -0.5XL, where XL is the reactance of the line. Although both

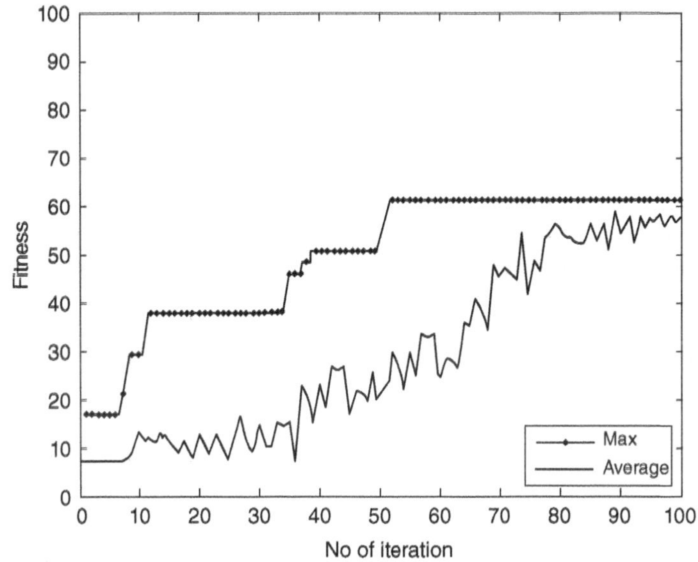

Figure 4.4.15 The convergence characteristic of fitness function with PSO to find solution.

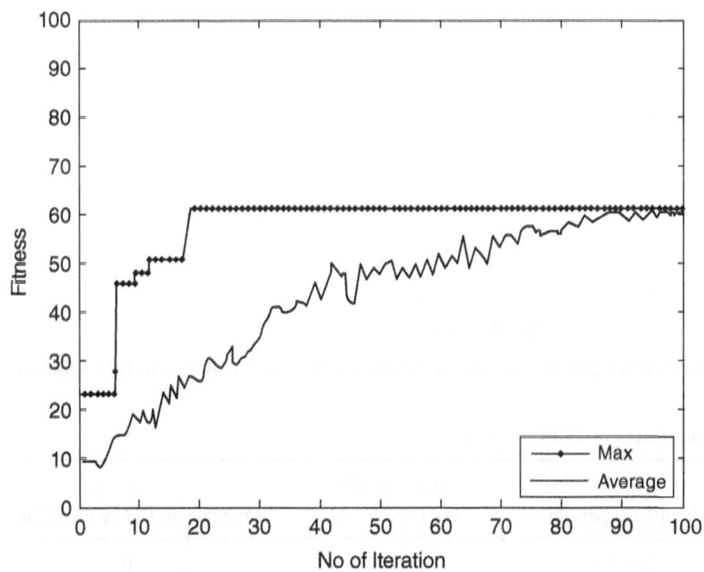

Figure 4.4.16 The convergence characteristic of fitness function with GCPSO to find solution.

versions of PSO find the same solution, due to having a complex search space, standard PSO does not converge properly while the convergence rate of GCPSO is much higher than standard PSO. The results are shown in Figures 4.4.15 and 4.4.16.

Figures 4.4.15 and 4.4.16 show that standard PSO finds the best solution very slowly (reaching to the Max), while GCPSO performs well and quickly finds the high-quality optimal solution.

To validate the results obtained, GA is applied to solve the problem. The number of chromosomes for a population is set to be 30. One point crossover is applied with the crossover probability $p_c = 0.9$ and the mutation probability is selected to be changed linearly from $p_m = 0.05$ to $p_m = 0.005$. Also, a weighted roulette wheel is used. As in the PSO, the number of iteration is considered to be 100.

The GA finds the same line and the same size for TCSC as GCPSO. The convergence characteristic of GA to find the solution is shown in Figure 4.4.17. The obtained result shows that GA performs similar to GCPSO. Table 4.4.2 shows the effects of TCSC on TTC.

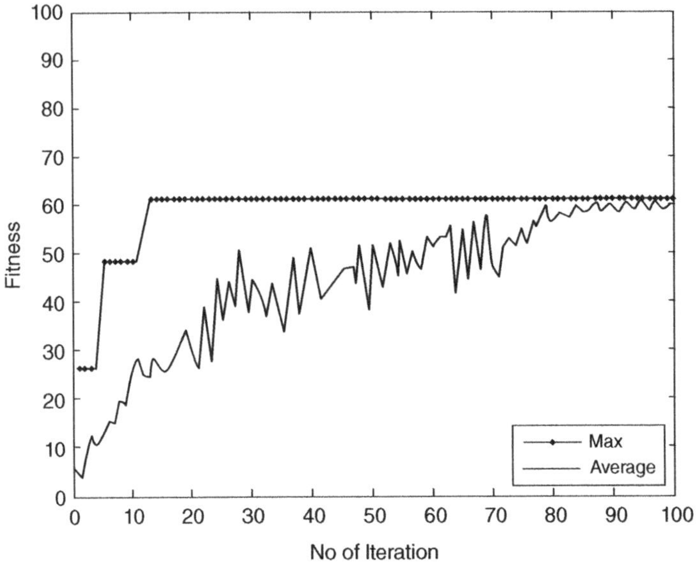

Figure 4.4.17 The convergence characteristic of fitness function with GA to find solution.

TABLE 4.4.2 The Effects of TCSC on TTC

TTC	Without Compensation	With TCSC On Line 52-42	Increase of TTC with TCSC
Area1	30.45%	30.45%	0%
Area3	46.69%	61.35%	14.67%
Area4	34.95%	44.95%	9.99%
Area5	57.85%	57.85%	0%
Total	169.94%	194.6%	24.66%

4.4.7 Conclusions

In this chapter, a hybrid of analytical and some heuristic techniques for application of FACTS devices are addressed. Analytical techniques can be best utilized in determining the location of FACTS devices for voltage improvement and in determining the most suitable supplementary input signals for FACTS devices for damping improvement. On the other hand, heuristic methods can be utilized in determining the location and size of FACTS devices. Heuristic methods have desirable characteristics as an optimization tool and offer significant advantages over traditional methods. They are inherently robust and have been shown to efficiently search the large solution space containing discrete or discontinuous parameters and nonlinear constraints, without being trapped in local minima. Therefore, to handle the complex problems in power system, a revolution from both the conventional control and the heuristic methods are required.

4.5 POWER SYSTEM AUTOMATION

E.M. Voumvoulakis and Nikos D. Hatziargyriou

National Technical University of Athens (NTUA), Athens, Greece

4.5.1 Introduction

Security is defined as the capability of maintaining the continuous operation of a power system under normal operation and following significant perturbations [120]. Power systems are forced to operate with narrower security margins since the increase in electric power demand outpaces the installation of new transmission and generation facilities. In this context, dynamic security assessment (DSA) functions that aim to assess in a fast and accurate way the dynamic performance of a power system for a number of critical contingencies are necessary. In the field of DSA much attention has been paid to preventive as well as corrective control [121, 122]. Preventive control refers to a set of actions that are applied when a potentially dangerous violation is detected through DSA. Corrective actions are applied to offset a security violation after the occurrence of a threatening contingency.

Power system security assessment has been traditionally performed by means of analytical methods, i.e. by performing numerical simulation of a contingency in a given state. However, the nonlinear nature of the models expressing physical phenomena and the growing complexity of modern power systems make online security assessment a very challenging task. Thus, for large power systems in which many contingencies must be assessed, computing, full simulation methods require times often unsatisfactory for online analysis. In this aspect, intelligent systems (IS) are seen to have features that can bring benefits in comparison to analytical methods [123]. Several AI techniques, like radial basis functions (RBF), artificial neural networks (ANNs), and support vector machines have been applied to DSA [124–129].

In this chapter, meta-heuristic-aided methods for DSA and corrective control determination are developed. The corrective control actions include load shedding, generation shedding, and system controlled islanding. In the first section, a PSO-aided automatic learning framework is developed for DSA and corrective control determination [130]. The corrective control is defined here as a nonlinear optimization problem that is solved by the method of PSO, while radial basis function neural networks (RBFNNs) are utilized to evaluate the objective function. The method is generic so that it can be applied for any kind of power system instability problems and requires from the user the definition of a set of disturbances, a set of eligible corrective controls, and a security criterion. These parameters are determined on the basis of engineering judgment and preliminary power system analysis. The method assesses online the security status of the system and proposes a corrective control action, out of the set of eligible control actions, when insecurity is detected. The method is applied for the Greek Interconnected power system, the IEEE-50 Generators system, and the autonomous power system of Cyprus.

The second section focuses on load shedding. Load shedding is one of the main corrective actions, because of its regulating capabilities with regard to the speed of voltage and frequency phenomena. In bibliography, corrective load shedding schemes to mitigate voltage or frequency collapse are mainly based on analytical methods [131–135]. Among IS, machine learning methods have been applied for the definition of remedial actions. In [136] a decision tree (DT) method is proposed for deriving corrective load shedding rules. This idea is further developed in [137] where an automatic learning hybrid method for corrective dynamic security is proposed, based on a self-organized map (SOM) that classifies the power system's security state according to its load profile. DTs nested in nodes with mixed security states are applied to investigate further their security status. In Section 4.5.3, the idea of DT method for the establishment of load shedding rules is enhanced by the use of GA, which defines the appropriate groups of loads that are eligible for load shedding.

The third section investigates the controlled islanding of a power section as a way to avoid a widespread blackout. The method proposes a hybrid algorithm which consists of a variable local topology SOM that gives a pseudo-geographical representation of power system buses onto a two-dimensional plane. The problem of finding the optimal power system islanding is then transformed to the problem of dividing the plane into a number of pieces, where each piece of the plane represents an island of the power system. The optimal solution is found by applying the PSO algorithm. The proposed method is applied on IEEE systems of different scales, and the results show its accuracy.

4.5.2 Application of PSO on Power System's Corrective Control

Corrective control can be formulated as a nonlinear optimization problem. The three major components of this problem are objective function, control measures, and security constraints, as defined in (4.5.1)

$$\min \quad f(\mathrm{OP},\mathrm{CCV}) \qquad \text{s.t.} \quad h(\mathrm{OP},\mathrm{CCV}) \ \rightarrow \text{Secure} \tag{4.5.1}$$

where *f*(OP, CCV) is the objective function and h(OP, CCV) the DSA function.

In general, the objective is to minimize control costs, by ensuring that the control action will drive the power system into the secure region of operation. In the proposed methodology an IS is trained to provide the system security constraints, while a PSO algorithm is employed to solve the optimization problem. Figure 4.5.1 shows the framework of the proposed methodology. Three main tasks are distinguished in the procedure:

- Knowledge base generation
- IS construction
- Online implementation

Knowledge Base Generation

Knowledge base (KB) comprises a large number of OPs covering all possible operating conditions of the system and is generated offline either by means of random sampling or on the basis of historical data. A set of predefined contingencies $D = \{D_1, \cdots, D_n\}$ is simulated for each OP with a time-domain simulation machine and the OPs are classified into secure and insecure, according to a security criterion related to the post-disturbance values of system variables. The security criterion takes the form of (4.5.2):

$$\begin{cases} g(y_t) \geq 0, & \text{OP is Secure} \\ g(y_t) < 0, & \text{OP is Insecure} \end{cases} \tag{4.5.2}$$

where y_t is a vector of post-disturbance values of system variables.

The user selects the appropriate security criterion g on the basis of engineering judgment and preliminary system analysis. This criterion may be related to any type of system instability, such as frequency, rotor angle, or voltage instability. Equation (4.5.2) is an example of security criterion of the under study case. For each insecure OP encountered during the procedure, a number of corrective control actions are created, represented by the corrective control vector (CCV), $\mathrm{CCV} = \{cc_1, \ldots, cc_n\}$. Each variable cc_i encodes a control action, such as load shedding, transformers tap changer blocking, etc. The corrective actions may be created either with the help of a deterministic technique that gives probable solutions or by random sampling. The latter has the disadvantage that generates a large number of ineffective corrective actions but on the other hand it provides better exploration of the search space, and finds solutions that may not be obvious by the deterministic approach.

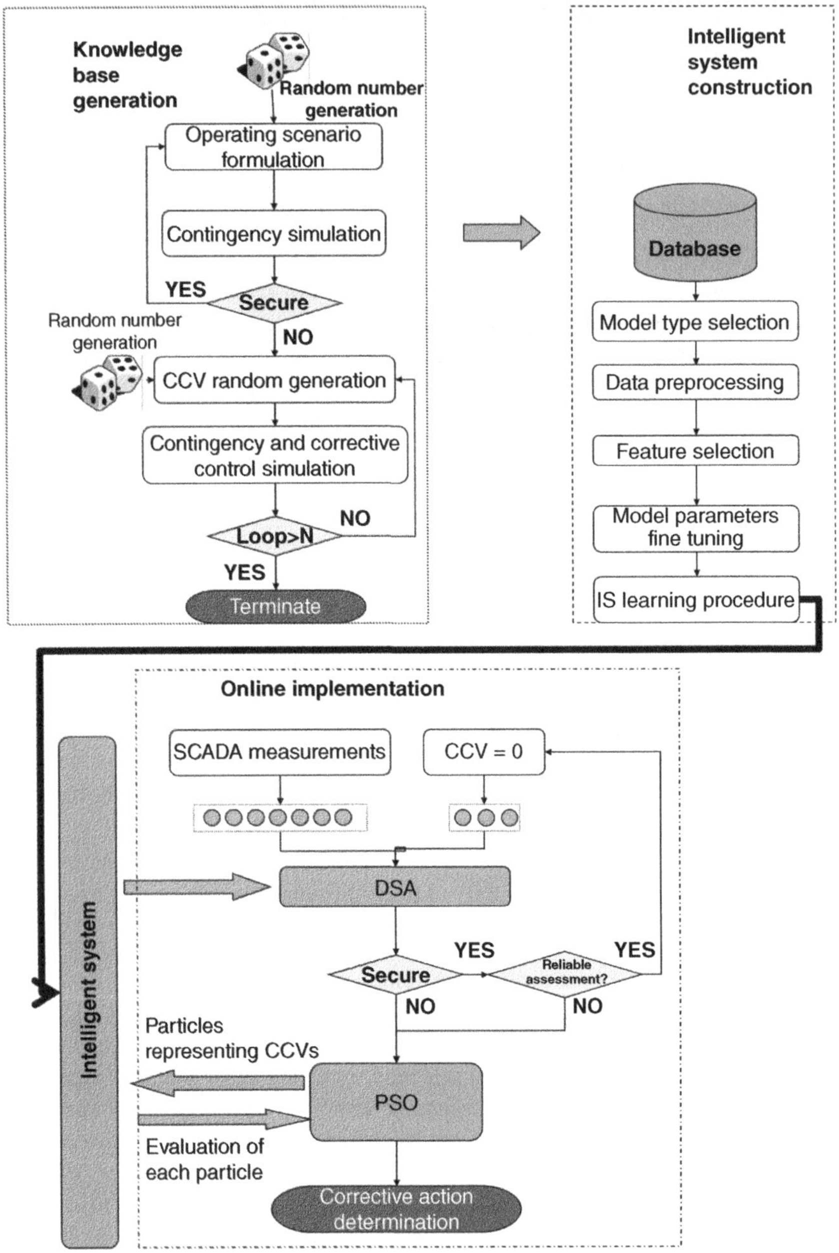

Figure 4.5.1 Framework of the proposed method.

The application of each corrective control is simulated in the time domain and each pair of "OP-CCV" is classified as successful or unsuccessful control, depending on the success of the control actions to bring the system into the secure region.

Intelligent System Construction

The IS is constructed and trained using the KB. The selection of the suitable machine learning model is essential for the efficiency of the method. The machine learning model employed here is RBFNN (see Section "Radial Basis Function Neural Networks"); however, any other machine learning algorithm could be used as well. A preprocessing step of the KB is required before training the IS. This step involves feature selection, i.e. algorithms that output a subset of the input feature set, where selected features characterize properly a variety of power system operating conditions. Two techniques are applied for feature selection, the first makes use of the concept of divergence [138], which is a measure of dissimilarity between two classes and is defined in (4.5.3):

$$J_{ij} = \frac{1}{2} t_r \left[(C_i - C_j)\left(C_j^{-1} - C_i^{-1}\right) \right] + \frac{1}{2} \left[\left(C_i^{-1} + C_j^{-1}\right)(m_i - m_j)(m_i - m_j)^t \right] \tag{4.5.3}$$

where

t_r is the trace of a matrix and is equal to sum of its eigenvalues,

C_i is the covariance matrix of class i of size [D x D],

m_i is the mean of class i of size [D x 1], and

D is the number of features.

Features, which give large divergence, are more important than others. There are many ways to search for an optimal feature combination, like backtrack method, forward sequential method, and backward sequential method [139]. An alternative to the divergence measure that can be used to avoid data repetitions is correlation coefficient (4.5.4):

$$C_{ij} = \frac{E(x_i x_j) - E(x_i)E(x_j)}{\sigma_i \sigma_j} \tag{4.5.4}$$

For each set of completely correlated features all but one are eliminated. A PSO technique is then used to select the optimal feature set from the uncorrelated features, using class separability index F. Each particle $x = \{x_1, \ldots, x_n\}$ represents a candidate feature combination, where if $x_i > 0.5$ feature i is included in the feature set, while if $x_i < 0.5$, feature is not included in the feature set. The fitness function of each particle is calculated using the measure of class separability index, which is defined in (4.5.5):

$$F = \frac{\left| m_i^{(1)} - m_i^{(2)} \right|}{\sigma_i^{(1)} - \sigma_i^{(2)}} \tag{4.5.5}$$

Training of RBFNN requires the determination of the design parameters, i.e. the number N of RBF nodes in the hidden layer and the value of their width σ. The method of S-fold cross validation is employed to attain the optimal network structure for better performance.

During the online implementation of the method, the IS takes as inputs the current OP, together with a zero-valued CCV in order to estimate the security status of the system. In case of an insecure OP, corrective control is necessary to drive the power system in a secure state. PSO (details in Section "Particle Swarm Optimization" and [45]) is employed to solve the optimization problem described in (4.5.1). The objective function is formulated so that it rewards low cost, effective, and reliable corrective control. The reliability of the estimation is provided by the extrapolation index of the RBFNN, calculated using (4.5.35). Values of the extrapolation index close to 1 indicate highly reliable estimations. Certain limits and constraints are imposed on the values of cc_i to account for control actions that may not be acceptable at a particular operating state of the system (e.g. maximum value for the load shedding factor). These limits can be modified during the online implementation according to the requirements of the operator.

The computation time of the optimization procedure depends on the dimension of the CCV and the population of the swarm. Since the dynamic behavior of the power system to the corrective control actions is represented by the offline constructed RBFNN, the scale of the power system does not affect the online required computation time of the optimization procedure (as long as the CCV dimension remains the same).

Intelligent System Construction

This section gives a brief description of the power system models that are used to exemplify the proposed methodology: i.e. the IEEE 50 generators test power system, The Hellenic interconnected power system, and the autonomous power system of Cyprus.

IEEE 50 Generators Test Power System The IEEE 50-generator system is derived from a representative model of a realistic power system in North America [140, 141]. It comprises 50 generators, 145 buses, 401 transmission lines, and 52 transformers. The 44 generators are represented using the classical model and the remaining 6 generators are represented using a two-axis model with an AC4IEEE exciter [142].

Hellenic Power System The second study case system is a model of the Hellenic mainland system corresponding to a stressed operating condition. The system comprises steam turbines, which produce the base load, hydro turbines which are used for peak load and frequency regulation and combined cycle units. The system is interconnected with the Balkan system with two 400 kV synchronous interconnections and with a DC interconnection to Italy. The peak load is approximately 10 GW. The derived model includes the generation and transmission system up to 20 kV buses. The total number of buses is 876, there are 846 lines and 206 transformers. The generating units are represented by 78 machines including their governors, exciters, and overexcitation limiters, as described below.

- Exciters. The IEEE type 1 is the excitation system used for the generators of the system.
- Steam turbine governor. The prime mover governor model is required to demonstrate long-term stability of the unit in response to frequency oscillations. The droop is high (1 p.u.), so that steam turbines do not participate in the frequency regulation. Only the combined cycle units have a droop around 7%.
- Hydro turbine governor. The hydro turbines are used for frequency regulation and their droop is around 4%.
- Overexcitation limiter. Overexcitation limiters are devices, which protect synchronous machine field windings from overheating [143]. The limiting action provided by these devices must offer proper protection while simultaneously allowing maximum field forcing for power system stability purposes. Typical operating characteristics attempt to mimic the field current short-time overload capability given in ANSI standard C50.13-1977.

The synchronous interconnections at the North of the country were modeled as synchronous generators with high nominal power (20 GW) and inertia coefficient (10 MW s/MVA). These generators have no overexcitation limiter and their droop is low, so that they can undertake the load, in case of a contingency. The ZIP model is used for the load assuming 25% constant power, 25% constant current, and 50% constant impedance.

Autonomous Power System of Cyprus The application system is based on a realistic model of the autonomous power system of Cyprus. There are four groups of generating units as illustrated in Table 4.5.1, with total installed capacity of 1118 MW.

TABLE 4.5.1 Generating Unit Groups

	Vasilikos Steam Units	Dhekelia Steam Units	Moni Steam Units	Moni and Vasilikos Gas Units
Number of Units	3	6	6	5
P_G^{min} (MW)	60	30	18	4
P_G^{max} (MW)	130	60	30	28
FOR	0.589%	3.04%	6.41%	9,788%
Forced outage duration (hours)	24	24	24	24
Number of forced outage events	2.13	11.08	23.4	35.72

Under unexpected disturbances, when the spinning reserve of the system fails to control frequency within certain bounds, load shedding automatons are triggered, according to the thresholds described in Table 4.5.2.

Application of the Method to the IEEE 50 Generators Test Power Systems

The creation of the data set starts from a base scenario available at [141]. The system was simulated for 4000 randomly generated operating conditions. In order to account for the effect of topology changes, the training patterns represent different power system topologies, generated by randomly removing a power system component at each OP. For each OP, the voltages, voltage angles, real and reactive loads and generation of the buses, as well as the power interchange of the branches of the system are recorded, formulating a KB of 1396 pre-disturbance steady-state variables (features) for each OP. A three-phase fault at bus #7 cleared by opening line between bus #7 and bus #6 after 200 ms is simulated. The transient stability of a power system at any time can be indicated by the distance of the rotor angles from the center of the inertia angle at that time [144]. A power system is always operated such that any generator rotor angle will not be greater than a threshold. If a generator's rotor angle is larger than such a threshold, the generator will be tripped by out-of-step relay to protect it from being damaged.

$$|\delta_i - \delta_{\mathrm{COI}}| \leq \delta_{\max} \tag{4.5.6}$$

where δ_{COI} is the center of inertia angle, M_i is the inertia constant of the ith generator, and $\delta_{\max}$ is taken equal to 120°. Table 4.5.3 illustrates the partition of the OPs to the secure and insecure class.

TABLE 4.5.2 Load Shedding Automatons Thresholds

Frequency Threshold	Ration of Load Shed to the Total Load of the System (%)
49.0	*10*
48.5	*10*
48.0	*15*

TABLE 4.5.3 Two-class Partition of the Operating Points

Data Set	Learning Set	Test Set
Number of Samples 4000	2000	2000
Class A (secure)	1194	1214
Class B (insecure)	806	786

Generation shedding and fast valving are the corrective actions proposed. Each CCV is defined as CCV=[cc_1...cc_{50}], where each element of the CCV is defined in Eq. (4.5.7):

$$cc_i = \begin{cases} 0, & \text{Perform no control at generator } i \\ 1, & \text{Trip Generator } i \\ 2, & \text{Fast Valving for Generator } i \end{cases} \tag{4.5.7}$$

Twenty CCVs are randomly generated for each one of the 1 592 insecure OPs of the data set, generating thus 31 840 "OP-CCV" pairs. The distribution of the values of variables cc_i is a function of the pre-disturbance angular deviation and generators inertia H_i so that generators with low inertia or high angular deviation are more probable to be shed or to apply fast valving. Table 4.5.4 illustrates the partition of "OP-CCV" pairs to classes B1 (successful corrective control) and B2 (unsuccessful corrective control).

The total generated data set comprises 1396 features, as discussed in the previous section, thus feature selection is necessary to reduce dimensionality. Two feature sets are derived by applying the methodologies described in Section "Intelligent System Construction." FS1 is the feature derived by the divergence-based method and comprises 25 features, while FS2 is the feature set derived by the PSO correlation-based method.

The RBFNN parameters are also tuned via S-fold cross validation. N is examined in the range of [1–500] while σ in range [0–10], deriving $N = 450$ and $\sigma = 3.5$. Table 4.5.5 presents the classification performance of the method.

TABLE 4.5.4 Two-Class Partition of Pairs "OP-CCV"

Data Set	Learning Set	Test Set
Number of samples 31840	16 120	15 720
Class B1 (successful control)	9 609	9 231
Class B2 (unsuccessful control)	6 511	6 489

TABLE 4.5.5 Classification of OPs and "OP-CCV" Pairs

Classification of OPs of Test Set by the RBFNN				
	Assessed as Secure		Assessed as Insecure	
	FS1	FS2	FS1	FS2
Secure (1 191)	1 154	1 178	60	36
Insecure (809)	57	43	729	743
Classification of "OP-CCV" pairs of the Test Set by the RBFNN				
16 180 Samples	**Assessed as Successful**		**Assessed as Unsuccessful**	
	FS1	**FS2**	**FS1**	**FS2**
Successful control (3 490)	8 532	8 768	699	463
Unsuccessful control (6 489)	423	254	6 066	6 235

TABLE 4.5.6 Evaluation of Classification Performance

Performance Evaluation					
Classification of OPs			Classification of "OP-CCV" pairs		
	FS1 (%)	FS2 (%)		FS1 (%)	FS2 (%)
Success rate	94.15	96.05	Success rate	93.14	95.43
False alarms	4.94	2.97	Successful assessed as unsuccessful	6.52	5.02
Missed alarms	7.25	5.47	Unsuccessful assessed as successful	7.57	3.91

TABLE 4.5.7 Generation Shedding and Fast Valving

	FS1	FS2
Accumulated GWs of shedded generation	3242.6	3202.1
Accumulated GWs of unnecessary shedded generation	249.5	187.5

The feature set FS2 derived from the PSO feature selection method improves the accuracy of the model leading to less false alarms (FA) and missed alarms (MA).

The RBFNN evaluates security of the power system (Table 4.5.6). In case it detects an insecure OP, the PSO algorithm is employed to find a CCV that leads to successful control. The objective function used is defined in (4.5.8),

$$\min\ f(\mathrm{OP}, \mathrm{CCV}) = \sum_{i=1}^{25} P_G^i(i : \mathrm{cc}_i = 1) + 0.2 \sum_{j=1}^{25} P_G^j(j : \mathrm{cc}_j = 2) \qquad (4.5.8)$$

$$\text{s.t}\quad (\mathrm{OP}, \mathrm{CCV}) \in \text{class } B1 \qquad \rho(\mathrm{OP}, \mathrm{CCV}) \geq \rho_{\text{threshold}}$$

where ρ is the extrapolation index defined in (4.5.35) and $\rho_{\text{threshold}}$ a threshold value required for the estimation of the RBFNN to be considered as reliable. The objective function penalizes generation shedding by the active production of the shed generator, and fast valving by 20% of generator's active production (Table 4.5.7). The constraints are evaluated by the RBFNN.

Application of the Method to the Hellenic Power System

The method described in Section 4.5.2 is applied on the Hellenic power system. The creation of the data set starts from a base scenario, which is a snapshot of the system at its maximum loading of the year 2005. A coefficient n is considered which takes values from 0.91 to 1.10 with a step of 0.01, and represents the ratio of the total load to the maximum total load of the year 2005. For each of the 20 values of the coefficient n, 200 OPs are created by varying randomly the load at each load busbar. This random distribution of the load to the buses ensures that the KB comprises several distribution patterns, since load distribution influences the security of the system. The load at each bus of the system is given by (4.5.9),

$$P_i^D = N\left(n \cdot P_{0,i}^D, \sigma\right) \tag{4.5.9}$$

where N represents the normal distribution, $P_{0,i}^D$ is the demand of bus i at the base scenario, and σ is the standard deviation ($\sigma = 4\%$). The generated active power of each unit P_i^G follows a normal distribution with mean value the production of the unit when the system operates at maximum load, multiplied with the coefficient n, as long as P_i^G is between its technical minimum and maximum.

$$P_i^G = \min\left(P_{i,\max}^G, N\left(n \cdot P_{0,i}^D, \sigma\right)\right) \tag{4.5.10}$$

where $P_{i,\max}^G$ is the technical maximum of the generator and $P_{0,i}^G$ is the generation of the unit at base case scenario.

The application of the method to one contingency, which is the loss of a combined cycle unit at Lavrio power station with 460 MW nominal active power, is presented here. This is a major contingency that can lead to dangerously low voltage levels, especially in the Athens region. In our application, monotonic voltage stability is considered. Bus voltages are recorded 150 seconds after the occurrence of the disturbance. In order to consider an OP as secure, the voltages at certain buses must be above a security threshold taken equal to 0.9, 150 seconds after the disturbance. This voltage threshold indicates that the power system survived the disturbance and avoided voltage collapse. Table 4.5.8 illustrates the partition of the OPs to class A (secure OPs) and class B (insecure OPs) for the learning and the test set.

If the OP is insecure, control action shall be applied to restore the power system to its normal state. 20 CCVs are randomly generated for each one of the 1 626 insecure OPs, generating thus 32 520 "OP-CCV" pairs. Each CCV is defined as $CCV = [cc_1 \ldots cc_8]$ where cc_i represents the percentage of load curtailed at area i (it is required that $cc_i < 20\%$). Table 4.5.9 illustrates the partition of "OP-CCV" pairs

TABLE 4.5.8 Two-class partition of the Operating Points

Data Set	Learning Set	Test Set
Number of samples 4000	2000	2000
Class A (secure)	1183	1191
Class B (insecure)	817	809

TABLE 4.5.9 Two-class Partition of Pairs "OP-CCV"

Date Set	Learning Set	Test Set
Total number of samples 32 520	16 340	16 180
Class B1 (successful control)	3 967	3 843
Class B2 (unsuccessful control)	12 373	12 337

to classes B1 (successful corrective control) and B2 (unsuccessful corrective control). Training is performed by random selection of the LS, while the TS is made of the rest of the data set. In this way, the learning rate is checked and the capability of the method to classify correctly unforeseen states can be evaluated on an objective basis.

The RBFNN is trained after applying the two feature selection techniques and S-folder cross validation. Parameter N is examined in the range of [1–300] while σ in range [0–50] and deriving values are $N = 250$ and $\sigma = 3$. Table 4.5.10 present the performance of the RBFNN in the classification of OPs and "OP-CCV" pairs of the test set.

The results derived from the two feature sets are practically equivalent. The corrective control applied under the event of a disturbance is determined via the PSO optimization method. Each particle represents a CCV=[$cc_1 \ldots cc_8$] and its fitness is evaluated by the objective function provided by (4.5.11) (Table 4.5.11).

TABLE 4.5.10 Classification of Ops and "OP-CCV" Pairs

Classification of OPs of Test Set by the RBFNN				
	Assessed as Secure		Assessed as Insecure	
	FS1	FS2	FS1	FS2
Secure (1 191)	1 185	1 180	6	11
Insecure (809)	9	13	800	796
Classification of "OP-CCV" pairs of the Test Set by the RBFNN				
16 180 Samples	**Assessed as Successful**		**Assessed as Unsuccessful**	
	FS1	**FS2**	**FS1**	**FS2**
Successful control (3 490)	3 204	380	286	240
Unsuccessful Control (6 489)	306	3 250	12 384	12 310

TABLE 4.5.11 Evaluation of Classification Performance

Performance Evaluation					
Classification of OPs			Classification of "OP-CCV" pairs		
	FS1 (%)	FS2 (%)		FS1 (%)	FS2 (%)
Success rate	99.25	98.80	Success rate	96.34	96.17
False alarms	0.50	0.90	Successful assessed as unsuccessful	8.02	6.88
Missed alarms	1.11	1.61	Unsuccessful assessed as successful	2.41	2.99

$$\min \quad f(\mathrm{CCV}) = \sum_{i=1}^{8} \mathrm{cc}_i \cdot P_L^i \cdot C(i)$$
$$\text{s.t} \quad (\mathrm{OP}-\mathrm{CCV}) \in \text{class } B1, \qquad \rho(\mathrm{OP}-\mathrm{CCV}) \geq \rho_{\text{thresshold}} \tag{4.5.11}$$

where $C(i)$ is the per unit value of load shed at area i, and P_L^i is the total load of area.

Here, the per unit value of load lost is considered to be the same for all areas of the power system, but the method allows for value differentiation, by choosing the suitable $C(i)$.

Comparison

In this section the corrective load shedding of the proposed method is compared with the results of the DT method proposed in [136], the SOM method, and the combined SOM-DT method. The above four methods have been applied on the OPs of the test set. Table 4.5.12 compares the performance indices of the proposed methodology for the OPs' partition task into the secure and the insecure classes.

The main difference between the methods is that the PSO-RBFNN method can freely select the appropriate inputs (voltages, power transfers, etc.), while SOMs and DTs are restricted to use mainly power-related inputs in order to be able to provide a load shedding scheme as explained in [136, 137]. Thus, the PSO-RBFNN method can exploit more knowledge, leading to better results (Table 4.5.13).

Table 4.5.14 shows the total load that needs to be curtailed in order to reach a secure state, as proposed by the four methods. Moreover, the total unnecessarily curtailed load, which is the total load suggested to be curtailed in secure cases, is shown. An insecure OP is simulated to show the effectiveness of the derived load shedding propositions.

Table 4.5.15 shows the load profile of the selected OP while Table 4.5.16 compares the load shedding strategy proposed by the four methods. Both PSO-RBFNN and SOM-DT propose a total load shedding of around 500 MW, but with

TABLE 4.5.12 Comparison of OP Classification by Each Method

	TS 2000 OPS			
	Secure		Insecure	
	1191		809	
	Secure OPs of the TS			
	DTs	**SOM**	**SOM-DTs**	**PSO-RBF**
Assessed as secure	1092	1086	1175	1185
Assessed as insecure	99	105	16	9
	Insecure OPs of the TS			
Assessed as secure	28	16	18	6
Assessed as insecure	781	793	791	800

TABLE 4.5.13 Comparison of Performance of Each Method

	Classification Performance Evaluation			
	DTs	SOM	SOM-DTs	PSO-RBF
Success rate	93.65% (1873/2000)	93.95% (1879/2000)	98.30% (1966/2,000)	99.25% (1985/2000)
False alarms	8.31% (99/1191)	8.81% (105/1191)	1.34% (16/1191)	0.50% (6/1191)
Missed alarms	3.46% (28/809)	1.98% (16/809)	2.22% (18/809)	1.11% (9/809)

TABLE 4.5.14 Load Shedding Performance for Each Method

—	DT	SOM	SOM-DT	PSO-RBF
Accumulated curtailed load	101	487 GW	423 GW	402 GW
Accumulated unnecessary curtailed load	18 GW	31.7 GW	37.6 GW	1.03 GW

TABLE 4.5.15 Load Profile of the OP Under Study

Parea1	Parea2	Parea3	Parea4	Parea5	Parea6	Parea7	Parea8
2250	1022	1140	1058	3690	1070	50	553

TABLE 4.5.16 Curtailed Load for Each Area

Area	1	2	3	4	5	6	7	8	Total
DT	164	75	83	77	269	78	4	40	790
SOM	176	92	91	77	305	95	0	47	884
SOM-DT	90	52	48	37	164	55	0	26	472
PSO-RBF	0	121	35	61	165	88	0	0	470

differences in the distribution of the curtailed load to the areas. In Figure 4.5.2 the effects of the suggested load shedding options on the voltage at a characteristic bus in the Attica region are shown and compared with the case of no corrective action leading the system to collapse. Even though SOM-DT and PSO-RBFNN methods propose the same total amount of load to be shed, the shedding scheme proposed by the PSO-RBFNN method is more effective in restoring the voltage level for the particular OP under study.

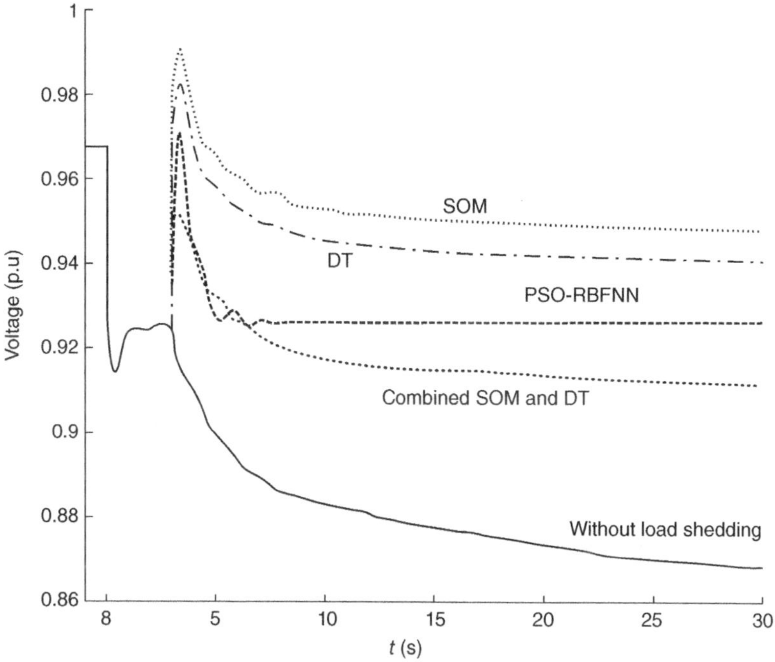

Figure 4.5.2 Simulation of load shedding strategies.

Frequency Stability – Cyprus

Reserve services are the services required for the control of system frequency within certain bounds in the presence of events. The primary reserves are necessary in order to enable the system to intercept runaway frequency after an unexpected disturbance and are typically provided by spinning units within a very short time range.

The classic approach for the determination of the necessary primary frequency control services is based on off-line stability analysis of selected extreme conditions. This approach does not take into consideration the varying behavior of the demand and the dynamic characteristics of the generation response [145–147]. Previous publications have faced this problem using online techniques based on energy functions [148–150] and AI [151–153]. A successful methodology is given in [153], which takes into account the effect of response time of the offered reserves from the generating units, but it suffers from long execution time. In [129], an efficient, preventive security assessment method, characterized from short execution times, is proposed, providing however suboptimal solutions. In [154], the principle of the method presented in [129] is extended into an ED algorithm for both active power and primary spinning reserves that includes security margins as constraints.

The objective of the proposed methodology is to determine the ED of the power generation and of the primary reserve services required to ensure the secure operating state of the system in the presence of preselected disturbances. The ED problem can be formulated as follows:

The objective function is described by Eq. (4.5.12),

$$\min \quad C_{\mathrm{S}}^{t} = \sum_{m=1}^{M} C_{G,m}^{t}\left(P_{G,m}^{t}\right) + \sum_{m=1}^{N} C_{\mathrm{PFC},m}^{t}\left(P_{\mathrm{PFC},m}^{t}\right) \tag{4.5.12}$$

subject to the constraint of security,
where

t time interval

N number of generation units biding in frequency control service markets

M number of generation units biding in energy market

C_{TC}^{t} total cost of generation and frequency control service

$C_{G,m}^{t}$ cost of active power provided by unit m

$C_{\mathrm{PFC},m}^{t}$ cost of spinning reserve power provided by unit m

$P_{G,m}^{t}$ active power generation of unit m

$P_{\mathrm{PFC},m}^{t}$ active power injection of unit m for frequency control

The optimization problem is similar to the problem defined in Eq. (4.5.1), and is solved according to the framework described in Figure 4.5.3. Extensive tests have shown that the development of different RBFNNs, each for a different load range (i.e. low load, medium load, and high load), ensure a better performance.

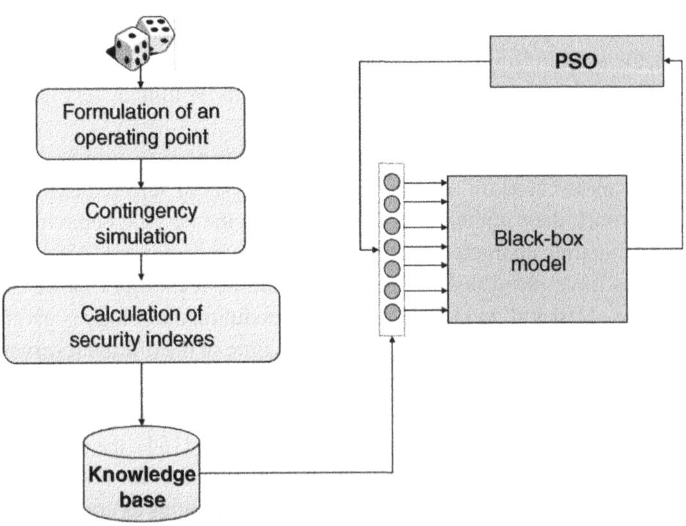

Figure 4.5.3 Flowchart of the proposed methodology.

The total load of the system is assumed to follow normal distribution around four operating profiles:

1. low load with a total load 300–450 MW
2. medium load with 450–600 MW
3. high load with 600–750 MW
4. peak load with 750–900 MW

The required spinning reserve of the system follows also normal distribution with mean value of the generation of the largest unit and standard deviation $\sigma = 20\%$. For the creation of the KB, a large number of initial OPs are obtained by varying randomly the load at each load busbar and the spinning reserve required. For each OP, the outage of the unit with the largest power production, which is a steam unit of Vasilikos station, is simulated using EUROSTAG [155]. The frequency deviation is recorded and the OPs are labeled according to the following security criterion (4.5.13):

$$IF\ \ f_{\min} < 48.9Hz\ \ THEN\ \ OP\ \ is\ \ SECURE\ \ ELSE\ \ OP\ \ is\ \ INSECURE \tag{4.5.13}$$

Each RBFNN is trained using the KB created above, and operates as a black box model. Its inputs are the parameters to be optimized as well as any other attribute that can contribute to the better security estimation. The output of the RBFNN is either the class of the OP, i.e. secure/insecure, or the minimum frequency. Using the trained RBFNN as a black box model to evaluate the security status of the OP under study, the PSO algorithm is applied in order to find the values of inputs that optimize the objective function (4.5.12).

The inputs of the RBFNN are chosen to be the power generation of the units, since these are the variables to be optimized in the objective function. Training of RBFNN requires the determination of the design parameters, i.e. the number N of RBF nodes in the hidden layer and the value of their width σ. The method of S-fold cross validation is employed to attain the optimal network structure for better performance [156]. Table 4.5.17 presents the regression performance of RBFNN

TABLE 4.5.17 RBFNN Regression Performance Evaluation for Testing Set

	Low Load	Medium Load	High Load	Peak Load
MAPE (%)	0.03	0.02	0.03	0.11
Max error (Hz)	0.191	0.141	0.435	0.794
MSE	0	0	0.001	0.007
SSE	1.229	0.697	1.784	20.95
Success rate	2998	3000	3000	2988
False alarms	2	0	0	6
Missed alarms	0	0	0	6

TABLE 4.5.18 Unit Commitment Patterns for the OP Under Study

	Vasilikos Steam Units	Dhekelia Steam Units	Moni Steam Units	Moni Gas Units	Vassilikos Gas Units
UC1	3	6	2	0	0
UC2	3	6	3	0	0
UC3	3	5	4	0	0

TABLE 4.5.19 Dispatch and Operational Cost for the Unit Commitment Patterns under Study

	Vasilikos Steam Units (Number of Units ∗MW)	Dhekelia Steam Units (Number of Units ∗MW)	Moni Steam Units (Number of Units ∗MW)	System Operational Cost £	Total Spinning Reserve
UC1	3∗109	6∗46	2∗23.3	18 945	156 MW
UC2	3∗108	6∗44	3∗20.15	19 134	184 MW
UC3	3∗105	5∗48.7	4∗23.1	19 221	152 MW

when applied for the four operating profiles under study. The performance indices used are the mean average percentage error (MAPE), maximum error, mean square error (MSE), sum square error (SSE), success rate, FA, and MA.

The PSO method is applied using the trained RBFNN in order to evaluate the security constraint of the optimization problem. Assuming a snapshot of the power system with total load 650 MW and three alternative unit commitment patterns (UC1, UC2, and UC3) as described in Table 4.5.18, the optimization method proposed is presented in Table 4.5.19.

Table 4.5.19 presents the ED as well as the respective operational and spinning reserve cost for each one of the commissioning patterns under study. According to Table 4.5.19, pattern UC1 is the indicated commitment pattern since it has the lowest cost and it satisfies the security constraints.

4.5.3 Genetic Algorithm-aided DTs for Load Shedding

In order to construct a DT (see Appendix Section "Decision Trees") to form a load shedding scheme, the candidate attributes that formulate the rules of the DT must be load-related and controllable variables. In [136], the DT is constructed using as candidate attributes the active load of each of the 9 areas to which the Greek power system is divided as illustrated in Figure 4.5.4, and the total load of the Greek power system (10 attributes). Once the DT is constructed, load shedding rules can be derived by inverse reading of it. Figure 4.5.4 illustrates the DT, where A6 is the total active load of Greek power system in MW and A7 is the total active load of region 6 in MW.

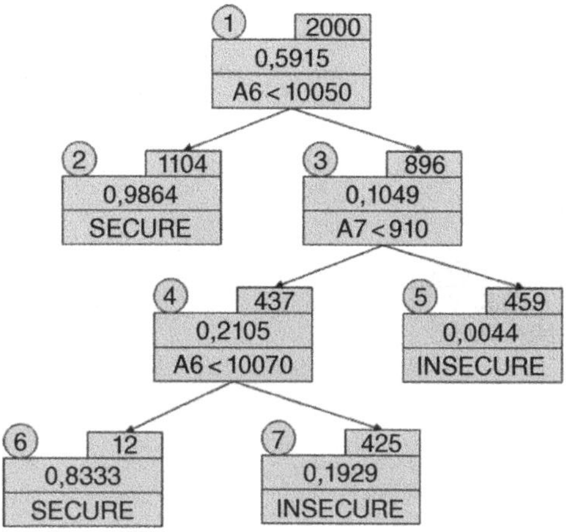

Figure 4.5.4 Decision tree for load shedding.

There are two options in order to transpose from the insecure state to a secure one: curtailment of the load such that the total load is less than 10 050 MW. This action equals to a transposition to leaf 2 of the DT by shedding the appropriate amount of load. Curtailment of the load, such that the load of Area 6 is less than 910 MW and the total load is less than 10 070 MW. This action is equivalent to the transposition from an insecure leaf to leaf 6. It should be noted, however, that the security index (SI) of leaf 6 is lower than the SI of leaf 5, which means that the first option is more reliable.

In this section the contribution of GA in the formulation of composite attributes that can be used as inputs to DT is investigated. More specific, instead of using the load of the TSO-defined areas of the power system, the aggregate load of suitable groups of load buses can be used to derive more accurate security rules. The objective is to use the aggregate load $\sum_{j \in A(i)} P_D^j$ of a group $A(i)$ of load buses as an attribute for the dichotomy of DT nodes.

The model of the Hellenic power system comprises 241 load buses. Each combination of these buses can formulate a group, which means that there are 2^{241} groups of load buses. The large number of possible solutions indicates that a heuristic search method shall be used to provide a suboptimal solution. A GA approach is used here to select appropriate groups of load buses and the process is executed in parallel with the construction of the DT.

The DT is constructed according to [136] with the difference that in this case attributes are not a-priori known, but searched via the GA method. An initial population of chromosomes is created, each of which represents a group $A(i)$ of load buses. Each chromosome is a vector $V_i : V_i(j) = \{0, 1\}$, $j = 1, \ldots, 241$. If $v(j) = 1$, then the bus j belongs to group $A(i)$, otherwise it does not. The chromosomes are evaluated according to the information gain they provide when the aggregate load $\sum_{j \in A(i)} P_D^j$ is used in a dichotomy test of a DT node.

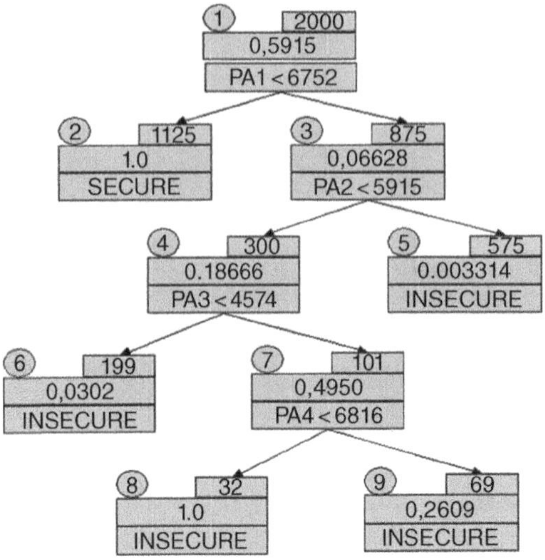

Figure 4.5.5 Genetic algorithm-aided DT.

TABLE 4.5.20 Attributes used by the GA-aided DT

PA1	Group comprising 157 load buses
PA2	Group comprising 135 load buses
PA3	Group comprising 108 load buses
PA4	Group comprising 159 load buses

After a number of iterations, the chromosome, which provides the maximum information gain, is chosen and an attribute is formulated. This attribute is used to split the node of the DT. The procedure is repeated recursively until the nodes of the DT cannot be further developed. The initial population comprises 482 chromosomes and the GA is repeated for 1000 iterations. The DT developed by the above procedure is illustrated in Figure 4.5.5 (Table 4.5.20).

4.5.4 Power System-Controlled Islanding

The method of VLT-SOM (details in Section "Variable Local Topology Self-Organized Maps (VLT-SOM)") is applied to map a power system onto a two-dimensional plane in a way that its topological features are preserved. The method is applied for the IEEE – 30 buses test system, the IEEE – 118 buses test system, and IEEE 17 generators test system. The matrix $[e_{ij}]$ is defined in such a way so as to represent the electrical distance between the two buses i and j. The Thevenin equivalent impedance between buses i and j can be used as a measure of electrical distance in the matrix $[e_{ij}]$. Another option is to consider the dissimilarity of the swing curves of the system, so that generators with different swing curves are assigned large values of e_{ij}. This can be done by introducing an index W'_{ij} which is defined by equation:

$$W'_{ij} = \sum_{t_k \in [0, T]} \left| \omega_i(t_k) - \omega_j(t_k) \right| \tag{4.5.14}$$

where $[0, T]$ is the under study time span. Index W'_{ij} W'_{ij} is normalized according to equation:

$$W_{ij} = \frac{W'_{ij}}{\max\left(W'_{ij}\right)} \tag{4.5.15}$$

A measure of coherence between generators i and j is formulated by equation:

$$C_{ij} = 1 - W_{ij} \tag{4.5.16}$$

Generators can be divided into clusters according to their coherence. For two generators i, j of the same coherency group $e_{ij} = C_{ij}$, while for two generators in different coherency groups $e_{ij} = -C_{ij}$.

After having mapped the power system onto the two-dimensional plane, the task of power system islanding consists in separating the plane into k segments. In order to achieve this, k points $p_1, p_2, \ldots, p_k$ are considered on the plane with coordinates $p_i = [x_i, y_i]$. These points divide the plane into k segments S_1, S_2, S_k, where segment i includes all points w, for which distance $\|p_i - w\|$ is minimized. Each segment represents an island of the power system comprising the buses that are mapped onto that segment. The task of finding the best positions for points p_1, $p_2, \ldots, p_k$ can be solved using a heuristic technique such as PSO [45], given a suitable objective function. This objective function F is defined in order to satisfy the constraints described in the following sections.

Generation and Load Balance

The load of the formulated islands must be equal to the generation. Quantity F_1 expresses load and generation imbalance:

$$F_1 = \sum_{i=1}^{k} \sum_{j=1}^{ni} \left(P_{ij}^G - P_{ij}^D\right) \tag{4.5.17}$$

where k is the number of islands, n_k is the number of buses of island i, P_{ij}^G is generation of bus j in island i, and P_{ij}^D is the demand of bus j in island i.

Uniform distribution of buses among the islands

It is often desirable to split the system into equal-sized islands. Quantity F_2 is a measure of uniform distribution of power system loads to the islands and is defined by equation:

$$F_2 = \sum_{i=1}^{k} \left| P_i^D - \overline{P}^D \right| \tag{4.5.18}$$

where P_i^D is the total demand of the ith island and $\overline{P}^D$ the average demand of all the islands.

Number of necessary switching

The number of branches that open in order to form the islands must be limited. Otherwise, the opening of a large number of branches can cause instability. Quantity F_3 is a measure of the number of opening branches:

$$F_2 = a{\cdot}n_{\text{lines}}^2 \tag{4.5.19}$$

where α is a weight factor and n_{lines} the number of opening branches.

Split the Power System to the Desirable Number of Islands

The number of islands that are created by the proposed methodology is not always equal to the number k of points used to segment the plane. This is because one of the k segments can contain two or more islands. Quantity F_4, which is a modification of F_1, expresses the requirement of creating the desirable number of islands.

$$F_4 = \left(1 + e^{\frac{(k-k_1)}{q}}\right) \sum_{i=1}^{k} \sum_{j=1}^{ni} \left(P_{ij}^G - P_{ij}^D\right) \tag{4.5.20}$$

where k_1 is the number of islands and q a weight coefficient. The objective function takes the form of equation:

$$F = c_1F_1 + c_2F_2 + c_3F_3 + c_4F_4 \tag{4.5.21}$$

where coefficients c_i are user defined. The method is applied for different test power systems and results are presented in the following sections.

4.5.5 Application of the method on the IEEE – 30 buses test system

The method is applied to the IEEE – 30 buses test system.

Figure 4.5.6 illustrates the initial random distribution of system buses onto the plane (left) and the distribution after the application of VLT-SOM method (right).

Table 4.5.21 presents islanding of the system in two islands, proposed by applying the PSO method to segment the plane of Figure 4.5.6.

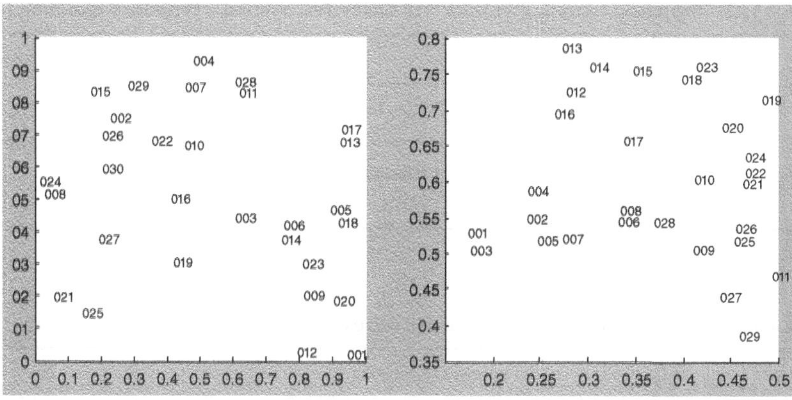

Figure 4.5.6 Mapping of system buses before and after the VLT-SOM training.

TABLE 4.5.21 Proposed Islanding

	Bus	Generation	Load
Island 1	6, 8, 9, 10, 28, 11, 21, 22, 27, 24, 25, 29, 30, 26	50	48,5
Island 2	1, 2, 3, 4, 5, 12, 7, 13, 14, 15, 16, 18, 23, 17, 19, 20	130	89
	Opened Branches		
	6-2, 6-4, 6-7, 10-17,10-20, 24-23		

4.5.6 Application of the method on the IEEE – 118 buses test system

IEEE – 118 buses test system comprises 179 lines, 93 load buses, and 15 generators Tables 4.5.22 and 4.5.23 present two options for islanding of the system in two islands, proposed by applying the PSO method.

4.5.7 Conclusions

In this section, meta-heuristic-aided methods for DSA and corrective control determination are developed. The methods proposed, perform online evaluation of the security status of a power system and provide the optimal corrective control when the system's security is endangered. The corrective control actions include load shedding, generation shedding, and system-controlled islanding. The framework proposed here is generic, so that it can be applied for any kind of power system instability problems and requires from the user the definition of a set of disturbances, a set of eligible corrective controls, and a security criterion.

In the first section a PSO-aided automatic learning framework is developed for DSA and corrective control determination. It is shown that compared to the results of other machine learning methods, the PSO framework leads to less

TABLE 4.5.22 Proposed Islanding of IEEE 118 Buses Test System – Proposed Islanding I

	Bus	Generation	Load
Island 1	23-25, 27-29, 31-32, 68-112, 114-116, 118	1964	1847
Island 2	1-22, 26, 30, 33-67, 113, 117	1823	1803
	Opened Branches		
	17-31, 22-23, 25-26, 47-69, 49-69, 65-68, 113-32		

TABLE 4.5.23 Proposed Islanding of IEEE 118 Buses Test System – Proposed Islanding II

	Proposed Islanding II		
	Bus	Generation	Load
Island 1	42, 45-68, 80, 81, 92-94, 98-112, 116	1636	1501
Island 2	1-41, 43-44, 69, 70, 72, 74-79, 82-91, 95-97, 113-115, 117-118	2151	2143
Disconnected buses	71, 73	0	6
	Opened Branches		
	38-65, 40-42, 41-42, 44-45, 69-47, 69-49, 69-68, 70-71, 72-71, 77-80, 79-80, 89-92, 91-92, 95-94, 96-80, 96-94, 97-80		

unnecessarily curtailed load and better estimation of insecure states. Another advantage of the method is that the RBFNN provides an extrapolation index for each estimation. This index gives a measure of reliability of DSA and proposed corrective control action.

In the second section the idea of DT method for the establishment of load shedding rules is enhanced by the use of GA, which defines the appropriate groups of loads that are eligible for load shedding. The results indicate that GA help to select the most critical loads as input to the DT.

The third section investigates the controlled islanding of a power section as a way to avoid a widespread blackout. The method proposes a hybrid algorithm which consists of a variable local topology SOM that gives a pseudo-geographical representation of power system buses onto a two-dimensional plane. The optimal solution is found by applying the PSO algorithm. The proposed method is applied on IEEE systems of different scales, and the results show its accuracy.

4.5.8 Appendix

Particle Swarm Optimization

PSO [45] is a stochastic, population-based evolutionary computer algorithm for problem solving inspired by social behavior of bird flocking or fish schooling. The swarm is typically modeled by particles in multidimensional space that have a position and a velocity, where each particle represents a candidate solution to the optimization problem. During the optimization procedure, particles communicate good positions to each other and adjust positions according to their history experience and the experience of neighboring particles.

The objective of the algorithm when applied to a minimization problem is to find a solution X^* out of a set $X \subseteq \Re^d$ such that: $X^* = \arg\min_{x \in X} f(x)$, where $f(x)$ is the objective function. The procedure is organized in the following sequence of steps:

1. A population of N particles is created uniformly distributed over X:

$$x_i^{(0)} = x_{\min} + \text{rand}(x_{\max} - x_{\min}) \tag{4.5.22}$$

$$u_i^{(t)} = \frac{x_{\min} + \text{rand}(x_{\max} - x_{\min})}{\Delta t} \tag{4.5.23}$$

   where $x_i^{(t)}$, $u_i^{(t)}$ are the position and velocity of particle i at iteration t, respectively, defined as $x_i^{(t)} = \left[x_{i,1}^{(t)} \cdots x_{i,d}^{(t)}\right]$ and $u_i^{(t)} = \left[u_{i,1}^{(t)} \cdots u_{i,d}^{(t)}\right]$, $x_{\min}$ and $x_{\max}$ are vectors of lower and upper limit values, respectively.

2. The position of each particle is evaluated according to the objective function $f(x)$ and each particle records its best previous position (local best, pbest) and the best previous position among all particles (global best, gbest):

$$\text{pbest}_i = \left(\text{pbest}_{i,1}\ \ \text{pbest}_{i,2} \cdots \text{pbest}_{i,d}\right) \tag{4.5.24}$$

$$\text{gbest} = \left(\text{gbest}_1\ \ \text{gbest}_2 \cdots \text{gbest}_d\right) \tag{4.5.25}$$

   The global best is known to all particles and immediately updated when a new best position is found.

3. Particles' position and velocity are updated according to Eqs. (4.5.26) and (4.5.27), respectively,

$$x_i^{(t+1)} = x_i^{(t)} + u_i^{(t+1)} \tag{4.5.26}$$

$$u_i^{(t+1)} = \omega u_i^{(t)} + c_1 \cdot \text{rand}_1\left(\text{pbest}_i - x_i^{(t)}\right) + c_2 \cdot \text{rand}_2\left(\text{gbest} - x_i^{(t)}\right) \tag{4.5.27}$$

   where

   ω inertial constant

   c_1, c_2 constants, which indicate how much the particles are directed toward the good positions

   rand_1, rand_2 uniform random value in the range [0,1]

4. Go to Step 2 until stopping criteria are satisfied. The stopping criterion set is usually the maximum change in best fitness to be smaller than a specified tolerance for a specified number of moves.

$$\left| f\left(\text{gbest}^{(t)}\right) - f\left(\text{gbest}^{(t-q)}\right) \right| \le \varepsilon \tag{4.5.28}$$

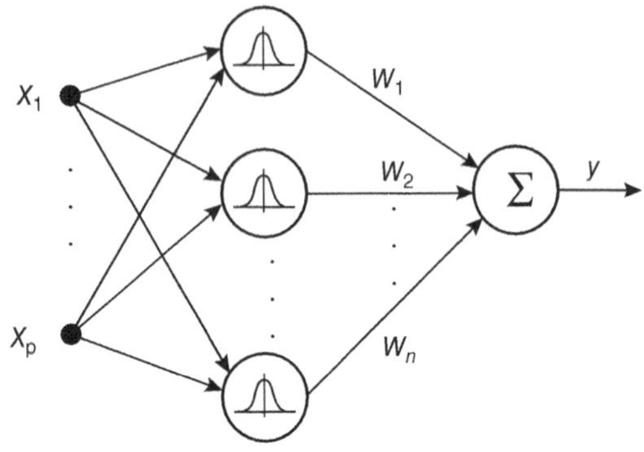

Figure 4.5.7 Radial basis function neural network.

The fitness function can be modified to include functional constraints, by imposing penalties when the constraints are not satisfied:

$$f(x) = f(x) + \sum_{i=1}^{Ncon} r_i \tag{4.5.29}$$

As the swarm iterates, the fitness of the global best solution improves. However, PSO iteration is a heuristic method and it is not able to guarantee the convergence to a global minimum but rather to give a good solution.

Radial Basis Function Neural Networks

RBFNNs is a well-known class of neural networks (NNs) emerged in the late 1980s [157, 158]. A typical RBFNN has a feedforward structure, as illustrated in Figure 4.5.7, and consists of three layers of nodes: the input layer, the hidden layer, and the output layer. Each of the p nodes of the input layer corresponds to a feature of the input pattern $X = [x_1, \ldots, x_p]$. The values of x are passed to the hidden layer. Each of the n nodes in the hidden layer is associated with a vector of centers with dimension equal to the number of inputs $C_n=[c_{n1},\ldots,c_{np}]$ and implements the RBF, which is a real-valued function defined by its center and distance scaling parameter σ.

The most commonly used RBF is the Gaussian function (4.5.30):

$$\varphi(z) = \exp\left(-\frac{z^2}{\sigma^2}\right) \tag{4.5.30}$$

Given an input X, the output of the nth hidden node is given in (4.5.31):

$$a_n(X) = \varphi(|X - C_n|) \tag{4.5.31}$$

where $|X - C_n|$ is the Euclidean distance between the input X and the center C_n.

Thus, the nth hidden node gives a maximum response to input vectors close to its center. The width σ determines how fast its response decreases with the distance of the center. When the width is small it decreases fast and when it is large the response of the RBF decreases slower. The output of the l-th output node is given in (4.5.32):

$$y_l = b_l + \sum_{n=1}^{N} w_{nl} \cdot \varphi(|X - C_n|) \tag{4.5.32}$$

where b_l is the bias of the lth output node.

The calculation of the RBF parameters takes place in three steps. First, the RBF node centers are calculated through a clustering algorithm. When the centers are determined, the width of each RBF node is selected with a nearest-neighbor method. Finally, weights and biases are computed using multiple regression techniques.

Extrapolation Index

RBFNNs can provide measures of reliability, such as an extrapolation index [159–161]. The extrapolation index determines whether there is enough training data in the vicinity of the test point, in order to consider the output of the RBFNN as reliable. The method of computation of extrapolation index is based on Parzen windows [162], which can be implemented by RBFs. At hidden node n, the Parzen density estimate based on the node width σn and node activation function $a_n(x)$ is

$$\rho_n = \rho(x_n) = \frac{1}{KV_n} \sum_{k=1}^{K} a_n(x_k) \tag{4.5.33}$$

where K is the number of training points and

$$V_n = \int_{-\infty}^{\infty} a_n(x) \mathrm{d}x \tag{4.5.34}$$

which means that ρ_n is the fraction of training patterns with high activation values per set volume. Since the nodes may overlap depending on their centers and widths, the extrapolation index is a weighted average as defined in (4.5.35):

$$\rho(x) = \frac{\sum_{n=1}^{N} a_n(x) \rho_n}{\sum_{n=1}^{N} a_n(x) + 1 - \max(a_n)} \tag{4.5.35}$$

Decision Trees

The DT is a tree, structured upside down, built on the basis of a KB consisting of a large number of OPs, covering all possible states of the under-study power system in order to ensure its representativity. The construction of a DT starts at the root node with the whole LS of two-class OPs (e.g. secure and insecure OPs). At each step, a tip-node of the growing tree is considered and the algorithm decides whether it will be a terminal node or should be further developed. To develop a node, an appropriate attribute is first identified, together with a dichotomy test on its values. The test T is defined as:

$$T : x_j \geq w_* \tag{4.5.36}$$

The selection of the optimal test is based on maximizing the additional information gained through the test. The information gain is based on the entropy of each subset E_n with respect to the class partition of its elements, defined as

$$H_c(E_n) = -(f_s \log f_s + f_i \log f_i) \tag{4.5.37}$$

where f_s and f_i are the relative frequencies of the secure and insecure OPs, respectively, in the subset. The selected test is applied to the LS of the node splitting it into two exclusive subsets, corresponding to the two successor nodes. Every subset (node) is characterized by its SI, defined as the percentage of secure OPs belonging to this subset. The optimal splitting rule is applied recursively to build the corresponding subtrees. In order to detect if one node is terminal, i.e. "sufficiently" class pure, the stop splitting rule is used, which checks whether the entropy of the node is lower than a preset minimum value. If it is, the node corresponds to a sufficiently pure subset (states belong to the same class) and is declared a leaf; otherwise, a test T is sought to further split the node. If the node cannot be further split in a statistically significant way, it is termed a dead-end, carrying the two-class probabilities estimated on the basis of the corresponding OPs subset. Figure 4.5.8 shows a sample DT: to infer the output information corresponding to given input attribute values, one traverses the tree, starting at the top-node, and applying sequentially the dichotomous tests encountered to select the appropriate successor. When a terminal node is reached, the output information stored is retrieved.

The nodes of the DT have the following scheme: there are four attributes. In the upper right side is the number of OPs, which belong to the node. In the upper left side is the label of the node. In the middle is the SI, which is the number of OPs belonging to one of the two classes divided to the total OPs of the node. Finally, in the bottom is the dichotomy test of the node or the characterization of the node as dead-end or leaf. In case there is a separation criterion, this is checked and if it is true the left node is accessed, otherwise the right.

Each leaf of a DT corresponds to a chain of rules, each rule having the form of an inequality (partition test). For example, leaf 4 of the sample DT illustrated in Figure 4.5.8 corresponds to the chain of rules:

IF (ATRR1 < 1968) AND (ATTR2 < 3376) THEN LEAF4

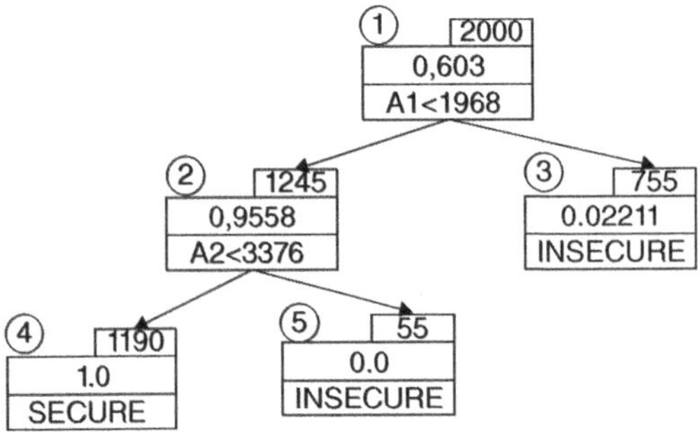

Figure 4.5.8 A sample decision tree.

DTs are evaluated using the testing set (TS). The most important evaluator of the DT reliability and performance is the rate of successful classifications, defined as the ratio of successfully classified OPs to the number of OPs tested.

$$\text{Success} \quad \text{Rate} = \frac{\text{OPs Successfully classified}}{\text{Total Number of OPs}}$$

For a two-class partition (Secure–Insecure), there can be distinguished two types of errors, depending on the actual class of the misclassified OP, FA, and MA:

$$\text{FA} = \frac{\text{OPs misclassified as Insecure}}{\text{Total Number of Secure OPs}}$$

$$\text{MA} = \frac{\text{OPs misclassified as Secure}}{\text{Total Number of Insecure OPs}}$$

In the above case only one input dimension was used at the dichotomy test (univariate tree). An alternative to this is a multivariate tree, where at a decision node, all input dimensions can be used and thus it is more general. When all inputs are numeric, a binary linear multivariate node is defined as

$$T : w_m^T x + w_{m0} \geq 0 \tag{4.5.38}$$

In the case of a multivariate DT, exhaustive search is no longer possible. Meta-heuristic methods can be used in order to fine tune the weights w_m. A special case is when weights are binary numbers. In that case the algorithm searches for the optimum aggregation between the candidate attributes, which provide the maximum information gain when it is used to split the node of the tree. In that case GA can be used to optimize the weights w_m.

Variable Local Topology Self-Organized Maps (VLT-SOM)

SOMs with variable local topology, proposed in [163], are shown to constitute a reasonably good heuristic to find approximate solutions to the NP-complete graph-partitioning problem where a weighted graph has to be partitioned in k subgraphs. The solution to this problem can be applied to power system-controlled islanding. The application of the variable local topology SOM for graph representation can be used as well for power system visualization purposes. In this section, the algorithm (thereafter called VLT-SOM for variable local topology SOM) is described. A nondirected graph G is characterized by a set of vertices V and a set of edges E connecting the vertices, which is represented by matrix [eij]:

$$V = \{u_i\}, \quad i = 1, \ldots, n \tag{4.5.39}$$

$$[e_{ij}] = \begin{cases} e_{ij} = 0, \text{ if edge } (u_i u_j) \text{ does not belong to } E \\ e_{ij} \neq 0, \text{ if edge } (u_i u_j) \text{ belongs to } E \end{cases} \tag{4.5.40}$$

The neighborhood of a vertex u_i is the set of vertices to which u_i is connected. Graph partitioning is an NP-complete problem and heuristic methods are proposed to solve it. In order to partition a graph $G = (V;E)$, vertices are assigned planar coordinates: let w_i denote the positional vector of vertex v_i in the plane.

The modified algorithm proposes moving the vertices' positional vectors appropriately, starting from an initially random spatial distribution of positional vector. In order to achieve this goal we use a NN algorithm, the SOM. The idea is that "training vectors" are drawn according to a certain spatial probability density and the graph tries to approximate the distribution of training vectors. Each time a training vector is presented, the closest positional vector is selected and its coordinates are modified so that it gets closer to the presented training vector. In addition, the coordinates of its neighbors are also modified to make them closer to the training vector. After presentation of sufficiently many training vectors, positional vectors should distribute themselves in space to approximate the distribution of training vectors.

1. Vertices of the graph are assigned random planar coordinates:

$$w_i = [w_{i1} \;\; w_{i2}] \quad \text{where } 0 \le w_{i1}, w_{i2} \le 1 \tag{4.5.41}$$

2. A vertice i is randomly selected.
3. For each vertice j with $a_{ij} = 1$, w vector is updated according to the equation:

$$w_j^{(t+1)} = w_j^{(t)} + \beta \frac{1}{e(i,j)} \left\| w_i^{(t)} - w_j^{(t)} \right\|. \tag{4.5.42}$$

where

$w_i^{(t)}$ is weight value at time step t;

β – learning rate;

$\left\| w_i^{(t)} - w_j^{(t)} \right\|$ – distance measure between i and j.

The planar can be either segmented beforehand in k rectangles each of them representing a partition of a graph. An alternative to this is the segmentation of the planar after the training procedure, using a heuristic algorithm such as PSO [45] and a suitable objective function.

4.6 POWER PLANT CONTROL

Xiao Wu[1] *and Kwang Y. Lee*[2]

[1]*Southeast University, Nanjing, China*
[2]*Baylor University, Waco, TX, USA*

4.6.1 Introduction

MHO recently has been widely studied and applied to solve the complex mathematical problems extracted from the modeling, control, and optimization issues. This chapter introduces several MHO's application in the fossil-fuel power plant (FFPP).

TABLE 4.6.1 Nomenclature

$q_{m,\text{air}}$ mass flow rate of inlet primary air	$q_{m,C}$ mass flow rate of inlet coal
t_{in} inlet primary air temperature	t_{out} mill outlet temperature
m_c the mass of raw coal in coal mill	m_{pf} the mass of pulverized coal in coal mill
$q_{m,pf}$ outlet pulverized coal flow of coal mill	θ_{CM} coal moisture
$W_{\text{free}}^{\text{water}}$ evaporation of coal moisture	P mill differential pressure
I current of coal mill	

$$\dot{m}_{pf} = K_{10}m_c - q_{m,pf} \tag{4.6.2}$$

$$q_{m,pf} = K_{11} \times \frac{22.4}{28.8} \times \frac{273 + t_{\text{in}}}{273} \times \left(\frac{q_{m,\text{air}}}{10}\right)^2 m_{pf} \tag{4.6.3}$$

$$\begin{aligned}\dot{t}_{\text{out}} = {} & (K_1 t_{\text{in}} + K_2)q_{m,\text{air}} + K_3 q_{m,C} - (K_4 t_{\text{out}} + K_5)\left(q_{m,\text{air}} + q_{m,C}\right) \\ & + K_9 I - K_{14} W_{\text{free}}^{\text{water}} + K_{12} t_{\text{out}}\end{aligned} \tag{4.6.4}$$

$$W_{\text{free}}^{\text{water}} = K_{13}\left(q_{m,C}\theta_{CM}\right)t_{\text{out}}\left(1 - e^{-q_{m,\text{air}}/K_{15}}\right) \tag{4.6.5}$$

$$I = K_6 m_{pf} + K_7 m_c + K_8 \tag{4.6.6}$$

$$P = (K_{16} + K_{17}M_c)q_{m,\text{air}}^2 + K_{18}M_{pf}q_{m,\text{air}} \tag{4.6.7}$$

where the input variables of the model are: the inlet coal flow rate $q_{m,\ C}$, inlet PA flow rate $q_{m,\ \text{air}}$, and inlet PA temperature t_{in}. The output variables are the current of the coal mill I, the pressure drop over the mill P, and the outlet temperature t_{out}. The states of the model are the mass of raw coal m_c, the mass of pulverized coal m_{pf}, and the outlet temperature t_{out}. K_1–K_{18} are model parameters to be identified. The nomenclature is presented in Table 4.6.1. The detailed modeling approach is well introduced in [167, 170], and is not repeated here.

For the nonlinear coal mills, identifying the parameters K_1–K_{18} is not easy, because a large-scale nonlinear optimization problem has to be solved to minimize the discrepancy between real plant output and model output. In this section, the GA is employed to fulfill this task on the basis of measurement data [173, 174].

The GA is an optimization algorithm which searches the solution by simulating the biological evolution and natural selection of the world. Because the aim of identification is to find a set of K_1–K_{18} that enables the model to predict the output of coal mill as accurately as possible, the discrepancy between model output and measurements is chosen as the optimization target.

The procedure of parameter identification using GA is given below.

Encoding Encoding is a mapping from feasible solution of the problem to the search space of GA. Different from the conventional binary encoding approach, it is more convenient to apply real-valued encoding in this modeling problem, because the model parameters are all real numbers. Note that, the coal moisture

θ_{CM} cannot be measured, and the initial value of the states $m_c^\theta(0), m_{pf}^\theta(0), t_{out}^\theta(0)$ should be known to make an output estimation, we have to include them in the search space θ:

$$\theta = \left(m_c^\theta(0), m_{pf}^\theta(0), t_{out}^\theta(0), \theta_{CM}, K_1, K_2, \ldots, K_{18}\right)$$

Setting GA Parameters The GA parameters to be set are generation gap, crossover rate, mutation rate, number of individuals, maximum number of generations, and the search range. Generation gap is the ratio between the amount of preserved individuals after selection and the amount of parental individuals before selection. A small generation gap is corresponding to a quick evolving speed; however, a too small generation gap will make the algorithm easily fall into a local minimum. The crossover is a gene segment exchange between parental chromosomes, and the mutation is a random change occurring at a certain gene segment of parental chromosome. Both the crossover and mutation can generate new genome (offspring) which can increase the diversity of the individuals, and improve the possibility of the algorithm in finding better solutions. The search range (θ_{min}, θ_{max}) has to be determined before the optimization, and will affect the performance and efficiency of the optimization greatly; however, because a priori knowledge for the model is very limited, determining the search range is time-consuming. One experience is that the parameter which has greater impact to the model dynamics should be determined with priority. The max number of generations gives a termination condition of the searching process.

Computing Fitness Value Fitness value evaluates the performance of individuals (candidate solutions). For modeling problem, the accuracy of output prediction is the most concerned issue. Therefore, the following quadratic output estimation error $V(\theta, z^N)$ is selected as an objective function to evaluate the model parameters:

$$V\left(\theta, z^N\right) = \frac{1}{N}\sum_{t=1}^{N} l(t, \theta, \varepsilon(t, \theta)) \tag{4.6.8}$$

$$l(t, \theta, \varepsilon) = \varepsilon^T \Lambda \varepsilon \tag{4.6.9}$$

$$\varepsilon(t, \theta) = z^N(t) - z(t, \theta) \tag{4.6.10}$$

where $\varepsilon(t, \theta)$ and $l(t, \theta, \varepsilon)$ are the prediction error and weighted quadratic prediction error at each sampling time, respectively, N is the amount of measurements, Λ is the weighting matrix, and $z^N(t)$ and $z(t, \theta)$ are the measured and predicted outputs, respectively.

Generally, individuals which have a larger fitness value will have a big possibility to survive. Therefore, an inverse form of $V(\theta, z^N)$ is used to ensure that model parameters with smaller prediction error have larger fitness values:

$$f\left(x_i^j\right) = \frac{1}{1 + V(\theta, z^N)} \tag{4.6.11}$$

where x_i^j is the ith individual of the jth generation.

Because the developed coal mill model is a continues model in the form of differential equations while the input measurements are discrete, model prediction is made with the assumption that input variables keep the value of last sample between two samples. Runge–Kutta method is then used to calculate the output prediction of the differential equation model.

At each generation, the fitness of all individuals are calculated. Then the crossover and mutation operations are performed and new offsprings are generated. According to the selection mechanism, parental individuals with lowest fitness are substituted by new offsprings and the individuals in the next generation are determined.

The GA will stop working when the maximum number of generations is attained or a satisfactory solution is found. However, sometimes GA can only find solutions that fit the identification data set well, but have poor performance for other groups of data. Therefore, model validation is necessary, and if the validation results have satisfactory prediction performance, the parameter identification is complete; or the GA parameters should be adjusted and a new search process should be started.

Model Validation

The proposed grey-box coal mill model is identified and verified using on-site data collected from Zhoushan power plant located in Ningbo, China. Operating data generated at three random days are downloaded from the distributed control system (DCS), during which the coal mill runs under various operating conditions. The model validation results are shown in Figure 4.6.2.

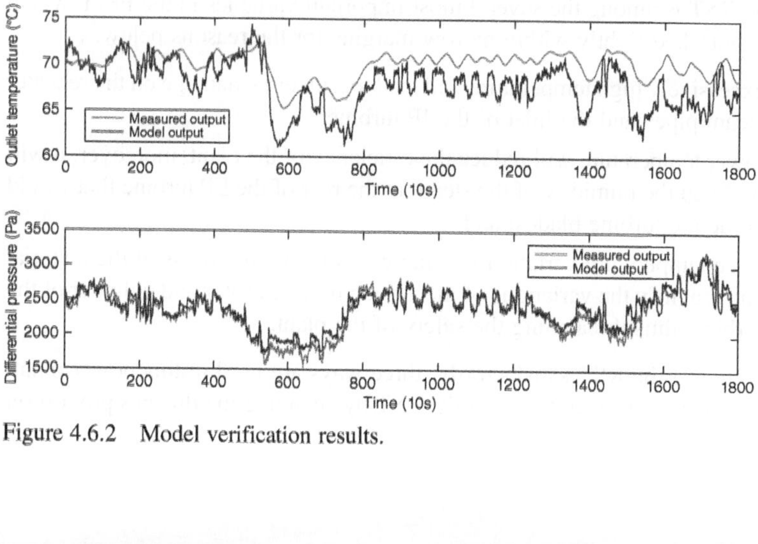

Figure 4.6.2 Model verification results.

The verification results show that dynamics of the identified model fits the real plant well. Although there is a little offset between model output and real output, the more cared dynamics of the coal mill, i.e. the trend and time constant, can be captured accurately, which makes the model suited for advanced control or fault diagnosis.

4.6.3 Nonlinear Model Predictive Control of Reheater Steam Temperature

The working process of FFPPs is a typical system with large inertia, large time-delay, strong nonlinearity, and strict input–output constraints. To overcome these issues and improve the operation performance of the plant, NMPCs have been extensively designed in recent years [164].

The main difficulty of NMPC design is online solving a nonlinear constrained optimization problem at every sampling time to find the optimal future control sequence. Due to the complexity of the FFPP, this calculation often costs considerable amount of effort and time, which will influence the online implementation of NMPC. Moreover, the conventional nonlinear optimization approaches lack robustness and easily fall into local minima problems. Therefore, it is a logical choice to take advantages of MHOs to develop efficient NMPCs with satisfactory performance.

This section introduces the PSO-based NMPC of reheater steam temperature (RST) system in a large-scale power plant [175].

Reheater Temperature System in the Power Plant

The power plant under investigation is an USC pulverized coal-fired power plant and the RST system of this power plant consists of the final and primary reheaters, which are arranged at the top of the boiler furnace and flue gas channel, absorbing the transfer heat from the flue gas and reheating the steam out of the HP turbine to a rated temperature (624 °C) before delivering it to the IP turbine. The sketch of the boiler unit is shown in Figure 4.6.3.

The RST is among the several most important variables in the FFPP which must be controlled tightly within narrow margins for the reasons below:

i. excessively high temperature will lead to a material damage on the reheater steam pipes and the inlet of the IP turbine;
ii. lower temperature will reduce the efficiency of the plant, moreover, it will build up the humidity of the steam in the rear of the LP turbine that would erode the turbine blades; and
iii. large temperature variation will increase the thermal stress of the material and magnify the variations of the air gap between rotor and cylinder of the turbine, thus threatening the safety of the plant.

To control the RST satisfactorily, three ways are used in this power plant: (i) a gas damper is installed to adjust the RST by changing the flue gas proportion

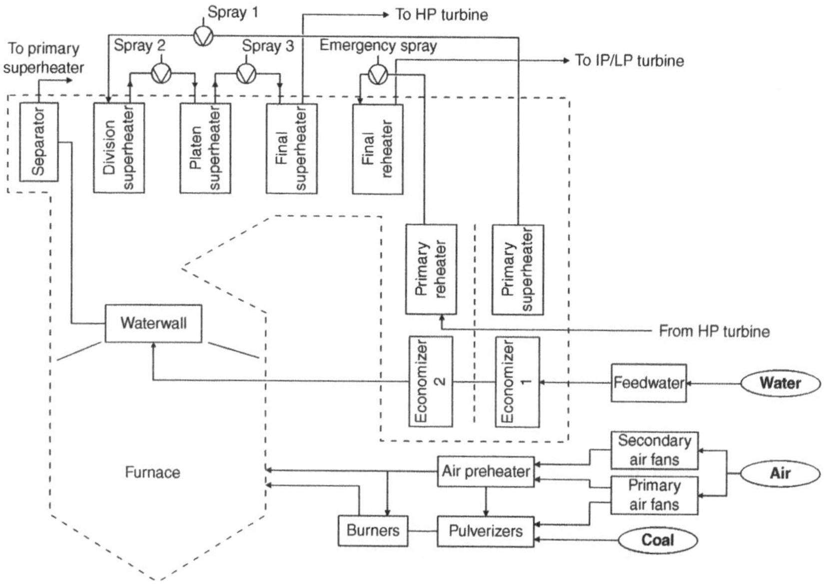

Figure 4.6.3 Sketch of the boiler of the investigated power plant.

flowing through the primary RH and primary SH. The gas damper–RST loop has a large time delay and big inertia due to slow heat transfer between gas and steam; (ii) a bypass line is installed to adjust the steam flow through the reheater, which can control the RST more quickly and accurately; and (iii) a water spray valve is installed before the final reheater; however, because the spray water will influence the efficiency of plant severely, it can only be used in emergency case, i.e. when the steam temperature is rising too fast and over 624 °C which endangers the structural safety. Under normal stable condition the emergency spray has very little opening or keeps fully closed. Therefore, the reheater damper and the bypass valve will be the two control objects with the NMPC scheme. However, the emergency spray valve still uses its original PID controller.

Nonlinear Model Predictive Control of the Reheater Temperature System Using PSO

The overall structure of the NMPC scheme for RST control is shown in Figure 4.6.4. The NMPC is composed of two parts: (i) a recurrent NN model is identified to capture the dynamic behavior of the RST, which receives the data from the power plant and evaluates the performance of different control sequences; and (ii) a PSO-based optimal search module, which will find best controls for the bypass valve and the reheater damper with iterative search.

The development of the recurrent NN model is well introduced in [175, 176], and thus is not repeated here. However, it is worth mentioning here that there are many variables which have big influences on the RST, such as air, feedwater, coal flow, reheater damper position, bypass valve opening, and emergency water-spray

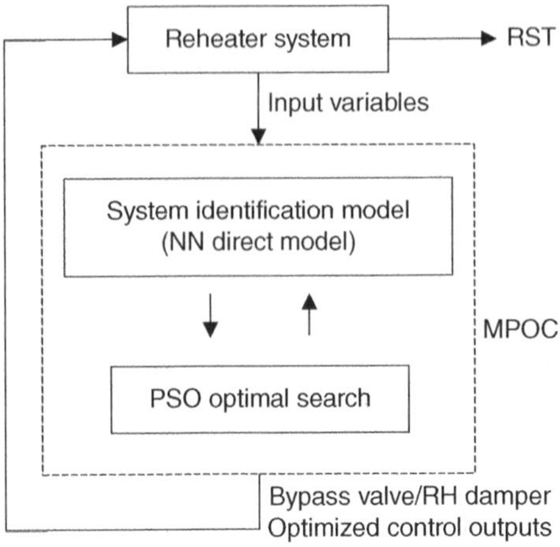

Figure 4.6.4 Structure of the NMPC for reheated steam temperature control.

TABLE 4.6.2 The Selected Input and Output Variables of the RST Model

	Inputs (9)
(1)	Boiler demand
(2)	Turbine demand
(3)	Forced draft fan demand
(4)	Primary air fan demand
(5)	Coal feeder demand
(6)	Feedwater pump demand
(7)	Emergency water-spray valve demand
(8)	Reheater damper demand
(9)	Bypass valve demand
	Output (1)
(1)	Final reheated steam temperature

attemperator. Therefore, the RST system model should consider all these important factors. The selected input and output variables of the RST model are shown in Table 4.6.2.

The PSO algorithm was created based on the simulation of birds flocking in two-dimensional space [177]. The position and velocity of each particle represents the position and velocity of a bird. The position and velocity vectors are both represented by *X–Y* coordinates. The flocking of birds around food is used as a model to represent the optimization of a function. Each particle (bird) saves its best value so far (called pbest) and its current position. Also, each particle knows the best pbest among the group (called gbest), which is the best value any particle in the group has had so far. By knowing its own best value (pbest), the particle knows its own personal experience. With the knowledge of the group's best (gbest), the particle knows the overall performance of the group. Each particle modifies its position by changing its velocity. The velocity of each particle can be updated by

$$v_i^{k+1} = wv_i^k + c_1 \times \text{rand}_1 \times \left(\text{pbest}_i - u_i^k\right) + c_2 \times \text{rand}_2 \times \left(\text{gbest}_i - u_i^k\right) \quad (4.6.12)$$

where v_i^k is velocity of particle i at iteration k, w is weighting function, c_1 and c_2 are weighting factors, rand_1 and rand_2 are random numbers between 0 and 1, u_i^k is current position of particle i at iteration k, pbest_i is the pbest of particle i, and gbest is the best value so far in the group among the pbests of all particles.

The weighting function in (4.6.12) is usually adjusted as:

$$w = w_{\max} - \left(w_{\max} - w_{\min}\right) \times \text{iter}/\text{iter}_{\max} \quad (4.6.13)$$

where $w_{\max}$ is the initial weight, $w_{\min}$ is the final weight, $\text{iter}_{\max}$ is the maximum iteration number, and iter is the current iteration number. Using the above equations, a certain velocity, which gradually brings the particles close to *pbest* and *gbest* can be calculated. The current position (search point in the solution space) can be modified by

$$u_i^{k+1} = u_i^k + v_i^{k+1} \quad (4.6.14)$$

The model using (4.6.12) is called the Gbest model. The model using (4.6.13) in (4.6.12) is called the IWA.

The above BPSO has some disadvantages, such as relapsing into a local extremum, slow convergence in velocity, and low convergence in precision in the later evolution. Recently, a simplified particle swarm optimization (sPSO) is proposed to overcome the disadvantages of the BPSO [178]. The sPSO discards the concept of particle velocity and reduces the BPSO from the second-order to the first-order difference equation. The evolutionary process is only controlled by the variables of the particles' position. The experimental results of some classic benchmark functions have shown that the sPSO greatly improves the convergence velocity and precision in the evolutionary optimization [178]. With this new sPSO the position of each particle is updated by:

$$u_i^{k+1} = wu_i^k + c_1 \times \text{rand}_1 \times \left(\text{pbest}_i - u_i^k\right) + c_2 \times \text{rand}_2 \times \left(\text{gbest}_i - u_i^k\right) \quad (4.6.15)$$

The sPSO algorithm will be used in search of the best controls for the reheater damper and the bypass valve of the boiler unit.

During the implementation of sPSO in NMPC of RST, the sPSO procedure runs 10 iterations with 10 particles before the optimal controls are sent to the reheater system [179, 180].

The candidate control actions are provided to the direct model, with which the RST of the boiler unit is estimated. The estimated output T_e is compared with the current set-point SP_RST in a cost function, the MSE between T_e and its set-point SP_RST, which is:

$$\text{Cost} = (T_e - \text{SP\_RST})^2 \quad (4.6.16)$$

To make further efforts to reduce the single-step RST prediction error of the neural direct model, each of the last-step error Δ between the network's predictive output T_e and the actual RST T_a is calculated and added to the cost function for current-step search of the best controls with PSO. The modified cost function with error compensation is

$$\text{Cost}' = (T_e + \Delta - \text{SP\_RST})^2 \quad (4.6.17)$$

In order for the PSO to find optimal solutions against time with less iteration, three approaches are used:

i. The search window of control output, i.e. maximum and minimum control values are determined at the PSO design stage. The unit load (power output) is used for establishing a control window for the reheater damper based on historical data. For the given large-scale boiler, the control window for the reheater damper during loading-down process is shown in Figure 4.6.5. As for the bypass valve, since it may change through 25–65% during the whole load scope according to its original control logic, we simply use 25 and 65% as its lower and upper limits during PSO control search.

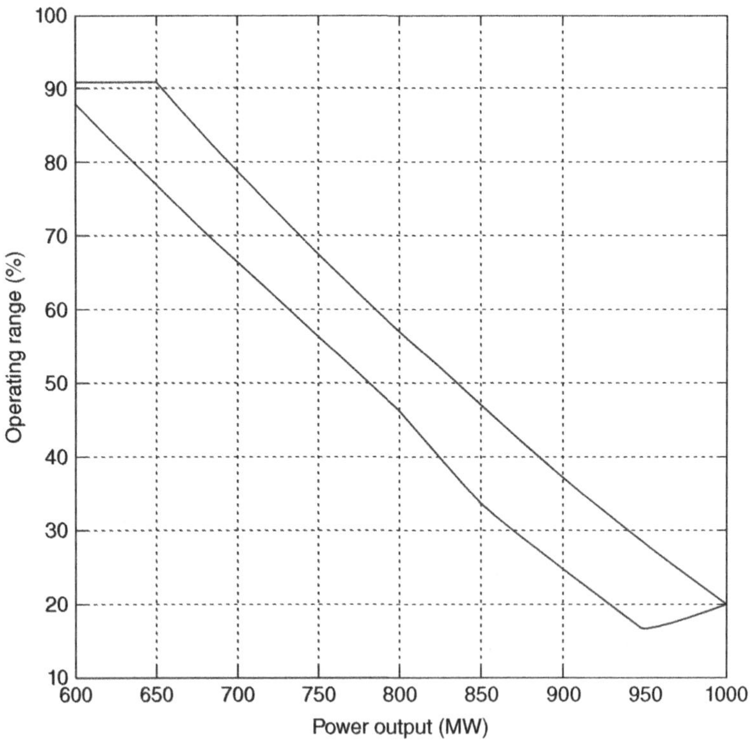

Figure 4.6.5 Reheater damper control search window for loading-down process.

ii. It is noted that the reheater damper and the bypass valve have very clear control direction in an actual power plant operation, which is based on the relationship between actual RST T_a and its set-point SP_RST. If $T_a >$ SP_RST, the damper will go down and the bypass will go up from the current position. On the other hand, if $T_a <$ SP_RST, the damper will go up and the bypass will go down from the current position. In fact, there are many combinations of controls for reheater damper and bypass valve in each step which can meet the temperature control demands. In order to avoid oscillation during the search, a special search direction control factor Dir can be added to (4.6.15) to obtain an optimal combination for the reheater bypass and the damper, which are the nearest to their last-step positions. With this new directional sPSO, the position of each particle is updated by

$$u_i^{k+1} = wu_i^k + c_1 \times \text{rand}_1 \times \text{Dir} \times \left(\text{pbest}_i - u_i^k\right) + c_2 \times \text{rand}_2 \times \text{Dir} \times \left(\text{gbest}_i - u_i^k\right) \quad (4.6.18)$$

where Dir is the directional control vector, whose value takes either $[1, -1]^{\text{T}}$ or $[-1, 1]^{\text{T}}$ under different cases.

Control Simulation Tests

In order to test the above NMPC scheme, some simulation tests have been made on a full-scope simulator for the large-scale power generating unit. The control results with the original PID control logic in the simulator and the proposed NMPC scheme are compared by changing the unit load from 100 to 75% load, with respective results shown in Figure 4.6.6.

From Figure 4.6.6, it can be seen that the NMPC scheme gives a better control result for the RST than the original control over time when the unit load is changed significantly, from 100 to 75%. This improvement is acquired by faster and optimized adjustments for the reheater damper and the bypass valve with the PSO optimal search.

4.6.4 Multi-objective Optimization of Boiler Combustion System

Fossil fuel-fired power plant is a place where electricity is generated, but also a place where energy is wasted and pollution is produced. Therefore, in the context of increasing power demand as well as the energy and environmental issues, the optimization of FFPPs is of great significance. From the operational level, the optimization of FFPPs does not refer to design advanced control strategy to improve the control level only, but more importantly, includes finding the optimal operation points for the plant. This section introduces MHO's application in finding the optimal operation points, and the most complex system of FFPP, boiler combustion system, is taken as an example.

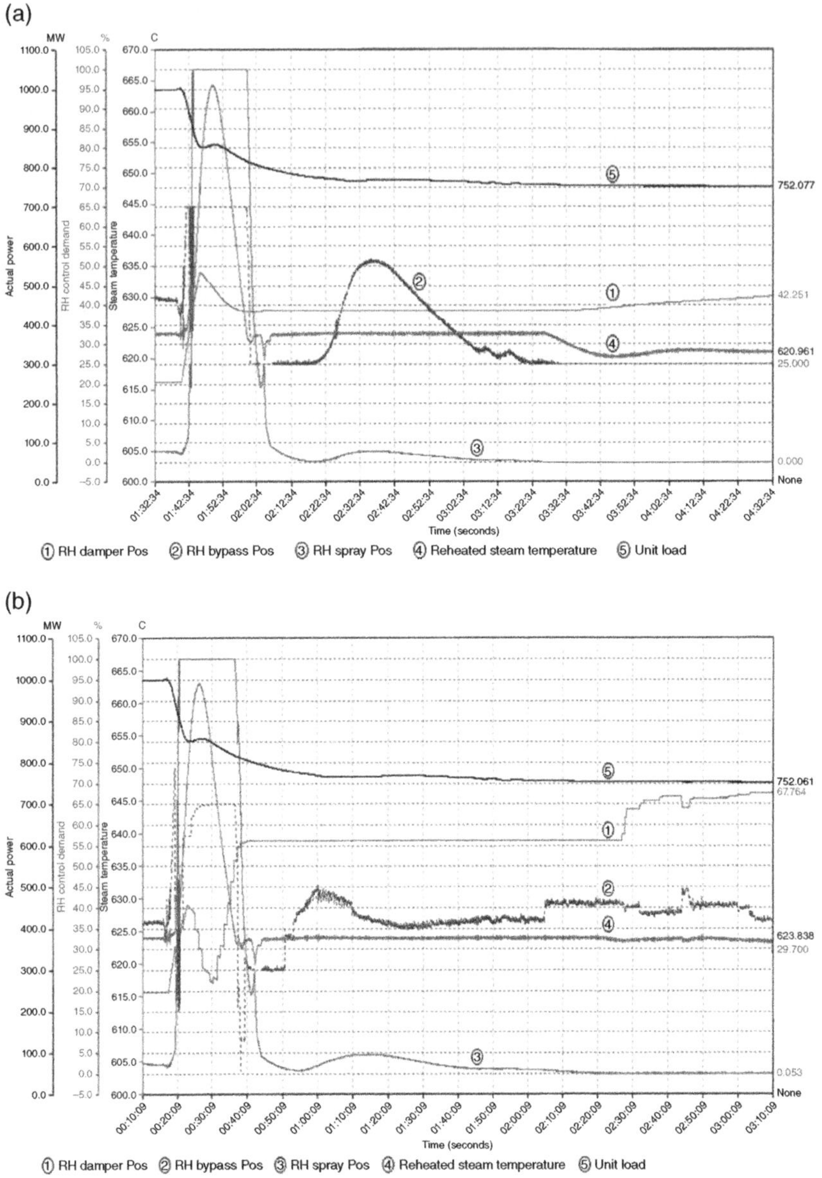

Figure 4.6.6 Damper, bypass, and RST responses due to load change from 100 to 75%. (a) The original PID control. (b) The NMPC control.

Boiler Combustion Optimization

For the modern FFPPs, the fuel cost accounts for more than 70% of the total operation cost. Generally speaking, if the boiler efficiency can be improved by 1%, the whole unit efficiency will be improved by 0.3–0.4%, and the standard fuel cost will be decreased by 3–4 g/kWh. Moreover, the emission of the nitrogen oxides (NO_x) during coal combustion in FFPPs accounts more than 35% of total NO_x emission according to the recent investigation in China [181]. For this reason, performing the combustion optimization for the boiler has a huge potential in energy saving and pollution reduction.

However, the boiler combustion optimization has great challenges:

1. The basic working principle of low NO_x combustion is to create a low oxygen and temperature environment in the furnace [182], which is generally achieved by adjusting the air distribution in the furnace and achieving a staged combustion of mixed fuel and air. However, the efficiency of the boiler is also degraded when the combustion temperature is decreased. Therefore, enhancing the boiler efficiency and reducing the NO_x generation are two conflict objectives, which cannot be attained simultaneously.
2. There are many factors which will influence the combustion condition, for example: the boiler load, PA and secondary air (SA) velocity, total air flow rate, speeds of mills, coal quality, and so on. The large number of optimized variables makes the optimization complex and computationally difficult.

Therefore, the boiler combustion optimization presents a large-scale multi-objective optimization problem, and MHOs which have quick convergence speed and good performance are highly desired. We will introduce an immune cell subsets-based multi-objective optimization algorithm (ICSMOA) to the boiler combustion optimization next [183].

Immune Cell Subsets-Based Multi-objective Optimization Algorithm

Generally, a multi-objective optimization problem can be described as follows:

$$\begin{aligned} &\min \quad \mathrm{f}(\boldsymbol{X}) = (\mathrm{f}_1(\boldsymbol{X}), \mathrm{f}_2(\boldsymbol{X}), \ldots, \mathrm{f}_n(\boldsymbol{X})) \\ &\text{s.t.} \begin{cases} g_p(\boldsymbol{X}) \geq 0, & p = 1, 2, \ldots, l \\ h_q(\boldsymbol{X}) = 0, & q = 1, 2, \ldots, m \end{cases} \end{aligned} \tag{4.6.19}$$

where $X=[x_1, x_2, \ldots, x_r] \in \Omega$ is the decision vector, Ω is the feasible region, $f_1(X), \ldots, f_n(X)$ are the multi-objectives, and $g_p(X)$ and $h_q(X)$ are the constraints of the system.

For the boiler combustion system, because the high efficiency and low NO_x emission objectives are in conflict, there is no absolutely optimal solution; therefore, the notion of "optimal" is transformed into "Pareto optimal," which is defined as:

Definition 1 $\forall X_i, X_j \in \Omega$, if $\forall t \in 1, 2, \ldots, n, f_t(X_i) \leq f_t(X_j)$; together with $\exists s \in 1, 2, \ldots, n, f_s(X_i) < f_s(X_j)$, we say that X_i dominates X_j, marked as $X_i \prec X_j$.

Definition 2 $\forall X_i, X_j \in \Omega$, if $\forall t \in 1, 2, \ldots, n, f_t(X_i) \leq f_t(X_j)$; together with $\exists s \in 1, 2, \ldots, n, s \neq t, f_s(X_i) > f_s(X_j)$, we say that there is no dominance relationship between X_i and X_i, marked as $X_i O X_i$.

Definition 3 $X_i \in \Omega$ is a Pareto optimal solution if $\neg\exists\, X_j \in \Omega$, s.t. $X_j \prec X_i$ ($\neg\exists$ means nonexistent).

The process of multi-objective optimization is continually updating the non-dominated solution set, and make it close to the real Pareto front.

The ICSMOA was created based on the simulation of immune cells which are the basic units participating in the immune response. The immune cells have to experience complex changes from generation to maturation, such as immunological tolerance, clone, mutation, memory, death, and others. Moreover, according to different functional characteristics, the immune cells can be divided into different subsets, for example B cells, T cells, and so on [184]. These features can be utilized in searching the Pareto front according to different preferences in the multi-objective optimization.

For convenience of description, some notations are given first: C is the abbreviation for cell, which means a candidate solution; CS is the abbreviation for cell subset; MC is the abbreviation for memory cell, which is used to represent a Pareto solution; MCS is the abbreviation for memory cell subset. The sizes of decision variables and objectives are denoted as des and obj, respectively.

We will introduce the operators of the artificial immune cell model first, which simulate the functions of immune cells, then the working procedure of the ICAMOA will be given.

Definition 4 (affinity operator). The affinity of immune cell-i represents the similarity between cell-i and other cells. It is defined as the number of immune cells within the neighborhood of an immune cell-i in the ICSMOA.

$$\mathrm{aff}(Ci) = \sum_j \mathrm{sig}(Ci, Cj) \tag{4.6.20}$$

$$\mathrm{sig}(Ci, Cj) = \begin{cases} 1 & \mathrm{dis}(Ci, Cj) < \mathrm{set\_limit\_R} \\ 0 & \mathrm{dis}(Ci, Cj) > = \mathrm{set\_limit\_R} \end{cases} \tag{4.6.21}$$

The bound of the neighborhood set_limit_R is adaptive during the implementation of the algorithm.

Definition 5 (subset division operator). Inspired by B cell and T cell subset theory, the subset division operator is designed to divide the immune cells into different subsets according to different preferences.

The total number of the subsets is obj +1. The first subset contains the cells which perform best at the most preferred objective, the second subset consists of the cells which have the best value of the second-most preferred objective. In this way, the first obj subsets can be determined and the remaining cells constitute the obj+1th subset.

The following two rules are satisfied by the subset division:

1. $CS1 + CS2 + \cdots + CS_k = \mathrm{cell\_set}$
2. $\forall i \neq j, i \le k, j \le k, CSi \cap CSj = \emptyset$

Definition 6 (non-dominated set solving operator). The non-dominated set solving operator is used to find the non-dominated solution set for each subset. The approach given in [179] is used here to achieve the goal.

Definition 7 (affinity maturation operator). The affinity maturation operator is used to determine whether the non-dominated solution of each subset is a non-dominated solution of the whole cell group.

$$\text{mature}(Ci) = \begin{cases} 1 & Ci \in \text{Pareto}(\text{Pareto}(\text{CS}_1) + \cdots + \text{Pareto}(\text{CS}_k)) \\ 0 & Ci \neg \in \text{Pareto}(\text{Pareto}(\text{CS}_1) + \cdots + \text{Pareto}(\text{CS}_k)) \end{cases} \tag{4.6.22}$$

If the immune cell C_i is judged affinity mature, it will be included in the memory cell set.

Theorem 1 Finding the non-dominated solutions directly from the original immune cell set is equivalent to finding the non-dominated solutions of the set, which is composed of non-dominated solutions of each subset [185], i.e.

$$\text{if cell\_set} = \text{CS}_1 + \text{CS}_2 + \cdots + CSk, andCSi \cap CSj = \emptyset, i \neq j, then$$
$$\text{Pareto(cell\_set)} = \text{Pareto}(\text{Pareto}(\text{CS}_1) + \cdots + \text{Pareto}(\text{CS}_k))$$

Definition 8 (immunological tolerance test operator). The process of immunological tolerance test operation is outlined in Figure 4.6.7. If the new immune cell can pass the immunological tolerance test, it will be regarded as a mature cell. According to the result in [186], most of the newly generated anti-self-reactive cells (the cells failed pass the immunological tolerance test) are not eliminated immediately; instead, a receptor editing operation occurs in most cases, through which the anti-self-reactive cells can be converted to non-self-reactive cells. Inspired by this mechanism, a random number is generated to determine whether the receptor editing should be performed on the anti-self-reactive cells.

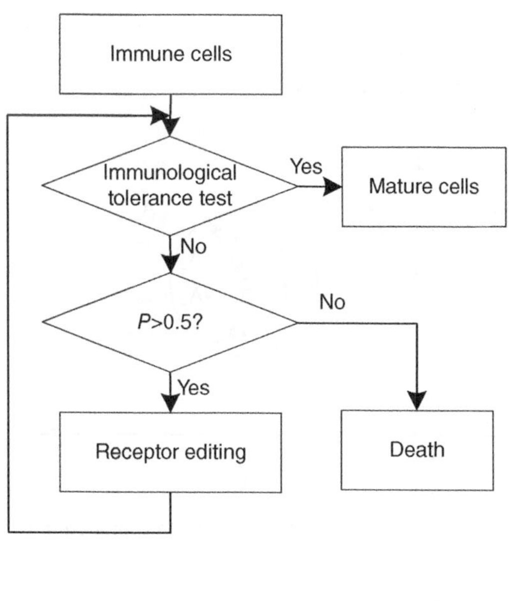

Figure 4.6.7 Process of immunological tolerance test.

The immunological tolerance operation includes the central and peripheral tolerance operation. The central tolerance operation is used for the newly generated immune cells which can be described as:

$$\text{central\_tole}(Ci) = \begin{cases} 1 & \text{neighbor\_count}(Ci) < \text{set\_limit} \\ 0 & \text{neighbor\_count}(Ci) > = \text{set\_limit} \end{cases} \quad (4.6.23)$$

If the number of the cells around the newly generated cell is under the set limit, the cell will pass the test. This operation guarantees a better distribution of the candidates.

The peripheral tolerance operation is used for the cells which have experienced mutation, and it can be described as

$$\text{peri\_tole}(C_i) = \begin{cases} 1 & C_i < C_{\text{source}}\ OR\ C_i O C_{\text{source}} \\ 0 & \text{others} \end{cases} \quad (4.6.24)$$

After the mutation, if the cell is not inferior to the clone source, it will pass the test. However, if it is dominated by the clone source, it is not necessary to preserve it in the next generation because it cannot become a non-dominated solution. This operation reduces the computational complexity in the next generation.

Definition 9 (receptor editing operator). For the cell which has not passed the tolerance test, a receptor editing operation is performed which maps the cell to the neighborhood of the cell which has the lowest affinity. The schematic diagram of receptor editing is shown in Figure 4.6.8 for a two-dimensional condition, where Ai is the cell need to be edited, B is the cell which has the lowest affinity, and A is the obtained new cell which is within a circle (B is the center and R is the radius).

For high-dimensional cases, suppose the coordinate of cell B is $(b_1, b_2, \ldots, b_d)$, the coordinate of A_i is $(a_{i1}, a_{i2}, \ldots, a_{id})$, the coordinate variation is limited to $(\min_j, \max_j), j \in \{1,2,\ldots,d\}$. Then, the coordinate of the new cell A_i' $(a_{i1}', a_{i2}', \ldots, a_{id}')$ can be calculated as

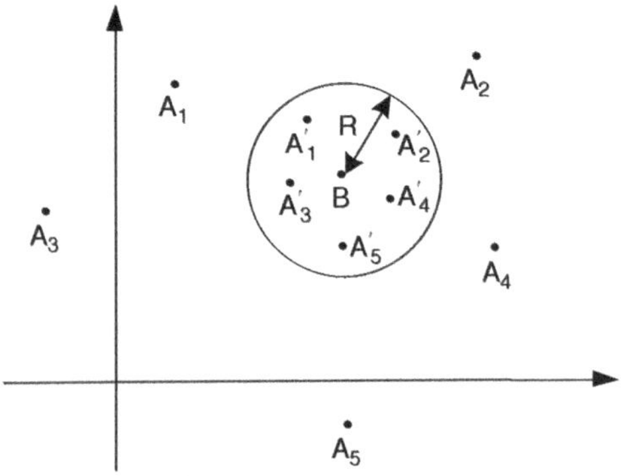

Figure 4.6.8 Schematic figure for receptor editing.

$$a'_{ij} = \begin{cases} b_j - \dfrac{b_j - a_{ij}}{b_j - \min_j} \times R & \text{if } a_{ij} < b_j \\ b_j + \dfrac{a_{ij} - b_j}{\max_j - b_j} \times R & \text{others} \end{cases} \tag{4.6.25}$$

Definition 10 (clone operator). As an asexual reproduction technique, clone can produce a group of identical cells from a single original ancestor. In the ICSMOA, the clone size of each source cell is determined by the affinity and the preference of the objectives. The cell with lower affinity and greater preference will be reproduced more times.

Suppose $(C_{11}, C_{12}, \ldots, C_{1i_1}; C_{21}, C_{22}, \ldots, C_{2i_2}; \ldots; C_{k1}, C_{k2}, \ldots, C_{ki_k})$ are the clone sources for subsets $1, 2, \ldots, k$, prefer(j) is the preference proportion for subset j, and M_j is the modified coefficient, then the clone size of cell C_{js} is determined by

$$Mjs = \text{int}\left(\frac{M_j \prod_{t=1}^{i_j} \text{aff}(Cjt)}{\text{aff}(Cjs)} \times \text{prefer}(j) \right) \tag{4.6.26}$$

After clone proliferation, a hyper mutation operation is imposed to every new cell, so that the diversity of the candidates can be improved.

Definition 11 (redundant memory cell removal operator). Redundant memory cell removal operation is executed in the case that memory pool size exceeds the set limit. The redundant cells are removed one by one according to the preference and the affinity.

Using the defined operators, the working principle of the ICSMOA is shown in Figure 4.6.9. Except the first generation, the immune cell set is composed of three parts: the new cells which have passed the central tolerance test, the mutant cells which have passed the peripheral tolerance test, and the memory cells obtained from the last generation. New cell sets are then divided into subsets, and the non-dominated set of each subset is regarded as clone source. The detailed procedures of processing memory cells are shown in Figure 4.6.10.

Optimization Results and Discussion

The boiler under consideration is a tangentially fired once-through boiler served in a 600MW supercritical (SC) power plant. Six layers of PA nozzles (A–F) and seven layers of SA nozzles (AA, AB, BC, CD, DE, EF, and FF) are installed in the main combustion zone. Two layers of compact overfire air (COFA) nozzles (CCOFA1 and 2) and five layers of separated overfire air (SOFA) nozzles (SOFA1–5) are installed over the furnace. The boiler is equipped with vertical spindle roller pressure mill direct-firing pulverizing system and six coal feeders (A–F) are used to provide pulverized coal to the boiler.

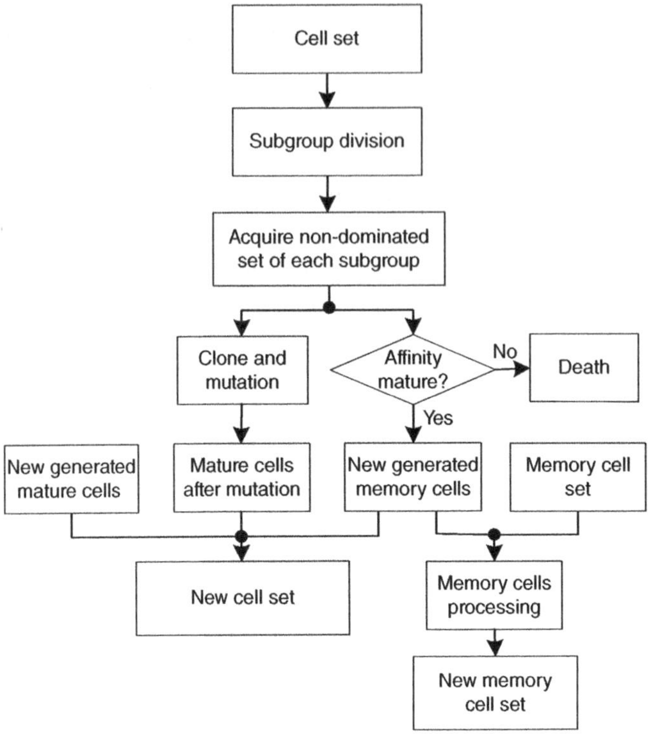

Figure 4.6.9 Working principle of ICSMOA.

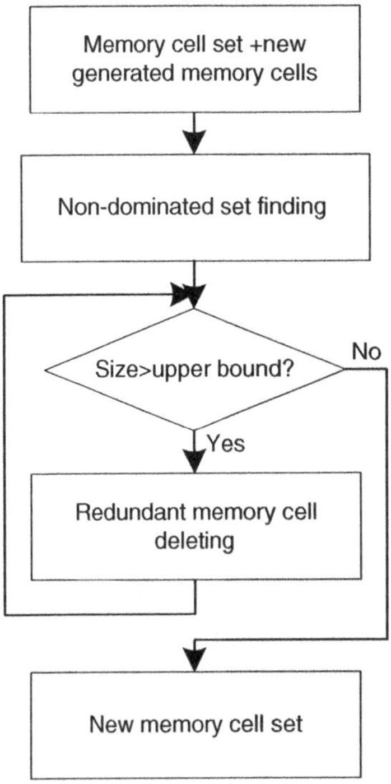

Figure 4.6.10 Detailed procedure of memory cells processing.

Historical operation data are downloaded from the DCS of the plant which can cover 360–550MW operating range of the plant. A steady-state model of the combustion system is then developed, where the load, coal feed rate, valve opening of six coal feeder, six PA, seven SA, two COFA, five SOFA, and the oxygen concentration at the economizer outlet are selected as the input; the boiler efficiency and NO_x concentration at the outlet of the furnace are selected as the output. The model is developed using the approach of least-square support vector machine (LSSVM) and then used in the combustion optimization. The detailed modeling approach can be found in [183], and not repeated here.

The ICSMOA is then used in the boiler combustion optimization, finding the optimal steady-state OP for different loading conditions. Because there are a huge number of variables which have influence on the boiler efficiency and NO_x generation, to improve the reliability of implementing the optimization and reduce the computational complexity, 10 key variables, the valve opening of 5 SA (AB, BC, CD, DE, EF) and 5 SOFA (SOFA1–5), are selected as the decision variables.

The limitations of the 10 decision variables are also given; for the SA valves, the opening range of sampling data is selected as the search region and the limitations of the SOFA valve are given in Table 4.6.3.

Here, 18 OPs along 360–550 MW region are optimized, and the optimization results for four typical OPs (OP 3, 8 12, and 17) are shown in Figure 4.6.11. Using the deigned air valve openings, these four points have the highest/lowest NO_x emission and boiler efficiency. The original NO_x emission and boiler efficiency of these four OPs are given in Table 4.6.4.

Considering the 100 mg/m^3 NO_x emission standard (GB 152 13223-2011) in China, and the operating cost of selective catalytic reduction (SCR) denitrification devices, the NO_x concentration at the outlet of the furnace is expected to be around 200 mg/m^3, which has been set as the preference during the optimization. Being different from the conventional multi-objective optimization algorithm which has an even solution distribution, we can see from Figure 4.6.11 that larger number of preferred solutions (NO_x emission around 200 mg/m^3) can be found by the ICS-MOA. This feature reduces the burden of computation and solution storage, and brings convenience in decision-making. Moreover, when the preference is

TABLE 4.6.3 The Limitations of the Separated Overfire Air Valve for OP 3,8,12,17

OP SOFA	3	8	12	17
SOFA1	0–100%	0–100%	0–100%	0–100%
SOFA2	0–90%	0–100%	0–100%	0–100%
SOFA3	0–80%	0–90%	0–100%	0–100%
SOFA4	0–40%	0–50%	0–70%	0–80%
SOFA5	0	0	0–30%	0–50%

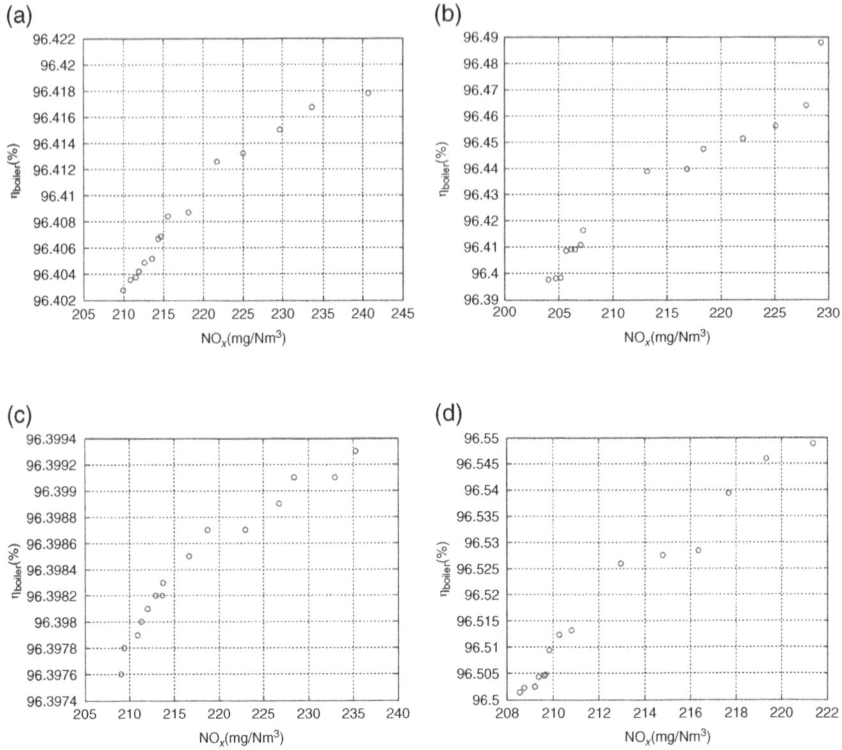

Figure 4.6.11 The optimization results for OP 3,8,12, and 17 using the ICSMOA. (a) *OP* 3. (b) *OP* 8. (c) *OP* 12. (d) *OP* 17.

TABLE 4.6.4 Original NO_x Emission and Boiler Efficiency for OP 3,8,12,17

OP	NO_x Emission (mg/NM^3)	Boiler Efficiency (%)
3	296.88	96.28
8	204.39	96.41
12	234.67	96.1
17	220.15	96.55

changed, the ICSMOA can respond rapidly, shifting the focus of search to the preferred region.

The results also show that, for OP 3 and 12, the optimization can improve the operating level greatly, reducing the NO_x generation and increasing the boiler efficiency. For OP 8 and 17, the original SA and SOFA valve openings are set well and we cannot find a solution which has higher efficiency and lower NO_x generation simultaneously. However, many alternative optimal solutions can be provided by the ICSMOA and the users can select one from them according to their preferences.

4.6.5 Conclusions

This chapter shows three different MHO techniques: the GA, PSO, and ICSMOA, as well as their applications in the FFPP.

Due to the complex system behavior, the main difficulty, which limits the implementation of advanced modeling, optimization, and control techniques in the FFPP, is solving the large-scale nonlinear optimization problem. For these reasons, MHOs that have quick convergence speed and good performance are highly desired, which can greatly improve the operating level of the plant.

The MHOs have been used extensively in both the academic study and engineering practice of the FFPP. Considering the benefits brought by the MHOs and the facilities provided by the wide use of DCS and fast development of computer technologies, it is safe to say that much more development in the intelligent operation of FFPPs is ahead of us, and the advanced modeling, optimization, and control techniques will replace the classical PI/PID controllers in a foreseeable future.

4.7 PREDICTIVE CONTROL IN LARGE-SCALE POWER PLANT

Liangyu Ma[1] and Kwang Y. Lee[2]

[1]North China Electric Power University, Baoding, China
[2]Baylor University, Waco, TX, USA

4.7.1 Introduction

SC and USC coal-fired power units have become the dominant energy conversion units in China and around the world. With the widespread implementation of AGC in regional power grids, these large-capacity power units are required to participate in peaking-load regulation frequently without exception and often operate under variable load conditions. The traditional coordinated control strategy cannot adapt well to the wide-scope loading situations, and often leads to slow load response speed and large main steam pressure fluctuations. Therefore, it is of great significance to improve the coordinated control quality of a SC/USC energy conversion unit with advanced intelligent control strategies.

With rapid development of modern control theory and computer technology, intelligent model predictive optimal control (MPOC) approach based on NN predictive model and swarm intelligence algorithm (such as PSO or GA) has drawn much attention and has been applied in many areas, including steam temperature control and coordinated control in coal-fired power units [179, 187–190]. With recognized strong nonlinear mapping capability, adaptive learning, parallel information processing ability, and good fault tolerance, ANNs are suitable for modeling and control of complex nonlinear dynamic systems [191–195]. PSO algorithm, with its advantages of high search efficiency, fast convergence speed, and less parameters to set, has also been successfully applied to complex system optimization and control of various industrial processes, including solving optimization problems, NN training, controller design, and optimization [45, 196–201].

To improve the coordinated control performance of a large-scale SC/USC power unit, an intelligent MPOC method using constrained PSO is proposed [187]. The proposed intelligent MPOC scheme is programmed with MATLAB software and implemented in the full-scope simulator of a 600 MW SC power unit to test the MPOC scheme through extensive control simulation tests.

4.7.2 Particle Swarm Optimization Algorithm

Mathematical Description of PSO Algorithm

PSO algorithm was first put forward in 1995 by Kennedy and Eberhart based on the simulation of birds flocking in two-dimensional space [45]. Since the PSO algorithm is concise, easy to implement with less parameters to adjust, and with no need to optimize the function of the gradient information, it has drawn wide interest and has been applied in various engineering optimization fields.

The expressions of the standard PSO algorithm, in which a colony is composed of m particles moving with a certain speed in a D-dimensional space, are given as follows [196]:

$$v_{id}^{k+1} = wv_{id}^{k} + c_1 r_1 \left(p_{id}^{k} - x_{id}^{k}\right) + c_2 r_2 \left(p_{gd}^{k} - x_{id}^{k}\right) \tag{4.7.1}$$

$$x_{id}^{k+1} = x_{id}^{k} + v_{id}^{k+1} \tag{4.7.2}$$

where, $i = 1, 2,\ldots,m$; $d = 1, 2,\ldots,D$; m and D are, respectively, the numbers of particles and dimension; and c_1 and c_2 are two non-negative acceleration constants. Random numbers r_1 and r_2 are between 0 and 1 to maintain the diversity of the group. The inertia weight w reflects the proportion of the local search and global search capability, which is a weight function rather than a fixed value, and usually adjusted by

$$w = w_{\max} - \frac{k(w_{\max} - w_{\min})}{k_{\max}} \tag{4.7.3}$$

where, $w_{\max}$ is initial weight, $w_{\min}$ is the final minimum weight, $k_{\max}$ is the maximum iteration number, and k is the current iteration count. This function makes the PSO explore larger area in the beginning and quickly locate the general area of the optimal solution.

There exist some disadvantages for the above standard PSO algorithm. Due to the stronger global search ability at the beginning, if the algorithm cannot find the best point in the initial stage, it is easy to fall into a local optimum with the decrease of w and focusing on local search.

An Efficient and Simpler PSO Algorithm

In overcoming the disadvantages of the standard PSO algorithm, such as relapsing into local extremum and slow convergence velocity in later iterations, there have been various improvements [199–201]. But, generally these measures lead the PSO algorithm to a more complex structure and make it not favorable for real-time online control of a complex industrial process.

Thus, a high-efficiency simplified PSO (sPSO) algorithm is proposed in [178], which can greatly improve the convergence speed and accuracy, and effectively overcome the weakness of the standard PSO algorithm. The sPSO cleverly skips the velocity calculation and the evolution is controlled only by the particle's position, thus reducing the PSO algorithm from the second-order to a first-order difference equation. The position of each particle in sPSO is updated by

$$x_{id}^{k+1} = wx_{id}^{k} + c_1 r_1 \left(p_{id}^{k} - x_{id}^{k}\right) + c_2 r_2 \left(p_{gd}^{k} - x_{id}^{k}\right) \tag{4.7.4}$$

To compare the performance of the given sPSO algorithm with the standard PSO, several typical benchmark functions are used for optimization solution tests. Detailed validating tests can be found in [178]. It is shown that the sPSO algorithm dramatically improves the convergence speed in the evolutionary optimization, which is in favor of the real-time performance for predictive optimal control. Therefore, the sPSO algorithm is used as the optimization solving algorithm for MPOC of the coordinated system of the SC power unit.

4.7.3 Performance Prediction Model Development Based on NARMA Model

Neural Network Modeling Method

With a traditional back-propagation (BP) network, a first-order time-delay nonlinear autoregressive moving average (NARMA) model for a nonlinear system with three inputs (u_1, u_2, and u_3) and two outputs (y_1, y_2) is constructed as shown in Figure 4.7.1, which takes the impact of the inputs at the kth and the $(k-1)$th steps and the outputs at the kth step on the outputs at the $(k+1)$th step into consideration. This model can realize nonlinear system identification based on dynamically updated historical data of the inputs and outputs with the advantages of less calculation and easy convergence.

Model Structure of Supercritical Power Unit

The simplified system of a SC power unit is shown in Figure 4.7.2. When a SC power unit works in the "once-through" stage, there are no clear demarcation points between steam and water. The feedwater pumped into the boiler is continuously heated, evaporated, and overheated in the waterwall and all stages of superheaters. The length of each stage changes with disturbances of fuel flow rate

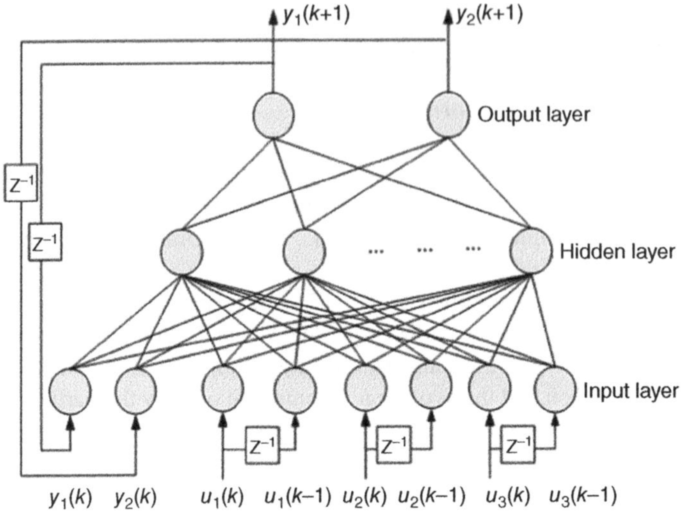

Figure 4.7.1 Simplified NARMA neural network model.

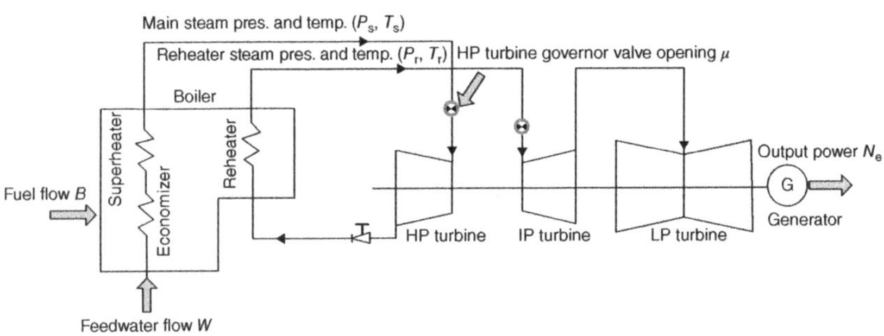

Figure 4.7.2 Schematic diagram for supercritical power unit.

B (kg/s), feedwater flow W (kg/s), and turbine governor valve opening μ (%), leading to changes of superheated steam temperature T_s (°C), main steam pressure P_t (MPa), and unit power N_e (MW) [204–206].

Thus, generally a SC/USC power unit's performance can be described as a model with three inputs and three outputs, revealing the strongly coupled nonlinear relationship between the fuel flow, feedwater flow, and turbine governor valve opening, and the unit's load, main steam pressure, and the intermediate point temperature (heat enthalpy). Considering that the modeling purpose in this work is mainly for coordinated control to improve the control quality of unit load and main steam pressure, the intermediate point temperature (heat enthalpy) can be omitted during modeling. Thus, the properties of the once-through boiler unit can be simplified into a nonlinear model with three inputs and two outputs, as shown in Figure 4.7.3.

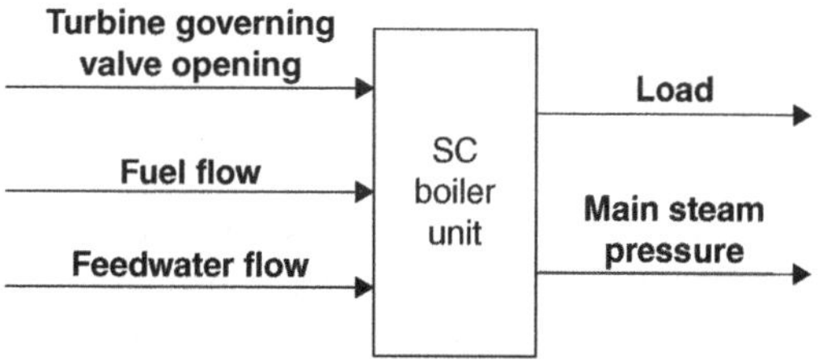

Figure 4.7.3 Model inputs and outputs for supercritical boiler unit.

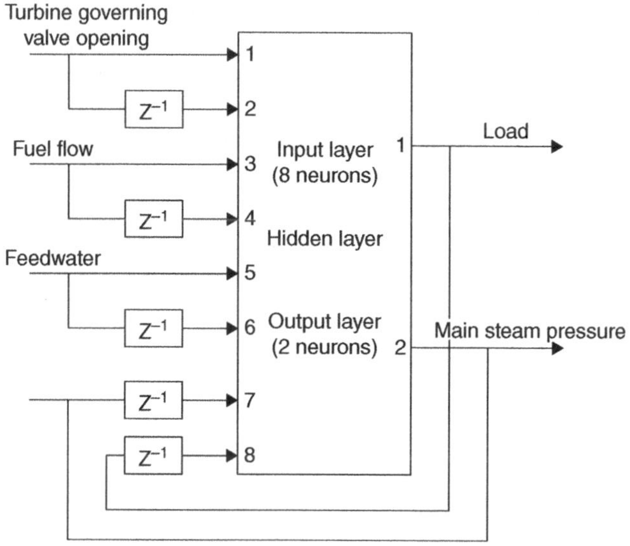

Figure 4.7.4 NARMA model for load and main steam pressure characteristics.

Then, a NARMA NN model structure, as shown in Figure 4.7.1, is utilized to develop the model of load and main steam pressure characteristics for the SC once-through power unit, as shown in Figure 4.7.4.

Model Training and Verification

The modeling object is a 600 MW SC boiler–turbine–generator unit. The boiler is of DG-1900/25.4-II type with single furnace, single reheater, manufactured by Dongfang Boiler Co. Ltd., China. In this work, the model's training data are obtained from a commercial-grade full-scope simulator of the given 600 MW SC power unit. During data acquisition, the simulator operates in the coordinated control mode, and the feedwater control, all levels of superheated steam temperature control, air flow control, etc., are all put into auto modes. To make the NN model reflect the dynamic and static characteristics of the controlled object comprehensively, the training data should have wide enough information to involve different conditions under which the model will be applied, containing different steady-state conditions and load-changing dynamic processes. In this work, 11 927 sets of data are collected with the sampling period of 2 seconds, including steady-state data at 600, 540, 480, 420 and 360 MW load levels, and the dynamic transient data between the above 5 load levels with the load ramping rate of 12 MW/min.

The NN model is established by using the MATLAB NN toolbox function, and is trained with modified Levenberg–Marquardt (L-M) algorithm. The training cycle is set for 1000 and the preset mean squared error (MSE) is 1e-6. With trial and error, the number of hidden layer neurons of the model are taken as 12. The model's MSE is less than 2e-7 after training.

To validate real-time predictive performance of the trained model, simulation test is executed by dropping the unit load demand from 540 to 360 MW with a ramping rate of 10 MW/min under coordinated control mode. The model's main steam pressure and unit load outputs are compared with those of actual plant in Figure 4.7.5. It can be seen that the model developed in this work has good

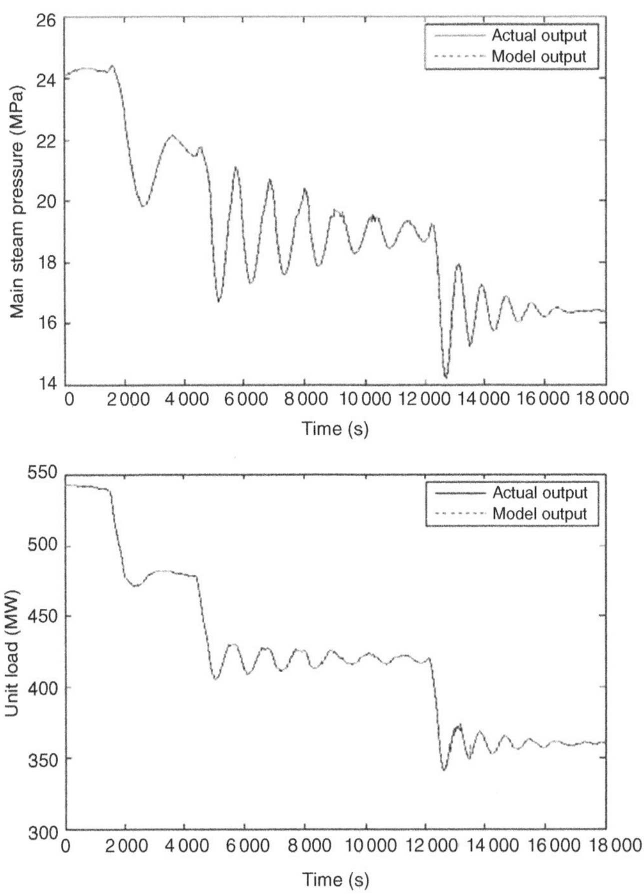

Figure 4.7.5 Model test under wide load-changing condition

prediction accuracy and strong generalization ability under wide load-changing conditions.

4.7.4 Design of Intelligent MPOC Scheme

Coordinated Control Modes Analysis

For a coal-fired SC energy conversion unit, its coordinated control system includes boiler master control (BMC), turbine master control (TMC), target load and load rate setting, target main steam pressure and pressure rate setting, and other function loops. According to whether BMC and TMC are put into automatic or not, there are several forms of coordinated control modes: manual mode, boiler-following (BF) mode, turbine-following (TF) mode, and boiler–turbine coordinated control [204–206].

If BMC and TMC are both in automatic modes, the unit works in boiler–turbine coordinated control mode. According to the principle of regulation, the coordinated mode is divided into "turbine-following based coordinated control" (TFCC) and "boiler-following based coordinated control" (BFCC). Under TFCC mode, the fuel flow is changed to adjust the unit's load and the valve opening is used to maintain main steam pressure. It means smaller steam pressure deviation, but slower load response and lower control precision because of the boiler's large delay and big inertia. Under BFCC mode, the load is controlled by changing valve opening when load demand changes, and fuel flow is responsible for maintaining the main steam pressure. It means faster load response and smaller load deviation, but relatively larger main steam pressure fluctuations due to the hysteresis of the boiler. The two different coordinated control modes are illustrated in Figure 4.7.6.

When a coal-fired power-generating unit is scheduled automatically through AGC by the regional grid load dispatch center, the power plant often puts the priority in meeting the power grid load demand to avoid additional penalty. Thus, the BFCC mode, with its fast load response, is the preferred coordinated control mode under AGC control, and it is adopted by most SC/USC power units. For the 600 MW SC power unit investigated in this work, the BFCC mode is also employed in its original control system.

Intelligent Predictive Optimal Control Scheme Design

MPOC is an advanced control algorithm which uses the dynamic model of the controlled object to predict its future outputs and calculate the best input sequences by means of optimizing the process outputs in the future period of time. Generally speaking, it can be divided into three main parts: predictive model, rolling optimization, and feedback correction [202, 203].

The role of the predictive model is to predict the future outputs based on historical information and future inputs of the object and to provide a priori knowledge for predictive optimal operation, so as to decide what control input sequence adopted to make prospective control output move toward the anticipated target. Predictive control is not carried out offline but online periodically with certain control period. In every epoch, there is an optimal PI based on which online rolling

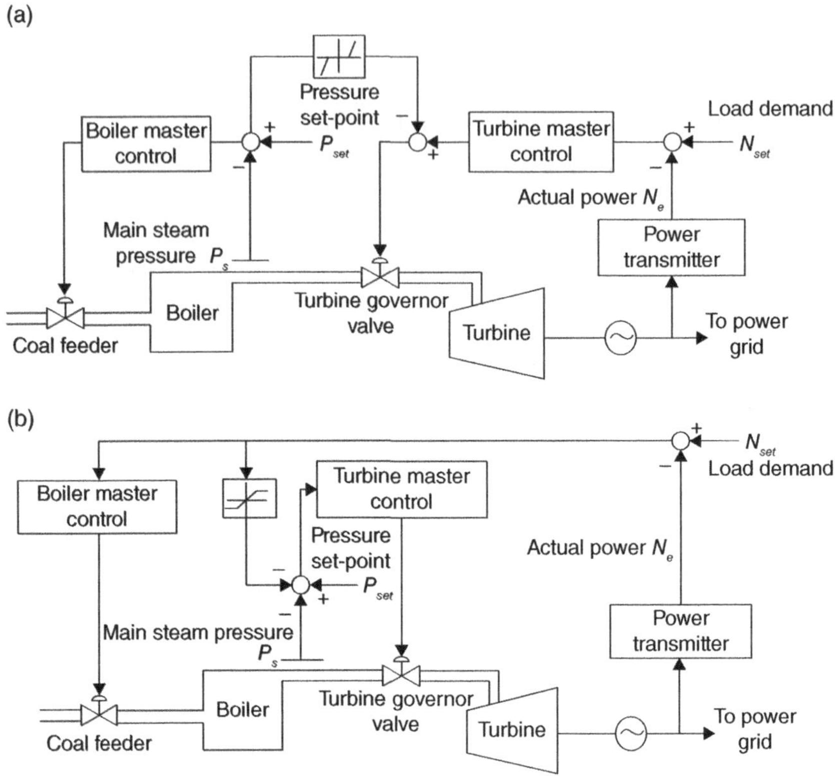

Figure 4.7.6 Schematics of two typical coordinated control modes. (a) Boiler-following-based coordinated control mode. (b) Turbine-following-based coordinated control mode.

optimization is executed to determine future control instruction. Considering the nonlinear, time-varying characteristics, model mismatch, and interference factors existing in the actual system, time-varying outputs-based predictive model could not completely yield the optimal performance. So, the actual output values of the plant are applied for feedback correction of the predictive model before a new optimization period starts.

For the coordinated control of a SC energy conversion unit, different NN MPOC schemes may be designed corresponding to different coordinated control modes. When the load of a coal-fired power-generating unit is scheduled automatically through AGC by the regional grid load dispatch center, a power plant often puts the unit operation priority in meeting the power grid load demand to avoid additional penalty. Thus, the BFCC mode, with its fast load response, is the preferred coordinated mode under AGC control. On the basis of the BFCC mode, the MPOC method based on NARMA NN predictive model with PSO algorithm is considered [189], as shown in Figure 4.7.7.

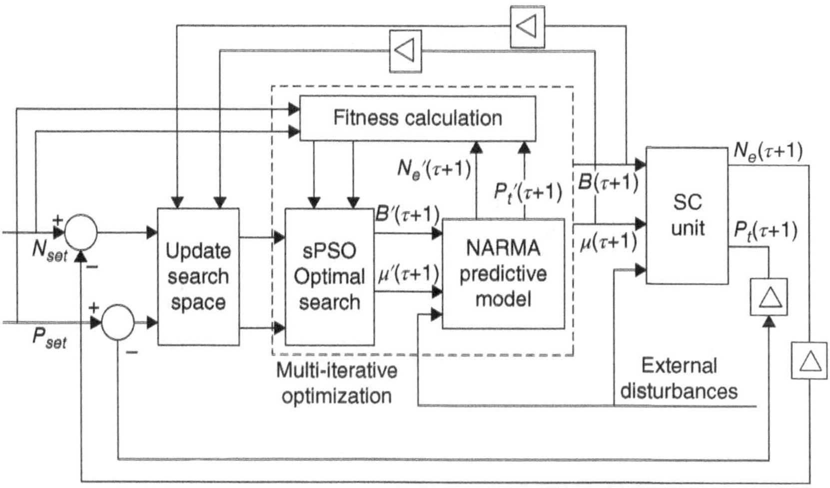

Figure 4.7.7 Intelligent coordinated predictive optimal control scheme.

For the coordinated control system with two main inputs and two outputs, when the PSO algorithm is used to optimize, the first thing is to determine the search spaces of control demand B (fuel flow) and μ (value opening). If the search spaces are too wide, it is easy to cause oscillation. On the contrary, if the search spaces are too narrow, it may miss the optimal control demands.

Therefore, a constrained dynamic search space updating method is put forward [189], with which the PSO search spaces are adjusted dynamically according to the real-time control errors between the system outputs N_e and P_t and their set points N_{set} and P_{set}. If the output error is small, the search space will narrow down; if the error is big, the search space will be enlarged. Based on this method, the elastic search spaces $\Omega_B = [B_{\min}, B_{\max}]$ and $\Omega_\mu = [\mu_{\min}, \mu_{\max}]$ are updated corresponding to the BFCC mode, as given below.

$$\text{If } N_e(\tau) < N_{\text{set}}: \quad \Omega_\mu = [\mu(\tau), \mu(\tau) + k_\mu \| N_{\text{set}} - N_e(\tau) \|] \tag{4.7.5}$$

$$\text{If } N_e(\tau) > N_{\text{set}}: \quad \Omega_\mu = [\mu(\tau) - k_\mu \| N_{\text{set}} - N_e(\tau) \|, \mu(\tau)] \tag{4.7.6}$$

$$\text{If } p_t(\tau) < P_{\text{set}}: \quad \Omega_B = [B_f(\tau), B_f(\tau) + k_b \| P_{\text{set}} - P_t(\tau) \|] \tag{4.7.7}$$

$$\text{If } p_t(\tau) > P_{\text{set}}: \quad \Omega_B = [B_f(\tau) - k_b \| P_{\text{set}} - P_t(\tau) \|, B_f(\tau)] \tag{4.7.8}$$

On the basis of the actual fuel flow command $B(\tau)$ at time τ, a fuel flow feedforward function $f(N_{\text{dmd}})$, relative to load demand $B_f(\tau)$, is added in (4.7.7) and (4.7.8) to make the main steam pressure track its set point much faster, which is defined as

$$B_f(\tau) = B(\tau) + k_f[f(N_{dmd}) - B(\tau)] \tag{4.7.9}$$

where k_f is the feedforward strength of fuel flow to load demand, taking a value between 0 and 1. The nonlinear function $f(N_{dmd})$is generally fitted in advance according to the historical operation data, and k_b and k_μ, are the zoom factors of the search windows for fuel and turbine valve opening, which should be optimized through tests. By introducing real-time tracking errors of the outputs and the feedforward fuel flow signal to set up constrained dynamic search spaces, the oscillation of MPOC during PSO search can be effectively eliminated and the PSO's searching speed can be greatly improved.

After the PSO's search spaces are determined, the sPSO algorithm will then be applied to search the optimal control demands. Thus, a fitness function should be defined to evaluate the control performance and to guide the search direction. Different fitness functions may lead to different optimization results, and the control effect also will be different. For the coordinated control problem in this work, the fitness function is defined as

$$\text{Fit} = R_1 \left\| \left(N'_e(\tau + 1) - N_{\text{set}}\right)/N_{\text{set}} \right\| + R_2 \left\| \left(P'_t(\tau + 1) - P_{\text{set}}\right)/P_{\text{set}} \right\| \tag{4.7.10}$$

where R_1 and R_2 are the weights for the control deviations of the unit load N_e and the main steam pressure P_t, respectively; $N'_e(\tau + 1)$ and $P'_t(\tau + 1)$ are the outputs of the NARMA predictive model.

4.7.5 Control Simulation Tests

After debugging, the sPSO parameters in the intelligent coordinated predictive optimal controllers are set as $c_1 = c_2 = 1.4$, $w_{\min} = 0.4$, $w_{\max} = 0.9$, $k_{\max} = 5$ and the population size takes 10. The weights in fitness function are taken as: $R_1 = 1$ and $R_2 = 0$, since load deviation is the most concerned PI under the AGC mode. After repeated simulation tests, the optimal zoom factors of the constrained PSO search spaces for the turbine governor valve opening and the fuel flow, and the feedforward strength factor for the fuel demand are determined as $k_\mu = 0.4$, $k_b = 21.5$, $k_f = 1$.

In the BFCC mode, the control tests are made for loading-down process from 600 to 540 MW, and from 540 to 480 MW with a load-changing rate of 12 MW/min. The control simulation results with the new intelligent MPOC scheme are compared with those of the original PID controllers in Figures 4.7.8 and 4.7.9.

It can be seen from Figures 4.7.8 and 4.7.9 that, with the intelligent MPOC scheme, the maximum load deviations during loading-down processes from 600 to 540 MW and from 540 to 480 MW are much smaller than those with the original control. Thus, the power unit can achieve faster load response and better meet the AGC requirement. This improvement comes from quicker and earlier coal flow and turbine valve opening actions with the new intelligent MPOC scheme.

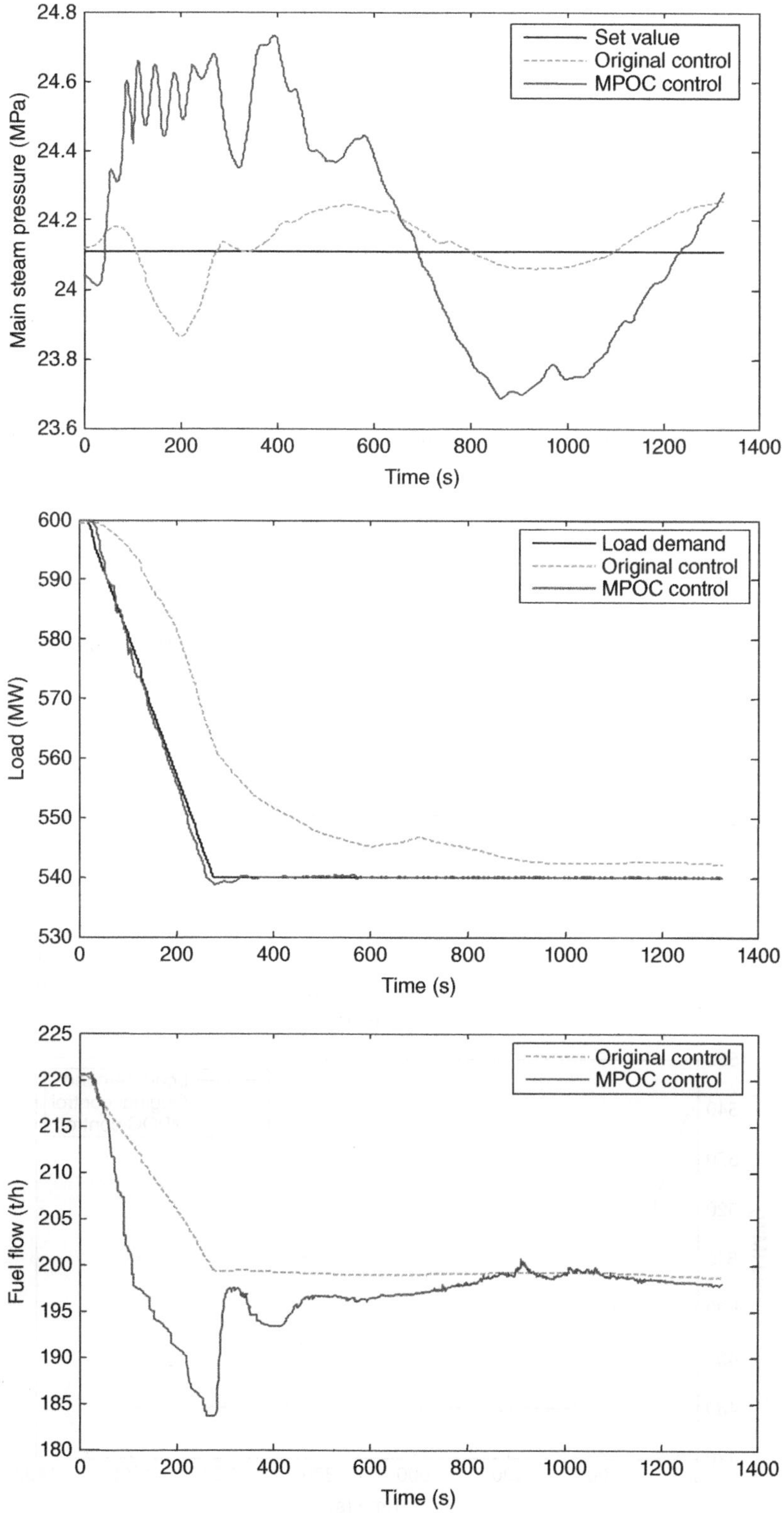

Figure 4.7.8 Test results when load drops from 600 to 540 MW.

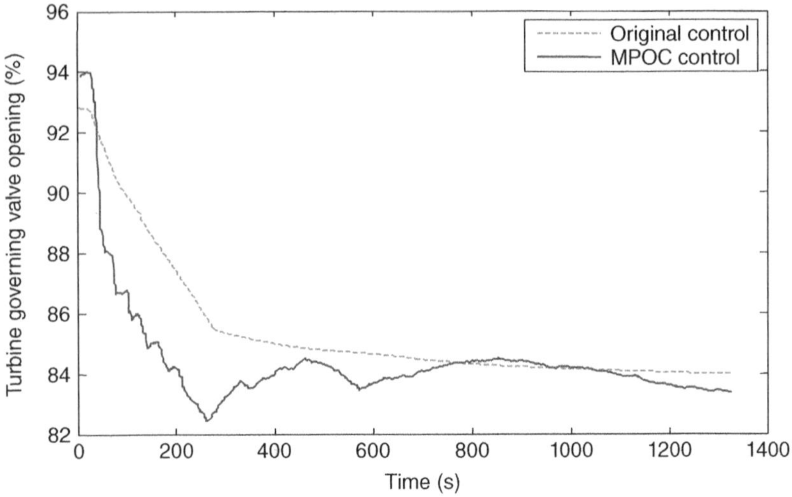

Figure 4.7.8 (Continued)

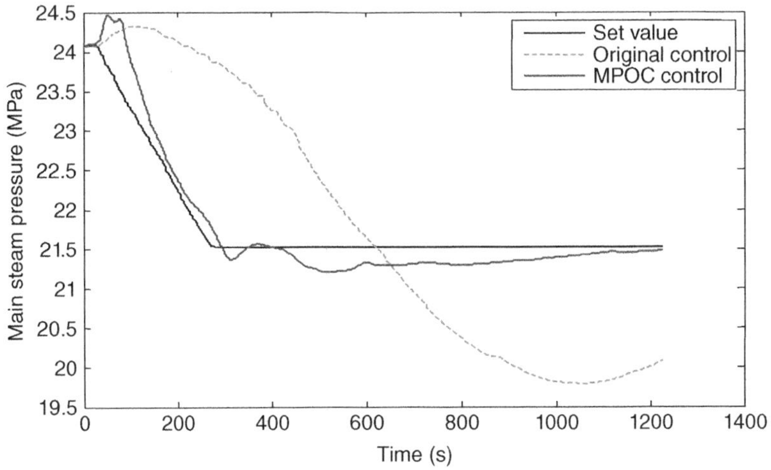

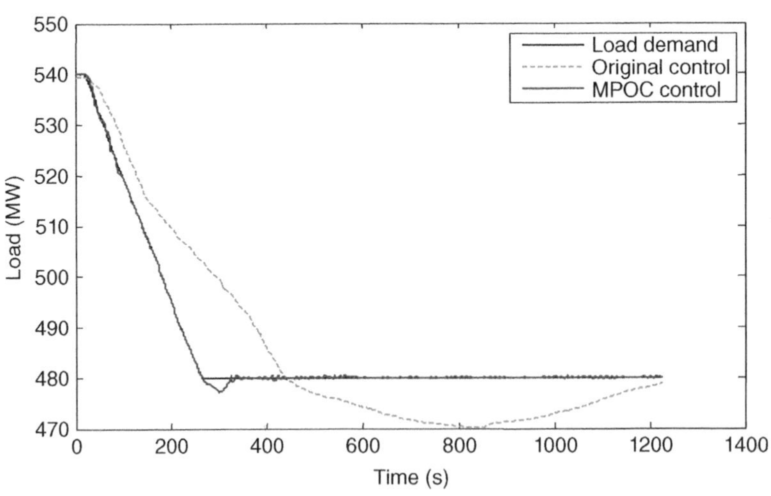

Figure 4.7.9 Test results when load drops from 540 to 480 MW.

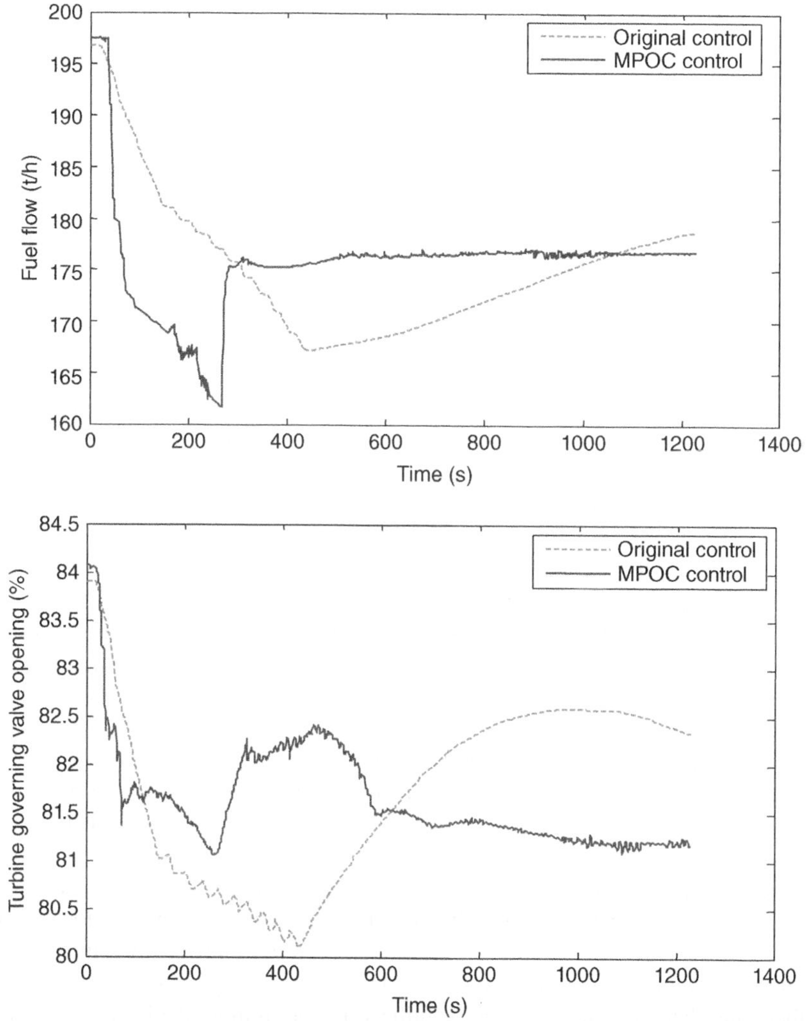

Figure 4.7.9 (Continued)

4.7.6 Conclusions

In this section, an intelligent MPOC approach based on NN modeling and constrained PSO algorithm is proposed to improve the load response speed of a large-scale SC power unit. The mathematical model for the load and the main steam pressure characteristics of a 600 MW SC power unit is built with a NARMA NN model and used as a dynamic predictive model. A new simpler PSO with fast searching speed and high convergence efficiency is adopted in the control scheme, which can meet the requirement of real-time control application. A constrained elastic search space is updated dynamically based on the load and steam pressure error feedbacks and coal demand feedforward signal to avoid oscillation during PSO search and to improve the load control precision and response speed.

Extensive control simulation tests with a commercial-grade full-scope simulator of the 600 MW SC power unit show that the MPOC scheme has better control effect than the original control by giving faster and optimized controls. The proposed MPOC method can achieve fast load response and at the same time keeping other key parameters, such as main steam pressure, within safety limits.

4.8 CONCLUSION

This chapter brought various aspects of power system and power plant operation in to light and has shown the application of meta-heuristic algorithms for reliable and efficient operation of a power system.

A PPSO technique for reactive power and voltage control has been presented. The studies show the fast computation by parallel computation along with dependability of PPSO for VVC. The application of PSO for gain tuning of secondary controllers for LFC applications has been demonstrated in this chapter. Various cost functions like ISE, ITSE, etc., have been presented and the effect of cost functions, optimized gains, and system parameters is highlighted particularly from stability perspective when communication delay plays a part in the LFC schemes.

As one of the objectives of power system operation is to minimize active and reactive power losses, it has been shown that FACTS devices can be helpful to meet this objective if these devices are properly included in the OPF problem. However, as these devices are quite costly, their proper location may improve the performance and the economic operation of the system. It is shown in this chapter that analytical techniques can be best utilized in determining the location of FACTS devices for voltage improvement and in determining the most suitable supplementary input signals for FACTS devices for damping improvement. On the other hand, heuristic methods can be utilized in determining the location and size of FACTS devices. The advantages of heuristic methods is well demonstrated.

DSA of the power system is presented using PSO. The corrective actions needed when the system security is under threat are optimized using intelligent techniques. It is shown that the PSO-RBF method presented in this chapter gives much more accuracy as compared to others.

Similar to the optimization of power system parameters, a power plant operation can also be optimized and the same is demonstrated in this chapter using GA, PSO, and ICSMOA. It is shown that such methods can help in optimizing various complex nonlinear problems associated with power plant operation, e.g. boiler combustion optimization, control of reheater temperature system. Finally, the chapter concludes by demonstrating the application of PSO for FEMS and it is shown that energy savings can be improved using these methods.

REFERENCES

1. Concordia, C. and Kirchmayer, L.K. (1953). Tie-line power & frequency control of electric power system. *AIEE Transactions, III-A* 72: 562–572.
2. Concordia, C. and Kirchmayer, L.K. (1954). Tie-line power & frequency control of electric power system: part II. *AIEE Transactions, III-A* 73: 133–146.
3. Elgerd, O.I. and Fosha, C.E. (1970). Optimum megawatt-frequency control of multiarea electric energy systems. *IEEE Transactions on Power Apparatus and Systems* PAS-89 (4): 556–563.
4. Fosha, C.E. and Elgerd, O.I. (1970). The megawatt-frequency control problem-A new approach via optimal control theory. *IEEE Transactions on Power Apparatus and Systems* PAS- 89 (4): 563–577.
5. Jaleel, N., VanSlyck, L.S., Ewart, D.N. et al. (1992). Understanding automatic generation control. *IEEE Transactions on Power Systems* 7 (3): 1106–1112.
6. Abdel-Magid, Y.L. and Abido, M.A. (2003). AGC tuning of interconnected reheat thermal systems with particle swarm optimization. *Proceedings of the 2003 10th IEEE International Conference on Electronics, Circuits and Systems, 2003*. ICECS 2003, Sharjah, United Arab Emirates (14–17 December 2003), vol. 1, pp. 376–379.
7. Nanda, J., Mishra, S., and Saikia, L.C. (2009). Maiden application of bacterial foraging-based optimization technique in multiarea automatic generation control. *IEEE Transactions on Power Systems* 24 (2): 602–609.
8. Beaufays, F., Abdel-Magid, Y., and Widrow, B. (1994). Application of neural networks to load-frequency control in power systems. *Neural Networks* 7 (1): 183–194.
9. Chaturvedi, D.K., Satsangi, P.S., and Kalra, P.K. (1999). Load frequency control: a generalised neural network approach. *International Journal of Electrical Power & Energy Systems* 21 (6): 405–415.
10. Chown, G.A. and Hartman, R.C. (1998). Design and experience with a fuzzy logic controller for automatic generation control (AGC). *IEEE Transactions on Power Systems* 13 (3): 965–970.
11. Panda, G., Panda, S., and Ardil, C. (2012). Automatic generation control of interconnected power system with generation rate constraints by hybrid neuro fuzzy approach. *International Journal of Electrical, Computer, Electronics and Communication Engineering* 6 (4): 27–32.
12. Yousef, H. (2015). Adaptive fuzzy logic load frequency control of multi-area power system. *International Journal of Electrical Power & Energy Systems* 68: 384–395.
13. Abdel-Magid, Y.L. and Abido, M.A. (1995). Tuning of agc of interconnected reheat thermal systems with Genetic algorithms. *IEEE International Conference on Systems, Man and Cybernetics, 1995. Intelligent Systems for the 21st Century* (22–25 October 1995), vol. 3, pp. 2622–2627.
14. Ghoshal, S.P. and Goswami, S.K. (2003). Application of GA based optimal integral gains in fuzzy based active power-frequency control of non-reheat and reheat thermal generating systems. *Electric Power Systems Research* 67: 79–88.
15. Ghoshal, S.P. (2004). Application of GA/GA-SA based fuzzy automatic generation control of a multi-area thermal generating system. *Electric Power Systems Research* 70: 115–127.
16. Panda, G., Panda, S., and Ardil, C. (2013). Automatic generation control of multi-area electric energy systems using modified GA. *International Journal of Electrical, Computer, Electronics and Communication Engineering* 7 (5): 308–316.

17. Rout, U.K., Sahu, R.K., and Panda, S. (2013). Design and analysis of differential evolution algorithm based automatic generation control for interconnected power system. *Ain Shams Engineering Journal* 4 (3): 409–421.
18. Dash, P., Saikia, L.C., and Sinha, N. (2014). AGC of a multi-area system using firefly optimized two degree of freedom PID controller. *Power and Energy Systems Conference: Towards Sustainable Energy* (13–15 March 2014), pp. 1–6.
19. Naidu, K., Mokhlis, H., Bakar, A.H.A. et al. (2014). Application of firefly algorithm with online wavelet filter in automatic generation control of an interconnected reheat thermal power system. *International Journal of Electrical Power & Energy Systems* 63: 401–413.
20. Sahu, R.K., Panda, S., and Padhan, S. (2015). A hybrid firefly algorithm and pattern search technique for automatic generation control of multi area power systems. *International Journal of Electrical Power & Energy Systems* 64: 9–23.
21. Debbarma, S., Saikia, L.C., and Sinha, N. (2014). Robust two-degree-of-freedom controller for automatic generation control of multi-area system. *International Journal of Electrical Power & Energy Systems* 63: 878–886.
22. Sathya, M.R. and Mohamed Thameem Ansari, M. (2015). Load frequency control using Bat inspired algorithm based dual mode gain scheduling of PI controllers for interconnected power system. *International Journal of Electrical Power & Energy Systems* 64: 365–375.
23. Dash, P., Saikia, L.C., and Sinha, N. (2015). Automatic generation control of multi area thermal system using Bat algorithm optimized PD–PID cascade controller. *International Journal of Electrical Power & Energy Systems* 68: 364–372.
24. Khooban, M.H. and Niknam, T. (2015). A new intelligent online fuzzy tuning approach for multi-area load frequency control: Self Adaptive Modified Bat Algorithm. *International Journal of Electrical Power & Energy Systems* 71: 254–261.
25. Nanda, J., Sreedhar, M., and Dasgupta, A. (2015). A new technique in hydro thermal interconnected automatic generation control system by using minority charge carrier inspired algorithm. *International Journal of Electrical Power & Energy Systems* 68: 259–268.
26. Barisal, A.K. (2015). Comparative performance analysis of teaching learning based optimization for automatic load frequency control of multi-source power systems. *International Journal of Electrical Power & Energy Systems* 66: 67–77.
27. Nanda, J. and Kaul, B.L. (1978). Automatic generation control of an interconnected power system. *Proceedings IEE* 125 (5): 385–390.
28. Kothari, M.L., Satsangi, P.S., and Nanda, J. (1981). Sampled data automatic generation control of interconnected reheat thermal system considering generation rate constraint. *IEEE Transactions on Power Apparatus and Systems* PAS- 100: 2334–2342.
29. Nanda, J., Kothari, M.L., and Satsangi, P.S. (1983). Automatic generation control of an interconnected hydrothermal system in continuous and discrete modes considering generation rate constraints. *Proceedings of the Institution of Electrical Engineering* 130, pt. D, no. 1: 17–27.
30. Hari, L., Kothari, M.L., and Nanda, J. (1991). Optimum selection of speed regularization parameter for automatic generation control in discrete mode considering generation rates constraint. *Proceedings of the Institution of Electrical Engineering* 138 (5): 401–406.
31. Nanda, J., Mangla, A., and Suri, S. (2006). Some new findings on automatic generation control of an interconnected hydrothermal system with conventional controllers. *IEEE Transactions on Energy Conversation.* 21 (1): 187–194.

32. Zare, K., Tarafdar Hagh, M., and Morsali, J. (2015). Effective oscillation damping of an interconnected multi-source power system with automatic generation control and TCSC. *International Journal of Electrical Power & Energy Systems* 65: 220–230.
33. Yu, B. and Li, W.-D. (2013). A simulation scheme for AGC relevant studies. *IEEE Transactions on Power Systems* 28 (4): 3621–3628.
34. Chavez, H., Baldick, R., and Matevosyan, J. (2014). CPS1 compliance-constrained AGC gain determination for a single-balancing authority. *IEEE Transactions on Power Systems* 29 (3): 1481–1488.
35. Chang-Chien, L.-R., Sun, C.-C., and Yeh, Y.-J. (2014). Modeling of wind farm participation in AGC. *IEEE Transactions on Power Systems* 29 (3): 1204–1211.
36. Leum, M. (1966). The development and field experience of a transistor electric governor for hydro turbines. *IEEE Transactions on Power Apparatus and Systems* PAS-85: 393–402.
37. Tripathy, S.C., Balasubramanian, R., and Nair, P.S.C. (1992). Effect of superconducting magnetic energy storage on automatic generation control considering governor deadband and boiler dynamics. *IEEE Transactions on Power Systems* 7 (3): 1266–1273.
38. Concordia, C., deMello, F.P., Kirchmayer, L.K., and Schulz, R.P. (1966). Prime-mover response and system dynamic performance. *IEEE Spectrum* 3 (10): 106–111.
39. deMello, F.P. (1991). Boiler models for system dynamic performance studies. *IEEE Transactions on Power Systems* 6 (1): 66–74.
40. Kundur, P. (2013). *Power System Stability and Control*. McGraw-Hill.
41. Elgerd, O.I. (1983). *Electric Energy Systems Theory An Introduction*. Tata McGraw-Hill.
42. Rowen, W.I. (1983). Simplified mathematical representations of heavy-duty gas turbines. *Journal of Engineering and Power* 105: 865–869.
43. Guda, S.R. (2005). Modeling and power management of a hybrid wind microturbine power generation system. M.S. Thesis, Montana State University (July 2005).
44. Nanda, J., Sharma, D., and Mishra, S. (2015). Performance analysis of automatic generation control of interconnected power systems with delayed mode operation of area control error. *IET The Journal of Engineering* 2015 (4): 164–173.
45. Kennedy, J. and Eberhart, R. (1995). Particle swarm optimization. *Proceedings of ICNN'95 – International Conference on Neural Networks*, Perth, WA, Australia (27 November–1 December 1995), vol. 4, pp. 1942–1948.
46. Wu, H., Tsakalis, K., and Heydt, G.T. (2004). Evaluation of time delay effects to wide-area power system stabilizer design. *IEEE Transactions on Power Apparatus and Systems* 19 (4): 1935–1941.
47. Yu, X. and Tomsovic, K. (2004). Application of linear matrix inequalities for load frequency control with communication delays. *IEEE Transactions on Power Apparatus and Systems* 19 (3): 1508–1515.
48. Liu, M., Yang, L., Gan, D. et al. (2007). The stability of AGC systems with commensurate delays. *European Transactions on Electrical Power* 17 (6): 615–627.
49. Gu, K., Kharitonov, V.L., and Chen, J. (2003). *Stability of Time Delay Systems*. Boston, MA, USA: Birkhauser.
50. Chen, J., Gu, G., and Nett, C.N. (1995). A new method for computing delay margins for stability of linear delay systems. *Systems and Control Letters* 26 (2): 107–117.
51. Rekasius, Z.V. (1980). A stability test for systems with delays. *Proceedings of the Joint Automatic Control Conference*, San Francisco, CA, USA (13–15 August 1980).
52. Olgac, N. and Sipahi, R. (2002). An exact method for the stability analysis of time-delayed linear time-invariant (LTI) systems. *IEEE Transactions on Automatic Control* 47 (5): 793–797.

53. Olgac, N. and Sipahi, R. (2004). A practical method for analyzing the stability of neutral type LTI-time delayed systems. *Automatica* 40 (5): 847–853.
54. He, Y., Wang, Q.G., Xie, L.H., and Lin, C. (2007). Further improvement of free-weighting matrices technique for systems with time-varying delay. *IEEE Transactions on Automatic Control* 52 (2): 293–299.
55. Wu, M., He, Y., She, J.H., and Liu, G.P. (2004). Delay-dependent criterion for robust stability of time-varying delay systems. *Automatica* 40 (8): 1435–1439.
56. Xu, S.Y. and Lam, J. (2007). On equivalence and efficiency of certain stability criteria for time-delay systems. *IEEE Transactions on Automatic Control* 52 (1): 95–101.
57. Sönmez, Ş., Ayasun, S., and Nwankpa, C.O. (2016). An exact method for computing delay margin for stability of load frequency control systems with constant communication delays. *IEEE Transactions on Power Systems* 31 (1): 370–377.
58. Hingorani, N.G. (1988). Power electronics in electric utilities: role of power electronics in future power systems. *Proceedings of the IEEE* 76 (4): 481–482.
59. Hingorani, N.G. and Gyugyi, L. (1998). *Understanding FACTS*. The Institution of Electrical and Electronics Engineers.
60. Song, Y.H. and Johns, A.T. (1998). *Flexible AC Transmission System*. IEE press.
61. Hingorani, N.G. (1993). Flexible AC transmission. *IEEE Spectrum* 30: 40–45.
62. Edris, A. (2000). FACTS technology development: an update. *IEEE Power Engineering Review* 20 (3): 4–11.
63. Hammons, T.J. and Lim, S.K. (1997). Flexible AC transmission systems (FACTS). *Electrical Machines and Power systems* 25: 73–85.
64. Gyugyi, L. (1992). A unified power flow control concept for flexible AC transmission systems. *IEE Proceedings, Part-C* 139 (4): 323–333.
65. Keri, A.J.F., Lombard, X., Edris, A.A. et al. (1999). UPFC: modeling and analysis. *IEEE Transactions on Power Delivery* 14 (2): 648–654.
66. Nabavi-Naiki, A. and Iravani, M.R. (1996). Steady state and dynamic models of unified power flow controller (UPFC) for power system studies. *IEEE Transactions on Power Systems* 11 (4): 1937–1945.
67. Rietman, T.R., Edris, A., Schauder, C.D. et al. (1995). The unified power flow controller: a new approach to power transmission control. *IEEE Transactions on Power Delivery* 10 (2): 1085–1093.
68. Rahman, M., Ahmad, M., Gutman, R. et al. (1997). UPFC application on the AEP System. *IEEE Transactions on Power Systems* 12 (4).
69. Kalyan, K.S. (1998). SSSC – static synchronous series compensator: theory, modeling and applications. *IEEE Transactions on Power Delivery* 13 (1): 241–246.
70. Pilotto, L.A.S., Ping, W.W., Carvalho, A.R. et al. (1997). Determination of needed FACTS Controller that increases asset utilization of power systems. *IEEE Transactions on Power Delivery* 12 (1): 364–371.
71. Lie, L.L. (2001). *Power System Restructuring and Deregulation*. UK: John Wiley and Sons.
72. Li, N., Xu, Y., and Chen, H. (2000). FACTS-based power flow control in interconnected power systems. *IEEE Transactions on Power Systems* 15 (1): 58–64.
73. Furete-Esquivel, C.R., Acha, E., and Ambriz-Perez, H. (2000). A thyristor controlled series compensator model for Power Flow solution of practical power network. *IEEE Transactions on Power Systems* 15 (1).
74. Liu, J.Y., Song, Y.H., and Mehta, P.A. (2000). Strategies for handling UPFC constraints in the steady state power flow and voltage control. *IEEE Transactions on Power Systems* 15 (2): 566–571.

75. Lambord, X. and Thernond, P.G. (1997). Control of unified power flow controller: comparison of methods on the basis of a detailed numerical model. *IEEE Transactions on Power Systems* 12 (2): 824–830.
76. Singh, S.N., Verma, K.S. and Gupta, H.O. (2001). Optimal power flow control using UPFC in open power market. *IEEE Summer Meeting*, Ohio Canada (July 2001).
77. Chung, T.S. and Li, Y.Z. (2001). A hybrid GA approach for OPF with consideration of FACTS devices. *IEEE Power Engineering Review* 21 (2): 47–50.
78. Shaoyun, G. (1999). Optimal power system operation and control incorporating FACTS devices. Thesis submitted for award of PhD Degree in Engineering. Hong Kong (August 1999).
79. Ooi, B.T., Kazerani, M., Marcowi, R., and Wolanskie, Z. (1997). Midpoint sitting of FACTS devices in transmission lines. *IEEE Transactions on Power Delivery* 12 (4): 1717–1722.
80. Rajaraman, R., Alvarado, F., Maniaci, A. et al. (1998). Determination of location and amount of series compensation to increase power transfer capability. *IEEE Tranactions on Power Systems* 13 (2): 294–299.
81. Paterni, P., Vitet, S., Bena, M., and Yokoyama, A. (1999). Optimal location of phase shifters in the French network by genetic algorithm. *IEEE Transactions on Power Systems* 14 (1): 37–42.
82. de Oliveira, E.J. and Lima, J.W.M. (1999). Allocation of FACTS devices in a competitive environment. *13th PSCC*, pp. 1184–1190.
83. Lie, T.T. and Deng, W. (1999). Optimal flexible AC transmission systems (FACTS) devices allocation. *International Journal of Electrical Power & Energy Systems* 19 (2): 125–134.
84. Fang, W.L. and Ngan, H.W. (1999). Optimising location of unified power flow controllers using the method of augmented lagrange multipliers. *Proceedings of the IEE Part-C* 146 (5): 428–434.
85. Verma, K.S., Singh, S.N., and Gupta, H.O. (2001). Optimal location of UPFC for congestion management. *Electric Power Systems Research* 58 (2): 89–96.
86. Mutale, J. and Strabac, G. (2000). Transmission network reinforcement Vs FACTS: an economic assessment. *IEEE Transactions on Power Systems* 15 (3): 961–967.
87. Singh, S.N. and David, A.K. (2001). Optimal location of FACTS devices for congestion management. *Electric Power Systems Research* 58 (2): 71–79.
88. Verma, K.S., Singh, S.N., Gupta, H.O. (2001). FACTS device location for enhancement of total transfer capability. *IEEE Power Winter Meeting*, Ohio, USA (28 January 2001).
89. Mihalic, R., Zunko, P., and Povh, D. (1996). Improvement of transient stability using unified power controller. *IEEE Power Delivery* 11 (1): 485–492.
90. Wang, H.F. (2000). A unified model for the analysis of FACTS devices in damping power system oscillations-part III: unified power flow controller. *IEEE Transactions on Power Delivery* 15 (3): 978–983.
91. Radu, D. and Besanger, Y. (2005). Blackout prevention by optimal insertion of FACTS devices in power systems. *Proceedings of the 2005 International Conference on Future Power Systems*, Amsterdam, Netherlands (November 2005), pp. 1–6.
92. da Silva, F.H.J.R., Guedes, R.B.L., Alberto, L.F.C., and Bretas, N.G. (2006). Parameter dependent control Lyapunov function for series and shunt FACTS devices considering uncertain power system model. *Proceedings of the IEEE Power Engineering Society General Meeting*, 2006, Montreal, Quebec, Canada (June 2006), pp. X004021–X004026.

93. Yong, C., Zheng, X., Gaihong, C., and Jiani, X. (2006). A novel SVC supplementary controller based on wide area signals. *Proceedings of the 2006 IEEE Power Engineering Society General Meeting,* 2006, Montreal, Quebec, Canada (June 2006).
94. Mhaskar, U.P. and Kulkarni, A.M. (2006). Power oscillation damping using FACTS devices: modal controllability, observability in local signals, and location of transfer function zeros. *IEEE Transactions on Power Systems* 21 (1): 285–294.
95. Sadikovic, R., Korba, P., and Andersson, G. (2006). Self-tuning controller for damping of power system oscillations with FACTS devices. *Proceedings of the 2006 IEEE Power Engineering Society General Meeting*, Montreal, Quebec, Canada (June 2006).
96. Haque, M.H. (2005). Use of series and shunt FACTS devices to improve first swing stability limit. *Proceedings of the 7th International Power Engineering Conference*, Singapore, Singapore (29 November–2 December 2005). IPEC 200, Vol. 2.
97. Haque, M.H. (2005). Stability improvement by FACTS devices: a comparison between STATCOM and SSSC. *Proceedings of the 2005 IEEE Power Engineering Society General Meeting*, San Francisco, CA, USA (June 2005), pp. 1708–1713.
98. Chatterjee, D. and Ghosh, A. (2005). Evaluation of transient stability margin of a power system containing multiple FACTS devices. *Proceedings of the 2005 IEEE Power Engineering Society General Meeting*, San Francisco, CA, USA (June 2005), pp. 1787–1794.
99. Wang, H.F. and Xu, H.Z. (2004). FACTS-based stabilizers to damp power system oscillations – a survey. *Proceedings of the 39th International Universities Power Engineering Conference, 2004.* UPEC 2004, Bristol, UK (September 2004), vol. 1, pp. 318–322.
100. Farsangi, M.M., Nezamabadi-Pour, H., Song, Y.H., and Lee, K.Y. (2007). Placement of SVCs and selection of stabilizing signals in power systems. *IEEE Transactions on Power Systems* 22 (3): 1061–1071.
101. Lee, K.Y. and El-Sharkawi, M.A. (eds.) (2002). Tutorial on modern heuristic optimization techniques with applications to power systems. *IEEE Power Engineering Society*, IEEE Catalog Number 02TP160, Piscataway, NJ, 2002.
102. S. Mollazei, M.M. Farsangi, H. Nezamabadi-Pour and K.Y. Lee (2007). Multiobjective optimization of power system performance with TCSC using the MOPSO algorithm. *IEEE Power Engineering Society General Meeting* (24–28 June 2007).
103. Lee, K.Y. (ed.) (2005). Tutorial on intelligent optimization and control of power systems. *Proceedings of the 13th International Conference on Intelligent Systems Application to Power Systems (ISAP)*, Arlington, VA, 2005.
104. Alabduljabbar, A.A. and Milanovic, J.V. (2006). Genetic algorithm based optimization for allocation of static VAr compensators. *Proceedings of the 8th IEE International Conference on AC and DC Power Transmission*, 2006 (March 2006), pp. 115–120.
105. Mori, H. and Maeda, Y. (2006). Application of two-layered tabu search to optimal allocation of UPFC for maximizing transmission capability. *Proceedings of the IEEE International Symposium on Circuits and Systems, 2006.* ISCAS 2006 (May 2006), p. 4.
106. Radu, D. and Besanger, Y. (2006). A multi-objective genetic algorithm approach to optimal allocation of multi-type FACTS devices for power systems security. *Proceeding of the IEEE Power Engineering Society General Meeting*, 2006 (June 2006).
107. Ongsakul, W. and Jirapong, P. (2005). Optimal allocation of FACTS devices to enhance total transfer capability using evolutionary programming. *Proceedings of the IEEE International Symposium on Circuits and Systems*, 2005. ISCAS 2005 (May 2005), vol. 5, pp. 4175–4178.

108. Hao, J., Shi, L.B., and Chen, C.H. (2004). Optimising location of unified power flow controllers by means of improved evolutionary programming. *IEE Proceedings Generation, Transmission and Distribution* 151 (6): 705–712.
109. Sudersan, A., Abdelrahman, M., and Radman, G. (2004). Contingency selection and static security enhancement in power systems using heuristics-based genetic algorithms. *Proceedings of the Thirty-Sixth Southeastern Symposium on System Theory*, 2004, pp. 556–560.
110. Khan, L., Rogers, G.J., and Lozano, C.A. (2004). Intelligent control of FACTS for damping power system low frequency oscillations. *Proceedings of INMIC 2004, 8th International Multitopic Conference* (December 2004), pp. 731–736.
111. Wong, K.P., Yuryevich, J., and Li, A. (2004). Evolutionary-programming-based load flow algorithm for systems containing unified power flow controllers. *IEE Proceedings Generation Transmission and Distribution* 150 (4): 441–446.
112. Bhasaputra, P. and Ongsakul, W. (2003). Optimal placement of multi-type FACTS devices by hybrid TS/SA approach. *Proceedings of the 2003 International Symposium on Circuits and Systems*, 2003, vol. 3, pp. III-375–III-378.
113. Ebrahimi, S., Farsangi, M.M., Nezamabadi-Pour, H., and Lee, K.Y. (2006). Optimal Allocation of STATIC VAR COMPENSATORS using modal analysis, simulated annealing and Tabu search. *Proceedings of the 2006 IFAC Symposium on Power Plants and Power Systems*, Calgary, Canada (July 2006).
114. Jeon, Y.-J., Kim, J.-C., Kim, J.-O. et al. (2002). An efficient simulated annealing algorithm for network reconfiguration in large-scale distribution systems. *IEEE Transactions on Power Delivery* 17 (4): 1070–1078.
115. Farsangi, M.M., Nezamabadi-Pour, H., and Lee, K.Y. (2006). Multi-objective VAr planning with SVC for a large power system using PSO and GA. *Proceedings of the 2006 IEEE PES Power Systems Conference and Exposition (PSCE)*, Atlanta, USA (29 October–1 November 2006).
116. Goldberg, D.E. (1989). *Genetic Algorithms in Search, Optimization, and Machine Learning*. New York: Addison-Wesley.
117. Van den Bergh, F. and Engelbrecht, A.P. (2002). A new locally convergent particle swarm optimizer. *Proceeedings of the IEEE Conference on Systems, Man and Cybernetics*, Hammamet, Tunisia (6–9 October 2002), vol. 3.
118. Khaleghi, M., Farsangi, M.M., Nezamabadi-Pour, H., and Lee, K.Y. (2011). Pareto-optimal design of damping controllers using modified artificial immune algorithm. *IEEE Transactions on Systems, Man, and Cybernetics, Part C: Applications and Reviews* 41 (2): 240–250.
119. Bijami, E., Askari, J., and Farsangi, M.M. Design of stabilising signals for power system damping using generalised predictive control optimised by a new hybrid shuffled frog leaping algorithm. *IET Generation Transmission and Distribution* 6 (10): 1036–1045.
120. CIGRE (2007). Technical Brochure, review of on-line dynamic security assessment: tools and techniques, SCC4. WG601 (January 2007).
121. Cutsem, T.V. and Vournas, C. (1998). *Voltage Stability of Electric Power Systems*. Boston: Kluwer Academics.
122. CIGRE T.F. 38.02.13 (1997). New trends and requirements for dynamic security assessment. Technical Report.
123. Morison, K. (2006). On-line dynamic security assessment using intelligent systems. Presented at the PES General Meeting, Tampa, Florida (June 2006).

124. El-Sharkawi, M. and Niebur, D. (1996). Artificial neural networks with applications to power systems. *IEEE-PES Special Publ. 96TP-112-0.*
125. Sobajic, D. and Pao, Y. (1989). Artificial neural-net based dynamic security assessment for electric power systems. *IEEE Transactions on Power Systems* 4: 220–228.
126. Gavoyiannis, A.E., Vogiatzis, D.G., and Hatziargyriou, N.D. (2001). Dynamic security classification using support vector machines. *Presented at the International Conference on Intelligent System Application to Power Systems*, Budapest, Hungury (June 2001).
127. Moulin, L., da Silva, A., El-Sharkawi, M., and Marks, R. (2001). Neural networks and support vector machines applied to power systems transient stability analysis. *Engineering Intelligent System Journal* 9 (4): 205–211.
128. Gavoyiannis, A.E., Voumvoulakis, E.M., and Hatziargyriou, N.D. (2005). Radial basis functions for dynamic security assessment of power systems with increased wind power penetration. *Presented at the 15th PSCC*, Liege, Belgium (August 2005).
129. Karapidakis, E. and Hatziargyriou, N. (2002). On-line preventive dynamic security of isolated power systems using decision trees. *IEEE Transactions on Power Systems* 17 (2): 297–305.
130. Voumvoulakis, E.M. and Hatziargyriou, N.D. (2010). A particle swarm optimization method for power system dynamic security control. *IEEE Transactions on Power Systems* 25 (2): 1032–1041.
131. Feng, Z., Ajjarapu, V., and Maratukulam, D. (1998). A practical minimum load shedding strategy to mitigate voltage collapse. *IEEE Transactions on Power Systems* 13 (4): 1285–1291.
132. Taylor, C. (1992). Concepts of under voltage load shedding for voltage stability. *IEEE Transactions on Power Delivery* 7 (2): 480–487.
133. Tuan, T., Fandino, J., Hadjsaid, N. et al. (1994). Emergency load shedding to avoid risk of voltage instability using indicators. *IEEE Transactions on Power Systems* 19 (1): 341–351.
134. Bijwe, P., Tare, R., and Kelapure, S. (1999). Anticipatory load shedding scheme for loadability enhancement. *IEE Proceedings on Generation, Transmission and Distribution* 146 (5): 483–490.
135. Enrico De Tuglie, M.D. and Scala, M.L. (2000). A corrective control for angle and voltage stability enhancement on the transient time-scale. *IEEE Transactions on Power Systems* 15 (4): 1345–1353.
136. Voumvoulakis, E., Gavoyiannis, A., and Hatziargyriou, N. (2006). Decision trees for dynamic security assessment and load shedding scheme. *Presented at the PES General Meeting*, Tampa, Florida (June 2006).
137. Voumvoulakis, E.M. and Hatziargyriou, N.D. (2008). Decision trees-aided self organized maps for corrective dynamic security. *IEEE Transactions on Power Systems* 23 (2): 622–630.
138. Niazi, K., Arora, C., and Surana, S. (2003). Power system security evaluation using ANN: feature selection using divergence. *Electric Power Systems Research* 69: 161–167.
139. Jensen, C., El-Sharkawi, M.A., and Marks, R.J. (2001). Power system security assessment using neural networks: feature selection using fisher discrimination. *IEEE Transactions on Power Systems* 16 (4): 757–763.
140. Committee Report, I.E.E.E. (1992). Transient stability test systems for direct stability methods. *IEEE Transactions on Power Systems* 7 (1): 37–44.

141. UW Power Systems Test Case Archive. Seattle, WA: University of Washington. http://www.ee.washington.edu/research/pstca (accessed 10 October 2019).
142. IEEE Committee Report (1981). Excitation models for power system stability studies. *IEEE Transactions on PAS* February: 494–509.
143. I. T. F. on Excitation Limiters (1995). Recommended models for overexcitation limiting devices. *IEEE Transactions on Energy Conversion* 10 (4): 706–713.
144. Gan, D., Thomas, R.J., and Zimmerman, R.D. (2000). Stability-constrained optimal power flow. *IEEE Transactions on Power Systems* 15 (2).
145. Brown, P. (1989). The dispatch of spinning reserve in multiple time scales on an island power system. *Proceedings of the University of Power Engineering Conference*, vol. 1, Belfast, Ireland, 1989, pp. 611–614.
146. Wood, A.J. and Wollenberg, B.F. (1996). *Power Generation, Operation, and Control*. New York: Wiley.
147. Bodo, D.R., Mauzy, D.M., and Trefny, F.J. (1994). Economic generation dispatch with response spinning reserve constraints. *IEEE Transactions on Power Systems* 9: 555–559.
148. Vaahedi, E., Mansour, Y., and Tse, E. (1998). A general purpose method for on-line dynamic security assessment. *IEEE Transactions on Power Systems* 13: 243–249.
149. Li, W. and Bose, A. (1998). A coherency based rescheduling method for dynamic security. *IEEE Transactions on Power Systems* 13: 810–815.
150. Bettiol, A.L., Wehenkel, L., and Pavella, M. (1999). Transient stability-constrained maximum allowable transfer. *IEEE Transactions on Power Systems* 14: 654–659.
151. El-Sharkawi, M. and Neibur, D. (1996). Artificial neural networks with applications to power systems, *IEEE Power Engineering Society, IEEE Neural Network Council, and IEEE Educational Activities, Handbook*, 1996.
152. Wehenkel, L. and Pavella, M. (1996). Special Issue on Automatic Learning Application to Power Systems of the Revue-E, SRBE, Belgium (December 1996).
153. Wehenkel, L. (1998). *Automatic Learning Techniques in Power Systems*. Norwell, MA: Kluwer.
154. Papadogiannis, K.A. and Hatziargyriou, N.D. (2004). Optimal allocation of primary reserve services in energy markets. *IEEE Transactions on Power Systems* 19 (1): 652–659.
155. EUROSTAG, Software for the Simulation of Power Systems Dynamics. Electricité de France, Tractebel, Package, User's Guide, release 4.4.
156. Devijver, P.A. and Kittler, J. (1982). *Pattern Recognition: A Statistical Approach*. London: Prentice-Hall.
157. Poggio, T. and Girosi, F. Networks for approximation and learning. *Proceedings of the IEEE* 78 (9): 1481–1497.
158. Haykin, S. (1994). *Neural Networks: A comprehensive Foundation*. Upper Saddle River, NJ: Prentice Hall.
159. Leonard, J.A., Kramer, M.A., and Unger, L.H. (1992). A neural network architecture that predicts its own reliability. *Computers in Chemical Engineering* 16: 819–835.
160. Papalexopoulos, A.D., Hubele, N.E., and Ranaweera, D.K. (1995). Application of radial basis function neural network model for short term load forecasting. *IEE Proceedings – Generation Transmission and Distribution* 142 (1): 45–49.
161. Yang, L., Kavli, T., Carlin, M. et al. (2000). An evaluation of confidence bound estimation methods for neural networks. *Proceedings of the 2000 ESIT*, Aachen, Germany, pp. 322–329.

162. Parzen, E. On estimation of a probability density function and mode. *Annals of Mathematical Statistics* 33: 1065–1076.
163. Bonabeau, E. and Hnaux, F. (1998). Self-organizing maps for drawing large graphs. *Information Processing Letters* 67: 177–184.
164. Wu, X., Shen, J., Li, Y., and Lee, K.Y. (2015). Steam power plant configuration, design and control. *WIREs Energy and Environment* 4 (6): 537–563.
165. Lee, K.Y. (2006). Intelligent techniques applied to power plant control. *2006 IEEE PES General Meeting*, Montreal, Canada (2006).
166. Flynn, D. (2003). *Thermal Power Plant Simulation and Control*. UK: IEE Press.
167. Cortinovis, A., Mercangoez, M., Mathur, T. et al. (2013). Nonlinear coal mill modeling and its application to model predictive control. *Control Engineering Practice* 21 (3): 308–320.
168. Agrawal, V., Panigrahi, B.K., and Subbarao, P.M.V. (2015). A unified thermo-mechanical model for coal mill operation. *Control Engineering Practice* 44: 157–171.
169. Zhang, Y.G., Wu, Q.H., Wang, J. et al. (2002). Coal mill modeling by machine learning based on onsite measurements. *IEEE Transactions on Energy Conversion* 17 (4): 549–555.
170. Zeng, D.L., Hu, Y., Gao, S., and Liu, J.Z. (2015). Modelling and control of pulverizing system considering coal moisture. *Energy* 80: 55–63.
171. Shin, H.B., Li, X.L., Jeong, I.Y. et al. (2009). Modeling and parameter identification of coal mill. *Journal of Power Electronics* 9 (5): 700–707.
172. Liang, X.F. (2016). Research on coal mill condition monitoring and fault detection based on first principle approach and on-site data. M.E. dissertation, School of Energy and Environment, Southeast University, Nanjing, Jiangsu, China (in Chinese).
173. Golberg, D.E. (1989). *Genetic Algorithms in Search, Optimization, and Machine Learning*. USA: Addison Wesley Publishing Company.
174. Whitley, D. (1994). A genetic algorithm tutorial. *Statistics and Computing* 4 (2): 65–85.
175. Lee, K.Y., Ma, L., Boo, C. et al. (2009). Intelligent modified predictive optimal control of reheater steam temperature in a large-scale boiler unit. *2009 IEEE PES General Meeting*, Calgary, Canada.
176. Lee, K.Y., Heo, J.S., Hoffman, J.A. et al. (2007). Neural network-based modeling for a large-scale power plant. *2007 IEEE PES General Meeting*, Tampa, USA.
177. Reynolds, C. (1987). Flocks, herds, & schools: a distributed behavioral model. *Computers and Graphics* 21 (4): 25–34.
178. Hu, W. and Li, Z. (2007). A simpler and more effective particle swarm optimization algorithm. *Journal of Software* 18 (4): 861–868.
179. Lee, K.Y., Heo, J.S., Hoffman, J.A. et al. (2007). Modified predictive optimal control using neural network-based combined model for large-scale power plants. *2007 IEEE PES General Meeting* (June 2007), pp. 1–8.
180. Heo, J.S. and Lee, K.Y. (2006). Multi-objective control of power plants using particle swarm optimization techniques. *IEEE Transsctions of Energy Conversion* 21 (2): 552–561.
181. Zhou, X. (2011). Immune multiobjective optimization algorithm and its application in boiler combustion optimization. PhD dissertation, School of Energy and Environment, Southeast University, Nanjing, Jiangsu, China (in Chinese).
182. Zeldovich, J.B. (1946). The oxidation of nitrogen in combustion and explosions. *ACTA Physicochemical URSS* 21 (4): 557–628.

183. Zhou, X. (2010). New immune multiobjective optimization algorithm and its application in boiler combustion optimization. *Journal of Southeast University (English Edition)* 26 (4): 563–568.
184. Gong, F. (2003). *Medical Immunology*. Beijing, China: Science Press (in Chinese).
185. Deb, K., Pratap, A., Agarwal, S., and Meyarivan, T. (2002). A fast and elitist multiobjective genetic algorithm: NSGA-II. *IEEE Transactions on Evolutionary Computation* 6: 182–197.
186. Li, T. (2004). *Computer Immunolgy*. Beijing: Publishing House of Electronics Industry (in Chinese).
187. Ma, L.Y., Cao, P.R., Gao, Z.Y., and Lee, K.Y. (2016). ANN and PSO based intelligent model predictive optimal control for large-scale supercritical power unit. *2016 IEEE International Conference on Information and Automation (ICIA)*, Ningbo (2016), pp. 690–695.
188. Piche, S., Sayyar-Rodsari, B., Johnson, D., and Gerules, M. (2000). Nonlinear model predictive control using neural networks. *IEEE Control Systems Magazine* 20 (3): 53–62.
189. Ma, L.Y., Lee, K.Y., and Ge, Y.P. (2012). An improved predictive optimal controller with elastic search space for steam temperature control of large-scale supercritical power unit. *2012 IEEE 51st Annual Conference on Decision and Control (CDC2012)*, Maui, USA, pp. 7024–7029.
190. Liu, X.J., Guan, P., and Chan, C.W. (2010). Nonlinear multivariable power plant coordinate control by constrained predictive scheme. *IEEE Transactions on Control Systems Technology* 18 (5): 1116–1125.
191. Irwin, G., Brown, M., Hogg, B., and Swidenbank, E. (1995). Neural network modeling of a 200MW boiler system. *IEE proceedings: Control Theory and Applications* 142 (6): 529–536.
192. Shen, B., Lang, W., and Hu, Z.G. (2009). Neural network modeling of the production process. *Control Theory & Applications* 28 (2): 1–3. (in Chinese).
193. Liu, X.J., Kong, X.B., Hou, G.L., and Wang, J.H. (2013). Modeling of a 1000 MW power plant ultra super-critical boiler system using fuzzy-neural network methods. *Energy Conversion and Management* 65: 518–527.
194. Niu, P.F., Guo, X.H., and Meng, F.D. (2010). Application of fuzzy neural network control to superheated steam systems. *Control Engineering of China* 17 (3): 293–296. (in Chinese).
195. Zhou, H., Zhu, H.B., and Zeng, T.H. (2002). Artificial neural network modeling on the unburned carbon in fly ash from utility boilers. *Proceedings of the CSEE* 22 (6): 96–100. (in Chinese).
196. Shi, Y. and Eberhart, R. (1998). A modified particle swarm optimizer. *IEEE International Conference on Evolutionary Computation Proceedings. IEEE World Congress on Computational Intelligence*, New York, USA, pp. 69–73.
197. Park, J.-B., Lee, K.-S., Shin, J.-R., and Lee, K.Y. (2005). A particle swarm optimization for economic dispatch with nonsmooth cost functions. *IEEE Transactions on Power Systems* 20 (1): 34–42.
198. Vlachogiannis, J.G. and Lee, K.Y. (2006). A comparative study on particle swarm optimization for optimal steady-state performance of power systems. *IEEE Transactions on Power Systems* 21 (4): 1718–1728.
199. Selvakumar, A.I. and Thanushkodi, K. (2007). A new particle swarm optimization solution to nonconvex economic dispatch problems. *IEEE Transactions on Power Systems* 22 (1): 42–51.

200. Wei, J.X. and Wang, Y.P. (2008). Smooth scheme and line search based particle swarm optimization for constrained optimization problems. *Systems Engineering and Electronics* 4 (30): 739–742. (in Chinese).
201. Fang, H.Q. and Shen, Z.Y. (2005). Optimal hydraulic turbogenerators PID governor tuning with an improved particle optimization algorithm. *Proceedings of the CSEE* 25 (20): 120–124. (in Chinese).
202. Lu, Z.G., Wu, S.C., and Yu, L.H. (2004). *Nonlinear Adaptive Inverse Control and Its Application*. Beijing: National Defence Industry Press.
203. Wei, D. (2007). *Neural Network Parameter Prediction and Control of Nonlinear System*. Beijing: China Machine Press.
204. Zhang, Z.Y., Li, W.H., and Song, Z.X. (2007). Control strategy of coordination control system of 600MW unit with supercritical monotube boiler. *North China Electric Power* 1: 24–28. (in Chinese).
205. Yang, L., Niu, Y.G., and Gao, Z.C. (2008). Analysis and optimization of coordinated control system of 2×600MW supercritical thermal power unit. *Cogeneration Power Technology* 3: 94–97. (in Chinese).
206. Ma, L.Y., Lee, K.Y., and Wang, Z.Y. (2016). Intelligent coordinated controller design for a 600MW supercritical boiler unit based on expanded-structure neural network inverse models. *Control Engineering Practice* 53: 194–201.

CHAPTER 5

DISTRIBUTION SYSTEM

5.1 INTRODUCTION

The active management of the increasing penetration of distributed generation (DG), mainly based on renewable energy sources (RES), has gradually converted the distribution networks to active distribution networks (ADNs). This transition has provided a new dimension and further complexity to the distribution network operation and planning problems. Indeed, in power distribution systems, most of the optimization problems are complex, large-scale, hard, nonlinear combinatorial problems of mixed integer nature. This means that the number of solutions to be evaluated grows exponentially with the system size, which means the problem is a non-polynomial time (NP-complete) with a large number of local optimal solutions, which makes the problem space a potentially high multimodal landscape.

This chapter presents meta-heuristic optimization techniques for the solution of complex power distribution system problems, including ADN planning, optimal selection of distribution system architecture, conservation voltage reduction (CVR) planning, dynamic distribution network expansion planning (DNEP) with demand side management (DSM), capacitor placement, distribution system reconfiguration and service restoration, and parameter identification of dynamic equivalents (DE) for ADNs.

This chapter is not meant to discuss all of the state-of-the-art models about optimal operation and planning of power distribution systems but rather to provide good examples of how the use of meta-heuristic optimization techniques has improved modeling, has allowed us to provide better results than with alternative models, or even has allowed us to obtain results that were not previously at hand with conventional models.

Applications of Modern Heuristic Optimization Methods in Power and Energy Systems,
First Edition. Edited by Kwang Y. Lee and Zita A. Vale.
© 2020 by The Institute of Electrical and Electronics Engineers, Inc.
Published 2020 by John Wiley & Sons, Inc.

5.2 ACTIVE DISTRIBUTION NETWORK PLANNING

Nikolaos C. Koutsoukis, Pavlos S. Georgilakis, and Nikos D. Hatziargyriou

National Technical University of Athens (NTUA), Athens, Greece

5.2.1 Introduction

The main goal of the DNP consists of determining the optimal network investment plan with the minimum cost in order to timely meet the load growth demand and guarantee the safe operation of the distribution network. The objective of the DNP is to minimize the investment and operational costs, subject to the network's technical and operational constraints. The investment costs include the costs for new distribution lines and substations. The operational costs include the cost of energy losses and the maintenance costs. DNP is traditionally a demand-led process mainly based on given load forecast values and considering occasional connection of DG units.

In recent years, distribution networks are facing the challenge to integrate large shares of DG, mostly based on RES, such as wind farms and photovoltaic (PV) panels [1]. The traditional approach, also called "fit and forget" approach, is to simply connect the DG sources to the network, which is in fact a passive distribution network. However, the recent advances in information and communication technologies (ICT) along with national and regional policies enable the control of DG units at the distribution level, the integration of distributed energy storage (ES), and the participation of consumers in the form of demand response (DR). As a result, the distribution networks are gradually converted from passive to ADNs. This transition demands new approaches for the DNP that fully exploit the capacity and control capabilities of distributed energy resources (DER) and provide cost-saving planning solutions. During the last decade, a wide variety of methods have been applied to the planning of ADNs [2, 3]. Incorporating DER control [4–7] along with DR schemes [8, 9] and online reconfiguration [10] into the DNP provide planning solutions with lower investment and operational costs than the passive network planning, in which there is no DER control and/or DR.

In Section 5.2.4, a genetic algorithm (GA) is applied to solve the optimal planning of ADNs that is formulated in Section 5.2.2. More specifically, the proposed GA-based method of Section 5.2.4 determines optimal distribution network reinforcement and expansion plan taking into account the control capability of the active and reactive power of the DG units. An overview of DNP models and methods is provided in Section 5.2.3.

5.2.2 Problem Formulation

Distribution networks are traditionally planned to cope with the increasing load demand along with the connection of new customers during a planning horizon. In case a DG unit is connected to the network, usually the "fit and forget" approach is adopted, which means that no control of DG is considered. High DG integration may cause feeder congestion or voltage rise issues leading to additional network reinforcement expenditures. Active network management (ANM) that involves control of the active/reactive output power of the DG units, instead of passive management (i.e. "fit and forget" approach), can potentially mitigate these feeder congestion or voltage rise issues. Thus, integrating DG and their active management into the DNP may lead to investment deferral and decreased operational costs.

The main goal of the DNP is to minimize the sum of the investment costs for network reinforcement and expansion and the operational costs associated with the energy losses of the distribution network, during the planning period. Since DG is integrated into the planning formulation, several load/generation scenarios should be taken into account. The DNP is a mixed integer nonlinear programming (MINLP) problem and it is formulated as follows:

$$\min f = \sum_{i,j \in N} x_{ij}^k \cdot C_k + \sum_{sc \in T_{sc}} \tau_{sc} \cdot P_{L,sc} \cdot C_L \tag{5.2.1}$$

subject to the following constraints:

$$P_{i,sc}^{SS} + P_{i,sc}^{DG} - P_{i,sc}^{d} = \sum_{j \in N} P_{ij,sc} \qquad \forall i \in N, \; sc \in T_{sc} \tag{5.2.2}$$

$$Q_{i,sc}^{SS} + Q_{i,sc}^{DG} - Q_{i,sc}^{d} = \sum_{j \in N} Q_{ij,sc} \qquad \forall i \in N, \; sc \in T_{sc} \tag{5.2.3}$$

$$P_{ij,sc}^2 + Q_{ij,sc}^2 \leq x_{ij}^k \cdot S_{k,\max}^2 \qquad \forall i,j \in N, \; sc \in T_{sc} \tag{5.2.4}$$

$$V_{\min} \leq V_{i,sc} \leq V_{\max} \qquad \forall i \in N, \; sc \in T_{sc} \tag{5.2.5}$$

$$\sum_{k \in \Phi_k} x_{ij}^k \leq 1 \qquad \forall i,j \in N \tag{5.2.6}$$

where N is the set of network buses; x_{ij}^k is the decision variable of the reinforcement or addition of the line that connects buses i and j using type k conductor; C_k is the investment cost of type k conductor; T_{sc} is the number of load/generation scenarios; τ_{sc} is the duration (hours) of scenario sc; $P_{L,sc}$ is the network's power loss during the sc load/generation scenario; C_L is the energy loss cost (\$/MWh); $P_{i,sc}^{SS}$ and $Q_{i,sc}^{SS}$ are the active and reactive power injected by the substation at bus i during the sc load/generation scenario, respectively; $P_{i,sc}^{DG}$ and $Q_{i,sc}^{DG}$ are the active and reactive power of the DG at bus i during the sc load/generation scenario, respectively; $P_{i,sc}^{d}$ and $Q_{i,sc}^{d}$ are the active and reactive load demand at bus i during the sc load/generation scenario, respectively; $P_{ij,sc}$ and $Q_{ij,sc}$ are the active and reactive power flow through line i–j during the sc load/generation scenario, respectively; $S_{k,\max}$ is the nominal capacity of type k conductor; $V_{i,sc}$ is the voltage magnitude at bus i during the sc load/generation scenario; $V_{\min}$ and $V_{\max}$ are the minimum and maximum bus voltage limit, respectively; and Φ_κ is the set with the available conductors.

The objective function of the DNP problem is given by (5.2.1). The first term of (5.2.1) corresponds to the total investment costs for the reinforcement or addition of distribution lines, while the second term of (5.2.1) refers to the total energy loss cost for all the different load/generation scenarios. The active and reactive power flow equations are given by (5.2.2) and (5.2.3), respectively. The limit of the apparent power flow for each distribution line is defined by (5.2.4). Equation (5.2.5) represents the bus voltage limits. To avoid more than one change in the conductor type of a distribution line, constraint (5.2.6) is applied.

The radial operation of the final planning solution should be ensured. Thus, the distribution network is represented as a spanning tree, which has as root the substation, regardless of the power flow direction. This is performed by introducing the two binary variables y_{ij} and y_{ji} for each i–j distribution line. The radiality constraints of the distribution network, expressed as spanning tree, are as follows [11]:

$$y_{ij,\mathrm{sc}} + y_{ji,\mathrm{sc}} = \mathrm{st}_{ij,\mathrm{sc}} \qquad \forall i,j \in N,\ \mathrm{sc} \in T_{\mathrm{sc}} \tag{5.2.7}$$

$$\sum_{j \in N} y_{ij,\mathrm{sc}} = 1, \qquad \forall i \in N \backslash N_{\mathrm{SS}},\ \mathrm{sc} \in T_{\mathrm{sc}} \tag{5.2.8}$$

$$y_{ij,\mathrm{sc}} = 0, \qquad \forall i \in N_{\mathrm{SS}},\ \mathrm{sc} \in T_{\mathrm{sc}} \tag{5.2.9}$$

$$0 \le \mathrm{st}_{ij,\mathrm{sc}} \le 1 \qquad \forall i,j \in N,\ \mathrm{sc} \in T_{\mathrm{sc}} \tag{5.2.10}$$

where N_{SS} is the set of substation buses. The binary variable $\mathrm{st}_{ij,\mathrm{sc}}$ is equal to 1, if the line i–j belongs to the spanning tree during scenario sc; otherwise $\mathrm{st}_{ij,\mathrm{sc}}$ is equal to 0, which means that the buses i and j are not connected. Equation (5.2.7) shows that if buses i and j are connected ($\mathrm{st}_{ij,\mathrm{sc}} = 1$), bus j is the parent of bus i ($y_{ji,\mathrm{sc}} = 1$) or bus i is the parent of bus j ($y_{ij,\mathrm{sc}} = 1$), during scenario sc. As shown in (5.2.8), each bus can have only one parent, except for the substation, which has no parents (5.2.9).

The DNP formulation given by (5.2.1)–(5.2.10) refers to the passive operation of DG units that includes no control of DG. In order to exploit the advantages of ANM for the solution of the DNP problem, the following constraints are added to the DNP formulation:

$$Q_{i,\mathrm{sc}}^{\mathrm{DG}} \ge -P_{i,\mathrm{sc}}^{\mathrm{DG}} \cdot \tan(\phi_{\min}) \qquad \forall i \in N, \mathrm{sc} \in T_{\mathrm{sc}} \tag{5.2.11}$$

$$Q_{i,\mathrm{sc}}^{\mathrm{DG}} \le P_{i,\mathrm{sc}}^{\mathrm{DG}} \cdot \tan(\phi_{\max}) \qquad \forall i \in N, \mathrm{sc} \in T_{\mathrm{sc}} \tag{5.2.12}$$

$$P_{i,\mathrm{sc}}^{\mathrm{DG}} = \mathrm{CF}_{i,\mathrm{sc}} \cdot P_{i,\mathrm{sc}}^{\mathrm{DG\_real}} \qquad \forall i \in N,\ \mathrm{sc} \in T_{\mathrm{sc}} \tag{5.2.13}$$

$$\mathrm{CF}_{\min} \le \mathrm{CF}_{i,\mathrm{sc}} \le 1 \qquad \forall i \in N,\ \mathrm{sc} \in T_{\mathrm{sc}} \tag{5.2.14}$$

The control of the reactive power output of the DG unit placed at bus i is given by (5.2.11) and (5.2.12). The reactive power output of the DG unit varies within specified limits given by the minimum ($\phi_{\min}$) and maximum ($\phi_{\max}$) phase angle of the DG power factor. The control of the active power output of the DG unit is given by (5.2.13). The curtailment factor ($CF_{i,sc}$) of the DG unit placed at bus i during the sc scenario, which has values between a minimum value (CF_{min}) and one, as (5.2.14) indicates, determines the active power of the DG unit ($P_{i,sc}^{DG\_real}$) at bus i during the sc scenario that is actually injected to the network.

5.2.3 Overview of the Solution Techniques for Distribution Network Planning

Various heuristic optimization techniques have been successfully applied to the solution of the DNP problem [2, 12]. Heuristic optimization methods are in general easy to implement and they allow the consideration of multiple objectives, such as network investment cost, reliability costs, loss minimization, voltage quality, and several technical aspects of the DNP problem. Following an iterative process, an optimal planning solution is determined by optimizing the value of an objective function, which is usually the sum of costs over the planning horizon.

The first attempt to solve the DNP problem with heuristic methods was in the early 1980s [13]. In 1990s, heuristic methods based on artificial intelligence emerged in distribution planning problems giving even better planning solutions and providing more flexibility in the DNP formulation [12]. In the last decade, meta-heuristic optimization techniques, such as GA, Tabu search (TS), and particle swarm optimization (PSO) are employed for the planning of ADNs.

In [14], a recombination-based evolutionary GA approach is adopted for the planning of ADNs in which automatic switching is enabled. DG integration is incorporated into the DNP and the optimization problem is solved by graph theory-based GA [15] and a hybrid GA-OPF (optimal power flow) method [16]. The use of GA is also adopted in [17] and [18] to solve the planning of ADNs in which DER control is enabled. GA is also adopted for the joint planning of network reinforcement, renewable DG, and smart-metering allocation [9]. Multi-objective TS [19] combined with GA and constructive heuristic approach (CHA) [20] solves the DNP problem considering automatic switching and reliability improvement. A DNP formulation that considers DG and uncertainties is solved by TS with an embedded Monte Carlo simulation method [21]. An evolutionary PSO examines the effect of dispatchable DG and load uncertainties in the solution of the DNP problem [22, 23]. The planning of DG, storage units, substations, and feeders are co-optimized by a modified PSO [24].

5.2.4 Genetic Algorithm Solution to Active Distribution Network Planning Problem

This section describes the application of GA to solve the planning problem of ADNs that was formulated in Section 5.2.2. The methodology involves the following steps: (i) data modeling, (ii) chromosome coding, (iii) creation of initial population, (iv) fitness function definition and chromosome evaluation, and (v) crossover, mutation, and next generations.

In the demand-led DNP methodologies, the planning solution was calculated based on the load forecasting values whereas DG was omitted. In the planning of ADNs, the stochastic behavior of DG and load demand over a time span, e.g. over a year, should be taken into account. The consideration of all possible values of load and generation over a year results in a large-scale computational simulation, which makes the planning process practically infeasible. Thus, a reasonable number of load and generation scenarios should be considered, which have similar stochastic behavior as the original complete list of scenarios (e.g. the complete yearly load and generation curves with a resolution of one hour, i.e. 8760 values per curve). The proposed method uses the *K*-means clustering method for grouping data with similar characteristics [25].

The *K*-means method divides a set of *m*-dimensional data into *k* clusters. The *k* clusters include data with similar stochastic behavior and the centroid of each cluster represents the average behavior of its included data.

Only integer variables, i.e. the investment variables (x_{ij}^k), are coded in the proposed method. Thus, each chromosome consists of two parts. The first part contains the candidate lines for reinforcement and the second part corresponds to the network expansion plan. The operation variables, such as power flow, voltage magnitudes, and angles, are determined by a nonlinear programming (NLP) solver. Considering the large number of line connections and their possible conductor type, the traditional GA binary coding would result in large chromosomes. Thus, the coding employed here uses integer numbers to represent the conductor type of each line [26]. Figure 5.2.1a shows a six-bus distribution network. In Figure 5.2.1a, the dashed lines depict the candidate distribution lines for the connection of new loads. Figure 5.2.1b presents a candidate planning solution for the six-bus distribution network of Figure 5.2.1a. Figure 5.2.1c illustrates the binary and integer coding of the candidate planning solution of Figure 5.2.1b. Moreover, Figure 5.2.1c shows the conductor type; e.g. line 2, which is the line between buses 2 and 3, is made of conductor type 5.

The random generation of initial population, as in the classic GA, is not efficient for radial networks, because most of the random candidate solutions would result in inconsistent or meshed networks. In order to generate initial radial networks, for a new load connection, first an available connection is randomly selected and then its conductor type is randomly chosen. For the existing distribution lines that are candidate to be reinforced, a conductor type with equal or higher capacity than the existing one is randomly selected. The optimization procedure terminates when the maximum number of generations is reached or the solution is not improved after a certain number of generations.

Once a candidate solution is selected, the integer variables, i.e. the investment variables (x_{ij}^k), are determined providing the network topology. Afterward, the feasibility of the candidate planning solution should be analyzed and evaluated. If a solution is not feasible, a penalty is assigned (f_{pen}), instead of simply rejecting

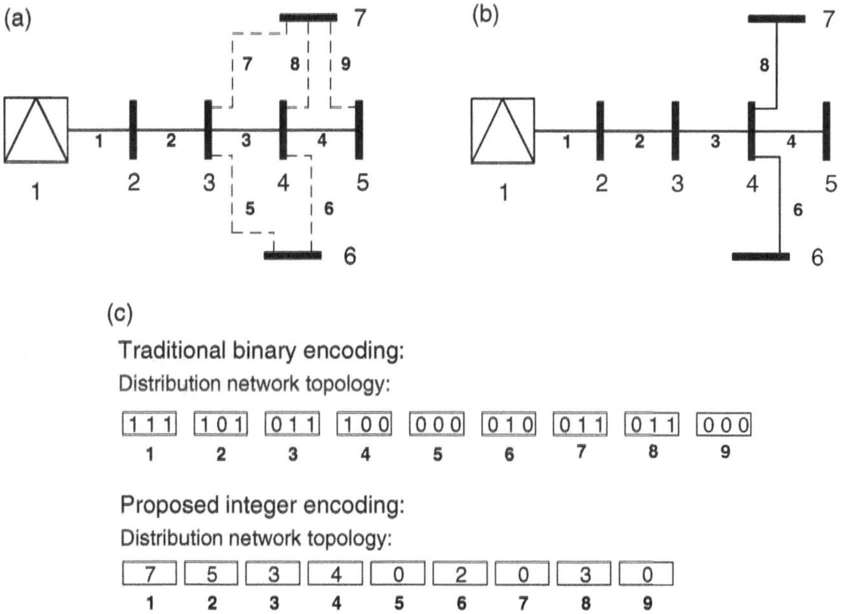

Figure 5.2.1 (a) Six-bus distribution network, (b) candidate planning solution, and (c) binary and integer chromosome coding.

the solution. The fitness function (ff) of the optimization procedure is given by (15), as follows:

$$\text{ff} = \sum_{i,j \in N} x_{ij}^{k} \cdot C_k + f_{\text{pen}} \tag{5.2.15}$$

In order to efficiently evaluate a planning solution, two artificial variables, which take positive or zero values, are added in (5.2.2) and produce (5.2.17), which expresses the balance of the active power flow. These two variables of (5.2.17) represent a fictitious load shedding ($r_{i,\text{sc}}$) and a fictitious generation shedding ($g_{i,\text{sc}}$) at each load bus i during scenario sc. For each candidate planning solution, the following NLP problem is solved:

$$\min f_{\text{pen}} = \sum_{\text{sc} \in T_{\text{sc}}} \tau_{\text{sc}} \cdot P_{\text{L,sc}} \cdot C_{\text{L}} + M \cdot \sum_{\text{sc} \in T_{\text{sc}}} \left(g_{i,\text{sc}} + r_{i,\text{sc}} \right) \tag{5.2.16}$$

subject to the following constraints:

$$P_{i,\text{sc}}^{\text{SS}} + P_{i,\text{sc}}^{\text{DG}} - P_{i,\text{sc}}^{\text{d}} - g_{i,\text{sc}} + r_{i,\text{sc}} = \sum_{j \in N} P_{ij,\text{sc}} \qquad \forall i \in N,\ \text{sc} \in T_{\text{sc}} \tag{5.2.17}$$

$$g_{i,\text{sc}}, r_{i,\text{sc}} \geq 0 \qquad \forall i \in N,\ \text{sc} \in T_{\text{sc}} \tag{5.2.18}$$

as well as the constraints (5.2.3)–(5.2.14), where M is a big number. Once the integer variables (x_{ij}^{k}) are determined by the structure of the chromosome, the problem given by (5.2.3)–(5.2.14) and (5.2.16)–(5.2.18) becomes nonlinear, whose solution indicates that the candidate planning solution is operationally feasible ($g_{i,\text{sc}} = 0$, $r_{i,\text{sc}} = 0$), or not ($g_{i,\text{sc}} \neq 0$, $r_{i,\text{sc}} \neq 0$).

A stochastic tournament selection operator is employed for the choice of next generation's population. First, two solutions p_i and p_j are randomly chosen from the initial population and the solution with the lower fitness function is selected. The aforementioned procedure is repeated until a mating pool is created. With this selection operator, the elitism of the solutions is preserved, while the bad solutions are eliminated. A two-point crossover operator is applied to the population of the mating pool to create an offspring population. The first cross point of two-point crossover operator is applied to the first part of the chromosome, which represents the candidate lines for reinforcement, while the second cross point is applied to the second part of the chromosome, which represents the expansion plan of the network. The mutation operator will be used in a percentage of the mating pool population. During the mutation operation, a gene is randomly chosen, and it is replaced by a conductor type with higher capacity.

5.2.5 Numerical Results

The GA was applied to the modified 21-bus distribution network [26], as shown in Figure 5.2.2. The 21-bus distribution network is a 13.8 kV network and it consists of 1 substation, 4 existing load buses, and 16 future load connections. The possible connections are 35 and they are depicted with dashed lines in Figure 5.2.2. The conductor type of the existing lines is given by the number next to them in Figure 5.2.2. The technical and economical characteristics of the available conductors are presented in Table 5.2.1. Two wind farms of 3 MW are connected to buses 15 and 21. The one-year hourly load profile of the network is presented in

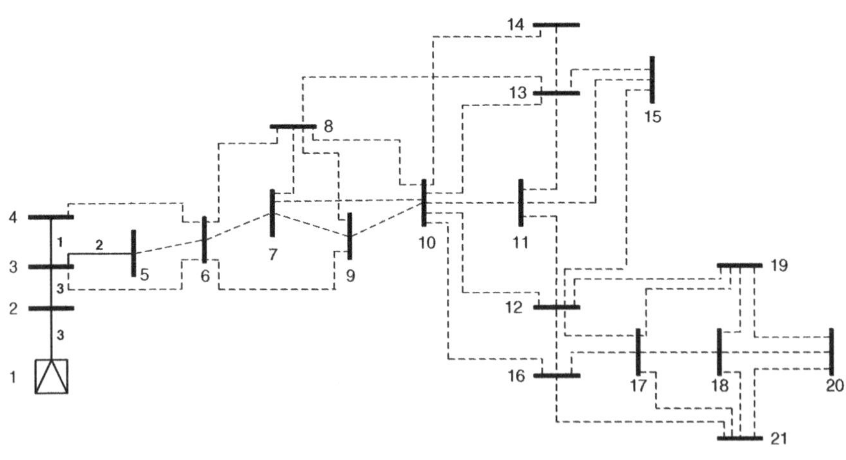

Figure 5.2.2 21-Bus distribution network.

TABLE 5.2.1 Technical and Economical Specification of the Available Conductors

Type	R (Ω/km)	X (Ω/km)	Ampacity (A)	Cost ($/km)
1	1.0145	0.4679	158	10 000
2	0.5205	0.4428	250	15 000
3	0.2006	0.4026	453	30 000

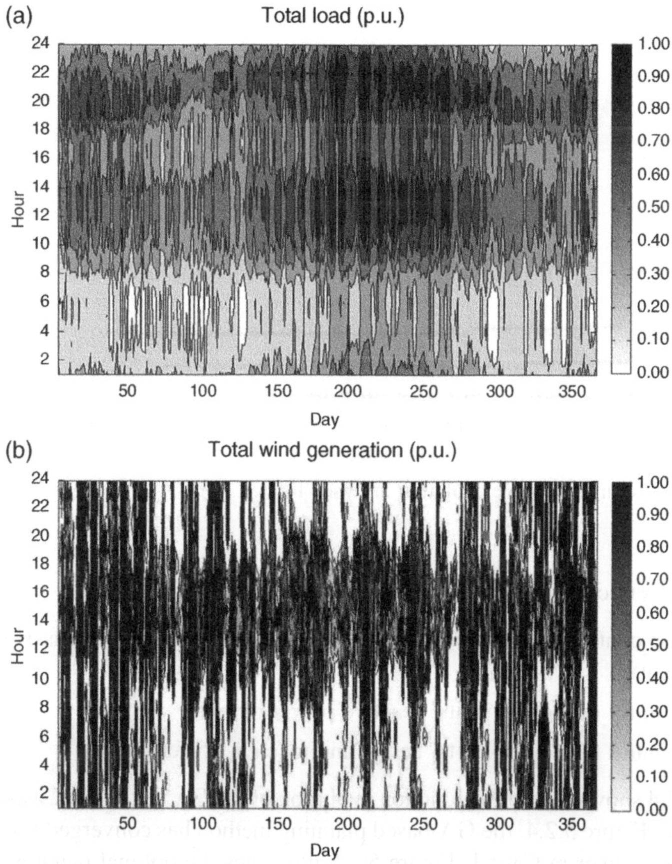

Figure 5.2.3 (a) Yearly load profile and (b) yearly wind generation profile.

Figure 5.2.3a. The one-year hourly wind generation profile at the network site is shown in Figure 5.2.3b. The data of Figure 5.2.3 were acquired from the case study of [27]. The data were grouped into 20 clusters using K-means method resulting in 20 load/generation scenarios. The energy loss cost is 40$/MWh. The GA parameters are determined by trial and error and are shown in Table 5.2.2. The results for the 21-bus distribution network are presented for one-year horizon.

TABLE 5.2.2 GA Parameters

Type	Value
Population size	200
Maximum generations	100
Maximum generations without improvement	50
Crossover rate	0.80
Mutation rate	0.05

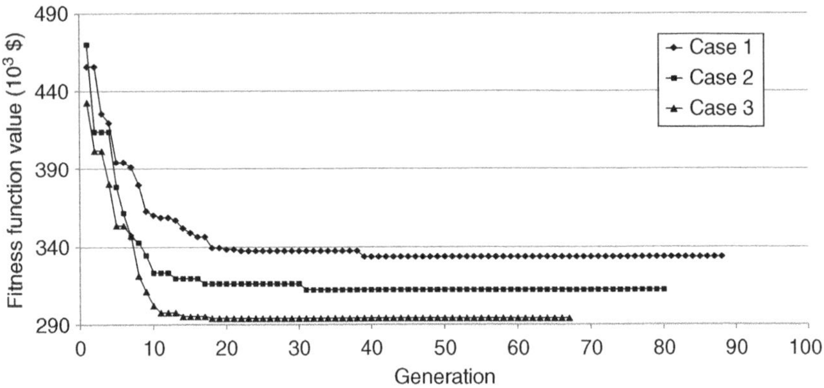

Figure 5.2.4 Fitness function value evolution in Cases 1–3.

Three different cases were examined:

- Case 1: No control of the DG units' output power is considered (passive management).
- Case 2: Control of the DG units' reactive power output.
- Case 3: Control of the DG units' active and reactive power output.

Figure 5.2.4 shows the fitness function evolution of the GA method in Cases 1–3. As shown in Figure 5.2.4, the GA-based planning method has converged faster in Case 3 and slower in Case 1. Figure 5.2.5 illustrates the optimal planning solutions in Cases 1–3. In Figure 5.2.5, the number next to the distribution lines represents their conductor type. The total cost of the planning solution in Case 1 is 334 000$, out of which the total network investment cost is equal to 293 500$ and the total energy loss cost is equal to 40 500$. The total cost of the planning solution obtained in Case 2 is 312 670$. The network investment cost in Case 2 is equal to 273 300$, which is 6.9% lower than the network investment cost in Case 1, while the energy loss cost is 40 400$, which is almost the same with Case 1. The planning solution with the lowest total cost is obtained in Case 3 and it is equal to 294 300$. The network investment cost for Case 3 is equal to 249 300$,

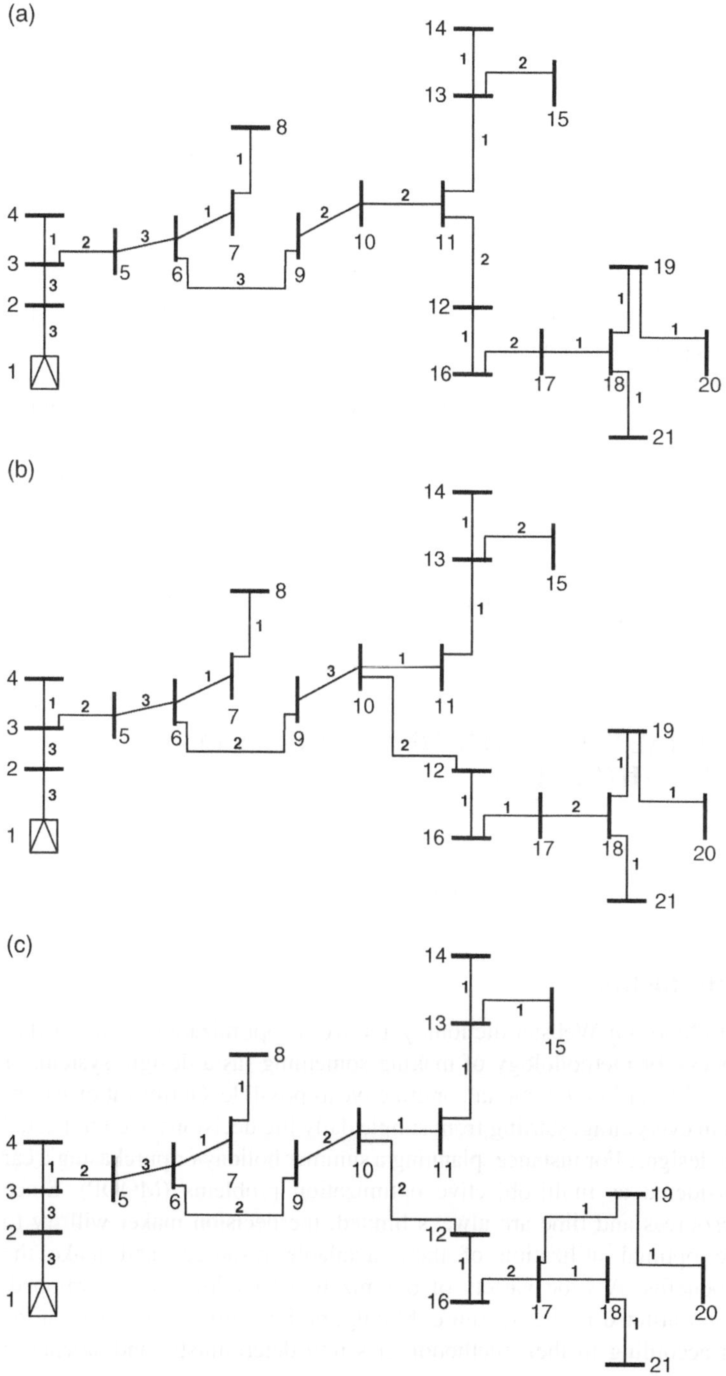

Figure 5.2.5 (a) Solution in Case 1, (b) solution in Case 2, and (c) solution in Case 3.

which is 15% lower than the investment cost in Case 1. It must be noted that in Case 3 the active power curtailment was enabled only in the load/generation scenarios where the actual DG power was greater than 90% of the nominal DG active power and CF_{min} was equal to 0.9. The computational time in Cases 1–3 was approximately 70 minutes using a PC with an Intel Core i7 CPU at 3.40 GHz and 4 GB of RAM.

5.2.6 Conclusions

The proposed GA method provides optimal planning solutions for large-scale network planning problems within reasonable computational time. The meta-heuristic optimization procedure enables the consideration of multiple load/generation scenarios and the formulation of different ANM schemes into the planning process providing new dimensions to the DNP problem. In fact, results show that cost-saving planning solutions are obtained when the control of DG units' active and reactive output power is considered highlighting the importance of ADNs. The proper tuning of the GA parameters, the chromosome integer encoding, and the efficient implementation of the crossover and mutation operators along with the creation of the initial population are the key features for the effective application of the GA in the planning of ADNs.

5.3 OPTIMAL SELECTION OF DISTRIBUTION SYSTEM ARCHITECTURE

Osama Mohammed and Ahmed Elsayed

Florida International University, Miami, FL, USA

5.3.1 Introduction

According to Merriam-Webster dictionary, the word "optimization" is defined as the act, process, or methodology of making something (as a design, system, or decision) as fully perfect, functional, or effective as possible. Optimization essentially exists in everything, starting from simple daily life decisions to complicated engineering designs. For instance, planning a summer holiday or purchasing a car can be considered as multi-objective optimization problems (MOOP). Since financial resources and time are always limited, the decision maker will try to achieve the optimal utilization of these available resources and make the maximum benefits. A wide variety of optimization techniques are developed, discussed, and utilized in the literature. Mainly, optimization techniques can be categorized according to their methodologies into deterministic and stochastic techniques.

5.3.2 Deterministic Optimization Techniques

Deterministic optimization, or sometimes referred to as mathematical programming, is the classical branch of optimization algorithms. These techniques involve algorithms which solely rely on linear algebra since they are commonly based on the computation of the gradient, and in some cases also of the Hessian matrices, of the response variables. Deterministic techniques are divided into two families depending on the problem either unconstrained or constrained [28]. The techniques for unconstrained optimization problems include:

- Simplex Method: Sometimes referred to as Polytope method, which was introduced in the early sixties by Spendley et al. [29]. This method involves starting from an initial polytope instead of one initial point. This initial polytope with $n + 1$ vertices (if n is the dimension of the parameter space) is set up. Hence, if $n = 2$, the polytope will take the shape of a triangle and quadrilateral for $n = 3$. At each iteration a new polytope will be generated by producing a new point to replace the "worst" point of the old polytope until the optimal point is reached.
- Newton's Method: Newton's method is the most classic and known optimization algorithm. It is also known as the second derivative method since it involves calculating the first and second derivatives of the objective function $f(x)$ (e.g. the gradient $g(x)$ and the Hessian matrix $H(x)$).
- Quasi-Newton Method: Quasi-Newton method is the simplified version of the Newton method which only requires the first derivative of the objective function to be computed. The method is superior when the Hessian matrix is not available, thus the most obvious thing to do is to approximate it by finite differences in the gradient vector. In other words, building up trend information as the iterations proceed, using the observed behavior of $f(x)$ and the gradient matrix $g(x)$ without explicitly forming the Hessian matrix.
- Other methods like Gauss–Newton and Levenberg–Marquardt methods are utilized.

The other family of deterministic techniques are used to solve constrained optimization problem, this family includes:

- Penalty and Barrier Function Method: When solving a general nonlinear optimization problem in which the constraints cannot be easily eliminated, it is necessary to balance the aims of reducing the objective function $f(x)$ and staying inside or at least very close to the feasible region. Thus, this method transforms the generic minimization problem to a minimization problem of an unconstrained function whose value is penalized in case the constraints of the original problem are not respected.
- Active Set Methods: Active set methods are methods for handling inequality constraints. The most common is the primal active set method. It is important to distinguish between constraints that hold exactly and those that do not.

At a feasible point x, the constraint $a_i^T x \leq b_i$ is said to be active or binding if $a_i^T x = b_i$ and inactive if $a_i^T x < b_i$. In this method, a correction $\delta^{(n)}$ is applied to a solution $x^{(n)}$ in iteration n. A line-search is performed along $\delta^{(n)}$ to find the best feasible point. If the search terminates at a point where an inactive constraint becomes active, the constraint is added to the active set. This process continues until the solution of the equality constraints problem yields $\delta^{(n)} = 0$, and no constraints are left to be removed from the active set, the optimization terminates and $x^{(n)}$ is flagged as the optimal solution.

- Lagrangian Methods: There are different Lagrangian methods, some of them are coupled with active set methods or Newton and Quasi-Newton methods.
- Quadratic Programming: Quadratic programming (QP) is a more direct approach than penalty and barrier function methods to NLP. It consists in iteratively solving subproblems in which the objective function is approximated to a quadratic function and the constraints functions are linearized.
- Mixed Integer Programming: Mixed integer programming is the study of optimization in cases where some of the variables are required to take integer or discrete values. For example, optimization of a certain manufacturing process, where a decision variable such as number of labors must be integer. Whereas, in the same problem, another variable which is the number of working hours is not restricted to be an integer. These kinds of problems are solved by the branch and bound method. In this method, the set of candidate solutions is thought of as a rooted tree with the full set at the root. Then, a subset of solutions is obtained by manipulating the rooted set. These subsets are called branches and the algorithm explores the branches of this tree [28].

Initially, it seems that the aforementioned deterministic techniques are appealing for practical problems, the main reason for this is that their convergence to a solution is much faster than the stochastic algorithms which involve random variables (please see the following section). Further, they involve a lower number of objective function evaluations. However, these methods are looking for stationary points in response for a certain variable, thus they are more susceptible to be trapped in local optimum points. Moreover, these methods are basically relying on the extensive mathematical modeling of the problem; when speaking about practical problems, finding an accurate mathematical model is not affordable. Therefore, if there is no or very little knowledge about the behavior of the objective function in the multidimensional search space, tackling the optimization problem with a stochastic method seems to be more practical. Hence, the deterministic techniques will not be discussed any further in this section, instead more focus will be dedicated to the stochastic techniques.

5.3.3 Stochastic Optimization Techniques

Stochastic optimization methods include the optimization methods in which randomness is present in the search procedure. As a matter of fact, stochastic methods choose their path through the parameter space from randomly selected initial configurations to the optimal solution using a "random factor." Unlike the case in deterministic methods, this random nature causes that different computation runs solving the same problem with the same conditions lead to quietly different results. Despite this fact, this random nature can be viewed from another angle as an advantage, since stochastic methods feature high robustness when dealing with noisy measurement data. Moreover, including random factors enables the stochastic methods to accept deterioration in the objective function (temporarily worse solutions) during the iteration process. This fact enables them to escape local solutions and find the region of the global optimum with a high probability. This process is guaranteed with a high probability regardless of the starting point location [30].

Another point of criticism for the stochastic optimization methods is fueled by the fact that they require a high number of objective function evaluations. With enough confidence, it can be said that this was a serious concern two to three decades ago, before the stunning advancements in the computation power. With the high availability of parallel computing systems, graphical processing units (GPUs), and powerful processors, several evaluations of complicated objective functions can be done in a reasonable time. This led to the high popularity of stochastic methods in solving real-world problems.

In the past, stochastic algorithms were often referred to as heuristic. Originally, the word "heuristic" is an ancient Greek word which means "to discover." Roughly speaking, heuristic means to discover by trial and error. First, heuristic was introduced as a psychological concept by Nobel Laureate Herbert A. Simon. Then, it was exploited in different science realms. As time progressing and with the introduction of more developed techniques, the word "meta" is attached to the word "heuristic" to form the new term "meta-heuristic." Here, "meta" means "beyond," so "meta-heuristic" refers to higher level algorithms.

When designing any meta-heuristic algorithms, the design should achieve a balance between two major concepts, namely intensification and diversification. Diversification means to generate diverse solutions so as to explore the entire search space on a wide scale (which consequently increases the algorithm robustness), while intensification means to focus the search in a local region knowing that a current good solution is found in this region. Designing an algorithm with more intensification will be more susceptible to be trapped in a local minimum, in contrast, an algorithm with more diversification will not converge in a reasonable time. Hence, a trade-off has to be made between the two concepts depending on the problem.

Many meta-heuristic algorithms are discussed in the literature, most of them are based on concepts from biology, metallurgy, and observations of nature or animal behaviors [30]. In this section, a brief overview on the existing meta-heuristic techniques will be provided.

Simulated Annealing

Simulated annealing (SA) is based on one of the metallurgy concepts which is the metal annealing. These techniques were introduced by Kirkpatrick et al. in 1983 [31]. The annealing is a technique which involves heating the metal to a very high temperature and slowly cooling it, to rearrange its crystals in a configuration where the minimum energy is reached. If the cooling is sufficiently slow, the system has time to explore many configurations and settles in the minimum energy one. The main purpose of this process is to increase the material machinability, the contrast of this process is quenching or hardening.

In SA, the actual search moves trace a piecewise path. With each move, an acceptance probability p is evaluated, which is based on Boltzmann probability distribution, given by:

$$p = e^{-\frac{\Delta E}{k_B \cdot T}} \tag{5.3.1}$$

where k_{B} is the Boltzmann's constant, T is the temperature for controlling the annealing process, and ΔE is the change in energy. This acceptance probability not only accepts changes that improve the objective function (a lower value in a minimization problem) but also keeps some changes that do not improve the objective (a larger objective value). This feature plays a significant role in helping the SA to achieve the global minimum. This transition probability is based on the Boltzmann distribution in statistical mechanics.

In SA optimization, a law defining how the temperature parameter decreases over successive iterations must be given. There are many ways to control the cooling rate or the temperature decrease. Two commonly used annealing schedules (or cooling schedules) are linear and geometric. The geometric cooling is the most widely used one, in which T is replaced by αT where α is the cooling rate in the range [0, 1]. The advantage of the geometric cooling is that it does not need to specify the maximum number of iterations. In order to make the cooling process slow enough to allow the system to explore more states and stabilize, α is selected in the range [0.7, 0.99]. However, more conservative researchers limit the practical range of α to be within [0.7, 0.85].

The objective function $f(x)$ to be minimized can be considered the energy of the system while the different combinations/values of the decision variables of the optimization are its atomic configurations. The probability that a particular configuration, even a worse one, is accepted is ruled by a Boltzmann probability is explained earlier. This particular acceptance criterion, which is the backbone of the method, allows some probability of accepting worse configurations. The probability of accepting worse solutions is dictated by the current temperature $T(n)$. As long as the temperature is very high, SA accepts every new solution, thus yielding random jumps through the search space. On the other hand, with a lower temperature, only improvements are accepted.

Particle Swarm Optimization

PSO based on swarm behavior is observed in nature such as in bird flock or fish schooling. PSO was developed by J. Kennedy and R.C. Eberhart in 1995 [32]; since then, PSO has attracted a lot of attention, and now forms one of the most exciting, ever-expanding stochastic techniques. It was found that when a flock of birds is looking for food, the individuals are following a leader who is the closest one to the food location. Then, each bird changes his location with a certain velocity in the direction of the leader. Thus, by imitating this concept, the food location will be the optimum solution in the search space and the flock leader is the fittest individual (the solution of the current iteration that achieves the minimum value of the objective function in a minimization problem).

PSO searches the space of an objective function by adjusting the trajectories of individual agents, called particles. When a particle finds a location that is better than any previously found locations, then it updates this location as the new current best for particle i. There is a current best for all particles at any time t at each iteration. The aim is to find the global best among all the current best solutions until the stopping criterion becomes true or after a certain number of iterations.

There are many algorithms that are derived from PSO, as well as hybrid algorithms obtained by combining PSO with other existing algorithms. One of the most noticeable improvements is probably introducing an inertia function $\theta(t)$ so that velocity of a particle i, v_i is replaced by $\theta(t) \cdot v_i$, where θ takes a value between 0 and 1. In the simplest case, the inertia function can be taken as a constant, typically in the range [0.5, 0.9]. This is equivalent to introducing a virtual mass to stabilize the motion of the particles, and thus the algorithm is expected to converge more quickly. Another improved PSO technique involves using a chaotic inertia weights and a crossover operator was introduced to solve the economic dispatch (ED) problem [33].

Differential Evolution

Differential evolution (DE) was developed by R. Storn and K. Price [34]. It is a vector-based evolutionary algorithm (EA), and can be considered as a further development of genetic algorithms (GAs). Both algorithms aim at simulating the evolution of a population through successive generations of better-performing individuals. Unlike GAs, DE carries out operations over each component (or each dimension of the solution). Almost everything is done in terms of vectors, so a mutant individual is represented by a vector. DE can be viewed as a self-organizing search, directed toward the optimum. Similar to GAs, DE consists of three main steps: mutation, crossover, and selection.

Ant Colony Optimization

Ant colony optimization (ACO) was pioneered by M. Dorigo [35] and is based on the co-operative behavior of social ants when looking for food. Ants are social insects and live together in well-organized colonies consisting of approximately 2–25 million individuals. When searching for food, a swarm of ants interact or communicate in their local environment. Each ant lays scent chemicals or pheromone to communicate with others. Each ant is also able to follow the route marked by the chemicals laid by other ants. When an ant finds a food source, it will mark it with these chemicals and also mark the path to and from it [36].

The process starts with an initial random route, where the individuals are dispersed over the search space. Since the pheromone (laid chemicals) concentration varies, the ants follow the route with higher pheromone concentration. In turn, the pheromone is enhanced by the increasing number of ants (more laid chemicals).

As more and more ants follow the same route, it becomes the favored path. Thus, some favorite routes emerge, often the shortest or more efficient ones leading to the optimal solution. In electrical engineering, ant colony was used in different subjects including generation planning, loss reductions of distribution systems, and scheduling of electric vehicle (EV) charging [37–41].

Bee Algorithms

Bee algorithms are another class of meta-heuristic algorithms, inspired by the foraging behavior of bees. A few derivatives of the classical algorithm exist in the literature, including honeybee algorithm (HBA), artificial bee colony (ABC), and virtual bee algorithm (VBA). These algorithms are pioneered by Pham et al. [42]. It was initially introduced in 2004–2005.

Honey bees construct colonies, where they live and store honey. The communication mean among colony members is pheromone and "waggle dance." When discovery bees find a good food source (flowers rich with nectar), they will communicate the location of the food source by performing what is called waggle dance as a signaling system. Such signaling dances vary from species to species, and it depends on the available amount of food. However, they are aimed at recruiting more bees by using directional dancing with varying strength so as to communicate the direction and distance of the food source. The same idea is utilized to develop the optimization algorithm where the optimal solution is referred to as the food source.

Cuckoo Search

Cuckoo search (CS) is one of the most recent nature-inspired meta-heuristic algorithms, developed by Xin-She Yang and Suash Deb [43]. Later on in 2010, the same developers of this algorithm enhanced it by incorporating the so-called Lévy flights. Recent studies show that CS is potentially more efficient than PSO and SA. Cuckoos are fascinating birds as they make adorable sounds; however, they have an aggressive parasitism strategy for their production. They remove other birds' eggs and lay theirs among the hosts' eggs to increase the hatching probability of their own eggs. The main steps of the algorithm are as the following:

a. Each cuckoo lays one egg at a time, and dumps it in a randomly chosen hosting nest.

b. The best nests with high-quality eggs are carried over to the next generations.

c. The number of available host nests is fixed, and the egg laid by a cuckoo is discovered by the host bird with a probability $p \in [0, 1]$, depending on the similarity of a cuckoo egg to its host eggs. In this case, the host bird can either get rid of the egg, or simply abandon the nest and build a completely new nest.

The CS is one of the promising optimization techniques to be used in MOOPs. It was shown that it has higher accuracy and good convergence capabilities.

Other Meta-Heuristic Algorithms

Some widely used techniques were selected to be briefly discussed in this book chapter; however, there are many other successful meta-heuristic techniques that cannot be covered in a single chapter or article. The interested reader is highly encouraged to read about the other optimization methods. For example, TS is one of the optimization techniques that explicitly uses memory and search history. Through the appropriate use of search history, revisiting the same solutions can be avoided. The algorithm records recently tried solutions in tabu lists, which the algorithm is prevented to go over again. Over a large number of iterations, these tabu lists could save a significant amount of computing time, leading to improvements in search efficiency. Another technique is the Firefly algorithm, which is based on the flashing (glowing) patterns of fireflies. Harmony search is a technique that is inspired by the improvisation process of a musician. Further, there is a technique called artificial immune system, which is inspired by the actions of the immune system. This algorithm uses memory and learning, which adds adaptive capabilities to its performance.

Genetic Algorithms

GAs are probably the most popular and successful evolutionary algorithms (EAs). Their success can be widely seen in their applicability, where they have been utilized in a diverse range of applications. Further, a considerable number of well-known and complicated optimization problems have been solved by GAs. These problems include graph coloring, pattern recognition, travelling salesman problem, aerospace engineering, financial markets, and ED. In addition, GAs are population-based and many modern EAs are directly based on, or have strong similarities to, GAs.

From historical point of view, GAs were developed more than half a century ago in the 1960s by J. Holland. The core of the GA is the abstraction of biological evolution based on Charles Darwin's theory of natural selection. The main steps of a GA are:

a. Initialize a population of m individuals $x_1^{(1)}$, $x_2^{(1)}$, …, $x_m^{(1)}$ on the discretized design space and evaluate the fitness function for each individual in the population. Each individual represents a solution and each solution is encoded as arrays of bits or character strings (chromosomes).

b. Each iteration produces a generation, which is a set of new individuals. For each generation n, these steps are done:

- Randomly select a pair of solutions. This process is called selection, which is the first one to take place. Although, the selection of the parents is random but the probability of being selected is not the same for each individual as in other EAs. Different rules have been used for the selection mechanism; the most common are the roulette-wheel selection and the tournament selection. The mechanism that will be used in this study is the tournament selection, so it will be briefly explained. In tournament selection, a few individuals are selected in a random manner to take part in a tournament. The individual with best fitness (achieves the minimum value of an objective function in a minimization problem) will be the winner of that tournament, which will be selected. The individuals are ranked according to their fitness, the best individual is selected with probability p_t, the second best with probability $p_t \cdot (1 - p_t)$, the third best with probability $p_t \cdot (1 - p_t)^2$, and so on. Typically, $p_t \in [0.5, 1.0]$. Allowing suboptimal solutions to be selected helps in maintaining diversity in the population and prevent premature convergence. As mentioned earlier in this section, the design should achieve a trade-off between diversification and intensification.

- Apply the crossover operator with probability p_c, giving birth to two children. If no crossover takes place, the two offsprings are exact copies of their parents. The crossover probability is generally quite high in the range [0.7, 0.99]. Some researchers reported successful using of values around 0.9. Simply, as implied from the process name, crossover involves interchanging portions from the parents and combining them together to produce children. The most common crossover techniques are the one-point crossover, two-point crossover, and uniform crossover. In one-point crossover, a single point along the parents' chromosomes is selected randomly and the genetic information beyond that point is swapped between the two parents in order to create the two children. While, in two-point crossover, two points along the parents' chromosomes are selected randomly and the genetic information in between the two points is swapped between the two parents. In this research work, the single point crossover is used.
- Apply the mutation operator to each one of the reproduced two offspring with probability p_m. The mutation probability is generally quite low ($p_m \in [0.005, 0.05]$). Thus, crossover has more contribution than mutation in the evolution process. Mutation operator acts simply swapping, from 0 to 1 and the other way around.

c. After finishing the abovementioned process for all the individual members, the new generation replaces the previous one and the fitness of their individuals is evaluated. If the number of individuals per generation is odd, then one child is discarded randomly.

Generally, the efficiency of GAs depends mainly on the accuracy of the fitness function and on the choice of the controlling parameters, such as p_c, p_m, and p_t. In addition, the choice of the right population size is also very important. If the population size is too small, there will not be enough evolution, and there is a risk for the whole population to converge prematurely. On the other hand, if the population is too large, more evaluations of the objective function are needed, which will require extensive computing time and effort.

5.3.4 Multi-Objective Optimization

Solving a multi-objective problem requires the determination of a set of points that all fit a predetermined definition of an optimum solution. Since more than an objective function is involved in the optimization process, there is no single ultimate solution. Actually, the multi-objective problems produce a set of solutions. This set of solutions is called Pareto front (PF). This concept of defining optimal solutions is called Pareto optimality. In multi-objective genetic algorithm (MOGA), a redefinition of the selection rules is needed. For instance, the probability of winning the tournament in tournament selection will be higher for the individuals belonging to the Pareto frontier. Then, less probability is assigned to the individuals at farther distance from the front. More complex selection operators, which aim at achieving a more uniform distribution of the solutions on the Pareto frontier, can also be defined. In this research work, it is proposed to use non-dominated sorting genetic algorithm II (NSGA-II) to generate non-dominated PF [44].

NSGA was a popular non-domination-based GA for multi-objective optimization (MOO). However, it is criticized for its computational complexity. Then, a modified version, NSGA-II, was developed, which has a better sorting algorithm and less computational complexity. First, let us breakdown the name; N for non-dominated: because it finds the non-dominated solution, non-dominated solution is the one achieving the minimum objectives (in a minimization problem). To relate it to our specific problem, the non-dominated solution is the one that is achieving the minimum voltage and frequency fluctuations in a certain distribution architecture in shipboard power systems. S for sorting, because it sorts the population based on a fitness function, puts the non-dominated individual in the front to form what is called PF or Pareto efficient solution. Furthermore, NSGA-II utilizes an elitism operator, which makes the most performing individual in the population to survive through the generations without mutation [44].

How it works: The population is initialized as usual. Once the population in initialized, the population is sorted based on non-domination into each front. The first front being completely non-dominant set in the current population and the second front being dominated by the individuals in the first front only and the front goes so on. Each individual in each front are assigned rank (fitness) values or based on front in which they belong to. Individuals in first front are given a fitness value of 1 and individuals in second are assigned fitness value as 2, and so on. Some examples in the literature showed that it is successful in avoiding the local minima/maxima or local traps [45]. As shown in Figure 5.3.1a, some functions exhibit nonlinear behavior, which can give a local minima and global minima. NSGA-II has the capability to avoid the local minima due to its sorting functionality. For further understanding of the concept of PF, please refer to Figure 5.3.1b. The PF is the set of solutions that meet the objective function criterion [45–48].

5.3.5 Mathematical Modeling for Power System Components

This section tends to be more problem specific and discusses how meta-heuristic techniques are applied to solve the problem understudy which is to determine the optimal distribution architecture [49]. The optimum architecture refers to the optimum states of the switches, type, location, and size of energy storage to minimize the voltage and frequency fluctuations during a disturbance. This disturbance can be the connection/disconnection of a high demanding load. Some of those high

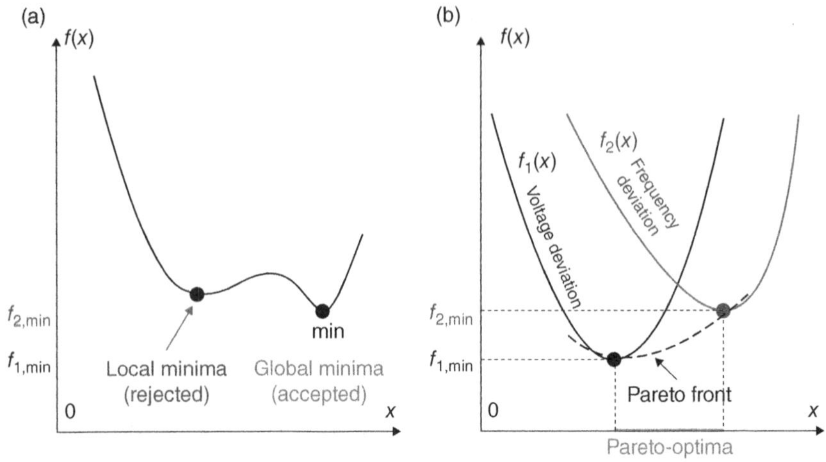

Figure 5.3.1 (a) Global minima and local minima. (b) Pareto front.

demanding loads are referred to as pulsed loads where the load profile takes a pulse shape. In another definition, pulsed loads can be defined as loads that draw very high current for short time in an intermittent fashion. Such current behavior can potentially cause the system voltage to drop in the entire system or in some of the buses. Also, high frequency regulation can occur, consequently, these disturbances can trip other sensitive loads or through some of the generation units out of the synchronized system. Examples of these loads are electromagnetic rail devices, launch systems, radar systems, free electron lasers, large group of EV during their charging process, or starting of a big machine [50–53]. According to standards (MIL-Std-1399, section 300b, 2008), pulsed load is defined as a load which demands frequent or regular repeated power input. A pulsed load is measured as the average power during the pulse interval minus the average power during the same interval immediately preceding the pulse. A pulse load lasts longer than 1 cycle at nominal frequency and less than 10 seconds.

The main objective of this study is to utilize the Pareto-based MOO to select the optimal design of the distribution architecture to reduce the negative effects of the pulsed loads when energized. The steps for solving this problem can be classified into two main categories as follows:

The Local Level

- Will be on the level of each architecture.
- Mathematical modeling for all power systems components.
- The power flow will be solved to determine the voltages at all the buses and the power injection of each generator.
- Using this power injection and given the generator excitation and governor models, we can determine the frequency oscillation.
- The deviation from the nominal voltage and frequency will be calculated.

The Global Level

- There is a Pareto-based optimization algorithm which will be controlling the decision variables and comparing the voltage and frequency fluctuations of each architecture.
- This process continues until the PF is formed. PF is the set of the solutions (architectures and configurations) that achieve the minimum fluctuations and meet the prespecified constraints [54].

As mentioned earlier, the efficiency of the GA is highly dependent of how accurate the objective function is formulated. Thus, mathematical modeling of each component in the distribution system must be carried out [55].

Voltage Source Converter

One of the challenging portions of this study is the modeling and representation of the voltage source converters (VSC) in power flow problems. The VSC can be modeled as an AC voltage source behind a phase reactance [56]. For more accuracy, the distribution line and transformer impedances are considered. Figure 5.3.2 shows the equivalent circuit for VSC in power flow studies. For more clarity, it is assumed that the AC system is divided into synchronized zones as shown in Figure 5.3.3. These zones are connected through a DC network. The equivalent injected active and reactive power from the inverter in zone z to the AC network is (at bus G):

$$P_G^z = -U_G^2 \cdot G_{\mathrm{tr}} + U_G \cdot U_{\mathrm{f}} \cdot [G_{\mathrm{tr}} \cos(\theta_G - \theta_{\mathrm{f}}) + B_{\mathrm{tr}} \sin(\theta_G - \theta_{\mathrm{f}})] \tag{5.3.2}$$

$$Q_G^z = U_G^2 \cdot B_{\mathrm{tr}} + U_G \cdot U_{\mathrm{f}} \cdot [G_{\mathrm{tr}} \sin(\theta_G - \theta_{\mathrm{f}}) - B_{\mathrm{tr}} \cos(\theta_G - \theta_{\mathrm{f}})] \tag{5.3.3}$$

where

Voltage at converter bus: $U_{\mathrm{c}}\angle\theta_{\mathrm{c}}$

Filter bus voltage: $U_{\mathrm{f}}\angle\theta_{\mathrm{f}}$

Network bus voltage: $U_G\angle\theta_G$

S_G is the injected apparent power to the AC network.

S_{c} is the injected apparent power from the VSC.

Q_{f} is the reactive power in the low-pass filter.

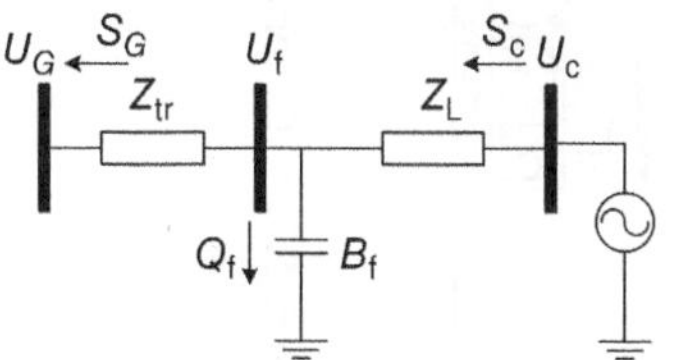

Figure 5.3.2 Equivalent circuit of VSC for power flow studies.

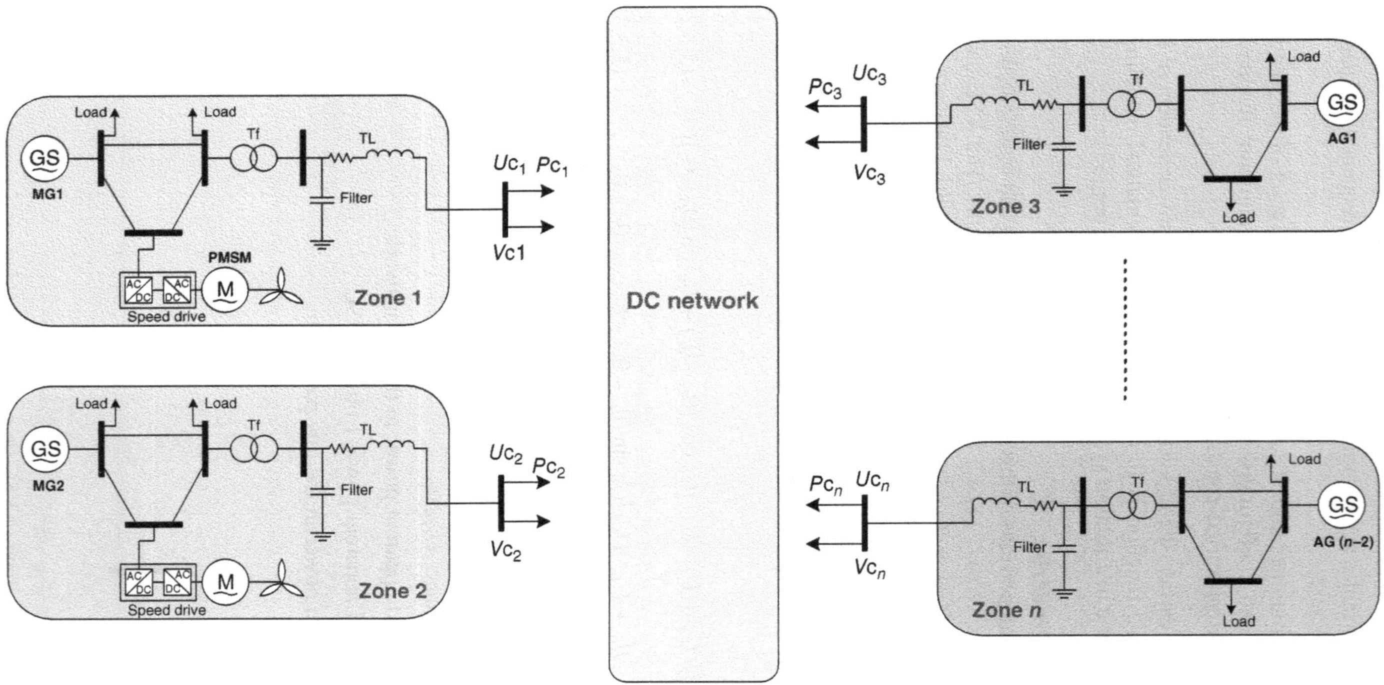

Figure 5.3.3 Generic distribution system with n asynchronized zones.

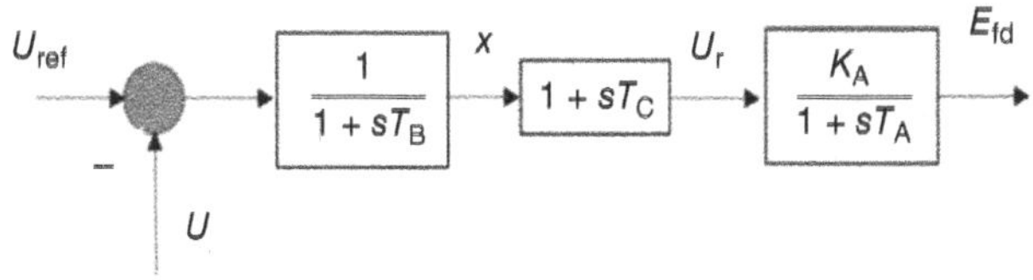

Figure 5.3.4 IEEE type AC4A excitation system.

Generator Excitation System

To determine the generator response upon sudden increase/decrease in the load, a mathematical model for the excitation systems should be derived. According to IEEE Std 421.5 (2005), the synchronous generator excitation systems can be classified as:

- DC – Direct Current Commutator Exciters; Types DC1A, DC2A, DC3A.
- AC – Alternator Supplied Rectifier Excitation Systems; Types AC1A, AC2A, AC3A, AC4A, AC5A, AC6A.
- ST – Static Excitation Systems; Types ST1A, ST2A, ST3A.

In this study, alternator supplied excitation system type AC4A is selected; however, the same procedure can be applied to represent any excitation type. The block diagram of the excitation system under study is shown in Figure 5.3.4. The differential equations of IEEE typeAC4A excitation system are given as follows:

$$\dot{x} = \frac{1}{T_B}(U_{ref} - U - x) \tag{5.3.4}$$

$$U_r = x + \dot{x}T_C \tag{5.3.5}$$

$$\dot{E}_{fd} = \frac{1}{T_A}(K_A U_r - E_{fd}) \tag{5.3.6}$$

$$\dot{E}_{fd} = \frac{1}{T_A}\left(K_A\left(x + \frac{T_C}{T_B}(U_{ref} - U - x)\right) - E_{fd}\right) \tag{5.3.7}$$

Distribution Line Models

This study is performed in the context of the distribution systems. Since the distribution lines in these systems are relatively short, they are represented by a series resistance and inductance (R_L, X_L).

5.3.6 AC/DC Power Flow in Hybrid Networks

To evaluate the effect and impact of pulsed load on system operation, the power flow problem should be solved. Solving the power flow for the hybrid AC/DC power system such as the shipboard power system is not straightforward but challenging. The power flow method that will be used here is Newton Raphson (NR) method, because it has a very high convergence rate. For example, for IEEE 30-bus system, the power flow using NR converges in three iterations, while using Gauss Siedel it takes 670 iterations to converge. However, the NR method is well studied for the legacy AC systems; applying this algorithm to hybrid AC/DC systems requires some manipulations. There are two methods adopted in the literature for solving such problem, the unified and sequential methods.

Unified Method

The AC and DC system equations are solved simultaneously, i.e. formulate a single Jacobian matrix for the entire system [57]. A disadvantage of this method is that the entire Jacobian matrix must be updated in each iteration.

Sequential Method

The AC system equations are solved and then their output parameters are passed to the DC system equations to be solved. The AC and DC set of equations are solved sequentially. For several considerations including computational simplicity, this method will be used in this study.

The main difference is in formulating the Jacobian matrix (matrix of active and reactive power derivatives). The unified approach solves the entire system in one time, i.e. formulate one Jacobian matrix for the entire system with its AC and DC sides. This matrix is massive and inversing it requires large computational effort [58]. The other approach is sequential approach, it solves the AC system first and then passes the required parameters to solve the DC, so this is why it is called sequential, i.e. it solves one side at a time [59]. So, to reduce the computational effort, the NR algorithm with sequential approach will be used in this study. Consider the simplified diagram shown in Figure 5.3.5 where the system is consolidated into two sections, AC and DC. The AC and DC sides of the system are interfaced through VSC. The steps of applying the sequential power flow algorithm to the system are depicted in the flow chart in Figure 5.3.6.

AC Power Flow

The AC power flow problem is well studied and investigated extensively in the literature, so it will be outlined briefly in this section. As depicted earlier in Figure 5.3.3, the AC system is divided into zones. Thus, for zone z, the bus admittance matrix Yz is formed. The injected active and reactive power at bus i are given through the following equations:

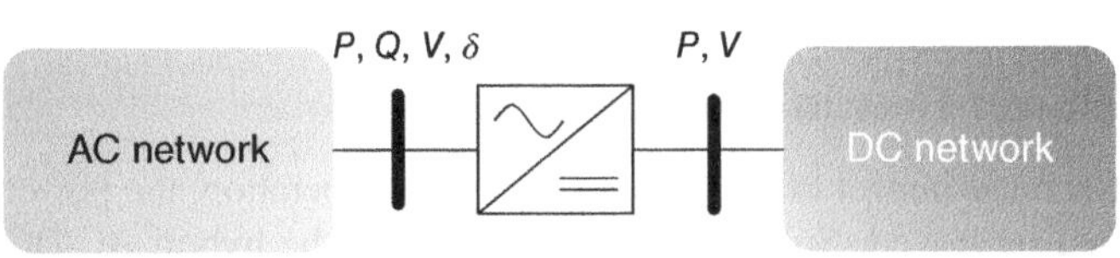

Figure 5.3.5 Simplified schematic for AC/DC power system.

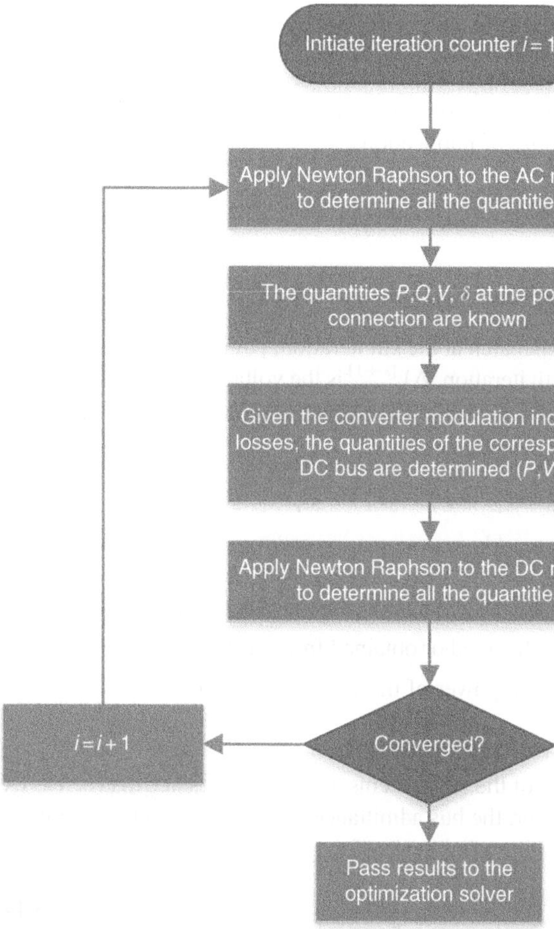

Figure 5.3.6 Flowchart for the sequential power flow approach.

$$P_i = U_i \sum_{\substack{j=1 \\ j \neq i}}^{M} U_j \left(G_{ij} \cos\left(\theta_i - \theta_j\right) + B_{ij} \sin\left(\theta_j - \theta_j\right) \right) \tag{5.3.8}$$

$$Q_i = U_i \sum_{\substack{j=1 \\ j \neq i}}^{M} U_j \left(G_{ij} \cos\left(\theta_i - \theta_j\right) - B_{ij} \sin\left(\theta_j - \theta_j\right) \right) \tag{5.3.9}$$

In this study, one of the classical methods is employed to solve the nonlinear AC power flow equations. This method is the NR method. Although the NR method requires higher computational efforts, it is highly preferred over the Gauss

Seidel method due to its fast convergence rate. Also, it should be highlighted here that the simplified version of NR method, which is the fast-decoupled method, cannot be used here. The main reason for this is because it neglects some terms which affect the accuracy of the load flow results. This claim coincides with the conclusions of a large number of publications.

$$0 = \left(S_i^{\text{sch}}\right)^* - \left(V_i^{[k]}\right)^* \sum_{j=1}^{n} Y_{ij} V_j^{[k]} + \varepsilon \tag{5.3.10}$$

$$\Delta V^{[k+1]} = \left[J\left(V^{[k]}\right)\right]^{-1} \cdot \Delta S^{[k]} \tag{5.3.11}$$

where $\Delta S^{[k]}$ is the power mismatch at the kth iteration, $[J(V^{[k]})]^{-1}$ is the inverse of the Jacobian matrix at the kth iteration. $\Delta V^{[k+1]}$ is the voltage incremental vector at the next iteration, which is $k+1$. S_i^{sch} is the scheduled apparent power. The power mismatch vector $\Delta S^{[k]}$ can be written as:

$$\Delta P_i^{[k]} = P_i^{\text{gen}} - P_i^{\text{dem}} - P_i^{[k]} \tag{5.3.12}$$

$$\Delta Q_i^{[k]} = Q_i^{\text{gen}} - Q_i^{\text{dem}} - Q_i^{[k]} \tag{5.3.13}$$

where P_i^{gen}, P_i^{dem}, Q_i^{gen}, Q_i^{dem} are the generated, demanded active and reactive power. These quantities are either scheduled or obtained from the previous iterations. $P_i^{[k]}$, $Q_i^{[k]}$ are the active and reactive power of the current iteration.

DC Load Flow

By taking the same notation of the AC systems, the bus admittance matrix Yz, DC for zone z is formed. However, the bus admittance matrix will be much simpler as all of its elements are real. The admittance of the line connecting bus i and j is:

$$y_{ij} = \frac{1}{R_{\text{dc},ij}} \tag{5.3.14}$$

The current injected at bus i in the DC side can be given as:

$$I_{\text{dc},i} = \sum_{\substack{j=1 \\ j \neq i}}^{n} Y_{\text{dc},ij}\left(U_{\text{dc},i} - U_{\text{dc},j}\right) \tag{5.3.15}$$

By assuming bipolar configuration, the transferred power over a transmission line is given as:

$$P_{\text{dc},i} = 2U_{\text{dc},i} \cdot I_{\text{dc},i} \tag{5.3.16}$$

By combining both equations, yields:

$$P_{\text{dc},i} = 2U_{\text{dc},i} \cdot \sum_{\substack{j=1 \\ j \neq i}}^{n} Y_{\text{dc},ij}\left(U_{\text{dc},i} - U_{\text{dc},j}\right) \tag{5.3.17}$$

The Jacobian matrix can be formed as follows:

$$\left(U_{\mathrm{dc},j}\frac{\partial P_{\mathrm{dc},i}}{\partial U_{\mathrm{dc},j}}\right)^{(j)} = -2U_{\mathrm{dc},i}^{(j)}Y_{\mathrm{dc},ij}U_{\mathrm{dc},j}^{(j)} \tag{5.3.18}$$

$$\left(U_{\mathrm{dc},i}\frac{\partial P_{\mathrm{dc},i}}{\partial U_{\mathrm{dc},i}}\right)^{(j)} = P_{\mathrm{dc},i}^{(j)} + 2U_{\mathrm{dc},i}^{(j)}U_{\mathrm{dc},i}^{(j)}\sum_{\substack{j=1\\ j\neq i}}^{n} Y_{\mathrm{dc},ij} \tag{5.3.19}$$

Similar to the AC systems, one of the converters will be responsible for regulating the DC bus voltage; this converter is called DC slack bus. The equations and terms corresponding to the slack bus are removed since its voltage is known prior to the DC network power flow. Since all the AC quantities are known from the previous iteration including $P_{\mathrm{ac}}^{(k)}$, the power flow from the DC side is given as:

$$P_{\mathrm{dc},i}^{[k]} = P_{\mathrm{ac},i}^{[k]} \pm P_{\mathrm{Loss\_conv}i} \tag{5.3.20}$$

where $P_{\mathrm{Loss\_conv}i}$ is the converter losses; the sign in the above equation is dependent on the power flow direction. It should be noticed that the P_{ac} is known from the AC power flow. In the next step, the DC power flow can be solved using the NR method.

$$\left(U_{\mathrm{dc}}\frac{\partial P_{\mathrm{dc}}}{\partial U_{\mathrm{dc}}}\right)^{(j)} \cdot \frac{\Delta U_{\mathrm{dc}}{}^{(j)}}{U_{\mathrm{dc}}} = \Delta P_{\mathrm{dc}}^{(j)} \tag{5.3.21}$$

$$\Delta P_{\mathrm{dc},i}^{(j)} = \begin{cases} P_{\mathrm{dc},i}^{(k)} - P_{\mathrm{dc},i}\left(U_{\mathrm{dc}}^{(j)}\right) & \forall i \leq k \\ -P_{\mathrm{dc},i}\left(U_{\mathrm{dc}}^{(j)}\right) & \forall k \leq i \leq n \end{cases} \tag{5.3.22}$$

where $\Delta P_{\mathrm{dc}}^{(j)}$ is the power mismatch vector for iteration j.

5.3.7 Pareto-Based Multi-Objective Optimization Problem

In this section, the formulation of the multi-objective problem will be covered and discussed in detail [60–64].

Problem Objectives

Minimize the Voltage Fluctuations (ΔV)

1. **For AC buses**

$$\Delta V_{\mathrm{ac},i} = |V_i| - V_{i,\mathrm{nominal}} \quad \forall i \in \{1,2,\ldots,N_i\} \tag{5.3.23}$$

where V_i is the AC voltage at bus i, N_i is the total number of AC buses in the system.

2. **For DC buses**

$$\Delta V_{\text{dc},j} = |V_j| - V_{j,\text{nominal}} \quad \forall j \in \{1,2,\ldots,N_j\} \tag{5.3.24}$$

where V_j is the AC voltage at bus j, N_j is the total number of DC buses in the system.

Overall voltage deviation is calculated as:

$$\Delta V = \max\{\Delta V_{\text{ac},i}, \Delta V_{\text{dc},j}\} \tag{5.3.25}$$

Minimize the Frequency Fluctuations (Δf)

$$\Delta f = |f - f_0| \qquad f_0 = 60, \text{assuming 60 Hz system} \tag{5.3.26}$$

Decision Variables

The decision variables are the variables that can be changed or controlled by the optimization algorithm to change the results and modify the architecture for the objectives to be met. In the literature, the decision variables are called control variables. The decision variables are the following:

Reconfiguration Switches A variable X_s is assigned to each switch s to represent its state, where:

$$X_s = \begin{cases} 0 & \text{if switch s is OFF} \\ 1 & \text{if switch s is ON} \end{cases} \tag{5.3.27}$$

Candidate Architecture A variable X_A is assigned to each architecture A, where there are n candidate architectures in the search space. For example:

$$X_A = \begin{cases} 1 \text{ for Architecture one} \\ 2 \text{ for Architecture two} \\ \vdots \\ n \text{ for Architecture } n \end{cases} \tag{5.3.28}$$

Energy Storage Sizing A variable y_j is assigned to each bus j to control the size of the energy storage connected to it, where $y_j = 0$ indicates no energy storage installed on this bus [65, 66].

$$E^j_{\text{ES}\_h} = y^h_j \times \text{certain step size } (h) \tag{5.3.29}$$

E^j_{ES} is the energy storage installed at bus j. h is indicating the type of the energy storage installed on this bus. Where:

$$h \in \{1,2,3\}$$

1. for lead-acid battery
2. for ultra-capacitors
3. for flywheel

It should be noticed that any other types of energy storage can be implemented given the model and constraints of that type. What is presented here is an optimization platform, which can accept any new parameters.

Constraints

In order to make sure that the optimal solution is within the acceptable operation and design limits for ship power systems, certain constraints are applied to avoid nonoptimal or nonpractical solutions. Further, it should be kept into consideration that the dynamic operation of the system should be within the limits specified by the standards.

Voltage Limits The voltage on all buses should not exceed the values specified by standards; in this study, MIL-Std 1399 section 300 is considered for AC components and section 390 is considered for DC components. Any other standard can be used. These constraints are the limits (boundaries) based on which the algorithm will accept or reject a solution. These constraints are set according to relevant standards such as MIL-Std 1399. Figure 5.3.7 depicts the boundaries specified for 440 V, 60 Hz systems by MIL-Std 1399.

$$V_{\min} \leq V_j \leq V_{\max} \qquad \forall j \in \{1, 2, \ldots, N\} \tag{5.3.30}$$

VSC Limits The most important constraints to be considered are the limits of the VSC, the operation point of the converter on the *P-Q* plane should be situated within the safe operation region. As shown in Figure 5.3.8, this region is defined by three boundaries [59, 67–70]. These boundaries are the following:

The first limiting factor is the maximum current through the insulated gate bipolar transistor (IGBT) valves. This leads to forming the maximum MVA circle in the *P-Q* plan.

$$|I_{\text{conv}}| \leq |I_{\text{conv\_max}}| \tag{5.3.31}$$

$$I_{\text{conv}} = \frac{\sqrt{P_{\text{conv}}^2 + Q_{\text{conv}}^2}}{\sqrt{3} \cdot U_{\text{conv}}} \tag{5.3.32}$$

From (5.3.32), the converter current can be calculated as a function of the active, reactive powers and the terminal voltage of the converter. $I_{\text{conv\_max}}$ is determined mainly based on the used IGBT characteristics. The converter active power is given by:

$$P_{\text{conv}} = P_{\text{dc}} - P_{\text{loss}} \tag{5.3.33}$$

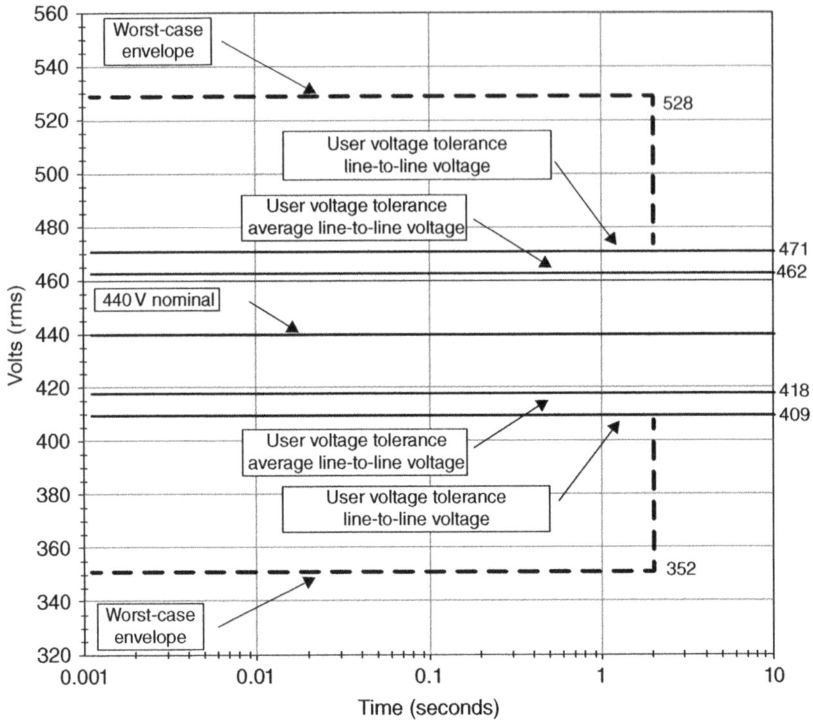

Figure 5.3.7 Voltage limits specified by MIL-Std 1399, section 300.

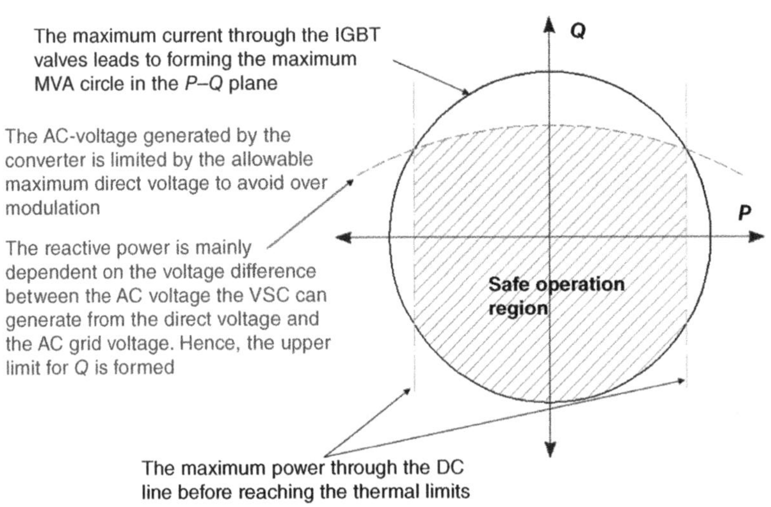

Figure 5.3.8 Safe operation region for VSC.

The losses in the converter are determined according to a generalized loss equation:

$$P_{\text{loss}} = a + b \cdot I_{\text{conv}} + c \cdot I_{\text{conv}}^2 \tag{5.3.34}$$

The second limiting factor is the maximum voltage. The AC-voltage generated by the converter is limited by the allowable maximum direct voltage. The reactive power is mainly dependent on the voltage difference between the AC voltage the VSC can generate from the direct voltage and the AC grid voltage (the dashed arc).

$$Q = \frac{|U_{\text{f}}|(|U_{\text{c}}|\cos(\delta) - |U_{\text{f}}|)}{X} \tag{5.3.35}$$

where X is Img (Z_{L}).

The third limit is the maximum current through the cable (the vertical dashed lines).

$$|I_{\text{dc},ij}| \leq |I_{\text{dc},ij\_\max}| \tag{5.3.36}$$

where $I_{\text{dc},ij}$ is the DC current between points i and j. $I_{\text{dc},ij\_\max}$ is the maximum current capacity. However, in some design practices the cable capacity is selected to be more than the converter capacity. Thus, these limits can be omitted. The working point must be placed within the P-Q capability chart of the converter, i.e. any solution yielding an operating point outside the shaded area will be rejected.

Frequency Limits The frequency of the AC side is considered as an important measure for the system stability. High frequency modulation can cause outage for generation units due to asynchronization. One of the major constraints applied is the frequency limitations. In this study, frequency envelopes specified by MIL-Std 1399 are followed. Figure 5.3.9 shows that the permitted frequency fluctuation is within ±0.5% which corresponds to ±2 Hz in 400 Hz systems or ±0.3 Hz in 60 Hz. However, the standard permits frequency fluctuations within ±1.5% but for 0.3 second only. Any fluctuations pertain longer than 0.3 second should be damped to be with the continuous range (±0.5%). So, the frequency of the system (f) should be:

$$f_{\min} \leq f \leq f_{\max} \tag{5.3.37}$$

Energy Storage Placement Constraints To avoid over distributing the energy storage all over the system, the number of the buses where energy storage can be installed should be limited [65].

$$\sum_i u_i^{\text{ES}} \leq U_{\text{ES}} \tag{5.3.38}$$

$$u_i^{\text{ES}} = \begin{cases} 0 \text{ No energy storage installed on bus } i \\ 1 \text{ Energy storage installed on bus } i \end{cases} \quad \forall i \in \text{SB} \tag{5.3.39}$$

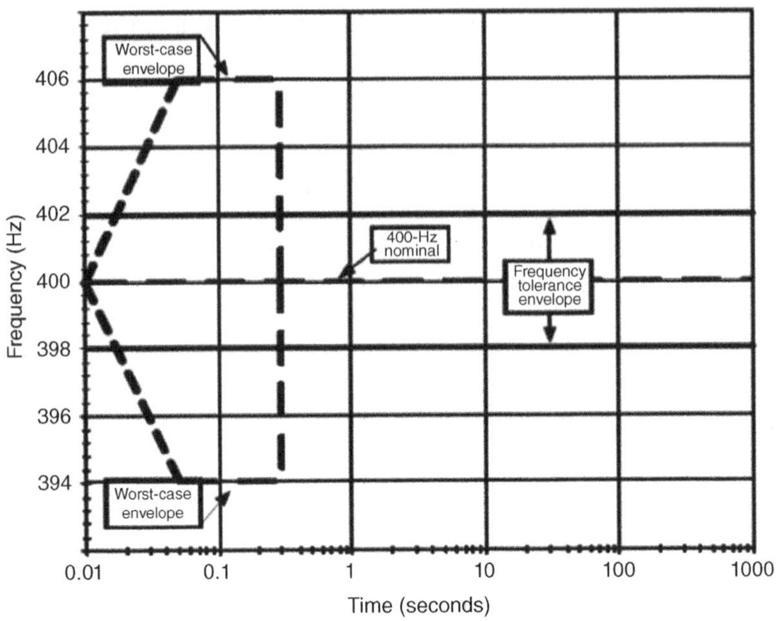

Figure 5.3.9 Frequency envelopes for the 400 Hz systems in marine applications.

where U_{ES} is the maximum number of buses allowed to install energy storage on. SB is the set of system buses. It is important to consider this constraint where the energy storage elements should not be dispersed over the entire system. Moreover, to avoid dispersing small energy storage units and installing very large units, the installed energy storage should be bounded [71–74]. So, the total power for energy storage installed on each bus should be constrained as follows:

$$S_i^{\mathrm{ES}} = x_i \times \text{certain discrete size} \tag{5.3.40}$$

$$u_i^{\mathrm{ES}} S_{i,\min}^{\mathrm{ES}} \le S_i^{\mathrm{ES}} \le u_i^{\mathrm{ES}} S_{i,\max}^{\mathrm{ES}} \tag{5.3.41}$$

The stored capacity of the energy storage can be constrained as:

$$E_i^{\mathrm{ES}} = y_i \times \text{certain discrete size} \tag{5.3.42}$$

$$u_i^{\mathrm{ES}} E_{i,\min}^{\mathrm{ES}} \le E_i^{\mathrm{ES}} \le u_i^{\mathrm{ES}} E_{i,\max}^{\mathrm{ES}} \tag{5.3.43}$$

It should be noted that x_i and y_i are controlled variables.

Energy Storage Sizing Constraints The total size of the installed energy storage in the entire system should not exceed certain limits.

$$\sum_i S_i^{\mathrm{ES}} \le S_{\max} \tag{5.3.44}$$

$$\sum_i E_i^{\mathrm{ES}} \leq E_{\max} \tag{5.3.45}$$

where $S_{\max}$ and $E_{\max}$ are the maximum power and energy of the installed energy storage in the entire system, respectively.

Energy Storage Operation Constraints (Battery) When sizing some types of energy storage, their operation limits should be taken into consideration. For example, some types of batteries are not allowed to be discharged below 20%, i.e. the State of Charge (SoC) should not go below 20%. When designing a battery system to supply pulsed loads, this factor should be considered to avoid high Depth of Discharge (DoD) over the pulsed load cycle. For the given pulsed load profile shown in Figure 5.3.10, t_{on} is the on pulse duration, t_{off} is the off pulse duration while $\Delta t = t_{\mathrm{on}} + t_{\mathrm{off}}$. When the ES is charging off the pulse ($P(t) > 0$):

$$\eta_{\mathrm{c}} P_i(t) t_{\mathrm{off}} \leq K_{\mathrm{c}} E_i^{\mathrm{ES}} \tag{5.3.46}$$

where η_{c} is the charging efficiency, K_{c} is a factor $(0 - 1)$ representing the maximum amount of energy that can be supplied to the ES element during a charging cycle. When the ES is discharging during the pulse ($P(t) < 0$):

$$\mathrm{SoC}(t_0 + t_{\mathrm{on}}) = \mathrm{SoC}(t_0) + P_i(t) t_{\mathrm{on}} / E_i^{\mathrm{ES}} \cdot \eta_{\mathrm{Dis}} \tag{5.3.47}$$

$$\mathrm{SoC}(t_0 + t_{\mathrm{on}}) \geq 25\% \tag{5.3.48}$$

$$-S_i^{\mathrm{ES}} \leq P_i(t) \leq S_i^{\mathrm{ES}} \tag{5.3.49}$$

Energy Storage Operation Constraints (Flywheel) The operation constraints for a flywheel differ from that of the other types since flywheel is characterized with high self-discharge rate. Thus, the self-discharge rate should be considered [75]. The stored energy in the flywheel is given by:

$$E_{\mathrm{fw}} = \frac{1}{2} \cdot J \cdot \omega_{\mathrm{fw}}^2 \tag{5.3.50}$$

where

E_{fw} is the stored energy in the flywheel

ω_{fw} is the flywheel speed

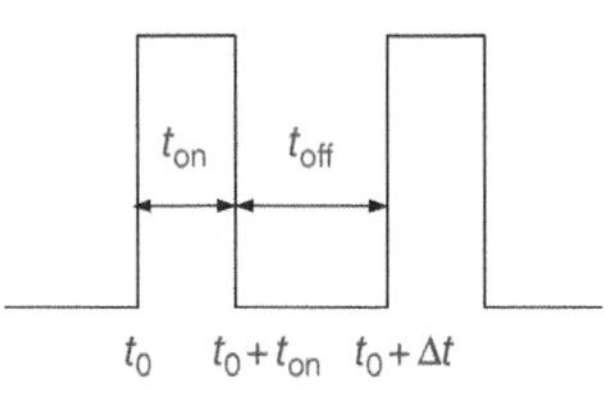

Figure 5.3.10 Pulsed load profile.

When the flywheel is discharging during the pulse ($P(t) < 0$):

$$E_{\text{fw}}(t_0 + t_{\text{on}}) = (1 - \sigma_{\text{dsr}})\, E_{\text{fw}}(t_0) + P(t) \cdot t_{\text{on}}/\eta_{\text{Dis}} \tag{5.3.51}$$

$$E_{\text{fw}}(t_0 + t_{\text{on}}) \geq K_{\text{d,fw}} \cdot E_{\text{fw}}^{\max} \tag{5.3.52}$$

where σ_{dsr} is the self-discharging rate, $P(t)$ is the pulsed load power, and η_{Dis} is the discharging efficiency for the flywheel. K is the lowest allowed SoC for the flywheel. E is the maximum energy that can be stored in the flywheel.

The power flow through each branch should not exceed the branch limits to avoid violation of thermal limits.

$$|S_i| < |S_{i\ \max}| \quad \forall i \in \{1, 2, 3, \ldots, N_{\text{L}}\} \tag{5.3.53}$$

where i is the branch index, N_L is the total number of transmission lines in the architecture. S_i is the calculated apparent power flow through branch i. $S_{i\max}$ is the maximum apparent power flow. $Si_{\max}$ is calculated based on the branch maximum thermal capacity as specified in *IEEE Recommended Practice for Electrical Installations on Shipboard*, IEEE Std-45 (2002).

Decision-Making Criterion

The complex MOOPs like the one under study are non-convex problems, i.e. the generated PF contains more than a single optimal solution. Some of these solutions are extreme which achieve a single objective and do not fully meet the other objectives. By referring to Figure 5.3.11, the extreme solutions are highlighted. The other type of solutions is the Pareto efficient solutions. These solutions are of interest and we target to select them. Thus, adopting an appropriate decision-making criterion is mandatory to select the most optimal solution. Since, the installed equipment weight is critical in marine environment. This decision criterion will be based on selecting the solution that achieves the minimum total weight of the installed energy storage units on the shipboard. After, solving the MOOP and generating the PF, the physical weight of the installed energy storage for each solution is evaluated.

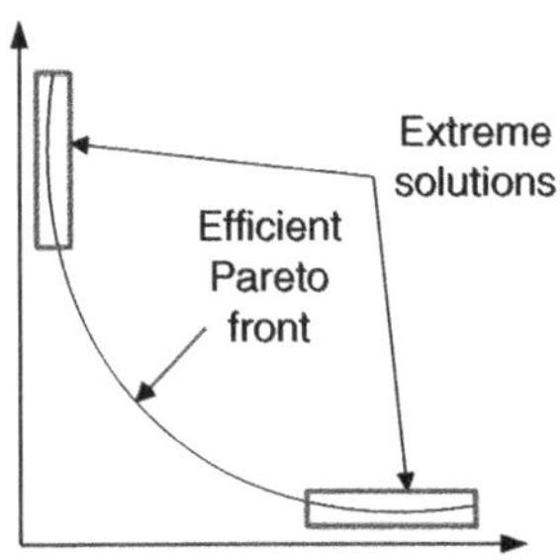

Figure 5.3.11 Pareto front and selection of efficient Pareto solutions.

$$W_{ES} = \sum_{h=1}^{3}\sum_{j=1}^{N_b} E^j_{ES\_h} \times \frac{1}{D_h} \tag{5.3.54}$$

where W_{ES} is the total weight of the installed energy storage units, h is representing the type of energy storage, and the total types are three: j is the bus number, N_b is the total number of buses, and D_h is the specific energy (energy density per unit mass). It is desired to minimize the weight of the installed energy storage, i.e. min(W_{ES}). This criterion can be flexibly used by the designer or the decision maker to select the most optimal solution that meets all the design perspectives for a certain applications or conditions. Figure 5.3.12 depicts a flow chart for the entire optimization procedure.

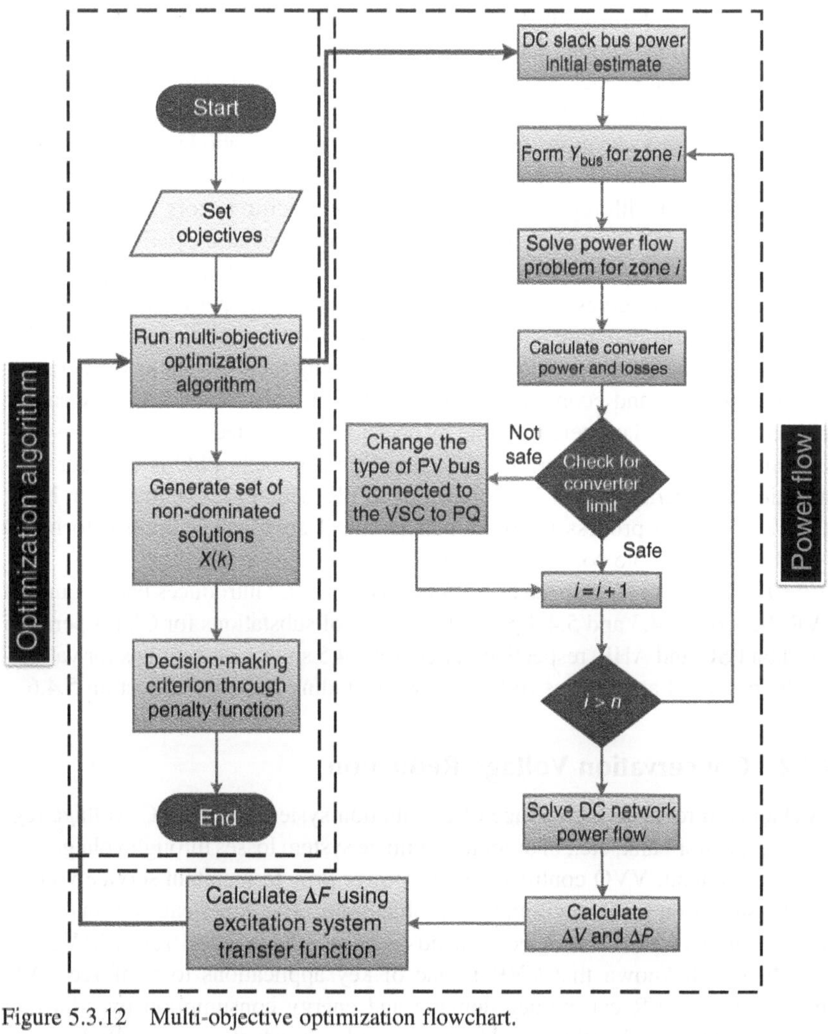

Figure 5.3.12 Multi-objective optimization flowchart.

5.4 CONSERVATION VOLTAGE REDUCTION PLANNING

Kyungsung An[1] *and Kyeon Hur*[2]
[1]*SK Telecom, Seoul, Korea*
[2]*Yeonsei University, Seoul, Korea*

5.4.1 Introduction

CVR is a technique to decrease peak demand and save energy by controlling the voltage level within acceptable voltage range on a distribution system [76]. Many utilities have pursued DSM to improve energy efficiency on a distribution system such as volt/var optimization (VVO), CVR, and DR [77, 78]. It is enough to operate CVR operation without additional equipment because it depends on load tap changer (LTC) or voltage regulator controls. Moreover, nonintrusive operation of CVR does not make uncomfortable experiences for consumer. Therefore, it is expected to execute more frequent CVR operation in modern electrical grid because of its numerous operational benefits. The impacts of large-scale CVR should be studied with regard to numerous utility-specific factors in order to provide more reliable and predictable CVR performance because existing practice of CVR study is usually based on CVR factor analysis at each substation or feeder.

This chapter serves a framework for evaluating the performance of large-scale CVR in grid planning and operations. This work supports decision-making for ranking substations based on multiple conflicting factors such as CVR factor, network analysis, and economic impact. Moreover, substations for CVR can be selected at a particular operating time by decision-making techniques such as heuristic and hierarchical approaches. As a heuristic approach, PSO is used for minimization of operation cost by connecting CVR with ED. On the other hand, analytic hierarchy process (AHP) for hierarchical approach is applied to MOO by balancing multiple conflicting criteria.

This chapter is organized as follows: Section 5.4.2 introduces background of CVR. Sections 5.4.3 and 5.4.4 present selection of substations for CVR operation based on PSO and AHP, respectively. Section 5.4.5 shows case studies for validating the proposed approaches followed by concluding remarks in Section 5.4.6.

5.4.2 Conservation Voltage Reduction

VVO aims to regulate the voltage of distribution system using LTC, voltage regulator, capacitor bank, etc., in order to minimize system losses through voltage profile improvement. VVO controls reactive power flow to maintain service voltage along distribution feeders within permissible limits. The American National Standards Institute (ANSI) C84.1 recommends service voltage range from 120 V ± 5% [79]. It is well known that CVR is one of key applications to reinforce VVO performances. CVR can reduce demand and energy consumption by adjusting the voltage level of distribution system. The voltage reduction is usually between

2 and 5% of the present voltage level. Many utilities have made efforts to execute CVR using real field test data so as to examine the amount of load reduction from lowering voltage level, and 0.3–1% load reduction can be achieved from 1% voltage reduction [80]. Recently, CVR operation has been more important because it is projected that demand and energy costs continue increasing, countries have been forced to reduce carbon dioxide emissions, and more operating reserve can be secured by CVR in power system with high penetration of renewable energy resources.

CVR Factor

The impact of CVR at each substation can be represented by conservation voltage reduction factor (CVR_f) as follows:

$$\text{CVR}_\text{f} = \frac{\Delta P}{\Delta V} = \frac{P_{\text{CVR\_off}} - P_{\text{CVR\_on}}}{V_{\text{CVR\_off}} - V_{\text{CVR\_on}}} \tag{5.4.1}$$

where ΔP and ΔV are the change in load consumption and voltage, respectively.

The CVR factor is not fixed because load behavior varies hourly and seasonally. ZIP load model, which consists of the sum of constant impedance (Z), constant current (I), and constant power (P) components, is generally used to represent load composition including voltage-dependency as follows:

$$\begin{aligned} P &= P_0\left(p_z\left(\frac{V}{V_0}\right)^2 + p_i\left(\frac{V}{V_0}\right) + p_p\right) \\ Q &= Q_0\left(q_z\left(\frac{V}{V_0}\right)^2 + q_i\left(\frac{V}{V_0}\right) + q_p\right) \end{aligned} \tag{5.4.2}$$

where P_0 and Q_0 are the active and reactive power at nominal voltage V_0, and p_z, p_i, p_p and q_z, q_i, q_p are coefficients for active power and reactive power, respectively.

Current CVR practices command accessible substations to reduce the voltage level based on operator's experience, during peak time or emergency situation. However, it is sometimes a radical measure because it could cause unforeseen (or adverse) problems and implementation costs due to sudden and considerable modification of network. Sufficient operating reserve can be secured by CVR operations at only several substations.

For successful CVR execution, operators should take account of operation conditions and uncertainty of grid through online energy management system (EMS). Online EMS conducts calculation of CVR factor and power system sensitivity analysis to reinforce flexibility and predictability of grid planning and operations. Then, substations for CVR operation to reduce desired number of loads can be determined through decision-making techniques such as heuristic or hierarchical approaches. The performance of CVR can be improved by using smart monitoring devices, communication and control schemes on the distribution system.

5.4.3 CVR Based on PSO

Background of PSO

PSO is one of the heuristic optimization techniques based on swarm intelligence developed by Eberhart and Kennedy in 1995, and inspired by the behaviors of bird flock and fish school [32]. Each particle's movement as conceptually illustrated in Figure 5.4.1 can be represented by the combination of self-experiences and social-experiences. PSO can handle nonlinear optimization problems with continuous and discrete variables. Moreover, PSO for mixed-integer nonlinear optimization problems presented for incorporation of continuous and discrete PSO [81]. The updates of each particle's velocity and position can be represented by the following equations:

$$v_i^{k+1} = wv_i^k + c_1r_1 \times \left(P_{i,\text{best}} - x_i^k\right) + c_2r_2 \times \left(G_{\text{best}} - x_i^k\right) \qquad (5.4.3)$$

$$x_i^{k+1} = x_i^k + v_i^{k+1} \qquad i = 1, \ldots, N \qquad (5.4.4)$$

$$w = w_{\max} - \frac{w_{\max} - w_{\min}}{k_{\max}} \times k \qquad (5.4.5)$$

where

v_i: the velocity of particle i

w: the inertia weight

c_1, c_2: the acceleration coefficients

r_1, r_2: the random numbers between 0 and 1

x_i: the position of particle i

N: the number of decision variables

The velocity update equation in (5.4.3) consists of three terms. The first term is the inertial velocity that is keeping the particle's movement in the same direction with previous velocity. The value of the inertia weight is generally between 0.4 and 0.9, which can be modified according to iteration k. High value of the inertia weight at the beginning of PSO process is expected to explore throughout search space, while low value at the end of the process is forced to move toward convergence of swarm's best position. The second term, which is referred to as self-experience, presents the movement to each particle's personal best. On the other hand, the third

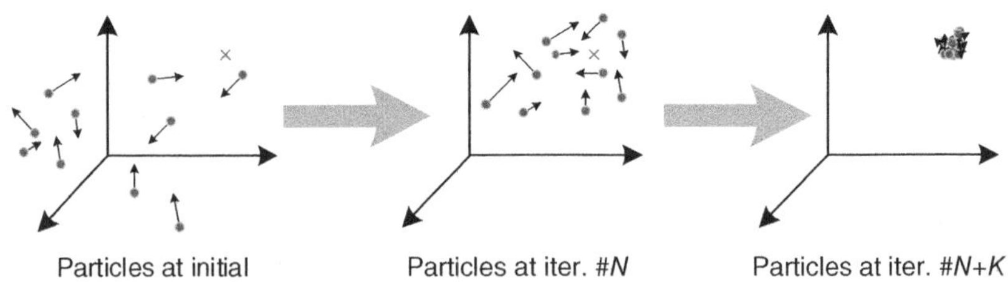

Figure 5.4.1 Concept of PSO.

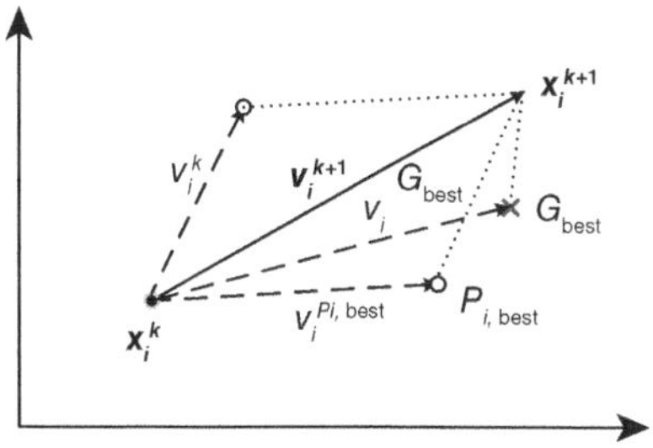

Figure 5.4.2 Modification of particle's velocity and position.

term, which is referred to as social-experience, encourages particles to fly toward swarm's global best so far. The value of acceleration coefficients are both 2.0, and it is validated to be pertinent in power system problems [82, 83]. Figure 5.4.2 shows a movement of particle i based on (5.4.3) and (5.4.4).

The process of PSO can be written as follows:

Step 1: Initialize particles with random position x_i and velocity v_i.

Step 2: Evaluate object function $J(x_i)$.

Step 3: Update particle's personal best and swarm's global best. If the $J(x_i)$ is better than current personal best $P_{i,\text{best}}$, the personal best $P_{i,\text{best}}$ can be changed to the current value $J(x_i)$. Then, if the current personal best $P_{i,\text{best}}$ is better than global best G_{best}, the global best G_{best} can be replaced by current personal best $P_{i,\text{best}}$.

Step 4: Modify particle's position x_i and velocity v_i using (5.4.3) and (5.4.4).

Step 5: Check the terminal condition. The process will continue finding the global solution until the terminal condition is satisfied.

Figure 5.4.3 shows the flowchart of PSO as described above.

Binary Particle Swarm Optimization

Discrete binary particle swarm optimization (BPSO) is developed for binary problems, which will make decisions such as yes or no, true or false, or on or off. In BPSO, the velocity of particle is updated by (5.4.3), and the position of particle is updated as follows:

$$x_i^{k+1} = \begin{cases} 1 & \text{if } \rho_i^{k+1} < \text{sig}\left(v_i^{k+1}\right) \\ 0 & \text{otherwise} \end{cases} \tag{5.4.6}$$

where ρ_i^{k+1} is uniform distribution random number in the range [0, 1], and $\text{sig}\left(v_i^{k+1}\right)$ is the sigmoid function, which indicates probabilities of particle's tendency toward one or zero, as follows:

$$\text{sig}\left(v_i^{k+1}\right) = \frac{1}{1 + \exp\left(-v_i^{k+1}\right)} \tag{5.4.7}$$

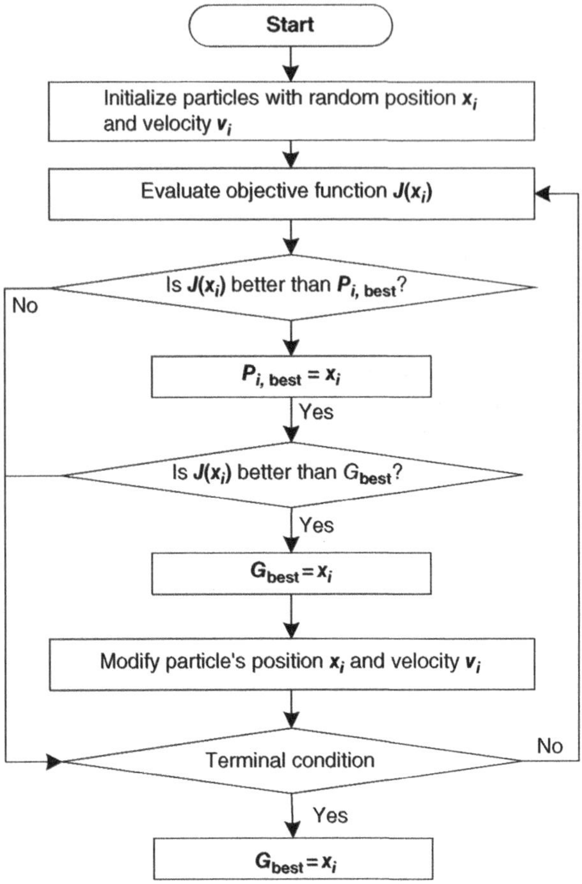

Figure 5.4.3 PSO flowchart.

CVR Using BPSO

CVR using BPSO finds best solution that minimizes operating cost by connecting CVR with ED in order to achieve the desired load reduction by voltage reduction. The state of each substation for CVR can be determined by BPSO. Substations with state 1 reduce their voltage level for load reduction, while substations with state 0 maintain their voltage level. The formulation of ED for CVR can be represented as following equations:

$$\min\ F_{\mathrm{T}} = \sum_{i=1}^{m} F_i(P_i) = \sum_{i=1}^{m} a_i + b_i P_i + c_i P_i^2 \tag{5.4.8}$$

such that:

$$\sum_{i=1}^{m} P_i = P_{\mathrm{D}}^{\mathrm{new}} + P_{\mathrm{L}} \tag{5.4.9}$$

$$\Delta P_{\text{target}} = P_{\text{D}}^{\text{new}} - P_D = \sum_{j=1}^{k} s_j \cdot P_{j,\text{CVR}} \tag{5.4.10}$$

$$\max\left(P_i^{\min}, P_i^0 - \text{DR}_i\right) \le P_i \le \min\left(P_i^{\max}, P_i^0 + \text{UR}_i\right) \tag{5.4.11}$$

$$|P_{j,k}| \le P_{j,k}^{\max}, \qquad V_i^{\min} \le V_i \le V_i^{\max} \tag{5.4.12}$$

where

P_i: generation of unit i

P_{D}: total system demand

P_{L}: total network losses

ΔP_{target}: the amount of desired target load reduction

$P_{\text{D}}^{\text{new}}$: new total demand after CVR

s_j: state variable of substation j

$P_{j,\text{CVR}}$: load reduction of substation j by CVR

$P_i^{\min}$, $P_i^{\max}$: power output limit of unit i

DR_i, UR_i: down/up ramp limit of unit i

$P_{j,k}$: real power flow of line j, k

$P_{j,k}^{\max}$: real power limit flow of line j, k

V_m: voltage at bus m

$V_i^{\min}$, $V_i^{\max}$: voltage limit at bus m

5.4.4 CVR Based on AHP

Background of AHP

AHP, developed by Saaty, is an effective technique for complex decision-making problems, and aids to make the best decision in multiple criteria [84]. The final ranking of alternatives can be determined by incorporating and analyzing each judgment matrix, which consists weights of alternatives through relative evaluation of operator's judgment and pairwise comparison. The AHP is a flexible and strong tool because it can handle not only quantitative problems but also qualitative problems. The process of the AHP can be written as follows [85]:

Step 1: Set up a hierarchy model including decision objective (H1), criteria (H2) for assessing alternatives, and alternatives (H3).

Step 2: Build a judgment matrix for each criterion. The value of elements in the judgment matrix reflects the relative importance between every pair of factors as shown in Table 5.4.1. The judgment matrix A can be written as follows:

TABLE 5.4.1 Intensity Scale of Importance

Intensity of Importance	Definition
1	Equal importance
3	Weak importance of one over another
5	Strong importance
7	Very strong importance
9	Absolute importance
2, 4, 6, 8	Median of both neighboring judgments

$$A = \begin{pmatrix} w_1/w_1 & w_1/w_2 & \cdots & w_1/w_N \\ w_2/w_1 & w_2/w_2 & \cdots & w_2/w_N \\ \vdots & \vdots & \ddots & \vdots \\ w_N/w_1 & w_N/w_2 & \cdots & w_N/w_N \end{pmatrix} \tag{5.4.13}$$

where w_i/w_j represents the relative importance of the ith alternative compared with the jth alternative.

Step 3: Calculate the eigenvector of each judgment matrix for weighting factors. It would be time-consuming to precisely calculate. Therefore, two approximate approaches such as root method and sum method can be adopted to determine weighting factors.

Step 4: Make the final hierarchy ranking.

CVR Using AHP

AHP for hierarchical approach is applied to MOO by balancing multiple conflicting criteria: CVR factor, power transfer distribution factor (PTDF), and voltage sensitivity factor (VSF). These analyses show influences on network after load reduction by CVR at each substation. The CVR factor analysis is related to the number of substations for CVR operation, while PTDF and VSF analyses explain the securement of operating reserve and the improvement of voltage profile according to location of CVR, respectively. Figure 5.4.4 shows the hierarchy model for CVR operation.

5.4.5 Case Studies for CVR in Korean Power System

CVR in Korean power system (2013 summer peak case) is implemented using PSO and AHP. In the case, total load is 77 577 MW and 29 653 Mvar. Industrial loads do not contribute to load reduction from voltage reduction because motor load is represented by constant power loads. Total load for CVR candidates is 56 731 MW/20 909 Mvar, and the number of substations is 655. The amount of potential load reduction by CVR is about 850 MW when all substations

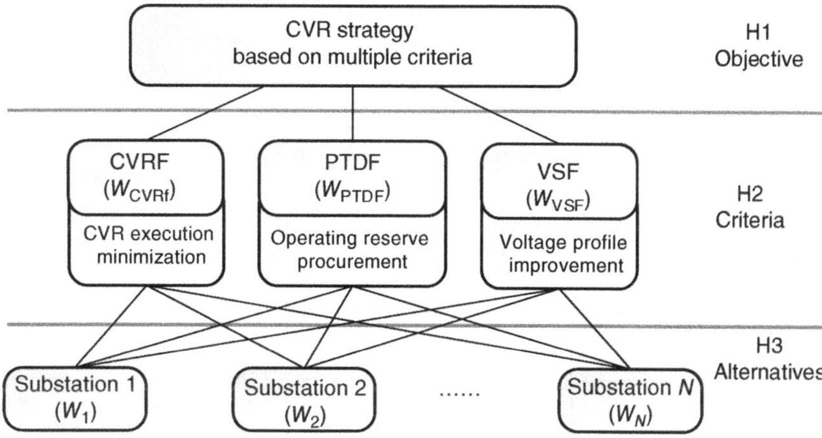

Figure 5.4.4 A hierarchy model for CVR.

execute CVR with 2.5% voltage reduction. It is assumed that target load reduction is 400 MW.

Simulation Result of CVR Using PSO

CVR using PSO not only minimize operation cost but also tends to reduce total generation and system loss. Swarms (CVR candidates) explore search space to find global solution (selection of substations) for CVR operation as shown in Figures 5.4.5 and 5.4.6. Table 5.4.2 shows the simulation results using PSO compared with base case.

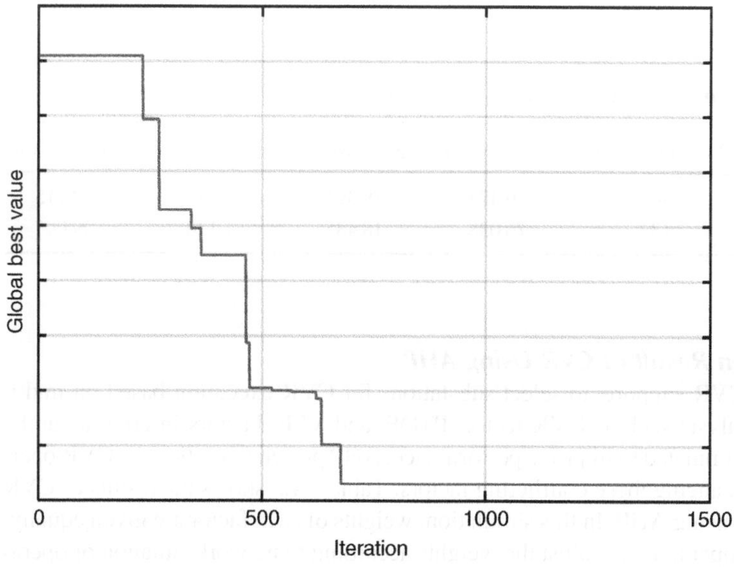

Figure 5.4.5 Global best value at each iteration.

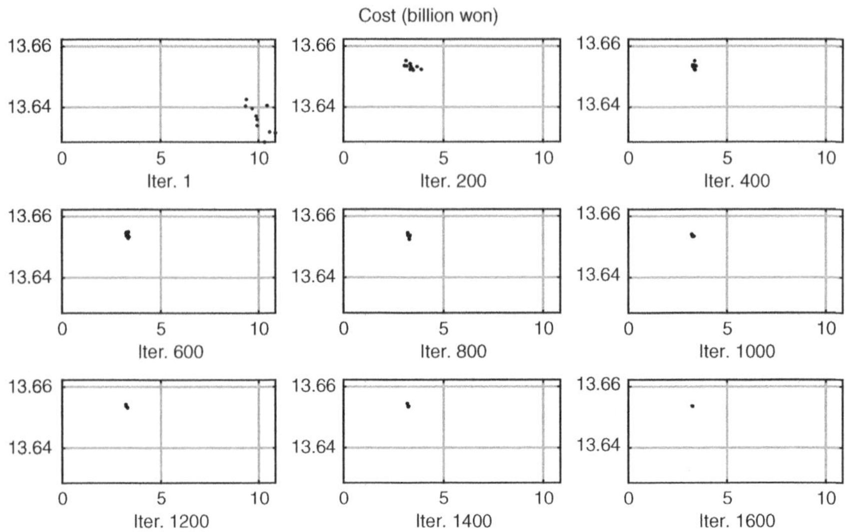

Figure 5.4.6 Particle's movement for CVR.

TABLE 5.4.2 Results of CVR Simulation Using PSO

	No. of CVR On	Operating Cost (×1000 Won)
Base case	—	13 730 320
PSO	142	13 651 340

TABLE 5.4.3 Results of CVR Simulation Using AHP

	No. of CVR On	Pgen (MW)	Qgen (Mvar)	Ploss (MW)	Qloss (Mvar)
Base case	—	76 475	19 391	1 155	29 715
AHP	133	76 038	18 485	1 120	29 097

Simulation Result of CVR Using AHP

AHP for CVR supports to select substations for CVR execution based on multi-criteria analysis such as CVR factor, PTOF, and VSF. Factors in criteria can be added or eliminated to improve performance. AHP presents solution for CVR operation by balancing these conflicting factors. Table 5.4.3 shows the results of CVR simulation using AHP. In this simulation, weights of each factor are given equally. However, operator can adjust the weights according to network situation or operator's intention.

5.4.6 Conclusion

This chapter introduced CVR planning and operation using PSO and AHP. The CVR using PSO finds best solution to minimize operation cost by incorporating the ED. CVR using AHP determines substations for CVR by incorporating multiple factors. It is expected that these approaches for CVR can improve operation flexibility and reduce uncertainty of large-scale CVR operation.

5.5 DYNAMIC DISTRIBUTION NETWORK EXPANSION PLANNING WITH DEMAND SIDE MANAGEMENT

Ngoc Hoang Luong, Peter A. N. Bosman, Marinus O. W. Grond, and Han La Poutré

Centrum Wiskunde and Informatica, Amsterdam, The Netherlands

5.5.1 Introduction

Static Planning and Dynamic Planning

DNEP [26, 86–88] is the central problem that distribution network operators (DNOs) have to solve to find the optimal plan for the construction and expansion of electricity grids. DNEP can be generally categorized into static planning and dynamic planning. In static DNEP, the problem formulation involves the questions *what* kinds of network reinforcements (i.e. physical asset installations) should be performed and *where* these enhancements should be done so that the distribution network can operate properly regarding the future power demand. In addition to these two issues, the dynamic DNEP formulation involves the question *when* these expansion activities should be carried out in multiple years toward a planning horizon. Solving the dynamic DNEP thus results in an expansion plan with a detailed multistage installation schedule, usually on a year-to-year (or stage-to-stage) basis.

Asset Investment and Demand Side Management

In the transition toward smart grids and renewable energy integration, DNOs face many challenges when tackling the DNEP problem. Traditionally, DNOs need to ensure that their network assets (e.g. cables, transformers) have enough capacity to support the required power flows that satisfy the power demands, especially in the worst-case scenario of peak loads. Situations of insufficient capacity can lead to overloads on network assets, which are harmful to network safety (e.g. cable failures) and the continuity of electricity supply (e.g. blackouts or load shedding). However, constructions of new facilities become increasingly expensive such that expansion plans like installing new underground cables in highly urbanized cities are often not favorable. Moreover, peak loads can become high but last for short durations compared to the lower base loads of much longer periods. This is even more the case with the future increase of RES. This fact means that if DNOs invest in physical assets just to satisfy (higher) peak loads, their networks will be severely underutilized for most of the time. Such overcapacity is uneconomic and undesirable. In this work, we assume that, in future sustainable energy systems, DNOs are allowed to directly be involved in DSM [89–92] activities so that consumers are incentivized to reduce their electricity consumption during peak-demand hours. The overall peak loads on the distribution network can then be brought toward or within the nominal capacities of network assets, aiming at avoiding overloads of network assets. DNOs can thus reduce or postpone costly physical investments in network capacity [91]. Therefore, in this scenario, DNOs have two different options to deal with the increasing load growth: network reinforcements or DSM contribution options. DSM options can be considered as part of operation cost of DNOs as we assume that consumers are financially compensated for changing the behavior to reduce peak loads. It is not necessarily that consumers directly participate in DSM programs. The DSM peak shaving effects can still be achieved through the applications of autonomous software agents in smart homes that can control electricity-consuming devices such that the electricity consumption cost is minimized without significantly compromising on the consumers' comfort [93].

Single-Objective and Multi-Objective Optimization

Beside investment costs, DNOs need to take into account other factors when solving DNEP: i.e. energy losses and electricity supply interruptions. Traditionally, expansion plans of low investment costs are often favored but, in transition toward a sustainable energy future, energy-efficient systems should be considered because of the increasing importance of environmental issues. It is also reported that outages due to faults on medium-voltage distribution (MV-D) networks have the largest contribution to the total System Average Interruption Duration Index (SAIDI) [94]. Such index reflects the reliability of the networks and also customer satisfaction, which are important for DNOs to pay attention to, especially in deregulated energy markets.

It can be seen that in order to draw out an appropriate solution plan for DNEP, DNOs have to consider many criteria, such as asset investment cost, DSM contribution cost, energy loss, and network reliability. One common approach to tackle this multi-criteria problem is to capitalize and aggregate all criteria into a single-objective cost function [26]. Available single-objective (SO) optimization algorithms can then be employed to minimize this integral cost. With this approach, DNOs are however limited to focus only on the financial perspective. Because SO optimizations can only return solutions of minimum cost, they cannot consider other alternatives if they would like to recognize what kinds of compromise on environmental issues and network reliability would need to be made for carrying out the cheapest expansion plan. Furthermore, it is not always possible to monetize nonfinancial terms in a sensible way. For example, conversion of the energy losses of an expansion plan into money requires assumptions about energy prices in the distant future and, in addition, the importance of the environment is hard to monetize. Similarly, capitalization of network reliability requires assumptions about penalty amounts that DNOs pay to consumers when electricity supply interruptions happen due to network failures. These assumptions are very likely to fluctuate during a long-term planning horizon and are difficult to be determined at the beginning. Moreover, it can be argued that even if a monetary conversion is possible, the loss of direct insights into the actual nonmonetary objective values and the trade-offs is undesirable. Instead of the aggregation approach by capitalization, we therefore argue that it is more beneficial to keep these criteria separately and treat them as different objectives when solving DNEP.

These objectives are often conflicting with each other. For example, focusing on investment cost reductions often results in lossy and less reliable systems [26]. Another example is the choice of asset installations or DSM contributions or some combinations of both of them to deal with peak load growth. Delaying assets installations (i.e. reducing investment costs) requires higher contributions of DNOs to DSM options (i.e. increasing operation costs) to keep peak loads within assets' nominal capacities. In financial terms, both are costs for DNOs, but in essence, they are two conflicting directions: physical asset investments versus policy investments. Note that one option can be favorable in one period (stage) but can be unfavorable in another period (stage) considering its long-term effect. An ideal solution that optimizes all these objectives at the same time does not exist. Instead, there exists a *set* of so-called Pareto-optimal solutions [95] which are optimal in the sense that no solution can be improved on some objective without diminishing other objectives. So, they can be seen as optimal trade-off alternatives. The corresponding set of objective value vectors of these Pareto-optimal solutions is called the PF. By inspecting and comparing these trade-offs (along the PF), DNOs can get insights in multiple aspects when making decisions and subsequently choose the Pareto-optimal solution that they prefer. For example, DNOs can immediately see how much energy will be lost during the planning period in case of the lowest-cost solutions and what the trade-offs between losses and cost are. DNOs may also recognize how many (extra) new assets must be installed to (further) improve the network reliability. It is beneficial if DNOs are exposed to these alternatives before choosing a concrete expansion plan.

Multi-Objective Optimization Algorithms

Classical Approaches Earlier approaches to find multiple alternatives in MO optimization share the same working mechanism: (1) using a conversion procedure to change the MOOP instance at hand to some SO optimization problem (SOOP) instance, (2) employing an available SO algorithm to solve that SOOP instance to obtain one solution, (3) updating the conversion procedure with new conversion parameters, (4) going back to Step 1 to begin a new optimization round with a new SOOP instance, or stop if having obtained enough solutions [95]. Two popular conversion procedures are: the weighted sum method and the ε-constraint method [95, 96]. The weighted sum method scales each objective with a coefficient and then adds all the scaled objectives into a composite objective function [95].

The conversion parameters of the weighted sum method are the coefficients. The weighted sum method cannot reach solutions on concave parts (if these exist) of the PF [95, 97]. The ε-constraint method keeps one objective as the master objective and converts other objectives into problem constraints bounded by an ε vector [95]. The conversion parameters of ε-constraint method is the selection of master objective and the ε vector. These classical approaches are time-consuming and inconvenient because multiple optimization processes must be executed [96]. It can be difficult for practitioners to properly choose good conversion parameters in order to obtain a good PF, which are problem-dependent and are often obtained from in-depth analysis. In modern MO optimizations, it is preferred that algorithms require parameter inputs from users as little as possible and are able to return a set of multiple Pareto-optimal solutions in a single optimization run that represent a good PF [95]. Applications of MO optimization in power and energy literature are currently indeed gradually evolving from classical approaches [98, 99] with single-objective reformulation to *true* multi-objective approaches [96], especially multi-objective evolutionary algorithms (MOEAs).

Multi-Objective Evolutionary Algorithms Approach EAs have been shown to be a favorable methodology for industrial optimization due to their relatively ease of implementation and excellent results for many applications. Moreover, EAs are population-based algorithms, in which a population of different candidate solutions is maintained and evolved. This makes EAs well-suited to meet the requirement of obtaining a diverse set of trade-off solutions in MO optimization [96]. Two exemplary MOEAs, NSGA-II [44] and Strength Pareto Evolutionary Algorithm 2 (SPEA2 [100]) have found their applications in numerous power system planning and operation problems [26, 101–104]. However, theoretical research in MO optimization [105] pointed out the main drawback of classical MOEAs (e.g. NSGA-II and SPEA2), namely scalability. An MOEA is regarded to be scalable if it can maintain its effectiveness and efficiency when the problem size increases, resulting in acceptable, i.e. low-order polynomial, increasing runtimes. In the context of DNEP, scalability requires that a good set of trade-off expansion plans for large networks can be obtained by MOEAs within a reasonable amount of computing time. MOEA literature [105–108] pinpoints essential components to construct such scalable MOEAs. Following these design guidelines, the multi-objective gene-pool optimal mixing evolutionary algorithm (MO-GOMEA [109]) is a recently developed scalable MOEA that was found to have superior performance in solving laboratory benchmarks [109] and to have very promising results when solving the real-world static DNEP problem [110] compared to the classical MOEA NSGA-II. Therefore, we choose the state-of-the-art MO-GOMEA to tackle the dynamic MO-DNEP problem.

Our Contributions

In this work, we consider the DNEP problem with multiple objectives and dynamic planning in stages. We incorporate the future options for DSM as an additional investment action for DNOs. We present efficient trade-off solutions for this

MO-DNEP problem and discuss their implications on DNEP decision-making. This work advances the state-of-the-art in MO-DNEP and contains several major extensions compared to our previous works [110, 111] in the following ways. First, the research in [110, 111] considered mainly facility constructions while this work includes both physical asset investments (i.e. traditional network reinforcements) and DSM options (i.e. policy alternatives). Second, in [110], we solved the MO-DNEP in a static manner, in which all expansion options are assumed to be done at the same time, i.e. in a single-stage expansion. Third, in [111], the DNEP problem was formulated in dynamic planning but was solved as a single-objective problem in which only the total cost was considered and minimized. In this work, we incorporate the asset-installation decomposition procedure that we developed for dynamic planning (in [111]) into our MO optimization framework (in [110]) to create a full-fledged MO-GOMEA dedicated for the dynamic MO-DNEP problem with DSM options. We will show that the algorithm can be employed to assist DNOs in designing effective expansion plans regarding the upcoming introductions of smart-grid technologies (e.g. DSM or storage systems) into the existing traditional electricity networks.

The remainder of the chapter is organized as follows. Section 5.5.2 introduces two types of expansion options: asset investment and DSM option. Section 5.5.3 formulates the MO-DNEP problem. Section 5.5.4 outlined the optimization algorithm MO-GOMEA. Section 5.5.5 shows and discusses the experimental results on several benchmark networks. Section 5.5.6 concludes the chapter.

5.5.2 Expansion Options

We assume that two categories of planning options are available to DNOs when solving DNEP: physical asset investment and DSM options. Asset investments (i.e. installing new facilities, upgrading existing equipment, or changing network topology) can literally increase the capacity of the distribution network. On the other hand, DSM policy contributions can be seen as improving the efficiency of network usage. DSM helps decrease the peak loads so that the current network capacity can still be adequate for the energy consumption.

Network Facilities (Assets)

In this work, network modeling is based on the specifications for MV-D networks in the Netherlands [86], which are also common for other highly urbanized regions/countries. A typical MV-D network contains lines/cables branching out of MV transmission substations (or HV/MV transformer substations) connecting MV nodes (i.e. MV/LV transformer substations or MV customer substations) into distribution rings or mesh structures. Some specific lines/cables, called normally open points (NOPs), are opened on one side and carry no power flow in the normal operation. Therefore, the physical topology of an MV-D network is a mesh grid but its operational topology is a radial grid. An MV-D network consists of a number of *feeders*. The commonly used term *feeder* is formally defined as follows. A feeder is a collection of lines/cables that corresponds to a maximal subnetwork that does not

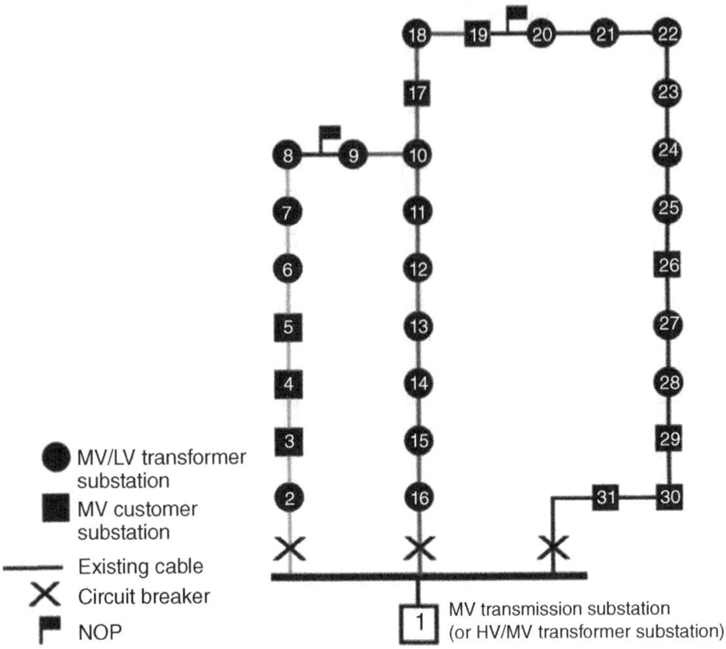

Figure 5.5.1 MV distribution network with three feeders.

contain any NOP or supply substation. Normally, *lines* refer to overhead connections, and *cables* refer to underground connections in the network. Here, without loss of generality, we consider MV-D networks containing only underground cables, which are typically found in urban areas. Figure 5.5.1 shows an example of an MV-D network.

Cables and MV/LV transformers are two typical types of assets in MV-D networks. They can be modeled by their equivalent impedances so that both can be abstractly seen as branches in a network [112]. In this study, we focus on MV-D cables as our main asset category but inclusions of transformers can be done without significant changes to the problem formulation. Each cable has a nominal capacity that defines the maximum apparent power that it can carry. Increasingly higher power demands require increasingly greater power flows, and bottlenecks happens when any cable is overloaded. The network should be enhanced beforehand to avoid such bottlenecks. Traditional enhancement options consist of activities to increase the network capacity: replacing legacy cables with higher-capacity cables, creating new feeders (i.e. installing a new cable to connect the MV transmission substation, or HV/MV transformer substation, with an MV node), or installing new cables to connect neighboring distribution rings [86]. Installations of new cables require additional placements of corresponding NOPs to keep the network operating radially [86].

Distribution Network Representation Let l_e denote the number of currently existing cables in an MV-D network. Let l_p denote the number of potential cable connections (i.e. pairs of nodes that are not directly connected at the moment but can be considered to be directly connected by a cable in the planning horizon). As the possible cable connections are numerous, expert knowledge should be applied here to disregard impractical or undesirable cable connections, such as connections between distant nodes. Let n_l denote the total number of branches (cable connections) that can be considered in the optimization process, i.e. $n_l = l_e + l_p$. The network configuration during a planning period from the beginning year t_0 until (and including) the final year t_{horizon} can be represented as a $n_y \times n_l$ matrix X where $n_y = t_{\text{horizon}} - t_0 + 1$ is the number of years. Let $\Omega(k)$ denote the set of cable types that can be installed at branch k. Each entry x_k^t of X indicates the status of branch k in year t as:

$x_k^t = \text{ID} > 0$: Active cable. A cable of type $\text{ID} \in \Omega(k)$ is installed at branch k.

$x_k^t = 0$: No cable connection. There is no cable at branch k.

$x_k^t = \text{ID} < 0$: A normally open point (NOP). A cable of type $\text{ID} \in \Omega(k)$ is installed at branch k but is out of normal operation.

The first row of X, $x^0 = \left(x_1^0, x_2^0, \ldots, x_{n_l}^0\right)$, is the vector representing the currently existing network. The last row of X, $x^{\text{horizon}} = \left(x_1^{\text{horizon}}, x_2^{\text{horizon}}, \ldots, x_{n_l}^{\text{horizon}}\right)$, is the vector indicating the network configuration at the final year t_{horizon}. With this matrix encoding, which is a common representation for dynamic planning in the literature, all the year-by-year changes in the network can be represented.

However, in real-world DNEP, a network does not change frequently nor arbitrarily given the current practice of DNOs which is as follows. First, DNOs may replace a legacy cable with a new cable of higher capacity but DNOs do not completely remove an existing cable connection because such removals will reduce network capacity. Second, network cables (as our main asset category in this study) have very long lifetimes (more than 40 years) while planning horizons of a too distant future can be regarded as impractical due to prohibitively high degrees of uncertainties in load growths and the emergence of new technologies. These facts suggest that, during a practical planning horizon (here, we assume planning period $\leq$30 years), a network branch requires reinforcement at most once. Expansion activities such as installing a thin cable first and then replacing it with a higher-capacity cable are regarded as impractical because construction costs are typically very high. Because each entry x_k^t of X corresponds with a decision variable in the optimization process, the matrix encoding thus contains a lot of redundancy. Instead, we can compare x^{horizon} with x^0 to know what new asset installations are required (i.e. if $x_k^{\text{horizon}} \neq x_k^0$). To find the asset installation schedule (i.e. the installation time of each new asset), we use the following decomposition heuristic.

Decomposition Heuristic for Dynamic Planning Given a forecast growth rate R of the peak power demand, the current network x^0, and the final-year network x^{horizon}, we derive an installation schedule for new assets in x^{horizon} in two phases as follows. In the first phase, based on R, we determine the first year t_X when the current network becomes infeasible (i.e. when any operation constraints is violated, see Section 5.5.3). Then, we create the base installation schedule by assuming all new assets are installed at the same time in the year t_X. We verify the feasibility and evaluate objective values of this base schedule (see Section 5.5.3). If the schedule violates any constraint, then the decomposition procedure is stopped because we do not want to decompose infeasible expansion plans. In the second phase, we loop repeatedly through the list of all new assets in a random order. We create a new schedule by delaying the installation of an asset a by one year. We evaluate the feasibility and objective values of this new schedule. If the new schedule is feasible and its objective values dominate those of the previous schedule (see Section 5.5.3), we then accept that postponement and a can be considered for another postponement again in a next loop. Otherwise, the installation of that asset a cannot be delayed any further. We continue this postponement checking until no asset can be postponed any more. Finally, we obtain a detailed year-by-year installation schedule for all new assets in an expansion plan and the concerned objectives can be evaluated accordingly. We argue that, based on the real-world practice of distribution network reinforcement in which a network branch generally requires enhancement at most once during a reasonable planning period (i.e. $\leq$30 years), this decomposition mechanism is sufficient for solving dynamic planning.

This asset-installation decomposition procedure was also used in our previous work [111] on solving the dynamic SO-DNEP. We note that the decomposition procedure that we use in this study is different from the decomposition algorithm that was proposed in [113]. The decomposition algorithm in [113] divides a multiyear DNEP into multiple one-year DNEPs, where each one-year DNEP can be solved independently, and the asset installations of these subproblems are coordinated through so-called forward/backward procedures in a recursive manner [113]. The decomposition algorithm in [113] can be seen as a framework to perform multiple single-year DNEPs and to synthesize the obtained results. In our study, the decomposition procedure is a heuristic that is embedded into the multiyear optimization process. During the optimization process, we evaluate multiple network configurations x^{horizon}'s, and for each of them, we need the decomposition procedure to derive an asset installation schedule of that network configuration.

Demand Side Management (Policies)

Parts of the power demand from the network are flexible loads, such as the use of dishwashers, washing machines, tumble dryers, or charging EV. DSM policies can motivate consumers to shift these flexible loads to different times out of the daily peak energy consumption hours by giving consumers, for example, financial compensations [114]. We assume that DNOs, as a stakeholder in energy markets, might be allowed (or required) to contribute parts of those compensations.

In this work, we assume that the flexible loads account for 10% of the peak loads at the beginning of a planning horizon, and linearly grow to 30% of peak loads at the end of the planning horizon. This assumption is based on the fact that the emergence of smart household appliances and EV gradually enlarges the magnitude of flexible loads. The range 10–30% of peak load can be flexible loads is determined based on the study in [115]. We also assume that DNOs can contribute to DSM policies through a financial means in EUR/year for each peak power demand reduction of 1 kW on an MV node in the whole year. Note that the numbers that we use here are simplified to set up a demonstrate case. Different and more fine-grained scenarios can be created by customizing these assumptions as input data. In this work, we take into account only the peak shaving effects of DSM because this is the most important aspect to DNEP while other related details, such as the actual mechanism of DSM, who administers DSM, and how consumers are incentivized to participate in DSM, are abstracted away.

Demand Side Management Representation Unlike (long lifetime) physical assets, DSM options can be seen as operational policies which can be changed from year to year (or stage to stage) during the planning horizon. In principle, in order to represent a DSM policy for a planning horizon of multiple years, we need a DSM decision variable for each year (or stage). However, we here use a DSM strategy which applies peak-shaving and only when it is necessary to prevent bottlenecks. Therefore, we only need DSM decision variables for the years from the first infeasible year t_X until (and including) t_{horizon}. We have $n_d = t_{\text{horizon}} - t_X + 1$ is the number of years in a DSM policy. We represent a DSM policy over n_d years as a vector of n_d non-negative integer elements.

$$y = \left(y^1, y^2, \ldots, y^{n_d}\right), \quad y^t \in \Omega(y^t), \quad t = 1, 2, \ldots, n_d \tag{5.5.1}$$

where $\Omega(y^t) \subset \mathbb{N}$ is the set of DSM option levels that can be used in a year t. The value of y^t indicates the chosen DSM level in year t. For the sake of simplicity in making decisions, DNOs are assumed to decide the amount of DSM contribution (i.e. corresponding with the desired amount of flexible load reduction) in discrete DSM levels, namely 0 (no DSM is needed), 1 (25% flexible load reduction), 2 (50% flexible load reduction), 3 (75% flexible load reduction), and 4 (100% flexible load reduction). Here, we take a holistic approach in which a single DSM level y^t is applied to all network nodes in each year t of the planning horizon. Our formulation can be easily extended to support a more fine-grained approach in which each network node i has its own DSM level per year y_i^t.

Effects of DSM on Peak Loads Let f^t denote the percentage of flexible loads in peak power demand in a year t. We assume that $f^0 = 10\%$ in the beginning year and $f^{\text{horizon}} = 30\%$ in the end year of the planning period. The values of f^t's, $0 < t < t_{\text{horizon}}$ can be calculated by linear interpolation.

For an MV-D network of n nodes, assuming that no DSM policy is used, the vector of peak active power demand at each node in year t is:

$$P^{t,0} = \left(P_1^{t,0}, P_2^{t,0}, \ldots, P_n^{t,0}\right) \tag{5.5.2}$$

The accompanying vector of peak reactive power is:

$$Q^{t,0} = \left(Q_1^{t,0}, Q_2^{t,0}, \ldots, Q_n^{t,0}\right) \tag{5.5.3}$$

In that year t, if we use a DSM policy at a level $y^t > 0$, $y^t = 25, 50, 75$, or 100%, then the vector of new peak active power demands will be:

$$P^{t,y^t} = \left(P_1^{t,y^t}, P_2^{t,y^t}, \ldots, P_n^{t,y^t}\right) \tag{5.5.4}$$

where $P_i^{t,y^t} = P_i^{t,0} \cdot (1 - y^t \cdot f^t)$. The corresponding vector of peak reactive power will be:

$$Q^{t,y^t} = \left(Q_1^{t,y^t}, Q_2^{t,y^t}, \ldots, Q_n^{t,y^t}\right) \tag{5.5.5}$$

where $Q_i^{t,y^t} = Q_i^{t,0} \cdot (1 - y^t \cdot f^t)$. Then, P^{t,y^t} and Q^{t,y^t} form the new load due to the peak shaving effect of the DSM level y^t in year t.

The total reduction in peak active power demands corresponds with employing the DSM level y^t in year t is:

$$P_{\text{total}}^{t,y^t} = \sum_{i=1}^{n} \left(P_i^{t,0} - P_i^{t,y^t}\right) \tag{5.5.6}$$

Solution Representation

A solution s of the DNEP problem for a network over a planning period consists of the network configuration x^{horizon} at the end of the planning period and the DSM policy y (i.e. the list of DSM levels that are applied on the network in each year). For the sake of convenience in representation, we shorten x^{horizon} as x.

$$s = (x, y) = \left(x_1, x_2, \ldots, x_{n_1}, y^1, y^2, \ldots, y^{n_d}\right) \tag{5.5.7}$$

The total number of decision variables is then $L = n_1 + n_d$. An installation schedule of new physical assets in x is determined by our installation decomposition approach. The DSM policy y indicates the DSM levels that affect the peak load profile P^{t,y^t} and Q^{t,y^t} in each year t, and will be used to verify constraint violations of the network indicated by x.

5.5.3 Problem Formulation

Constraints

We can represent a DNEP case as a graph $G = (V,E)$, where V is the set of all vertices/nodes (i.e. substations) and E is the set of all edges (i.e. existing and potential cable connections). We have $V = V_s \cup V_c$, where V_s is the set of supplying nodes (i.e. MV transmission substations or HV/MV transformer substations) and V_c is the set of consuming nodes (i.e. MV customer substations or MV/LV transformer substations). We have $E = E_e \cup E_p$, where E_e is the set of all existing cable connections and E_p is the set of all potential cable connections.

A network configuration **x** consists of: $E_x \subseteq E$, the set of all cables installed in **x**, and $E_x^a \subset E_x$, the set of all active cables in **x**. In each year t during the planning horizon, given the forecast peak consumption at every node and the peak shaving effect of the chosen DSM policy y^t in that year, we can determine the peak load profile P^{t,y^t} and Q^{t,y^t}. The following constraints must be satisfied by **x** regarding P^{t,y^t} and Q^{t,y^t}:

1. Connectivity: All nodes should be connected to the HV network.

   For all $u \in V_c$ there exists a $w \in V_s$ such that there is a path $p = (p_1 = w, p_2, \ldots, p_{|p|} = u)$ through the graph G that connects u and w, and all edges on the path are active, i.e. for all $i \in \{1, 2, \ldots, |p| - 1\}, (p_i, p_{i+1}) \in E_x^a$.
2. Normal operation constraints: During normal operation, the voltage at each node is within allowable ranges and the magnitude of the power flow through each cable is within the nominal capacity of that cable.

   Let V_i be the voltage at node $v_i \in V$ and V_i^{nom} is the nominal voltage at node v_i.

   $$\text{For all } v_i \in V, 0.9 \cdot V_i^{nom} \leq V_i \leq 1.1 \cdot V_i^{nom}$$

   Let S_i be the magnitude of the power flow through each active cable $s_i \in E_x^a$ and S_i^{nom} is the nominal capacity of cable s_i.

   $$\text{For all } s_i \in E_x^a, \ |\ S_i\ |\ \leq S_i^{nom}$$
3. Radiality constraint: NOPs are placed in such a way that the network can operate radially.

   For all $u \in V_c$ there exists exactly one supplying node $w \in V_s$ and one path $p = (p_1 = w, p_2, \ldots, p_{|p|} = u)$ through the graph G that connect u with w, and all edges on the path are active, i.e. for all $i \in \{1, 2, \ldots, |p| - 1\}, (p_i, p_{i+1}) \in E_x^a$.
4. Reconfigurability constraint: When an active cable fails, the feeder containing that failed cable (i.e. from the supplying MV transmission substation or HV/MV transformer substation to the corresponding NOPs) will be out of service. The DNO isolates the failed cable and then reconfigures the network by closing some NOPs to bring back operation. While maintenance activities take place, a mild overload, i.e. 130% of nominal capacity, is allowed and the radiality constraint can be compromised.

   Suppose an active cable c fails, it will be temporarily isolated from the network. During repair, some NOPs are closed so that the connectivity constraint (constraint 1) is satisfied and the network operation can be restored.

   For all $c \in E_x^a$, during repair, $E_x^{c,fail} = E_x \backslash \{c\}$, there exists a $C \subseteq E_x \backslash E_x^a$ such that for all $s_i \in E_x^{repair} = E_x^{c,fail} \cup C$, $|\ S_i\ |\ \leq 1.3 \cdot S_i^{nom}$ and E_x^{repair} satisfies the connectivity constraint.
5. Substation capacity constraint: We assume that each MV transmission substation (or HV/MV transformer substation) has limited physical space to install at most three new outgoing cables.

Let E_0^u be the set of outgoing cables at supplying substation $u \in V_s$ in the currently existing network. Let E_x^u be the set of outgoing cables at supplying substation $u \in V_s$ in the network configuration $\mathbf{x}$.

$$\text{For all } u \in V_s, 0 \le |E_x^u| - |E_0^u| \le 3$$

Constraints 1, 3, and 5 can be verified by checking the topology of the network in each stage. Constraints 2 and 4 require computationally expensive alternating-current power flow calculations (AC PFCs [116]) for the peak load profile P^{t,y^t} and Q^{t,y^t}. Note that the simpler DC model is not accurate enough for PFCs in distribution networks since the condition that branch resistance is negligible compared to reactance ([117]) does not hold for distribution networks. However, a complete verification of the reconfigurability constraint, which requires at least n_l AC PFCs for a network of n_l active cables, is too cumbersome. Here, we employ the line outage distribution factor (LODF [118]) method to verify this constraint. LODFs provide rapid assessment of multiple branch outage impacts and require only one (pre-contingency) PFC. LODFs in combination with an AC power flow was shown to require much less computing time but to have an acceptable accuracy for the capacity evaluation of MV-D networks [118].

To compare different solutions of a constrained optimization problem like DNEP, we employ the constraint-domination principle [119]. It is important to note that because PFCs can only be performed for connected networks, the connectivity constraint needs to be satisfied so that constraints 2 and 4 can be properly verified. Therefore, we first need to check the connectivity of the networks encoded in the solutions that we want to compare. A connected network is always better than an unconnected network. If both networks are unconnected, then the one that differs the most in terms of network topology from the currently existing network is the worse one. If both networks are connected, we then compare them by using the total violation of other constraints (i.e. constraints 2–5). A feasible solution has a total violation of value 0. A feasible solution is always better than an infeasible solution. If both solutions are infeasible, then the one that has greater total constraint violations is the worse one. If both solutions are feasible, we then evaluate their objective values and use the Pareto domination principle to compare them.

Objectives

Pareto Domination A feasible solution s^0 of an MOOP is said to *Pareto dominate* (i.e. to be multi-objectively better than) another feasible solution s^1 (denoted $s^0 \succ s^1$) if s^0 is strictly better than s^1 in at least one objective and s^0 is no worse than s^1 in all the other objectives. A Pareto set $\mathcal{P}$ is a set of solutions in which no solution dominates any other solutions that also exist in $\mathcal{P}$. A solution s^0 is said to be *Pareto optimal* if and only if there exists no other solution s^1 that dominates s^0. The optimal Pareto set $\mathcal{P}_S$ is the set of all possible Pareto-optimal solutions. The optimal Pareto front $\mathcal{P}_F$ is the image set of $\mathcal{P}_S$ in the objective space. Because the size of $\mathcal{P}_S$ is usually very large or even infinite, it is very hard to find all Pareto-optimal solutions. MO optimization normally aims to obtain a good Pareto set $\mathcal{P}$ whose front (i.e. the image set in the objective space) approximates $\mathcal{P}_F$. The approximation is considered in the objective space because we generally want a diverse set of good solutions with respect to all objectives.

In the following section, we introduce the typical objectives that are used for the MO-DNEP problem.

Physical Asset Installation Cost – Capital Expenditure (CAPEX) We use the annuity method [120] to convert the expenditure on each new asset a into a series of uniform annual payments, called annuities. Assuming the length of this series to be equal to the economic lifetime of the asset t_a^{life}, the annuity AN_a of an asset a with a discount rate $i = 4.5\%$ (as in [86]) can be computed as:

$$AN_a = Price_a \cdot \frac{i}{1-(1+i)^{-t_a^{life}}} \tag{5.5.8}$$

where $Price_a$ is the installation cost of a, including both acquisition and construction cost. CAPEX for the asset a in a year t can be calculated as:

$$CAPEX_a(t) = \begin{cases} AN_a, & \text{if } t_a^{inst} \le t < t_a^{inst} + t_a^{life} \\ 0, & \text{else} \end{cases} \tag{5.5.9}$$

with t_a^{inst} is the year when the asset a is firstly installed. The total CAPEX in a year t over the whole network is defined as:

$$CAPEX(t) = \sum_{\substack{\text{new asset } a \text{ in} \\ [t_0,\, t_{horizon}]}} CAPEX_a(t) \tag{5.5.10}$$

We minimize the net present value (NPV) of the total CAPEX over the planning period with a discount rate i:

$$CAPEX_{NPV} = \sum_{t=t_X}^{t_{horizon}} \frac{CAPEX(t)}{(1+i)^{t-t_0}} \tag{5.5.11}$$

We assume that $t_a^{life} = 30$ years for all new assets a's. Because physical assets have a very long lifetime, it is possible that $t_a^{inst} + t_a^{life} > t_{horizon}$, which means a part of the investment cost of an asset a would not be included in this objective value. However, this method is favored for comparing multiple scenarios of load growths [86, 120].

Demand Side Management Cost

From Eq. (5.5.6), we can obtain P_{total}^{t,y^t}, i.e. the total reduction of peak active power demand due to the application of the DSM level y^t on the network in year t. The cost that the DNO spends on the DSM level y^t in year t is computed based on this reduction as:

$$\mathrm{DSM}(t) = \mathrm{Price}_{\mathrm{unit}}^{\mathrm{DSM}} \cdot P_{\mathrm{total}}^{t,y^t} \tag{5.5.12}$$

where $\mathrm{Price}_{\mathrm{unit}}^{\mathrm{DSM}} = 1$ EUR is the unit price for power demand reduction per kW. For the purpose of demonstration, we here choose only a single DSM price but this price can be easily customized as input data to create different scenarios of DSM costs.

We minimize the NPV of the total DSM policy cost over the planning period with a discount rate i.

$$\mathrm{DSM}_{\mathrm{NPV}} = \sum_{t=t_X}^{t_{\mathrm{horizon}}} \frac{\mathrm{DSM}(t)}{(1+i)^{t-t_0}} \tag{5.5.13}$$

We here assume that the DNO would start to apply DSM options from the first year t_X that the current network would become infeasible. From t_0 until t_X, the current network can still operate properly, and the DNO has little practical incentive to employ DSM earlier than needed.

Energy Loss

Since the network cable is our main asset category in this work, the energy loss of the network in a year t can be taken as:

$$E_{\mathrm{loss}}(t) = P_{\mathrm{peak\ loss}}(t) \cdot T_{\mathrm{loss}}(t) \tag{5.5.14}$$

where $P_{\mathrm{peak\ loss}}(t)$ is the total network peak loss which can be obtained by performing PFC regarding peak demands P^{t,y^t} and Q^{t,y^t}, i.e. the peak load profile in year t with respect to a chosen DSM level y^t. $T_{\mathrm{loss}}(t)$ is the service time of peak loss for year t, defined as the ratio of the area under the normalized yearly loss profile shape over the normalized peak loss value [120, 121]. The exact values of $T_{\mathrm{loss}}(t)$ depend on the nature of each specific network and its power demands. Here, we take $T_{\mathrm{loss}}(t) = 2000$ hours for all t, which has been reported to be a realistic value for real-world MV distribution cables [121]. Note that this value was calculated based on traditional DNEP [121], but for the purpose of demonstration in this study, this value is sufficient. More accurate and dedicated models to calculate T_{loss} in the presence of smart-grid technologies, when available, can be employed and be straightforwardly considered as input data.

We want to minimize the total energy losses on the network during the planning period:

$$E_{\mathrm{loss}} = \sum_{t=t_0}^{t_{\mathrm{horizon}}} E_{\mathrm{loss}}(t) \tag{5.5.15}$$

Customer Minutes Lost (CML) Per Year We choose to evaluate the total number of customer minutes lost (CML [110]) per year to quantify the reliability of an MV-D network. Alternatively, we can also measure the network reliability by employing the SAIDI, which can be easily obtained by dividing the total CML per year over the number of customers in the network [94]. When a network cable fails, the feeder containing that cable (i.e. from an MV transmission substation, or HV/MV transformer substation, with circuit breakers to corresponding NOPs) is put out of operation because the corresponding circuit breaker will be triggered. Customers connected to all the nodes along this feeder suffer a temporary power outage. The DNO needs to find out and isolate the failed cable along the feeder. After the failed cable is isolated, the DNO can close the corresponding NOP and the circuit breaker of the feeder so that the electricity supply is resumed. The number of failures NF_k over a cable k per year can be estimated as:

$$\mathrm{NF}_k = F_k \cdot L_k \tag{5.5.16}$$

where F_k is the annual failure rate of cable k per kilometer, and L_k is the length of cable k. The restoration time T_{res} can be defined as the duration between failure occurrence and power supply restoration. This duration depends on the number of nodes connected to the feeder. The average restoration time (in minutes) [122] when a cable k fails can be taken as:

$$T_{\mathrm{res}}(k) = 75 + \frac{\mathrm{NS}(\mathrm{Feeder}(k))}{2} \cdot 10 \tag{5.5.17}$$

where Feeder(k) denotes the feeder containing the cable k and NS(Feeder(k)) is the number of MV/LV transformer substations and MV customer substations connected to the feeder Feeder(k). CML for a cable k per year can now be defined as follows:

$$\mathrm{CML}_k = \mathrm{NF}_k \cdot \mathrm{NC}(\mathrm{Feeder}(k)) \cdot T_{\mathrm{res}}(k) \tag{5.5.18}$$

where NC(Feeder(k)) is the total number of customers connected along the feeder Feeder(k). In this study, when conducting experiments, we assume the number of customers stay the same during the planning period. More accurate prediction models about the number of customers, if available, can be straightforwardly employed to provide new input data.

By the application of the asset installation decomposition procedure (see Section 5.4.2) on the final network configuration x^{horizon}, we obtain an installation schedule from which we can derive the network configuration x^t in each year t. The CML of a network x^t in a year t can be computed as:

$$\mathrm{CML}(x^t) = \sum_{\substack{k=0 \\ x_k^t > 0}}^{n_l} \mathrm{CML}_k \tag{5.5.19}$$

where n_l is the total number of cable connections, and we compute CMLs only for active branches ($x_k^t > 0$) as there is no power flow through a cable with an NOP ($x_k^t < 0$). The total CML during the planning period is:

$$\mathrm{CML}_{\mathrm{total}} = \sum_{t=t_0}^{t_{\mathrm{horizon}}} \mathrm{CML}(x^t) \tag{5.5.20}$$

We want to minimize the averaged CML:

$$\text{CML}_{\text{averaged}} = \frac{\text{CML}_{\text{total}}}{n_{\text{y}}} \tag{5.5.21}$$

where $n_{\text{y}} = t_{\text{horizon}} - t_0 + 1$ is the number of years in the planning period.

Solution Evaluation

The objectives CAPEX, DSM cost, and averaged CML per year for a feasible solution can be efficiently evaluated by checking the network topology in each year along the planning horizon. The total energy loss objective, and the normal operation constraint and the reconfigurability constraint, however require computationally heavy PFCs. Evaluating solutions for every year during a planning period of 30 years takes a long time regarding the available resource and time for the authors in doing this study (also see Section 5.5.5). Moreover, in the current practice of real-world DNEP, DNOs hardly ever add a new cable to a network every year. Adding a new cable can increase the network capacity by a great amount and is normally very expensive. Therefore, instead of evaluating solution constraints and performing the decomposition procedure (see Section 5.5.2) in a fine-grained year-by-year manner as presented above, we run experiment with a stage-by-stage approach, where each stage covers a period of at most 5 years and at least 3 years. Thus, the decomposition procedure will try to delay each asset installation by one stage of multiple years later. The peak load of the final year in each stage is considered to be the peak load of the stage (because peak loads normally increase monotonically) which is used in constraint evaluations of the network configuration in that stage. For the sake of simplicity and computing time reasons, we also assume that DNOs can make decisions about DSM options per stage, instead of per year as in the formulation in Section 5.5.2. Therefore, the number of DSM variables n_{d} is equal to the number of stages. For every year in a stage, the network is applied the same DSM level of that stage. Note that the same DSM level can have different costs per year, depending on the amount of peak load reduction in that year. Although we perform stage-by-stage planning, the objective values of CAPEX, DSM cost, and CML can still be evaluated in a year-by-year manner. As we perform PFCs only for the final year of each stage, we can obtain the accurate value of peak loss only in that year (to be used to compute energy loss). For the remaining years in stage, peak losses can be estimated by taking the assumption that the peak loss also has a growth rate related to the load growth R as follows [86]:

$$P_{\text{peak loss}}(t) = P_{\text{peak loss}}(t-1) \cdot (1 + R)^2 \tag{5.5.22}$$

5.5.4 Optimization Algorithm

This section outlines the MO-GOMEA, the optimization algorithm that we use to solve our MO-DNEP problem.

Multi-Objective Gene-pool Optimal Mixing Evolutionary Algorithm

MO-GOMEA is a state-of-the-art scalable MOEA which was recently developed by combining well-established researches in evolutionary multi-objective (MO) optimization [105–108] with research on single-objective (SO) GOMEA [123] to bring the superior performance of (SO) GOMEA into the MO realm. MO-GOMEA has been employed to solve the static MO-DNEP problem with promising results [110]. In-depth descriptions of MO-GOMEA can be found in the literature [109]. Here, we present the main components of MO-GOMEA and our adaptations in the context of the dynamic MO-DNEP problem.

Elitist Archive

During the optimization process, MO-GOMEA maintains an elitist archive $\mathcal{A}$ of all non-dominated solutions [124], i.e. solutions that are so far not (Pareto-)dominated by any solutions. Every time a new candidate solution s is evaluated, we check if it can enter $\mathcal{A}$ by comparing s with all solutions $s^{\mathcal{A}} \in \mathcal{A}$ and employing the constraint-Pareto domination principle (see Sections 5.5.1 and 5.5.2). If s is dominated by any $s^{\mathcal{A}}$, then s is not accepted into $\mathcal{A}$. If s is not dominated by any $s^{\mathcal{A}}$, then s can enter $\mathcal{A}$ and all $s^{\mathcal{A}}$'s that are dominated by s need to be removed. The archive thus keeps track of all best solutions that have been found so far, hence the name elitist archive. When optimization terminates, the elitist archive is considered as the result of MO optimization, i.e. the approximation Pareto set for the true optimal Pareto set $\mathcal{P}_S$.

Initialization

MO-GOMEA can normally start with a population of randomly generated solutions. However, for a constrained optimization problem like DNEP, random solutions often have many constraint violations. We consider the following mechanism (also described in [111]) which is based on engineering rules [86] to generate candidate network configurations to give MO-GOMEA a good start. First, it can be observed that currently existing cables should be left intact or replaced with higher-capacity cables. DNOs do not remove existing connections or install lower-capacity cables because these activities decrease the capacity of the network. Thus, every asset decision variable x_k in a candidate solution can only be randomly initialized with a non-negative value (i.e. we do not place NOPs at this step) as long as it does not downgrade the currently existing capacity (i.e. $x_k \geq | x_k^0 |$). This principle also ensures that the candidate network configuration is a connected topology since the current network is connected. Second, we check if the number of new cables branching out from each MV transmission substation (or HV/MV transformer substation) is more than the remaining capacity of that substation (i.e. whether constraint 5 is violated). New outgoing cables are then randomly deleted until constraint 5 is satisfied. Third, we loop through the list of all asset decision variables that have positive values (i.e. active cables) in a random order. For each cable $x_k > 0$, we try to place an NOP on that cable by negating its value. If the network is still connected, then the NOP can be placed; otherwise, we undo this operation. This procedure, called radialization, creates a network of radial topology with random placements of NOPs (i.e. constraint 3 is satisfied). Note that we do not repair solutions to meet the normal operation and reconfigurability constraints because they require computationally heavy PFCs. Finally, the DSM decision variable y^t can be randomly generated for each year t from t_X until (and including) t_{horizon}.

Population Clustering The optimal Pareto set $\mathcal{P}_S$ can have a very large optimal Pareto front $\mathcal{P}_F$ that consists of many different regions. It is possible that optimization algorithms can undesirably focus on a particular region, especially the middle region of $\mathcal{P}_F$ due to the pressure of Pareto domination. A good approximation set must be diverse enough to equally cover all parts of $\mathcal{P}_F$. Besides, variation operators of EAs may create undesirable candidate solutions by incidentally performing fruitless recombinations of very different solutions that are located in distant regions along the PF. For example, it is not sensible to recombine the most economical solution with the most reliable solution because these two solutions typically have very different network configurations (e.g. see Figure 5.5.1). It is usually beneficial if we can restrict recombinations among solutions pertaining to the same regions. To this end, MO-GOMEA divides its working population into clusters that have the same number of candidate solutions by using the balanced k-leader-means clustering algorithm [107], partitioning the population as observed in the objective space. Each cluster can then be seen as separately approaching a different part of the optimal Pareto front $\mathcal{P}_F$, ensuring the search efforts and resources are distributed equally along the PF. Note that the number of clusters k is thus a control parameter of the optimization algorithm MO-GOMEA which needs to be properly determined to achieve desirable results. In this chapter, we present a mechanism that automatically adapts k in conjunction with another important parameter, i.e. the population size of MO-GOMEA, so that the burden of parameter settings is alleviated.

Linkage Learning

MO-GOMEA differs from traditional MOEAs like NSGA-II and SPEA2 in the use of a linkage learning (LL) procedure during the optimization process [109]. In essence, LL tries to detect from the population of candidate solutions possible dependencies between decision variables, also known as linkage information. This information can be used by the optimization algorithms to juxtapose parts of different existing solutions to generate new promising candidate solutions while disrupting strong and important linkages as little as possible [123]. This is not guaranteed if traditional variation operators, such as uniform (UX), one-point (1X), or two-point (2X) crossovers, are employed, potentially resulting in poor optimization performance [125]. The complicated structures of distribution networks suggest that LL can be essential to the success of MOEAs in solving DNEP. Possible linkage knowledge in DNEP, for example, can be groups of cables on the same feeder or the fact that the addition of a new cable connection requires an appropriate placement of an NOP. For each cluster in the working population,

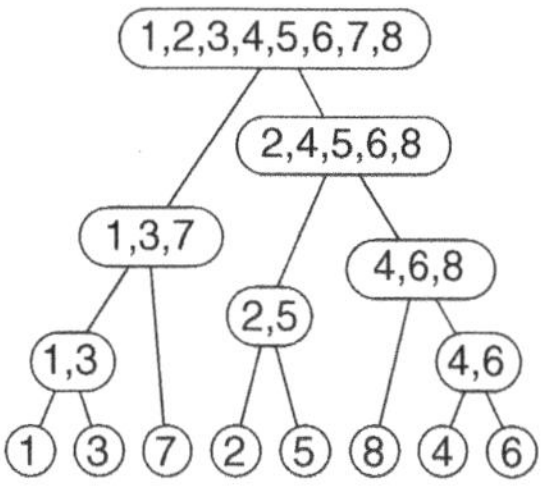

Figure 5.5.2 An example linkage tree.

MO-GOMEA performs LL separately to learn the linkage knowledge that is specific to solutions in that cluster.

LL detects and encodes the learned linkage knowledge into a hierarchical data structure called the linkage tree (LT) [123]. First, all L decision variables are assumed to be totally independent from each other. We have L singleton linkage sets, and each linkage set contains one unique decision variable. All these univariate linkage sets are put at the lowest level (i.e. leaf nodes) of the LT. At higher levels of the LT are branch nodes, which are multivariate linkage sets. A multivariate linkage set contains decision variables that are dependent on each other to some degree. A multivariate linkage set can be formed by merging a pair of linkage sets located at lower levels. In the LT structure, each linkage set is allowed to be merged once. Pairs of linkage sets that are considered to have stronger degrees of dependency are merged first. As a measure of dependency, mutual information (MI) can be employed here because MI was shown to be a good metric to build the LT [126]. Following this bottom-up principle, linkage sets are iteratively merged, forming higher-level multivariate linkage sets, until the root node, which is a set containing all decision variables, is created. An example of the LT can be seen in Figure 5.5.2. Given a set of L decision variables and a population of N candidate solutions, an LT can be constructed in $\mathcal{O}(NL^2)$ time by the unweighted pair grouping method with arithmetic mean (UPGMA) [127].

Optimal Mixing

The LT in each cluster is then used to generate new candidate solutions for that cluster. For each existing solution, called a parent p, in a cluster, MO-GOMEA employs a variation operator, called optimal mixing (OM [109]), to convert p into a new solution, called an offspring o, in a stepwise manner. First, o and a backup b are entirely cloned from p. Then, for each linkage set in the LT, a donor solution d is randomly selected from the same cluster containing p. The values of the decision variables indicated by the linkage set are copied together from d into o. Because these decision variables are in the same linkage set, they are considered to have some dependencies and should be treated jointly like a substructure. The partially altered solution o is evaluated for constraint violations and objective values and is compared against the backup b. If o dominates b or if o is not dominated by any solutions in the current elitist archive $\mathcal{A}$, then the changes are accepted and b is updated accordingly. Otherwise, the changes are undone and o rolls back to the backup b. After all linkage sets in the LT are traversed, a new solution, i.e. offspring o, is said to be fully constructed. In essence, OM exploits the linkage knowledge to recombine p with other solutions d's in the same cluster in such a way that important substructures are maintained and mixed to effectively create a new solution o that is no worse than p (i.e. o has the same objective values as p, o dominates p, or o is not dominated by $\mathcal{A}$).

It may happen that all the mixing steps of OM cannot create a new solution o. In this case, an additional procedure called forced improvement (FI) is invoked [109]. FI is, in essence, a second round of OM in which the donors are randomly selected from the elitist archive $\mathcal{A}$ instead of the cluster containing p. FI only accepts a mixing event if such mixing results in a strict improvement over the parent (i.e. o dominates p) or a true Pareto-front improvement (i.e. $o \notin \mathcal{A}$ and o is not dominated by $\mathcal{A}$). FI stops as soon as the first mixing is accepted. Besides, it can also happen that there exist many solutions having different decision variable values but the same objective values. In such cases, OM keeps changing solutions back and forth without truly improving the PF and FI is thus never invoked. To overcome this problem, when the number of consecutive iterations that the PF of the elitist archive $\mathcal{A}$ does not change exceeds the threshold $1 + \lfloor \log_{10} N \rfloor$, FI is invoked, where N is the population size [109]. If FI does not succeed in creating a new solution o either, we randomly choose a solution from the elitist archive as the new solution o. OM (together with FI) is applied to all existing solutions in the current working population to generate a population of new solutions. If the computing time budget (i.e. the allowed amount of time for the optimization process) has not been fully spent, we take the new population as the working population and go back to the Population Clustering step to begin a new iteration.

The above OM variant, called MO-OM, is employed for clusters in the middle regions. For extreme clusters (i.e. a cluster that has the best cluster objective mean with respect to one objective), we employ the OM version that was originally developed for SO optimization [123]. The SO-OM accepts a mixing event (i.e.

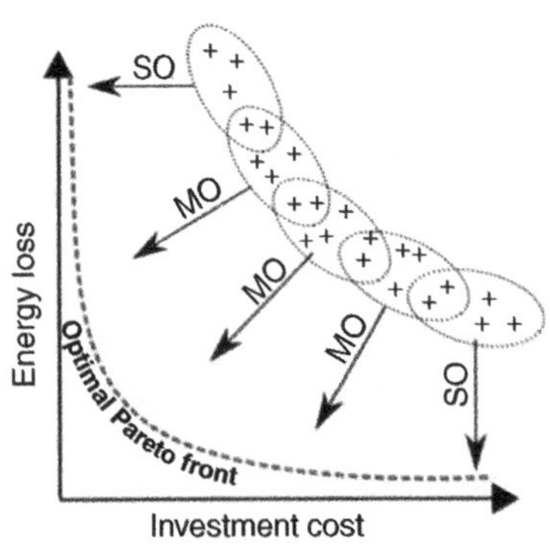

Figure 5.5.3 Different clusters approach different parts of the optimal Pareto front.

copying values of a linkage set from a donor d into o) if such mixing results in an equal or better solution with respect to that one objective. The SO-FI procedure of SO-OM also only accepts mixing events that improve the parent solution regarding the single objective of the corresponding cluster. Theoretical research on MO optimization showed that it is more efficient if the extreme regions (i.e. regions contain solutions that skew toward one objective) of the optimal PF are approached by SO optimizations [107, 109]. The working principle of multiple clusters with their corresponding OM variants (i.e. MO-OM or SO-OM) approaching different parts of the optimal PF are illustrated in Figure 5.5.3.

Adaptions of the Optimal Mixing Operator

It is generally not straightforward to recombine solutions for a highly constrained optimization problem like DNEP. New network configurations can violate many constraints, in particular the connectivity constraint, which is the crucial constraint that needs to be satisfied before other constraints can be evaluated (see Section 5.5.3). During the construction process of new solutions, in order to maintain its non-deteriorating principle, the OM variation operator needs to evaluate and compare many different intermediate solutions. Therefore, in other to effectively traverse the search space, OM must be adapted to handle this connectivity constraint properly. In this section we present two practical DNEP-related adaptations for the OM variation operator. Note that the ideas of these adaptations can be reused for other MOEAs to solve DNEP as well.

Disconnectivity Quantification This adaptation can be considered as the closest to the out-of-the-box version of MO-GOMEA because it requires the least modification. If the network configuration part x of a solution s is unconnected, we need to check how much its topology differs from the currently existing network x^0 (i.e. at time t_0). We go through all the asset decision variables (i.e. x_k's) and count the number of branches where the existing network x^0 has an active cable (i.e. $x_k^0 > 0$) but the solution network x has an NOP (i.e. $x_k < 0$), or where the existing network x^0 has an NOP or no cable connection (i.e. $x_k^0 \leq 0$) but the solution network x has an active cable (i.e. $x_k > 0$). This number is called the disconnectivity value. Note that connected networks do not require disconnectivity quantification (DQ) and are assumed to have the disconnectivity value 0. OM can then simply use the constraint-domination principle (see subsection "Constraint" in Section "5.5.3") during its construction process of new solutions. We call this adaptation DQ. DQ was also used in our previous work [111] on solving the single-objective DNEP.

Branch Exchange Because feasible solutions must be connected networks, it is beneficial if we can directly traverse the search space of connected topologies. Because all solutions in the initial population are generated such that they are all connected and radial networks, the adaptation we present here can help OM generate new connected and radial topologies during its mixing process. At every mixing event in OM corresponding with a linkage set in LT, the values of decision variables indicated by the linkage set are copied from a donor d to the current solution o. For asset decision variables i where both o_i and d_i are active cables (i.e. $o_i > 0$, $d_i > 0$) or both are NOPs or have no cable connections (i.e. $o_i \leq 0$, $d_i \leq 0$), values of such d_i's can be copied into o because they have the same network topology. For asset variables i where o_i is an active branch and d_i is an NOP or has no cable connections (i.e. $o_i > 0$, $d_i \leq 0$), we need to find another asset variable j such that o_j is an NOP or has no cable connection and d_j is an active cable (i.e. $o_j \leq 0$, $d_j > 0$). In other words, if an active cable is removed at branch i, then an active cable must be brought back at branch j so that the connectivity constraint is maintained. Similarly, for asset variables i where o_i is an NOP or has no cable connection and d_i is an active cable (i.e. $o_i \leq 0$, $d_i > 0$), we need to find another asset variable j such that o_j is an active cable and d_j is an NOP or has no cable connections (i.e. $o_j > 0$, $d_j \leq 0$). In other words, if an active cable is added at branch i, then an NOP must be added at branch j so that the network is still radial. Then, both d_i and d_j are copied into o_i and o_j if and only if such copying results in a connected and radial topology. To enhance the exploration capability of the search, during every mixing event, we allow a small probability of $p_m = 1/L$ (L is the total number of decision variables) that each copied variable value can be randomly mutated as long as such mutation does not violation the connectivity and radiality constraints. This adaptation can be seen as bringing DNEP problem-specific knowledge about connectivity and radiality into the variation operator OM, making MO-GOMEA a DNEP-dedicated optimization algorithm. We call this adaptation branch exchange with mutation (BX-M).

Note that in our previous work [111], the search for the variable j was limited to the current linkage set. The pairs of variables i and j correspond to pairs of an active cable and an NOP, and they should be copied together when performing mixing. However, one NOP can correspond with many active cables, and the LT, having $2L - 1$ linkage sets, might not capture all these pairs. It can happen that, for some pairs of active cable and NOP, i and j are not in the same linkage set, strongly reducing the potential of BX for small linkage sets. Therefore, in this study, we update BX by allowing j to be searched among all asset decision variables. We also note that there exists the *branch exchange algorithm* in the literature [128] which is similar to the BX adaptation in the idea of constructing connected and radial networks but is different in the purpose of usage. In [128], the branch exchange is an optimization algorithm for DNEP to construct the expansion plan that optimizes the concerned objective. In our study, BX is an adaptation for the OM operator of MO-GOMEA to generate new connected and radial networks while the optimization part is handled by MO-GOMEA.

Removing the Population Size Parameter and the Number of Clusters Parameter from MO-GOMEA

For population-based optimization algorithms like MO-GOMEA or NSGA-II, the size of the working population is a crucial parameter. If the population size is too small, the algorithm cannot solve the problem properly or, at best, will converge prematurely to local optimal solutions. If the population size is too large, the search effort becomes overly diverse, and it would take a lot of solution evaluations before approaching the desired regions (i.e. the optimal PF). In real-world optimization like DNEP, since evaluating solution quality usually takes a long time, too large populations may use up the allowed budget of computing time before the desired solutions are reached. However, it is almost impossible to determine the optimal population size before actually executing optimization because it depends not only on the problem size and the problem structure of the specific problem instance at hand but also the specific optimization algorithm that is used. Here, we overcome this issue by running MO-GOMEA following a scheme that gets rid of the population size setting. Note that the same scheme was also used and showed its effectiveness in our previous work [110] for solving the static MO-DNEP.

This scheme was originally developed for single-objective GAs [129]. Here, we adapt it into the context of MO optimization. The principal idea is that, instead of a single working population, we run MO-GOMEA with multiple populations of different sizes in an interleaved fashion where populations with large sizes have slower iteration cycles. Specifically, the first population P_1 of some small size n_1 begins the optimization process. Each next population P_i is twice as large as the previous one, i.e. $n_i = 2 \cdot n_{i-1}$ for $i > 1$. All populations are scheduled such that for every two iterations of population P_i, one iteration of population P_{i+1} is run (or initialized if this is the first iteration for P_{i+1}). An example schedule is shown in Table 5.5.1.

The same elitist archive is shared between all populations. Note that in its original implementation for SO optimization [129], populations of small sizes are stopped if they have converged or their averaged objective values are worse than those of larger population sizes. In the context of MO optimization, due to the goal of finding a diverse approximation Pareto set, its populations do not converge to a single solution and it is also not sensible to compute the average value of different conflicting objectives. Therefore, all populations of MO-GOMEA are kept running during the optimization process. We do not set a maximum population size for MO-GOMEA. The algorithm runs and grows the population size until the allowed computing time is over.

TABLE 5.5.1 Population Sizing-Free Scheme

ID	P_1	P_2	P_3	P_4	...
1	x				
2	x	x			
3	x				
4	x	x	x		
5	x				
6	x	x			
7	x				
8	x	x	x	x	
...	...	...	...	...	...

Research on MO-GOMEA [109] showed that the number of clusters parameter k is also an important control parameter that affects the quality of the resulting PF approximation. However, compared to the more troublesome population size parameter, the number of clusters k can be intuitively set such that for each objective we have one cluster approaching the extreme region on the PF corresponding with that objective and that we have at least one cluster approaching the middle regions. In our previous work [110], we fixed the number of clusters as $k = 5$ for every population. Here, we propose a more flexible approach that adapts k according to the above population sizing-free scheme. The initial population P_1 has the number of clusters $k_1 = m + 1$ (m is the number of objectives). Subsequent populations P_i's ($i = 2, 3, \ldots$) have the number of clusters $k_i = k_{i-1} + 1$.

5.5.5 Case Studies

Benchmark Networks

We consider three benchmark networks that are adapted from real data of a Dutch DNO. Cable-type data can be found in Appendix A.1. General data about each network are shown in Table 5.5.2 and network topologies are shown in Figure 5.5.4 while details about the current peak loads of the three networks are listed in Appendices A.2, A.3, and A.4. Based on the forecast peak load growth rate (see Table 5.5.3), we calculate the first year t_X that each network becomes infeasible. The values of t_X are needed for the decomposition mechanism to determine *when* each new asset should be installed along the planning horizon (i.e. from t_X until t_{horizon}). Note that as mentioned in Section 5.5.3, due to computing time reasons, instead of the more fine-grained year-by-year approach, we perform the expansion planning in a stage-by-stage manner, where each stage covers a period of maximum 5 years and minimum 3 years, which are reasonable stage lengths in DNEP practice. Details about the number of stages and the beginning year and end year of each stage are shown in Table 5.5.3.

We use two MO-GOMEA variants with two variation operators DQ and BX to solve the MO-DNEP for these three benchmark networks. We run MO optimization with respect to various combinations of two out of four objectives presented in Section 5.5.3: CAPEX versus DSM, total cost (CAPEX + DSM) versus total energy losses, and total cost (CAPEX + DSM) versus CML per year. Note that MO optimizations with three or four objectives are also possible but would not necessarily yield more illustrative results. Each MO-GOMEA variant is run

TABLE 5.5.2 Benchmark Network Size

ID	No. of Nodes	No. of HV/MV Substations	No. of Asset Variables n_l
1	10	1	17
2	31	1	59
3	51	4	190

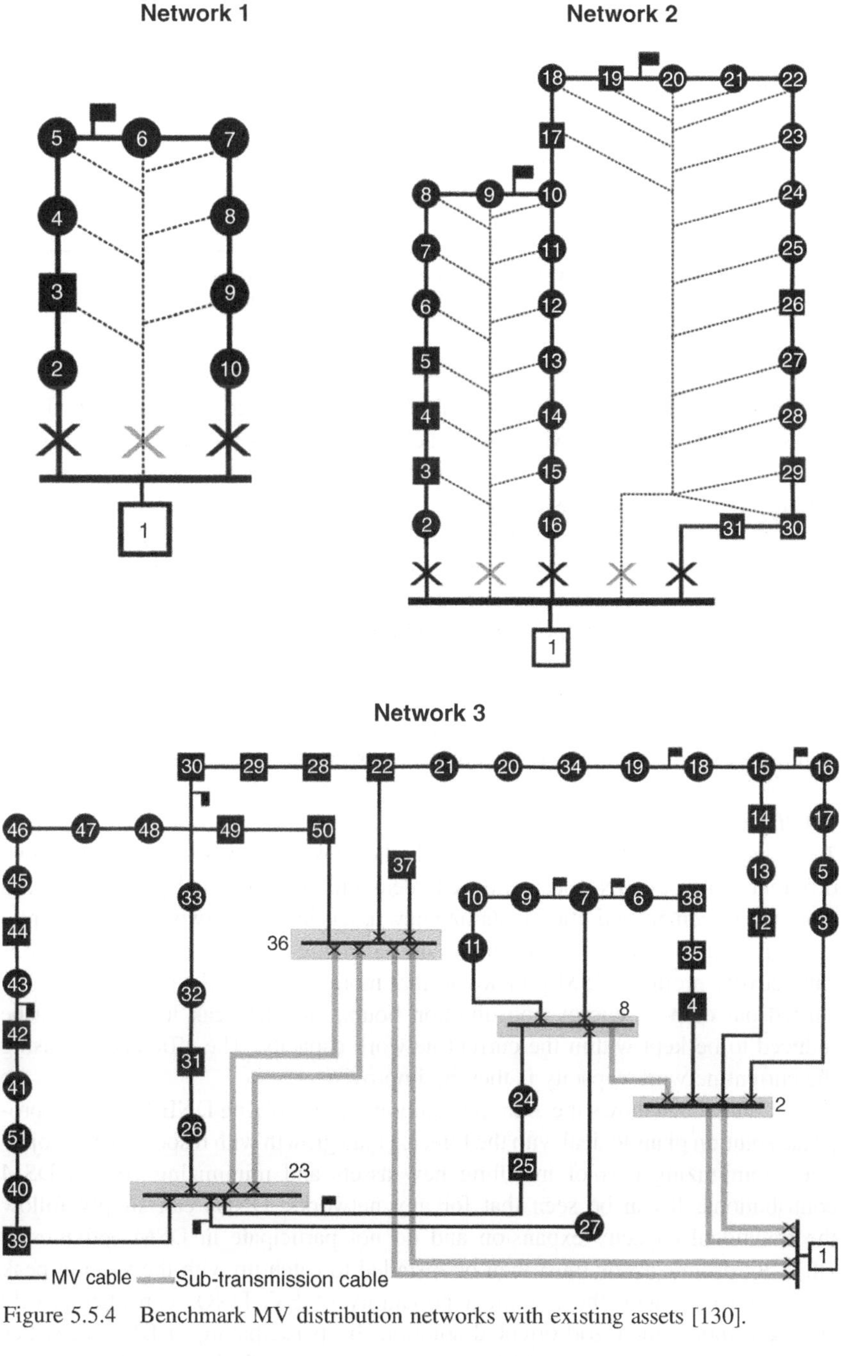

Figure 5.5.4 Benchmark MV distribution networks with existing assets [130].

TABLE 5.5.3 Planning Periods of Networks 1, 2, and 3

ID	t_{horizon}	R (%)	t_X	Stages $S_i = [t_{\text{begin}} - t_{\text{end}}]$	n_{d}
1	10	2	8	$S_1 = [8-10]$, $S_2 = [11-15]$, $S_3 = [16-20]$, $S_4 = [21-25]$, $S_5 = [26-30]$	5
2	31	2	10	$S_1 = [10-12]$, $S_2 = [13-15]$, $S_3 = [16-20]$, $S_4 = [21-25]$, $S_5 = [26-30]$	5
3	51	3	3	$S_1 = [3-5]$, $S_2 = [6-10]$, $S_3 = [11-15]$	3

t_{horizon} is the horizon year. R is the peak load growth rate. t_X is the first year in which the network becomes infeasible. Stage S_i is the duration of each planning stage during $[t_X - t_{\text{horizon}}]$. n_{d} is the number of DSM variables.

10 times on each network, and the final PFs are assembled by combining all the results of 10 runs. The maximum number of evaluations for each run is 100 000 evaluations for network 1, 200 000 for network 2, and 300 000 for network 3. Each solution evaluation consists of: one AC PFC to verify the normal operation constraint and one LODF to verify the reconfigurability constraint of network configuration in the final year (see Section 5.5.3), and if the solution is feasible, then the decomposition procedure is invoked to find an asset installation schedule (see Section 5.5.2). The decomposition procedure also requires AC PFCs and LODFs to perform constraint verifications for network configuration in each year t in the period $[t_X, t_{\text{horizon}}]$. The allowed number of solution evaluations are set as above, given the available time for the authors in this study. If we can afford more time for running experiments, better results may be obtained.

CAPEX vs. DSM

To manage the MV-D networks to handle the peak load growth, DNOs have two different strategies. DNOs can choose to deal with the situation by traditional network reinforcement activities: installing new assets into the networks along the planning period. The network capacity is then increased. On the other hand, DNOs can actively promote DSM policies so that more and more flexible loads can be shifted out of peak energy consumption hours, and the peak load can then be reduced to be kept within the current network capacity. The efficiency of using the current network capacity is thereby improved.

Table 5.5.4 shows the extreme solutions when solving DNEP to find appropriate solution plans to deal with the forecast load growth with respect to two objectives: minimizing cost of installing new assets and minimizing cost of DSM contributions. It can be seen that for any networks, DNOs can simply follow the traditional capacity expansion and do not participate in DSM activities at all. A lot of new assets must then be installed to catch up with the surging peak load, especially with the imminent popularity of EV. DNOs can alternatively choose a more smart-grid-oriented solution: by participating in DSM activities to ensure that peak load would be kept under control. Table 5.5.4 shows that if DNOs invest in DSM policies, a great deal of new asset installation cost can be

TABLE 5.5.4 Solutions of DNEP Considering CAPEX vs. DSM

Extreme Solution	CAPEX (EUR)	DSM (EUR)
Network 1		
Most Asset-pro	40 177.08	0
Most DSM-pro	12 437.85	5 697.25
Network 2		
Most Asset-pro	23 285.49	0
Most DSM-pro	533.74	19 290.71
Network 3		
Most Asset-pro	35 465.06	0
Most DSM-pro	0	25 410.58

saved by being able to delay expensive network reinforcements to many years later. The combined cost of CAPEX + DSM of the most DSM-pro solutions is shown to be significantly less than the CAPEX of the traditional physical asset investment alone. Note that the costs of DSM options here are based on our assumption about the DSM price and different DSM price levels can be used to create multiple scenarios. We here use a single DSM price for the purpose of demonstration only. Interestingly, for network 3 (i.e. the largest network considered in this study), with a good DSM policy investment, the network does not require any new asset installation during its planning period of 15 years. Only the NOPs need to be relocated to reconfigure the power flow paths and the cost of NOP relocations is generally negligible. Indeed, Figure 5.5.5 shows three example solution plans for network 3. First, the DNO can actively stimulate DSM policies to shift all flexible loads and the network can be left intact for 15 years. Second, the DNO can choose a combination of new asset installations and DSM options but the asset installations are delayed until later years in the planning period. We observed in the obtained outcome data: in the first stage, the network can be reconfigured by changing the locations of NOPs and new assets are only needed to be added in the last stage. Third, the peak load growth is not controlled and the DNO needs to install one new cable connection and replaces three other legacy cables. Note that it is not necessary that a DSM-pro solution is better than an asset-pro solution. Which solution will be chosen to be carried out depends on specific situations.

Solving DNEP while considering CAPEX versus DSM results in many interesting alternatives (see Figure 5.5.6). Although both asset investment and DSM option can be considered as costs for DNOs, they are in essence two different strategies: adding more network capacity versus increasing efficiency in the usage of current network capacity by decreasing peak power demands. It is important that DNOs are informed about all these alternatives before making a decision, and it is crucial that these alternatives must be (approximately) optimal trade-off solutions so that the best-informed decision can be made. Figure 5.5.6 shows the approximate PF (i.e. the image set of the best-found non-dominated solutions) found

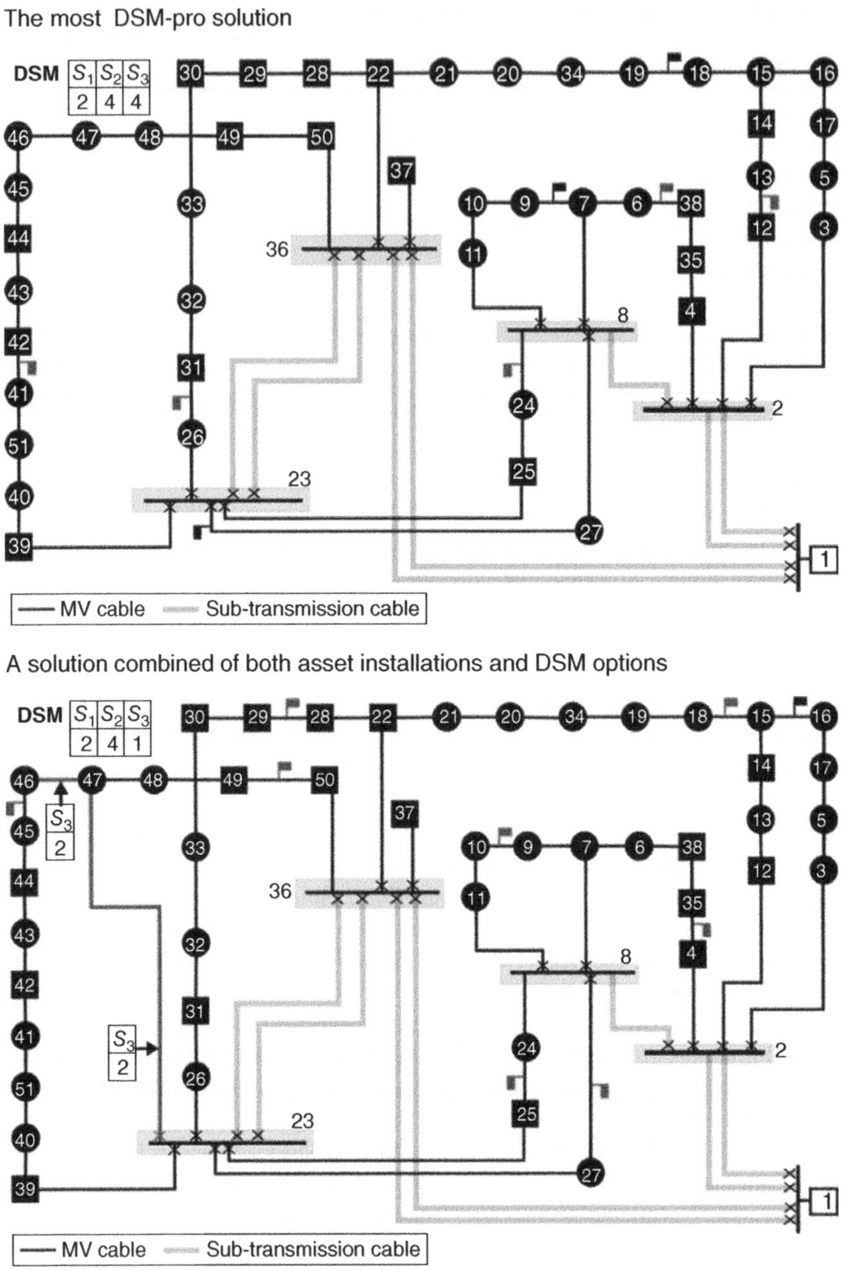

Figure 5.5.5 Network 3: CAPEX vs. DSM.

The most asset investment-pro solution

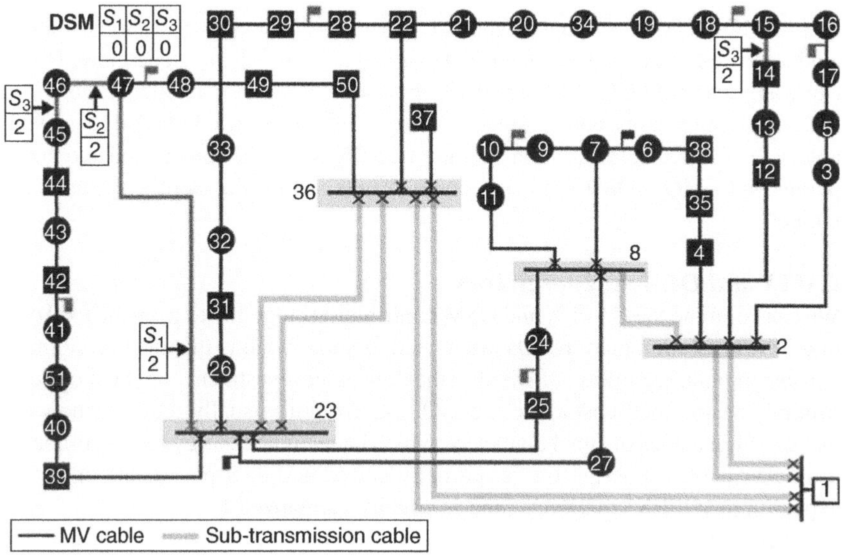

Figure 5.5.5 (Continued)

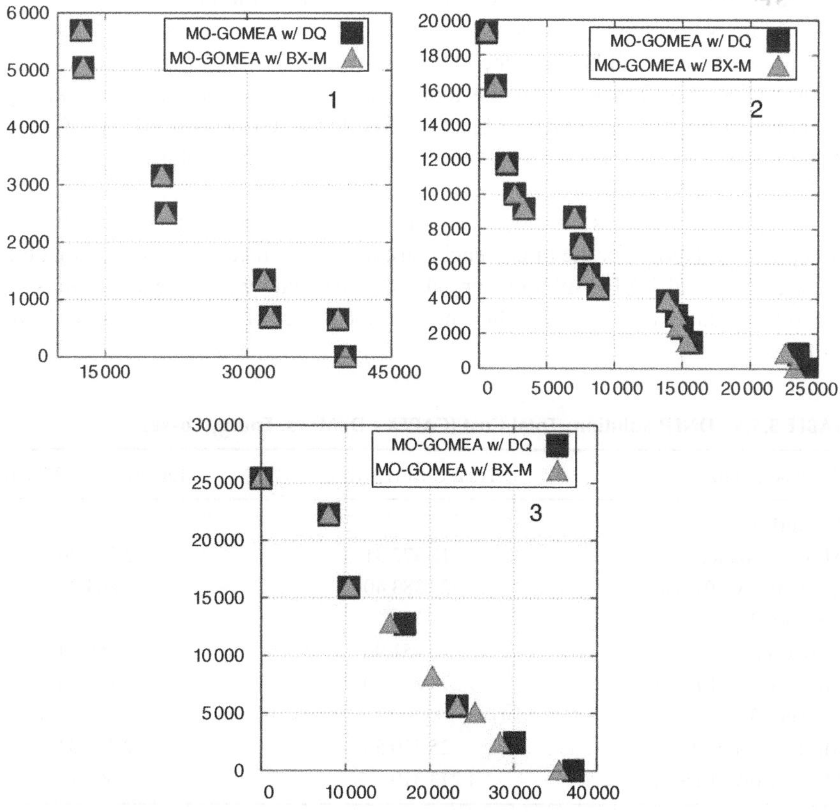

Figure 5.5.6 Pareto fronts of MO optimizations for three networks.

by two MO-GOMEA variants DQ and BX-M. Both MO-GOMEA variants obtain similar results while the PF of the BX-M variant is slightly better. Therefore, DNOs can easily use MO-GOMEA out-of-the-box with some minor modifications and still obtain reasonably good results. If more problem-specific knowledge is available, a DNEP-dedicate variation operator like BX-M can be constructed and incorporated into MO-GOMEA to find better results within the same computational budget.

CAPEX and DSM vs. Energy Loss

We can combine the CAPEX and DSM contributions together to form the total cost of a DNEP solution plan. DNOs often need to know what is the most economical expansion plan, regardless of whether they are asset investments or DSM options. Minimizing the total cost alone (i.e. SO optimization) usually returns a network that uses thin cables of small diameters, which have higher energy losses compared to thicker cables. Energy loss is normally considered as a part of the operating expenditure (OPEX) of DNOs, and is usually capitalized to be aggregated with CAPEX so that SO optimization can be employed to find the solution that has the minimum lump sum cost.

Being informed about only one solution plan, DNOs will find it difficult to reduce energy losses (in the most efficient way) if they would like to do so. Moreover, in transition toward a greener and environment-friendlier energy consumption future, DNOs will be motivated to consider energy-efficient networks. Here, we retain energy losses out of the total cost and try to optimize it separately by considering it as a separate objective. We have run MO-GOMEA DQ and MO-GOMEA BX-M for the two objectives: total cost and total energy losses. The results are as follows.

Table 5.5.5 shows the most economical and the most energy-efficient solution plans for DNEP for the three benchmark networks. It can be observed that the most economical DNEP solutions are also the least energy-efficient solutions. If we do not take into account the specific locations of NOPs (because NOP

TABLE 5.5.5 DNEP Solution: Total Cost (CAPEX + DSM) vs. Energy Losses

Extreme Solution	Total Cost (EUR)	Energy Loss (MWh)
Network 1		
Most economical	18 372.31	2 333.70
Most energy-efficient	327 383.40	804.20
Network 2		
Most economical	12 451.90	4 369.34
Most energy-efficient	279 079.80	2 105.16
Network 3		
Most economical	25 410.58	2 263.04
Most energy-efficient	1 273 310	663.76

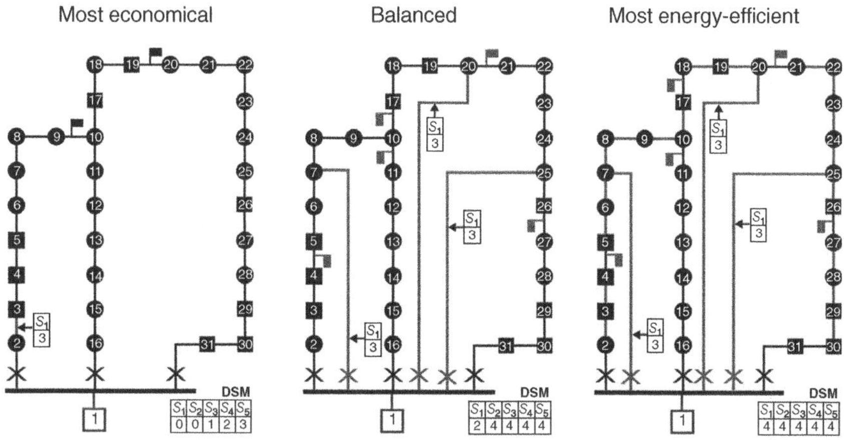

Figure 5.5.7 Network 2: total cost (CAPEX + DSM) vs. energy loss.

relocations do not have a cost in our model), then, for network 3, the most economical solution is the same as the most DSM-pro solution (compare Tables 5.5.4 and 5.5.5). Because network 3 has many legacy cables (i.e. cables of very small diameters) and there is no cable needed to be replaced in the most DSM-pro solution, its total energy loss is therefore the highest. In order to reduce energy losses, many of these legacy cables must be replaced with new cables of bigger sizes, bringing up the total investment cost. Table 5.5.5 also indicates that the cost of improving energy efficiency is very high. Therefore, instead of choosing the most efficient solution, it might be more reasonable for DNOs to look into solutions that balance cost and efficiency (i.e. the middle regions of the PFs, see Figure 5.5.8).

Figure 5.5.7 shows three example solutions for network 2. We can see that the most energy-efficient solution (i.e. the solution with the least energy losses) requires a lot of new cable installations and replacements because new cables of higher capacities generally have less branch resistance. This solution also makes use of the highest level of DSM options during the planning period because reducing peak load can also reduce energy losses as well. The most economical solution replaces only one cable and has little participation in DSM contributions; this solution suffers the highest total energy losses. A balanced solution might be a favorable combination of installing new cables and actively contributing to DSM options (with respect to our assumption about DSM price). All three solutions in Figure 5.5.7 indicate that all new asset installations should be done in the first stage, which is understandable as the sooner we replace legacy cables with new and more efficient cables, the less energy will be lost.

Figure 5.5.8 shows the PFs of MO-GOMEA DQ and MO-GOMEA BX-M solving DNEP with respect to the total cost (CAPEX + DSM) and the total energy loss. For network 1, the obtained results of the two variants are quite similar. For network 2, MO-GOMEA BX-M found a slightly better front than MO-GOMEA DQ. However, these differences are not critical regarding the numerous amount of solutions on the PF

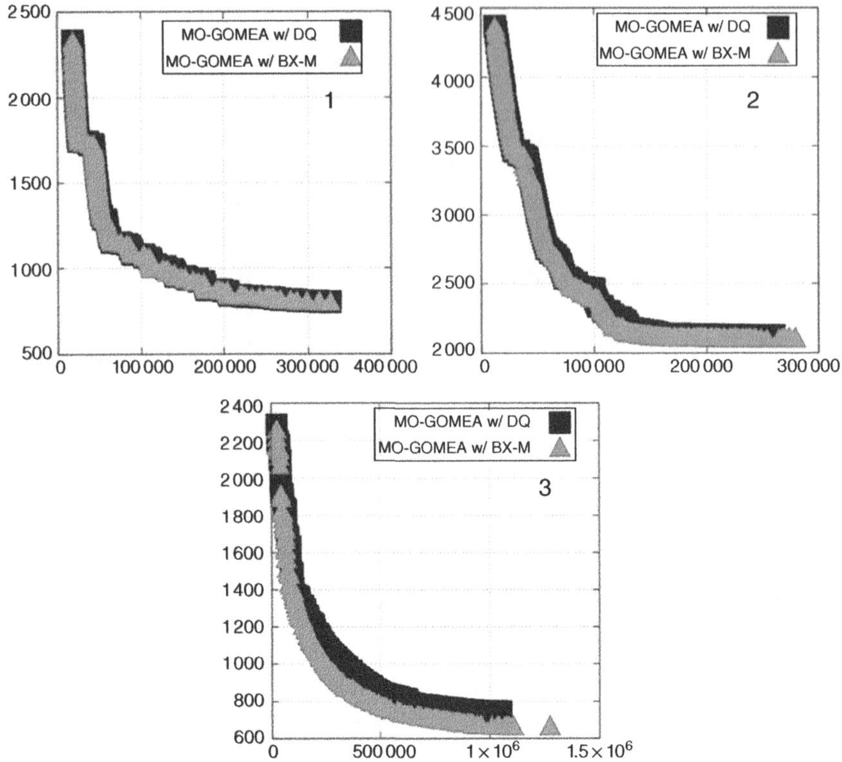

Figure 5.5.8 Fronts of MO optimization (CAPEX + DSM) vs. energy loss for three networks.

(see Figure 5.5.8), which is different from the much sparser PFs when considering CAPEX vs. DSM (see Figure 5.5.6). For network 3, it can be seen more clearly that the solutions found by MO-GOMEA BX-M (Pareto) dominate those found by MO-GOMEA DQ. However, the gap between the DQ's PF and BX-M's PF is not significantly big. If we run MO-GOMEA DQ longer, we may obtain results that are similar to MO-GOMEA BX-M's results, but within the same amount of computing time, BX-M gives MO-GOMEA some slight advantages over DQ. Compared to the more DNEP problem-dedicated MO-GOMEA BX-M, the performance of the out-of-the-box MO-GOMEA DQ can be considered as quite robust. This suggests that the LL capability of MO-GOMEA is powerful enough to help the algorithm reach acceptably good solutions even when problem-specific knowledge (e.g. the way the connectivity constraint should be handled in BX-M) is not available.

CAPEX and DSM vs. CML

The average CML per year during the planning period is regarded as the measure for network reliability in this study. CML in our model depends on the distribution of customers along each feeder (i.e. the fewer customers connect to a feeder, the fewer customers would be affected if an outage occurs), on the number of MV nodes on that feeder (i.e. the fewer MV nodes, the less time is required to localize and isolate the failed cable), and also on the total length of the feeder (i.e. the shorter the feeder is, the smaller the failure rate is). Minimizing CML can add more new cable connections to the network (i.e. creating new feeders) so that each feeder connects fewer nodes and fewer customers, which increase the cost of asset installations.

On the other hand, to minimize CAPEX, it is more economical to try to relocate existing NOPs first (rather than adding new cables immediately) to reconfigure the network so that parts of power flows are rerouted through different paths, avoiding heavily loaded cables. To this end, NOPs are usually located at locations that make the power flow equally distributed for each feeder. However, these positions might not be optimal locations to minimize CML. Minimizing CML tends to relocate NOPs so that the number of customers are distributed equally per feeder.

An MV customer substation node and an MV/LV transformer substation node might have the same power demand, but a customer substation is counted as 1 customer (i.e. 1 company) while a transformer substation can be counted as many customers (i.e. many households). Therefore, minimizing CAPEX and minimizing CML are two conflicting objectives.

Here, we also combine CAPEX and DSM together and regard that as the total cost of a solution plan. We have run MO-GOMEA DQ and MO-GOMEA BX-M for the two objectives: total cost and CML per year. The results are as follows. Table 5.5.6 shows the extreme solutions for each benchmark network: the most economical solution (i.e. the least total cost) and the most reliable solution (i.e. the least CML). From Table 5.5.6 and Figure 5.5.10, it can be seen that themost economical solution is also the least reliable one and vice versa. For network 3, we observed that we again obtain the most economical solution that is similar to the most DSM-pro solution as in Figure 5.5.5 (not taking into account the NOP locations in each case). It can be inferred that due to the peak shaving effects of DSM options and the relocation of NOPs, no new asset installation is

TABLE 5.5.6 DNEP Solutions: Total Cost (CAPEX + DSM) vs. CML per Year

Extreme Solution	Total Cost (EUR)	CML per Year (min)
Network 1		
Most economical	29 670.62	5 634
Most reliable	157 581	4 005
Network 2		
Most economical	12 451.90	22 178
Most reliable	95 198.62	14 221
Network 3		
Most economical	25 410.58	22 212
Most reliable	411 415.30	12 032

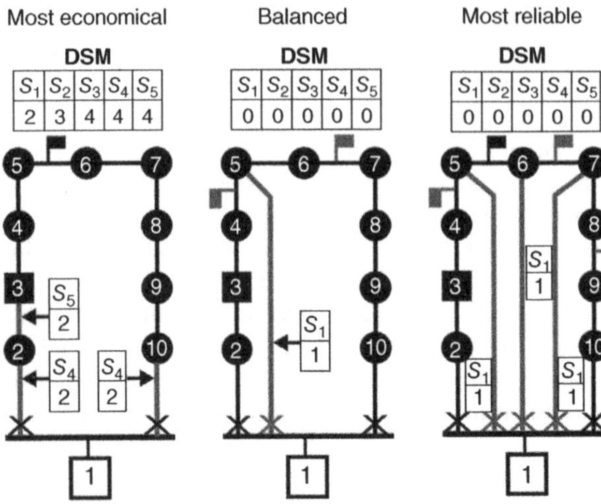

Figure 5.5.9 Network 1: total cost (CAPEX + DSM) vs. CML.

required during the planning period. However, those NOP locations are not the favorable positions to reduce CML.

Figure 5.5.9 shows three examples for network 1. The most economical solution promotes DSM options and replaces three overload cables at much later stages of the planning period. The most reliable solution adds three new cable connections in the first infeasible year t_X to create three new feeders and to reduce the number of customers per feeder. A balanced solution adds only one new cable connections in the year t_X.

Figure 5.5.10 shows the PF of solving DNEP regarding the total cost (CAPEX + DSM) versus the average CML. For networks 1 and 2, both MO-GOMEA variants DQ and BX-M obtain similar results. For network 3, MO-GOMEA BX-M obtains a better PF than MO-GOMEA DQ. However, the performance of the (out-of-the-box) MO-GOMEA DQ is still reasonably good. This confirms the reliability of the basic variant of MO-GOMEA. The effects of DQ and BX-M here conform quite well with one of our previous studies [111] in the case of solving the single-objective DNEP with (SO) GOMEA. EAs that make use of linkage learning (LLEAs) are usually robust solvers, which can be used out-of-the-box or with minimum modifications to solve complicated problems like MO-DNEP or SO-DNEP. Customizing LLEAs with expert knowledge to make them problem-dedicated solvers can help to further improve the obtained results.

5.5.6 Conclusions

In this work, for the first time, we considered a multi-objective DNEP problem formulation that includes both asset investments and DSM policies as expansion options in combination with an efficient decomposition heuristic that can convert

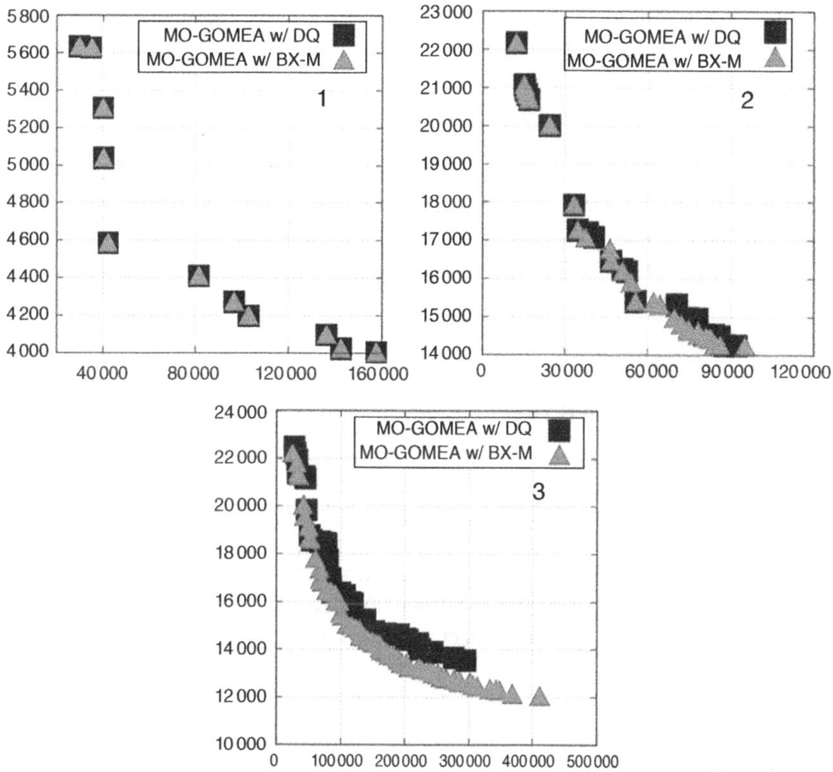

Figure 5.5.10 Fronts of MO optimization (CAPEX + DSM) vs. CML for three networks.

static (single-stage) network configurations into feasible dynamic (stage-by-stage) installation schedules while regarding practical problem constraints. We argued that DNEP is a true MOOP because trade-offs between many conflicting criteria must be taken into account before deciding to carry out a specific expansion plan. We considered the MO-DNEP problem with various combinations of two out of four different objectives: minimizing the physical asset investment cost CAPEX, minimizing the cost of using DSM options, minimizing the total energy loss, and minimizing the CML per year. Note that more than two objectives can be used in our framework, but for the purpose of demonstration in this study, considering two objectives at a time is sufficient. We employed the state-of-the-art MO-GOMEA together with some problem-dedicated adaptations to solve the complicated MO-DNEP for several benchmark networks.

Based on the experimental results, we conclude that solving DNEP in an MO optimization manner returns a far richer set of valuable results and alternatives for DNOs to consider than when solving the SO-DNEP. Being exposed to different possible (optimal) trade-offs, DNOs can make well-informed decisions for each specific situation. The PFs found by MO optimization can be used as a visualization tool to effectively exhibit how much energy efficiency and network reliability are compromised as DNOs reduce investment costs and also by how much investments must be increased to improve these two objectives as well. We also showed that by using DSM options to deal with the peak load growth, DNOs can indeed postpone costly asset installations. This can be an incentive for DNOs to actively participate in DSM research, development, and deployment. Finally, MO-GOMEA is shown to be a robust MO solver that can tackle a complicated problem like MO-DNEP well, making it a promising framework to be considered for solving other power and energy optimization problems in the future.

ACKNOWLEDGEMENT

The research conducted for this work is part of the "Computational Capacity Planning in Electricity Networks (COCAPLEN)" projects. COCAPLEN is a joint project between the Intelligent System Groups at Centrum Wiskunde & Informatica (CWI) (The Dutch National Research Institute for Mathematics and Computer Science) and the Electrical Energy System Group at Eindhoven University of Technology. The project is funded by the Netherlands Organisation for Scientific Research (NWO).

APPENDIX A

A.1. MV Cable Types

ID	Type (mm^2)	I_{nom} (A)	R (Ω/km)	X (Ω/km)	C (μF/km)	Cost (EUR/km)
1	120	215	0.257	0.085	0.38	50 000
2	150	295	0.208 58	0.095 92	0.383 3	59 000
3	240	370	0.135 17	0.108 23	0.435 53	62 000
4	400	475	0.080 77	0.099 72	0.534 4	N/A
5	630	605	0.051 1	0.092 72	0.641 03	N/A
6	N/A	135	0.532 53	0.097 77	0.270 72	N/A
7	N/A	160	0.373 7	0.093 67	0.306 71	N/A
8	N/A	195	0.267 56	0.089 95	0.345 05	N/A
9	N/A	225	0.328 29	0.101 34	0.326 67	N/A
10	N/A	320	0.130 79	0.077 57	0.531 09	N/A
11	N/A	350	0.101 93	0.080 04	0.484 85	N/A
12[a]	N/A	N/A	N/A	N/A	N/A	N/A

[a] Type 12 is sub-transmission cable and is not considered in MV DNEP. Types 4–11 are currently existing cable types but are not used for new cable installations.

A.2. Network 1 Data

Node Information				Cable Information				
	Load			Existing			Potential	
ID	P (kW)	Q (kvar)	Customers No.	Branch	Length (m)	Type	Branch	Length (m)
1	0	0	0	1–2	654	1	1–3	1235
2	271	168	131	1–10	710	1	1–4	1259
3	924	573	1	2–3	610	1	1–5	1323
4	394	244	190	3–4	163	1	1–6	1711
5	409	253	197	4–5	511	1	1–7	1904
6	394	244	190	5–6	496	1	1–8	1976
7	370	229	179	6–7	420	1	1–9	1781
8	117	72	57	7–8	297			
9	259	160	125	8–9	336			
10	431	267	208	9–10	690			

A.3. Network 2 Data

Node Information				Cable Information				
	Load			Existing			Potential	
ID	P (kW)	Q (kvar)	Customers No.	Branch	Length (m)	Type	Branch	Length (m)
1	0	0	0	1–2	481	3	1–3	526
2	35	17	25	1–16	246	3	1–4	469
3	1113	539	1	1–31	761	3	1–5	989
4	348	216	1	2–3	96	2	1–6	2062
5	871	286	1	3–4	48	2	1–7	738
6	332	109	232	4–5	498	2	1–8	2227
7	132	82	92	5–6	86	2	1–9	2307
8	170	82	119	6–7	288	2	1–10	2633
9	22	14	15	7–8	935	2	1–11	3041
10	202	98	141	8–9	200	2	1–12	3041
11	120	0	84	9–10	470	2	1–13	1395
12	88	55	62	10–11	851	2	1–14	1194
13	284	137	199	10–17	736	2	1–15	923

(continued)

Node Information				Cable Information				
	Load			Existing			Potential	
ID	P (kW)	Q (kvar)	Customers No.	Branch	Length (m)	Type	Branch	Length (m)
14	219	136	153	11–12	220	3	1–17	2808
15	314	152	220	12–13	300	3	1–18	2760
16	185	90	130	13–14	284	3	1–19	2653
17	127	79	1	14–15	479	3	1–20	1275
18	17	8	12	15–16	846	3	1–21	1205
19	896	434	1	17–18	101	2	1–22	1136
20	314	152	220	18–19	154	2	1–23	1131
21	125	77	88	19–20	283	2	1–24	1041
22	248	120	174	20–21	308	2	1–25	950
23	85	41	60	21–22	133	2	1–26	966
24	123	76	86	22–23	132	2	1–27	900
25	209	130	146	23–24	138	2	1–28	804
26	566	274	1	24–25	140	2	1–29	677
27	266	129	186	25–26	103	2	1–30	801
28	126	61	88	26–27	215	2		
29	360	174	1	27–28	139	2		
30	273	169	191	28–29	218	2		
31	263	163	1	29–30	136	2		
				30–31	160	3		

A.4. Network 3 Data

Node Information				Cable Information						
	Load			Existing			Potential			
ID	P (kW)	Q (kvar)	Customers No.	Branch	Length (m)	Type	Branch	Length (m)	Branch	Length (m)
1	0	0	0	1–2	1	12	2–5	1335	8–38	1900
2	67	32	22	1–36	1	12	2–6	1357	8–40	2754
3	185	90	108	2–3	670	6	2–9	1256	8–41	3542
4	112	54	1	2–4	280	7	2–10	1040	8–42	3252
5	194	94	92	2–8	1	12	2–13	1840	8–43	2847

Node Information				Cable Information						
	Load			Existing			Potential			
ID	P (kW)	Q (kvar)	Customers No.	Branch	Length (m)	Type	Branch	Length (m)	Branch	Length (m)
6	61	30	14	2–12	1820	7	2–14	1987	8–44	2696
7	152	73	216	3–5	570	6	2–15	2123	8–45	2510
8	158	76	143	4–35	380	9	2–16	1948	8–46	2148
9	282	136	317	5–17	570	6	2–17	1541	8–47	1900
10	193	94	153	6–7	1300	6	2–18	2507	8–48	1781
11	165	54	39	6–38	759	9	2–19	2437	8–49	1725
12	148	72	1	7–8	1421	6	2–20	3102	8–51	3405
13	91	44	47	7–9	610	7	2–21	2885	9–23	1840
14	119	57	1	8–11	360	7	2–28	3605	9–36	1693
15	311	150	186	8–24	570	8	2–29	3160	10–23	1981
16	314	152	295	8–27	570	8	2–30	3950	10–36	1890
17	333	161	245	9–10	250	7	2–31	3560	13–23	1496
18	351	170	397	10–11	340	7	2–32	3855	13–36	1189
19	236	114	167	12–13	320	7	2–33	3769	14–23	1822
20	297	144	351	13–14	380	7	2–34	2437	14–36	1388
21	253	122	264	14–15	150	7	2–35	372	15–23	1978
22	355	172	9	15–16	800	7	2–38	372	15–36	1496
23	492	238	208	15–18	510	7	2–40	4483	16–23	2726
24	152	74	34	16–17	570	7	2–41	5283	16–36	2277
25	156	75	1	18–19	280	7	2–42	4969	17–23	3063
26	186	90	41	19–34	510	7	2–43	4632	17–36	2704
27	310	150	211	20–21	300	7	2–44	4487	18–23	1887
28	292	141	4	20–34	510	7	2–45	4291	18–36	1321
29	11	6	1	21–22	530	7	2–46	3972	19–23	1524
30	230	111	8	22–28	955	8	2–47	3726	19–36	983
31	136	66	1	22–36	313	9	2–48	3593	20–23	1563
32	287	139	232	23–25	400	7	2–49	3546	20–36	920
33	298	144	186	23–26	350	7	2–51	5159	21–23	1219
34	174	84	167	23–27	350	7	5–8	1919	21–36	583
35	180	87	1	23–36	1	12	5–23	3038	23–28	1951
36	0	0	0	23–39	590	5	5–36	2718	23–29	2217
37	806	500	1	24–25	365	7	6–8	3086	23–30	2029
38	156	75	1	26–31	785	7	6–23	4250	23–31	900
39	259	126	134	28–29	465	8	6–36	4122	23–32	1194

(*continued*)

Node Information				Cable Information						
	Load			Existing			Potential			
ID	P (kW)	Q (kvar)	Customers No.	Branch	Length (m)	Type	Branch	Length (m)	Branch	Length (m)
40	281	136	76	29–30	740	8	8–9	671	23–33	1515
41	310	150	69	30–33	685	8	8–10	819	23–34	1524
42	217	105	2	31–32	272	7	8–13	530	23–35	3061
43	153	74	34	32–33	671	7	8–14	974	23–38	3061
44	137	66	1	35–38	426	9	8–15	1190	23–40	1593
45	259	126	62	36–37	94	9	8–16	1772	23–41	2374
46	261	126	121	36–50	450	7	8–17	1977	23–42	2092
47	226	110	50	39–40	640	7	8–18	1342	23–43	1678
48	269	130	139	40–51	1251	7	8–19	1036	23–44	1528
49	218	106	1	41–42	430	7	8–20	1557	23–45	1340
50	136	66	5	41–51	229	9	8–21	1231	23–46	1017
51	240	116	53	42–43	387	7	8–28	2088	23–47	778
				43–44	130	7	8–29	1947	23–48	629
				44–45	520	6	8–30	2354	23–49	592
				45–46	350	6	8–31	1731	23–51	2235
				46–47	233	6	8–32	2033	28–36	1328
				47–48	180	6	8–33	2039	29–36	1581
				48–49	150	7	8–34	1036	30–36	1456
				49–50	435	7	8–35	1900	31–36	632
									32–36	928
									33–36	1009
									34–36	983
									35–36	2945
									36–38	2945
									36–40	2006
									36–41	2721
									36–42	2489
									36–43	1973
									36–44	1818
									36–45	1672
									36–46	1204
									36–47	971
									36–48	936
									36–49	850
									36–51	2567

5.6 GA-GUIDED TRUST-TECH METHODOLOGY FOR CAPACITOR PLACEMENT IN DISTRIBUTION SYSTEMS

Hsiao-Dong Chiang[1], *Jinda Cui*[2], *and Tianshi Xu*[3]

[1]*Cornell University, Ithaca, NY, USA*
[2]*Lehigh University, Bethlehem, PA, USA*
[3]*Tianjin University, Tianjin, PRC*

5.6.1 Introduction

For most practical applications of optimization technologies, the underlying objective functions are often nonlinear and depend on a large number of variables, making the task of searching the solution space to find the globally optimal solution very challenging. The primary challenge is that, in addition to the high-dimension solution space, there are many local optimal solutions in the solution space in which a local optimal solution is optimal in a local region of the solution space, but not in the global solution space. The globally optimal solution is just one solution and yet both the globally optimal solution and locally optimal solutions share the same local properties. In general, the number of local optimal solutions is unknown, and it can be quite large. Furthermore, the values of an objective function at the local optimal solutions and at the global optimal solution may differ significantly. Hence, there are strong motivations to develop effective methods for finding the global optimal solution.

One popular method for solving nonlinear optimization problems is to use an iterative local improvement search procedure. Local improvement search methods usually get trapped at local optimal solutions and are unable to escape from them. In fact, the great majority of existing nonlinear optimization methods for solving optimization problems usually produce local optimal solutions but not the globally optimal solution. The drawback of iterative local improvement search methods has motivated the development of more sophisticated local search methods, designed to find better solutions via introducing some mechanisms that allow the search process to escape from local optimal solutions. The underlying "escape" mechanisms use certain search strategies that accept a cost-deteriorating neighborhood to make an escape from a local optimal solution possible. These sophisticated local search algorithms include SA, GA, TS, evolutionary programming (EP), and PSO methods. However, these sophisticated local search methods, among other problems, require intensive computational effort and usually cannot find the globally optimal solution.

Significant efforts have been made to develop hybrid search methods which combine a stochastic method and an iterative local improvement search method. For instance, a class of hybrid methods is formed by the combination of GAs and local gradient methods (or the sensitivity-based methods). But these hybrid methods still suffer from several difficulties in the computational effort and the quality of optimal solutions found. To overcome the difficulties encountered by the majority of existing optimization methods, the following two important and challenging issues in the course of searching for multiple high-quality optimal solutions need to be fully addressed:

C1. How to effectively move (escape) from a local optimal solution and move on toward another local optimal solution; and

C2. How to avoid revisiting local optimal solutions which are already known.

In the past, significant effort has been directed toward attempting to address these two issues, but without much success. Issue (C1) is difficult to solve and the existing methods all encounter this difficulty. Issue (C2), related to computational efficiency during the course of the search, is also difficult to solve and again, the majority of the existing methods encounter this difficulty. Issue (C2) is a common problem, which degrades the performance of many existing methods in searching for the globally optimal solution. Revisitation of the same local optimal solution several times indeed wastes computing resources without gaining new information regarding the location of the globally optimal solution. From the computational viewpoint, it is important to avoid revisiting the same local optimal solution in order to maintain a high level of efficiency.

The purpose of this chapter is to present a GA-guided Trust-Tech method. This method belongs to the class of hybrid methods; however, it is not the traditional one which combines a stochastic method such as GA and an iterative local search method. This new method, a combination of GA and the Trust-Tech method, has several distinguishing features to be described in the next section and can address the two challenging issues: (C1) and (C2). The GA-guided Trust-Tech method takes the global search capability of GA and the tier-by-tier search capability of the Trust-Tech method. In this combination, the role of GA is to identify promising regions in the solution space and guide the Trust-Tech method to deterministically find all of the local optimal solutions located in the promising regions.

We illustrate the effectiveness of the GA-guided Trust-Tech method in solving the capacitor placement problem in distribution networks. Capacitor placement/replacement is a practical way to minimize losses as well as operating costs in power distribution systems. It also benefits the network with a better voltage profile, power factor correction, and system delivery capacity enhancement. The extent of the improvements usually depends on the placement scheme which consists of the locations, sizes, types, states, and quantity of capacitors. An adverse placement may lead to loss increment or damage in other operating indices.

Considerable research efforts have been devoted to this area during the past years. A variety of algorithms have been developed and attempted for a better solution of this problem, such as GA [131–135], PSO [136, 137], TS [138], Fuzzy theory [139], and Memetic Algorithm [140]. Research efforts were also directed to the consideration of DG [141], combining network reconfigurations with capacitor placement [142, 143]. It is, however, still important to improve traditional algorithms due to their intrinsic deficiencies [144]. Meta-heuristic methods lack the fine-tuning capability to zoom-in local optimal solutions and are stochastic in nature to find solutions.

The two-stage algorithm according to the GA-guided Trust-Tech method for capacitor placement consists of the following steps:

I. Identify the regions of solution space that contain high-quality optimal solutions by using the global search capability of GA.

II. Exploit the promising regions to compute the local optimal using the Trust-Tech method.

It will be numerically shown that the two-stage algorithm according to the GA-guided Trust-Tech method achieves better solutions in a much more efficient way as compared with the GA alone.

5.6.2 Overview of the Trust-Tech Method

The only reliable way to find the globally optimal solution of a nonlinear optimization problem is to first find all the local optimal solutions and then, from them, find the globally optimal solution. To this end, we will present in this section a systematic method, which is deterministic in nature, to locate all the local optimal solutions of nonlinear optimization problems. This method is based on the following transformations:

- The transformation of a local optimal solution of a nonlinear optimization problem into a stable equilibrium point of a continuous nonlinear dynamical system.
- The transformation of the search space of nonlinear optimization problems into the union of the closure of stability regions of stable equilibrium points.

Hence, the optimization problem (i.e. the problem of finding local optimal solutions) is transformed into a problem of finding stable equilibrium points, and it will become clear that the stability regions of stable equilibrium points play an important role in finding these local optimal solutions. This methodology is termed Trust-Tech, which stands for Transformation Under Stability-reTaining Equilibria Characterization [145–152].

One distinguishing feature of the Trust-Tech methodology is that it systematically searches all the local optimal solutions of nonlinear optimization problems in a deterministic manner. Another important feature of the Trust-Tech method is that it finds local optimal solutions in a tier-by-tier manner, starting from a local optimal solution, finds the nearby first-tier local optimal solutions, then the second-tier local optimal solutions, and so on.

Trust-Tech-based methods are dynamical methods for obtaining a set of local optimal solutions of general optimization problems. We believe that an effective approach to search for high-quality optimal solutions of general optimization problems is the one that first finds multiple, if not all, local optimal solutions and then selects the best solution from them. We explain the Trust-Tech framework in solving the following unconstrained NLP problem:

$$\begin{gathered}\min C(x)\\ x \in R^n\end{gathered} \tag{5.6.1}$$

where $C: R^n \rightarrow R$ is a function bounded below and possesses only finite local optimal solutions. Our focus for this problem is to locate all or multiple local optimal solutions of $C(x)$.

Instead of solving the unconstrained optimization problem, described by (5.6.1) directly, we consider the corresponding dynamical system:

$$\dot{x}(t) = -\nabla C(x) \tag{5.6.2}$$

where $x \in R^n$. It will be shown that each local optimal solution of problem (5.6.1) corresponds to a stable equilibrium point of the gradient system (5.6.2). This transformation allows us to locate each local optimal solution of problem (5.6.1) via each stable equilibrium point of the gradient system (5.6.2). We next derive several geometrical and topological properties of the gradient system (5.6.2), which will be useful in our development of a systematic method for obtaining several local optimal solutions.

Theorem 5.6.1: (Completely Stable) [148] If there exist an ε and σ such that $\|\nabla C(x)\| > \varepsilon$ unless $x \in B_\delta(\widetilde{x})$, where $\widetilde{x}$ is an equilibrium point, then the gradient system (5.6.2) is completely stable and $C(x)$ is an associated Liapunov function.

For a completely stable system, every trajectory converges to one of its equilibrium points. Hence, the state space is the union of the closure of the stability regions; see, for example, Figure 5.6.1. In other words, every trajectory converges to one of its stable equilibrium points or to one of its unstable equilibrium points.

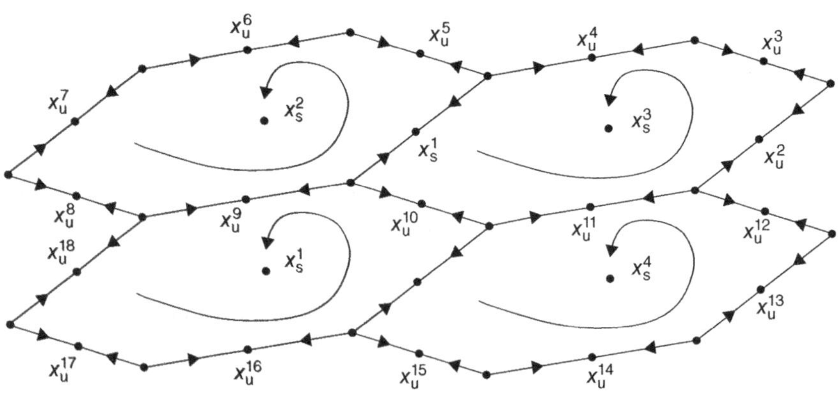

Figure 5.6.1 Convergence for equilibrium points.

From a numerical simulation viewpoint, every trajectory does converge to one of its stable equilibrium points.

Next is presented a complete characterization of the stability boundary of the gradient system (5.6.2). This characterization is expressed in terms of the stable manifolds of the equilibrium points lying on the stability boundary.

Proposition 5.6.2: (Characterization of the Stability Boundary) [148] Suppose that all the equilibrium points of the gradient system (5.6.2) are hyperbolic. Let x_i, $i = 1, 2, \ldots$ be the equilibrium points on the stability boundary $\partial A(x_s)$ of a stable equilibrium point, say x_s. Then, the stability boundary is contained in the union of the stable manifolds of the equilibrium points on the stability boundary; in other words,

$$\partial A(x_s) \subseteq \bigcup_{\sigma_i \epsilon \partial A} W^s(\sigma_i) \tag{5.6.3}$$

Theorem 5.6.3: (Equilibrium Points and Local Optimal Solutions) [148] If $\bar{x}$ is a hyperbolic equilibrium point of gradient system (5.6.2), then $\bar{x}$ is a stable equilibrium point of system (5.6.2), if and only if $C(x)$ has an isolated minimum of the optimization problem (5.6.1) at $\bar{x}$.

Theorem 5.6.3 characterizes the relationship between the optimal solutions of the unconstrained optimization problem (5.6.1) and the stable equilibrium points of its corresponding dynamical system (5.6.2). Hence, if x_s is a stable equilibrium point of (5.6.2), then it is a local optimal solution of the unconstrained optimization problem (5.6.1). Conversely, if x_s is a local optimal solution of (5.6.1), then it is a stable equilibrium point of the gradient system (5.6.2). Our efforts are hence focused on developing effective algorithms to locate stable equilibrium points of (5.6.2), based on several topological and geometrical properties of stability boundaries of the gradient system (5.6.2). Because of such correspondence, the problem of computing multiple local optimal solutions of the optimization problem is then transformed to finding multiple stability regions in the defined dynamical system, each of which contains a distinct stable equilibrium point (SEP). A SEP can be computed with the trajectory method or using a local method with a trajectory point in its stability region as the initial point.

Recall the concept of the stability region of equilibrium point x_s:

$$A(x_s) = \{x \in R^n : \lim_{t \to \infty} \emptyset\,(x, t) = x_s\} \tag{5.6.4}$$

The boundary of $A(x_s)$ is called the stability boundary and is denoted by $\partial A(x_s)$. The concept of qausi-stability region $A_q(x_s)$ is derived from the concept of a stability region $A(x_s)$ and is defined as quasi-stability region $A_q(x_s) = \mathrm{int}\left(\overline{A(x_s)}\right)$ [152], where $\overline{A(x_s)}$ represents the closure of $A(x_s)$ and int(:) represents the interior of the set. Hence, the quasi-stability region is the interior of the closure of the stability region. Its boundary is defined as the quasi-stability boundary and denoted by $\partial A_q(x_s)$. It has been shown that the quasi-stability region eliminates the complex portion of the stability boundary lying inside the interior of $\overline{A(x_s)}$ and is more suitable for practical applications.

We shall call a type-one equilibrium point lying on a quasi-stability boundary $\partial A_q(x_s)$ a decomposition point (with respect to x_s). Decomposition points and their stable manifolds will be used to characterize the structure of a quasi-stability boundary. The searching procedure of Trust-Tech will be based on such a characterization.

The decomposition point x_d distinguishes itself from other type-one equilibrium points in the following way: as the Liapunov (or energy) function value increases from $c(x_d) - \varepsilon$ to $c(x_d) + \varepsilon$, the number of path components of $S_{c(x_d)}(x)$ decreases by one, i.e. $S_{c(x_d)+\varepsilon}(x) = S_{c(x_d)-\varepsilon}(x) - 1$. On the other hand, if x_1 is a type-one equilibrium point other than a decomposition point, the number of path components remains the same as the Liapunov function value increases from $c(x_d) - \varepsilon$ to $c(x_d) + \varepsilon$.

Before describing the procedure to locate multiple local optimal solutions, we need to justify that decomposition points can serve as the bridge linking two stable equilibrium points. The theorem below provides such a justification.

Theorem 5.6.4: (Decomposition Point and Stable Equilibrium Points) Suppose that every equilibrium points of dynamical system (5.6.2) is hyperbolic and its stable and unstable manifolds satisfy the transversally condition. If x_s^1 is a stable equilibrium point of dynamical system (5.6.2) and x_d is a decomposition point on its quasi-stability boundary, then there exists another stable equilibrium point x_s^2 to which the one-dimensional unstable manifold of x_d converges.

We note that the decomposition point must be type-one, the unstable manifold $W^u(x_d)$ is a one-dimensional manifold. Remove x_d from the unstable manifold to form two half-manifolds, $W^{u+}(x_d)$ and $W^{u-}(x_d)$. Each half-manifold is itself an invariant manifold composed of a single trajectory. These two trajectories must converge to two stable equilibrium points. These characterizations can be used to devise a mechanism to escape from one stable equilibrium point to its adjacent stable equilibrium points. The basic idea is described as follows: the entire state space is decomposed into the closure of all the stability regions. Two adjacent stability regions are separated by the intersection of their stability boundaries which is the stable manifold of a decomposition point. To identify the adjacent stable equilibrium point from the initial stable equilibrium point, it suffices to devise a mechanism to cross the stability boundary from one stability region to reach its adjacent stability region.

5.6.3 Computing Tier-One Local Optimal Solutions

We present a Trust-Tech-based dynamic decomposition point (DDP) method for locating another local optimal solution, i.e. tier-one local optimal solution, from a local optimal solution of the unconstrained optimization problem (5.6.1). The local optimal solution, say x_s, can be found by an optimization method, such as an interior point method, a gradient-based method, or a sequential quadratic programming (SQP) method. Starting from an initial stable equilibrium point x_s^0 (i.e. a local optimal solution), we follow a specified search direction which is one of the eigenvectors of the Jacobian matrix of the gradient system at the stable equilibrium point and progressively sample a sequence of points $\{x_i\}$ with a small distance ε. By monitoring the value of the objective function $C(x)$ starting from the local optimal solution, it is clear that the value along the sequence will initially increase, since the sequence starts from a local optimal solution. The values along the sequence will continue to increase until the sequence reaches a point at which the value of objective function $C(x)$ decreases. This point is termed an approximated exit point. This approximated exit point should be very close to the exact exit point, which lies on the stable manifold of a DDP and is the intersection between the stability boundary and the curve connecting the sequence (see Figure 5.6.2).

The exit point serves as the initial point of a trajectory lying on the stability boundary to locate the DDP. This procedure computes the DDP via moving along the stable manifold of DDP. Due to the digital simulation of the trajectory, this stability-boundary-following procedure constructs a sequence of points, starting from the exit point, moving along the stability boundary of x_s leading to a point which is close to the corresponding DDP. This stability-boundary-following procedure will generate a sequence of points $\{z_i\}$ which is close to the stability boundary. Since each point in sequence $\{z_i\}$ is close to the stability boundary, the behavior of the sequence will approximately follow the behavior of a trajectory located on the stability boundary. Due to the structure of the stability boundary, the point in the sequence $\{z_i\}$ having the minimal value of the norm of the vector field is the closest point to the decomposition point x_d. The point at the end of this procedure is termed the minimum gradient point (MGP). From a geometric viewpoint, the MGP is close to the desired DDP. Using the MGP as an initial guess and applying a robust nonlinear algebraic solver, one can obtain the DDP x_d.

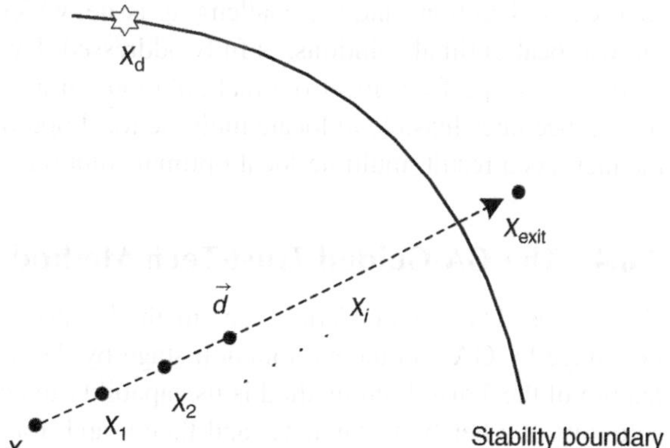

Figure 5.6.2 x_{exit} should be very close to the exact exit point.

Now, with the information of a known local optimal solution x_s^0 and a DDP x_d located on the stability boundary, this step seeks to locate a point lying inside the stability region of another stable equilibrium point (i.e. another local optimal solution). It asserts that the unstable manifolds of DDP x_d connect the initial stable equilibrium point x_s^0 and another SEP, say x_s^1.

We present a method to numerically implement this step. The method is to move outward along the direction $x_s^0x_d$, say a step-size of 10% to obtain a point, say y_0 (i.e. $y_0 = x_s{}^0x_d + 0.1x_s^0x_d$). It is clear that point y_0 is located inside the stability region of x_s^1, provided the step-size is not large. We use point y_0 as an initial condition to numerically integrate the dynamical system. This numerical integration will generate a trajectory converging to the stable equilibrium point x_s^1. Hence, another local optimal solution is located.

One can monitor the ensuing trajectory until it reaches a point close to the targeted stable equilibrium point x_s^1 (i.e. the norm of the vector field is close to zero); one may stop the numerical integration and resort to a local optimization solver. The ending point of the numerical integration serves as an initial condition of the solver to locate another local optimal solution x_s^1. The efficiency of searching another local optimal solution can be improved by using a hybrid of the proposed method and an efficient local optimizer.

The Trust-Tech-based DDP method requires the stability-boundary-following procedure involving numerical integration, functional evaluation (for closeness to the stability boundary), and a nonlinear algebraic solver (for locating the DDP). This method has a nice feature of not repeatedly finding the same local optimal solution starting from a known local optimal solution. The theoretical basis of this feature is explained as follows. The unstable manifold of decomposition point x_d connects the initial stable equilibrium point x_s^0 and the targeted stable equilibrium point x_s^1. It is obvious that different search directions from the initial stable equilibrium point x_s^0 will lead to different exit points on the stability boundary of x_s^0. If multiple different exit points move along the same stable manifold of a DDP, then one locates the same decomposition point from these multiple different exit points. It gives us the fact that the same local optimal solution can be found from the two search directions and the challenging issue, which is how to avoid revisiting known local optimal solutions, is fully addressed. By exploring the Trust-Tech's ability to escape from local optimal solutions in a systematic and deterministic way, it becomes feasible to locate multiple local optimal solutions in a tier-by-tier manner. As a result, multiple local optimal solutions can be obtained.

5.6.4 The GA-Guided Trust-Tech Method

The proposed GA-guided Trust-Tech method consists of two stages: the exploration stage by GA and the exploitation stage by the Trust-Tech method. The key feature of the Trust-Tech method is its capability to compute all the local optimal solutions in a tier-by-tier manner and then search for the global optimum among them. If the initial point is not close to the global optimal solution, then the task of finding the global optimal solution may take several tiers of local optimal solutions computation. Hence, the goal of Stage I is to reduce the number of tiers to be computed by the Trust-Tech method.

Stage I: Exploration Stage

The basic idea of GA is to try to mimic a simple picture of natural selection in order to find a good algorithm. The first step is to mutate, or randomly vary, a given collection of sample programs. The second step is a selection step, which is often done through measuring against a fitness function. The process is repeated until a suitable solution is found. There are a large number of different types of GAs.

One advantage of a GA is that it does not require the fitness function to be very smooth, since a random search is done instead of following the path of a greedy search.

Outline of the Genetic Algorithm

Step 1: Generate the random population of n chromosomes (suitable solutions for the problem).

Step 2: Evaluate the fitness $f(x)$ of each chromosome x in the population.

Step 3: Create a new population by repeating the following steps until the new population is complete.

[Selection] Select two parent chromosomes from a population according to their fitness (the better the fitness, the bigger the chance to be selected).

[Crossover] With a crossover probability, crossover the parents to form a new offspring (children). If no crossover was performed, the offspring is an exact copy of the parents.

[Mutation] With a mutation probability, mutate new offspring at each locus (position in chromosome).

[Accepting] Place new offspring in a new population.

Step 4: If the end condition is satisfied, **stop**, and return the best solution in the current population; otherwise, go to **Step 2**.

Termination of Stage I

The GA calculation is terminated when its speed of convergence to a better solution slows down. A criterion is proposed here to judge the performance of slow down, as described below. We observe that the objective function value drops sharply in early stages of GA and becomes relatively flat (i.e. very slowly convergent) as GA calculation proceeds [153]. If we can determine the "dropping" speed of each generation, the termination criterion can be easily made based on that value. However, due to the random and discrete nature of GAs, simply combining the "speed" property with a single generation is insufficient. For example, the objective value may remain flat in several generations of GA, but after a few generations, changes in the objective function value (better/worse) could be dramatic. In order to capture these changes in the objective function value, we divide the obtained generations into several segments and combine the "speed" property with each segment.

However, it should be noticed that the "dropping speed" of a specific segment may not always be a positive term; the objective function value may increase during a segment of generations but produce better solutions in the following segments, making it difficult to prompt a termination condition. In GA, the diversity of population is of great importance to its global search capability. If we express this capability as "vitality," the objective value changes among generations in a segment can reflect the diversity, and hence it can be a measure of the vitality of the calculation. We define the vitality of the *i*th segment as follows:

$$V(i) = \frac{\left(F^i_{\max} - F^i_{\min}\right)}{F^i_{mean}} \tag{5.6.5}$$

where $F^i_{\max}$, $F^i_{\min}$, F^i_{mean} are the maximum, minimum, and mean of the objective function value of the generations within the segment.

The segment width (number of generations within a segment) of the *i*th segment increases with *i*:

$$\text{Gen}(i) = r \cdot \text{Gen}(i-1) \tag{5.6.6}$$

where *r* is the selected increment ratio. Given Gen(0), we can get the corresponding *V*(0) and then further calculate each *V*(*i*) as GA proceeds. We compute the ratio *V*(*i*)/*V*(0) at each segment, and predefine a threshold value, which we call the "termination ratio." GA calculation is continued until *V*(*i*)/*V*(0) is less than the threshold value.

Stage II: Exploitation Stage

The Trust-Tech method plays a key role in this stage and exploits all the local optimal solutions in each region in a tier-by-tier manner:

1. From an obtained local optimal solution, the Trust-Tech method intelligently moves away from the local optimal solution and approach, together with the local method, another local optimal solution in a tier-by-tier manner.
2. After finding the set of tier-1 local optimal solutions, the Trust-Tech method continues to find the set of tier-2 local optimal solutions if necessary.

We present a Trust-Tech-based DDP method for locating another local optimal solution from a local optimal solution of the unconstrained optimization problem (5.6.1). The local optimal solution, say x_s, can be found by an optimization method, such as an interior point method, a gradient-based method, or a SQP method.

Given: an unconstrained optimization problem (5.6.1) and a local optimal solution, say x_s^1.

Goal: find a tier-one local optimal solution

Step 0. Construct the corresponding gradient system (5.6.2).

Step 1. Construct a path moving away from the local optimal solution, x_s^1 (which is a stable equilibrium point of the gradient system (5.6.2) and toward the stability boundary $\partial A(x_s^1)$).

Step 2. Identify the exit point at which the constructed path intersects with the stability boundary $\partial A(x_s^1)$.

Step 3. If the exit point exists, say x_e, then another stable equilibrium point (i.e. another local optimal solution) exists. (According to Theorem 5.6.3, this exit point must lie on the stable manifold of a DDP).

Step 4. Starting from the exit point x_e and integrating the gradient system (5.6.2) results in a trajectory moving along the stable manifold of the DDP until the point whose norm of the vector field of the gradient system (5.6.2) is zero (or close to zero).

Step 5. Apply the point obtained from Step 4 to compute the DDP which separates the initial stable equilibrium point x_s^1 and a corresponding stable equilibrium point whose stability boundary contains the joint stability boundary of $\partial A(x_s^1)$.

Step 6. From the found DDP, generate one point, which is a vector lying inside the stability region of the corresponding stable equilibrium point, as stated in Step 5.

Step 7. Starting from the generated point at Step 6 and integrating the gradient system (5.6.2) results in a trajectory, which will converge to the corresponding stable equilibrium point (i.e. another local optimal solution).

The Trust-Tech-based DDP method requires the stability-boundary-following procedure involving numerical integration, functional evaluation (for closeness to the stability boundary), and a nonlinear algebraic solver (for locating the DDP). This method has a nice feature of non-repeatedly finding the same local optimal solution starting from a known local optimal solution. The theoretical basis of this feature is explained as follows.

From Theorem 5.6.4, it follows that the unstable manifold of decomposition point x_d connects the initial stable equilibrium point x_s^1 and the targeted stable equilibrium point x_s^2. It is obvious that different search directions from the initial stable equilibrium point x_s^1 will lead to different exit points on the stability boundary of x_s^1. If multiple different exit points move along the same stable manifold of a DDP, then one locates the same decomposition point from these multiple different exit points. It follows from Theorem *N*-4 that the same stable equilibrium point (i.e. the same local optimal solution) will be found if one applies Steps 5 and 6. Therefore, under this circumstance, these two steps are not necessary and the repetition of finding the same local optimal solution from two different search directions can be avoided. Hence, the challenging issue (C2) is fully addressed.

Theoretically speaking, Trust-Tech can continue to find the set of third-tier or higher-tier local optimal solutions at the expense of considerable computational efforts. Since the GA of Stage I provides solution points around the global optimal solution, the Trust-Tech method of Stage II searches fewer numbers of tiers to obtain high-quality local optimal solutions. It has been our experience that of the set of first-tier local optimal solutions, there usually exists a very high-quality local optimal solution, if it is not the global optimal solution. Hence, the exploitation process is terminated after finding all the tier-1 local optimal solutions.

5.6.5 Applications to Capacitor Placement Problems

In this section, we present the application of the proposed GA-guided Trust-Tech Methodology to the capacitor placement problem in distribution systems.

Problem Formulation

To determine the optimal capacitor placement schemes in a realistic distribution system, we formulate a nonlinear constraint optimization problem as follows:

$$\begin{aligned} &\min f(x) \\ \text{s.t. } &h(x) = 0 \\ &g(x) \leq 0 \end{aligned} \tag{5.6.7}$$

where $f(x)$ is the objective function. The x here represents a set of state and control vectors, $x = [x_s, x_c]$, where $x_s = [V_1, V_2, \ldots, V_n, \delta_1, \delta_2, \ldots, \delta_n]$ is the state vector that-consists of the voltage amplitude and angle of each phase on each bus. $x_c = [t_{1,\,1}, t_{1,\,2}, \cdots, t_{n,\,c}]$ is the control vector that consists of the existence information of the size-c capacitor on bus-n, 1 = exist, and 0 = non-exist. We can obtain the control vector which represents the capacitor placement scheme by minimizing the objective function subject to both inequality and equality constraints, $h(x)$ and $g(x)$.

Objective Function The effectiveness of the optimal capacitor placement is measured by the system total cost over an operation period, which consists of the real power loss cost, the capacitor installation cost, and capacitor replacement cost. The objective function or the cost function summarizing costs over each system component can be expressed as:

$$f(x) = C_{\text{loss}} + C_{\text{cap}} \tag{5.6.8}$$

where

$C_{\text{loss}} = c_{\text{rp}} \sum_{\text{ll}=1}^{N_{\text{ll}}} D_{\text{ll}} P_{\text{L}}^{\text{ll}}(x_{\text{ll}})$ is the cost of real power loss within a time period

$C_{\text{cap}} = \sum_{n=1}^{N_{\text{bus}}} \sum_{\text{p}=\text{a}}^{\text{c}} \left(c_{\text{new}}\left(x_{\text{c},n}^{\text{new,p}}\right) + c_{\text{rep}}\left(x_{\text{c},n}^{\text{rep,p}}\right)\right)$ represents the cost of capacitors

N_{ll} = no. of load levels, P_{L}^{ll} = total power loss of load level ll, c_{rp} = \$ cost of real power loss per kWh, D_{ll} = duration of the load level ll, N_{bus} = total no. of buses, $x_{\text{c},n}^{\text{new,p}}$ = the size of new installing capacitors, at bus n, phase p. $x_{\text{c},n}^{\text{rep,p}}$ = the size of replaced capacitors, at bus n, phase p.

$c_{\text{new}}\left(x_{\text{c},n}^{\text{new,p}}\right)$ = cost of the new installing capacitor and $c_{\text{rep}}\left(x_{\text{c},n}^{\text{rep,p}}\right)$ = cost of the replaced capacitor

Equality Constraints

In capacitor placement problems, the power flow equations are incorporated as the equality constraints. We developed three-phase unbalanced power system models such as three-phase distribution lines, varied transformers/regulators, grounded/ungrounded loads, etc., to truly reflect the targeted system as well as yield practical capacitor placement schemes.

Inequality Constraints

Subject to power system equipment limitations, economic considerations, and operating regulations, the optimal capacitor placement should yield solutions that keep the system states/variables within all the inequality constraints. Usually these constraints consist of voltage constraints of each bus, thermal limitation at each branch, and minimum/maximum number of capacitors that can be installed.

1. Voltage Constraints

   The voltage constraints at each bus can be expressed as:

$$V_n^{\mathrm{p,l}} \leq V_n^{\mathrm{p}} \leq V_n^{\mathrm{p,u}} \tag{5.6.9}$$

   where $V_n^{\mathrm{p,l}}$ and $V_n^{\mathrm{p,u}}$ represent the lower and upper bound of voltage magnitude, respectively, of phase p on bus n. Practically, $V_1^{\mathrm{p,l}} = V_2^{\mathrm{p,l}} = \cdots = V_n^{\mathrm{p,l}}$ and $V_1^{\mathrm{p,u}} = V_2^{\mathrm{p,u}} = \cdots = V_n^{\mathrm{p,u}}$.

2. Thermal Limitations

   For phase p of a specific branch i, the thermal limitation is defined as:

$$I_i^{\mathrm{p,l}} \leq I_i^{\mathrm{p}} \leq I_i^{\mathrm{p,u}} \tag{5.6.10}$$

   Similar to voltage constraints, $I_i^{\mathrm{p,l}}$ and $I_i^{\mathrm{p,u}}$ represent the lower and upper bound of current magnitude, respectively. Usually, $I_1^{\mathrm{p,l}} = I_2^{\mathrm{p,l}} = \cdots = I_n^{\mathrm{p,l}}$ and $I_1^{\mathrm{p,u}} = I_2^{\mathrm{p,u}} = \cdots = I_n^{\mathrm{p,u}}$.

3. Capacitor Number Limits

   Giving a range of capacitor numbers can shrink the searching space and limit the calculations in a realistic scenario. The capacitor number limit is expressed as:

$$\mathrm{Cap}_{\min} \leq \sum_{l=1}^{k \times c} t_l \leq \mathrm{Cap}_{\max} \tag{5.6.11}$$

   where $\mathrm{Cap}_{\min}$ and $\mathrm{Cap}_{\max}$ represent the lower and upper limitation of the total number of all kinds and each kind of capacitors, respectively. k is the total bus number, c is the total number of available capacitor sizes, and t is the control variable mentioned before.

The Two-Stage Algorithm

As illustrated by Figure 5.6.3, we apply the proposed GA-guided Trust-Tech method to solve the capacitor placement problem via the two stages.

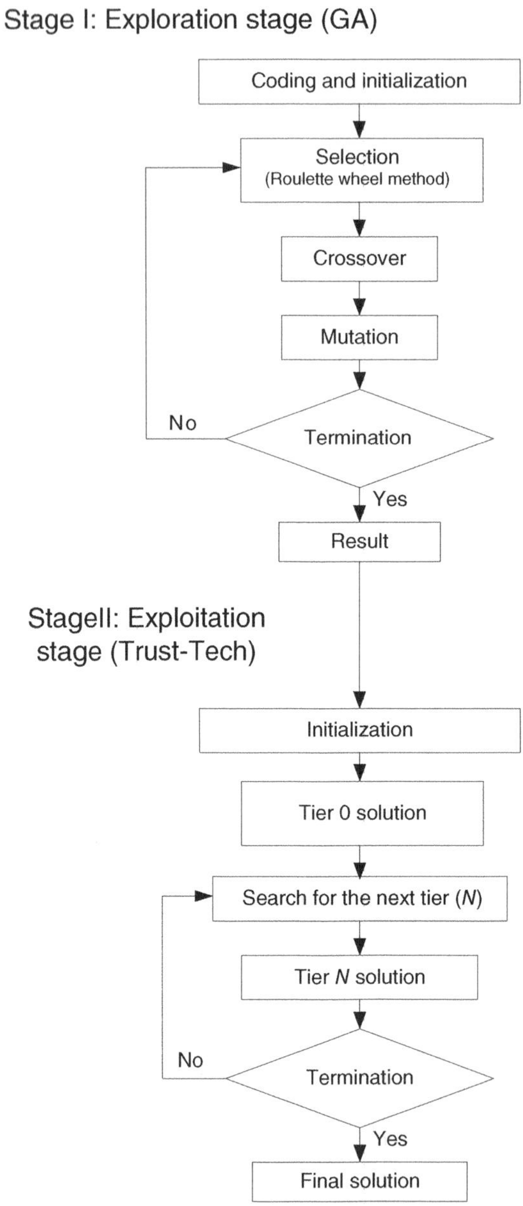

Figure 5.6.3 Flowchart of the GA-guided Trust-Tech.

The implementation of the two stages follows the procedures described in Section 5.6.4. For the capacitor placement problem, the following properties need to be mentioned:

In Stage I, each individual in the population is represented by a code string consisting of the binary information of all the locations and sizes of the candidate capacitors. The fitness function is defined as:

$$\text{Fitness}_i = f_{\max} - f_i \tag{5.6.12}$$

where

Fitness_i is the fitness value of the ith individual in the population,

$f_{\max}$ equals to the maximum among all the objective function values in the population,

f_i is the objective function value of the ith individual in the population.

The roulette wheel method is used for the selection during each iteration.

In Stage II, the optimization problem (5.6.8) is formulated into the corresponding nonlinear dynamic system using the solution by Stage I as the initial guess. The search process is terminated after finding all the tier-1 local optimal solutions, or it has reached the maximum number of search directions or the maximum number of local optimal solutions.

5.6.6 Numerical Study

The numerical study of the proposed two-stage algorithm was performed with a real network, which is a practical distribution system. Its basic statistics are summarized in Table 5.6.1. The stage-I algorithm (GA) has been implemented in C language and its solutions are then used as initial conditions for the Trust-Tech, i.e. Stage-II calculation.

Simulation Parameters

The parameter settings of this simulation are summarized in the following tables. For system parameters, please see Table 5.6.2. For capacitor costs, please refer to Table 5.6.3. Table 5.6.4 includes the GA and Trust-Tech settings.

TABLE 5.6.1 System Summary

Practical Distribution System					
No. of Buses	394	No. of Branches	343	No. of Capacitors	0
Total Loads					
Real power (kW)		36 436.50	Reactive power (kvar)	19 140.36	
Total Loss					
Total loss (kW)		1 594.182 085	Total loss (%)	4.3%	

TABLE 5.6.2 System Parameters

Real Power Cost	Planning Period	Loading Levels	0.7 × Base Case Load	1.0 × Base Case Load	1.3 × Base Case Load
0.06$/kWh	8 yr	Duration	2 yr	4 yr	2 yr

TABLE 5.6.3 Capacitor Cost

Phases	Type	Conn	Size (kvar)	Nom (kV)	Total New Cost ($)	Total Rep Cost ($)
3	F	Wye	50	12.47	2416	889
3	F	Wye	100	12.47	2704	1177
3	F	Wye	133.33	12.47	2838	1312
3	F	Wye	150	12.47	2907	1380
3	F	Wye	200	12.47	3186	1659
3	F	Wye	300	12.47	4315	2761
3	F	Wye	400	12.47	4874	3320

TABLE 5.6.4 GA and Trust-Tech Settings

GA Settings					
Population Size	Initial Segment Width	r	Termination Ratio	P (Crossover)	P (Mutation)
20	10 generations	1.484 65	20%	0.8	0.05
Trust-Tech Settings					
	Max no. of search directions			Max no. of local optimal solutions	
	50			50	
	Gradient tolerance			Solution tolerance	
	1.0E−8			1.0E−8	

Stage I: Exploration Stage (GA)

The proposed GA capacitor placement method was applied to the 394-bus system with the above settings. For comparison purposes, we output the best solution obtained at the segment in which the GA termination has been triggered, but kept the GA running. The objective function was optimized through 82 generations (shown in Figure 5.6.4) until it reached the termination ratio (20%). The first five segments of generations are presented in Table 5.6.5. The termination segment, Seg. 4, is highlighted.

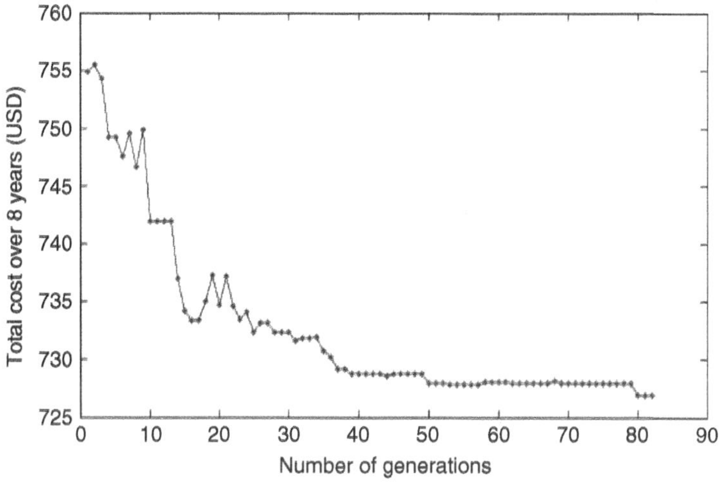

Figure 5.6.4 GA cost function value (Seg. 4, up to 82 generations).

TABLE 5.6.5 Segments of GA Generations

Segment	Segment Ending (Generation)	Vitality Value	Ratio (%)
Seg. 1	11	0.011 811	100
Seg. 2	26	0.007 605	64.39
Seg. 3	49	0.004 419	37.43
Seg. 4	82	0.001 979	16.76
Seg. 5	131	0.002 600	22.01

After 82 generations, GA has achieved a 7.127% improvement with the total cost of 6 408 540.61 USD compared to the initial state's 6 900 305.48 USD. Its capacitor placement scheme is summarized in Table 5.6.6.

We also obtained the optimal capacitor placement schemes from the two reference segments, Segs. 3 and 5, which are neighboring segments to Seg. 4.

In Seg. 3's optimal solution, GA has a 6.961% improvement with the total cost of 6 419 983.06 USD compared to the initial state's 6 900 305.48 USD. Its capacitor placement scheme is summarized in Table 5.6.7.

In Seg. 5's optimal solution, GA has a 7.368% improvement with the total cost of 6 391 889.93 USD compared to the initial state's 6 900 305.48 USD. Its capacitor placement scheme is summarized in Table 5.6.8.

The GA process has been kept running for 3500 generations. As shown in Figure 5.6.5, its calculation has entered a steady state since approximately the 1000th generation, where it has little chance to make further improvements.

TABLE 5.6.6 GA Optimal Capacitor Placement (Seg. 4, Up to 82 Generations)

Bus Name	Capacitor Size (kvar)	Bus Name	Capacitor Size (kvar)
1198	300	1075	300
1228	300	1148	400
1092	200	1107	400
1146	300	1024	300
1368	400	1030	150
1332	400	1195	400
1023	400	1261	400
1027	300	1356	300
1376	300	1031	300
1046	400	1316	300

TABLE 5.6.7 GA Optimal Capacitor Placement (Seg. 3, Up to 49 Generations)

Bus Name	Capacitor Size (kvar)	Bus Name	Capacitor Size (kvar)
1204	200	1075	300
1228	300	1148	400
1092	200	1107	400
1141	300	1024	300
1369	400	1030	150
1332	400	1194	400
1023	400	1260	400
1032	300	1356	300
1181	200	1031	300
1046	400	1316	300

TABLE 5.6.8 GA Optimal Capacitor Placement (Seg. 5, Up to 131 Generations)

Bus Name	Capacitor Size (kvar)	Bus Name	Capacitor Size (kvar)
1198	300	1107	300
1025	300	1268	400
1092	200	1107	400
1145	300	1024	300
1368	400	1030	150
1332	400	1195	400
1023	400	1261	400
1027	300	1356	300
1354	300	1031	300
1046	400	1316	300

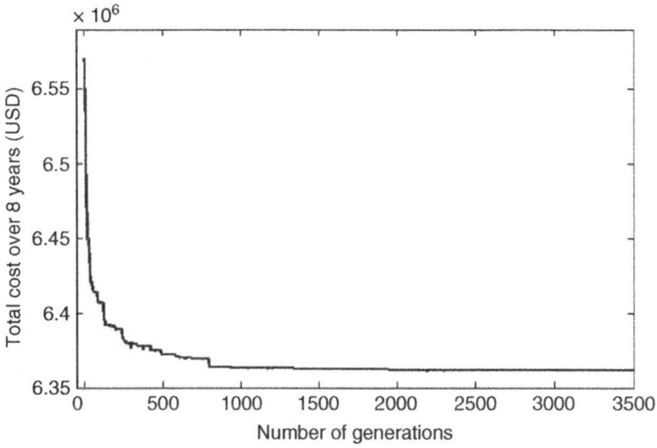

Figure 5.6.5 GA cost function value (3500 generations).

TABLE 5.6.9 GA Optimal Capacitor Placement (3500 Generations)

Bus Name	Capacitor Size (kvar)	Bus Name	Capacitor Size (kvar)
1208	300	1029	300
1016	300	1279	400
1147	300	1109	400
1137	300	1004	300
1368	400	1013	300
1329	400	1196	400
1067	400	1260	400
1026	300	1358	300
1358	300	1026	300
1067	400	1319	300

At the 3500th generation, GA has achieved a 7.809% improvement with the total cost of 6 361 468.57 USD compared to the initial state's 6 900 305.48 USD. Its placement scheme is shown in Table 5.6.9.

Stage II: Exploitation Stage (Trust-Tech)

From Stage I, we obtained a GA solution based on the termination criterion and two reference GA solutions from its neighboring segments. Using them as starting point we performed Trust-Tech optimization. As presented below, the results show significant improvements over the Stage I solutions.

The Trust-Tech method found the tier-0 optimal solution using Stage I's optimal capacitor placement scheme as its initial guess, and successfully found tier-1 local optimal solutions. The obtained solutions and their corresponding capacitor placement schemes are summarized below. From Tables 5.6.10 to 5.6.15, the Trust-Tech's tier-by-tier local optimal solutions based on the GA solution (Seg. 4) and two reference GA solutions (Segs. 3 and 5) are presented. Their final optimal solutions are highlighted.

GA Solution

TABLE 5.6.10 Trust-Tech Solutions (Seg. 4, Up to 82 Generations)

Local Optimal Solutions		Tier 0	Tier 1 (1)	Tier 1 (2)
Kw loss (kW)	Level 0	715.723	715.257	715.980
	Level 1	1 432.377	1 431.995	1 432.397
	Level 2	2 348.252	2 348.025	2 348.038
Loss cost ($)		6 232 279.13	6 230 747.79	6 232 366.54
Cap cost ($)		94 685.00	94 685.00	94 685.00
Total cost ($)		6 326 964.13	6 325 432.79	6 327 051.54
Improved (based on initial state)		$573 341.35	$574 872.69	$573 253.94
Improved (improvements over Stage I)		16.589%	16.900%	16.571%

TABLE 5.6.11 Trust-Tech Optimal Capacitor Placement (Seg. 4, Up to 82 Generations)

Bus Name	Capacitor Size (kvar)	Bus Name	Capacitor Size (kvar)
1004	300	1137	300
1004	400	1137	400
1028	400	1208	400
1026	300	1227	400
1026	400	1279	400
1013	300	1320	400
1013	400	1335	400
1022	400	1361	400
1062	400	1362	300
1064	400	1362	400

Reference GA Solution 1

TABLE 5.6.12 Trust-Tech Solutions (Seg. 3, Up to 49 Generations)

Local Optimal Solutions		Tier 0	Tier 1 (1)
Kw loss (kW)	Level 0	715.699	715.116
	Level 1	1 432.185	1 431.563
	Level 2	2 347.796	2 347.406
Loss cost ($)		6 231 372.30	6 229 040.89
Cap cost ($)		94 685.00	94 685.00
Total cost ($)		6 326 057.30	6 323 725.89
Improved (based on initial state)		$574 248.18	$576 579.59
Improved (improvements over Stage I)		19.555%	20.040%

TABLE 5.6.13 Trust-Tech Optimal Capacitor Placement (Seg. 3, Up to 49 Generations)

Bus Name	Capacitor Size (kvar)	Bus Name	Capacitor Size (kvar)
1004	300	1136	400
1004	400	1137	400
1028	400	1207	400
1026	300	1208	300
1026	400	1279	400
1013	300	1320	400
1013	400	1335	400
1022	400	1361	400
1064	400	1362	300
1067	400	1362	400

Reference GA Solution 2

TABLE 5.6.14 Trust-Tech Solutions (Seg. 5, Up to 131 Generations)

Local Optimal Solutions		Tier 0	Tier 1 (1)
Kw loss (kW)	Level 0	714.617	715.471
	Level 1	1 429.759	1 432.012
	Level 2	2 344.294	2 347.660
Loss cost ($)		6 221 452.95	6 230 623.99
Cap cost ($)		95 803.00	94 685.00
Total cost ($)		6 317 255.95	6 325 308.99
Improved (based on initial state)		$583 049.53	$574 996.49
Improved (improvements over Stage I)		14.680%	13.096%

TABLE 5.6.15 Trust-Tech Optimal Capacitor Placement (Seg. 5, Up to 131 Generations)

Bus Name	Capacitor Size (kvar)	Bus Name	Capacitor Size (kvar)
1004	400	1136	400
1028	400	1137	400
1024	400	1207	400
1025	400	1279	400
1012	400	1319	300
1026	400	1320	300
1013	400	1335	400
1022	400	1361	400
1064	400	1362	300
1067	400	1362	400

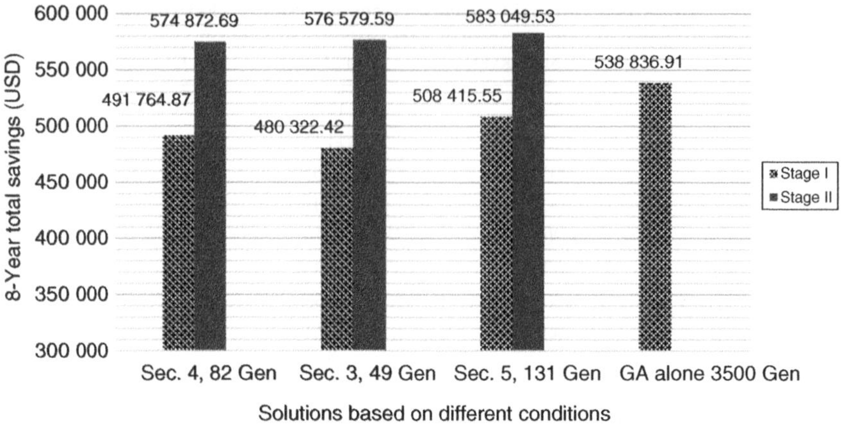

Figure 5.6.6 Solution comparison.

A comparison of the obtained solutions by GA alone, Stage I, and Stage 2 is presented below:

From Figure 5.6.6, we have the following observations:

The Stage II-Trust-Tech routine consistently improves the solutions obtained by the Stage I-GA. The improvements can be significant.

Terminating Stage I calculations at different segments has little impact on the results of Stage II.

After 3500 generations, the GA's solution is still inferior to the ones obtained by the proposed two-stage algorithm.

These observations reveal that the proposed GA-guided Trust-Tech method can assist GA to gain significant computational efficiency in finding better, if not much better solutions.

5.6.7 Conclusions

We present a GA-guided Trust-Tech method for nonlinear optimization with the goal of achieving better, if not much better solutions more efficiently, as compared with the GA. The proposed GA-guided Trust-Tech method consists of two stages: the exploration stage by GA and the exploitation stage by the Trust-Tech method. The key feature of the Trust-Tech method is its capability to compute all the local optimal solutions in a tier-by-tier manner and then search the global optimum among them. If the initial point is not close to the global optimal solution, then the task of finding the global optimal solution may take several tiers of local optimal solutions computation.

Theoretically speaking, the Trust-Tech can continue to find the set of second-tier, third-tier, or higher-tier local optimal solutions at the expense of considerable computational efforts. Since GA of Stage I provides solution points close to high-quality optimal solution, it makes the Trust-Tech method of Stage II search less number of tiers and obtain higher-quality optimal solutions. It has been our experience that of the set of first-tier local optimal solutions, there usually exists a very high-quality local optimal solution, if it is not the global optimal solution. Hence, the exploitation process is terminated after finding all the tier-1 local optimal solutions.

Another feature of the proposed GA-guided Trust-Tech method is its deterministic property of computing solutions, rather than the relatively random ones from GA. We also develop and integrate into the GA-guided Trust-Tech method an efficient method, termed the Trust-Tech-based DDP method for fast locating another local optimal solution from a known local optimal solution.

5.7 NETWORK RECONFIGURATION

Eduardo N. Asada[1], João Bosco A. London Jr[1], and Filipe O. Saraiva[2]

[1]University of São Paulo, São Carlos, Brazil
[2]Federal University of Pará, Belém, Brazil

5.7.1 Introduction

This chapter addresses two related problems in distribution systems that involve changes in system topology. One is the well-known topology reconfiguration problem and the other one is the restoration of system service. Both deal with changes in the system topology to achieve the best solution under given operation conditions. In terms of complexity, however, both present differences due to the final purpose. The optimal reconfiguration problem is frequently considered as a planning tool, or as a correction on the present system topology, which provides the best topology to a specific situation. On the other hand, in the restoration problem a sequence of actions must be performed to restore part of the system, which is out of service. In the context of advanced distribution systems, both problems can be integrated as part of the automatic actions designed as part of smart-grid functionality. Despite the difference in the concepts, both present elevated complexity concerning the modeling for optimal solution due to the nonlinearity and presence of discrete actions (switching actions) which modifies the problem structurally. This complexity is easily noticed when the simple assessment on the performance of the system is done. In this case, the information regarding the system configuration is always assumed correct and the major consequence for the occurrence of topology errors is the observation of a set of inconsistencies between measurements and the status of the controlled variables. The optimized operation that takes into account topology changes can assume significantly different solutions due to this fact and the resolution of this problem is frequently carried out by classical or approximate algorithms such as heuristic algorithms or branch-and-bound algorithms that present an incremental or constructive strategy of the solutions.

The development and deployment of modern distribution systems aim at the correct and automated response of the system to occurrences such as faults, operations that hinder the reliability of the system, power delivery capability issues, impact of DG, mobile loads, to name a few. The features and tools that have been developed for this environment is mainly based on an increasing monitoring, starting from the substation toward the consumers. Based on the information collected by sensors and meters, the diagnosis of the performance and the corresponding control actions are applied in centralized or decentralized way. In order to describe the two features of this topic, we will present the context in which both can be applied as part of smart-grid functions.

5.7.2 Modern Distribution Systems: A Concept

The research on power distribution systems has gained momentum in recent years. The application of state-of-art technology to improve the quality of service and quality of product involves the entire power industry in terms of investment and research, which can assume a multidisciplinary nature. Different areas such as computer science, communication, distributed computation systems, artificial intelligence, economy, and social sciences should be considered in this context. In the last decades, research on distribution systems has been carried out to provide modern systems with elevated level of monitoring and automation. For these systems, the term "smart grids" has been recently coined to summarize a modern automated system with operation capabilities in the presence of different generation sources. Some of these features have been presented in [154, 155]. They are:

- System auto-recovery or auto-healing
- High-quality energy delivery
- Resilience against cyberattacks
- Presence of massive DG
- Management of energy storage
- Detailed monitoring of consumer energy consumption
- Monitoring of consumers by distribution companies
- Reconfiguration of distribution systems from real-time measurements
- Increased energy efficiency in consumer level
- Possibility for the low-voltage consumer to sell or buy energy
- Real-time monitoring of energy price
- Energy management of electrical vehicles
- Minimization of operation and maintenance costs, among others

For the features mentioned earlier, it is possible to observe for most of them, the presence of reconfiguration of the network, which involves switching actions, which can be performed by different equipments. For this, the application of knowledge-based or intelligent decision-making is possible because in a smart-grid environment, advanced measurement and control systems are responsible for acquiring analog data, setting status of system components and sending data to other devices in the system. From the data collected, it is possible, according to a decentralized or centralized control structure, to execute more effective actions. The data can also feed special database to be used by expert systems to perform specific interventions on the system and plays a critical role in this environment. From this perspective, the smart grid can be described by layers of functionalities with the data flowing from each layer. This is illustrated in Figure 5.7.1 where the consumer layer, the communication layer, and the distribution layer are shown. The flow of information is from the consumer to the distribution system through communication network. Each layer has its own technology and consequently its own specification. As it can be observed, the information is distributed and the decisions can be distributed as well, for this it is interesting to visit the concept of distributed systems.

Distributed Systems and Smart Grid

Distributed systems are systems composed of hardware or software in networked computer system that communicate and coordinate their actions by message passing [156]. The concept of distributed systems has been well explored in computer science, telecommunication, distributed computing, etc. The main motivation for the application of distributed systems is the sharing of computing resources and the scheduling of tasks that deals with an elevated number of requisitions, such as web servers, where various computers process and respond to the request of some specific task. Another application is for computer architectures where the components are spatially dispersed. The smart

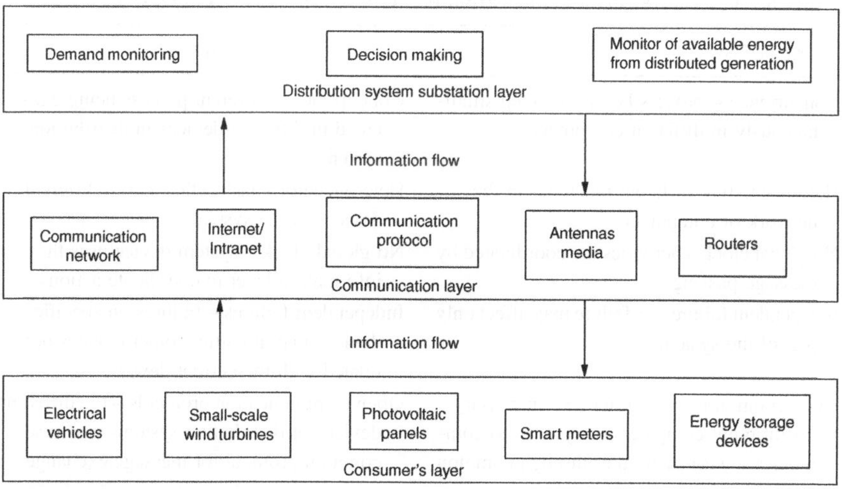

Figure 5.7.1 Smart-grid in layers.

grids can be placed in the latter area of application in distributed systems. The similarity with distributed systems can be observed by the classification presented in [156], which describes the following features of distributed systems.

- Concurrency – Tasks are processed concurrently. When a given task has the full prerequisites and available resources, it will be executed in a computer. If other task is ready for execution, it can be deployed to another computer in the system.
- No global clock – whenever the programs need cooperating with each other to solve a task, it is done by message passing. The task synchronization based on time is difficult to be accomplished in networked computers.
- Independent failures – some machines may fail during their tasks. However, others will continue to operate. It is possible to create fault-tolerant techniques to detect and correct this situation.
- Heterogeneity – the distributed system may be composed of different machines that present different characteristics which may cooperate to achieve a common goal.
- Open communication protocols – required for communication of devices of different manufacturers.
- Transparency – the distributed systems do not appear to be formed by various interconnected machines. The system manages automatically the division of the tasks.

After presenting the characteristics of distributed systems, a comparison between the features desired in smart grid and those present in computational distributed system can be summarized in Table 5.7.1.

TABLE 5.7.1 Features Present in Distributed Systems and Smart-Grid Devices

Distributed Systems	Smart Grids
Concurrence – process being executed simultaneously in different computers	Concurrence – different process being executed in different devices in distribution system
Heterogeneity – different machines in the network of computers	Heterogeneity – tasks allocated to different devices of the system
No global clock – activities are coordinated by message passing	No global clock – system device uses the information layer to coordinate actions
Independent failures – a failure may affect only part of the system	Independent failures – failures on specific devices may not affect other devices not related with the current device
Open communication protocols – necessary for different computers or systems to communicate with each other through common protocol	Open communication protocols – the different devices of distribution systems must use common protocol for message exchange

As a conclusion of comparison of those similarities, functions that involve topology reconfiguration can be applied coordinated with other actions based on real-time and forecast data within smart grid. In order to propose such type of application, we will present initially the fundamentals of two problems: the reconfiguration of distribution systems and the distribution service restoration, and then, an example of multi-agent systems applied to reconfiguration of distribution systems in smart-grid environment is presented.

5.7.3 Distribution System Reconfiguration

Distribution systems consist of several feeders (electrical distribution circuits) that are fed from a single source at the substation. They are planned to serve all loads in the area of operation of a particular substation. Although it can be connected as a meshed circuit, normally they operate in radial form. The switches allow changing the network topology by modifying their status (open/close). For a highly configurable system, it is expected to have a considerable number of tie switches and sectionalizing switches. The maneuverable switches determine the number of possible topologies, which characterizes a combinatorial problem. The action of modifying the topology results in redistribution of loads between substations and also in parts of the feeders, consequently affecting the voltage profile and the losses; moreover, these changes are nonlinear. Therefore, it is possible to describe the problem of finding the topology (or a radial topology) of the distribution network that results in minimum losses as a mixed-integer nonlinear problem with combinatorial nature. Another way to formulate this problem is by using the graph theoretic approach, in which the objective is to find the tree of the graph that will provide the lowest level of loss during the operation. As mentioned, the loss minimization might be the main objective; however, it is possible to consider other objectives simultaneously, as a multi-objective problem.

The reconfiguration of distribution systems has been analyzed since the 1970s [157]. This problem has been mathematically modeled as mixed integer nonlinear problem [130, 158]. The objective function is the loss minimization subject to power flow equations, operational constraints, and radial operation condition. As mentioned earlier, this problem has combinatorial nature and the mathematical representation of the radiality operation is nontrivial. Due to the natural difficulty in modeling and also solving with more exact formulation (AC model) due to nonlinearity, heuristic algorithms have appeared as the earliest solution proposals. Since then, the application of meta-heuristics appeared naturally as in [159], and many other research papers. Recent advances in classical mathematical modeling have been observed; however, the application of meta-heuristics is still very active and the tendency is toward the modeling of more complex objectives and constraints, such as application of various different controls, e.g. simultaneous control of capacitor banks, voltage regulators, and transformer taps. Multi-objective models have also been proposed as many population-based meta-heuristics are available nowadays. In terms of constraints, the daily load variation should be taken into account with detail, as well as more precise load models.

Challenges

The very basic model for the reconfiguration problem is:

$$\min f = \min\left(P_{\text{loss}}\right) \tag{5.7.1}$$

where P_{loss} is the total system loss for a certain topology. This optimization problem is subject to operating limits such as voltage magnitude limits, current magnitude limits at the feeders, and power flow equations subject to certain load pattern. Modern formulations should also consider the following aspects:

- Detailed modeling of operating conditions
- Environmental issues
- Modeling and interaction of new devices (voltage controls, power factor control, DG, protection system, etc.)
- Optimization of various objectives
- Variable demand at each bus
- Efficient fast solvers

Efficient heuristic methods have been proposed so far, however, in order to cope with the challenges mentioned earlier.

Meta-Heuristics for Distribution Network Reconfiguration

In terms of solution methods, it is possible to classify three classes of optimization methods: (i) classical optimization such as branch and bound and mathematical decomposition, (ii) heuristic methods, and (iii) meta-heuristics. One of the favorable points in using meta-heuristics is the fact that the optimal solution can be found with the search starting from infeasible solutions. In this context, the formal mathematical modeling may assume a secondary position. Meta-heuristics can solve problems that are hard to model mathematically. Another important feature is the low degradation of the solution quality as the complexity of the mathematical model increases. These features are due to the fact that the meta-heuristics are not conceived based on concepts such as differentiability, convexity, continuity, etc., which are required in classical optimization techniques. The major shortcoming for this is not being able to ensure that a certain solution is an optimal solution (local or global).

For the reconfiguration problem, many different solution methods with meta-heuristics have been proposed in the technical literature [37, 159–169]. All these solution proposals share similar ways of modeling the problem to the meta-heuristic method. Most of the methods encode only part of the variables of problem, i.e. usually they consider only the topological information in a solution vector (only the radial configurations are stored). Therefore, in this case, the representation of a solution proposal must define the encoding that maps (stores) radial topology in a vector. The advantage of this type of coding is to let the meta-heuristic to control indirectly the radiality of the solutions (this is the most difficult constraint to be considered in the problem). An alternative and simple coding is to only encode the switches' status as binary vector [160].

In the latter case, a candidate solution is represented only by binary values, which represents open and closed switches. The main drawback in this case is the lack of information in the coding vector regarding the remaining topology of the system and the radial operation condition can only be verified by an auxiliary function, or by verifying the occurrence of infeasibility in the constraints that model the radiality.

Once a solution proposal has been chosen, one must evaluate its quality and the feasibility of the new topology. This evaluation is carried out by solving the radial power flow that considers the Kirchhoff's laws for the AC model of the system without taking into account operating limits constraints. Suppose there are 100 topologies that must have their quality assessed. In this case, 100 power flow calculations are required. From the solution it is possible to calculate the limits violations, such as current limits, voltage bus limits and also to calculate the losses for each topology, which will define the value of the objective function. These basic steps are carried out by most of the technical papers, thus the differences among different proposals reside in the different ways to manage the solutions generated by the meta-heuristics, i.e.

- How to generate the initial set of solutions
- How to generate new solutions
- How to deal with unfeasible solutions
- How to control the radiality of new topologies generated by the algorithm

The reconfiguration of distribution systems has already been considered in planning problems. For example, in distribution system expansion planning, one of the conditions to consider, besides the capacity expansion is the ability of the expanded system to perform the reconfiguration. In other words, the reconfiguration can be considered as part or condition for the new system. This has been shown in [170] where a multi-objective model and solution method based on NSGA-II is proposed. In that paper, the system planning considers $n-1$ security condition, i.e. the planning should avoid configurations not able to withstand contingencies by corrective actions. In order to ensure this condition, the reconfiguration is also considered. In other words, expansion plans should consider the ability of the system to change the topology to deal with contingencies. This impacts the expansion routes and the switching branches to be placed accordingly to satisfy best the costs and the system reliability.

Another challenge regarding system operation is to consider the reconfiguration as an autonomous function, which allows changing the topology according to the system changes. Differently from some proposals in the literature, this operation can be modeled as a set of autonomous functions based on multi-agent system, as will be shown in the following sections. In the next section, the distribution system service restoration will be addressed and the importance of meta-heuristics will be highlighted.

5.7.4 Distribution System Service Restoration

This section addresses the problem of service restoration following outages in electric distribution systems (EDSs). Outage caused by steady-state faults or overloaded network equipment has a radical influence on power system operating objectives, since it degrades the most important function of an electrical system, which is supplying the customers [171]. Therefore, the problem of service restoration is one of the most relevant topics related to the efficient operation of EDSs [172–174].

Service restoration problem is computationally complex (NP-complete complexity class) [171, 175], which in practice means that there is no computer algorithm known that can guarantee the optimal solution in a tractable run-time. Therefore, no proposals for the computer approach has solved the service restoration problem for relatively large EDS networks using traditional optimization techniques [173], and several approaches based on heuristics [176], expert systems [177], database dedicated [178], as well as meta-heuristics [172, 174, 175, 179–182] have been proposed.

Distribution Systems

Although EDS networks are structured in meshes, they are usually operated in radial configurations for a better coordination of their protective schemes and reduction in the fault level [171, 183]. Therefore, several EDS problems, such as the service restoration problem, involve network reconfiguration procedures [175, 184].

As mentioned in the earlier section, network reconfiguration is the process of altering the topological structure of EDSs by opening sectionalizing (normally closed [NC]) switches and closing tie (normally open [NO]) switches. For problems involving network reconfiguration, the EDS network can be viewed as a set of sectors[1] interconnected by switches. Consequently, for service restoration problem, the distribution system is usually represented by a graph, i.e. a set of nodes connected by edges [171, 174, 175].

Figure 5.7.2a illustrates a typical EDS with three feeders. Figure 5.7.2b presents the corresponding graph representation, where the nodes represent the sectors, edges in solid lines correspond to NC switches, and edges in dashed lines symbolize NO switches. Note that the three feeders of the EDS illustrated in Figure 5.7.2a are represented by three graph trees[2] formed by the solid lines in Figure 5.7.2b. Nodes S1, S2, and S3 in the graph are the root nodes of trees 1, 2, and 3. These nodes correspond to buses S1, S2, and S3, which are in substations 1, 2, and 3 respectively.

Service Restoration Problem

Service restoration problem emerges after the faulted sectors have been identified and isolated by switching operations. Since EDSs are usually operated in radial

[1] A sector (also called section or block) is a set of buses and conductors connected by lines without switches.

[2] A graph tree is a connected and acyclic subgraph of a graph.

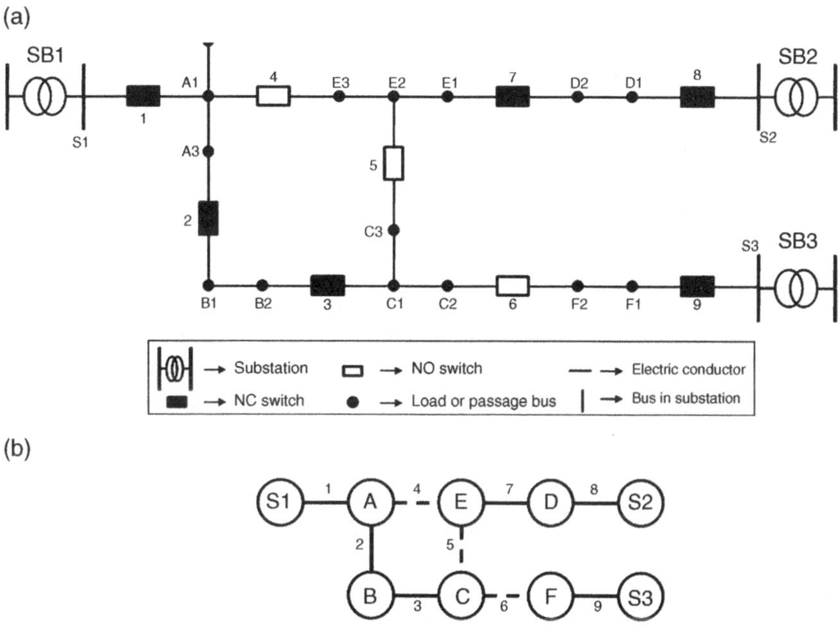

Figure 5.7.2 EDS and its representation by graph. (a) Example of a typical EDS with three feeders. (b) Graph representation.

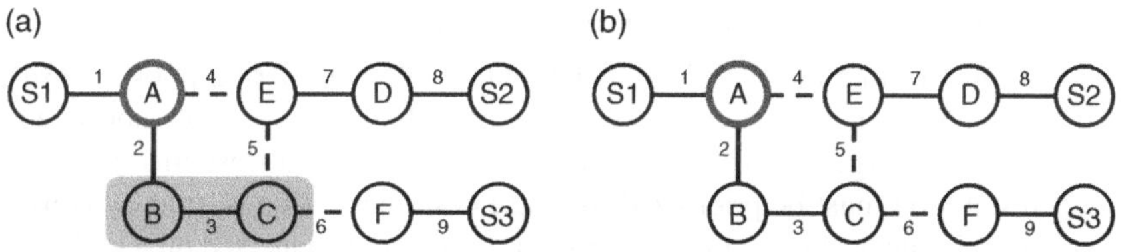

Figure 5.7.3 Example of service restoration. (a) Section in fault. (b) New configuration.

configurations, when the faulted sectors are identified and isolated for maintenance, disconnected loads (without energy supply) are those in the faulted sectors and also in their downstream sectors (the healthy out-of-service areas).

Figure 5.7.3 shows an illustrative example of service restoration. The graph in Figure 5.7.3a represents an EDS configuration with a permanent fault in sector A. This fault forces the system protection to open the NC switch 1 (the protection switch closest to the failure in the path to the feeder substation) disconnecting sectors A, B, and C. Once the faulted sector is identified (sector A), it is isolated by the opening of NC switch 2 (for maintenance) and the sectors B and C become healthy out-of-service sectors (shaded area in Figure 5.7.3a). One way to restore the service for these sectors is closing switch 6, as shown in Figure 5.7.3b. If the obtained radial configuration satisfies the operational constraints (bounds for node voltage, network loading, and substation loading), the restoration process is finished. Otherwise, additional switching operations must be investigated.

In typical EDSs there are several combinations of switching operations that could restore the energy for the healthy out-of-service areas or at least for some of them. Optimizing this process means reconnecting the system in an optimal (or near optimal) way, satisfying the radiality and operational constraints.

The service restoration problem has two fundamental objectives [171, 173, 174]: to minimize both the number of healthy out-of-service areas and the time to implement the restoration process. Since the highest the number of switching operations, the higher the time spent on the implementation of the restoration process; service restoration problem consists in determining the minimal number of switching operations that results in a configuration with minimal number of healthy out-of-service areas without violating the radiality and operational constraints [172]. Therefore, network reconfiguration for service restoration is a computationally complex problem, since it is [172, 174] (i) highly combinatorial, due to the large number of switching elements; (ii) nonlinear, since the equations governing the electrical system are in general nonlinear; (iii) non-differentiable, since a switch status change may result in crisp variations of values in objectives and constraints; (iv) constrained, due the radiality and operational restrictions; and (v) multi-objective, since the service restoration should minimize the number of healthy out-of-service areas and the number of switching operations. Moreover, a practical and effective service restoration plan should meet the following requirements:

- It must prioritize operation in remotely controlled switches (RCSs) instead of manually controlled switches (MCSs), since the time taken for operating a MCS is significantly longer than that for operating a RCS.
- It must restore as much load as possible while considering priority customers. Since real EDSs have customers with higher priority of supply (e.g. hospitals, large industries, traffic lights, and other critical costumers), service restoration must be prioritized for these customers. That is, priority consumers in the healthy out-of-service-areas must be reconnected on a preferential basis, and those in sectors not affected by the fault must not have swapped their feeders.
- It must provide an admissible sequence of switching operations to reach the final configuration from the initial configuration (the configuration with the faulted sectors identified and isolated). That is, a switching operation sequence that generates only radial intermediate configurations [179].
- It must consider the possibility of load curtailment of in-service customers. That is, it must allow, for instance, for a non-priority customer in a sector not affected by the fault to be turned off in order to enable the reconnection of priority consumers in sectors affected by the fault (sectors in the healthy out-of-service-areas).

Literature Review

The service restoration problem has remained a fertile and contentious research area since the eighties. The review of the key publications up to 1995 presented in [171] discusses the main characteristics of the problem.

According to the characteristics mentioned in the previous section, service restoration is a computationally complex problem (NP-complete). Therefore, there is no computer algorithm known that can solve this problem for relatively large networks using traditional optimization techniques in a tractable run-time [171, 174]. In the context of small networks, the mathematical model presented in [173] (based on a transformation of a MINLP problem into a mixed integer second-order cone programming problem) can be solved using commercial solvers based on the efficient branch and bound optimization technique.

Several methods based on heuristics [176], expert systems [177], and databases dedicated [178] have been proposed to solve the service restoration problem for relatively large networks. They essentially attempt to reproduce the procedures applied by the EDS operators during the process of finding the restoration solution. Therefore, they are dependent on the acquisition of operators' knowledge that is often a very difficult task [171, 179]. The methods based on meta-heuristics do not have the aforementioned dependencies.

In general, a meta-heuristic is a search algorithm that starts from an initial point (initial solution), or a set of initial points, through the search space of a problem and guided by strategies to escape from local optimal solutions. The results obtained by methods based on meta-heuristics, mainly those based on GA [181, 185] and MOEAs [172, 174, 175, 179], have been encouraging. Usually the GA-based methods employ a composite objective function that weights the multiple objectives and penalizes the violation of constraints. This strategy suffers from various disadvantages [95, 179, 186]. By other side, the MOEA-based methods retain the multi-objective nature of the problem without the need of any tunable weights or parameters [172, 175, 179, 186].

The MOEA-based methods can work with relatively large networks; however, most of them demand long running time when applied to large-scale EDS (with thousands of buses and switches) modeled without any simplifications (considering all switches and buses) [165, 172]. According to the literature, there are two critical factors for the performance of GA-based and MOEA-based methods applied to ESD problems involving network reconfiguration [165, 166, 175, 179, 187, 188]: the tree encoding (i.e. the data structure to computationally represent the EDS topology) and the genetic operators that are implemented. Generally, these operators do not guarantee the generation of only radial configurations and need to analyze the topological structure of each generated configuration (candidate solution) which consumes additional computing time. To overcome these hurdles, several tree encodings, with the corresponding genetic operators, have been proposed [165, 166, 172, 174, 187].

The combination of GA-based and MOEA-based methods, with the tree encoding named node-depth encoding (NDE) proposed in [189], has proved able to efficiently generate adequate service restoration plans (the solution of the service restoration problem) for large-scale EDSs (from EDS with 632 switches and 3860 buses to EDS with 5166 switches and 30,880 buses).

NDE and its operators improve the performance of GA-based and MOEA-based methods in service restoration problem because of the following NDE properties [172, 175]: (i) NDE operators produce exclusively feasible configurations, i.e. radial configurations able to supply energy for the whole reconnectable system.[3] Therefore, a specific routine to verify and correct unfeasible configurations is not necessary; (ii) NDE operators can generate more feasible configurations in comparison to other encodings in the same running time since its average time complexity is $O(\sqrt{n})$, where n is the number of graph nodes (each graph node corresponds to a EDS sector); and (iii) NDE enables a more efficient forward–backward sweep load-flow algorithm (SLFA) for EDSs. Observe that once a feasible configuration is guaranteed, the objective function and the network operational constraints of the service restoration problem can be analyzed solving a radial load flow. Typically, this kind of load flow applied to radial networks requires a routine to sort network buses into the terminal-substation order (TSO) before calculating the bus voltages [190]. Fortunately, the buses of each configuration produced by NDE operators are naturally arranged in the TSO, so that the SLFA can be significantly improved by using NDE and its operators.

The first interesting result of the combination of NDE and meta-heuristic approaches to solve the service restoration problem was presented in [191]. The proposed method combined NDE with a GA-based method. Afterward, two methods have been developed combining NDE with two MOEAs: one of them, named NSGA-N, combines a modified version of the NSGA-II with NDE [192]; the other, named MEAN [175], uses a technique of MOEA based on subpopulation tables with NDE. Each table stores the found solutions that better have met an objective or a constraint of the problem. These MOEA-based methods have proved able to efficiently generate adequate service restoration plans for large-scale networks (networks modeling EDSs with thousands of buses and switches). However, the service restoration plans for larger networks have not been as good as the ones obtained for smaller networks in terms of number of switching operations and margins for operational constraints. Moreover, the performances of those methods are degraded when applied to multiple faults, that is, the service restoration plans obtained for multiple faults are not as good as the ones obtained for a single fault. To overcome those drawbacks, the method proposed in [172] combines the main characteristics of both MEAN and NSGA-N. The method is based on the idea of subpopulation tables, as MEAN. However, it has new subpopulation tables to store the found solutions based on their dominance, according to the Pareto-optimality criteria (the evaluation criteria used by NAGA-II) [44]. Moreover, to manage single and multiple faults in an efficient way, the method proposed in [172] uses an alarming heuristic that focuses switching operations on alarming zones of the network according to technical constraints.

As a practical implementation, in [193] the alarming heuristic method was confirmed on several tests performed on the real and large-scale EDS of Londrina city (Brazil), which has 21 642 buses, 3196 NC switches, 357 NO switches, 7 substations, 86 feeders, 3133 sectors, and around 231 000 customers.

[3] The term "reconnectable system" means all out-of-service areas having at least one switch (NC or NO) linking them to energized areas. Some out-of-service areas may not have any switch to reconnect them to the remaining energized areas.

Other combinations of NDE with MOEA-based methods have also been proposed. The method proposed in [188] combines NSGA-II with a modified version of NDE in order to solve the service restoration problem considering the possibility of load curtailment. In order to improve the capacity of MEAN [175] to explore both the search and objective spaces, the method proposed in [186], as well as that one proposed in [194], combines MEAN with the DE algorithm [34]. It is important to highlight that the DE algorithm has received increased interest from the Evolutionary Computation community, since DE has shown better performance over other well-known meta-heuristics, like GA and PSO [195]. MOEA using subproblem Decomposition (MOEA/D) and NDE was also investigated in [194].

5.7.5 Multi-Agent System for Distribution System Reconfiguration

The smart grids represent the state-of-the-art of advance in distribution systems in terms of quality of service, power, and market opportunities. The infrastructure behind combines the advances in computation, communication, and artificial intelligence to provide services based on real-time monitoring, management of distributed generators, control, among others. There are two key points for the performance evaluation of distribution systems: the continuity of service and efficiency regarding the power delivery. Both consider the current topology and the ability in changing it while maintaining the quality of service. With this scenario in mind, the application of multi-agent systems technology can bring benefits regarding the improvement of quality of service.

The changes observed in the power system industry regarding the market and system management affect all the grid levels in both planning and operation. The advanced distribution network is expected to provide, for example, from the operator side, automatic reconfiguration of the grid, management of large-scale distributed energy sources (DG), network "self-healing," to name a few; from the consumer side, the real-time monitoring of energy price, monitoring of the consumption of home appliances, energy produced by micro-generators, and so on. One of the demands for changes in power systems came from the environmental concern, increasing the investments in renewable energy, and the deregulation of electricity market [154].

The deployment of earlier new features requires a series of new devices and expert systems located along the grid. These devices require sensors to provide measurements throughout the grid, and the exchange of the information collected with other devices. The information can be used, for instance, to decide whether reconfiguration should be carried on or not in order to achieve the expected behavior or to optimize some objective function, providing the new features planned. These characteristics – the perception of network state, the decision-making, and the autonomous reconfiguration of the system topology – are expected features of a smart grid. Essentially, the smart grids can be considered as distributed systems [156] as we have mentioned in the beginning of the chapter and, consequently, the most fitted solutions reside in the class of distributed methods and algorithms. From this assumption, multi-agent systems [196] emerges as an interesting tool for application in autonomous distribution system and the literature about this theme reports a set of papers following this path [197–200].

Topological reconfiguration of distribution systems is a classical problem that can be adapted to smart-grid scenario. In this problem there is a set of switches with open or closed state that defines the power flow path. Due to components' constraints and parameters, it has been commonly modeled as a problem to determine the configuration of switches in order to optimize the performance of the system [158].

When sensors and metering devices exist, it is possible to elaborate responsive methods such as based on topological reconfiguration, where the reconfiguration actions are performed according with some attribute monitored by the sensors. For example, from the monitoring of consumer's load profiles, the smart grid can reconfigure its topology in order to perform efficient power balance of the system by rerouting the power flow according to the changes in the demand, generation, or contingencies along the day.

Dynamic Reconfiguration of Power Distribution Systems

As observed from previous sections regarding network reconfiguration and restoration, the topological reconfiguration of power systems can be used as a tool to a wide set of applications. For example, load balancing [201], system restoration [202], voltage stability [203], minimization of electrical losses [130], among others.

There are several research papers that apply classical mathematical approaches for power distribution network reconfiguration. One of the earliest mathematical modeling of the problem was presented in [157]. In [130], the authors have proposed a branch exchange heuristic method for the problem. In [158], an improvement of the previous method using power flow model approximation is presented.

There are also some approaches using artificial intelligence methods for the problem. In [204], a GA is utilized, and in [205], an ACO algorithm is proposed. In [205], a hybrid method combining ACO and EP is proposed.

Finally, there are few papers using distributed computation for this problem. In [197, 206, 207], the application of multi-agent systems is proposed, but only for static scenarios. For dynamic operation scenarios, papers about this type of problem are rarely found, most of the papers on multi-agent systems are designed for market operation aiming economic gains. The dynamic characteristic is described by system changes in energy consumption and the operational changes regarding contingencies caused by natural or abnormal situations.

Design of Multi-Agent System to Reconfiguration of Distribution Systems

An agent system consists of an entity that perceives the environment and acts on it to achieve an objective or to reproduce a desired behavior [208]. The agent can be modeled as software or as an autonomous device. The main characteristic is the autonomous behavior. An intelligent agent presents similar characteristics of intelligent systems: a knowledge base that will provide support to solve a problem and search and database manipulation tools for finding information that will provide support for decision-making. The agent is composed of two parts that allow its interaction with the environment. The sensors are tools that allow the agent to perceive the environment; the actuators are responsible to execute the action.

Agent is a concept related with an entity, which can be software such as a bot that performs a task or hardware like a robot, which perceives the environment around and performs actions on the environment to achieve the stated objectives. Agents are autonomous and react to changes (they change their own status and modify the environment according to changes observed on them), they can also be configured to being proactive, i.e. they can start actions to achieve their goals.

The problems can be solved with a single agent or modeled to be solved with a set of agents. In case of a set of agents, it is normal to have agents to perform different tasks. In this case, the coordination of agents is critical to obtain good solutions, especially in cases of nonlinear problems.

In case of smart-grid applications, the multidisciplinary nature is especially visible. It is explained by the necessity to involve areas such as computer science, social science, economics, and management science to improve the technical decisions and automatic controls. The multi-agent system is in multidisciplinary area and utilizes all theories and concepts from the areas mentioned earlier and it is also part of distributed artificial intelligence [209]. As a conclusion, the multi-agent system is a natural option for modeling and simulation of smart grids. Here, we present a conceptual design of smart-grid devices modeled as agents, which allows simulating behaviors, data exchange, and decision-making in multi-agent systems.

The multi-agent system for topological reconfiguration requires at least four types of agents. One agent monitoring the loads, which informs its status to other agents; another agent is the substation agent, located at substations, and this agent has a broader view of the system; the agent located with switching devices, such as breakers and switches; and one agent to control the demand or DG. The coordination between agents depends on the current topology and the current load demand. The intelligence behind the coordination can be distributed or centralized. For instance, the agent in the substation can receive the information regarding the demands from the agents located at the loads. This information together with the system parameters and topology allows the calculation of power flow, which may guide the actions to be sent to the switch agents to modify the topology.

In this context, two concepts can be used in for agent management. The first one is a type of coalition, representing the set of load agents located in the same region limited by the same switching agents. Each coalition defines a leader for making the necessary calculus of the respective coalition. For example, Figure 5.7.4 shows an EDS where load agents are represented by small black circles, and switching agents are the small squares. The figure shows three different coalitions, where coalition 1 is formed by 2, 3, and 4; coalition 2 is 5, 6, and 7; and coalition 3 is 8, 9, and 10.

The second concept is the electrical path. Electrical path is the minimal electrical length between a substation agent and any load agent of a coalition, the length

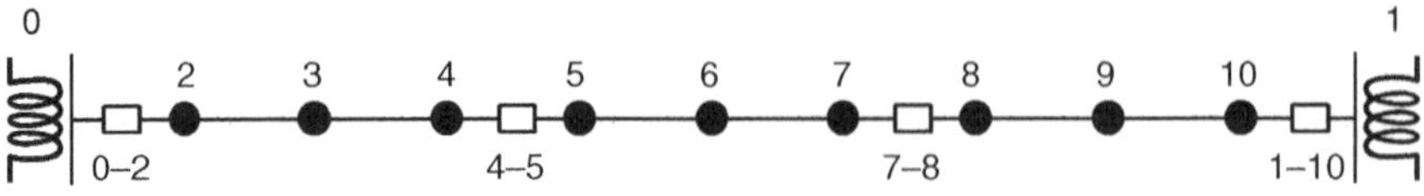

Figure 5.7.4 Example of distribution system with three coalitions: coalition 1 has agents located at 2, 3, and 4; coalition 2 has agents located at 5, 6, and 7; coalition 3 has agents located at 8, 9, and 10.

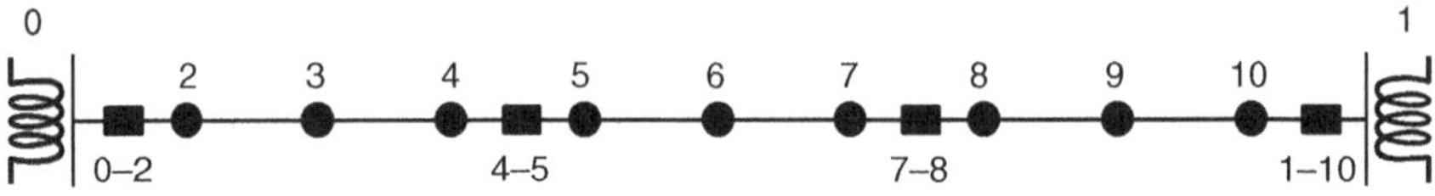

Figure 5.7.5 Example of electrical path between substation 0 and coalition 2.

is the sum of impedances of lines that forms the path. Figure 5.7.5 presents the electrical path between coalition 2 and substation 0, where the substation agent is located. The switch agent in lines 0–2 and 4–5 are closed, consequently, there is power flow on these lines.

The interaction between agents can be briefly explained as follows. Considering the initial state is known, each substation agent sends messages to switching agents asking for their current state. The load agents send the status of the demand to the substation agents. The message also reports the electrical paths and the current demand in each coalition. After receiving messages from all coalition leaders, the substation is aware of the demand of each coalition and it allows the calculation of power flow and consequently the energy losses of current topology. In this case the forward–backward sweep-based methods for distribution systems [210] can be applied.

The current topology may result in constraint violation. Consequently, the substation agent will remove the farthest coalition of the model, i.e. the coalition with greater electrical distance and then perform the power flow calculation again with the remaining coalitions. After achieving the convergence of power flow, the power losses for each coalition can be obtained.

Each substation agent sends messages to load coalition's leaders informing the power losses for the coalition in case of energy that flows to the coalition comes from the corresponding substation agent. Load coalition's leaders will decide which substation agent will be responsible for sending the energy to the coalition and will inform all substation agents about the decision.

After receiving replies from all load coalition's leaders, each substation agent will verify the switch that must be closed in order to the power reach the coalition. Substation agents will send messages to switching agents requesting the action "close switch." At this point, substation agents will suspend their actions until next iteration, when they receive message from load coalition's leaders again. In order to clarify the workings, Figure 5.7.6 presents the actions of the substation agent SAg in a flowchart.

The load agents receive one system identifier corresponding to the position in the power system, which is represented by an integer number (like a label of a

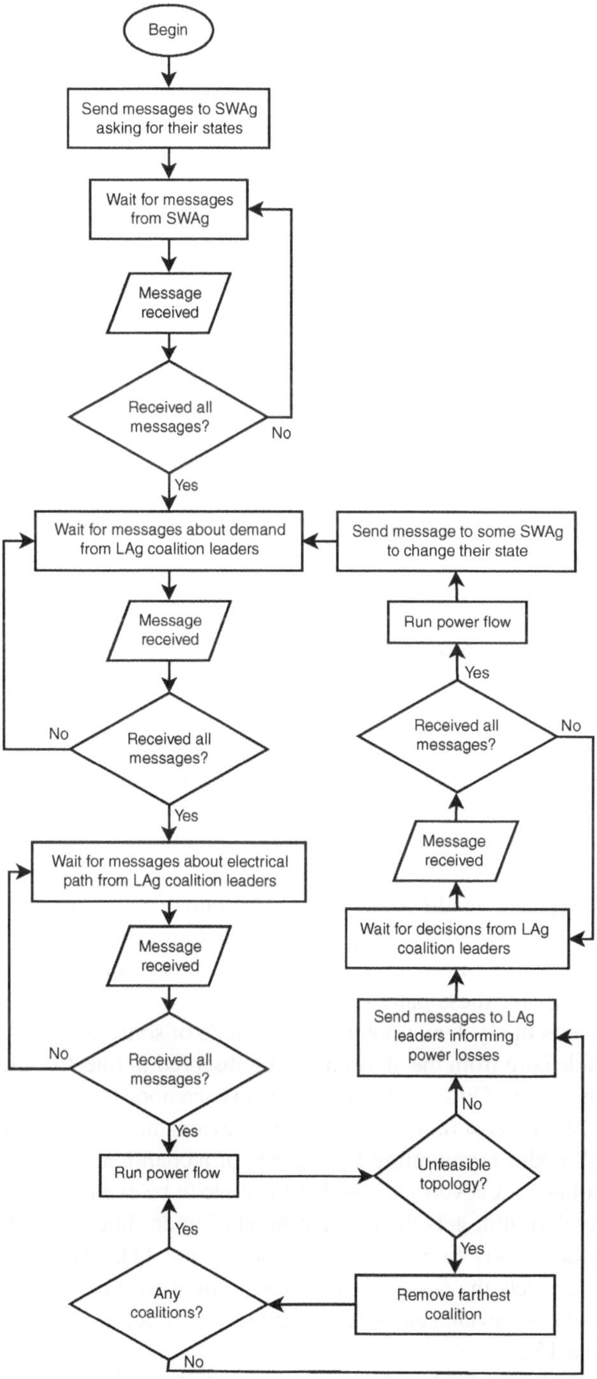

Figure 5.7.6 Substation agent (SAg) actions in a flowchart.

Algorithm 1 Depht-first search algorithm

Require: G, v, SW, Z
1: **for all** $z \in neighborhood(v)$ **do**
2: **if** $(z \notin Z)$ and $(v, z) \notin SW$ **then**
3: $Z = Z \cup \{z\}$
4: $Z = depht - search(G, z, SW, Z)$
5: **end if**
6: **end for**
7: **return** Z

Figure 5.7.7 Depth-first search algorithm for the creation of coalitions.

node in the graph), and the value of the monitored electrical load. Therefore, the distribution system is represented by a graph G.

In the algorithm of Figure 5.7.7, G is the entire graph of the system, v is the node where the load agent is located, SW is the set of lines with switching agents, and Z is the set of load agents of a coalition (it is empty in the initialization). In line 1, the algorithm gets one node neighbor of node v and in line 2, it is verified whether this node is in Z and if the edge (v,z) does not have a switching agent. In case those conditions are satisfied, z is added to Z and depth-first search method is executed again to verify the neighborhood of z. This function is called recursively for each neighbor node, and in the end of the process, the coalition load agents will be formed.

The criterion to define the leader of the coalition is the agent with highest identifier. According to the demand informed by other load agents, the leader will collect the information and inform the total for each substation agent.

The load agent leaders will search on the graph G to find the minimal electrical distance between each substation agent and their corresponding load agent coalition, using, for example, a modified Dijkstra algorithm [211], shown in Figure 5.7.8. The resulting path is the electrical path of the coalition. The load agent leaders send the electrical path's information to substation agents and wait for replies with the power losses for these paths.

The algorithm requires the system graph G and the node of some substation agent to be the source node (*sag* from the algorithm). As observed in line 1 of the algorithm, it initializes the costs of linking the variable Q to each node i with value ∞, except the source node with cost 0. A_i represents the previous node. F defines the set of end nodes – all nodes representing load agents of a coalition.

In the algorithm at line 9, it is verified whether the node to be evaluated is a final node. Case not, the algorithm gets the neighborhood of node (line 10), performs calculation of linkage costs to each neighbor node (line 11), marks the node as examined (line 12), gets the non-visited node with minimal linkage cost (line 14), and puts the previous evaluated node as the previous node of the next node to be evaluated (line 15).

When the method finds one final node, in line 18 a set of verified edges is registered and at line 20 the subgraph with the corresponding edges and the evaluated nodes is created.

Algorithm 2 Dijkstra algorithm with various end nodes

Require: G, sag

$Q_i = costs - initialization(G), \forall i$
$A_i = previous - node - initialization(G), \forall i$
$F = final - node - definition()$
$S = \emptyset$
$R = \emptyset$
$E = \emptyset$
$temp = 0$
$node = sag$
while $node \notin F$ **do**
  $N = neighborhood(G, node)$
  $Q_i = relative - cost - calculation(G, N_i), \forall i$
  $S = S \cup \{node\}$
  $temp = node$
  $node = Min((V - S), Q)$
  $A_{node} = temp$
end while
for all $v \in S$ **do**
  $E = E \cup (v, A_v)$
end for
$R = (S, E)$
return R

Figure 5.7.8 Modified Dijkstra algorithm for finding the minimal path between the substation agent (SAg) and the load agents.

The subgraph requires a processing to remove any unnecessary edge or node other than that of the path between the source node (with substation agent) and the final node found (the load agent). The resulting path is the electrical path of the coalition. Load agent leaders will send the information regarding the formed electrical paths to the substation agent and will wait for message replies carrying information of power losses for these paths.

After receiving all replies, each load agent leader selects one substation agent that will be responsible to send energy to its coalition. The selected substation agent is the one that presents minimal losses and the load leader informs all substation agents regarding the decision. The load agents will monitor the demand, if large variations occur, all previous actions will be performed again. Figure 5.7.9 presents the flowchart of load agent actions.

The switching agents are reactive, and all actions require requests from substation agents. This agent has only two actions: to inform its current state (open/close), and to change its own state.

The flowchart of switching agent is presented in Figure 5.7.10.

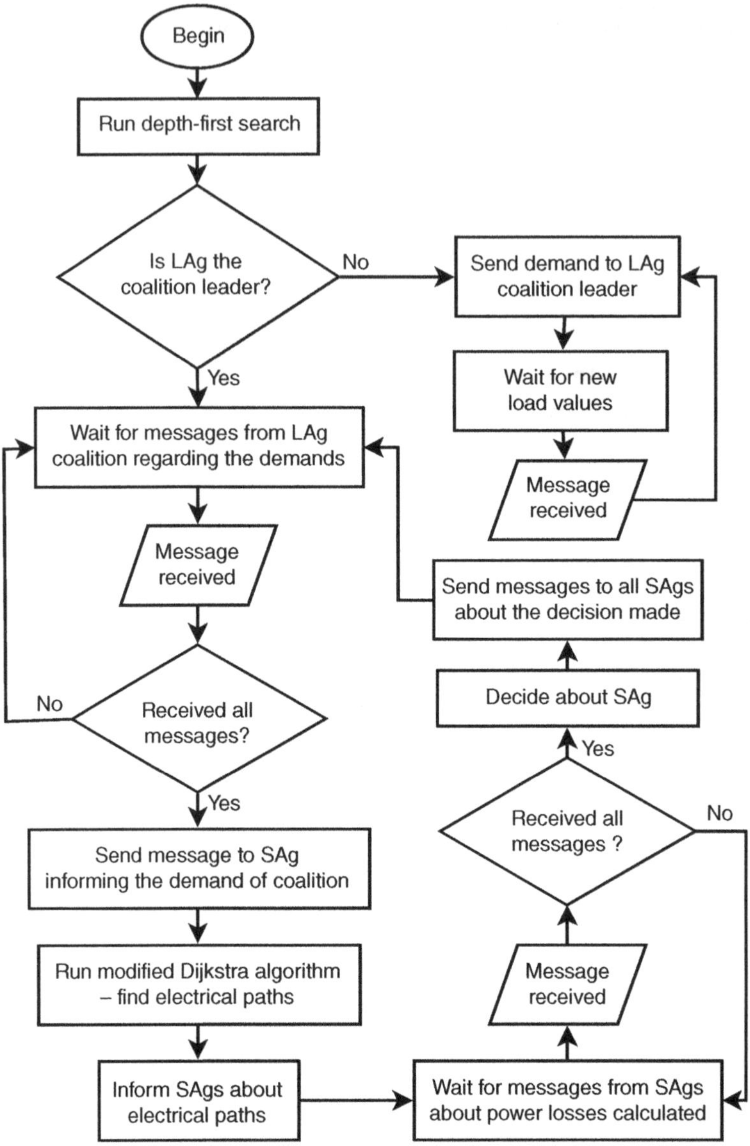

Figure 5.7.9 Load agent (LAg) actions in a flowchart.

The demand agent is a special agent utilized to modify the demands of load agents. The demand agents only send messages to specific load agent informing their new demands. Each load agent will update its demand and will inform the load agent coalition leader about the new demand. Figure 5.7.11 presents a sequence chart of the conceptual multi-agent system, showing the interactions between the different agents.

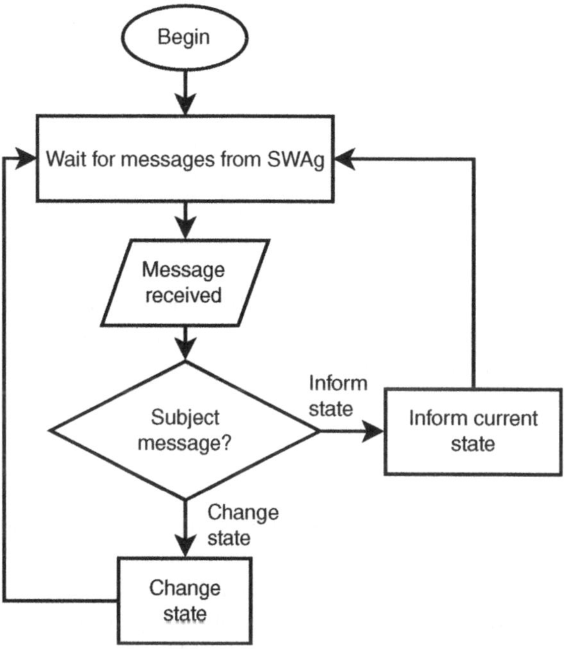

Figure 5.7.10 Flowchart of switching agent (SWAg).

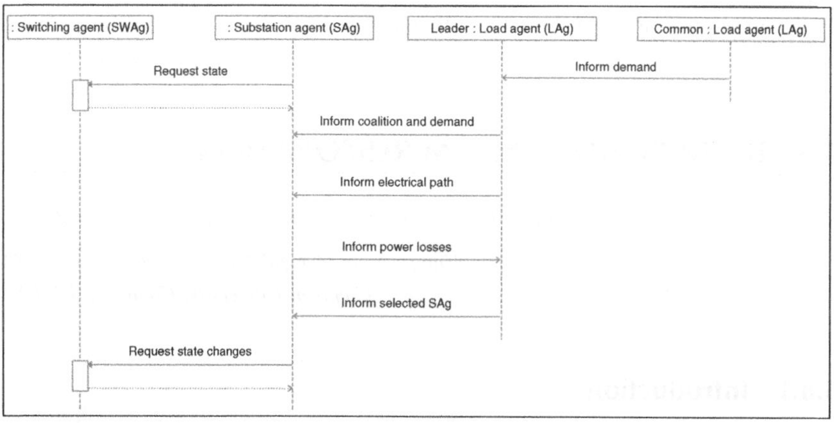

Figure 5.7.11 Sequence chart of the multi-agent system.

Development Framework

One possibility for the implementation of the proposed multi-agent system is the use of programming framework, such as JADE (*Java Agent DEvelopment*) framework [212]. This framework enables the use of a set of classes to develop multi-agent systems with easy access to multi-threading programming, computer network, message exchanging, creation of agents and behaviors, and protocol definitions. The messages can be implemented using Gson, a Google library to convert Java objects in JSON (Javascript objects), reducing the size of messages and making easy the operations on it.

The dynamic nature of systems refers to the varying demand and natural occurrences that can lead to a fault. Therefore, there are events in different and varying time frames. For instance, the occurrence of a fault may require fast actions; on the other hand, variations on demand are considered slow events. The speed of response varies according with the event and this fact also affects the data processing by the agents. When simulating in a computing framework, such as JADE, the situations must also be taken into account in order to bring the simulation of agents closer to reality. In this case, an interaction between the JADE infrastructure with distribution system real-time simulators might be essential.

5.7.6 Conclusions

This chapter addressed the application of meta-heuristics and distributed computation to problems that involve changes in the distribution system topology. The changes in topology to achieve an improvement of system performance or to restore the services represent a hard-combinatorial problem when the best topology is sought. In special, the restoration of services also implies on observing the priorities and the sequence of required actions which mainly involves switching on/off of specific parts of the network. It is expected that both problems can be solved autonomously without human intervention considering an advanced distribution system. In this context, the multi-agent systems can play an important role, based on distributed computation and distributed intelligence. In this sense, the extension of meta-heuristics as "distributed meta-heuristic" should be an interesting research topic.

5.8 DISTRIBUTION SYSTEM RESTORATION

Wei Sun[1], Kumar Venayagamoorthy[2], and Qun Zhou[1]

[1]University of Central Florida, Orlando, FL, USA
[2]Clemson University, Clemson, SC, USA

5.8.1 Introduction

Power system restoration (PSR) following a blackout of power grids is one of the most important tasks for power system planning and operation. The restoration process returns the system to a normal operating condition. Guided by restoration plans prepared off line, dispatchers assess system conditions, start blackstart (BS) units, establish the transmission paths to crank non-blackstart (NBS) generating units, pick up the necessary loads to stabilize the power system, and synchronize the electrical islands [213, 214]. System restoration is a complex problem involving a large number of generation, transmission and distribution, and load constraints. The general requirements of PSR are defined by a series of standards developed by North American Electric Reliability Corporation (NERC). There are various PSR strategies for systems with different characteristics. However, most bulk power systems share some common characteristics and different restoration strategies share a number of common guidelines [215].

Power system dispatchers are likely to face extreme emergencies threatening the system stability [216]. They need to be aware of the situation and adapt to the changing system conditions during system restoration. Therefore, utilities in the NERC Reliability Council regions conduct system restoration drills to train dispatchers in restoring the system following a possible major disturbance. To better support dispatchers in the decision-making process, several approaches and analytical tools have been proposed for system restoration strategies. There are simulation-based training tools; for example, EPRI-OTS and PowerSimulator offer training on system restoration for control center dispatchers [217, 218]. The system restoration problem can be formulated as a multi-objective and multistage nonlinear constrained optimization problem. The combinatorial nature of the problem presents challenges to dispatchers and makes it difficult to apply restoration plan system-wide. Modern heuristic optimization techniques have been applied in different aspects of system restoration problem. The application of seven heuristic optimization techniques in PSR will be introduced in the following sections.

5.8.2 Power System Restoration Process

The restoration process is divided into three stages: preparation, system restoration, and load restoration [219]. According to these restoration stages, PSR strategies can be categorized into six types [213], i.e. *Build-Upward*, *Build-Downward*, *Build-Inward*, *Build-Outward*, *Build-Together*, and *Serve-Critical*. There are different restoration actions in each restoration stage, explained as follows [220]:

1. In the Preparation stage, there are three actions for preparing restoration plans:
   - **Action 1**: Determine the generator start-up sequence. BS generating units will be started first to provide cranking power to start up NBS generators in an optimal way that total system generation capability can be maximized.
   - **Action 2**: Determine the transmission path to deliver the cranking power. Based on the available information of system status, the energization time of each bus and line will be determined to minimize the restoration time.
   - **Action 3**: Determine the critical load pickup sequence. The critical load needs to be picked up to maintain system stability, and the sequence will be based on first two actions.

   By completing these three actions, the restoration plan will proceed to the next stage.

2. In the System Restoration stage, there are three actions needed to prepare restoration plans:
   - **Action 4**: Determine the transmission energization path to energize and build the skeleton of the transmission system. The critical restoration

actions, such as energization of high-voltage lines and switching actions, need to follow the transmission path search.

- **Action 5**: Determine the dispatchable load pickup sequence. Sufficient loads need to be restored to stabilize generation and voltage. Larger or base-load units are prepared for load restoration in the next stage.
- **Action 6**: Check the resynchronization of electrical islands. Many system parameters, such as voltage stability, VAR balance, and voltage/frequency response, need to be checked and monitored to synchronize islands in a reliable way.

3. In the Load Restoration stage, there is one action required for restoration plans:

- **Action 7**: Determine the load pickup sequence. This is different from load pickup in the previous two stages that are aimed at stabilizing the power system. The objective in this stage is to minimize the unserved load according to the system total generation capability.

The seven actions in three restoration stages help to prepare the restoration plan, as illustrated in Figure 5.8.1.

A comprehensive strategy to facilitate system restoration is proposed in [221] to develop computational modules for the generation, transmission, and distribution subsystems. The primary modules include generation capability optimization module (GCOM), transmission path search module (TPSM), distribution restoration module (DRM), and constraint checking module (CCM). These developed optimization modules interact with each other to develop a PSR plan that incorporates generation, transmission, distribution, and load constraints. Specifically, GCOM first provides an optimal generator starting sequence, and TPSM identifies

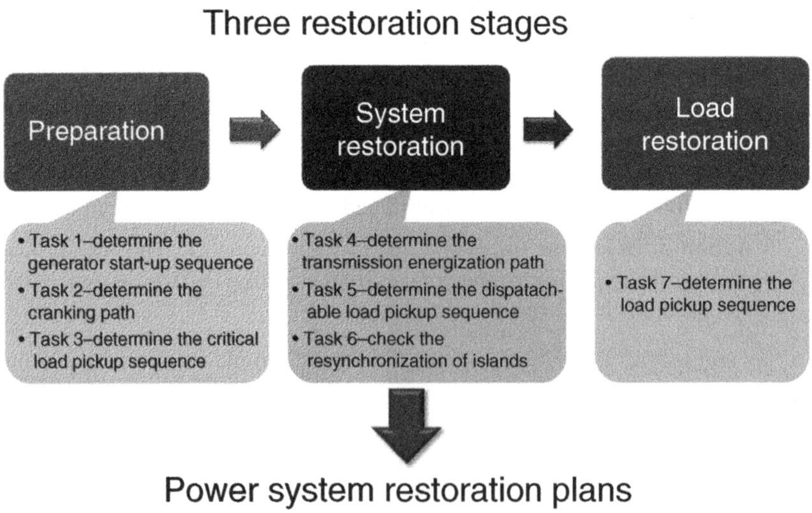

Figure 5.8.1 Actions and stages to prepare power system restoration plans.

the paths for the cranking sequence from GCOM and energizes the transmission network [222]. Then, DRM provides the load pickup sequence to maintain the system stability and minimize the unserved load. The sequences from GCOM and DRM and the transmission paths from TPSM are checked by CCM using power system simulation software tools to ensure that various constraints are met. They interact with each other by receiving the input from and passing the output to other modules. Eventually, this will lead to the successful restoration of a power system. The problem formulation in each subsystem will be explained in the next section.

Many modern heuristic optimization algorithms have been mainly applied in distribution system restoration. In the next section, the application of seven heuristic optimization techniques, including GA, EA, PSO, ant colony search (ACS), TS, SA, and fuzzy systems (FS), in distribution system restoration will be reviewed.

Distribution System Restoration

Load restoration, also known as distribution system restoration, is performed to re-establish the electric service and minimize the impact of the outage by finding the optimal sequence of feeders that need to be restored. Before distribution system restoration, some generation and load have been connected to the system, part of transmission system has been restored, and a certain degree of stability has been achieved. Therefore, voltage and frequency control will not be the primary goal in this stage. The objective is to restore as much load as possible or minimize the unserved load.

The distribution restoration problem presents several challenges to the operator. For example, system operators have to handle a large scale of system and face the simultaneous restoration of many distribution circuits. The restoration guidelines prepared off-line cannot be applied to each scenario. The combinatorial nature of the problem and the high level of detailed models may lead to suboptimal solutions due to the large solution space of the restoration problem. To better support the dispatchers in the decision-making process, different modern heuristic optimization techniques have been proposed for distribution system restoration strategies.

Problem Formulation ***Objective function:*** The general distribution restoration problem includes the restoration of load as seen from the distribution primary feeder and the service restoration by system reconfiguration. There are many different objectives, e.g. to minimize the time to restore all or a given percentage of system loads, to minimize switching operations, or to minimize the time to restore loads with high priority. In [223, 224], the objective is to minimize the cost of unserved energy, as shown in the following:

$$\min \sum_{i=1}^{N_F} C_i(t)P_i(t)t_i \tag{5.8.1}$$

where $C_i(t)$ is the unserved energy cost function of the ith feeder, $P_i(t)$ is the expected load of the ith feeder, t_i the restoration time of the ith load, and N_F is the total number of feeders or substations.

Constraints: There are different constraints in different formulations, while many share some common constraints. In [223], the basic constraints are summarized in the following:

Active Power Balance

$$P_G \geq P_L \tag{5.8.2}$$

where P_G is the available active power and P_L is the active power demand. It is assumed that the generation and the total load are closely matched at any time during the restoration process.

Reactive Power Requirement

$$Q_G \geq Q_L \tag{5.8.3}$$

where Q_G is the reactive power and Q_L is the reactive power demand.

Frequency Limits

$$f_{\min} \leq f \leq f_{\max} \tag{5.8.4}$$

where $f_{\min}$ and $f_{\max}$ are the minimum and maximum frequency permitted by the system. Each type of generating units has unique frequency response rate of load pickup. Generally, hydro, combustion turbine, and steam turbine units can only pick up the load with 15, 10, and 5% of each unit's capacity, respectively [225]. These numbers represent the maximum load each generator can pick up with acceptable frequency dip due to the sudden change of load. PJM uses these numbers to maintain the frequency within 60 ± 0.5 Hz [226].

Voltage Limit

$$|V|_{\min} \leq |V_i| \leq |V|_{\max} \tag{5.8.5}$$

where $|V_i|$ is the magnitude of the voltage at bus i, $|V|_{\min}$ and $|V|_{\max}$ are the minimum and maximum voltage magnitudes needed to guarantee acceptable system operation, respectively.

Single Switching Constraint

$$u_i(t^-) \leq u_i(t^+) \tag{5.8.6}$$

where u is the status of the ith feeder. The status of the feeder is a binary variable, where 1 represents energized feeders and 0 represents de-energized feeders. t^- and t^+ represent the time before and after energization of the feeder. This constraint guarantees that once a feeder is online, it must remain online for the remainder of the restoration process.

Review of Solution Methodologies

Genetic Algorithm

General Genetic Algorithm GA is a search mechanism based on the principle of nature selection and population genetics. GA uses a set of chromosomes to represent solution variables. Those chromosomes can be regarded as search points to find the optimal solution. The basic operation of GA includes reproduction, crossover, and mutation, which perform the tasks of copying strings, exchanging position of strings, and changing some bits of string. Finally, the string with the largest or smallest value of fitness function is found and decoded from the last pool of mature string [227]. The general procedure is summarized as follows:

1. Problem representation using strings: Chromosomes can be represented by a series of binary, decimal, and floating point number, namely string [228].
2. Generation of initial population: GA carries out evolutionary operations on groups and sets the initial population.
3. String evaluation and selection: The fitness function is used to evaluate strings and decide the probability of each string to be inherited to the next generation. Generally, the fitness function can be represented by the objective function. According to certain rules or genetic models, strings with higher fitness values can be selected to the next generation with higher probability.
4. String operation: The string operation includes crossover and mutation. Crossover exchanges partial of current strings and generates new generation based on certain probabilities. Mutation can maintain genetic diversity from one generation to the next by changing one or several genetic values of a string with low probability.

Genetic Algorithm in Distribution System Restoration GA is able to find global optimal solutions of large-scale combinatorial optimization problems. Service restoration problem is a multi-objective, combinatorial, and nonlinear constrained optimization problem. GA has been proved to be efficient in service restoration problem [229].

1. *The representation of the problem using strings*
   Generally, there are two string encoding methods. In [228], the length of a string equals to the number of loads in the out-of-service area, and each string position represents the section number of upstream load or power source. However, this string encoding method is difficult in application. In [229], since the topology of a distribution network can be uniquely defined by the status of all available switches, the solution can be encoded as a function of the controllable switch states of the network.

   The most common encoding scheme is the binary coding, which uses one binary bit to represent the state of switches, 1 means close and 0 means open. For example, in the network of Figure 5.8.2, switches S3 and S5 are open, and the rest are closed. Therefore, the corresponding string for this configuration is $S = \{1\ 1\ 0\ 1\ 0\ 1\}$, where the indices of this string represent the switch number.

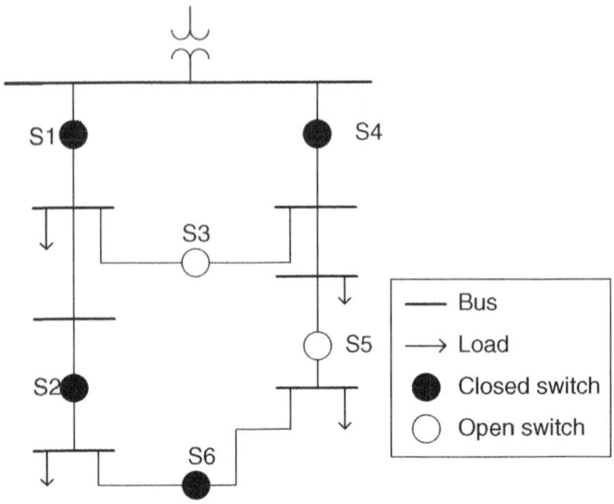

Figure 5.8.2 Example of six-switch system.

2. *The generation of initial population*

   Initial network configurations can be generated by various methods, such as random and problem-dependent methods [230]. In the random method, the population initialization procedure is to assign the value of 0 or 1 to each switch randomly, without any knowledge about the network. In the problem-dependent method, the probability for any given switch to be open or closed is determined by the proportion of open and closed switches in the original network. Also, if the prefault configuration is known, the initial strings can be directly set as the strings representing this configuration, which may accelerate the speed of convergence. In [230], the problem-dependent method is used for faster convergence to the global optimal solution.

3. *String evaluation and selection*

   i. String evaluation

   String evaluation can be obtained by their fitness to the environment. In [230], the fitness function is the inverse of the sum of objective functions and constraints, as defined in the following:

$$F = \frac{1}{\left\{\alpha\left(\sum_{i=1}^{b1} L_i - \sum_{i \in B} L_i\right) + \beta\left(\sum_{j=1}^{N_m} |\mathrm{SW}_j - \mathrm{SWR}_j|\right) + \mu_1 \max\left[0, (|I_i| - I_{i\max})^2\right] + \mu_2 \max\left[0, (|V_i| - V_{i\max})^2\right] + \mu_3 \max\left[0, (|V_i| - V_{i\min})^2\right]\right\}} \tag{5.8.7}$$

   where α, β, μ_1, μ_2, μ_3 are the weighting factors. This fitness function shows that the MOOP is converted to a SOOP. The values of weighting factors depend on the importance of the objective function, as well as on the scaling of objective functions and constraints [179].

   ii. String selection

   - The procedure of string selection can be summarized as follows [229]:
   - Sum up the fitness values of each string ($\sum F_i$).
   - Calculate the following probability: $\mathrm{pselect}_i = F_i / \sum F_i$.

- The expected number of strings (e_i) can be calculated using: $e_i = \text{pselect}_i \cdot gn_s$, where n_s is the number of strings.
- Each string is reproduced using the integer part of e_i.
- Each string is reproduced again by using the fraction part of e_i as the weight of roulette wheel selection. The procedure is repeated until the number of strings reaches n_s.

4. *String operation*

   Crossover operation exchanges partial service restoration strategies at the boundary of the load at the crossover point. This indicates an exchange of the direction of power source [229].

Comparison of Different Applications of Genetic Algorithm In [227], a combination of fuzzy and GA method is proposed to solve service restoration problem. It uses the interactive fuzzy satisfying to evaluate the imprecise nature of five objective functions' priority and then solve the optimization problem by GA. In [179], a technique based on NSGA-II is developed to solve service restoration problem. While the conventional GAs need to convert multi-objective functions into a single-objective function with weighting factors, this approach is able to retain the multi-objective nature without any weighting factors. In [228], a coarse-grain parallel genetic algorithm (PGA) is developed for service restoration. PGA uses several GA procedures in parallel to search various solution spaces, which can get fast convergence. However, its string encoding method makes the approach difficult to put into practice. In [229], a GA-based method is proposed to decide the service restoration and optimal load shedding strategy. This method forms the service restoration problem as the well-known "travelling salesman problem." It uses an "integer permutation" encoding scheme to represent the indices of switches and decides the state of each switch according to the graph theory, subject to the radial network constraint. In [231], a two-stage GA is proposed for the minimization problem of expected energy not supply (EENS) during distribution system restoration. In the first stage, GA creates radial network configuration. In the second stage, GA searches for an optimal sequence of switching operations that minimize EENS. Although this GA method can improve the small system reliability, it is difficult to apply to large-scale system considering the computation time. Then, in [232], a hybrid GA is developed to improve the performance using three speed-up strategies to replace GA in the second stage.

Evolutionary Algorithm

General Evolutionary Algorithm EA is a stochastic optimization technique based on the principles of natural evolution of "best is live" [233]. EA involves three operations: (i) selection, (ii) crossover, and (iii) mutation. First, the population is generated randomly. Second, the fitness value of individual members in the population is evaluated and selected. Then, the crossover operation combines parts of two parent chromosomes to produce a new one, which inherits characteristics from both parents. The mutation operation changes the arrangement of genes in the chromosome at specific mutation rate, which brings some variations in the population. The iteration process continues until the best solution is obtained.

Evolutionary Algorithm in Distribution System Restoration In [165, 191, 234], EA has been applied in the feeder reconfiguration problem in distribution system restoration. The objective is to minimize the number of unserved loads and switching operations. The genes represent the sectionalizing switches in distribution system, and its state is denoted by binary number of 0 or 1. Chromosome represents the configuration of switches to restore service. Initially, random chromosomes are generated to form population. Individual chromosome is evaluated based on the fitness value. Better switching configurations that minimize the objective function and satisfy constraints are selected. EA operations generate new population from better switching configurations and select the best configuration as the optimal solution.

In [191], EA is integrated with new data structure (Node Depth Representation) to increase convergence of the algorithm. Simulation results demonstrate the effectives in large systems. Traditional GA is limited to find the optimal solution in the radial distribution system. In [165], a modified EP algorithm is proposed to overcome this limitation. In [235, 236], service restoration problem is formulated as a multi-objective and multi-constraint problem. The computation time increases with the number of objectives in the mathematical model. Simulation results in [163] demonstrate that EP algorithm is applicable in online computation.

Particle Swarm Optimization

General Particle Swarm Optimization PSO is inspired by the social behavior of animals, such as bird flocking or fish schooling [237]. This technique has been applied in different areas, including adjustment of controllers' parameters in power electronic devices, fuel cost minimization, voltage profile improvement, loss minimization, optimal reactive power dispatch, etc. In PSO searching algorithm, first, a set of particles is randomly generated. Then, each particle is ranked by a chosen cost function, which is an indication of its suitability as the optimal solution of the problem. Each particle searches the space with a given velocity. This velocity is dynamically modified according to the flying experience of this and other particles. Classical PSO uses the following iteration:

$$X_i(K+1) = X_i(K) + V_i(K+1) \tag{5.8.8}$$

where $V_i(K)$ is the velocity of particles at time K and $X_i(K)$ denotes particle i location at time k.

In the each iteration, the velocity of each particle can be modified according to the best position for the particle itself and the best position in the neighborhood, by the following equations [238]:

$$\begin{aligned} V_i(K+1) = {} & W(K) \times V_i(K) + C_1 \times \text{rand}(0,1) \times (g_{\text{best}}(K) - X_i(K)) \\ & + C_2 \times \text{rand}(0,1) \times (P_{\text{best}_i}(K) - X_i(K)) \qquad i = 1, 2, \ldots, N \end{aligned} \tag{5.8.9}$$

where $W(K)$ is inertia weight in the velocity equation, C_1 and C_2 are learning factors, g_{best} is the best position of all particles, $P_{\text{best}i}$ is the best local position of particle i, and N indicates the number of particles. At each time instant, each particle searches for the best position with updated velocity.

Particle Swarm Optimization in Distribution System Restoration In [239], the interconnected power system with multiple generation units and transmission loops is considered.

Objective function: The objective is to minimize the number of unserved loads at each stage of restoration and expressed as follows:

$$f_s = \min\left\{ -A_L \sum_{L=1}^{N_L} (S_L \times M_L \times X_L) + A_s \times \Delta S^2 + \sum_{g=1}^{N_g} \left(A_{PG_g} \times \Delta P_{PG}^2\right) + \sum_{g=1}^{N_g} \left(A_{QG_g} \times \Delta Q_{gG}^2\right) + \sum_{K=1}^{N_K} \left(A_{PF_K} \times \Delta PF_K^2\right) \right\} \tag{5.8.10}$$

where f_s is objective function, s represents stage index, S_L denotes load capacity (apparent power), and M_L is percentage of load L being restored. X_L is a binary value, where 1 means load L is restored, otherwise X_L is 0. In this approach, penalty functions A_s, A_{QG_g}, A_L, A_{PF_K}, and A_{PG_g} are defined to force variables remain inside their allowable limits.

Constraints: Three groups of constraints were defined in [239]:

- Switching scenario constraints allows only a load, a transformer, a transmission line, a generator unit, or a load to be switched on at each stage of restoration.
- Power balance constraints based on load flow equations that guarantee load demand will be met.
- The operational constraints including active power, reactive power, line power flow and voltage limits of each bus, and line in the power system which guarantee a safe operation and stability of the system.

Algorithm: In PSO algorithm, each particle swarm is a candidate solution to the problem. Those particles are distributed into a space of n dimensions (problem domain). On the other hand, each particle includes several control variables, based on the size of the system to be restored. For a nine-bus system, 23 control variables were defined. Discreet binary variables are handled by BPSO module, shown as follows [239]:

- State of generation units: one if the generation unit is turned on, otherwise zero.
- State of loads: one if load is in restoration path, otherwise zero.
- Status of transmission lines and transformers: one if it is in the restoration, otherwise zero.

The continuous control variables, including voltage magnitude of generation units, active power generation of each source, and load percentage of each load, are adjusted in the limit range by continuous PSO module. At each stage of restoration, once the PSO algorithm converges, the optimal solution is fixed for the next step of the process. This algorithm can be applied to both total and partial blackout [239].

Comparison of Different PSO Applications

1. The comparison of PSO and GA methods in distribution system restoration is discussed in [240]. The objective function is to maximize the number of energized load using reconfiguration scheme. Two algorithms are used sequentially:
   - First algorithm creates a list of the status of each node with or without energy. The objective is to maximize the amount of power demand supplied to the system and to minimize the number of operating NO switches. Besides, it detects the overloaded line by running load flow algorithm. Once it detects an overloaded line, that case will be saved and moves to the next step until finding the best solution.
   - Second algorithm is activated when overloaded line is detected in first algorithm. This algorithm will open the NC switches to relieve the overload, by minimizing the number of operating NC switches. Then, the program returns to first algorithm to close another NO switches.
     In both algorithms, each particle represents a feasible solution of the optimization problem [240].
2. All particles in the swarm learn from the gbest point, even if the current g_{best} is far from the global optimum. Therefore, particles may be easily attracted to one area and trapped in a local optimum. Chaos particle swarm optimization (CPSO) algorithm proposed in [241] improves the search efficiency. The following is an example using chaos mapping method [242]:

$$g_{\text{best}}(K+1) = g_{\text{best}}(K) + \sigma r_1 X(K+1) - \sigma r_2 X(K) \tag{5.8.11}$$

$$X(K+1) = r X(K)(1 - X(K)) \tag{5.8.12}$$

$$(x) = \begin{cases} 1 & r_3 < P_{\text{m}} \\ 0 & r_3 \geq P_{\text{m}} \end{cases} \tag{5.8.13}$$

where r, r_1, r_2, and r_3 are random numbers between 0 and 1, and P_{m} is the probability constant between 0.05 and 0.5 [242]. CPSO has been used for distribution planning [241] and reconfiguration. This approach can be applied in other PSR problems.

1. During cold load pick up (CLPU), the optimal sizing and location of DG to reduce transformers and transmission lines load is discussed in [243]. CLPU may cause excessive overloading in substation transformers and feeders and voltage drop through the feeder. DG units are able to provide the local service and help energizing grid to reduce the outage time. CLPU can be modeled as a delayed exponential model. The objective function is defined as:

$$\text{Min} f = W_{\text{load}} \left[\frac{S_{\text{total}} - S_{\text{supplied}}}{S_{\text{total}}} \right] + W_{\text{T}}\text{LOL} + W_{\text{IV}} I_{\text{V}} + W_{\text{VV}} V_{\text{V}} + W_{\text{OC}}\text{OC} \tag{5.8.14}$$

where W represents penalty factor, LOL denotes transformer loss of life, I_V and V_V are line voltage and line flows, and operating cost is represented by OC.

Each swarm represents the candidate bus for DG installation. First, the load served by each DG is determined. Second, load flow is performed for both transmission and distribution networks. Then, the initial rollback time is computed and constraints are verified. At each step, transformer's loss of life and the objective function are calculated until the stopping criteria are met. Based on simulation results in [243], PSO outperforms GA in the objective function value and rollback time.

2. A skeleton-network reconfiguration strategy is proposed in [244]. A skeleton-network is composed of power sources, important loads, and transmission lines. Skeleton-network reconfiguration is a typical multi-constraint, multi-objective, and multistep nonlinear optimization problem that can be solved by discrete PSO algorithm. This paper investigates the restoration process from network topological structure point of view. The purpose is to find restoration target instead of switching sequence. First, node importance degree is calculated. Then, network reconfiguration efficiency is defined as the objective function to evaluate the reconfiguration effect. The outcomes are optimal restoration path sequences that help system operators to make decisions. Simulation results demonstrate that all power sources and some important loads are taken as the restoration target in the first stage. After the formation of the skeleton network, generator loading ability has been improved, and loads can be restored quickly in the second stage.

Ant Colony Algorithm

General Ant Colony Algorithm ACO is a swarm intelligence strategy to find an optimal solution by representing the problems as an undirected graph. ACO starts with a random search enumerating the entire possible paths from the source to destination. Then, ants lay pheromones on the path during the search of food. These pheromones evaporate quickly, and increase the probability of other ants following the path with highest intensity of pheromones. The shortest path between the source and destination will have the strongest intensity of pheromones. This process determines the optimal path for ants to search food. The probability that an ant follows a path depends on the attractiveness to the pheromones and the visibility. The number of optimal search is reduced by eliminating the state of the agent and enlisting the visited state to a tabu list. The probability that the ant/agent moves from one state to other is presented by the following equations [245]:

$$P_{ij}^{k} = \frac{\tau_{ij}^{\alpha} \cdot \eta_{ij}^{\beta}}{\sum\limits_{s \in J_i^k} \tau_{is}^{\alpha} \cdot \eta_{is}^{\beta}} \tag{5.8.15}$$

$$\tau_{ij}(t) = (1-\psi)\tau_{ij}(t) + \psi\tau_0(t) \tag{5.8.16}$$

where τ_{ij} is the pheromone trail, η_{ij} is a given heuristic value, α and β are parameters used to tune the relative importance of τ and η, J_k is the set of all possible states of agent/ant k at state i, ψ is the constant governing the pheromone decay process, and τ_0 is the initial pheromone level.

Both the pheromones level and attractiveness of the pheromones (τ) and the specific visibility function (η) are updated during each iteration. The global solution is compared with the local solution to choose the global optimal solution.

Ant Colony Algorithm in Distribution System Restoration In the literature, there are different objectives, such as loss minimization, minimization of unserved energy, and minimization of the number of switching operations in the distribution system restoration [246–249]. For example, in [247], two objectives are to maximize the capacity of served loads and to minimize the number of switching operations. Besides the constraints of (5.8.2)–(5.8.6), two constraints, including the current limit of transmission lines to minimize the loss and the guarantee of radial topology of distribution network, are added.

The general steps of applying ACO to determine the optimal restoration strategy following a fault that happened in a distribution system are summarized as follows:

- The buses are represented as nodes and the connecting lines represented as arcs, which makes the distribution system as an undirected graph.
- Each ant makes a sequence of switches, either NC or NO switches. During the restoration process, the NO breakers must be closed and the NC breakers must be opened to maintain the radial topology.
- The agents/ants move through the arcs to different states determining the switches to remain closed or open to minimize the loss.
- Each network line is assigned an initial pheromone concentration and the ants generated or positioned randomly at different buses. When an agent/ant is on ith node, it chooses to move to jth neighboring node based on the pheromone concentration.
- At node j, the concentration is updated. The choice of the node depends upon the pheromones associated with switches. The higher the values, they are more likely to be involved in the construction of solution set.
- The objective function is calculated when all nodes have been visited. The process continues to expand until the route expands out of local optimization. A new optimal path is generated after each generation search and compared with the global optimal path. The better one replaces the global solution.

The pheromone concentration on the local optimal solution is increased to a global updating. After certain number of iterations, when the population of ants in a particular path increases, the best path or global solution is found by the colony. These developed algorithms determine the optimal re-energization path and build the restorative topology in order to restore the unserved load while minimizing the restoration time and cost.

Combination of ACO and Other Algorithms In [247], ACO is combined with the multi-agent technique to formulate a MOOP. The outage area is divided into several intelligent agents. The interaction of intelligent agents with the heuristic search methods of ACO results in an optimal restoration solution which is robust and flexible.

In [250], ACO is used to determine the radial configuration topology while minimizing the system loss. The agents move through adjacent nodes and select the switches to remain closed based on the attractiveness of the pheromones deposited. The simulation results demonstrate the faster convergence by using ACO than GA or SA algorithms.

In [251], ACO is combined with a path tracing algorithm to determine the BS sequence. The path tracing algorithm helps form a directed graph based on the adjacency matrix. The ACO algorithm is able to accelerate the operation procedure by choosing the candidates with high potential, and brings a faster visualization of the optimal BS path. This hybrid algorithm improves the solution optimality by evaluating the low susceptance path and achieves the objective of the minimization of number of switching operations.

A hybrid fuzzy control method is combined with ACO algorithm in [252]. Based on the active and reactive power from supervisory control and data acquisition (SCADA), the fuzzy control initiates ACO for pheromones production. The advantage of using fuzzy control is to reduce the number of iterations for ACO to reach the pheromone level and accelerate the process. This hybrid algorithm provides more accurate topographical solutions and restores maximum loads in the faulted area.

Two algorithms, the absolute switch position ACO (ASP-ACO) algorithm and the relative switch position ACO (RSP-ACO) algorithm are compared in [253]. Simulation results demonstrate that the RSP-ACO algorithm converges more quickly than the other algorithm due to the pheromones reinforcement.

The advantage of ACO is the convergence to a solution, but the computational time may be long. ACO is able to only find radial solutions, which excludes the use of separated algorithm to test the radial nature of the system [250]. Also, the algorithm structure can easily accommodate and embed other algorithms for state transition and parameters update, which increases the flexibility and system adaptability [251]. The decisions in each step are not independent but are sequential. The combination of ACO with other algorithms results in a faster convergence and more robust solutions due to the changing probability distribution with each iteration.

Tabu Search

General Tabu Search TS algorithm is based on the hill climbing method (HCM). It first generates a potential solution, and then checks the neighbor of this solution. Based on the tabu criteria, solutions with better objective function value will replace the original solution. This algorithm can avoid retracing previous steps by enlisting the local solution into one or more tabu lists. It maintains two lists, Elite list (long-term memory) and tabu list (short-term memory). Best solutions are stored in the Elite list, while solutions previously visited within a short-term period or solutions that do not meet the constraints will be included in the tabu list. In each iteration, the new solution will be compared with all stored solutions and update both lists. After a certain number of iterations, the algorithm arrives at global or near global optimum.

Tabu Search in Distribution System Restoration TS has been applied in PSR to solve the MOOP [254–256]. The objective function includes the minimization of restoration time, power loss of the system, and the switching operation. General steps of applying TS in distribution system restoration are shown in the following:

- Step 1: Isolate the de-energized buses by tripping all connected switches, connecting the possible switches to healthy section of the network, and redefining the topology using load-flow analysis.
- Step 2: Consider the newly constructed set as the initial solution set for TS algorithm and put the current state in the tabu list.
- Step 3: Initialize the tabu length. The tabu length depends upon the number of nodes or number of loads in the affected area.
- Step 4: Generate neighboring states based on the selection of loads that can change the direction of power source in the current network configuration. These states act as candidates for next states.
- Step 5: Move the current state to the next state, update tabu list, and check the objective function value.
- Step 6: Go back to Step 2 until meeting the convergence criterion.

Compared to other algorithms that use the history information to determine future search, TS algorithm uses the tabu length as one of many parameters to accelerate the process. According to [254], TS has a better solution quality and faster convergence compared to SA or GA. However, in case of very large networks, TS requires more computational time to create solution candidates. There are different modifications of TS and integration of TS with other algorithms to improve the performance.

In [257], the length of tabu is fine-tuned with the feedback-based mechanism, named as reactive tabu search (RTS) algorithm. In RTS, the tabu length is not fixed, which brings much faster computation [230]. RTS algorithm uses an escape procedure from the long search in TS by checking whether the current solution has been found already. Another modification of parallel tabu search (PTS) is employed in various works [256]. PTS algorithm divides the solution space into equal divisions with different tabu lengths. Multiple tabu lengths allow the evaluation of solution in different search processes. A probabilistic sampling approach is developed in [255]. Starting with a random selection as the initial solution set, solution candidates in the neighborhood will be given certain probabilities. The probabilistic sampling is able to improve the computational time with the same solution quality. Artificial intelligence has been incorporated with TS algorithm to develop a more robust and efficient restoration procedure in [258]. The hybrid algorithm first closes all the switches converting the radial network to a mesh network. A new set of switches is opened to regain the radial network forming the individual. A large number of individuals compose the population for the process. Based on the fitness calculation of each individual, the mutation is carried out to select the best solution for tabu list. Multi agent technology (MAT) is integrated with TS in [259] for a new distribution model of PSR. The intelligent agents represent the subdivision of the affected area. The TS algorithm searches the solution globally based on the initial set of solution provided by the MAT. Local optimal solutions are selected by intelligent agents, and a faster and more reliable restoration process is achieved.

Simulated Annealing Algorithm

General Simulated Annealing Algorithm SA is driven by an analogy to annealing of metal treatment. It is a stochastic search method to find global optimal solution for a constraint optimization problem. SA starts with a valid initial solution and randomly generates potential solutions during the process [260]. The accepted potential solution replaces the existing solution, and the worse potential solutions are evaluated based on the swap probability function. The selection of potential solution and updating probability functions depend on a control parameter known as temperature parameter, whose value reduces gradually as the algorithm proceeds. The process of searching and updating optimal solution continues until optimal solution is found or given criteria is met.

Simulated Annealing Algorithm in Distribution System Restoration When applying SA in distribution system restoration, a build-up restoration strategy is considered. The entire system is divided into islands, and each island is restored using distributed generations (DGs) within the islands. After islands are restored to the stable state, they are synchronized together. The following assumptions are considered: (i) each island has at least one BS unit; (ii) total installed capacity of DG and load is known; (iii) secure voltage limits are specified; (iv) line current capacity is specified; (v) loads are picked up based on the priority; and (vi) the number of islands has been determined.

The objective is to minimize the unserved load in each island. The unserved load can be represented as the energy in the annealing process of metal treatment. The algorithm starts with the schedule of DGs and load pickup. DG's schedule is changed randomly by changing the state of DGs. Then, the resulted change of energy in the annealing process is evaluated by the objective function. The new schedule of DGs with less energy is accepted and replaces the existing schedule. Otherwise, if the energy of new DGs schedule is higher, it is accepted based on the probability function P given by Boltzmann distribution [261].

$$P(\Delta E) = e^{-(\Delta E/kT)} \tag{5.8.17}$$

where ΔE is the change of energy between the two consecutive DGs schedule, T is the existing system temperature, and k is Boltzmann's constant.

A certain number of iterations are executed at each temperature, and the temperature is reduced in subsequent steps. The process is repeated until temperature is reduced to a specified value, or a specified number of iteration has been executed. After two or more islands become stable, they can be synchronized together to form a large island. This process continues until all systems are restored.

In the literature, SA or modified SA has been applied in distribution system restoration. In [230], the parallel SA was used to optimize service restoration in distribution systems. The comparison of algorithm effectiveness and computational time is performed for four heuristic optimization algorithms, including GA, TS, RTS, and SA. Simulation results demonstrate that RTS is the most practical for combinatorial optimization problem. In [262], a modified SA optimization technique is developed for network reconfiguration after a fault to maximize customer service with minimum switching operations. In [263], SA is used to reconfigure radial distribution networks for loss reduction and service restoration. A MOOP, including the minimization of switching operations, is developed and tested for a 52-bus distribution network to verify the effectiveness of SA.

Fuzzy System

General Fuzzy System Fuzzy logic (FL) is the pioneering work of Lotfi A. Zadeh who invented the theory of fuzzy sets [264, 265]. By definition, if X is a collection of objects denoted generically by x, then a fuzzy set A in X is defined as a set of ordered pairs:

$$A = \{(x, \mu_A(x)) \mid x \epsilon X\} \tag{5.8.18}$$

where $\mu_A(x)$ is called the membership function for the fuzzy set A. The membership function maps each element of X (the universe of discourse) to a membership grade between 0 and 1. A higher membership value indicates that an element more closely matches the characteristic feature of the set.

The application of FL in power systems is extensive with more than 100 archival publications in the 1995 survey [266]. Many research works have investigated the application of FS in PSR, as reviewed in the next section.

Fuzzy System in Distribution System Restoration

1. In [227], a combination of fuzzy-GA method is developed to solve the service restoration problem.

   Objective functions: There are five different objective functions, including the minimization of out-of-service area, number of switching operation, current of feeders, division of the bus voltage, and load of transformers. The objective functions are modeled with fuzzy sets to evaluate their imprecise nature.

   - Minimize the out-of-state service area

   $$\text{Min} \quad f_1(\overline{X}) \tag{5.8.19}$$

   where $\overline{X}$ denotes the switch state vector, including each switch status (0 and 1 represent opened and closed status, respectively). $f_1(\overline{X})$ denotes the number of non-faulty out-of-service area under the state $\overline{X}$.

- Minimize the number of switching operations

$$\text{Min} \quad f_2(\overline{X}) = \sum_{i=1}^{N_s} |S_i - S_{0i}| \tag{5.8.20}$$

where $f_2(\overline{X})$ denotes the number of switching operations under state $\overline{X}$, and S_{0i} represents the original state of switch i (after fault isolation).

- Minimize the deviations of bus voltage

$$\text{Min} \quad f_3(\overline{X}) = \max |V_i - 1.0| \quad i = 1, 2, \ldots, N_b \tag{5.8.21}$$

where N_b is the number of buses, V_i denotes the voltage on bus i with p.u. input. $f_3(\overline{X})$ represents the maximal deviation of bus voltage in the considered system.

- Minimize the line current

$$\text{Min} \quad f_4(\overline{X}) = \max \left\{ \frac{I_{i\text{Load}}}{I_{i\text{Rate}}} \right\} \quad i = 1, 2, \ldots, N_L \tag{5.8.22}$$

where N_L is the total number of the feeder lines, $I_{i\text{Load}}$ and $I_{i\text{Rate}}$ denote the load current and the rated current of branch i, respectively. $f_4(\overline{X})$ represents the maximal normalized line current in the considered system.

- Minimize the loading of the transformer

$$\text{Min} \quad f_5(\overline{X}) = \max \left\{ \frac{\text{tr}_{i\text{Load}}}{\text{tr}_{i\text{Rate}}} \right\} \quad i = 1, 2, \ldots, N_t \tag{5.8.23}$$

where N_t is the total number of the transformers, $\text{tr}_{i\text{Load}}$ and $\text{tr}_{i\text{Rate}}$ denote the load current and the rated current of transformer i, respectively. $f_5(\overline{X})$ represents the maximal normalized line current in the considered system.

Constraints: A radial network structure must remain after service restoration. Considering the imprecise nature of each objective function, it can be formulated as fuzzy sets. To elicit a membership function for each objective function, $\mu_{fi}(\overline{X})$, authors in [227] defined a strictly monotonically decreasing and continuous function together with lower and upper bands as follows:

$$\mu_{fi}(\overline{X}) = \begin{cases} 1 & \text{if } f_i(\overline{X}) < f_i^{\min} \\ h_i(f_i(\overline{X})) & \text{if } f_i^{\min} < f_i(\overline{X}) < f_i^{\max} \\ 0 & \text{if } f_i(\overline{X}) < f_i^{\max} \end{cases} \tag{5.8.24}$$

Then, the service system restoration problem is solved through the following steps:

- Input the location of faulted buses.
- Identify the out-of-service areas.

- Determine the membership function, $\mu_{fi}(x)$, of each objective function.
- Select the initial expected membership value of each objective function, $\overline{\mu_{fi}^{(0)}}$.
- Apply GA to solve the below min–max problem, $\min\limits_{X \in S} \left\{ \max_{i=1,2,\ldots,n_0} \left[\overline{\mu_{fi}^{(0)}} - \mu_{fi}(\overline{X}) \right] \right\}$, where S denotes the vector space of $\overline{X}$ and n_0 represents the number of total objective functions.
- Check the feasible solution after stopping criteria satisfaction, otherwise choose new expected value, $\overline{\mu_{fi}^{(m)}}$, and redo the GA algorithm.
- Select the most satisfactory feasible solutions, $\overline{X}, f_i(\overline{X})$, and $\mu_{fi}(\overline{X})$ for $i = 1, 2, \ldots, n_0$.

2. In [267], fuzzy set is applied to solve the multi-objective service restoration problem. The main objective is to restore the electricity service to the interrupted customers outside the faulted zone after fault isolation. The following requirements must be satisfied: restore as much load within the out-of-service area as possible, operate minimal number of switches, and avoid component overloading. A modified fuzzy reasoning approach by reformulating the crisp objective function was proposed to avoid unbalanced loading conditions, where some lateral are loaded very close to their limits. The fuzzy reasoning approach can be expressed as follows:

$$\min \widetilde{Z} = 2\sum_{i=1}^{8} x_i + 1 \tag{5.8.25}$$

where x_i is the status of switches at each lateral (in this case, totally eight switches needs to be operated).

A high membership function value indicates a desirable situation. In this case, switching operation less than three is desirable. For each restoration plan, the fuzzy objective function, $\widetilde{Z}$, is solved, and the corresponding membership functions μ_z is calculated. Then, the minimum membership function is selected for each plan. The overall membership value of each specific restoration plan is computed, and the maximum value among those plans is selected, which represents the best restoration plan.

3. In [268], the FL technique is applied to find the most preferable restoration plan. First, the candidate set of feasible restoration plans is listed. Then, the fuzzy evaluation is used to evaluate the preference of each plan. Four fuzzy criterion and membership functions are considered: number of switching, maximum loading of the backup feeder, load transfer, and contingency preparedness. The weighted sum of each evaluation result is calculated for each plan to determine the most preferable plan.

4. Quantitative analysis of the preference ranking of various restoration plans has been discussed in [269]. The fuzzy-grey relational analysis is used to help operators evaluate each restoration plan and make decisions. The objective functions include the minimization of the number of switching operations, the maximum loading among supported feeders, the maximum loading among supported laterals, and the unbalance loading index of feeders after switching operations. No overload equipment and retaining radial structure of distribution system are assumed constraints.
5. In [270], based on grey system theory, the grey relational analysis (GRA) is used to measure the relationship between two sequences by calculating their correlative degree. Considering a reference sequence $x_0 = (x_0(1), x_0(2), \dots x_0(n))$ and m comparative sequence $x_i = (x_i(1), x_i(2), \dots, x_i(n))$, $i = 1, 2, \dots, m$, the grey relational coefficient (GRC) of x_i with respect to x_0 at the kth entry is defined as follows [230]:

$$\gamma(x_0(k), x_i(k)) \equiv \frac{\Delta_{\max} - \Delta_{0i}(k)}{\Delta_{\max} - \Delta_{\min}} \tag{5.8.26}$$

where $\Delta_{\max} \equiv \text{Max}_{\forall i}\text{Max}_{\forall k}|x_0(k) - x_i(k)|$, $\Delta_{\min} \equiv \text{Min}_{\forall i}\text{Min}_{\forall k}|x_0(k) - x_i(k)|$, and $\Delta_{0i}(k) \equiv |x_0(k) - x_i(k)|$. The GRCs between each comparative sequence x_i and the reference sequence x_0 can be derived from the average of GRC:

$$\Gamma_{0i} = \sum_{k=1}^{n} \frac{1}{n} \gamma(x_0(k) - x_i(k)) \tag{5.8.27}$$

where Γ_{0i} represents the degree of relation between each comparative sequence and reference sequence. The higher value shows that the comparative sequence is more similar to reference sequence. The steps of the proposed approach are summarized as follows:

- Generate all feasible restoration plans according to the operation constraints.
- Compute the value of objective function of all possible restoration plans.
- Use the fuzzy multi-criteria evaluation to evaluate the fitness degree of each objective function.
- Rank the restoration plans in preference order according to their GRCs.
- Minimize unbalance loading index of laterals after switching operations.

Simulation results demonstrate that this algorithm assists system operators in making restoration decisions by providing alternative restoration paths.
6. Compared to [269], authors in [271] take the priority of each objective function into consideration. AHP is utilized to assess the weighting factors of each objective function. AHP can be applied to service restoration to

derive numerical weight of each objective function through the following steps:

- Create pairwise comparison matrix. Let $f_1, f_2, \ldots, f_n$ be the set of objective functions, the pairwise comparison matrix A can be represented by the following n-by-n matrix:

$$A = \begin{bmatrix} 1 & a_{12} & \cdots & a_{1n} \\ \dfrac{1}{a_{12}} & 1 & \ldots & a_{2n} \\ \vdots & \vdots & \ddots & \vdots \\ \dfrac{1}{a_{a1n}} & \dfrac{1}{a_{2n}} & \ldots & 1 \end{bmatrix} \tag{5.8.28}$$

  where the element a_{ij} represents quantified judgment on a pair of objectives (f_i, f_j) and is defined as relative importance of two objectives [84].
- Calculate the largest eigenvalue and the corresponding normalized eigenvector.
- Verify the consistency of comparison matrix by the consistency ratio (CR). The CR is the ratio of consistency index (CI) and corresponding random index (RI). CI is the deviation of the maximum eigenvalue from the number of criteria used in the comparison process.

Finally, the results derived from fuzzy multi-criteria evaluation and AHP can be used as an input of GRA to calculate the performance index of each restoration plan. Each restoration plan can be evaluated in a quantitative fashion for operators to choose.

7. Restoration index in distribution system and its application to system operation has been discussed in [272]. The evaluation indices indicate the restoration capability of distribution feeders. These indices would help operators recognize a potential problem regarding restoration before any fault occurs.
8. An integer linear programming technique is developed in [273]. The objective function includes the minimization of the number of switching operations and load balancing. FL is used to select the best solution from all candidates that are able to relieve overloads of distribution lines. The best solution is able to bring lowest overload for upstream feeders with the minimum number of switching operations.

ACKNOWLEDGEMENT

This work is supported in part by SDSU Research/Scholarship Support Fund. The authors acknowledge the contribution of A. Golshani, N. Kadel, D. Chaudhary, and S. Ma.

5.9 GROUP-BASED PSO FOR SYSTEM RESTORATION

Hsiao-Dong Chiang[1], *Shuo Wang*[2], *and Yong-Feng Zhang*[3]

[1]*Cornell University, Ithaca, NY, USA*

[2]*China Electric Power Planning & Engineering Institute, Beijing, China*

[3]*University of Jinan, Jinan, China*

5.9.1 Introduction

Particle swarm optimization (PSO) is a population-based global optimization technique that has been widely applied to various fields [32]. However, PSO has some drawbacks in searching for the global optimal solution. One drawback, a common one, is that PSO is not guaranteed to converge to the global optimum and can easily converge to a local optimum [274]. Another drawback is that PSO can have a slow convergence rate [275]. In addition, PSO lacks the scalability in solving large-scale problems [276].

A large number of improved PSOs have been proposed. These improvements can be categorized into three groups. The first group adjusts parameters of the mathematical model of PSO [277, 278]. The second group proposes improved mathematical models of PSO for speed and convergence [279–281]. In the third group, the PSO is combined with different strategies with the aim to enhance the performance of the PSO. The methods of this category include the hybrid PSOs, a novel chaotic PSO combined with an implicit filtering local search [282], a hybrid of PSO with a modified Broyden–Fletcher–Goldfarb–Shanno method [283], a new PSO with moderate-random-search [284], a comprehensive learning particle swarm optimizer which maintains the particles' diversity by using a learning strategy to update a particle's velocity [275], a PSO-based memetic algorithm [285], and a hybrid of the population utilization strategy and PSO [286].

Many proposed variations of PSO have contributed to the improvement on the convergence of the population-based methods. In view of PSO, one important issue is how to prevent particles from being prematurely trapped in a local optimal solution. In addition, how to reduce the running time needed for PSO in solving a large-scale optimization problem is another key consideration.

In this chapter, we present a novel group-based PSO method, which can be classified into the second and third groups described above, for solving constrained global optimization problems. This novel method is proposed to overcome the challenges faced by PSO in issues such as extensive computation efforts (i.e. lengthy computation time) and scalability using the following three stages:

- (exploration and grouping) Stage 1: Apply the PSO method and group all the particles using the condition that all the particles reach a consensus instead of the maximal iteration numbers, which is the traditional criterion.
- (selection) Stage 2: From each group of particles, select the top particles in terms of quality of particles.

- (exploitation and fine-tune) Stage 3: Starting from the multiple solutions provided by the group-based PSO method, apply a local method to find multiple optimal solutions and compose the top solutions, i.e. high-quality optimal solutions.

Stage 1 uses the PSO to explore the search space and apply a grouping scheme to stop the PSO iterations when all the populations have reached a certain level of consensus. This new stopping criterion is designed to overcome the slow convergence of PSO. Stages 2 and 3 seek to find high-quality optimal solutions starting from those obtained by the proposed group-based PSO.

The premise for the proposed group-based PSO method to find the global optimal solution is the following:

- All the particles of the PSO have reached a high level of consensus by forming into several groups. Each group contains a number of particles (large or small) that lie close to each other in the search space.
- Each group of particles reveals that high-quality local optimal solutions, even the global optimal solution, are located in the region "covered" by the particles and are close to the particles.
- From the high-quality solutions obtained by the PSO algorithm, a local method can efficiently find a high-quality local optimal solution.

The proposed group-based PSO method is applied to solve large-scale service restoration problems in distribution networks which are nonlinear, discrete, and constrained optimization problems. Several studies on utilities' experience suggest that customer satisfaction is closely correlated with service interruption frequency and duration. Service interruption duration can be significantly decreased via effective service restoration procedures. Service restoration is achieved by changing the topological structure of a power system, which means altering the status of both tie switches (NO switches) and sectionalizing switches (NC switches) in a network. The main objective in service restoration procedures is to restore as much load as possible by transferring de-energized loads in the out-of-service areas via network reconfiguration to other supporting feeders without violating operation and electrical constraints.

Service restoration is a significant issue in power system operation as it may seriously affect the system reliability, including the outage duration and the number of out-of-service customers. Due to the ever-increasing demand of electric power, power systems are operated closer to their limits, being more likely to cause fault and overload events. Additionally, the increase of complexity and size in distribution systems also increases the risk of major power faults. Fast and complete service restoration has multifold benefits. For example, it reduces the inconvenience and the cost of the outage to customers; it enables the utility to resume earning revenue for energy sales; and it enables the utility to provide enhanced service to priority customers such as hospitals, police stations, fire departments, etc.

In practice, distribution dispatchers need to restore service to the outage areas as quickly as possible with a minimal number of switching operations. A minimal number of switch operations is required because of switch life expectancy concerns and manpower limitations since not all switches in the network are currently automated. From a computational viewpoint, the service restoration problem is an NP-complete problem. NP-complete problems typically require time-consuming combinatorial optimization algorithms to solve for global optimal solutions. However, for service restoration, there may exist many solutions to restoring power to the out-of-service areas.

In the past, there has been considerable effort addressing service restoration. Heuristic techniques and expert systems have been developed for quickly determining restoration plans [184, 287–293]. In general, these methods are very effective in a real-time environment, but the performance of the solutions obtained from those methods may be degraded because of the effectiveness of the heuristic rules. Meta-heuristic methods and optimization-based methods can provide effective solutions for addressing service restoration [230, 294–298]. Analytical methods for service restoration can be found, for example, in [299–303].

In the second part of this chapter, a three-stage solution algorithm based on the group-based PSO is developed for solving service restoration problems. This three-stage solution algorithm has several distinguishing features:

- it can find multiple local optimal solutions, and then allow operators and users to choose from or find the global optimal solution among them
- it explores the global search capability of PSO but overcome the blindness or randomness introduced by the PSO algorithm
- priority customer considerations
- voltage, current, and feeder capacity constraints
- multiple fault and fault location considerations
- comprehensive system modeling with three-phase, two-phase, and single-phase representation
- applicability to large-scale distribution systems

This proposed solution algorithm determines multiple representative optimal solutions based on the solution set provided by the group-based PSO algorithm. It performs especially well under heavy loading conditions and allows the operator to select their preferable solution based on his/her engineering judgment and operational experience. The proposed solution algorithm is evaluated in the standard IEEE 123-node test case system and an 1101-node practical distribution network. We use the K-means clustering approach to group the particles in the group-based PSO method. Test results show the feasibility and effectiveness of the proposed group-based PSO algorithm in solving the service restoration problem.

5.9.2 Group-Based PSO Method

We feel that PSO has an excellent capability of "global" search while it lacks the local fine-tuning capability and suffers the slow convergence to find optimal solutions.

Stage I: exploration and consensus stage

The first stage of the proposed methodology is designed to take advantage of the capability of PSO in exploring the promising regions in the search space. PSO is a computational intelligence-based technique that seems to be not affected by the size and nonlinearity of the study problem, and can converge to the optimal solution in many problems where most analytical methods fail to converge.

Traditional PSO

There are several variants of PSO methods to which the proposed methodology is applicable. As an illustration, we use the traditional PSO for continuous optimization in the following presentation. In the initialization phase of PSO, the positions and velocities of all particles are randomly initialized. Fitness value which is the objective function value, is calculated at each random position. The fitness values are, respectively, *pbests* of each particle which implies the optimal fitness of each particle so far. Among these fitness values, the best one is the initial *gbest* which is the optimal fitness value among all the particles so far. In each step, PSO relies on the exchange of information between particles of the swarm. This process includes updating the velocity of a particle and then its position. The former is accomplished by the following equation:

$$v_i^{k+1} = \omega v_i^k + c_1 r_1 \left(p_{i\text{best}} - x_i^k\right) + c_2 r_2 \left(g_{\text{best}} - x_i^k\right) \tag{5.9.1}$$

where v_i^k is the velocity of the ith particle at the k step. x_i^k denotes the position of the ith particle at the k step. ω is the inertia weight which is used for seeking a balance between the exploitation and exploration ability of particles. Typically, these are both set to a value of 2.0. r_1 and r_2 are elements from two uniform random sequences in the range (0,1).

The equation describing the PSO procedure consists of three parts and the relationship among them is described in Figure 5.9.1. The first section presents the inertia of a particle itself. The second section presents in the next step that each particle should move toward its own previous best position. The third section implies that each particle should move toward the best position of all particles so far.

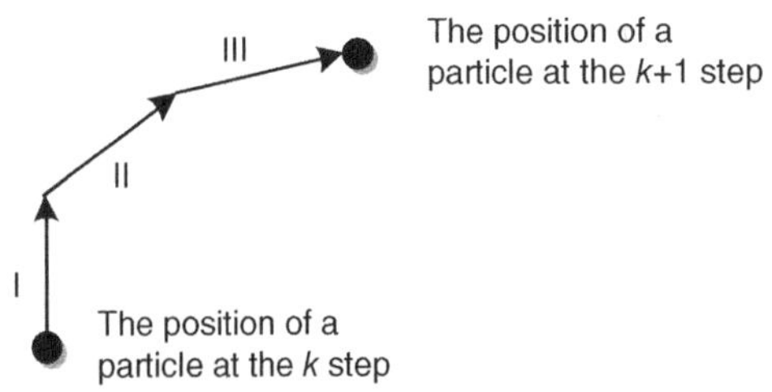

Figure 5.9.1 An illustration of three steps involved in the traditional PSO procedure.

Binary PSO

In the BPSO, trajectories are changes in the probability that coordinates will take on a zero or one value. The moving velocity is defined in terms of changes of probabilities that a dimension will be in one state or the other. Thus, a particle moves in a state space restricted to 0 and 1 for each dimension. Each individual of the population has to make a binary decision, either Yes or No. In that sense, according to the social approach of PSO, the probability of an individual to decide Yes or No can be modeled as,

$$P(X_{id}=1)=f\left(X_{id}(t-1),V_{id}(t-1),p_{id},p_{gd}\right) \tag{5.9.2}$$

In this model, the probability that the ith individual chooses 1 for the dth dimension in the particle, i.e. $P(X_{id}=1)$, is a function of the previous state of that dimension, i.e. $X_{id}(t-1)$, and V_{id}, the measure of the individual's predisposition to choose 1 or 0. This predisposition is derived based on individual and group performance. Therefore, the probability $P(X_{id}=1)$ implicitly depends on p_{id} and g_{id}. The former is the best individual state found so far; it is 1 if the best individual success occurred when X_{id} was 1, and 0 otherwise. The latter corresponds to the neighborhood best; this parameter is 1 if the best of any member of the neighborhood occurred when X_{id} was 1, and 0 otherwise. Mathematically, V_{id} determines a threshold in the probability function, and therefore should be bounded in the range [0.0, 1.0]. This threshold can be modeled with the well-known sigmoidal function

$$S(V_{id})=\frac{1}{1+\exp\left(-V_{id}\right)} \tag{5.9.3}$$

Applying (5.9.3), the state of the dth position in the particle for the ith individual at time t, X_{id} can be expressed as,

$$\begin{aligned} X_{id}(t)&=1 \text{ if } \rho_{id}<s(V_{id})\\ X_{id}(t)&=0 \text{ otherwise}\end{aligned} \tag{5.9.4}$$

where ρ_{id} is a random number with a uniform distribution in the range [0.0, 1.0]. This procedure is repeatedly iterated over each dimension (d: 1, …, N) and each individual (i: 1, …, N), testing if the current value $X_{id}(t)$ outperforms p_{id}. In that case, the value of X_{id}(t) will be stored as the best individual state. Similar to the case of a real number space, the velocity update can be expressed as,

$$\begin{aligned} V_{id}(t)=V_{id}(t-1)+\varphi_1\cdot \text{rand}_1\cdot\left(p_{id}-X_{id}(t-1)\right)\ldots+\varphi_2\cdot\text{rand}_2\\ \cdot\left(p_{gd}-X_{id}(t-1)\right)\end{aligned} \tag{5.9.5}$$

where φ_1, φ_2 are two positive numbers and rand_1, rand_2 are two random numbers with uniform distribution in the range [0.0, 1.0].

Consensus Stage

Since each particle of PSO has the ability to exchange information with each other, the information can guide each particle to promising regions that may contain the global optimal solution. However, it is possible that each particle has different information

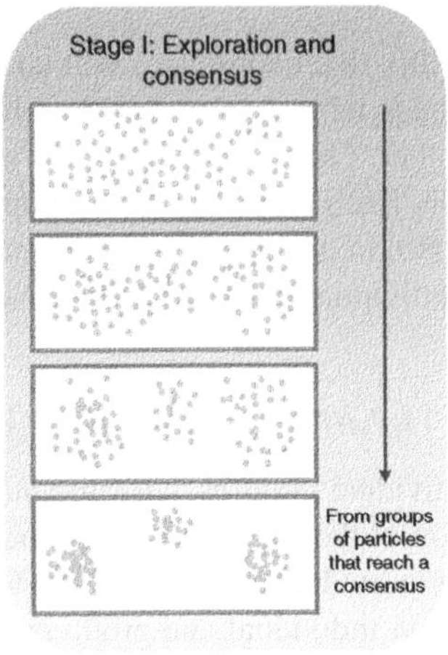

Figure 5.9.2 The procedure of Stage I.

regarding the location of the global optimal solution. Hence, they hold different views of the locations and all the particles may gather at several different regions of the state space (see Figure 5.9.2). These particles start to form groups of particles as they progress. We propose the following "equilibrium state" for consensus:

The number of groups of particles is not changed.

The members in each group do not alter.

The consensus condition serves to stop the PSO search procedure. Our observations have indicated that all the particles settle down to different locations which form several different groups in the search space. All the particles do not form only one group. It should be noted that a group containing a large number of particles does not indicate that the region in which members of particles of this group settle down contains the global optimal solution. It can occur that distinct particles with outstanding performance can move toward the region that contains the global optimal solution. We also observe that the size of particles in each group and the quality of the fitness value that each member possesses do not reveal information regarding the quality of local optimal solutions lying in the region. Consequently, the region in which each group of particles settles down should be exploited in order to find high-quality local optimal solutions. Therefore, we should explore all groups to locate the high-quality local optimal solution, if not the global optimum.

To make PSOs more efficient in solving continuous optimization problems, all the particles are clustered using Iterative Self-Organizing Data Analysis Techniques Algorithm (ISODATA) to identify the groups after certain iterations, say every 50 iterations. In view of the results of clustering, the stopping criterion (i.e. the consensus condition) of Stage I is reached if all the particles have reached a consensus. If not, the PSO method continues the exploration stage. We note that ISODATA is an unsupervised classification method and the user needs to provide threshold values which are used to determine the groups and their members. If the members in each group do not alter and the number of groups from ISODATA is not changed, this indicates that all the particles have reached a consensus. These search regions around the clusters are further explored for searching high-quality optimal solutions. As previously stated, the quantity and quality of cluster cannot determine the location of the global optimal solution. So we should consider all the obtained clusters.

To make PSOs more efficient in solving discrete optimization problems, all the particles are clustered using the *K*-means to identify the groups after a certain number of iterations, say every 50 iterations. An effective clustering algorithm can be incorporated into the group-based PSO method, aiming to group the set of particles provided by PSO into a user-specified number of clusters such that similar particles are assigned into one cluster. It groups the particles in PSO iteration to check whether a consensus condition among particles has been reached. In this paper, the *K*-means algorithm is used for the clustering task for the service restoration problem. The *K*-means algorithm is an unsupervised classification technique which requires the number of clusters to be prespecified. It attempts to find the cluster centers such that the sum of the squared distances of each data point to its nearest cluster center is minimized. When used on a set of particles, it helps to identify some inherent property presented in the objects by classifying them into subsets that have some practical applications (see Figure 5.9.3).

The *K*-means clustering scheme proceeds as follows:

Step 1: Fix the number of clusters k and initialize the cluster centers CP^1, $CP^2,\ldots,CP^k$ randomly from the population $\{CP_1, CP_2, \ldots, CP_{NP}\}$.

Step 2: Assign CP_n, $n = 1, \ldots, N_P$ to cluster C_j, $j \in 1, 2, \ldots, k$ if and only if $\|CP_n - CP^j\| < \|CP_n - CP^p\|$, $p = 1, 2, \ldots, k$ and $j \neq p$ where $\|CP_n - CP^j\|$ is the distance between CP_n and CP^j.

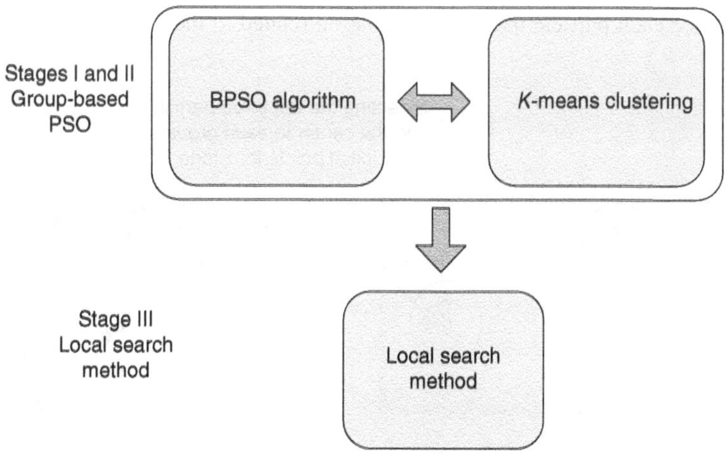

Figure 5.9.3 Procedure of the three-stage group-based PSO.

Step 3: Compute new cluster centers $CP^{1,*}$, $CP^{2,*}$, ..., $CP^{k,*}$ as follows:

$$CP^{j,*} = \frac{1}{M_j} \sum_{CP_n \in C_j} CP_n, \quad j = 1, 2, \ldots, k \tag{5.9.6}$$

where M_j is the number of elements belonging to cluster C_j.

Step 4: If $CP^{j,*} = CP^j, j = 1, 2, \ldots, k$, the process will be terminated. CP^1, CP^2, ..., CP^k are chosen as the cluster centers; otherwise, assign each CP^j with $CP^{j,*}$, $j = 1, 2, \ldots, k$, and go to Step 2 will be started again.

Stage II: Selection Stage

The PSO algorithm was shown to successfully converge during the initial stage of a global search to a region surrounding the global optimal solution; the search process will become very slow after the initial stage. The K-means algorithm, on the contrary, can evaluate the particles distribution and identify representative particles of each cluster. Stage I is designed to explore the merits of traditional PSO and the clustering method for efficient and diverse searches.

There are multiple particles contained in each cluster of particles. This stage selects the top particles in terms of objective function values and the center particle from each cluster for further refinements. Hence, Stage II sets up several "good" initial points for further improvement to be executed in Stage III. We found that particle clusters are useful in service restoration procedure because they can be used to identify representative feasible solutions in less computation time as well as help to exploit more optimal solutions.

An effective clustering scheme is utilized during the PSO algorithm to judge whether the particles have come to a consensus. We group the particles obtained in the PSO stage into a certain number of clusters. Then, the best solutions of each cluster are selected to form a user preference table for the users. The user can choose their preferable one for further refinement via a local search (see Figure 5.9.4). There are a few advantages for the clustering process. First of all, this method makes use of the high efficiency of PSO in maintaining the diversity of swarms, since each particle uses the information related to the most successful

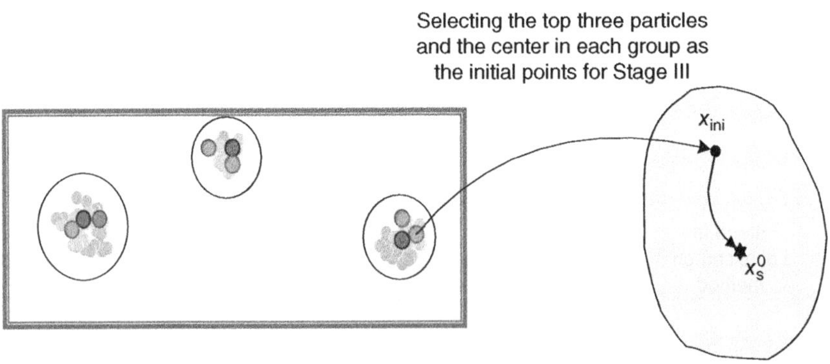

Figure 5.9.4 The top three solution points in each group.

particle in order to improve itself. Second, the group-based PSO algorithm can overcome the weak fine-tuning capability of the PSO algorithm. Third, the clustering process can greatly improve the searching efficiency of PSO to avoid time-consuming computation near local optimal solutions. Above all, the cluster process incorporated into the PSO algorithm can identify the inherent characteristics of different particles during iteration of the global search. Therefore, it is able to group similar particles and to select representative solutions. This provides a mechanism to explore more feasible optimal solutions through the group-based PSO.

Stage III: Local Search

Stage II provides several "good" initial regions for further exploration by a local method. Basically, the selection of a local method depends largely on the problem structure. Hence, depending on the study problem, a local method, whether it is a general-purpose local method such as the Interior Point Method, the Trust-Region Method, or a tailor-made local method that fully explores the problem structure, should be appropriately applied.

Next, we will describe algorithmic aspects of the three stages in detail.

Stage I (see Figure 5.9.5):

Step I-1: Set CheckNum = 100, initialize all parameters of the PSO-based method, and the maximum iteration number is N_{max}.

Step I-2: Set $i = 1$, while the condition of the iteration number of the PSO-based method $i \leq N_{max}$ is satisfied.

Step I-3: Solve the optimization problem by a PSO-based method.

Step I-4: When the iteration number of PSO i is the multiple of CheckNum, the Cluster operator will cluster all populations and check whether the stopping condition of the PSO-based method is met. If the stopping condition is met and all the populations have reached a consensus, go to Stage II. Otherwise, $i = i + 1$ and go back to Step I-3.

Stage II (see Figure 5.9.4):

Step II-1: The top three particles and the center point in each group are broadcasted to a local method.

Stage III: (local search)

Step III-1: The local method searches for the corresponding local optimal solution x_s^0. The number local optimal solutions is n which is equal to or less than four times of the number of groups obtained from Stage II.

Step III-2: A set of high-quality optimal solutions and the best local optimal solution can be identified from all the obtained local optimal solutions.

5.9.3 Overview of the Service Restoration Problem

Developing effective service restoration procedures is a cost-effective approach to improve service reliability and, consequently, enhance customer satisfaction. Service restoration through network reconfiguration is the process of altering the

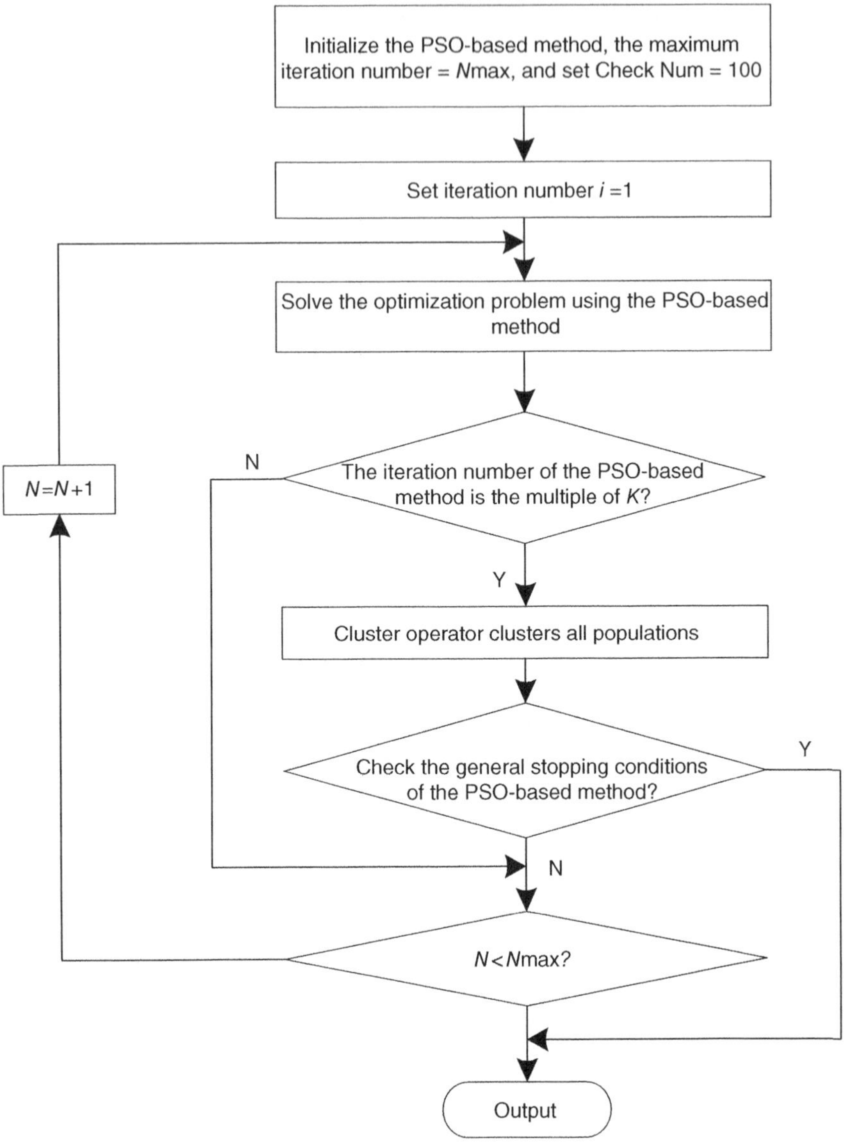

Figure 5.9.5 The flowchart of Stage 1 for the group-based PSO method.

topological structures of distribution feeders by changing the open/closed status of existing network switches and the network capacitor settings such that service is restored after a fault while electrical and operational constraints are met. Service restoration plans are to assist distribution personnel in making timely decisions on how to restore as much load as possible in the out-of-service areas. Service restoration also demands the utmost consideration of crew and public safety issues.

Consequently, we believe a practical and effective service restoration plan should meet the following requirements [300]:

1. it must be reached in a very short time
2. it must restore as much load as possible while considering priority customers
3. it must not cause violations in either engineering or operating constraints
4. it must outline a feasible sequence of operations to reach the final configuration
5. it must use a minimal number of switching operations
6. it must retain the radial network structure. This requirement is based on the feeder design for ease in fault location, isolation, and protective device coordination.

Service restoration via network reconfiguration is generally formulated as a constrained nonlinear binary optimization problem. The objective of this optimization problem can be expressed as follows:

$$\max_{g_k \in G_R} f_1 = \sum_{k \in N_R} |P_{L,k}| \tag{5.9.7}$$

where g_k is selected network configuration; G_R is the set of possible radial configurations; N_R is the set of restored buses; and $P_{L,k}$ is the total real power injected to bus k. Additionally, operating (topological, switch requirement, and loading capabilities) and power quality constraints should be respected, which are subject to:

Bus voltage constraint:

$$V_{k,\min} \le |V_k| \le V_{k,\max} \; \forall k \in N \tag{5.9.8}$$

Line current constraint:

$$I_{k,\min} \le |I_k| \le I_{k,\max} \; \forall k \in N \tag{5.9.9}$$

Power source limit constraint:

$$P_l^2 + Q_l^2 \le \left(S_l^{\max}\right)^2 \quad \forall l \in F \tag{5.9.10}$$

Network topology constraint:

$$G \in G_r \tag{5.9.11}$$

Unbalanced power flow constraints are satisfied:

$$f(V, g_k) = 0 \tag{5.9.12}$$

where V is the vector of node voltages; F is the set of all feeders; $S_l = \sqrt{P_l^2 + Q_l^2}$ is the total apparent power entering feeder l; V_k^p is the voltage at k bus; I_k^p is the current in the kth line; and N is the set of all buses. G presents the topology produced by the algorithm, while G_r presents the set that contains all the radial network topologies of the distribution system.

It is important for service restoration in a distribution power system to keep the radial structure during the iteration process. Once the state of a switch changes from open to closed, it is likely to form a loop, breaking the constraint of the radial structure. Thus, in each iteration, we need to check whether a loop is formed and we need to open a sectionalizing switch on the loop to maintain a radial structure.

The service restoration problem is a constrained NP-hard combinatorial problem. To illustrate the combinatorial nature, let us take a distribution network with 10 tie switches, we then have 10! (over 3.6 million) different possible sequences [304]. It is true that some sequences may be pruned out because they are not valid. However, distribution networks generally have a three-figure number of switches [305]. For this reason, the task of going through all combinations while optimizing the goals under constraints is prohibitive. There are several methods which are not mutually exclusive: heuristic rules, meta-heuristic technique, and the learning method. Heuristic rules are not based on a rigorous physical and mathematical proof but rather on operator experience and beliefs. For this reason, they are also referred to as knowledge-based techniques or expert systems. Meta-heuristic techniques are random searches (explorations) in combination with heuristics (exploitations). These methods belong to the class of stochastic optimization methods. The designer of the algorithm has to find good heuristics, set the initial parameters, and fine-tune empirically the parameters with the hope that they can adapt well to new situations. In this chapter, a three-stage methodology based on the combination of a group-based PSO algorithm and local heuristic methods is applied to the service restoration problem. We take advantage of the global capability of PSO and generate a set of "good" solution vectors located in several promising subregions in the search space. Starting from each solution vector generated from PSO, the local search method finds a set of the corresponding local optimal solutions. General principles to be considered in the restoration procedures and clustering scheme are combined with the meta-heuristic method proposed to support the operators' decisions.

5.9.4 Application to the Service Restoration Problem

The application of the proposed three-stage group-based PSO method to the service restoration problem (see Figure 5.9.6) is described in this section and is numerically evaluated in the next section.

Particle Representation

As the configuration of a network is represented by the status of all switches in the network, the particle used in the BPSO implementation of the service restoration problem represents the status of all the switches in the system. The dimension of each particle is equal to the number of switches in the system. In order to avoid the low searching efficiency in high-dimension distribution networks, we divided the search vector into two parts: the status of tie switches and the status of sectionalizing switches.

Clustering Scheme

The K-means algorithm is incorporated with the PSO algorithm. For the service restoration problem, we represent the entire distribution network configuration

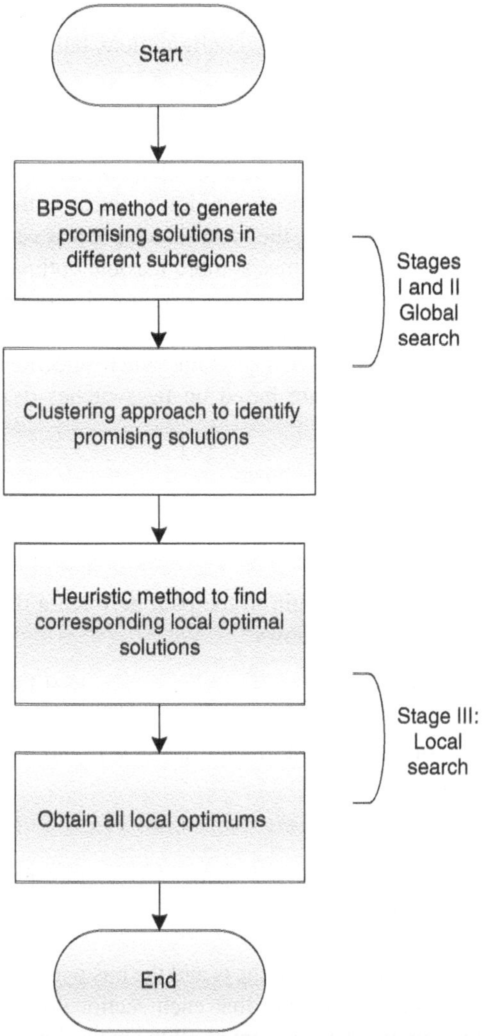

Figure 5.9.6 Group-based PSO algorithm flowchart for the service restoration.

with each particle representing "a network configuration." According to the characteristic of the data set composed of all particles, we choose the status of all switches, and the total power loss of the network to be the input features for clustering. In this case, a data point is composed of three attributes:

1. Tie switches: 0–1 variable, which has M dimensions
2. Sectionalizing switches: 0–1 variable, which has N dimensions
3. System loss: continuous variable, which has two dimensions representing real power loss and reactive power loss, respectively

We address the task of clustering all of the particles to find representative solutions. The particles in a cluster have similar network configurations but with different value of objectives.

Local Search

A ranking-based switch search method is utilized as the local search method in this paper, which has shown to be effective for the service restoration problem in the literature [275, 285]. As we will see in Section 5.9.5, the local search process can help exploit more fully than restore schemes, which means more feasible optimal solutions can be located by the proposed method than by a traditional service restoration method. This ranking-based switch search method conducts the restoration procedures based on pre-established rules. The main switch selection indices list below is adapted from [285], which are based on tie switches (ts) and sectionalizing switches (ss):

Z_{path}: the electrical distance of each ts to other buses

I_{ss}: the amount of load each ss can transfer to a ts

I_{M}: the spare capacity of each ts

For each ts, I_{M} is defined as the smallest margin in the path between a tie switch and the substation, which is expressed as follows:

$$I_{\text{M}} = \min_{k \in N_{\text{u}}, p} \left(I_k^{\max} - \left| I_k^p \right| \right) \tag{5.9.13}$$

where N_{u} is the set of upstream buses on the path between a tie switch and the substation.

The electrical distance Z_{path} is given by

$$Z_{\text{path}} = \frac{V_{\text{ts}}}{\sum_{k \in N_{\text{v}}} I_{\text{L},k}} \tag{5.9.14}$$

where N_{v} is the set of all buses lying in the path between the ts and the bus with the voltage violation. I_{ss} is defined as the amount of load that each sectionalizing switch can transfer to a tie switch. It is computed as the sum of all the loads downstream from the sectionalizing switch:

$$I_{\text{ss}} = \sum_{C_{\text{d}}} I_{\text{L}} \tag{5.9.15}$$

where C_{d} is the set of all buses lying downstream from the secondary side of the sectionalizing switch. The transferrable load is also used as an index to distinguish switch pairs in order to handle current flow and feeder capacity constraints.

The above three indices are used to develop ranking-based switch search methods for finding local optimal solutions as follows [299, 300]:

Step 3-1: Check the unrestored outage. Update the outage area.

Step 3-2: Build the candidate tie and sectionalizing switch list for Tier 1.

Step 3-3: Close the tie switch with the largest spare capacity nearest to the outage area to attempt full restoration of an out-of-service area. If unrestored loads still exist, go to the next step.

Step 3-4: Select and operate the candidate switch pairs for full restoration of the out-of-service area. If not fully restored, go to the next step; otherwise, stop.

Step 3-5: Determine which loads should not be restored and execute the load shedding operation.

The best solution suggested by the proposed method needs to be tempered with engineering judgment, and thus, there are advantages to developing a high-performance method which can offer diverse optimal solutions for users to choose from.

5.9.5 Numerical Results

We evaluate the proposed three-stage group-based PSO method on two test systems: the IEEE 123-node test feeder (see Figure 5.9.7) and an 1101-node practical distribution network.

IEEE 123-Node Test Feeder

The group-based PSO algorithm for service restoration is evaluated on the IEEE 123-node test feeder (see Table 5.9.1). In this test case, the normal total load of the system is 4930 kW and 2970 kvar. The bus voltage magnitude is limited to

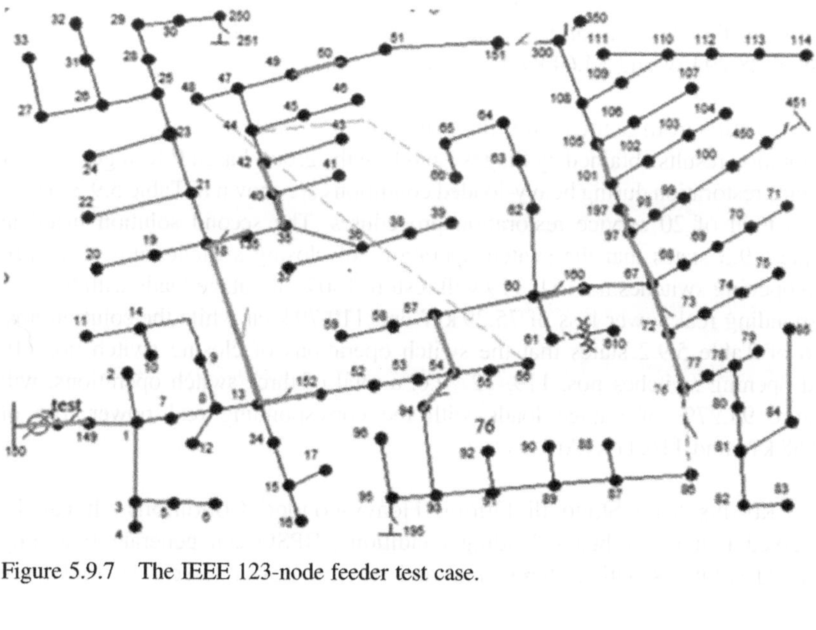

Figure 5.9.7 The IEEE 123-node feeder test case.

TABLE 5.9.1 123-Node Network Component List

Component Type	No. of Each
Tie switches	7
Sectionalizing switches	15
Balanced loads	4
Unbalanced loads	81

110–130 kV. All transformers and lines immediately before or after a transformer are protected by circuit breakers. Faults occurring on these branches can be isolated by operating these devices. Faults on other branches are isolated by operating the upstream and downstream sectionalizing switches or breakers. We have considered heavy loading conditions by varying load from their base case value to 200% of that value.

Results by from Stages I and II During Light-Loaded Conditions Service restoration results obtained by Stages I and II of the group-based PSO algorithm for light-loaded conditions are shown in Table 5.9.2. When the loading condition is light, the proposed three-stage algorithm is fast to locate 17 feasible restoration solutions for the fault occurring at the branch between nodes 160 and 167. For instance, solution no. 1 of Table 5.9.2 states that the switch operations of closing switch no. 113, opening switches nos. 51, 119 will restore 58.53% of outage loads with the corresponding real power loss 20.49 kW and 18.45 kvar while the solutions nos. 3–6 of Table 5.9.2 states that the switch operations of closing switch no. 116 and opening switches no. 119, i.e. a total of two switch operations, will restore 100% (full restoration) of outage loads with the corresponding real power loss of 28.29 kW and 34.64 kvar.

Results from Stages I and II During Heavy-Loaded Conditions Service restoration results obtained by Stages I and II of the group-based PSO algorithm for service restoration during heavy-loaded conditions are shown in Table 5.9.3. There are a total of 20 service restoration procedures. The second solution listed in Table 5.9.3 states that the switch operations of closing switches nos. 113, 116 and opening switches nos. 51, 119 will restore 100% of outage loads with the corresponding real power loss of 75.29 kW and 119.79 kvar while the solution nos. 3–6 of Table 5.9.2 states that the switch operations of closing switch no. 116 and opening switches nos. 119, 127, i.e. a total of three switch operations, will restore 94.27% of outage loads with the corresponding real power loss of 68.58 kW and 116.41 kvAr.

Results from Stage III During Heavy-Loaded Conditions It can be observed that under heavy loading conditions, BPSO can generate five fully restored solutions with different real and reactive power losses. To find more

TABLE 5.9.2 Solutions Obtained from PSO on Fault 160–167 at Light-Loaded Level

No.	Switch Operation	Loss	Switch No.	Ratio (%)	No.	Switch Operation	Loss	Switch No.	Ratio (%)
1	Close 113	20.49	3	58.53	11	Close 113, 114, 115, 116, 124, 125	27.98	11	100
	Open 51, 119	18.45				Open 118, 119, 122, 123, 131	34.23		
2	Close 113, 114, 116	28.34	7	96.66	12	Close 113	20.49	3	58.53
	Open 51, 117, 119, 131	34.88				Open 51, 119	18.45		
3	Close 116	28.29	2	100	13	Close 116	28.29	2	100
	Open 119	34.64				Open 119	34.64		
4	Close 116	28.29	2	100	14	Close 113	20.49	3	58.53
	Open 119	34.64				Open 51, 119	18.45		
5	Close 116	28.29	2	100	15	Close 113, 116	28.05	4	100
	Open 119	34.64				Open 51, 119	34.16		
6	Close 116	28.29	2	100	16	Close 116	28.29	2	100
	Open 119	34.64				Open 119	34.64		
7	Close 116, 125	27.99	4	100	17	Close 113, 114, 125	29.28	8	90.54
	Open 119, 127	34.02				Open 117, 118, 119, 127, 131	36.67		
8	Close 116	28.29	2	100	18				
	Open 119	34.64							
9	Close 116	28.29	2	100	19				
	Open 119	34.64							
10	Close 125	24.18	3	75.77	20				
	Open 119, 131	28.30							

TABLE 5.9.3 PSO on Fault 160–167 at Heavy-Loaded Conditions

No.	Switch Operation	Loss	Switch No.	Ratio (%)	No.	Switch Operation	Loss	Switch No.	Ratio (%)
1	Close 113, 116, 125, 130	75.37	8	100	11	Close 125	94.75	3	98.43
	Open 51, 119, 126, 131	130.15				Open 51, 119	169.85		
2	Close 113, 116	75.29	4	100	12	Close 113	84.07	3	94.27
	Open 51, 119	129.79				Open 119, 127	147.61		
3	Close 116	68.58	3	94.27	13	Close 113, 116	75.29	5	100
	Open 119, 127	116.41				Open 51, 119, 131	129.79		
4	Close 113, 116	75.29	4	94.27	14	Close 113	84.98	3	94.27
	Open 119, 131	129.79				Open 118, 119	149.28		
5	Close 113, 116	75.29	4	94.27	15	Close 113, 116	75.29	4	100
	Open 119, 131	129.79				Open 119, 131	129.79		
6	Close 116	75.19	3	98.43	16	Close 116	65.54	3	91.67
	Open 51, 119	130.04				Open 119.128	110.19		
7	Close 125	94.75	3	98.43	17	Close 116	70.13	3	94.27
	Open 51, 119	169.85				Open 118, 119	119.47		
8	Close 116	75.19	3	98.43	18	Close 125	94.75	3	98.43
	Open 51, 119	130.04				Open 51, 119	169.85		
9	Close 116	75.19	3	98.43	19	Close 116	65.54	3	91.67
	Open 51, 119	130.04				Open 119, 128	110.19		
10	Close 116, 125	75.64	4	100	20	Close 113	84.98	3	94.27
	Open 119, 127	130.74				Open 118, 119	149.28		

TABLE 5.9.4 ***K*-means (Set the Number of Clusters to 4)**

No.	Switch Operation	Loss	Switch No.	Ratio (%)
1	Close 116 (16, 19)	65.54	3	91.67
	Open 119, 128	110.19		
	Close 116 (17)	70.13	3	94.27
	Open 118, 119	119.47		
2	Close 116 (3)	68.58	3	94.27
	Open 119, 127	116.41		
	Close 116, 125 (10)	75.64	4	100
	Open 119, 127	130.74		
3	Close 113, 116	75.29	4	100
	Open 51, 119 (2)	129.79		
	Close 116 (6, 8, 9)	75.19	3	98.43
	Open 51, 119	130.04		
4	Close 125 (7, 18)	94.75	3	98.43
	Open 51, 119	169.85		
	Close 113 (12)	84.07	3	94.27
	Open 119, 127	147.61		
	Close 113 (14, 20)	84.98	3	94.27
	Open 118, 119	149.28		
	Close 113, 116, 125, 130	75.37	8	100
	Open 51, 119, 126, 131 (1)	130.15		
	Close 113, 116	75.29	4	100
	Open 119, 131 (4, 5, 15)	129.79		
	Close 113, 116 (13)	75.29	5	100
	Open 51, 119, 131	129.79		

restoration schemes with 100% load restoration, we cluster all the particles into four clusters by the *K*-means method. As shown in Table 5.9.4, there are four clusters of solutions.

After applying Stage 3 of the proposed method, the ranking-based switch search procedure, we are able to find three new solutions from the top solutions of each cluster (see Table 5.9.5). For instance, the 94.27% restoration ratio solution, which is a three-operation, close no. 116 and open nos. 119, 127 with 68.58 kW and 116.41 kvar losses, is improved by Stage III into a 100% full restoration solution, which is a four-operation, close nos. 116, 125 and open nos. 119, 127 with 75.64 kW and 130.74 kvar losses. Another improvement is on the 94.27% restoration ratio solution, which is a three-operation, close no. 113 and open nos. 118, 119 with 84.98 kW and 149.28 kvar losses. Now, this solution is improved by Stage III into a 100% full restoration solution, which is a four-operation, close nos. 113, 115 and open nos. 118, 119 with 96.65 kW and 173.34 kvar losses.

TABLE 5.9.5 Improvement in Local Search Stage

No.	Stage	Switch Operation	Loss	Switch No.	Ratio (%)
7	PSO	Close 125	94.75	3	98.43
		Open 51, 119	169.85		
	Local	Close 125, 113	98.12	4	100
		Open 51, 119	176.77		
17	PSO	Close 116	70.13	3	94.27
		Open 118, 119	119.47		
	Local	Close 116, 125	98.12	4	100
		Open 118, 119	176.77		
3	PSO	Close 116	68.58	3	94.27
		Open 119, 127	116.41		
	Local	Close 116, 125	76.63	4	100
		Open 119, 127	133.01		
20	PSO	Close 113	84.98	3	94.27
		Open 118, 119	149.28		
	Local	Close 113, 115	96.65	4	100
		Open 118, 119	173.34		

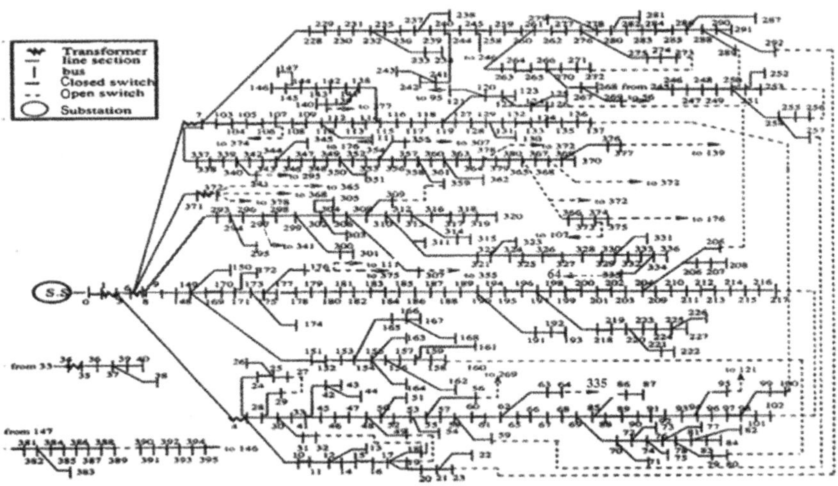

Figure 5.9.8 One-line diagram of the 394-bus, 1101-node test system.

The 1101-Node Practical System

The proposed three-stage method is also evaluated on a 394-bus, 1101-node practical unbalanced network (see Figure 5.9.8). Detailed information about the network is displayed in Table 5.9.6. The acceptable voltage range used is 110–130 V.

When a fault occurs on line 297–298, Stage I generates multiple solutions to restore the out-of-service load. Table 5.9.7 shows the top particles, obtained by Stage I of the group-based PSO method and selected by users based on his/her

TABLE 5.9.6 1101-Node Network Component List

Components Type	No. of Each
No. of lines	343
No. of transformers	7
Load condition (total)	26.03 MW, 13.48 Mvar
No. of tie switches	27
No. of sectionalizing switches	46

TABLE 5.9.7 Solutions Obtained in Group-based PSO on Fault 297–298

		Two Objectives		
No.	User Preferred Solutions	Load Served	No. of Switch Operations	Ratio (%)
2, 5, 7	Open 296–297, open 298–299, close 64–335	2 628 655 + j1 369 295	3	99.97
14	Open 296–297, open 298–299, close 355–307	2 628 655 + j1 369 295	3	99.97
15, 8	Open 296–297, open 298–299, close 359–309	2 628 655 + j1 369 295	3	99.97
4, 10	Open 296–297, open 298–299, close 341–297	2 145 226 + j1 134 749	3	81.86

TABLE 5.9.8 Solutions Obtained by Stage III: Local Stage

			Two Objectives		
Solution		User Preferred Solutions	Load Served	No. of Switch Operations	Improvement
8	Stages I and II	Open 296–297, open 298–299, close 341–297	2 145 226 + j1 134 749	3	22.1%
	Stage III	Open 296–297, open 298–299, close 355–307	2 628 655 + j1 369 295	3	

preference. It is observed that three different feasible solutions are found by the proposed method.

We select the solution with an 81.86% restoration ratio for further improvement. We apply the rank-based local method with the solution as the initial condition. It converges to a near fully restored solution which is the same as particle 14 (see Table 5.9.8). It appears that particle 14 is the local optimal solution in the subregion near the selected one. This study illustrates that multiple effective solutions can be obtained by the proposed method.

5.9.6 Conclusions

PSO has an excellent capability of "global" search while it lacks the local fine-tuning capability and suffers from slow convergence in finding optimal solutions. PSO suffers from the following two problems: (i) premature converging to a local optimum and (ii) a slow convergence rate when searching for a local optimum. A group-based PSO method is proposed to overcome these two problems. In this chapter, we present a novel group-based PSO method to overcome the challenges faced by PSO in issues such as extensive computation efforts (i.e. lengthy computation time), scalability, and lack of fine-tuning capability. In addition, the group-based PSO method provides a general framework for improving the various versions of PSO methods. Stage 1 explores the global capability of PSO and serves to check our proposed stopping criterion for PSO when all the populations have reached a certain level of consensus. We propose the following "equilibrium state" for consensus:

- The number of groups of particles is not changed
- The members in each group do not alter

The consensus condition serves to stop the PSO search procedure. Stages 2 and 3 combine the global capability with the fine-tuning capability of local methods to find multiple high-quality optimal solutions.

Our observations have indicated that all the particles settle down to different locations which form several different groups in the search space. All the particles do not form only one group. It should be noted that a group containing a large number of particles does not indicate that the region where members of particles of this group settle down contains the global optimal solution. It can occur that distinct particles with outstanding performance can move toward the region that contains the global optimal solution. We also observe that the size of particles in each group and the quality of fitness value that each member possesses do not reveal information regarding the quality of local optimal solutions lying in the region. Consequently, the region that where each group of particles settles down should be exploited in order to find high-quality local optimal solutions. Therefore, we should explore all groups to locate high-quality local optimal solutions, if not the global optimum.

Service restoration plans are to assist distribution personnel in making timely decisions on how to restore as much load as possible in the out-of-service areas. The proposed three-stage group-based PSO method is applied to the service restoration problem in distribution power systems with very promising results. This application also demonstrates the key function of each stage of the proposed method in finding optimal service restoration plans. Since the requirement of maintaining radial structure in a distribution network is rigid, the searching space is dynamic with different switch operations. It brings about computational challenges for an efficient searching process of the PSO algorithm.

5.10 MVMO FOR PARAMETER IDENTIFICATION OF DYNAMIC EQUIVALENTS FOR ACTIVE DISTRIBUTION NETWORKS

Digvijay Gusain[1], Jose Rueda[1], Jens C. Boemer[2], and Peter Palensky[1]

[1]TU Delft, Delft, The Netherlands

[2]Electric Power Research Institute, Seattle, USA

5.10.1 Introduction

Power system dynamic studies rely on model-based simulations. With increasing penetration of DG, modeling of a network that is not only geographically diverse but is also technologically mixed entails great efforts and accuracy for such studies. Despite the availability of highly powerful simulation tools and increased processing power, performing time domain simulations on such interconnected and complex system remains a great hurdle. Reducing the complexity of the model using aggregation can help with these problems. The purpose of this chapter is twofold. First, we go through the concept of network aggregation and introduce a white box DE suitable for system studies with high penetration of distributed PV. Next, we introduce a sophisticated approach for solving the nonlinear optimization problem of parameter identification of the developed DE. This approach is based on a new optimization technique called mean-variance mapping optimization (MVMO).

The parameters of the DE are identified by measuring the response of DE to certain disturbances and comparing them with reference signal sets. By using the proposed approach, we determine the parameters of our suggested DE that entails high accuracy. The accuracy of the DE is measured by calculating the root mean square error (RMSE) value. The time required to simulate faults and perform time domain simulations using a DE is considerably less than for a full-scale model. This has significance for system planners and researchers who want to analyze the effect of increasing penetration of DE into distribution grid on the system as a whole with regard to strategy development.

5.10.2 Active Distribution System

All around the world, the share of RES in the grid is continuously increasing. The lowering costs of renewable energy, as well as growing consciousness to tackle the threat of global warming, is the driving force for this increase in penetration of RES. Most of this RES integration into the grid has been on the distribution level where the traditional consumers of electricity also installing small-scale PV or wind generators. The introduction of DG into the grid, while certainly helps with the adoption of the concept of clean energy, also presents some interesting technical and operational challenges [306]. The system operators need to address these challenges adequately to ensure the reliability of power system.

The increasing participation of DG in LV/MV networks is changing the present network infrastructure definition; the distribution network is being transformed

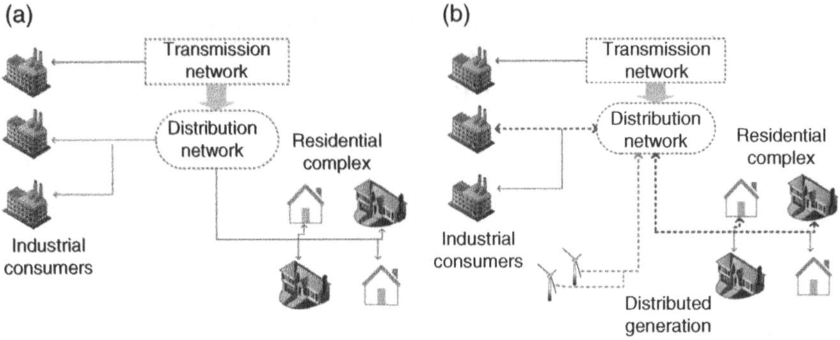

Figure 5.10.1 Distribution network configurations. (a) Passive network. (b) Active network.

from a passive network to an active network. DG being variable generation is dependent on continuously changing environmental conditions. The network operating conditions, therefore, also vary accordingly. On a day with high winds and sun, the wind turbine generators (WTG) and PV can produce enough electricity to meet the local load demands. Sometimes the generation can be so high that power may flow in an unconventional pattern from LV to MV grids [307]. This makes the LV grid and active participant in system studies. It is for this reason that such a network with an increased penetration of DG is referred to as an ADN. Hence, power flow in an ADN is bidirectional. The consumers not only consume power but are also providing power to the grid when their local generation exceeds their local demand. Figure 5.10.1a shows the passive network structure while Figure 5.10.1b shows the active network structure.

The changes in the existing electrical network structure require utilities to define a new set of rules for the safe and reliable operation of the electrical power systems. These rules and regulations are referred to as Grid Code Requirements (GCR). The GCR have to be strictly adhered to by all the generating stations. Earlier, GCR was to be followed by only large generating stations, mainly synchronous generators. The insignificant amount of DG connected was seen as a load element instead of a generation station and thus required to disconnect from the network in case of voltage dips at the point of common coupling (PCC) below the threshold voltage (usually 0.8 p.u.) [308]. This connect-and-forget approach was satisfactory when the DG-based generation was insignificant. However, with large penetration of DG, disconnecting this generation represents a significant active power loss, potentially leading to system instability.

5.10.3 Need for Aggregation and the Concept of Dynamic Equivalents

As already mentioned, the increasing penetration of DG is complicating the existing network infrastructure. Additionally, the distribution grid is playing an increasing role in determining system stability and dynamics due to the presence of increasing amounts of generation. This requires planners to accurately model such system for analysis of contingencies using a simulation-based approach. However, there are two main impediments to this requirement. First, modeling the entire detailed network infrastructure is a complex task. This is because each DG employs a unique control mechanism that satisfies the GCR at its PCC. However, the utilities keep updating the GCR to maintain system reliability according to the network configuration at that time period. Hence, different DGs installed in different time periods have different control schemes that adhere to GCR specified at that time. Second, to simulate this diverse and interconnected system for disturbances poses a significant computational overhead.

To overcome these challenges, a simplified model is developed. This technique to develop simplified models is known as dynamic equivalence. DE allows for representing the detailed network model in a simplified fashion which results in a reduction of the number of equations to be solved during simulations. The determination of DE requires defining a model structure to represent detailed network dynamics and identifying the parameters to increase the accuracy of DE [309]. With the exceptional advances being made in communication and sensory technologies, such model simplification would help us to not only make a lot of savings on computational expenses but also use available sensory information from grid in real time (using phasor measurement units, PMUs) for more efficient system performance (like predictive loadings, undesirable operating conditions, optimal remedial actions, etc.). Moreover, the motivation for modeling accurate DE is in the fact that detailed dynamic models and their associated control structures are protected by the intellectual property of the manufacturers and not readily and freely available to the general research community. The basic idea for simplification of detailed models to DE is to place representative generation and load behind an equivalent impedance as is shown in Figure 5.10.2. It must be kept in mind that

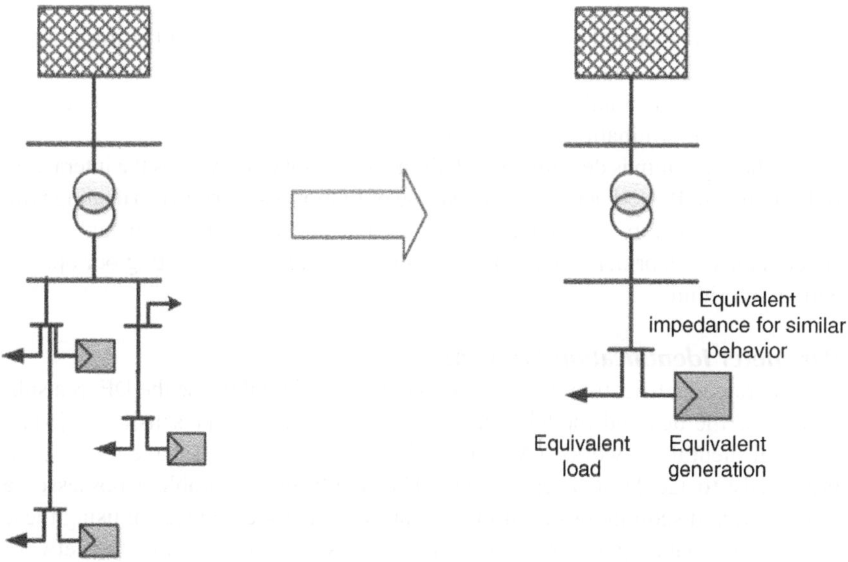

Figure 5.10.2 Concept of aggregation.

such representations are situation dependent and vary according to the detailed network configuration and method of aggregation.

The first stage of aggregation is identification and selection of generating stations that have similar dynamic behaviors. Once a group of generators is identified, they can be aggregated into a simplified representative model by one of the many aggregation techniques available in the literature [310].

5.10.4 Proposed Approach with MVMO

The proposed approach for the problem introduced is highlighted in Figure 5.10.3. A brief explanation to the approach follows.

After the dynamic equivalent model is defined, a set of multiple disturbances at the PCC to the external grid for which our DE should be identified is established. For our test case, we define two disturbances (3φ bolted fault with retained voltages of 0.3 and 0.65 p.u.). Initially, the solution vector containing optimization variables (like aggregated model's control system parameters, equivalent impedance, etc.) is randomly initialized. The search range for all optimization parameters is then normalized from [min, max] to [0, 1]. This is one of the uniqueness of MVMO. Normalization of search space is required as a precondition for transformation in later stages. This also has the advantage that the generated offspring is never beyond the searching boundaries. The solution vector containing normalized solutions is de-normalized to obtain real-valued parameters that are fed into models and used to perform time domain simulations. The desired signals from this model are then fed into the fitness evaluation test along with a set of reference signals. The value of objective function after being subjected to constraints forms the fitness value. If the objective function violates any of the constraints, the value is given a penalty. Such an optimization forms a constrained optimization problem. All the acceptable solutions from the process are stored to an external csv file. The best solutions are stored and according to a mapping function, a new solution set is created that is again de-normalized and fed to models for simulations. This process is then continued. The process is stopped when termination criterion is satisfied.

The algorithm is developed in Python 3.4. This work involves the interaction of Python with PowerFactory. We make use of Python's superiority in dealing with matrices and vectors as compared to PowerFactory's inbuilt vector and matrix functionality as a principle reason to use this software for formulating our optimization algorithm.

Parameter Identification Problem

An accurate identification of DE's parameters is essential to use the DE as a substitute for the detailed model. Techniques such as trajectory sensitivity [311], artificial neural networks (ANN) [312], and ACO [313] have been used previously to tackle such challenges. The identification problem possesses a non-convex, discontinuous multimodal nature. The disadvantages of using these methods are that they require a lot of time and historical data to train the network

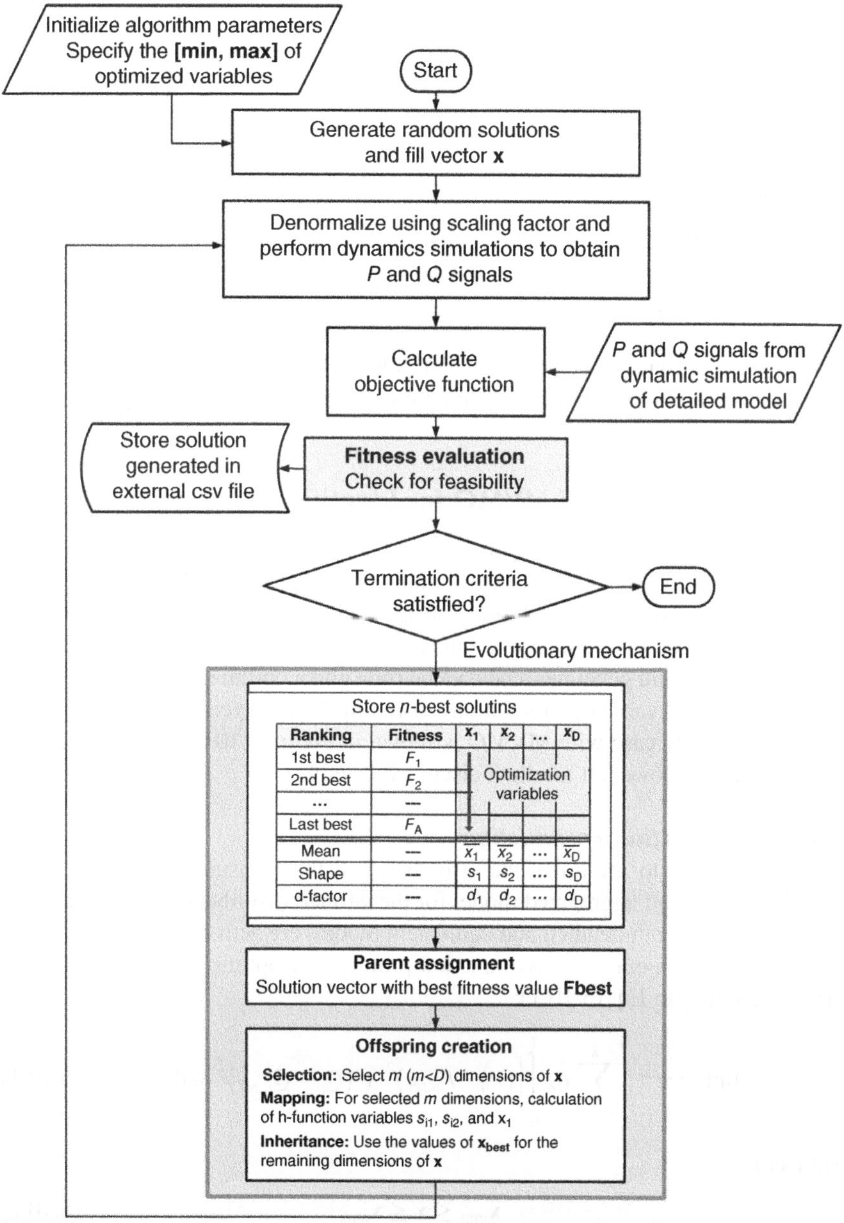

Figure 5.10.3 Flowchart for identification of MVMO parameters.

to be able to provide us reliable results (ANN) or output variable equation in terms of system variables (trajectory sensitivity). Additionally, many of these techniques get easily trapped into local optimum, since their searching performance depends on its appropriate parameter settings. Premature convergence and local stagnation is also a problem that these techniques face, especially for problems that are non-convex or multi-objective. Thus, classical optimization problems are not very useful to find solutions. Nonetheless, it is exactly the kind of problems that heuristic optimization techniques excel at. Driven by the excellent performance of MVMO algorithm to other power systems problems, including an exploratory study in a similar field of parameter identification for DE [314, 315], the presented work uses similar ideas of MVMO and attempts to estimate the parameters of the equivalent model. The MVMO is enhanced to increase its searching performance by using dynamic parameter settings that change as the optimization progresses.

MVMO has a reduced risk of premature convergence due to its mapping function evolutionary operations, which allows adaptively changing the search focus between exploration and exploitation [316].

5.10.5 Adaptation of MVMO for Identification Problem

MVMO evolves a candidate solution based on the statistics of the successful solutions recorded in the solution archive. This basically consists of creating a new solution based on the best solution found so far (firstly ranked in the archive), which is modified by the mapping function operator. This operator is the uniqueness of MVMO and generates a new value for a given optimization variable, based on the mean and variance of solutions stored in the archive [317]. The approach used in our study case with MVMO is shown in Figure 5.10.3. The various components of this flowchart are then discussed briefly.

Objective Function Formulation

Our objective is to minimize the error between the measured responses of the detailed model and aggregated model for the defined disturbances. The quantities compared from both detailed and equivalent models are active and reactive power at the PCC. Thus, our objective function problem for optimization can be formulated according to Eq. (5.10.1):

$$\text{Minimize:}\quad \sum_{n=1}^{k} \alpha_n \int_0^t \left[\left(P_n - P_{n,\text{ref}}\right)^2 + \left(Q_n - Q_{n,\text{ref}}\right)^2 \right] \cdot \mathrm{d}t \qquad (5.10.1)$$

subject to:

$$\mathbf{x}_{\min} \le \mathbf{x} \le \mathbf{x}_{\max} \qquad (5.10.2)$$

where $\mathbf{x}$ is the D-dimensional solution vector filled with the parameters of DE to be optimized while $\mathbf{x}_{\min}$ and $\mathbf{x}_{\max}$ are the bounds of each element of $\mathbf{x}$. The values of these bounds are generally ±10% of their standard value. P_n and Q_n are the responses of DE at any time t, while $P_{n,\text{ ref}}$ and $Q_{n,\text{ ref}}$ are the corresponding response from detailed models at the same time. α_n is the probability of the nth disturbance.

Fitness Evaluation

For a candidate solution generated in each function evaluation, the fitness value is the objective function value after application of constraint conditions. Thus, the fitness value is the sum of squares of errors of P and Q curves between DE and detailed models for a set of disturbances simulated in PowerFactory over a period of time. The following strategy for fitness evaluation is given in (5.10.3):

$$f' = \begin{cases} \text{OF} & \text{if solution within limits} \\ \rho & \text{otherwise} \end{cases} \tag{5.10.3}$$

The value of ρ is set to a high value (1e50) in our case.

Solution Archive

One of the strengths of MVMO lies in the fact that it has a knowledge archive that is used to derive constants that shape the transformation function, thus acting as a search guide. Moreover, this archive is dynamically updated as the algorithm finds better solutions. The archive stores n best solutions in a descending order of fitness, as seen in Figure 5.10.4. The archive size is fixed at the start of the procedure (four in our case). An update to the archive is made only if a solution with a fitness value better than stored solutions is found.

Offspring Generation

Offspring creation in MVMO is a three-step process. This is the part, where the uniqueness of MVMO from other EAs becomes evident. The steps are highlighted in Figure 5.10.5 and described in detail in the following sections.

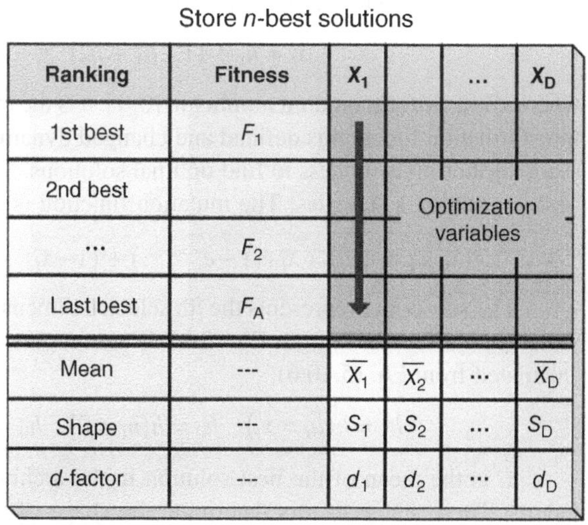

Figure 5.10.4 MVMO archive where solutions are stored.

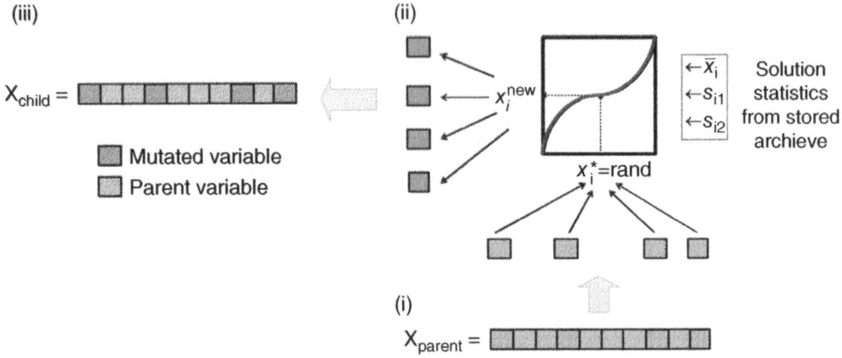

Figure 5.10.5 Procedure of mutation of selected genes.

The Parent of a New Solution MVMO adopts a single parent–offspring approach. As the name suggests, there is only a single parent involved in the creation of child vector. Parent selection can be done in multiple ways:

- Best-fit solution from archive
- A combination of best solution with another solution from archive

For this work, we chose to adopt the methodology to select the best-fit solution as the parent. Thus, the first entry from solution archive is taken as the parent to create offspring.

Evolution of Parent Based on Mapping Function This step forms the essence of MVMO's working. The step is illustrated in Figure 5.10.5(ii). From the parent vector, m $(m < D)$ variables are randomly selected. These variables are then given to the transformation function. The new value of each selected dimension is computed using Eq. (5.10.4):

$$x_i = h_x + (1 - h_1 + h_0) \cdot x_i' - h_0 \tag{5.10.4}$$

where x_i' is a uniform random number in [0, 1]. h is the transformation function. The transformation function is defined and changed dynamically as the algorithm proceeds further in its process to find optimal solutions. The function is a function of best-fit solution's statistics. The mutation function is given by Eq. (5.10.5):

$$h(\bar{x}_i, s_{i1}, s_{i2}, u_i) = \bar{x}_i \cdot \left(1 - e^{-u_i \cdot s_{i1}}\right) + (1 - \bar{x}_i) \cdot e^{-(1-u_i) \cdot s_{i2}} \tag{5.10.5}$$

The subscript i represents the ith selected dimension from parent vector. The variables h_x, h_1, and h_0 from Eq. (5.10.4) are outputs of transformation functions calculated from Eq. (5.10.6):

$$h_x = h\left(u_i = x_i'\right), \quad h_0 = h(u_i = 0), \quad h_0 = h(u_i = 1) \tag{5.10.6}$$

$\bar{x}_i$ is the mean of the best solution in the archive. s_{i1} and s_{i2} are the shape factors. These shape factors determine the shape of the transformation function as is seen in Figure 5.10.5(ii). These shape factors are determined from the variance of the stored solutions. This relationship between shape factors and variance is a logarithmic one as is shown in Eq. (5.10.7):

$$s_r = -\ln(v_r) \cdot f_s \tag{5.10.7}$$

where f_s is the user input scaling factor, generally set between [0.9, 10]. The factors s_1 and s_2 are assigned with the scheme shown in (5.10.8):

$$
\begin{aligned}
& s_1 = s_2 = s_i = -\ln(v_r) \cdot f_s \\
& \text{If } s_i > 0 \text{ then} \\
& \quad \Delta d = (1 + \Delta d_0) + 2 \cdot (\text{rand} - 0.5) \\
& \quad \text{if } s_i > d_i \\
& \quad\quad d_i = d_i \cdot \Delta d \\
& \quad \text{else} \\
& \quad\quad d_i = \frac{d_i}{\Delta d} \\
& \quad \text{end if} \\
& \quad \text{if rand} \geq 0.5 \\
& \quad\quad s_1 = s_i;\, s_2 = d_i \\
& \quad \text{else} \\
& \quad\quad s_2 = s_i;\, s_1 = d_i \\
& \quad \text{end if} \\
& \text{end if}
\end{aligned}
\tag{5.10.8}
$$

Starting from the first function evaluation, the first value of x_i will be the first randomly initialized value. The variance is set to 1.0, therefore $s_i = 0$. But with the continuation of the optimization process, these values are recalculated and updated to obtain a dynamically changing transformation function. The variable d_i is a parameter that determines the limits of variation of s_1 and s_2. Randomly varying s_1 and s_2 helps in fully exploiting transformation function's characteristics leading to a good balance between exploitation and exploration. On a more general note, the algorithm's ability to focus either on exploitation or exploration can be set using the shape factor (f_s). Setting f_s to a large value (≥1) focuses the algorithm toward exploration, while a low value (≤1) means the focus is more on exploitation.

The newly mutated variables from the above process are crossed with the parent vector to produce new solution vector $\mathbf{x}^{\text{new}}$ for the next function evaluation. The newly generated solution, hence, preserves features of the chosen parent solution and new generated values of the selected dimensions. The process is highlighted in Figure 5.10.5(iii). The *D-m* variables from parents are combined with the *m* mutated variables. This new vector has the parameters that are de-normalized and fed into the PowerFactory model for time domain simulations.

5.10.6 Case Study

Detailed Test System: IEEE 34-Bus Feeder

Our detailed test case is modeled in DigSILENT PowerFactory and is a modification of IEEE 34-bus feeder system that has three PV stations providing a combined generation of 1.5 MW. This is shown in Figure 5.10.6. Each of the PV station is modeled as a current source with a shunt. This is connected to a transformer via a dc–ac converter module as shown in Figure 5.10.7. The PV station also has a voltage source connected via a reactor to the dc side of the converter. However, we disable the reactor–voltage source combination as it was noticed the voltage source dynamics do not contribute significantly to system dynamics, while also speeding up the simulation process. One of the three generating stations is equipped with low voltage ride through (LVRT) capability (connected at bus 828) while the other two are not. They disconnect from the network as the voltage drops below 0.8 p.u. at the PCC. This follows in line with our challenge highlighted in Section 5.9.3 about DG being equipped with different control schemes over a certain time period to adhere to continuously updating GCRs, ensuring topological diversity. The PV stations are also placed on different buses that are separated by quite some distance physically in the feeder. This is another aspect that helps achieve a geographical diverseness in the system. Such representation, thus, ensures we have an as-real-as-possible scenario for distribution feeder. The feeder has two types of loads: 6 spot loads (0.7 MW) and 19 distributed loads (0.8 MW). The total active power requirement is equal to 1.5 MW. Thus, in steady-state, pre-disturbance period, there is no exchange of active or reactive power with the external grid. The

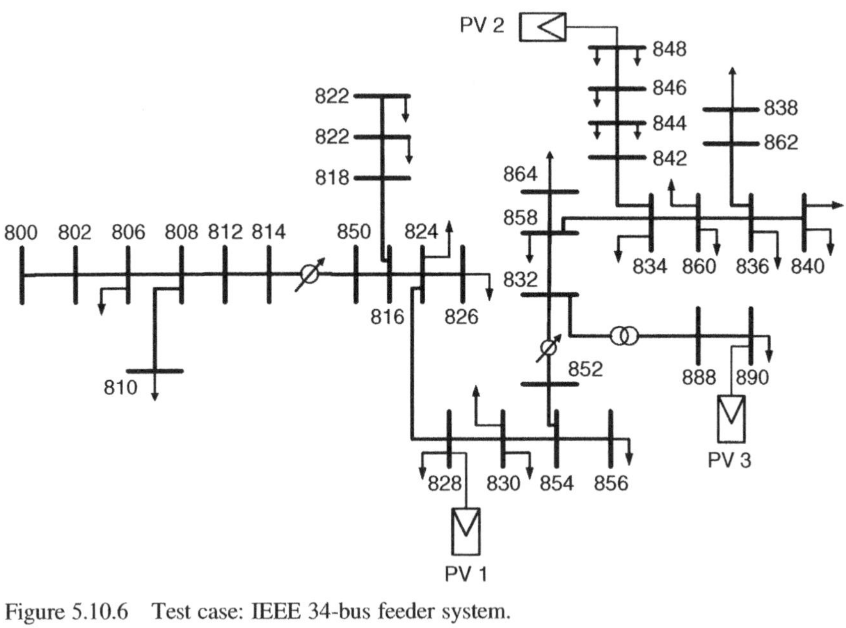

Figure 5.10.6 Test case: IEEE 34-bus feeder system.

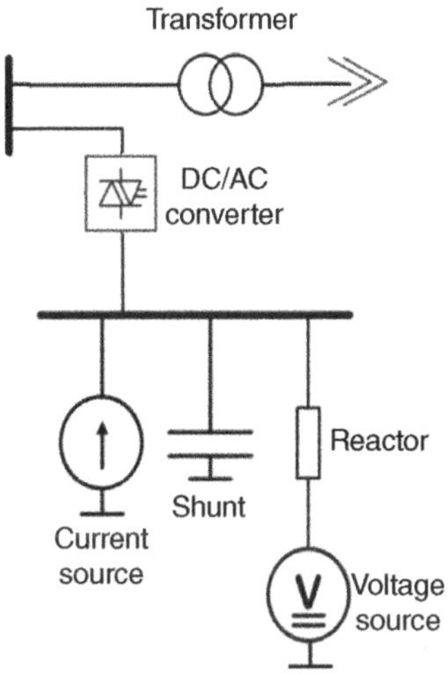

Figure 5.10.7 PowerFactory representation of PV station.

connecting lines are all overhead lines carries defined as *TypTow* elements in PowerFactory. The feeder also consists of two voltage regulators and two 2-winding three-phase transformers. It is tried that the detailed system represents an actual feeder as closely as possible. Recommendations from Electric Power Research Institute and the University of Washington [318] were taken into consideration for detailed system modeling. For standard detailed network data for other elements, the reader is referred to [319].

Aggregated System: The PVD1 Model

The aggregated DE is a distributed PV (now on referred to as PVD1) model, developed by Western Electric Coordination Council (WECC) Renewable Energy Modelling Task Force (REMTF). This model is developed with an aim to represent an aggregation of smaller, distribution-level connected systems that comprise a portion of composite distribution load model [320]. We use PVD1 model because of two main reasons. It fits in perfectly with our challenges of modeling a diverse repertoire of DG in the grid. The PVD1 model has a protection scheme that can capture the effect of diverse terminal conditions on aggregated generation. Also, it is nonproprietary and has been made freely available for research purposes.

Layout The aggregated generation in the form of PVD1 model is placed alongside an aggregated load behind equivalent impedance. Equivalent impedance is needed because the current PVD1 model with aggregated load does not include circuit impedance by itself. The aggregated generation is the sum of generation from PV stations in the detailed model. It is thus set to output active power of 1.5 MW. The aggregated load is roughly the sum of all loads present in the system. This arrangement is then connected to the external grid via a step-up transformer as is shown in Figure 5.10.8.

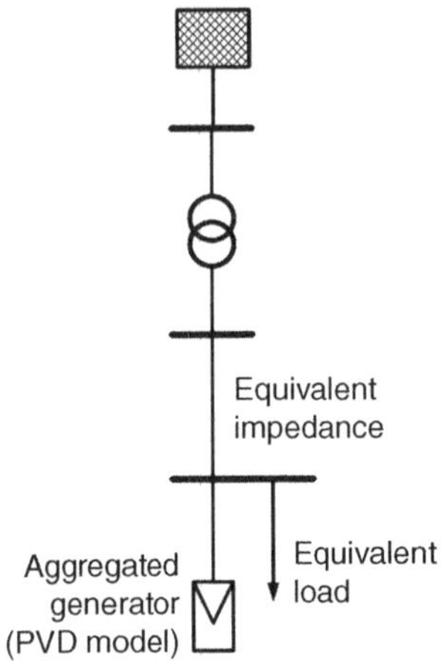

Figure 5.10.8 DE for test case shown.

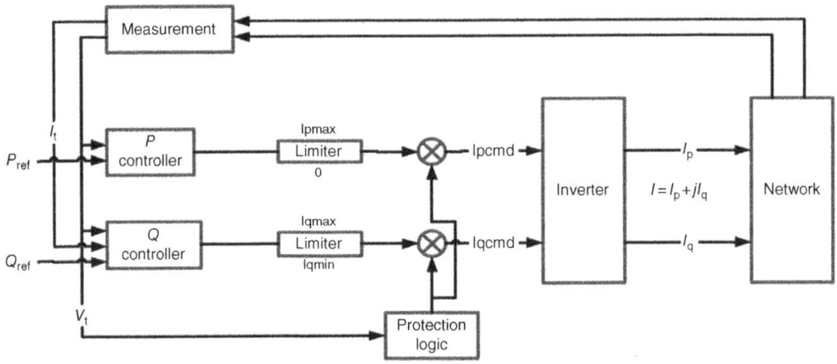

Figure 5.10.9 Basic block diagram of PVD1 model.

In our test case, we have three PV stations that are interfaced to the grid via a power electronic converter system as is described in Section 5.10.6. Since all generation is power electronic based with different topology, we can modify our control parameters of the PVD1 model to adequately represent the detail test model. The generic framework of the PVD1 model is as shown in Figure 5.10.9. The network solution provides terminal voltages and currents (V_t, I_t) that are used by the active power controller, reactive controller, and the protection logic block.

The main components, as can be seen from Figure 5.10.9, are:

- Active power controller unit
- Reactive power controller unit
- Protection logic unit

These blocks are briefly explained in the following sections.

Active Power Control This subsystem provides active current injection, which is subject to current limiting, to the network solution. It takes input reference active power (P_{ref}) and terminal voltage (V_t) from load flow solution. The active power reference is taken as initial power flow value from load flow solution in this work. The output of this block is an active power current command (Ipcmd).

Reactive Power Control Similarly, to the active power controller, the reactive power controller issues a reactive current command to the network. Current limiting mechanism is applied to this block too. The block inputs a reference reactive power (Q_{ref}), load flow obtained solutions of terminal current (I_t), and terminal voltage (V_t). The reactive power reference command is the sum of initial reactive power from network load flow solution and droop signal derived from voltage deviation at the said bus. The model can switch between active or reactive current injection modes in the event of a disturbance by setting internal parameter (Pqflag). This flag parameter can be accessed via model definition and can be set to either 0 (active power injection) or 1 (reactive power injection). The setting of this parameter allows the model to set limits for current limiter model which determines the cap on the active (Ipmax) and reactive current (Iqmax, Iqmin) output of the model. The derivations are in Table 5.10.1, where Imax refers to maximum allowable current through the inverter (between 1.0 and 1.3 p.u. on mbase, set to 1.2 as a default value).

Protection Logic This block performs the most important task of capturing diverse fault ride through (FRT) criteria among all the installed DGs in the system. This is based on modeling the converter considering partial dropout of generation according to [321]. Since this block provides the most important function to working on our aggregated model, we will describe it in more detail. The protection and partial dropout are implemented by this block by providing two outputs in accordance with the terminal voltage and frequency obtained from network solution. The internal parameter settings of the block determine if the generation recovery occurs when voltage and/or frequency disturbances reverse, and if so, then in what proportion.

In this work, we consider the effect of voltage deviations on the generation and thus only the four internal parameters are considered. These are V_{t0}, V_{t1}, V_{t2}, V_{t3}. These are the per unit voltage values that are constant for a given model definition and determine the amount of generation that is disconnected and/or

TABLE 5.10.1 Current Limits According to Pqflag Priority

$Q_{priority}$ (Pqflag = 0)	$P_{priority}$ (Pqflag = 1)
Iqmax = Imax	Ipmax = Imax
Iqmin = − Iqmax	Iqmax = $(\text{Imax}^2 - \text{Ipcmd}^2)^{1/2}$
Ipmax = $(\text{Imax}^2 - \text{Iqcmd}^2)^{1/2}$	Iqmin = − Iqmax

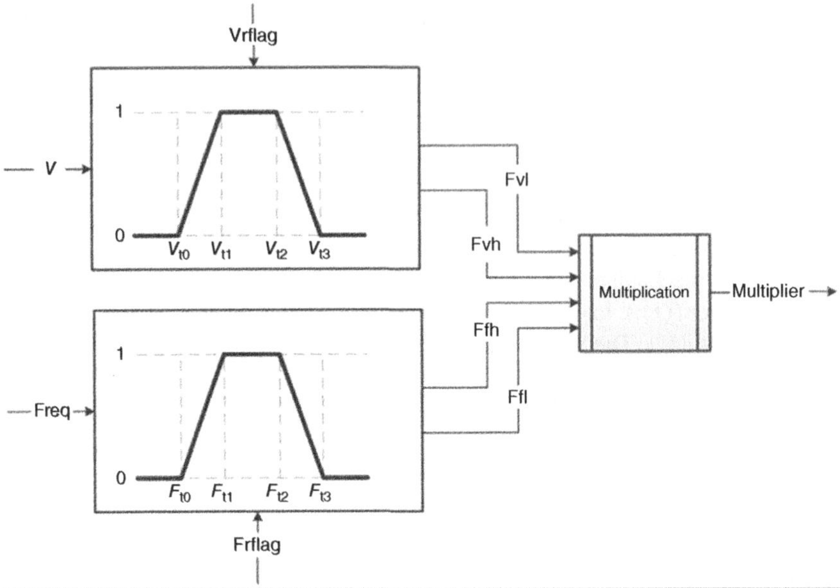

Figure 5.10.10 Protection logic behind PVD1 model.

TABLE 5.10.2 Derivation of Low-Voltage Tripping Logic

Condition	Fvl Value	Remarks
$V_t < V_{t0}$	0.0	V_t below minimum voltage
$V_{t0} < V_t < V_{t1}$	$\frac{V_{min} - V_{t0}}{V_{t1} - V_{t0}}$	Decreasing terminal voltage
$V_{t0} < V_t < V_{t1}$	$\frac{(V_{min} - V_{t0}) + Vrflag \cdot (V_t - V_{min})}{V_{t1} - V_{t0}}$	Recovering, partial reconnection
$V_t > V_{t1}$	1.0	Normal disturbance free operation
$V_t < V_{t1}$	$\frac{(V_{min} - V_{t0}) + Vrflag \cdot (V_t - V_{min})}{V_{t1} - V_{t0}}$	V_t was below V_{t1}, but has recovered now

reconnected following a disturbance in the network. Figure 5.10.10 gives a block diagram representation of this logic.

The Fvl (low-voltage tripping logic), Fvh (overvoltage tripping logic), Ffl (underfrequency tripping logic), and Ffh (over-frequency tripping logic) are the dropout factors that are multiplied by the current commands as shown in Figure 5.10.9. All the factors follow a similar logic. Determination of undervoltage dropout is explained in Table 5.10.2. The other three factors can be similarly derived. Reader is referred to [319] for detailed derivations of dropout factors.

The parameters described in Table 5.10.2 are defined as follows:

- V_t: Terminal voltage.
- V_{t0}: Lower threshold below which all generation will trip.
- V_{t1}: Lower limit of disconnection free operation.
- V_{t2}, V_{t3}: Similar to V_{t0} and V_{t1}, but for overvoltage disconnections.
- V_{min}: Tracks the lowest voltage through simulation but not below V_{t0}.
- Vrflag: Internal flag setting which determines the percentage of reconnection or the percentage of active power recovered after a disturbance.

Reference Signals and Detailed Network Dynamics

The reference signals (active and reactive power flows) to be used for identification of DE are generated from the detailed test network, modeled in Section 5.10.6, at the HV side of the interfacing transformer to the external grid. The two faults are simulated at different locations inside the external grid by varying the fault impedance at the HV side of the transformer:

- Fault 1: Three-phase short circuit at the HV side of the transformer with $X_f = 1.02\ \Omega$. This corresponds to retained voltage of 0.30 p.u. at the terminal. The fault is impressed on at 0.1 second for duration of 200 ms.
- Fault 2: Three-phase short circuit at the same location, but with fault impedance $X_f = 4.42\ \Omega$. This corresponds to retained voltage of 0.65 p.u. The fault is again impressed on at 0.1 second for duration of 200 ms.

The effect of applying these faults for such disturbance is different for different PV stations. As already highlighted in Section 5.10.6, only one of the PV station, located on bus 828, is capable of riding through the fault. The other two generators are disconnected from the grid when the voltage drops below 0.8 p.u. Thus, by using these fault conditions for identification, we automatically consider the case of network dynamics due to generation disconnection as well.

Sensitivity Analysis

Sensitivity analysis is performed on a simulation model to determine which parameters influence the required output of the model significantly. In our case, the model is checked on the basis of its active and reactive power curves. Hence, we determine suitable parameters from the aggregated model derived in Figure 5.10.8. These parameters are found to be V_{t0}, V_{t1} from the PVD1 model protection block, the PVD1 model inverter time constant, the dropout factor Vrflag, the aggregated line length, and the load active and reactive power values. We consider only the undervoltage V_t constants for optimization problem because we focus entirely on this part of system dynamics. It is assumed that there are no significant overvoltage dynamics in the system and so V_{t2}, V_{t3} are set to fixed values (1.1, 1.2). The dropout factor is optimized in a very narrow range around the expected value of 0.33 since only one out of three generators will be able to ride through the fault. The line length (dline) is optimized to obtain equivalent R, X values that form the equivalent impedance in Figure 5.10.8. Similarly, load active power (plini) and reactive power (qlini) are varied in a narrow ±5% range around the aggregated sum of all loads in the detailed system. This is done to take into account the unknown system losses that occur.

Once the parameters to be optimized have been defined, we revisit our optimization objective function equation (5.10.1). The values of which are the probability factor for the established faults are $\alpha_1 = 1$ and $\alpha_2 = 1$. Due to unavailability of standard probability factors for the network, we assume that both the faults are equally probable and hence the factors are set to 1. The optimization vector for the formulated objective function is shown in (5.10.9):

$$\mathbf{x} = [V_{t0}, V_{t1}, \mathrm{Tg}, \mathrm{Vrflag}, \mathrm{dline}, \mathrm{plini}, \mathrm{qlini}] \tag{5.10.9}$$

with limits

$$\mathbf{x}_{\min} = [0.45, 0.88, 0.01, 0.3, 10, 1.25, 0.8] \tag{5.10.10}$$

$$\mathbf{x}_{\max} = [0.5, 0.92, 0.5, 0.35, 100, 1.50, 0.93] \tag{5.10.11}$$

5.10.7 Application to Test Case

We now use the developed algorithm adapted to the test case and apply it for results. Before application, a few settings need to be configured. These include specifying the algorithm internal parameters, parameters from test case to be optimized, etc. After several discussions, it was decided to keep the internal parameters of the optimization constant. The scaling factor, however, was varied as the optimization proceeded. Initially, it is important to have a greater focus on exploration, and hence, during the first 100 function evaluations, the scaling factor (f_s) was set to 3. After 100 function evaluations, though, the scaling factor was set to 0.9 to focus more on the exploitation of already achieved solutions. The internal parameter settings of MVMO are described in Table 5.10.3.

In Table 5.10.3, MaxEval corresponds to maximum function evaluations. Randomly denotes the number of variables randomly chosen from the optimization vector. ρ is the penalty factor that is imposed on the objective function value if the solution does not obey the fitness criteria specified by the equation.

TABLE 5.10.3 MVMO Internal Variable Settings

Parameter	Value
MaxEval	1000
Nrandomly	3
fs	$3 \rightarrow 0.9$
ρ	10^{50}

5.10.8 Analysis

The optimization problem in our test case contains seven optimization variables which have been listed along with their search ranges in Eq. (5.10.9). The optimization problem and the objective function are recalled again at this point from Eq. (5.10.1). We minimize the sum of squares of ΔP and ΔQ where $\Delta P(Q)$ is the point-wise difference between the active (reactive) power curves of detailed and aggregated network. The analysis for this test case is done for 1000 function evaluations. The CPU time of MVMO execution for this optimization problem is approximately 28 minutes.

The convergence graph is shown in Figure 5.10.11. It can be seen from the convergence graph that the MVMO possesses a fast convergence characteristic. This demonstrates that MVMO is a fast and powerful heuristic optimization algorithm. This convergence is reflected in high resemblance of DE's response to detailed network's response as is visible in Figures 5.10.12 and 5.10.13; the faults for which DE was identified. To validate our DE, we measure its response against fault scenarios which were not used for defining DE's parameters. A fault level of 0.50 and 0.79 p.u. were randomly selected and analyzed. The results are presented in Figures 5.10.14 and 5.10.15. Since the DER cutoff limit was specified at 0.8 p.u., most erroneous results are expected to be around this value. It is, therefore, interesting to note the behavior of the network and DE at this level.

It is seen that the DE is able to catch the behavior of the detailed network with high accuracy even in such a zone, implying that the DE developed can be accepted as a good simplified representation of the detailed network. Additionally, it must be noted that the time to simulate the DE for similar disturbance as defined for the

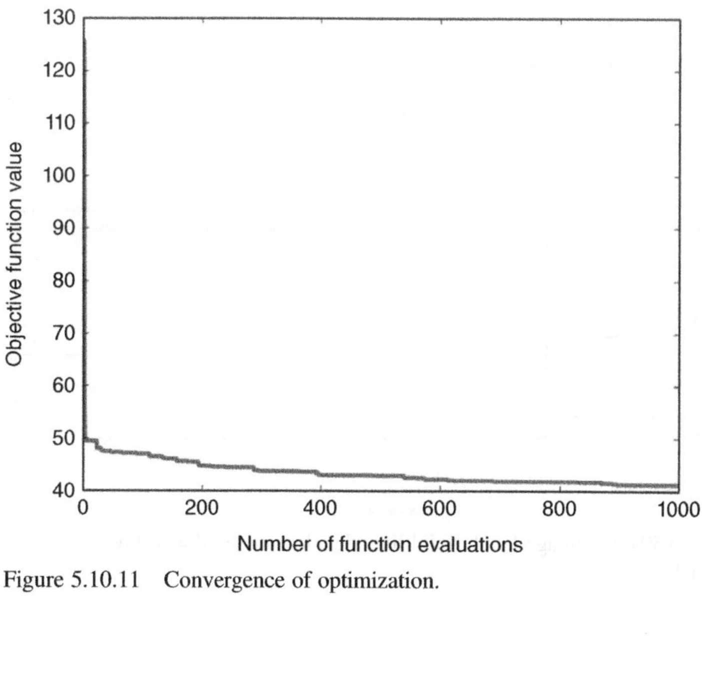

Figure 5.10.11 Convergence of optimization.

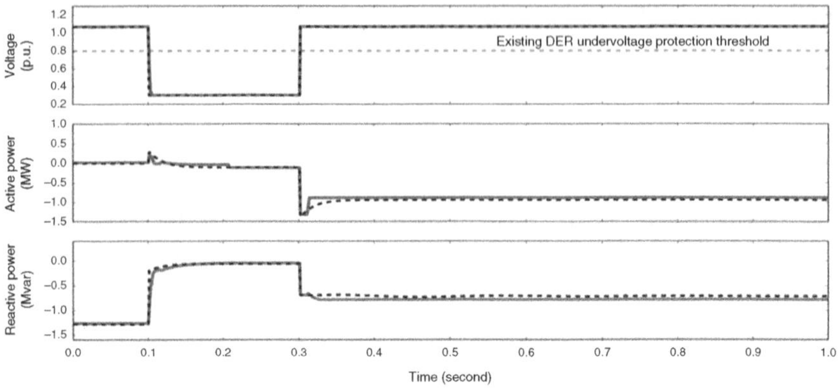

Figure 5.10.12 0.30 p.u. voltage level. Solid line: detailed model; dotted line: aggregated model.

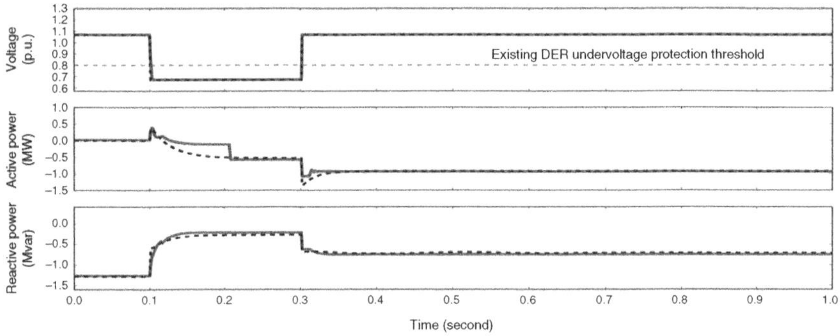

Figure 5.10.13 0.65 p.u. voltage level. Solid line: detailed model; dotted line: aggregated model.

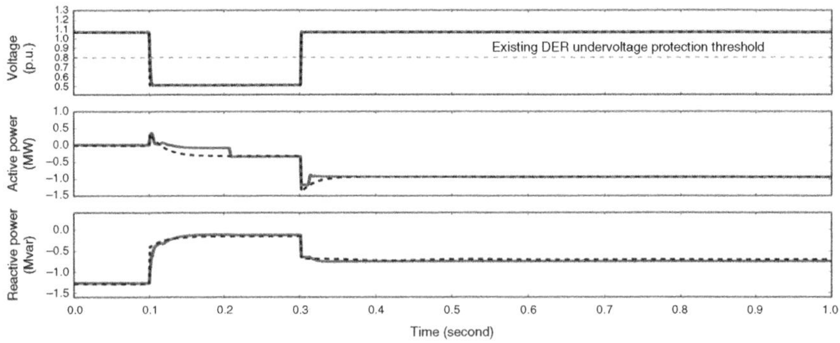

Figure 5.10.14 0.50 p.u. voltage level. Solid line: detailed model; dotted line: aggregated model.

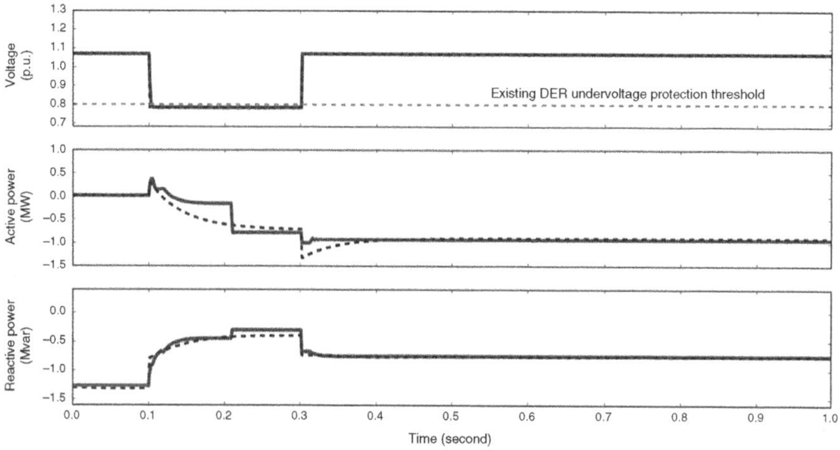

Figure 5.10.15 0.75 p.u. voltage level. Solid line: detailed model; dotted line: aggregated model.

TABLE 5.10.4 Optimized Parameters for Sub Test Case 1

Parameters		Values
PVD1 model	V_{t0}	0.870 0
	V_{t1}	0.898 57
	Tg	0.041 0
	Vrflag	0.303 9
Load model	Active power	1.269 3
	Reactive power	0.842 5
Line	Length	95.9

detailed network is also reduced. An average of five iterations showed that to perform a 2-second simulation on the detailed network, the system required 2.73 seconds while a similar simulation required only 0.26 second in DE-based network, representing a 90% reduction in simulation time. This is a tremendous improvement in computational speed and it can be safely assumed that this improvement in computational speed would be even significant for a much more detailed network than the one presented here. Table 5.10.4 lists the optimized parameters' values after performing the optimization.

It is important to bring to attention again that the analysis is based on producing equivalents that have satisfactory post-fault behavior. The performance of DE is measured by comparing the two curves. This is done by calculating the RMSE of the two curves. The RMSE values are calculated in Table 5.10.5.

Low RMSE values attest to the fact that the DE performs competently and can be used as a viable alternative for detailed networks in bulk system stability studies.

TABLE 5.10.5 Root Mean Square Values for Active and Reactive Power Curves

Uret	RMSE (*P*)	RMSE (*Q*)
0.30	0.130	0.0830
0.50	0.110	0.0700
0.65	0.108	0.0800
0.75	0.125	0.0645

5.10.9 Reflections

The PVD1 model was developed to be a representative model for large distribution feeders with increased penetration of PV that could provide an accurate post-fault behavior for the grid. This is used for bulk system stability studies, where the equivalent model shown in Figure 5.10.9 is plugged into the HV/eHV network for transmission system studies. Although a great DE, the PVD1 model has its own limitations. For cases where the detailed network model includes DG which are capable of not only riding through fault but also support voltage in the grid by providing an additional reactive current injection (aRCI), the PVD1 model is not a very good DE. There is significant mismatch of *P* and *Q* curves during the fault part, which is attributed to the fact that in the current version of PVD1 model, reactive current injection is not modeled. Although a control parameter exists, it is deactivated. The next version of PVD1 model is expected to overcome this deficiency and provide a better representation of generators with aRCI.

The PVD1 model has a great potential as an equivalent DE, especially for feeders with high penetration of distributed PV. This is possible if its parameters are accurately tuned. Tuning of the parameters requires data from the grid or from a detailed model simulation. Going into the future, DE can be tuned in real time for simulation studies if the data are directly available from the grid. The distribution system operator has to play a great role in providing these data by installing PMUs. Providing data with PMUs ensures enough observability of phenomenon in real time, enabling the identified DE to be relevant to local operating conditions.

5.10.10 Conclusions

The chapter presents an approach to tackle the challenge of parameter identification and distributed DG aggregation by presenting a heuristic optimization technique MVMO. It also presents a promising DE in WECC PVD1 model. The identification is done by obtaining a set of reference measurements for predetermined disturbances and then allowing DE optimization algorithm to determine appropriate parameters for mimicking a similar response from DE. The optimization is enhanced by adopting a dynamic MVMO procedure whereby the algorithm focuses initially on exploring the solution and after 100 function evaluations, changes it focus on the exploitation of already gathered best solutions. This is done by changing the algorithm's internal parameter settings. The DE is introduced and identified for an IEEE test feeder which contains three PV stations. The control technology and placement of the stations differ from one another to represent a diverse topological and geographical characteristic of the detailed network. Results obtained from analysis confirm MVMO's ability to quickly and efficiently determine the results of our identification problem. The time taken to complete the optimization process is significant. This can be attributed to the fact that current optimization technique uses only a single core of the four cores available for computing. Parallel and distributed processing can significantly reduce the time taken by the algorithm to identify the parameters of DE and even help to apply MVMO as a suitable candidate to perform parameter identification in real time. Future work can be on the lines of abovementioned shortcomings.

5.11 PARAMETER ESTIMATION OF CIRCUIT MODEL FOR DISTRIBUTION TRANSFORMERS

David L. Alvarez and Sergio Rivera

Universidad Nacional de Colombia, Bogotá, Colombia

5.11.1 Introduction

The frequency response analysis (FRA) is a test developed by Dick and Erven [322] as a diagnostic tool in transformers, given its sensitivity to mechanical deformations and the condition of the insulation. This test is based on making a frequency sweep or injecting a pulse [323] between two terminals of a transformer and calculating the transfer function (TF) between measurements. This TF depends on the material properties and the geometry of the transformer; thus, any changes in these parameters are reflected in the TF [322, 323], it is the principle of FRA.

Due to the effectiveness of FRA in the detection of failures in the core, windings, and insulation [324], different standards [325–327] for the measurement and analysis of the TF have arisen. These standards have defined two methodologies for the interpretation and analysis of the results. The first one consisting in comparisons between test transformer (they have a suspected condition) and records of similar transformers (measurements in different phases or previous tests where has certainty of the transformer status) [328]. The second methodology consists in comparing test results with the frequency response calculated by an equivalent circuit. In this last method, it is possible to associate the changes in the frequency response to one or more circuit parameters, allowing making an analysis and interpretation in more detail, without need for prior records because sometimes they may not be available. Different statistical indicators [329, 330] are applied independent of the methodology used, in order to assess the condition of the transformer. They can be applied in the frequency domain or wavelet transform spaces [331].

Since FRA was proposed, different RLC circuits for modeling transformers have been used. The first ones consisted of constant elements with the frequency for modeling transformer windings [322], subsequently, it proceeded to model each resonance and antiresonance through circuits groups which were coupled [332]. With the need to improve the performance of the circuit, it was necessary to model the frequency dependence of the RLC elements, given its influence on the frequency range of the test [333]. These values are generally calculated analytically through Wilcox expressions [334], but with increasing processing capacity of computers, the finite element method (FEM) emerged as a tool for calculating circuit parameters due to their ability for modeling the magnetic field distribution, and the skin and proximity effects in complex geometries such as transformers [335–337]. As a result of high influence of stray flux in FRA [338], which is affected by the use of symmetries when 2D models are used [339], it is necessary to optimize the circuit parameters or 3D transformer model. This last option requires large computational resources.

In this chapter the frequency response of transformer windings is optimized using different statistical indicators through a heuristic algorithm, whereby the values of the parameters of an equivalent circuit, representing the frequency variation of R and L through constants parameters, are fitted, using FEM and the vector fitting (VF) algorithm. The content of this chapter is as follows: in Section 5.11.2, the equivalent circuit used to model the frequency response with the calculation method for RL parameters is described. In Section 5.11.3, the main statistical indicators used for the interpretation of FRA tests in transformers are analyzed. In Section 5.11.4, differential evolutionary particle swarm optimization (DEEPSO) heuristic optimization algorithm is described, which is applied in a case study in Section 5.11.5.

5.11.2 Transformer Winding Equivalent Circuit

The equivalent circuit of Figure 5.11.1 is used to model the behavior of windings in transformers within the frequency ranges where FRA is carried out. This circuit is characterized by a strong magnetic coupling between all elements of the circuit due to proximity effect, and by the variation of the circuit parameters in function of the

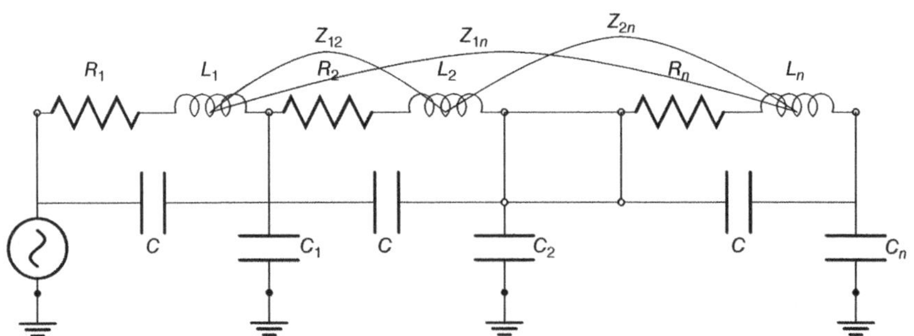

Figure 5.11.1 Simplified equivalent circuit of a transformer winding.

frequency as a result of skin effect. These two phenomena are mathematically modeled by the impedance matrix described in (5.11.1), which is composed of elements that vary with frequency.

$$Z = \begin{bmatrix} Z_{11}(w) & Z_{12}(w) & \cdots & Z_{1n}(w) \\ Z_{21}(w) & Z_{22}(w) & \cdots & Z_{2n}(w) \\ \vdots & \vdots & \ddots & \vdots \\ Z_{n1}(w) & Z_{n2}(w) & \cdots & Z_{nn}(w) \end{bmatrix} \tag{5.11.1}$$

In order to model the frequency dependence of R and L in transformer windings through the constant elements, the equivalent Mombello circuit (MC) can be used [333]. The latter consists in adding to each section of the circuit loops magnetically coupled, resulting in the circuit of Figure 5.11.2.

In MC, the behavior of self and mutual impedance are approximated by rational functions of the form of (5.10.2), where r is the number of additional loops, $k_0 = R_{1ii}$, $k_\infty = L_{1ii}$, $k_l = -M_{r\,ij}^2/L_{1\,ii}$. Finally, $\lambda_l = R_{2ij}/L_{2ij}$. Given that the number of equations is less than the number of unknowns, arbitrarily is imposed that $L_{2ij} = L_{1ii}$.

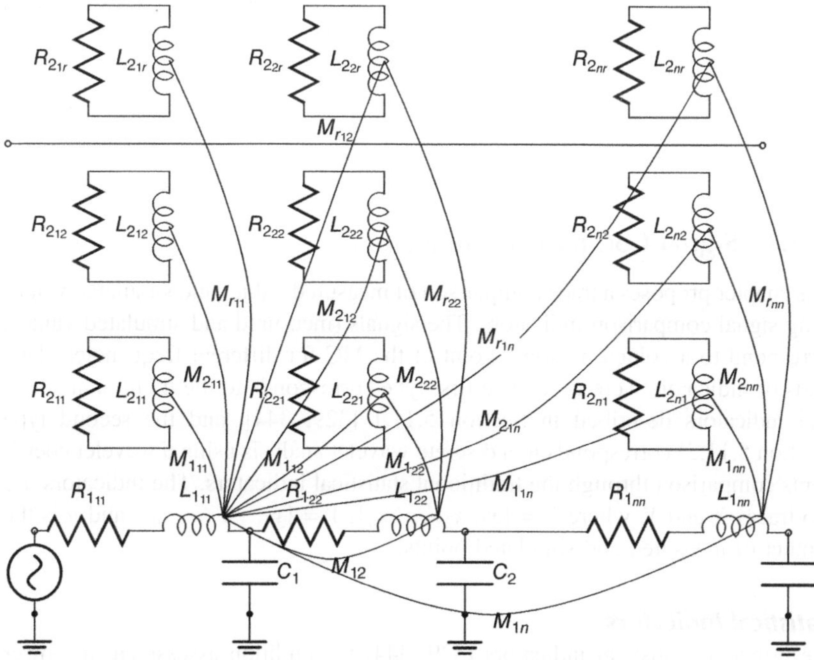

Figure 5.11.2 Equivalent circuit for multiple coupling windings.

$$Z(s) = k_0 + K_\infty s + s^2 \sum_{l=1}^{r} \frac{k_l}{s + \lambda_l} \tag{5.11.2}$$

In (5.11.3), the matrix of MC is described, by which it is possible to model the different physical phenomena that occur in the windings. Sub-matrices R_1 and L_1 (coefficients k_0 and k_∞) represent the self and mutual impedances at low frequencies or DC of each winding section; the R_2 and L_2 sub-matrices are associated with the additional loops and described by eddy currents that are generated product of skin and proximity effect. Finally, the magnetic fields that generated these currents are represented by the impedances M coupled magnetically between each loop and winding section.

$$Z = \begin{bmatrix} R_1 + sL_1 & sM_1 & sM_2 & \cdots & sM_r \\ M_1^T & R_{2_1} + sL_{2_1} & 0 & \cdots & 0 \\ M_2^T & 0 & R_{2_2} + sL_{2_2} & \cdots & 0 \\ \vdots & \vdots & \vdots & \ddots & \vdots \\ M_r^T & 0 & 0 & \cdots & R_{2_r} + sL_{2_r} \end{bmatrix} \tag{5.11.3}$$

In this chapter, the behavior of self and mutual impedances in (5.11.1) are calculated from FEM using the proposal of [340], and described by rational functions of the type (5.11.4), which are estimated by VF algorithm proposed in [341]. From the knowledge of the coefficients of VF, it is possible to calculate the coefficients of 5.11.2 and therefore MC parameters as shown in [342, 343].

$$Z(s) = b_0 + b_\infty s + \sum_{i=1}^{r} \frac{b_i}{s + \lambda_i} \tag{5.11.4}$$

5.11.3 Signal Comparison Indicators

This chapter proposes a trace comparison of measured values and simulated values using signal comparison indicators. The signals (measured and simulated values) correspond to a voltage in some point of the MC for different frequencies. Two kinds of indicators were used. The first type corresponds to the traditional statistical indicators described in Section 5.11.1 [329, 344], and the second type (Section 5.11.2) corresponds to a discrete wavelet analysis using a wavelet coefficients comparison through the traditional statistical indicators. The indicators use two traces X and Y, where $X = \{x_1, x_2, \ldots, x_n\}$, $Y = \{y_1, y_2, \ldots, y_n\}$, and n is the number of measured and simulated points.

Statistical Indicators

The principal statistical indicators [329, 344] to condition assessment in power transformers are described in Eqs. (5.11.5)–(5.11.11):

Correlation Coefficient

$$\rho = \frac{\sum_{i=1}^{N} x_i y_i}{\sqrt{\sum_{i=1}^{N} x_i^2 \sum_{i=1}^{N} y_i^2}} \tag{5.11.5}$$

Relative Error

$$\eta = \frac{\sum_{i=1}^{N} \left| \frac{x_i - y_i}{x_i} \right|}{N} \tag{5.11.6}$$

Relative Factor

$$r = \frac{\sum_{i=1}^{N} (x_i - \overline{x})(y_i - \overline{y})}{\sqrt{\sum_{i=1}^{N} (x_i - \overline{x})^2 \sum_{i=1}^{N} (y - \overline{y})^2}} \tag{5.11.7}$$

where $\overline{x}$ and $\overline{y}$ are the mean of X and Y, respectively.

Min-Max

$$\text{MM} = \frac{\sum_{i=1}^{N} \min(|x_i|, |y_i|)}{\sum_{i=1}^{N} \max(|x_i|, |y_i|)} \tag{5.11.8}$$

DABS

$$\text{DABS} = \frac{\sum_{i=1}^{N} |y_i - x_i|}{N} \tag{5.11.9}$$

ASLE

$$\text{ASLE} = \frac{\sum_{i=1}^{N} |20 \log_{10} y_i - 20 \log_{10} x_i|}{N} \tag{5.11.10}$$

Spectrum Deviation

$$\sigma = \frac{1}{N} \sum_{i=1}^{N} \sqrt{\frac{x_i - \frac{(x_i + y_i)}{2}}{\frac{(x_i + y_i)}{2}} + \frac{y_i - \frac{(x_i + y_i)}{2}}{\frac{(x_i + y_i)}{2}}} \tag{5.11.11}$$

In [344], it is reported that MIN-MAX and ASLE are very sensitive to shorted turns and axial displacements; and ASLE, DABS, MM, and ρ to disc deformations. In [336], ρ has a better performance in comparison with ASLE, it shows high sensitivity to conductors till and bending. In [345] are compared ρ vs σ, showing that ρ has a better performance and σ has troubles analyzing small changes in product of normal variations and not of damages. In [330], it is concluded that ASLE is the most pertinent indicator for FRA.

The different statistical indicators are sensitive to different transformer damages, for this reason it has not been possible to select one for assessment transformer through FRA.

Wavelet Coefficients Comparison

The wavelet transform is a mathematical tool developed to map a signal to the translation and scale representation based on a reference providing multiresolution analysis with dilated windows [331]. The data function is processed in every scale and different positions with a reference function called "mother wavelet." In this chapter, it is used a discrete wavelet transform (DWT), where the wavelets are discretely sampled. It captures both frequency and location information. In this way, we get a wavelet decomposition (wavelet coefficients) for the simulated and measured signals; these decompositions are compared using the traditional trace comparison statistical indicators (Section 15.10.3). Thus, the wavelet signal indicators are W_η, W_ρ, W_r, W_{MM}, W_{DABS}, W_{ASLE}, W_σ corresponding to the wavelet coefficients comparison indicators using the relative error, correlation coefficient, relative factor, MIN-MAX, DABS, average signal Liapunov exponent, and spectrum deviation, respectively.

5.11.4 Coefficients Estimation Using Heuristic Optimization

This section presents the proposed coefficients estimation approach. Additionally, it is presented an outline of heuristic optimization and the heuristic search strategy used in order to solve the optimization problem.

Different heuristic algorithms have been used to optimize RLC parameters in circuits that model frequency response. In [346], ABC is applied for synthesizing a mutually coupled lumped-parameter; here, the error was defined as the deviation in the max and min peak frequencies. In [347], an improved PSO is used, fitting the least square error between measurements and equivalent circuit response. In [348], a similar analysis was carried out, seeking to minimize the sum of the differences between measurements and the circuit response. In previous researches, the RLC parameters are considered constants with the frequency [346–348].

Coefficients Estimation Approach

The proposed approach is outlined in Figure 5.11.3. The required input data are the transformer geometry and the measured voltage in a determined winding for different frequencies ($v(s)$). There is an initial coefficients estimation through the blocks: FEM, VF, and MC. Using the transformer geometry, the behavior of self and mutual impedances is calculated through FEM (previous work has been presented in [338, 340, 349, 350]). The impedances can be described by rational functions of the type (5.11.4), using the proposal of VF algorithm [341]. From the

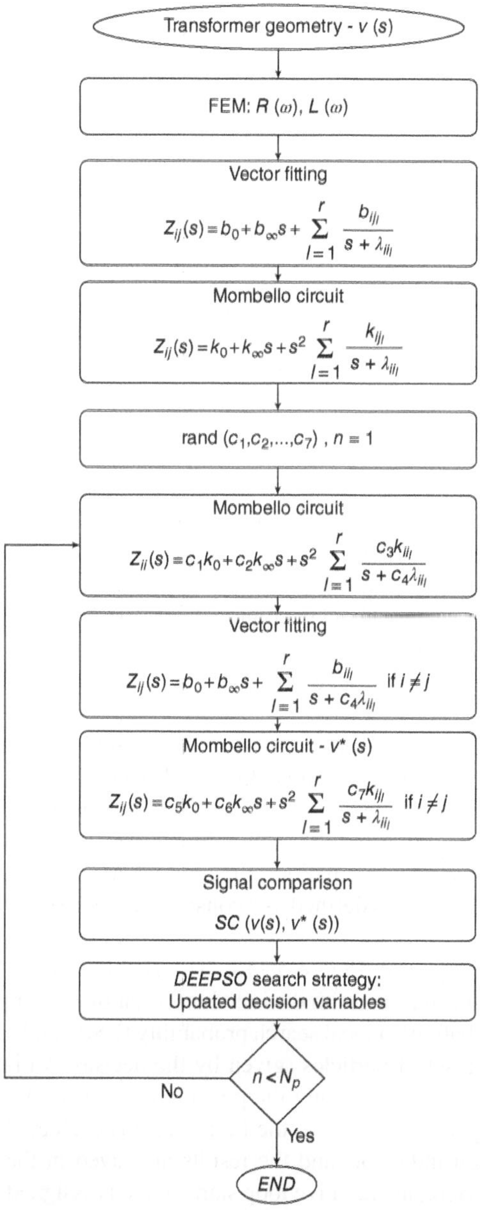

Figure 5.11.3 Coefficients estimation approach.

knowledge of the coefficients of VF, it is possible to calculate the coefficients of (5.11.2) and therefore MC parameters (previous work has been presented in [338, 342, 343]).

In order to improve the initial estimation, the approach selects (randomly) the decision variables (c_1, c_2, ..., c_7) that multiply the MC parameters. Here, the iterative loop starts, with the new MC; the voltage simulated ($v^*(s)$) for a certain set of decision variables is calculated. The signal comparison is carried out using one of the indicators presented in Section 5.11.1. The target is to minimize the signal comparison indicator using heuristic optimization, through the DEEPSO method (described in the next section).

Differential Evolutionary Particle Swarm Optimization

Heuristic techniques are optimization algorithms based on rules that can follow different type of procedures like natural processes [351]. Population-based heuristic techniques use a population of solutions and a search strategy in order to get a better value of the target function and modify the population, in each iteration [352]. In this chapter, the heuristic search strategy mentioned in Figure 5.11.3 follows the DEEPSO rules; nevertheless, it can be used for different heuristic search strategies.

DEEPSO is an enhancement of evolutionary particle swarm optimization (EPSO) method [353–356]. EPSO is a mixture algorithm that combines a PSO algorithm with an EP approach through a self-adaptive recombination operator. EP algorithms simulate the individual evolution in the populations based on processes such as selection, mutation, and recombination; which permits the subsistence of individuals with better performances from generation to generation [353, 355]. Alternatively, PSO is based on social behavior of animal swarms, in which each particle (given by the decision variables) moves in the search space according to a rule (given by a velocity) that combine three different parameters: inertia, memory, and cooperation [353].

The hybrid DEEPSO retains the self-adaptive properties of EPSO but borrows the concept of rough gradient from DE algorithms [355, 357]. The basic idea of DE is to produce a new solution from an existing individual by adding some fraction of the difference between two other points, randomly selected from the current population. After generating a new population, a recombination process ensures more diversity and a new population is defined as a consequence of a selection procedure [355–357].

The details of DEEPSO are outlined in Figure 5.11.4. The DEEPSO parameters are set through an initial weights matrix (with the initial information of inertia, memory, cooperation, and perturbation), a local search probability (LSP), and a mutation rate. The initial swarm is a set of particles (given by the decision variables). Each particle (a set of c_1, c_2, …, c_7 indicating the position of the particle) has associated a velocity and a weights matrix. Next, the first swarm is evaluated (calculation of the signal comparison indicator) and the results are saved in the memory of the DEEPSO algorithm. Here, the iterative loop starts, there is assigned a copied swarm, and the current and copied swarm are modified according to the following two rules: (i) if a random number is bigger than the LSP, a process of replication, mutation, and reproduction modifies the current and copied swarms. The modification calculates the new current swarm considering the velocity and the weights matrix, after that the weights is mutated and a new copied swarm is calculated considering the copied velocity and the new weights matrix. (ii) If

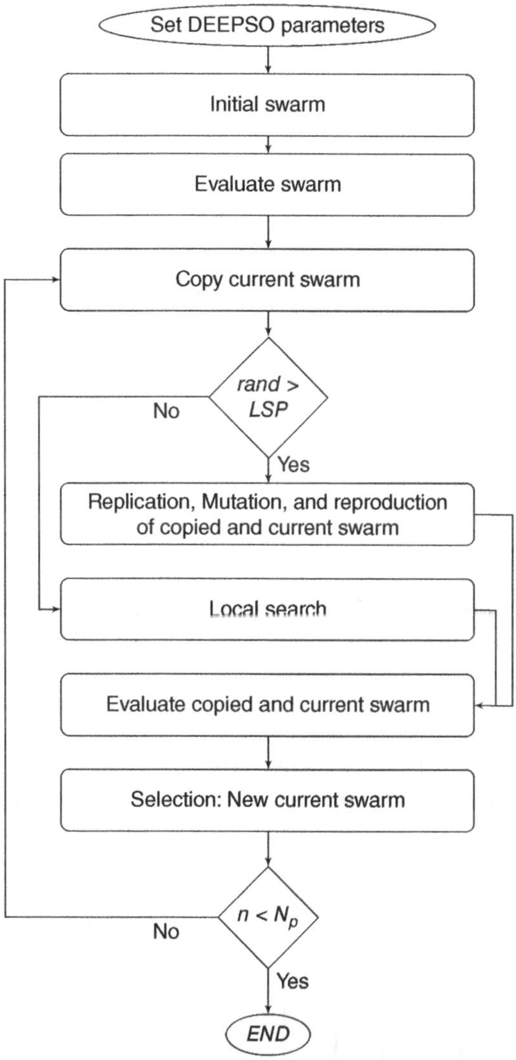

Figure 5.11.4 DEEPSO flowchart.

the random number is less than the LSP, a local search is carried out, computing a new current swarm that considers the mutation of some decision variables using a certain probability.

Then, the new current and copied swarms are evaluated (calculation of the signal comparison indicator) and it is selected a new current swarm given by the selection of particles with better signal indicators between the current and copied swarms. After a determined number of iterations, the DEEPSO search strategy ends, choosing the decision variables with the best signal comparison indicator.

5.11.5 Coefficients Estimation Results and Conclusion

The coefficients estimation approach was tested in a real transformer winding. "This winding has 12 coils, each one with 10 turns. The conductor dimensions are 2.5 × 7.5 [mm], made of copper; the core diameter is 90 [mm]. Due to the small size of the transformer, it was necessary to connect capacitors between each coil and ground in order to generate resonances falling in the range of power transformer resonances." (For more details about the characteristics of the transformer, see [333].)

Applying the formulation of the Transformer Geometry, FEM, VF, and MC blocks of Figure 5.11.3 over the mentioned real transformer, it is possible to get the FEM simulation of a voltage signal in a point of the MC in order to contrast this signal with a real measurement signal. In this way, in Figure 5.11.5 is shown the mentioned signals (measured and FEM simulation) using the proposed formulation in [338, 340, 342, 343, 349, 350].

The FEM simulated results are very good as an initial approximation, indeed with the MC parameters obtained with this simulation, it is possible to get diagnostics in real transformers [338, 340, 342, 343, 349, 350]. Although the two signals are similar in shape, the different peaks and zero values are presented in different frequencies, and the peaks are not the same in magnitude. Additionally, there is a more pronounced mismatch between the measured and simulated signals for high frequencies. In this way, this chapter proposes to improve the initial estimation using the iterative loop presented in Figure 5.10.3, which uses a heuristic search strategy.

In order to realize the improvements with the proposed method (Figure 5.11.3), in Figures 5.11.6–5.11.12, it is possible to get the range for each statistical indicator after the heuristic optimization when the different indicators are used as target function. In these figures, there are two curves, one when the statistical indicators are applied directly to the measured and simulated frequency data,

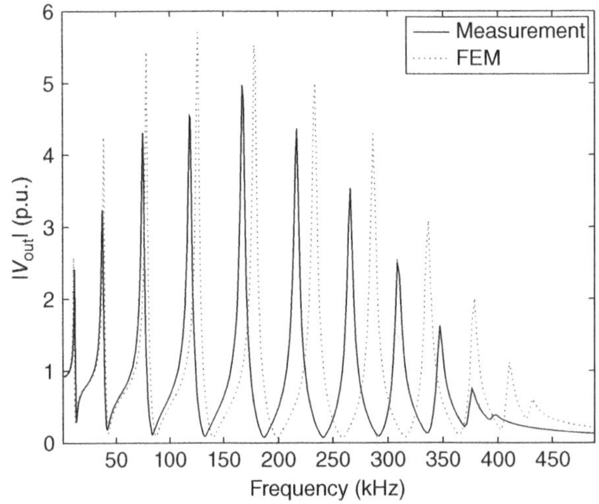

Figure 5.11.5 Measured and FEM simulated signal in a real transformer.

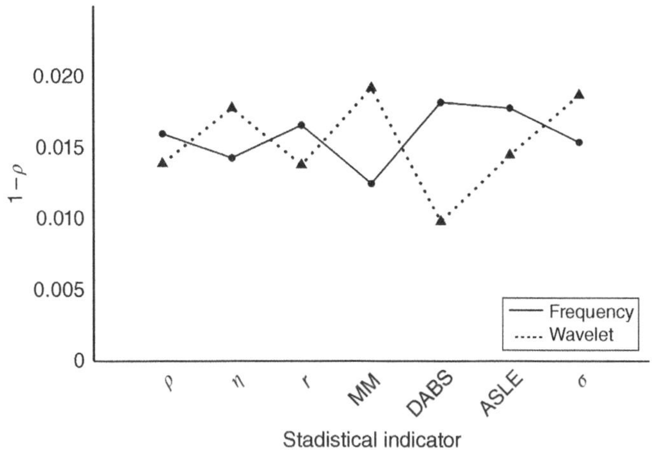

Figure 5.11.6 Correlation coefficient indicator $(1-\rho)$ for each target function.

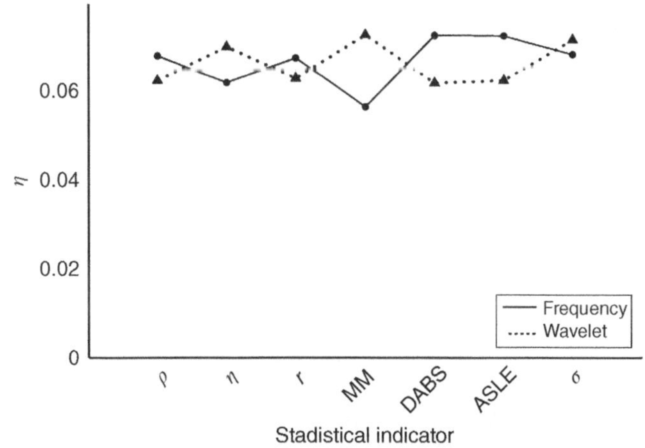

Figure 5.11.7 Relative error indicator (η) for each target function.

and the second one when statistical indicators are applied to the wavelet coefficients of the measured and simulated frequency data.

The lower the statistical indicator, more similar the measured and simulated signals. In this way, the correlation coefficient indicator $(1-\rho)$ before the optimization, that is to say from Figure 5.10.5, was 0.2626; and after the heuristic optimization, the range for the different target function is 0.0098–0.0192. In the same way, the relative error indicator (η) before the optimization was 0.7768, and after the heuristic optimization the range is 0.0621–0.0729. The relative factor indicator (r) before the optimization was 0.7575, and after the heuristic optimization the range is 0.0299–0.0601. The MIN-MAX indicator (1 – MM) before the optimization was

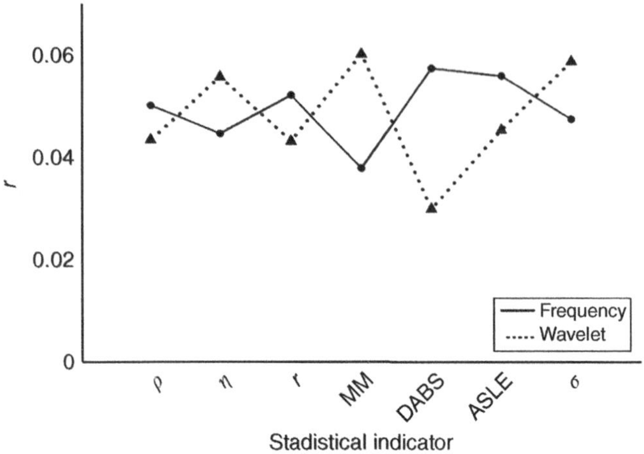

Figure 5.11.8 Relative factor indicator (r) for each target function.

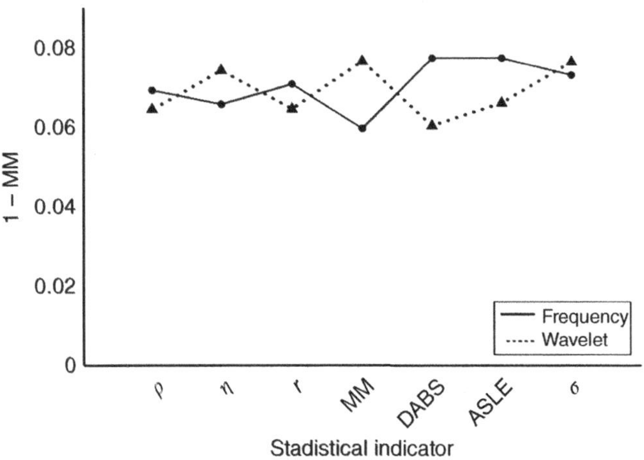

Figure 5.11.9 MIN-MAX indicator (1 – MM) for each target function.

0.3302, and after the heuristic optimization the range is 0.0596–0.0765. The DABS indicator (*DABS*) before the optimization was 0.3810, and after the heuristic optimization the range is 0.0579–0.0759. The ASLE indicator (*ASLE*) before the optimization was 3.5873, and after the heuristic optimization the range is 0.4813–0.6986. The spectrum deviation indicator (σ) before the optimization was 0.2456, and after the heuristic optimization the range is 0.0389–0.0547.

After the heuristic optimization, it is possible to get the simulations of a voltage signal in a point of the MC in two cases. The first case is using the statistical indicator to optimize directly to the measured and simulated frequency data, and the second one using the statistical indicator to optimize over the wavelet

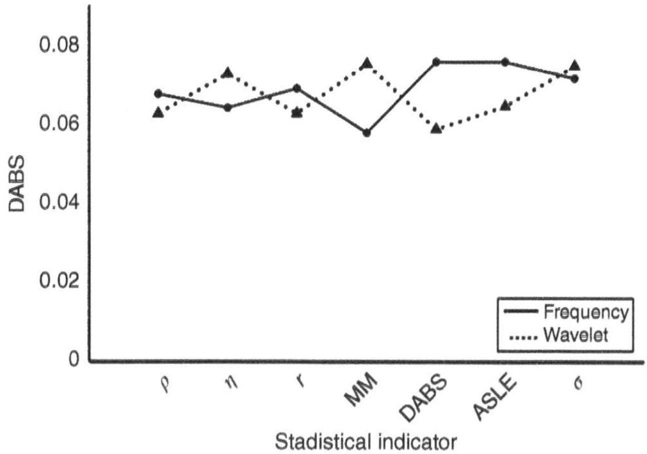

Figure 5.11.10 DABS indicator for each target function.

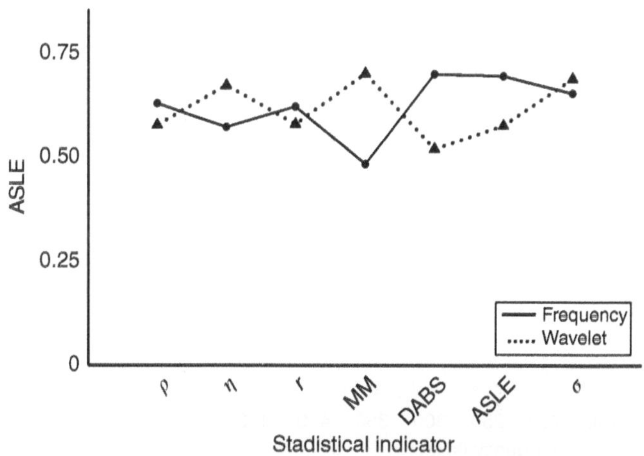

Figure 5.11.11 ASLE indicator for each target function.

coefficients of the measured and simulated frequency data, in order to contrast these signals with a real measurement signal. In Figures 5.11.13–5.11.19 are these traces comparison. It is possible to figure out the improvement with regard to the initial estimation (Figure 5.11.5). The frequencies of the peaks are very similar with regard to the measured data. Regarding the two mentioned cases, the statistical indicators applied to the wavelet coefficients give better results. The transformers windings coefficients estimation method proposed in this chapter, using heuristic optimization, is a convenient way to get the MC parameters.

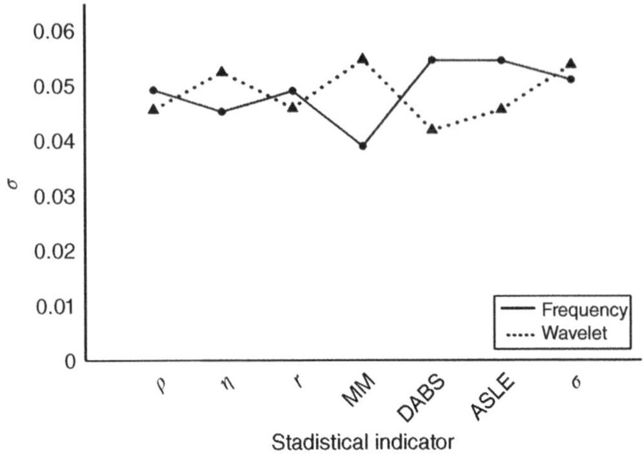

Figure 5.11.12 Spectrum deviation indicator (σ) for each target function.

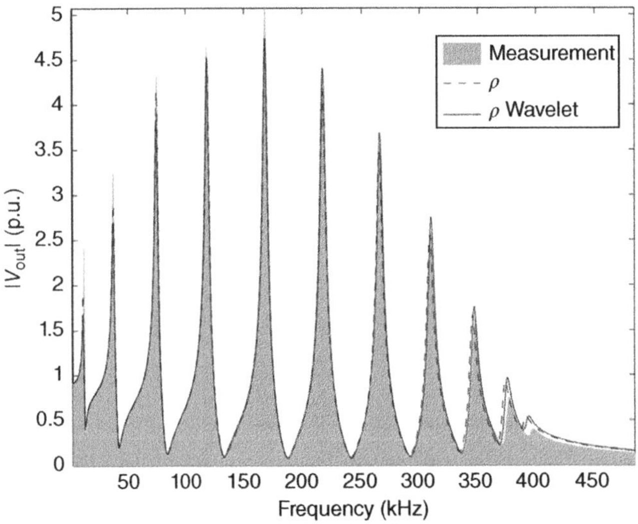

Figure 5.11.13 Measured and simulated signals with correlation coefficient.

5.11.6 Conclusions

The application of meta-heuristic, especially evolutionary, algorithms in distribution system planning is tempting due to the complexity and nonlinearity of the problems, which typically involve an evaluation of investment and an assessment of operational costs. The dimension of the problems is always very large, because in many cases planning implies an evolution in the time domain. Besides, many of the options available are of discrete nature and the problem becomes mixed-integer with a combinatorial characteristic.

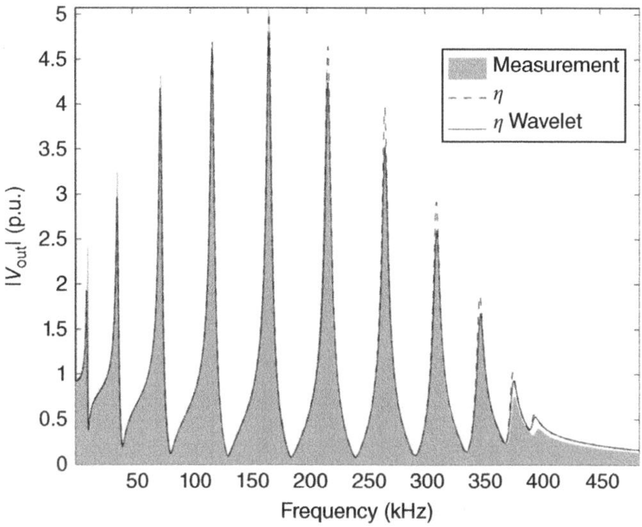

Figure 5.11.14 Measured and simulated signals with relative error.

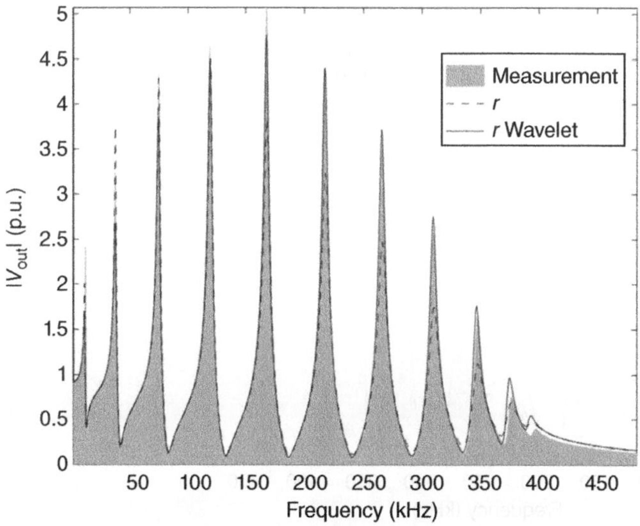

Figure 5.11.15 Measured and simulated signals with relative factor indicator.

We have visited in this chapter models that have application in the optimal operation and planning of power distribution systems. Examples include GA for ADN planning and capacitor placement, PSO methodology for distribution system service restoration and CVR, and multi-agent system with distributed intelligence for distribution system reconfiguration. Their authors claim a reasonable amount of success and also leave open the avenue for further improvement.

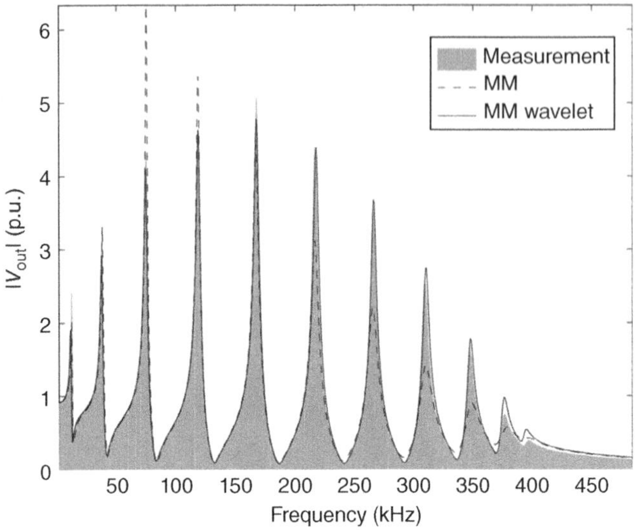

Figure 5.11.16 Measured and simulated signals with MIN-MAX indicator.

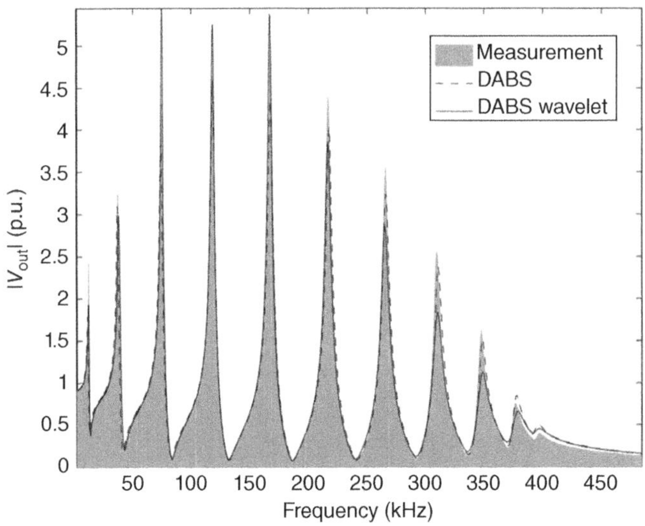

Figure 5.11.17 Measured and simulated signals with DABS indicator.

As a general comment, we may state that dynamic expansion planning models are a challenge, whereas for static expansion planning quite satisfactory experiences have been reported. The objective of the chapter was not to be exhaustive in the description of all the models proposed in the literature (certainly

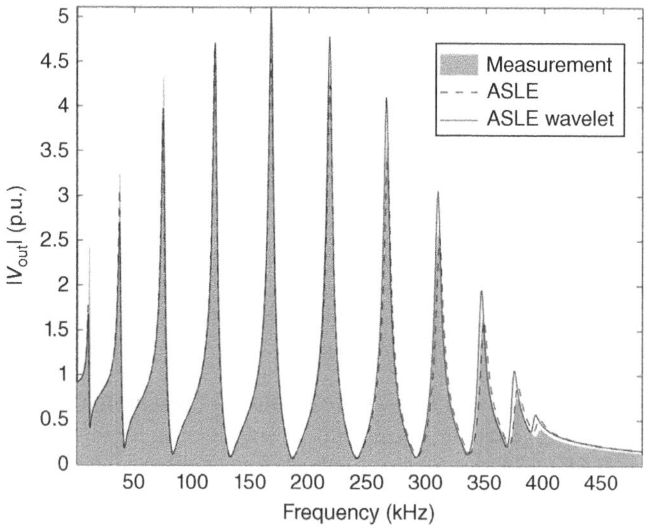

Figure 5.11.18 Measured and simulated signals with ASLE indicator.

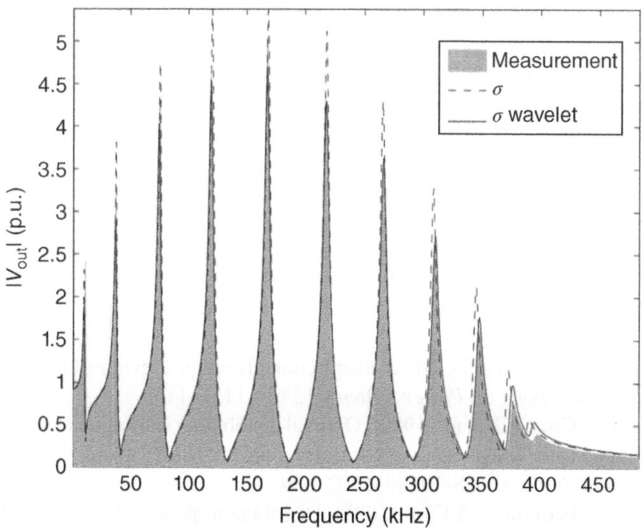

Figure 5.11.19 Measured and simulated signals with spectrum deviation indicator.

some other proposals with their own merits have come to light) but rather to provide enough evidence that the meta-heuristic approach works and should be seriously considered in practical distribution system operation and planning applications.

REFERENCES

1. Peças Lopes, J.A., Hatziargyriou, N.D., Mutale, J. et al. (2007). Integrating distributed generation into electric power systems: a review of drivers, challenges and opportunities. *Electric Power Systems Research* 77 (9): 1189–1203.
2. Georgilakis, P.S. and Hatziargyriou, N.D. (2015). A review of power distribution planning in the modern power systems era: models, methods and future research. *Electric Power Systems Research* 121: 89–100.
3. CIGRE Working Group C6.19 (2013). Planning and optimization methods for active distribution systems. In: *Technical Brochure*, vol. 591. Paris: CIGRE.
4. Celli, G, Mocci, S, Pilo, F, et al. (2007). Multi-year optimal planning of active distribution networks. *Proceedings of CIRED – International Conference on Electric Distribution*, Vienna (21–24 May 2007).
5. Haesen, E., Alarcon-Rodriguez, A.D., Driesen, J., and Belmans, R. (2009). Opportunities for active DER management in deferral of distribution system reinforcements. In: *Proceedings of IEEE PES Power Systems Conference and Exposition*, 1–8. IEEE.
6. Pilo, F., Celli, G., Mocci, S., and Soma, G.G. (2010). Multi-objective programming for optimal DG integration in active distribution systems. In: *Proceedings of the IEEE PES General Meeting*, 1–7. IEEE.
7. Ziari, I., Ledwich, G., Ghosh, A., and Platt, G. (2013). Optimal distribution network reinforcement considering load growth, line loss, and reliability. *IEEE Transactions on Power Systems* 28 (2): 587–597.
8. Martins, V.F. and Borges, C.L.T. (2011). Active distribution network integrated planning incorporating distributed generation and load response uncertainties. *IEEE Transactions on Power Systems* 26 (4): 2164–2172.
9. Zeng, B., Zhang, J., Yang, X., and Wang, J. (2014). Integrated planning for transition to low-carbon distribution system with renewable energy generation and demand response. *IEEE Transactions on Power Systems* 29 (3): 1153–1165.
10. Celli, G., Pilo, F., Pisano, G., and Soma, G.G. (2008). Optimal planning of active networks. In: *Proceedings of 16th Power Systems Computation Conference (PSCC 2008 Glasgow)*, 1362–1369. PSCC.
11. Jabr, R.A., Singh, R., and Pal, B.C. (2012). Minimum loss network reconfiguration using mixed-integer convex programming. *IEEE Transactions on Power Systems* 27 (2): 1106–1115.
12. Khator, S.K. and Leung, L.C. (1997). Power distribution planning: a review of models and issues. *IEEE Transactions on Power Systems* 12 (3): 1151–1159.
13. Sun, D.I., Farris, D.R., Cote, P.J. et al. (1982). Optimal distribution substation and primary feeder planning via the fixed charge network formulation. *IEEE Transactions on Power Apparatus and Systems* PAS-101 (3): 602–609.
14. Carvalho, P.M.S. and Ferreira, L.A.F.M. (2005). Distribution quality of service and reliability optimal design: individual standards and regulation effectiveness. *IEEE Transactions on Power Systems* 20 (4): 2086–2092.
15. Ouyanga, W., Chenga, H., Zhanga, X. et al. (2010). Distribution network planning considering distributed generation by genetic algorithm combined with graph theory. *Electric Power Components & Systems* 38 (3): 325–339.
16. Falaghia, H., Singh, C., Haghifam, M.R., and Ramezani, M. (2011). DG integrated multistage distribution system expansion planning. *International Journal of Electrical Power & Energy Systems* 33 (8): 1489–1497.

17. Nazar, M.S. and Haghifam, M.R. (2009). Multiobjective electric distribution system expansion planning using hybrid energy hub concept. *Electric Power Systems Research* 79 (6): 899–911.
18. Chen, T.H., Lin, E.H., Yang, N.C., and Hsieh, T.Y. (2013). Multi-objective optimization for upgrading primary feeders with distributed generators from normally closed loop to mesh arrangement. *International Journal of Electrical Power & Energy Systems* 45 (1): 413–419.
19. Junior, B.R.P., Cossi, A.M., Contreras, J., and Mantovani, J.R.S. (2014). Multiobjective multistage distribution system planning using tabu search. *IET Generation, Transmission and Distribution* 8 (1): 35–45.
20. Cossi, A.M., Da Silva, L.G.W., Lázaro, R.A.R., and Mantovani, J.R.S. (2012). Primary power distribution systems planning taking into account reliability, operation and expansion costs. *IET Generation, Transmission and Distribution* 6 (3): 274–284.
21. Koutsoukis, N.C., Georgilakis, P.S., and Hatziargyriou, N.D. (2014). A Tabu search method for distribution network planning considering distributed generation and uncertainties. In: *Proceedings of IEEE Probabilistic Methods Applied to Power Systems (PMAPS)*, 1–6. IEEE.
22. Samper, M.E. and Vargas, A. (2013). Investment decisions in distribution networks under uncertainty with distributed generation – part I: model formulation. *IEEE Transactions on Power Systems* 28 (3): 2331–2340.
23. Samper, M.E. and Vargas, A. (2013). Investment decisions in distribution networks under uncertainty with distributed generation – part II: implementation and results. *IEEE Transactions on Power Systems* 28 (3): 2341–2351.
24. Sedghi, M., Aliakbar-Golkar, M., and Haghifam, M.R. (2013). Distribution network expansion considering distributed generation and storage units using modified PSO algorithm. *International Journal of Electrical Power & Energy Systems* 52: 221–230.
25. Jain, A.K. (2010). Data clustering: 50 years beyond K-means. *Pattern Recognition Letters* 31 (8): 651–666.
26. Carrano, E.G., Soares, L.A., Takahashi, R.H. et al. (2006). Electric distribution network multiobjective design using a problem-specific genetic algorithm. *IEEE Transactions on Power Delivery* 21 (2): 995–1005.
27. Katsigiannis, Y.A., Georgilakis, P.S., and Karapidakis, E.S. (2012). Hybrid simulated annealing–tabu search method for optimal sizing of autonomous power systems with renewables. *IEEE Transactions on Sustainable Energy* 3 (3): 330–338.
28. Baumgartner, U., Ebner, T., and Magele, C. *Optimization in Electrical Engineering*. Austria: Technical University of Graz.
29. Spendley, W., Hext, G.R., and Himsworth, F.R. (1962). Sequential application of simplex designs in optimization and evolutionary operation. *Technometrics* 4: 441.
30. Cavazzuti, M. (2013). *Optimization Methods: From Theory to Design*. Springer.
31. Kirkpatrick, S., Gelatt, C.D. Jr., and Vecchi, M.P. (1983). Optimization by simulated annealing. *Science* 220 (4598): 671–680.
32. Kennedy, J. and Eberhart, R. (1995). Particle swarm optimization. In: *Proceedings of ICNN'95 – IEEE International Conference on Neural Networks*, Perth, Australia (November 1995), 1942–1948. IEEE.
33. Park, J.-B., Jeong, Y.-W., Shin, J.-R., and Lee, K.Y. (2010). An improved particle swarm optimization for nonconvex economic dispatch problems. *IEEE Transactions on Power Systems* 25 (1): 156–166.

34. Storn, R. and Price, K. (1997). Differential evolution: a simple and efficient heuristic for global optimization over continuous spaces. *Journal of Global Optimization* 11 (4): 341–359. https://doi.org/10.1023/A:1008202821328.
35. Dorigo, M. (1992). Optimization learning and natural algorithm. PhD dissertation, Politecnico di Milano.
36. Saber, A.Y. and Senjyu, T. (2007). Memory-bounded ant colony optimization with dynamic programming and A^* local search for generator planning. *IEEE Transactions on Power Systems* 22 (4): 1965–1973.
37. Chang, C.F. (2008). Reconfiguration and capacitor placement for loss reduction of distribution systems by ant colony search algorithm. *IEEE Transactions on Power Systems* 23 (4): 1747–1755.
38. Vlachogiannis, J.G., Hatziargyriou, N.D., and Lee, K.Y. (2005). Ant colony system-based algorithm for constrained load flow problem. *IEEE Transactions on Power Systems* 20 (3): 1241–1249.
39. Favuzza, S., Graditi, G., Ippolito, M.G., and Sanseverino, E.R. (2007). Optimal electrical distribution systems reinforcement planning using gas micro turbines by dynamic ant colony search algorithm. *IEEE Transactions on Power Systems* 22 (2): 580–587.
40. Yang, S., Wu, M., Yao, X., and Jiang, J. (2015). Load modeling and identification based on ant colony algorithms for EV charging stations. *IEEE Transactions on Power Systems* 30 (4): 1997–2003.
41. Tsai, M. and Chu, C. (2011). Applications of hybrid ep-aco for power distribution system loss minimization under load variations. In: *16th International Conference on Intelligent System Application to Power Systems*, 1–7. IEEE.
42. Pham, D.T., Ghanbarzadeh, A., Koc, E. et al. (2006). The bees algorithm: a novel tool for complex optimisation problems. *Proceedings of the IPROMS2006 Conference*, pp. 454–461.
43. Yang, X.-S. and Deb, S. (2009). Cuckoo search via Lévy flights. In: *2009 World Congress on Nature & Biologically Inspired Computing (NaBIC 2009)*. IEEE.
44. Deb, K., Pratap, A., Agarwal, S., and Meyarivan, T. (2002). A fast and elitist multi-objective genetic algorithm: NSGA-II. *IEEE Transactions on Evolutionary Computation* 6 (2): 182–197. https://doi.org/10.1109/4235.996017.
45. Li, H. and Zhang, Q. (2009). Multiobjective optimization problems with complicated Pareto sets, MOEA/D and NSGA-II. *IEEE Transactions on Evolutionary Computation* 13 (2): 284–302.
46. Li, Y., Lu, X., and Kar, N.C. (2014). Rule-based control strategy with novel parameters optimization using NSGA-II for power-split PHEV operation cost minimization. *IEEE Transactions on Vehicular Technology* 63 (7): 3051–3061.
47. Ye, C.-J. and Huang, M.-X. (2015). Multi-objective optimal power flow considering transient stability based on parallel NSGA-II. *IEEE Transactions on Power Systems* 30 (2): 857–866.
48. Kannan, S., Baskar, S., McCalley, J.D., and Murugan, P. (2009). Application of NSGA-II algorithm to generation expansion planning. *IEEE Transactions on Power Systems* 24 (1): 454–461.
49. Elsayed, A.T., Mohamed, A.A., and Mohammed, O.A. (2015). DC microgrids and distribution systems: an overview. *Electric Power Systems Research Journal* 119: 407–417.
50. Baldwin, T.L. and Lewis, S.A. (2004). Distribution load flow methods for shipboard power systems. *IEEE Transactions on Industry Applications* 40 (5): 1183–1190.

51. Jun, K., Lijun, F., Fan, M. et al. (2011). Some subjects of load flow calculation in shipboard integrated power systems. In: *2011 International Conference on Electrical Machines and Systems (ICEMS) (20–23 August 2011)*, 1–5. IEEE.
52. Medina, M.M., Qi, L., and Butler-Purry, K.L. (2003). A three phase load flow algorithm for shipboard power systems (SPS). In: *2003 IEEE PES Transmission and Distribution Conference and Exposition (7–12 September 2003)*, vol. 1, 227–233. IEEE.
53. Huang, J., Zhang, X., and Ye, Z. (2009). A combined load flow method for large shipboard power systems. In: *2009 Asia-Pacific Power and Energy Engineering Conference (APPEEC) (27–31 March 2009)*, 1–6. IEEE.
54. Mostafa, H.A., El-Shatshat, R., and Salama, M.M.A. (2013). Multi-objective optimization for the operation of an electric distribution system with a large number of single phase solar generators. *IEEE Transactions on Smart Grid* 4 (2): 1038–1047.
55. Cherkaoui, R. and Paolone, M. (2014). Optimal allocation of dispersed energy storage systems in active distribution networks for energy balance and grid support. *IEEE Transactions on Power Systems* 29 (5): 2300–2310.
56. Beerten, J., Cole, S., and Belmans, R. (2012). Generalized steady-state VSC MTDC model for sequential AC/DC power flow algorithms. *IEEE Transactions on Power Systems* 27 (2): 821–829.
57. Baradar, M. and Ghandhari, M. (2013). A multi-option unified power flow approach for hybrid AC/DC grids incorporating multi-terminal VSC-HVDC. *IEEE Transactions on Power Systems* 28 (3): 2376–2383.
58. Narayanan, K.N. and Mitra, P. (2013). A comparative study of a sequential and simultaneous AC-DC power flow algorithms for a multi-terminal VSC-HVDC system. In: *2013 IEEE Innovative Smart Grid Technologies – Asia (ISGT Asia)*, 1–6. IEEE.
59. Chaudhuri, N.R., Chaudhuri, B., Majumder, R., and Yazdani, A. (2014). *Multi-Terminal Direct-Current Grids; Modeling, Analysis, and Control.* Hoboken, NJ: Wiley.
60. Yu, H., Liu, H., and Liu, J. (2011). Service restoration reconfiguration in distribution networks based on a multi-objective optimization algorithm. In: *2011 International Conference on Electrical and Control Engineering (ICECE) (16–18 September 2011)*, 3624–3627. IEEE.
61. Akorede, M.F., Hizam, H., Aris, I., and Ab Kadir, M.Z.A. (2011). Effective method for optimal allocation of distributed generation units in meshed electric power systems. *IET Generation, Transmission & Distribution* 5 (2): 276–287.
62. Inglis, S., Ault, G.W., Alarcon-Rodriguez, A., and Galloway, S.J. (2010). Multi-objective network planning tool for networks containing high penetrations of DER. In: *2010 45th International Universities Power Engineering Conference (UPEC), (31 August–3 September 2010)*, 1–6. IEEE.
63. Chen, C., Duan, S., Cai, T. et al. (2011). Optimal allocation and economic analysis of energy storage system in microgrids. *IEEE Transactions on Power Electronics* 26 (10): 2762–2773.
64. Butler, K.L., Sarma, N.D.R., and Ragendra Prasad, V. (2001). Network reconfiguration for service restoration in shipboard power distribution systems. *IEEE Transactions on Power Systems* 16 (4): 653–661.
65. Awad, A.S.A., EL-Fouly, T.H.M., and Salama, M.M.A. (2014). Optimal ESS allocation and load shedding for improving distribution system reliability. *IEEE Transactions on Smart Grid* 5 (5): 2339–2349.

66. Awad, A.S.A., EL-Fouly, T.H.M., and Salama, M.M.A. (2015). Optimal ESS allocation for load management application. *IEEE Transactions on Power Systems* 30 (1): 327–336.
67. Cao, J., Du, W., Wang, H.F., and Bu, S.Q. (2013). Minimization of transmission loss in meshed AC/DC grids with VSC-MTDC networks. *IEEE Transactions on Power Systems* 28 (3): 3047–3055.
68. Bahrman, M.P. and Johnson, B.K. (2007). The ABCs of HVDC transmission technologies. *IEEE Power and Energy Magazine* 5 (2): 32–44.
69. Cole, S., Beerten, J., and Belmans, R. (2010). Generalized dynamic VSC MTDC model for power system stability studies. *IEEE Transactions on Power Systems* 25 (3): 1655–1662.
70. Guan, M. and Xu, Z. (2012). Modeling and control of a modular multilevel converter-based HVDC system under unbalanced grid conditions. *IEEE Transactions on Power Electronics* 27 (12): 4858–4867.
71. Atwa, Y.M., El-Saadany, E.F., Salama, M.M.A., and Seethapathy, R. (2010). Optimal renewable resources mix for distribution system energy loss minimization. *IEEE Transactions on Power Systems* 25 (1): 360–370.
72. Levron, Y., Guerrero, J.M., and Beck, Y. (2013). Optimal power flow in microgrids with energy storage. *IEEE Transactions on Power Systems* 28 (3): 3226–3234.
73. Honda, T., Cucci, F., and Yang, M.C. (2009). Achieving pareto optimality in decentralized design environment. *2009 International Conference on Engineering Design (ICED'09)*, Palo Alto, CA (24–27 August 2009).
74. Mostafa, H.A., El Shatshat, R., and Salama, M.M.A. (2014). Optimal distribution systems operation using smart matching scheme (SMS) for smart grid applications. *IEEE Transactions on Smart Grid* 5 (4): 1938–1948.
75. Rudraraju, S.R., Srivastava, A.K., Srivastava, S.C., and Schulz, N.N. (2009). Small signal stability analysis of a shipboard MVDC power system. In: *Electric Ship Technologies Symposium, 2009 (ESTS 2009) (20–22 April 2009)*, 135–141. IEEE.
76. Scalley, B. and Kasten, D. (1981). The effects of distribution voltage reduction on power and energy consumption. *IEEE Transactions on Education* 24 (3): 210–216.
77. Diaz-Aguilo, M., Sandraz, J., Macwan, R. et al. (2013). Field-validated load model for the analysis of CVR in distribution secondary networks: energy conservation. *IEEE Transactions on Power Delivery* 28 (4): 2428–2436.
78. Pinney, D. (2014). *Costs and Benefits of Conservation Voltage Reduction*, DE-OE0000222. Arlington, VA: NRECA.
79. American National Standard Institute (2011). *American National Standard for Electric Power Systems and Equipment-Voltage Ratings (60 Hertz)*, ANSI C84.1-2011. ANSI, Inc.
80. Wang, Z. and Wang, J. (2014). Review on implementation and assessment of conservation voltage reduction. *IEEE Transactions on Power Systems* 29 (3): 1306–1315.
81. Lee, K.Y. and El-Sharkawi, M.A. (eds.) (2002). *Modern Heuristic Optimization Techniques with Applications to Power Systems*. IEEE Power Engineering Society (02TP160).
82. Yoshida, H., Kawata, K., Fukuyama, Y. et al. (2000). A particle swarm optimization for reactive power and voltage control considering voltage security assessment. *IEEE Transactions on Power Systems* 15 (4): 1232–1239.
83. Naka, S., Genji, T., Yura, T., and Fukuyama, Y. (2003). A hybrid particle swarm optimization for distribution state estimation. *IEEE Transactions on Power Systems* 18 (1): 60–68.

84. Saaty, T.L. (1980). *The Analytic Hierarchy Process*. New York: McGraw Hill.
85. Zhu, J.Z. (2009). *Optimization of Power System Operation*. Hoboken, NJ: Wiley.
86. Grond, M.O.W., Luong, N.H., Morren, J. et al. (2014). Practice-oriented optimization of distribution network planning using metaheuristic algorithms. In: *Proceedings of the Power Systems Computation Conference (PSCC 2014)*, 1–8. IEEE.
87. Lavorato, M., Rider, M.J., Garcia, A.V., and Romero, R. (2010). A constructive heuristic algorithm for distribution system planning. *IEEE Transactions on Power Systems* 25: 1734–1742.
88. Ganguly, S., Sahoo, N., and Das, D. (2013). Multi-objective planning of electrical distribution systems using dynamic programming. *International Journal of Electrical Power & Energy Systems* 46: 65–78.
89. Finn, P., O'Connell, M., and Fitzpatrick, C. (2013). Demand side management of a domestic dishwasher: wind energy gains, financial savings and peak-time load reduction. *Applied Energy* 101: 678–685.
90. Gelazanskas, L. and Gamage, K.A. (2014). Demand side management in smart grid: a review and proposals for future direction. *Sustainable Cities and Society* 11: 22–30.
91. Zhang, N., Ochoa, L.F., and Kirschen, D.S. (2011). Investigating the impact of demand side management on residential customers. In: *2011 2nd IEEE PES International Conference and Exhibition on Innovative Smart Grid Technologies*, 1–6. IEEE.
92. Momoh, J. (2012). *Smart Grid: Fundamentals of Design and Analysis*. Wiley-IEEE Press.
93. Ramchurn, S.D., Vytelingum, P., Rogers, A., and Jennings, N.R. (2011). Agent-based control for decentralised demand side management in the smart grid. *10th International Conference on Autonomous Agents and Multiagent Systems (AAMAS 2011)*, Taipei, Taiwan (2-6 May 2011), vol. 1–3, pp. 5–12.
94. Slootweg, J.G. and Van Oirsouw, P. (2005). Incorporating reliability calculations in routine network planning: theory and practice. In: *18th International Conference and Exhibition on Electricity Distribution, CIRED 2005*, 1–5. IET.
95. Deb, K. and Kalyanmoy, D. (2001). *Multi-Objective Optimization Using Evolutionary Algorithms*. New York: Wiley.
96. Alarcon-Rodriguez, A., Ault, G., and Galloway, S. (2010). Multi-objective planning of distributed energy resources: a review of the state-of-the-art. *Renewable and Sustainable Energy Reviews* 14: 1353–1366.
97. Das, I. and Dennis, J.E. (1997). A closer look at drawbacks of minimizing weighted sums of objectives for Pareto set generation in multicriteria optimization problems. *Structural Optimization* 14: 63–69.
98. Celli, G., Ghiani, E., Mocci, S., and Pilo, F. (2005). A multiobjective evolutionary algorithm for the sizing and siting of distributed generation. *IEEE Transactions on Power Systems* 20: 750–757.
99. Ochoa, L., Padilha-Feltrin, A., and Harrison, G. (2006). Evaluating distributed generation impacts with a multiobjective index. *IEEE Transactions on Power Delivery* 21: 1452–1458.
100. Zitzler, E., Laumanns, M., and Thiele, L. (2001). SPEA2: Improving the Strength Pareto Evolutionary Algorithm. Technical report, Swiss Federal Institute of Technology.
101. Ahmadi, M.H., Hosseinzade, H., Sayyaadi, H. et al. (2013). Application of the multi-objective optimization method for designing a powered Stirling heat engine: design with maximized power, thermal efficiency and minimized pressure loss. *Renewable Energy* 60: 313–322.

102. Sousa, A.S. and Asada, E.N. (2015). Long-term transmission system expansion planning with multi-objective evolutionary algorithm. *Electric Power Systems Research* 119: 149–156.
103. Ramesh, S., Kannan, S., and Baskar, S. (2012). Application of modified NSGA-II algorithm to multi-objective reactive power planning. *Applied Soft Computing* 12: 741–753.
104. Murugan, P., Kannan, S., and Baskar, S. (2009). NSGA-II algorithm for multi-objective generation expansion planning problem. *Electric Power Systems Research* 79: 622–628.
105. Pelikan, M., Sastry, K., and Goldberg, D.E. (2005). Multiobjective hBOA, clustering, and scalability. In: *Proceedings of the 2005 Conference on Genetic and Evolutionary Computation – GECCO '05*, 663. New York, NY: ACM Press.
106. Bosman, P.A.N. and Thierens, D. (2003). The balance between proximity and diversity in multiobjective evolutionary algorithms. *IEEE Transactions on Evolutionary Computation* 7: 174–188.
107. Bosman, P.A.N. (2010). The anticipated mean shift and cluster registration in mixture-based EDAs for multi-objective optimization. In: *Proceedings of the 12th Annual Conference on Genetic and Evolutionary Computation – GECCO '10*, 351–358. New York, NY: ACM Press.
108. Knowles, J. and Corne, D. (2003). Properties of an adaptive archiving algorithm for storing nondominated vectors. *IEEE Transactions on Evolutionary Computation* 7: 100–116.
109. Luong, N.H., La Poutré, H., and Bosman, P.A.N. (2014). Multi-objective gene-pool optimal mixing evolutionary algorithms. In: *Proceedings of the 2014 Conference on Genetic and Evolutionary Computation – GECCO '14*, 357–364. New York, NY: ACM Press.
110. Luong, N.H., Grond, M.O.W., La Poutré, H., and Bosman, P.A.N. (2015). Scalable and practical multi-objective distribution network expansion planning. In: *2015 Power & Energy Society General Meeting*. IEEE.
111. Luong, N.H., La Poutré, H., and Bosman, P.A.N. (2015). Exploiting linkage information and problem-specific knowledge in evolutionary distribution network expansion planning. In: *The Genetic and Evolutionary Computation Conference (GECCO 2015)*, 1231–1238. New York: ACM.
112. da Rocha, M.C. and Saraiva, J.A.T. (2013). A discrete evolutionary PSO based approach to the multiyear transmission expansion planning problem considering demand uncertainties. *International Journal of Electrical Power & Energy Systems* 45: 427–442.
113. Nara, K., Satoh, T., Aoki, K., and Kitagawa, M. (1991). Multi-year expansion planning for distribution systems. *IEEE Transactions on Power Systems* 6: 952–958.
114. Darby, S.J. and McKenna, E. (2012). Social implications of residential demand response in cool temperate climates. *Energy Policy* 49: 759–769.
115. Veldman, E., Gibescu, M., Slootweg, J.G., and Kling, W.L. (2013). Scenario-based modelling of future residential electricity demands and assessing their impact on distribution grids. *Energy Policy* 56: 233–247.
116. Grainger, J.J. and Stevenson, W.D. (1994). *Power System Analysis*. McGraw-Hill.
117. Van Hertem, D., Verboomen, J., Purchala, K. et al. (2006). Usefulness of DC power flow for active power flow analysis with flow controlling devices. In: *The 8th IEE International Conference on AC and DC Power Transmission (ACDC 2006)*, 58–62. IET.

118. Grond, M.O.W., Pouw, J.I.P., Morren, J., and Slootweg, J.G. (2014). Applicability of line outage distribution factors to evaluate distribution network expansion options. In: *2014 49th International Universities Power Engineering Conference (UPEC)*, 1–6. IEEE.
119. Deb, K. (2000). An efficient constraint handling method for genetic algorithms. *Computer Methods in Applied Mechanics and Engineering* 186: 311–338.
120. Verzijlbergh, R.A., Grond, M.O.W., Lukszo, Z. et al. (2012). Network impacts and cost savings of controlled EV charging. *IEEE Transactions on Smart Grid* 3: 1203–1212.
121. Grond, M.O.W. (2011). Impact of future residential loads on medium voltage networks. Master thesis. Delft University of Technology.
122. de Groot, R., Morren, J., and Slootweg, J.G. (2012). Smart integration of distribution automation applications. In: *2012 3rd IEEE PES Innovative Smart Grid Technologies Europe (ISGT Europe)*, 1–7. IEEE.
123. Thierens, D. and Bosman, P.A.N. (2011). Optimal mixing evolutionary algorithms. In: *Proceedings of the 13th Annual Conference on Genetic and Evolutionary Computation – GECCO '11*, 617. New York, NY: ACM Press.
124. Luong, N.H. and Bosman, P.A.N. (2012). Elitist archiving for multi-objective evolutionary algorithms: to adaptor not to adapt. In: *Parallel Problem Solving from Nature – PPSN XII*, Lecture Notes in Computer Science, vol. 7492 (eds. C.A.C. Coello, V. Cutello, K. Deb, et al.), 72–81. Berlin, Heidelberg: Springer Berlin Heidelberg.
125. Thierens, D. and Goldberg, D.E. (1993). Mixing in Genetic Algorithms. In: *5th International Conference on Genetic Algorithms, ICGA 1993*, 38–47. Morgan Kaufmann.
126. Bosman, P.A.N. and Thierens, D. (2012). On measures to build linkage tree in LTGA. In: *Parallel Problem Solving from Nature – PPSN XII*, Lecture Notes in Computer Science, vol. 7491 (eds. C.A.C. Coello, V. Cutello, K. Deb, et al.), 276–285. Berlin, Heidelberg: Springer Berlin Heidelberg.
127. Gronau, I. and Moran, S. (2007). Optimal implementations of UPGMA and other common clustering algorithms. *Information Processing Letters* 104: 205–210.
128. Nara, K., Satoh, T., Kuwabara, H. et al. (1992). Distribution systems expansion planning by multi-stage branch exchange. *IEEE Transactions on Power Systems* 7: 208–214.
129. Harik, G.R. and Lobo, F.G. (1999). A parameter-less genetic algorithm. In: *Proceedings of the Genetic and Evolutionary Computation Conference 1999 (GECCO '99)*, 258–267. San Francisco, CA: Morgan Kaufmann.
130. Civanlar, S., Grainger, J., Yin, H. et al. (1988). Distribution feeder reconfiguration for loss reduction. *IEEE Transactions on Power Delivery* 3 (3): 1217–1223.
131. Masoum, M.A.S., Ladjevardi, M., Jafarian, A., and Fuchs, E.F. (2004). Optimal placement, replacement and sizing of capacitor banks in distorted distribution networks by genetic algorithms. *IEEE Transactions on Power Delivery* 19: 1794–1801.
132. Haghifam, M.-R. and Malik, O.P. (2007). Genetic algorithm-based approach for fixed and switchable capacitors placement in distribution systems with uncertainty and time varying loads. *IET Generation, Transmission and Distribution* 1: 244–252.
133. Kalantar, M., Dashti, R., and Dashti, R. (2006). Combination of network reconfiguration and capacitor placement for loss reduction in distribution system with based genetic algorithm. In: *Proceedings of the 41st International Universities Power Engineering Conference*, vol. 1, 308–312.

134. Boone, G. and Chiang, H.D. (1993). Optimal capacitor placement in distribution systems by genetic algorithm. *International Journal of Electrical Power & Energy Systems* 15 (3): 155–162.
135. Haibo, Z., Lizi, Z., and Fanling, M. (1998). Reactive power optimization based on genetic algorithm. *International Power Conference on Power System Technology* 2: 1448–1453.
136. Sailaja Kumari, M., Priyanka, G., and Sydulu, M. (2007). Comparison of genetic algorithms and particle swarm optimization for optimal power flow including FACTS devices. In: *Power Tech*, 1105–1110. IEEE.
137. Yu, J.X. and Zhao, B. (2005). Improved particle swam optimization algorithm for optimal power flow problems. *Proceedings of the CSU-EPSA* 17 (4): 83–88.
138. Yang, H.T., Huang, Y.C., and Huang, C.L. (1995). Solution to capacitor placement in a radial distribution system using tabu search method. *Proceedings of International Conference on Energy Management and Power Delivery* 1: 388–393.
139. Masoum, M.A.S., Jafarian, A., Ladjevardi, M. et al. (2004). Fuzzy approach for optimal placement and sizing of capacitor banks in the presence of harmonics. *IEEE Transactions on Power Delivery* 19: 822–829.
140. Medeiros, M.F. (2011). A memetic algorithm based on mixed ant colony optimization and genetic algorithm for optimal capacitor placement. In: *2011 16th International Conference on Intelligent System Application to Power Systems (ISAP)*. IEEE.
141. Haghdar, K. and Shayanfar, H.A. (2010). Optimal placement and sizing of DG and capacitor for the loss reduction via methods of generalized pattern search and genetic algorithm. In: *2010 Asia-Pacific Power and Energy Engineering Conference (APPEEC)*. IEEE.
142. Guimaraes, M.A.N. and Castro, C.A. (2011). An efficient method for distribution systems reconfiguration and capacitor placement using a Chu-Beasley based genetic algorithm. In: *2011 IEEE Trondheim Power Tech*. IEEE.
143. Madeiro, S., Galvão, E., Cavellucci, C. et al. (2011). Simultaneous capacitor placement and reconfiguration for loss reduction in distribution networks by a hybrid genetic algorithm. In: *2011 IEEE Congress on Evolutionary Computation (CEC)*. IEEE.
144. Huang, S.J. and Liu, X.Z. (2012). A plant growth-based optimization approach applied to capacitor placement in power systems. *IEEE Transactions on Power Systems* 27 (4): 2138–2145.
145. Lee, J. and Chiang, H.D. (2002). Theory of stability regions for a class of nonhyperbolic dynamical systems and its application to constraint satisfaction problem. *IEEE Transactions on Circuits and Systems I: Fundamental Theory and Applications* 49 (2): 196–209.
146. Chiang, H.-D. and Lee, J. (2008). A dynamical trajectory-based hybrid method for computing high-quality optimal solutions: method and theory. In: *Modern Heuristic Optimization Techniques: Theory and Applications to Power Systems* (eds. K. Lee and M.A. El-Sharkawi). IEEE Press.
147. Lee, J. and Chiang, H.-D. (2004). A dynamical trajectory-based methodology for systematically computing multiple optimal solutions of general nonlinear programming problems. *IEEE Transactions on Automatic Control (Regular paper)* 49 (6): 888–899.
148. Chiang, H.-D. and Chu, C.-C. (1996). Systematic search method for obtaining multiple local optimal solutions of nonlinear programming problems. *IEEE Transactions on Circuits and Systems I: Fundamental Theory and Applications* 13: 99–109.

149. Wang, B. and Chiang, H.-D. (2011). ELITE: ensemble of optimal, input-pruned neural networks using trust-tech. *IEEE Transactions on Neural Networks* 22 (1): 96–109.
150. Chiang, H.-D., Wang, B., and Jiang, Q.-Y. (2009). Applications of trust-tech methodology in optimal power flow of power systems. In: *Optimal Operations of Energy Systems*, International Series in Operations Research and Management Science. Springer.
151. Chiang, H.-D., Chen, J.-H., and Reddy, C. (2009). Trust-tech-based global optimal methodology for nonlinear programming. In: *Recent Advances in Global Optimization Methodology*, The Fields Institute Communication Series. American Mathematical Society (Invited paper, one book Chapter).
152. Chiang, H.-D. and Fekih-Ahmed, L. (1996). Quasi-stability regions of nonlinear dynamical systems: optimal estimation. *IEEE Transactions on Circuits and Systems I: Fundamental Theory and Applications* 43 (82): 636–642.
153. Miu, K.N., Chiang, H.D., and Darling, G. (1997). Capacitor placement, replacement and control in large-scale distribution systems by a GA-based two-stage algorithm. *IEEE Transactions on Power Systems* 12: 1160–1166.
154. Brown, R. (2008). Impact of smart grid on distribution system design. In: *Power and Energy Society General Meeting – Conversion and Delivery of Electrical Energy in the 21st Century*, 1–4. Pittsburgh: IEEE.
155. Wei, X., Yu-Hui, Z., and Jie-Lin, Z. (2009). Energy-efficient distribution in smart grid. In: *2009 International Conference on Sustainable Power Generation and Supply: SUPERGEN'09*, 1–6. [S.l.]: IEEE.
156. Coulouris, G., Dollimore, J., and Kindberg, T. (2005). *Distributed Systems: Concepts and Design*. [S.l.]: Addison-Wesley Longman.
157. Merlin, A. and Back, G. (1975). Search for minimum-loss operational spanning tree configuration for an urban power distribution system. In: *Proceedings of the Fifth Power System Conference (PSCC)*, Cambridge, 1–18. Department of Electrical & Electronic Engineering, Queen Mary College, University of London.
158. Baran, M.E. and Wu, F.F. (1989). Network reconfiguration in distribution systems for loss reduction and load balancing. *IEEE Transactions on Power Delivery* 4 (2): 1401–1407.
159. Nara, K., Shiose, A., Kitagawa, M., and Ishihara, T. (1992). Implementation of genetic algorithm for distribution systems loss minimim reconfiguration. *IEEE Transactions on Power Systems* 7 (3): 1044–1051.
160. Lin, W.M., Cheng, F.S., and Tsay, M.T. (2000). Distribution feeder reconfiguration with refined genetic algorithm. *IEE Proceedings – Generation, Transmission and Distribution* 147 (6): 349–354.
161. Chiou, J.P., Chang, C.F., and Su, C.T. (2005). Variable scaling hybrid differential evolution for solving network reconfiguration of distribution systems. *IEEE Transactions on Power Systems* 20 (2): 668–674.
162. Hong, Y.Y. and Ho, S.Y. (2005). Determination of network configuration considering multiobjective in distribution systems using genetic algorithms. *IEEE Transactions on Power Systems* 20 (2): 1062–1069.
163. Hsiao, Y.T. (2004). Multiobjective evolution programming method for feeder reconfiguration. *IEEE Transactions on Power Systems* 19 (1): 594–599.
164. Augugliaro, A., Dusonchet, L., Favuzza, S., and Riva, E. (2004). Voltage regulation and power losses minimization in automated distribution networks by an evolutionary multiobjective approach. *IEEE Transactions on Power Systems* 19 (3): 1516–1527.

165. Delbem, A.C.B., Carvalho, A.C.P.L.F., and Bretas, N.G. (2005). Main chain representation for evolutionary algorithms applied to distribution system reconfiguration. *IEEE Transactions on Power Systems* 20 (1): 425–436. https://doi.org/10.1109/TPWRS.2004.840442.
166. Mendoza, J., Lopez, R., Morales, D. et al. (2006). Minimal loss reconfiguration using genetic algorithms with restricted population and addressed operators: real application. *IEEE Transactions on Power Systems* 21 (2): 948–954. https://doi.org/10.1109/TPWRS.2006.873124.
167. Carreño, E.M., Romero, R., and Padilha-Feltrin, A. (2008). An efficient reconfiguration to solve distribution network reconfiguration for loss reduction problem. *IEEE Transactions on Power Systems* 23 (4): 1542–1551.
168. Raju, G.K.V. and Bijwe, R.R. (2008). An efficient algorithm for minimum loss reconfiguration of distribution system based on sensitivity and heuristics. *IEEE Transactions on Power Systems* 23 (3): 1280–1287.
169. Wang, C. and Cheng, H.Z. (2004). Optimization of network reconfiguration in large distribution system using plant growth simulation algorithm. *IEEE Transactions on Power Systems* 23 (1): 119–126.
170. Almeida, E.M. and Asada, E.N. (2015). NSGA-II applied to the multi-objective distribution system expansion planning problem. *18th Intelligent Systems Applications to Power Systems Conference and Debate*, Porto, Portugal (11–17 September 2015).
171. Curcic, S., Ozveren, C.S., Crowe, L., and Lo, P.K.L. (1995). Electric power distribution network restoration: a survey of papers and a review of the restoration problem. *Electric Power Systems Research* 35 (2): 73–86. https://doi.org/10.1016/0378-7796(95)00991-4.
172. Sanches, D.S., London, J.B.A. Jr., and Delbem, A.C.B. (2014). Multi-objective evolutionary algorithm for single and multiple fault service restoration in large-scale distribution systems. *Electric Power Systems Research* 110: 144–153. https://doi.org/10.1016/j.epsr.2014.01.017.
173. Romero, R., Franco, J.F., Leão, F.B. et al. (2015). A new mathematical model for the restoration problem in balanced radial distribution systems. *IEEE Transactions on Power Systems* 31 (2): 1259–1268. https://doi.org/10.1109/TPWRS.2015.2418160.
174. Carrano, E.G., da Silva, G.P., Cardoso, E.P., and Takahashi, R.H.C. (2016). Subpermutation-based evolutionary multiobjective algorithm for load restoration in power distribution networks. *IEEE Transactions on Evolutionary Computation* 20 (4): 546–562. https://doi.org/10.1109/TEVC.2015.2497361.
175. Santos, A.C., Delbem, A.C.B., London, J.B.A. Jr., and Bretas, N.G. (2010). Node-depth encoding and multiobjective evolutionary algorithm applied to large-scale distribution system reconfiguration. *IEEE Transactions on Power Systems* 25 (3): 1254–1265.
176. Gholami, M., Moshtagh, J., and Rashidi, L. (2015). Service restoration for unbalanced distribution networks using a combination two heuristic methods. *International Journal of Electrical Power & Energy Systems* 67: 222–229. https://doi.org/10.1016/j.ijepes.2014.11.024.
177. Solanki, J.M., Khushalani, S., and Schulz, N.N. (2007). A multi-agent solution to distribution systems restoration. *IEEE Transactions on Power Systems* 22 (3): 1026–1034. https://doi.org/10.1109/TPWRS.2007.901280.
178. Zhigang, L. and Yuxiang, D. (2007). Service restoration strategy for the distribution system with DGs. *Automation of Electric Power Systems* 31 (1): 89–92.

179. Kumar, Y., Das, B., and Sharma, J. (2008). Multiobjective, multiconstraint service restoration of electric power distribution system with priority customers. *IEEE Transactions on Power Delivery* 23 (1): 261–270. https://doi.org/10.1109/TPWRD.2007.905412.
180. Jorge, M.B., Hector, V.O., Miguel, L.G., and Hector, P.D. (2014). Multi-fault service restoration in distribution networks considering the operating mode of distributed generation. *Electric Power Systems Research* 116: 67–76. https://doi.org/10.1016/j.epsr.2014.05.013.
181. Zidan, A. and El-Saadany, E.F. (2015). Incorporating customers' reliability requirements and interruption characteristics in service restoration plans for distribution systems. *Energy* 87: 192–200. https://doi.org/10.1016/j.energy.2015.04.067.
182. Chong, Z., Dai, Z., Wang, S. et al. (2014). The application of binary particle swarm optimization in power restoration. In: *Proceedings of the IEEE 10th International Conference on Natural Computation (ICNC)*, 349–353. IEEE https://doi.org/10.1109/ICNC.2014.6975860.
183. Swarnkar, A., Gupta, N., and Niazi, K.R. (2012). Distribution network reconfiguration using population-based AI techniques: a comparative analysis. In: *Proceedings of the IEEE PES General Meeting*, 1–6. IEEE https://doi.org/10.1109/PESGM.2012.6345013.
184. Aoki, K., Nara, K., Itoh, M. et al. (1989). A new algorithm for service restoration in distribution systems. *IEEE Transactions on Power Delivery* 4 (3): 1832–1839. https://doi.org/10.1109/61.32679.
185. Junlakarn, S. and Ilic, M. (2014). Distribution system reliability options and utility liability. *IEEE Transactions on Smart Grid* 5 (5): 2227–2234. https://doi.org/10.1109/TSG.2014.2316021.
186. Prado, R.S., Silva, R.C.P., Neto, O.M. et al. (2014). Differential evolution using ancestor tree for service restoration in power distribution systems. *Applied Soft Computing* 23: 498–508. https://doi.org/10.1016/j.asoc.2014.06.005.
187. Carreno, E.M., Romero, R., and Padilha-Feltrin, A. (2008). An efficient codification to solve distribution network reconfiguration for loss reduction problem. *IEEE Transactions on Power Systems* 23 (4): 1542–1551. https://doi.org/10.1109/TPWRS.2008.2002178.
188. Huang, X., Taylor, G., Irving, M. et al. (2013). Service restoration of distribution systems using enhanced NDE with LCO. In: *Proceedings of the IEEE PES General Meeting*, 1–5. IEEE https://doi.org/10.1109/PESMG.2013.6672400.
189. Delbem, A.C.B., de Carvalho, A.C.P.L.F., Policastro, C.A. et al. (2004). Node-depth encoding for evolutionary algorithms applied to network design. In: *Genetic and Evolutionary Computation (GECCO 2004)*, Lecture Notes in Computer Science, vol. 3101, 678–687. http://link.springer.com/chapter/10.1007%2F978-3-540-24854-5_70 (accessed 12 October 2019).
190. Shirmohammadi, D., Hong, H.W., Semlyen, A., and Luo, G.X.O. (May 1988). A compensation-based power flow method for weakly meshed distribution and transmission networks. *IEEE Transactions on Power Systems* 3 (2): 753–762. https://doi.org/10.1109/59.192932.
191. Santos, A.C., Delbem, A.C.B., and Bretas, N.G. (2008). Energy restoration for large-scale distribution system using EA and a new data structure. In: *2008 IEEE Power and Energy Society General Meeting – Conversion and Delivery of Electrical Energy in the 21st Century*, 1–8. IEEE https://doi.org/10.1109/PES.2008.4596388.
192. Mansour, M.R., Santos, A.C., London, J.B.A. Jr. et al. (2009). Energy restoration in distribution systems using multi-objective evolutionary algorithm and an efficient data structure. In: *2009 IEEE PowerTech Conference*, Bucharest. IEEE.

193. Camillo, M.H.M., Romero, M.E.V., Fanucchi, R.Z. et al. (2014). Validation of a methodology for service restoration on a real Brazilian distribution system. In: *Proceedings of the IEEE PES Transmission & Distribution Conferences and Exposition – Latin America*, 1–6. IEEE https://doi.org/10.1109/TDC-LA.2014.6955192.
194. Sanches, D.S., London, J.B.A. Jr., Delbem, A.C.B. et al. (2014). Multiobjective evolutionary algorithm with a discrete differential mutation operator developed for service restoration in distribution systems. *International Journal of Electrical Power & Energy Systems* 62: 700–711. https://doi.org/10.1016/j.ijepes.2014.05.008.
195. Das, S., Abraham, A., Chakraborty, U.K., and Konar, A. (2009). Differential evolution using neighborhood-based mutation operator. *IEEE Transactions on Evolutionary Computation* 13 (3): 526–553. https://doi.org/10.1109/TEVC.2008.2009457.
196. Wooldridge, M. (2009). *An Introduction to Multiagent Systems*. Chichester, UK: Wiley.
197. Rumley, S., Kagi, E., Rudnick, H., and Germond, A. (2008). Multi-agent approach to electrical distribution networks control. In: *32nd Annual IEEE International Computer Software and Applications Conference, COMPSAC'08*. IEEE.
198. Pipattanasomporn, M., Feroze, H., and Rahman, S. (2009). Multi-agent systems in a distributed smart grid: design and implementation. *Power Systems Conference and Exposition, PSCE'09*, Phoenix, AZ, USA (18 March 2009).
199. Veit, A., Xu, Y., Zheng, R. et al. (2013). Multiagent coordination for energy consumption scheduling in consumer cooperatives. *Proceedings of 27th AAAI Conference on Artificial Intelligence*, Bellevue, WA, USA (14–18 July 2013).
200. Issicaba, D. (2013). Block-oriented agent-based architecture to support the power distribution system operation. PhD thesis. Universidade do Porto.
201. Ke, Y.-L. (2004). Distribution feeder reconfiguration for load balancing and service restoration by using g-nets inference mechanism. *IEEE Transactions on Power Delivery* 19 (3): 1426–1433.
202. Botea, A., Rintanen, J., and Banerjee, D. (2012). Optimal reconfiguration for supply restoration with informed a∗ search. *IEEE Transactions on Smart Grid* 3 (2): 583–593.
203. Kayal, P., Chanda, S., and Chanda, C. (2011). An ann based network reconfiguration approach for voltage stability improvement of distribution network. In: *2011 International Conference on Power and Energy Systems*. IEEE.
204. Ravibabu, P., Ramya, M., Sandeep, R. et al. (2010). Implementation of improved genetic algorithm in distribution system with feeder recon guration to minimize real power losses. In: *2010 International Conference on Computer Engineering and Technology (ICCET)*. IEEE.
205. Ahuja, A. and Pahwa, A. (2005). Using ant colony optimization for loss minimization in distribution networks. In: *Proceedings of the 37th Annual North American Power Symposium*, 470–474. IEEE.
206. Saraiva, F.O. and Asada, E.N. (2013). Decision making in intelligent electrical systems using distributed artificial intelligence and heuristic methods. In: *2013 IEEE PES Conference on Innovative Smart Grid Technologies Latin America (ISGT LA)*, 1–7. IEEE.
207. Saraiva, F.O. and Asada, E.N. (2014). Multi-agent systems applied to topological reconfiguration of smart power distribution systems. In: *2014 International Joint Conference on Neural Networks (IJCNN)*, 2812–2819. IEEE.
208. Russel, S. and Norvig, P. (2009). *Artificial Intelligence: A Modern Approach*, 3e. Pearson.

209. Weiss, G. (1999). *Multiagent Systems: A Modern Approach to Distributed Artificial Intelligence*. [S.l.]: The MIT Press.
210. Kersting, W. (2006). *Distribution System Modeling and Analysis*. [S.l.]: CRC.
211. Dijkstra, E.W. (1959). A note on two problems in connection with graphs. *Numerical Mathematics* 1: 269–271.
212. Bellifemine, F.L., Caire, G., and Greenwood, D. (2007). *Developing Multi-Agent Systems with JADE*. Wiley.
213. Adibi, M.M., Clelland, P., Fink, L. et al. (1987). Power system restoration – a task force report. *IEEE Transactions on Power Systems* 2 (2): 271–277.
214. Adibi, M.M., Borkoski, J.N., and Kafaka, R.J. (1987). Power system restoration – the second task force report. *IEEE Transactions on Power Systems* 2 (4): 927–932.
215. Fink, L.H., Liou, K.L., and Liu, C.C. (1995). From generic restoration actions to specific restoration strategies. *IEEE Transactions on Power Systems* 10 (2): 745–751.
216. Fountas, N.A., Hatziargyriou, N.D., Orfanogiannis, C., and Tasoulis, A. (1997). Interactive long-term simulation for power system restoration planning. *IEEE Transactions on Power Systems* 12 (1): 61–68.
217. Sun, W., Liu, C.C., and Zhang, L. (2011). Optimal generator start-up strategy for bulk power system restoration. *IEEE Transactions on Power Systems* 26 (3): 1357–1366.
218. Sun, W., Liu, C.C., and Chu, R.F. (2009). Optimal generator start-up strategy for power system restoration. In: *2009 15th International Conference on Intelligent System Applications to Power Systems*, Brazil. IEEE.
219. Liu, C.C., Liou, K.L., Chu, R.F., and Holen, A.T. (1993). Generation capability dispatch for bulk power system restoration: a knowledge-based approach. *IEEE Transactions on Power Systems* 8 (1): 316–325.
220. Sun, W., Liu, C.C., and Liu, S. (2011). Black start capability assessment in power system restoration. In: *2011 IEEE Power and Energy Society General Meeting*, 1–7. IEEE.
221. Liu, C.C., Vittal, V., Heydt, G.T., and Tomsovic, K. *Development and Evaluation of System Restoration Strategies from a Blackout*, PSERC Publication 09-08. Tempe, AZ: Power Systems Engineering Research Center, Arizona State University.
222. Sun, W. and Liu, C.C. (2013). Optimal transmission path search in power system restoration. In: *2013 IREP Symposium Bulk Power System Dynamics and Control – IX Optimization, Security and Control of the Emerging Power Grid (IREP)*, 1–5. IEEE.
223. Pérez-Guerrero, R.E. and Heydt, G.T. (2008). Distribution system restoration via subgradient based lagrangian relaxation. *IEEE Transactions on Power Systems* 23 (3): 1162–1169.
224. Pérez-Guerrero, R.E., Heydt, G.T., Jack, N.J. et al. (2008). Optimal restoration of distribution systems using dynamic programming. *IEEE Transactions on Power Delivery* 23 (3): 1589–1596.
225. Adibi, M.M., Borkoski, J.N., Kafka, R.J., and Volkmann, T.L. (1999). Frequency response of prime movers during restoration. *IEEE Transactions on Power Systems* 14 (2): 751–756.
226. PJM (2019). PJM Manual 36: System Restoration. http://www.pjm.com/~/media/documents/manuals/m36.ashx (9 September 2019).
227. Hsiao, Y.-T. and Chien, C.-Y. (Nov. 2000). Enhancement of restoration service in distribution systems using a combination fuzzy-GA method. *IEEE Transactions on Power Systems* 15 (4): 1394–1400.

228. Fukuyama, Y. and Chiang, H.-D. (1995). A parallel genetic algorithm for service restoration in electric power distribution systems. In: *Proceedings of 1995 IEEE International Conference on Fuzzy Systems*, vol. 1, 275–282. IEEE.
229. Luan, W.P., Irving, M.R., and Daniel, J.S. (2002). Genetic algorithm for supply restoration and optimal load shedding in power system distribution networks. *IEE Proceedings – Generation, Transmission and Distribution* 149 (2): 145–151.
230. Toune, S., Fudo, H., Genji, T. et al. (2002). Comparative study of modern heuristic algorithms to service restoration in distribution systems. *IEEE Transactions on Power Delivery* 17 (1): 173–181.
231. Watanabe, I. and Nodu, M. (2004). A genetic algorithm for optimizing switching sequence of service restoration in distribution systems. In: *Proceedings of the 2004 Congress on Evolutionary Computation*, vol. 2, 1683–1690. IEEE.
232. Watanabe, I., Kurihara, I., and Nakachi, Y. A hybrid genetic algorithm for service restoration problems in power distribution systems. In: *2006 IEEE International Conference on Evolutionary Computation*, 3250–3257. IEEE.
233. Akduman, B., Türkay, B., Uyar, A.S. et al. (2010). Service restoration in distribution systems using an evolutionary algorithm. In: *7th Mediterranean Conference and Exhibition on Power Generation, Transmission, Distribution and Energy Conversion*. IEEE.
234. Csatar, J., Dán, A., Vokony, I. et al. (2012). Evolutionary algorithm application to medium voltage power system restoration. ICREPQ, Santiago de Compostela, Spain (28–30 March 2012).
235. Hsu, F.-Y. and Tsai, M.-S. (2006). A Multi-Objective Evolution Programming Method for Feeder Reconfiguration of Power Distribution System. In: *ISAP*. IEEE.
236. Lambert-Torres, G., Martins, H.G., Pereira Coutinho, M. et al. (2008). Genetic algorithm to system restoration. *The Online Journal on Electronics and Electrical Engineering OJEEE* 1 (1): 9–14.
237. Kennedy, J. and Eberhart, R. (2001). *Swarm Intelligence*. Morgan Kaufmann Publishers.
238. Golshani, A., Alizadeh Bidgoli, M., and Bathaee, S.M.T. (2013). Design of optimized sliding mode control to improve the dynamic behavior of PMSG wind turbine with NPC back-to-back converter. *International Review of Electrical Engineering (IREE)* 8 (4): 1170–1180.
239. Singh, A. and Swarnkar, K.K. (2011). Power system restoration using particle swarm optimization. *International Journal of Computers and Applications* 30 (2): 25–32.
240. Lambert-Torres, G., Martins, H.G., Coutinho, M.P. et al. (2009). Comparison between PSO and GA in system restoration solution. In: *2009 15th International Conference on Intelligent System Applications to Power Systems*, 1–6. IEEE.
241. Sheng, S.-Q., Cao, Y., and Yu, Y. (2009). Distribution network reconfiguration based on particle swarm optimization and chaos searching. In: *2009 Asia-Pacific Power and Energy Engineering Conference*, 1–4. IEEE.
242. Dong, S.-f., Dong, Z.-c., Jun-jian, M.A., and Chen, K.-n. (2010). Improved PSO algorithm based on chaos theory and its application to design flood hydrograph. *Water Science and Engineering* 3 (2): 156–165.
243. El-Zonkoly, A.M. (2012). Distributed generation approach for single step system restoration during cold load pickup. In: *IEEE PES Transmission and Distribution Conference and Exposition (T&D)*, Orlando. IEEE.
244. Liu, Y. and Xueping, G. (2007). Skeleton-network reconfiguration based on topological characteristics of scale-free networks and discrete particle swarm optimization. *IEEE Transactions on Power Systems* 22 (3): 1267–1274.

245. Mitra, P. and Venayagamoorthy, G.K. (2009). Implementation of an intelligent reconfiguration algorithm for an electric ship power system. In: *2009 Industry Applications Society Annual Meeting*, 1–8. IEEE.
246. Watanabe, I. (2005). An ACO algorithm for service restoration in power distribution systems. *2005 IEEE Congress on Evolutionary Computation* 3: 2864–2871.
247. Li, L., Liao, Z., Huang, S., and Wang, G. (2005). A distributed model for power system restoration based on ant colony optimization algorithm. In: *2005 IEEE/PES Transmission and Distribution Conference and Exhibition: Asia and Pacific*, 1–5. IEEE.
248. da Silva, L.E., Lambert-Torres, G., Martins, H.G. et al. (2010). An application of ACO in system reconfiguration. In: *2010 IEEE PES Transmission and Distribution Conference and Exposition*, 1–6. IEEE.
249. Ahuja, A., Das, S., and Pahwa, A. (2007). An AIS-ACO hybrid approach for multi-objective distribution system reconfiguration. *IEEE Transactions on Power Systems* 22: 1101–1111.
250. Pereira, F.S., Vittori, K., and Da Costa, G.R.M. (2006). Distribution system reconfiguration for loss reduction based on ant colony behavior. In: *2006 IEEE/PESTransmission & Distribution Conference and Exposition: Latin America*, 1–5. IEEE.
251. Yi-Ting, C., Chih-Wen, L., Yi-Jen, W. et al. (2013). Development of a black start decision supporting system for isolated power systems. *IEEE Transactions on Power Systems* 28: 2202–2210.
252. Abd El-Hamed, M.Z., El-Khattam, W., and El-Sharkawy, R. (2013). Self-healing restoration of a distribution system using hybrid Fuzzy Control/Ant-Colony Optimization Algorithm. In: *2013 3rd International Conference on Electric Power and Energy Conversion Systems*, 1–6. IEEE.
253. Mohanty, I., Kalita, J., Das, S. et al. (2003). Ant algorithms for the optimal restoration of distribution feeders during cold load pickup. In: *Proceedings of the 2003 IEEE Swarm Intelligence Symposium. SIS '03*, 132–137. IEEE.
254. Mori, H. and Furuta, A. (2005). A new approach of hierarchical optimization to distribution system service restoration. In: *2005 IEEE International Symposium on Circuits and Systems*, vol. 5, 4751–4754. IEEE.
255. Mori, H. and Muroi, T. (2011). Application of probabilistic Tabu Search to distribution system service restoration. In: *2011 IEEE International Symposium on Circuits and Systems (ISCAS)*, 1037–1040. IEEE.
256. Mori, H. and Ogita, Y. (2002). A parallel tabu search based approach to optimal network reconfigurations for service restoration in distribution systems. *Proceedings of the 2002 International Conference on Control Applications* 2: 814–819.
257. Fukuyama, Y. (2000). Reactive tabu search for distribution load transfer operation. In: *Power Engineering Society Winter Meeting, 2000*, vol. 2, 1301–1306. IEEE.
258. Ting-Chia, O., Whei-Min, L., Cong-Hui, H., and Fu-Sheng, C. (2009). A hybrid programming for distribution reconfiguration of dc microgrid. In: *2009 IEEE PES/IAS Conference on Sustainable Alternative Energy (SAE)*, 1–7. IEEE.
259. Xiangzhen, H., Zhiwei, L., Wenxin, G. et al. (2008). A multi-objective model for transmission network restoration based on multi-agent and Tabu Search. In: *2008 Third International Conference on Electric Utility Deregulation and Restructuring and Power Technologies*, 2392–2397.
260. Ahmed, W. and Sheta, A.F. (2008). Optimization of electric power distribution using hybrid simulated annealing approach. *American Journal of Applied Sciences* 5 (5): 559–564.

261. Gandomkar, M. and Toulabi, H.B. (2010). Investigation of simulated annealing, ant-colony and genetic algorithms for distribution network expansion planning with distributed generation. In: *Proceedings of the 9th WSEAS International Conference on Instrumentation, Measurement, Circuits and Systems*. Stevens Point, WI: World Scientific and Engineering Academy and Society (WSEAS).
262. Chiang, H.-D. et al. (1990). Optimal network reconfigurations in distribution systems: part 2: solution algorithms and numerical results. *IEEE Transactions on Power Delivery* 5 (3): 1568–1574.
263. Matos, M.A. and Melo, P. (1999). Multi-objective reconfiguration for loss reduction and service restoration using simulated annealing. In: *Proceedings of IEEE Power Tech'99, Budapest, Hungary (29 August–2 September 1999)*. IEEE.
264. Zadeh, L.A. (1972). A rationale for fuzzy control. *Journal of Dynamic Systems, Measurement, and Control* 94 (series G): 3–4.
265. Zadeh, L.A. (1973). Outline of a new approach to the analysis of complex systems and decision processes. *IEEE Transactions on Systems, Man, and Cybernetics* SMC-3 (1): 28–44.
266. Momoh, J., Ma, X., and Tomsovic, K. (1995). Overview and literature survey of fuzzy set theory in power systems. *IEEE Transactions on Power Systems* 10 (3): 1676–1690.
267. Hsu, Y.-Y. and Kuo, H.-C. (1994). A Heuristic based fuzzy reasoning approach for distribution system service restoration. *IEEE Transactions on Power Delivery* 9 (2): 948–953.
268. Lee, S.-J., Lim, S.-I., and Ahn, B.-S. (1998). Service restoration of primary distribution systems based on fuzzy evaluation of multi-criteria. *IEEE Transactions on Power Systems* 13 (3): 1156–1163.
269. Chen, W.-H., Tsai, M.-S., and Kuo, H.-L. (2005). Distribution system restoration using the hybrid fuzzy-grey method. *IEEE Transactions on Power Systems* 20 (1): 199–205.
270. Deng, J.L. (1982). Control problems of grey systems. *Systems and Control Letters* 1: 288–294.
271. Chen, W.-H. (2010). Quantitative decision-making model for distribution system restoration. *IEEE Transactions on Power Systems* 25 (1): 313–321.
272. Lim, S.-I., Lee, S.-J., Choi, M.-S. et al. (2006). Restoration index in distribution systems and its application to system operation. *IEEE Transactions on Power Systems* 21 (4): 1966–1971.
273. Zhou, Q., Shirmohammadi, D., and Liu, W.H.E. (1997). Distribution feeder reconfiguration for service restoration. *IEEE Transactions on Power Systems* 12 (2): 724–729.
274. van Den Bergh, F. and Engelbrecht, A.P. (2006). A study of particle swarm optimization particle trajectories. *Information Sciences* 176 (8): 937–971.
275. Liang, J.J., Qin, A.K., Suganthan, P.N., and Baskar, S. (2006). Comprehensive learning particle swarm optimizer for global optimization of multimodal functions. *IEEE Transactions on Evolutionary Computation* 10 (3): 281–295.
276. Bergh, F.V.D. and Engelbrecht, A.P. (2004). A cooperative approach to particle swarm optimization. *IEEE Transactions on Evolutionary Computation* 81 (3): 225–239.
277. Corne, D., Dorigo, M., and Glover, F. (1999, ch. 25). *New Ideas in Optimization (Advanced Topics in Computer Science)*, 379–387. New York: McGraw-Hill.
278. Clerc, M. and Kennedy, J. (2002). The particle swarm: explosion, stability, and convergence in a multi-dimensional complex space. *IEEE Transactions on Evolutionary Computation* 6: 58–73.

279. Kennedy, J. (2000). Stereotyping: improving particle swarm performance with cluster analysis. In: *Proceedings of the 2000 Congress on Evolutionary Computing*, 1507–1512. IEEE.
280. Sun, J. and Xu, W.B. (2004). A global search strategy of quantum-behaved particle swarm optimization. *Proceedings of IEEE International Conference on Cybernatics and Intelligent Systems* 1: 111–116.
281. Kennedy, J. and Mendes, R. (2002). Population structure and particle swarm performance. In: *Proceedings of the 2002 World Congress on Computation. CEC'02 (Cat. No.02TH8600)*, 1671–1676. Honolulu, HI: IEEE.
282. Coelho, L.S. and Mariani, V.C. (2009). A novel chaotic particle swarm optimization approach using Hénon map and implicit filtering local search for economic load dispatch. *Chaos, Solitons and Fractals* 39 (2): 510–518.
283. Li, S., Tan, M., Tsang, I.W., and Kwok, J.T.-Y. (2011). A hybrid PSO-BFGS strategy for global optimization of multimodal functions. *IEEE Transactions on Systems, Man, and Cybernetics. Part B, Cybernetics* 41 (4): 1003–1014.
284. Gao, H. and Xu, W. (2011). A new particle swarm algorithm and its globally convergent modifications. *IEEE Transactions on Systems, Man, and Cybernetics. Part B, Cybernetics* 41 (5): 1334–1351.
285. Liu, B., Wang, L., and Jin, Y.H. (2007). An effective PSO-based memetic algorithm for flow shop scheduling. *IEEE Transactions on Systems, Man, and Cybernetics. Part B, Cybernetics* 37 (1): 18–27.
286. Hsieh, S.T., Sun, T.Y., Liu, C.C., and Tsai, S.J. (2009). Efficient population utilization strategy for particle swarm optimizer. *IEEE Transactions on Systems, Man, and Cybernetics. Part B, Cybernetics* 39 (2): 444–456.
287. Moradzadeh, B. and Tomsovic, K. (2011). Optimal distribution system reconfiguration and restoration without heuristic. *17th PSCC*, Stockholm, Sweden (22–26 August 2011).
288. Liu, C.C., Lee, S.J., and Venkata, S.S. (1988). An expert system operational aid for restoration and loss reduction of distribution system. *IEEE Transactions on Power Systems* 3: 619–626.
289. Tsai, M.-S. (2008). Development of an object-oriented service restoration expert system with load variations. *IEEE Transactions on Power Systems* 23: 219–225.
290. Gomes, F.V., Carneiro, S. Jr., Pereira, J.L.R. et al. (2005). A new heuristic reconfiguration algorithm for large distribution systems. *IEEE Transactions on Power Systems* 20: 1373–1378.
291. Gomes, F.V., Carneiro, S. Jr., Pereira, J.L.R. et al. (2006). A new distribution system reconfiguration approach using optimum power flow and sensitivity analysis for loss reduction. *IEEE Transactions on Power Systems* 21: 1616–1623.
292. Das, D. (2006). A fuzzy multiobjective approach for network reconfiguration of distribution systems. *IEEE Transactions on Power Delivery* 21: 202–209.
293. Castro, C.H., Bunch, J.B., and Topka, T.M. (1980). Generalized algorithms for distribution feeder deployment and sectionalizing variations. *IEEE Transactions on Power Apparatus and Systems* PAS-99: 549–557.
294. Jiang, Y., Jiang, J., and Zhang, Y. (2012). A novel fuzzy multiobjective model using adaptive genetic algorithm based on cloud theory for service restoration of shipboard power systems. *IEEE Transactions on Power Systems* 27: 612–620.
295. Shirmohammadi, D. (1994). Service restoration in distribution networks via network reconfiguration. *IEEE Transactions on Power Delivery* 7 (2).

296. Kumar, Y. and Das, B. (2008). Multiobjective, multiconstraint service restoration of electric power distribution system with priority customers. *IEEE Transactions on Power Delivery* 23: 261–270.
297. Enacheanu, B., Raison, B., Caire, R. et al. (2008). Radial network reconfiguration using genetic algorithm based on the matroid theory. *IEEE Transactions on Power Systems* 23: 186–195.
298. Fukuyama, Y., Chiang, H.D., and Miu, K.N. (1996). Parallel genetic algorithm for service restoration in electric power distribution systems. *Electrical Power and Energy Systems* 18 (2): 111–119.
299. Miu, K.N., Chiang, H.-D., Yuan, B.B., and Darling, G. (1998). Fast service restoration for large-scale distribution systems with priority customers and constraints. *IEEE Transactions on Power Systems* 13 (3): 789–795.
300. Miu, K.N., Chiang, H.-D., and McNulty, R.J. (2000). Multi-tier service restoration through network reconfiguration and capacitor control for large scale radial distribution system. *IEEE Transactions on Power Systems* 15: 1001–1007.
301. del Valle, Y., Venayagamoorthy, G., Mohagheghi, S. et al. (2008). Particle swarm optimization: basic concepts, variants and applications in power systems. *IEEE Transactions on Evolutionary Computation* 12 (2): 171–195.
302. Sarma, N.D.R., Ghosh, S., Rao, K.S.P., and Srinivas, M. (1994). Real time service restoration in distribution networks. *IEEE Transactions on Power Delivery* 9 (4): 2064–2070.
303. Ucak, C. and Pahwa, A. (1994). An analytical approach for step-by-step restoration of distribution systems following extended outages. *IEEE Transactions on Power Delivery* 9 (3): 1717–1723.
304. Da Silva Filho, J.I. and Rocco, A. (2008). Power systems outage possibilities analysis by paraconsistent logic. In: *2008 IEEE Power and Energy Society General Meeting – Conversion and Delivery of Electrical Energy in the 21st Century*. IEEE.
305. Ahuja, A., Das, S., and Pahwa, A. (2007). An AIS-ACO hybrid approach for multi-objective distribution system restoration. *IEEE Transactions on Power Systems* 22 (3): 1101–1111.
306. Richardson, D.B. (2013). Electric vehicles and the electric grid: a review of modeling approaches, impacts, and renewable energy integration. *Renewable and Sustainable Energy Reviews* 19: 247–254.
307. Cipcigan, L. and Taylor, P. (2007). Investigation of the reverse power flow requirements of high penetrations of small-scale embedded generation. *IET Renewable Power Generation* 1 (3): 160–166.
308. Van Ruitenbeek, E. (2014). Impact of large amounts of photovoltaic installations in low-voltage networks on power system stability during transmission system faults: a german case study. PhD dissertation. TU Delft, Delft University of Technology.
309. Cepeda, J.C., Rueda, J.L., and Erlich, I. (2012). Identification of dynamic equivalents based on heuristic optimization for smart grid applications. In: *2012 IEEE Congress on Evolutionary Computation*, 1–8. IEEE.
310. Braun, M. and Strauss, P. (2008). A review on aggregation approaches of controllable distributed energy units in electrical power systems. *International Journal of Distributed Energy Resources* 4 (4): 297–319.
311. Cari, E.P.T., Alberto, L.F.C., and Erlich, I. (2015). Assessment of model parameters to identify an equivalent wind power plant. In: *2015 IEEE Eindhoven PowerTech*, 1–5. IEEE.

312. Jongepier, A.G. (1996). Artificial neural networks applied to power systems. PhD dissertation. TU Delft, Delft University of Technology.
313. Sun, L., Qu, P., Huang, Q., and Ju, P. (2007). Parameter identification of synchronous generator by using ant colony optimization algorithm. In: *2007 2nd IEEE Conference on Industrial Electronics and Applications*, 2834–2838. IEEE.
314. Rueda, J.L., Cepeda, J., Erlich, I. et al. (2015). Heuristic optimization based approach for identification of power system dynamic equivalents. *International Journal of Electrical Power & Energy Systems* 64: 185–193.
315. Cepeda, J.C., Rueda, J.L., and Erlich, I. (2013). MVMO^{s}-based approach for identification of dynamic equivalents from PMU measurements. In: *2013 IEEE Grenoble Conference*, 1–6. IEEE.
316. Rueda, J.L. and Erlich, I. (2013). Evaluation of the mean-variance mapping optimization for solving multimodal problems. In: *2013 IEEE Symposium on Swarm Intelligence (SIS)*, 7–14. IEEE.
317. Erlich, I., Venayagamoorthy, G.K., and Worawat, N. (2010). A mean-variance optimization algorithm. In: *IEEE Congress on Evolutionary Computation*, 1–6. IEEE.
318. Edwards, A. (2016). Modeling of distributed generation under next generation interconnection requirements. Master thesis. University of Washington.
319. Kersting, W.H. (2001). Radial distribution test feeders. In: *Power Engineering Society Winter Meeting, 2001*, vol. 2, 908–912. IEEE.
320. Ellis, A., Behnke, M., and Elliott, R. (2013). *Generic Solar Photovoltaic System Dynamic Simulation Model Specification*. Tech. Rep. Albuquerque, NM: Sandia National Laboratories (SNL-NM).
321. Sugita, S., Kataoka, Y., Naoi, S. et al. (2010). Study of modeling method of distributed generators considering partial dropout for trunk transmission system. In: *The 2010 International Power Electronics Conference (IPEC)*, 975–980. IEEE.
322. Dick, E.P. and Erven, C.C. (1978). Transformer diagnostic testing by frequency response analysis. *IEEE Transactions on Power Apparatus and Systems* PAS-97 (6): 2144–2153.
323. Gomez Luna, E. (2015). Experiences with non-intrusive monitoring of distribution transformers based on the on-line frequency response. *Ingeniería e Investigación* 35 (1): 55–59.
324. Abeywickrama, K.G.N.B., Serdyuk, Y.V., and Gubanski, S.M. (2006). Exploring possibilities for characterization of power transformer insulation by frequency response analysis (FRA). *IEEE Transactions on Power Delivery* 21 (3): 1375–1382.
325. Ji, T.Y., Tang, W.H., and Wu, Q.H. (2011). Frequency response analysis of power transformer winding deformation based on a hybrid model. In: *2011 IEEE Power Engineering and Automation Conference, Wuhan*, 144–147. IEEE.
326. IEC 60076-18 (2012). *Power Transformers – Part 18: Measurement of Frequency Response*, 1–94. IEC.
327. IEEE Power and Energy Society (2013). *IEEE Std C57.149-2012, Guide for the Application and Interpretation of Frequency Response Analysis for Oil-Immersed Transformers*, 1–72. IEEE.
328. C. international des grands reseaux electriques (2008). *Comite d' etudes A2, Mechanical Condition Assessment of Transformer Windings Using Frequency Response Analysis, FRA*. CIGRE.
329. Behjat, V. and Mahvi, M. (2015). Statistical approach for interpretation of power transformers frequency response analysis results. *IET Science, Measurement and Technology* 9 (3): 367–375.

330. Kim, J.W., Park, B., Jeong, S. et al. (2005). Fault diagnosis of a power transformer using an improved frequency-response analysis. *IEEE Transactions on Power Delivery* 20 (1): 169–178.
331. Gonzales Arispe, J.C. and Mombello, E.E. (2014). Detection of failures within transformers by FRA using multiresolution decomposition. *IEEE Transactions on Power Delivery* 29 (3): 1127–1137.
332. Pleite, J., Olias, E., Barrado, A. et al. (2002). Transformer modeling for FRA techniques. *IEEE/PES Transmission and Distribution Conference and Exhibition 2002: Asia Pacific* 1: 317–321.
333. Mombello, E.E. and Moller, K. (2000). New power transformer model for the calculation of electromagnetic resonant transient phenomena including frequency-dependent losses. *IEEE Transactions on Power Delivery* 15 (1): 167–174.
334. Wilcox, D.J., Conlon, M., and Hurley, W.G. (1988). Calculation of self and mutual impedances for coils on ferromagnetic cores. *IEE Proceedings A – Physical Science, Measurement and Instrumentation, Management and Education – Reviews* 135 (7): 470–476.
335. Bjerkan, E. (2005). *High Frequency Modeling of Power Transformers: Stresses and Diagnostics*, Doktoravhandlinger ved NTNU. Norwegian University of Science and Technology, Faculty of Information Technology, Mathematics and Electrical Engineering, Department of Electrical Power Engineering.
336. Yousof, M.F.M., Ekanayake, C., and Saha, T.K. (2015). Frequency response analysis to investigate deformation of transformer winding. *IEEE Transactions on Dielectrics and Electrical Insulation* 22 (4): 2359–2367.
337. Zhang, Z.W., Tang, W.H., Ji, T.Y., and Wu, Q.H. (2014). Finite-element modeling for analysis of radial deformations within transformer windings. *IEEE Transactions on Power Delivery* 29 (5): 2297–2305.
338. Alvarez Alvarez, D.L., Rosero Garcia, J.A., and Mombello, E.E. (2015). Core influence on the frequency response analysis (FRA) of power transformers through the finite element method. *Ingeniería e Investigación* 35 (1): 110–117.
339. Martinez Velasco, J.A. and Martinez-Velasco, J.A. (2010). *Power System Transients: Parameter Determination*. Taylor & Francis.
340. Alvarez, D.L., Rosero, J.A., and Mombello, E.E. (2014). Analysis of impedance matrix in transformer windings through the Finite Element Method (FEM). In: *2014 IEEE PES Transmission Distribution Conference and Exposition – Latin America (PES T D-LA)*, 1–7.
341. Gustavsen, B. and Semlyen, A. (1999). Rational approximation of frequency domain responses by vector fitting. *IEEE Transactions on Power Delivery* 14 (3): 1052–1061.
342. Alvarez, D.L., Rosero, J.A., and Mombello, E.E. (2013). Circuit model of transformers windings using vector fitting, for frequency response analysis (FRA). In: *2013 Workshop on Power Electronics and Power Quality Applications (PEPQA)*, 1–6. IEEE.
343. Alvarez, D.L. (2013). Circuit model of transformers windings using vector fitting, for frequency respon3se analysis (FRA) part II: core influence. In: *2013 Workshop on Power Electronics and Power Quality Applications (PEPQA)*, 1–5.
344. Firoozi, H. and Shishehchian, M. (2009). Frequency response analysis condition assessment of power transformers using mathematical and statistical criteria. In: *2009 IEEE 9th International Conference on the Properties and Applications of Dielectric Materials*, 253–256. IEEE.
345. Ryder, S. (2002). Methods for comparing frequency response analysis measurements. In: *Conference Record of the the 2002 IEEE International Symposium on Electrical Insulation (Cat. No.02CH37316)*, 187–190. IEEE.

346. Mukherjee, P. and Satish, L. (2012). Construction of equivalent circuit of a single and isolated transformer winding from FRA data using the ABC algorithm. *IEEE Transactions on Power Delivery* 27 (2): 963–970.
347. Rashtchi, V., Rahimpour, E., and Fotoohabadi, H. (2011). Parameter identification of transformer detailed model based on chaos optimisation algorithm. *IET Electric Power Applications* 5 (2): 238–246. https://doi.org/10.1049/iet-epa.2010.0147.
348. Arivamudhan, M. and Santhi, S. (2013). Model based approach for fault detection in power transformers using particle swarm intelligence. In: *Recent Advancements in System Modelling Applications*, Lecture Notes in Electrical Engineering, vol. 188 (eds. R. Malathi and J. Krishnan), 287–300. Springer India.
349. Betancourt, L., Martinez, G., Alvarez, D., and Rosero, J. (2013). Losses characterization on distribution transformer windings in frequency domain by means of finite element method (fem): part I. In: *2013 Fourth International Conference on Power Engineering, Energy and Electrical Drives (POWERENG)*, 103–108. IEEE.
350. Betancourt, L., Martinez, G., Alvarez, D., and Rosero, J. (2013). Losses characterization on distribution transformer windings in frequency domain by mean of finite element method (fem): part ii. In: *2013 Fourth International Conference on Power Engineering, Energy and Electrical Drives (POWERENG)*, 109–114.
351. Varvarigos, E. and Christodoulopoulos, K. (2014). Algorithmic aspects in planning fixed and flexible optical networks with emphasis on linear optimization and heuristic techniques. *Journal of Lightwave Technology* 32 (4): 681–693.
352. Peng, F., Tang, K., Chen, G., and Yao, X. (2010). Population-based algorithm portfolios for numerical optimization. *IEEE Transactions on Evolutionary Computation* 14 (5): 782–800.
353. Miranda, V. and Alves, R. (2013). Differential evolutionary particle swarm optimization (DEEPSO): a successful hybrid. In: *2013 BRICS Congress on Computational Intelligence and 11th Brazilian Congress on Computational Intelligence (BRICS-CCI CBIC)*, 368–374. IEEE.
354. Carvalho, L., Loureiro, F., Sumaili, J. et al. (2015). Statistical tuning of DEEPSO soft constraints in the security constrained optimal power flow problem. In: *2015 18th International Conference on Intelligent System Application to Power Systems (ISAP)*, 1–7. IEEE.
355. Pinto, R., Carvalho, L., Sumaili, J. et al. (2015). Coping with wind power uncertainty in unit commitment: a robust approach using the new hybrid metaheuristic DEEPSO. In: *2015 IEEE Eindhoven PowerTech*, 1–6. IEEE.
356. Miranda, V. and Alves, R. (2014). PAR/PST location and sizing in power grids with wind power uncertainty. In: *2014 International Conference on Probabilistic Methods Applied to Power Systems (PMAPS)*, 1–6. IEEE.
357. Seyedmahmoudian, M., Rahmani, R., Mekhilef, S. et al. (2015). Simulation and hardware implementation of new maximum power point tracking technique for partially shaded pv system using hybrid depso method. *IEEE Transactions on Sustainable Energy* 6 (3): 850–862.
358. Mohammed, O.A. (2015). Ship power system optimal distribution architectures and energy storage allocation under dynamic operational scenarios. Presentation at NSWC Carderock (20 July 2015).
359. Elsayed, A. and Mohammed, O.A. (2015). A comparative study on the optimal combination of hybrid energy storage system for ship power systems. In: *2015 IEEE Electric Ship Technology Symposium (ESTS) (21–24 June 2015)*. IEEE.

CHAPTER 6

INTEGRATION OF RENEWABLE ENERGY IN SMART GRID

6.1 INTRODUCTION

The present chapter focuses on two main topics – Renewable energy and Smart Grid. These topics are very distinct but at the same time very complementary; in this chapter, these will be addressed both separately and together, covering operation and control aspects of different sources, namely reactive power control in the scope of wind power integration. The different sources will be treated in an operation context as working all together, but also specific discussion will be devoted to wind power, photovoltaic (PV) generation control, and forecasting.

On the demand side, demand response (DR) will be discussed as a tool to optimally integrate distributed generation (DG), namely the one based on renewables. Load forecasting methods are presented in comparison with the generation forecast.

Finally, storage means and electric vehicles (EVs) are addressed as a consumption/generation resource with interesting characteristics for the integration of renewables.

6.2 RENEWABLE ENERGY SOURCES

H. Morais[1], Zita A. Vale[2], João Soares[2], and T. Sousa[3]

[1]INESC-ID / University of Lisbon, Lisbon, Portugal
[2]Polytechnic of Porto, Porto, Portugal
[3]Technical University of Denmark, Lyngby, Denmark

6.2.1 Renewable Energy Sources Management Overview

The development of renewable energy sources (RES) is of main importance in the way toward a sustainable development, as a form to limit greenhouse gasses

Applications of Modern Heuristic Optimization Methods in Power and Energy Systems, First Edition. Edited by Kwang Y. Lee and Zita A. Vale.
© 2020 by The Institute of Electrical and Electronics Engineers, Inc.
Published 2020 by John Wiley & Sons, Inc.

(GHG) emissions. The intensive use of RES has been established as a target by the energy policy makers, namely in the European Union (EU) and in the United States. Presently, power systems use a diversity of energy resources, with the increasing use of DG units, with the ones based on RES being already very significant in some countries. The integration of these resources in power systems and their operation in the scope of liberalized and competitive markets according to the smart grid concept require significant changes in operation methods and business models.

The main difficulties with RES are the continuity and reliability problems associated with the variability and unpredictable nature of the primary natural energy sources. The output of some renewable generation, such as wind generators and PV systems, is determined by the climate and weather conditions and operating patterns will therefore be constrained by these natural conditions. The increasing number of generation units based on RES makes their dispatch a large-dimension problem for which several meta-heuristic approaches have been proposed.

A review of the optimization methods that have been applied to the renewable energy field is presented in [1], which includes a large reference set concerning the use of meta-heuristic methods.

An optimization model that uses a genetic algorithm (GA) technique for the calculation of the additional cost of electricity due to the penetration of RES is presented in [2]. This model also analyzes the required increase in consumer tariffs, and the overall cost increase to fund RES, as well as the optimum feed-in-tariff to be offered to RES-based producers.

A methodology for solving generation planning unit commitment (UC) problem for thermal units integrated with wind and solar energy systems using an improved particle swarm optimization (PSO) combined with GA operators is presented in [3].

A multi-objective energy dispatch method that considers environment and fuel cost under large wind energy is proposed in [4] and uses the strength Pareto evolutionary algorithm (SPEA). The integration of renewable-based generation in electricity markets with virtual power producers is addressed in [5].

The problem of minimizing costs and emissions in a smart grid considering the use of RES and gridable vehicles is addressed in [6] using a PSO approach. An optimization method to maximize the utilization of RES in order to reduce both cost and emissions to an optimum level, considering the use of gridable vehicles, using PSO is presented in [7]. A simulated annealing (SA) and modified PSO approaches to energy resource management considering vehicle-to-grid in the context of a distribution network with intensive use of RES are proposed in [8, 9]. A PSO approach in [10] has been adapted to integrate DR programs in EVs for handling with the energy resource management.

A distribution state estimation (DSE) including RESs based on the combination of Nelder–Mead simplex search and PSO algorithms is presented in [11]. The proposed algorithm can estimate the load and the RES output values. The DSE methodology is compared with other evolutionary optimization algorithms, such as original PSO, honey bee mating optimization (HBMO), neural networks (NNs), and ant colony, in order to determine its effectiveness and efficiency.

6.2.2 Energy Resource Scheduling – Problem Formulation

This section presents the energy resource scheduling mathematical formulation considering DG units, storage systems, EVs, direct load control (DLC) DR programs, and the acquisition of energy to external suppliers. The problem can be solved as a mixed-integer nonlinear programming problem. The objective function (6.2.1) intends to determine the resources scheduling considering the minimum operation cost. Therefore, the costs associated with each type of distributed energy resource (DER) are considered.

DR programs provide the system operator with additional capacity that it can use to balance the system demand/generation. In the model presented here, the implemented DR contracts correspond to DLC DR programs that consist of a reduction of power consumption. Two different types of DLC programs are considered. The first one (DR_A) is applied to loads with the possibility of continuous regulation whereas DR_B is applied to discrete loads (on–off).

The energy resource management problem, as previously defined, can be solved with the following objective function (6.2.1):

$$\text{Minimize } f = \sum_{t=1}^{T} \left(\begin{array}{l} \sum_{SP=1}^{N_{SP}} P_{SP(SP,t)} \times c_{SP(SP,t)} + \sum_{DG=1}^{N_{DG}} P_{DG(DG,t)} \times c_{DG(DG,t)} \\ + P_{GCP(DG,t)} \times c_{GCP(DG,t)} \\ + \sum_{ST=1}^{N_{ST}} P_{Dch(ST,t)} \times c_{Dch(ST,t)} - P_{Ch(ST,t)} \times c_{Ch(ST,t)} \\ + \sum_{EV=1}^{N_{EV}} P_{Dch(EV,t)} \times c_{Dch(EV,t)} - P_{Ch(EV,t)} \times c_{Ch(EV,t)} \\ + \sum_{L=1}^{N_L} P_{DR\_A(L,t)} \times c_{DR\_A(L,t)} + P_{DR\_B(L,t)} \times c_{DR\_B(L,t)} \\ + P_{NSD(L,t)} \times c_{NSD(L,t)} \end{array} \right) \tag{6.2.1}$$

Equations (6.2.4)–(6.2.23) in Section 6.2.6 refer to the considered constraints. Equations (6.2.4) and (6.2.5) refer to the active and reactive power balance constraints, in each bus b, respectively. The power flow constraints, based on bus power balance, are expressed in (6.2.6) and (6.2.7). Equations (6.2.8) and (6.2.9) represent the bus voltage magnitude and bus voltage angle limits, respectively. Equation (6.2.10) represents the line thermal limits considering AC power flow and the shunt impedance in the power line.

Equations (6.2.11)–(6.2.23) represent the constraints concerning the maximum capacity of the considered resources, corresponding to suppliers (6.2.11) and (6.2.12), generation (6.2.13a), (6.2.13b), (6.2.13c), and (6.2.14), the DG ramp rate constraints (6.2.15) and (6.2.16), the DR programs (6.2.17) and (6.2.18), and EVs and storage units (6.2.19)–(6.2.23).

DG limits depend on the used technology. When combined heat and power (CHP) units are used, they can be modeled by Eq. (6.2.13a), considering the minimum and the maximum generation limits. For technologies based on natural resources (6.2.13b), the minimum generation is zero and the forecast processes determine the maximum limit. Equation (6.2.13c) represents the DG units with "take or pay" contracts. In these cases, it is necessary to include the variable to penalize the generation curtailment of these units.

DR programs are represented by (6.2.17) and (6.2.18). Equation (6.2.17) is used for DR programs applied to loads with continuous regulation. Equation (6.2.18) is used for discrete loads (on–off). The use of binary variable is important to define the DR application to each load L with this type of program.

EVs and the storage systems based on batteries can be formulated with the same constraints. The main difference is that the storage systems are permanently connected to the network and the energy spent in the trips is zero. In Section 6.2.6 it is only presented the formulation for the EVs.

The energy stored in the EVs and storage system batteries (6.2.19) is limited by their maximum capacity. The minimum value of stored energy can be equal to zero. However, to avoid the fast degradation of the batteries, it is recommended to impose a minimum around 15% of the maximum capacity.

The power discharge rate, for each EV and storage unit, is considered in Eq. (6.2.18) and the power charge rate in (6.2.21). In each period, each storage unit can only be charged or discharged as imposed by binary variables and in (6.2.20) and (6.2.21). The binary parameter gives information about the EVs' state (1 when the EV is connected to the electric grid, 0 when the EV is disconnected).

Finally, the EV batteries' state is obtained considering the initial stored energy, the charge and the discharge in each time period, and the energy spent during the trips (6.2.23), corresponding, respectively, to the discharge and charge efficiency of the batteries. These efficiency values depend on each battery technology and also on the type of network connection.

Particle Swarm Optimization

The PSO was proposed in [12–14] trying to simulate simple social systems like flocks of birds or schools of fish. The PSO algorithm starts determining with a random initial individuals population, representing solutions of a problem, to which are assigned random velocities. The individuals, called particles, evolve throughout the problem space searching for the optimal solution for the specific problem. A fitness function is used in each PSO iteration to determine the one that offers the best solution for the proposed problem. However, each particle also keeps track of its own best. Both values, the global best and the own particle best, are used by the particles in the searching process. Three main vectors govern the particle velocities: particle's inertia, the attraction toward its best position so far, and the attraction to the best global position. In the traditional PSO, the particle's velocity is clamped to avoid overshooting. The weights of the velocity equation should be tuned according to the problem characteristics. For instance, the inertia weight value carries a strong influence on the evolution of the particle, determining to a certain extent whether it will fall into a local optimum, converge to a global maximum, or simply overshoot. It is therefore common to apply to this component a function that decreases as it converges to the global solution, but even the decreasing rate of this function must be carefully defined. This method is also complemented with the clamping of the particle's velocity to maximum and minimum allowed values [15]. The setting of these values is another externally defined operation, which is critical to obtain accurate results: if the velocity is too high, the particle risks pass beyond a good solution, but if it is too low, it is probable that it will get stuck in a local optimum.

6.2.3 Energy Resources Scheduling – Particle Swarm Optimization

The proposed PSO algorithm has been developed to address the problem of energy resources scheduling (ERS) considering high quantities of EVs. The method uses a centralized approach in which an entity is responsible for the management of all distributed resources connected in a distribution network. The main advantage of PSO-based algorithms is their simplicity, while being capable of delivering accurate results in a consistent manner. PSO-based algorithms are fast and very flexible, being applicable to a wide range of problems, with limited computational requirements [12].

Evolutionary Particle Swarm Optimization

To overcome the PSO limitations, several variant methodologies have been proposed in recent years. One of the most interesting approaches is the inclusion of evolutionary algorithms with traditional PSO [16]. The evolutionary particle swarm optimization (EPSO) algorithm is proposed in [17]. EPSO can be seen as a self-adaptive evolutionary algorithm where the recombination is replaced by an operation called particle movement. It does not rely on the external definition of weights or other PSO crucial parameters. The basic idea of this method can be summarized as follows:

1. Every particle is replicated a certain number of times (one time in the case study).
2. Every particle's weights are mutated.
3. A movement rule is applied to each mutated particle.
4. Each new particle is evaluated according to the problem-specific fitness function.
5. The best particles are picked to form the new generation.

Modified Particle Swarm Optimization

Other variant of the traditional PSO is proposed in [9] considering the requirements of the ERS problem. The modified particle swarm optimization (MoPSO) uses some concepts traditionally used in the GAs in the PSO process. MoPSO makes use of mutation in the definition of the inertial weight but discards the recombination and selection steps. This mutation is governed by a Gaussian distribution [16]. One of the key points of the proposed method is the definition of the velocity limits boundaries [18]. The proposed method changes the velocity limits dynamically [19] during the search process using an intelligent mechanism, indirectly skewing some of the variables toward the desired outcomes. This mechanism is applied in two moments: evaluation and premovement phase. In the evaluation phase, after power flow evaluation, the mechanism will check for constraint violations.

Two-Step Particle Swarm Optimization

The two-step particle swarm optimization (2sPSO) uses the same model as MoPSO algorithm presented in the previous section. However, in the 2sPSO, the initial solution is performed by a deterministic approach using a relaxation of the problem, namely excluding the AC power flow constraints. The deterministic model in the case of 2sPSO uses a DC power flow model. This simplification reduces dramatically the execution time of the nonlinear deterministic technique, given an initial solution near by the optimal one.

6.2.4 Energy Resources Scheduling – Simulated Annealing

The proposed SA algorithm has been developed to deal with the day-ahead optimal resource scheduling considering large amounts of DERs. A robust AC power flow model is included in the proposed methodology to guarantee the feasibility of the solutions during the SA process.

Simulated Annealing Principles

In 1983, Kirkpatrick et al. [20, 21] presented a meta-heuristic technique to deal with hard and complex optimization problem that is called SA algorithm. The SA algorithm has been successfully adapted to deal with hard and complex combinatorial optimization problems in many applications [8, 22–24]. SA is based on the cooling process seen in metallurgy, in which the algorithm simulates the heat process of a metal, followed by the cooling process that reduces the temperature until the metal achieves a crystallized state [25]. The SA algorithm starts with an initial solution, and then a neighborhood mechanism is developed to obtain the neighbor solutions. If the neighbor solution is better than the initial one, then the neighbor will be used for the next iteration. Otherwise, the neighbor can be also accepted based on the Boltzmann probability (6.2.2) [20]. If this probability is higher than a random value, then the neighbor will be used in the next iteration.

$$p = e^{-\frac{\left|F(x_{K+1}) - F(x_{K+1})\right|}{T_K}} \tag{6.2.2}$$

The Boltzmann probability will enable the SA algorithm to accept a worse solution in order to avoid local optima. The temperature parameter controls the behavior of the Boltzmann probability during the iterative process. Typically, the SA algorithm accepts worse neighbor solutions in first's iterations avoiding these solutions in the last iterations, due to the reduction of the temperature value. This temperature reduction is achieved from a cooling scheduling that decreases the temperature value during the iterative process. Typically, the geometric scheme is used to reduce the temperature value during the iterations (6.2.3). The SA algorithm will stop when the maximum number of iterations was reached or when the temperature reaches a minimum value.

$$T_{K+1} = \alpha \times T_K \tag{6.2.3}$$

Simulated Annealing Algorithm

The implemented SA algorithm was proposed in [8, 22] in order to determine the day-ahead optimal resource scheduling including DG units, external suppliers, DR programs, storage system units, and EVs. The generic version of the SA is used in this methodology with a couple of changes in the neighborhood scheme, in the fitness function, and in the constraints control strategy [22]. The initial solution is randomly generated. During the iterative process, a single solution is generated in the neighborhood scheme in order to compare with the previous one. In the neighborhood scheme, an intelligent allocation heuristic to obtain a solution for the EVs and storage system is used. The main objective of this heuristic is to schedule the EVs and storage charge in off-peak periods and to schedule their discharge in peak periods. Moreover, the order of merit heuristic is applied to schedule the DG units, external suppliers, and DR based on the resources with a low operation cost. The fitness function is composed by the function (6.2.1) and a set of penalty factors to avoid solutions with constraints violations. The penalty factors are applied to EVs' constraints violation, storage's constraints violation, voltage magnitude and angle limits violation, and line thermal limits violation. The temperature value is achieved in each iteration using the geometric cooling scheduling proposed in [25]. The SA algorithm stops when the maximum number of iterations is reached, or when the fitness function does not improve during a certain number of iterations.

Energy Resources Scheduling Simulated Annealing

The energy resources scheduling simulated annealing (ERS2A) algorithm is adapted from the previously described SA algorithm [8, 22]. To improve the method efficiency and effectiveness, a heuristic mechanism is implemented to generate the initial solution without violation of any constraint. This heuristics preschedules the EVs and storage charge in off-peak periods and the EVs and storage discharge in the peaks. This prescheduling process improves the fitness function value of the initial solution. Furthermore, the other DERs are sorted by merit order according to the problem goal (the operation cost in the present case). The initial heuristic preschedules these units balancing the generation and the demand in each period. This approach helps the ERS2A to provide better solution when compared with the SA approach presented in section "Simulated Annealing Principles" that uses a random initial solution. Other advantage of this method is the lower computational effort (time and memory) required to the scheduling process.

In the iterative process, the neighborhood solutions are found using the mechanism proposed in [8, 22]. This mechanism tests each neighbor solution avoiding solutions with violations in the DER operation constraints (6.2.11)–(6.2.23). Additionally, a robust radial AC power-flow [26] for each neighbor solution runs in each iteration to ensure that the network operation limits are respected (6.2.4)–(6.2.10). If any constraint has constraints violations, the neighborhood mechanism adjusts the DER scheduling providing a feasible solution. This methodology proposed in [27, 28] can be designed as an implicit handling approach. The use of the implicit handling approach avoids the use of penalty factors and direct repair solutions helping the ERS2A algorithm to focus more on the fitness function improvement instead to spend most of the time repairing infeasible solutions. To guarantee the bus voltage magnitude limits (6.2.8) and also the lines thermal limits (6.2.10), the algorithm proposed in [9] is adopted. This algorithm increases the DER generation or decreases the power demand in downstream buses when it is possible relaxing the congested lines. A similar process is used to avoid the voltage magnitude violations, but in this case changing the initial reactive power scheduling. The temperature value is achieved in each iteration using the geometric cooling scheduling proposed in [25]. The ERS2A methodology stops when the maximum number of iterations is reached, or if the fitness function does not present any improvement during a certain number of iterations.

Hybrid Simulated Annealing and Deterministic Technique

The hybrid simulated annealing and deterministic technique (SADT) uses the same model as the ERS2A algorithm presented in the previous section. The main difference between the two approaches is the method to obtain the initial solution. In the ERS2A, a heuristic mechanism considering the resources' merit order is used to obtain the initial solution. In the SADT, the initial solution is obtained through a relaxed deterministic solution of the problem. The relaxed formulation of the energy resource scheduling problem does not consider the network constraints, namely the active (6.2.6) and reactive (6.2.7) power flow, the voltage magnitude (6.2.8) and angle (6.2.9), and the line thermal limit (6.2.10) constraints. Additionally, the active (6.2.4) and reactive (6.2.5) power balance constraints do not consider power losses. This simplification reduces dramatically the execution time of the deterministic technique. This simplified mathematical formulation is used to obtain a good solution in the initial step in order to reduce the effort for the SA to search in the feasible space.

As in the ERS2A, the neighborhood scheme uses the mechanisms proposed in [8, 22] and a robust radial AC power-flow guarantees the feasibility of each solution considering the network constraints (6.2.4)–(6.2.10). The SADT methodology stops when the maximum number of iterations is reached, or if the fitness function does not present any improvement during a certain number of iterations.

6.2.5 Practical Case Study

In this section, the described methodologies are applied to a realistic case study of a distribution network with 37 buses and the obtained results for each of them are compared. The results for the operation cost and the execution time are discussed in detail and the energy resource scheduling solution is evaluated for each methodology. The proposed network considers an energy-mix scenario for the year 2050 proposed in [29].

Case Study Description

The presented case study considers the distribution network presented in [30] which is shown in Figure 6.2.1. The distribution network is composed by 37 buses

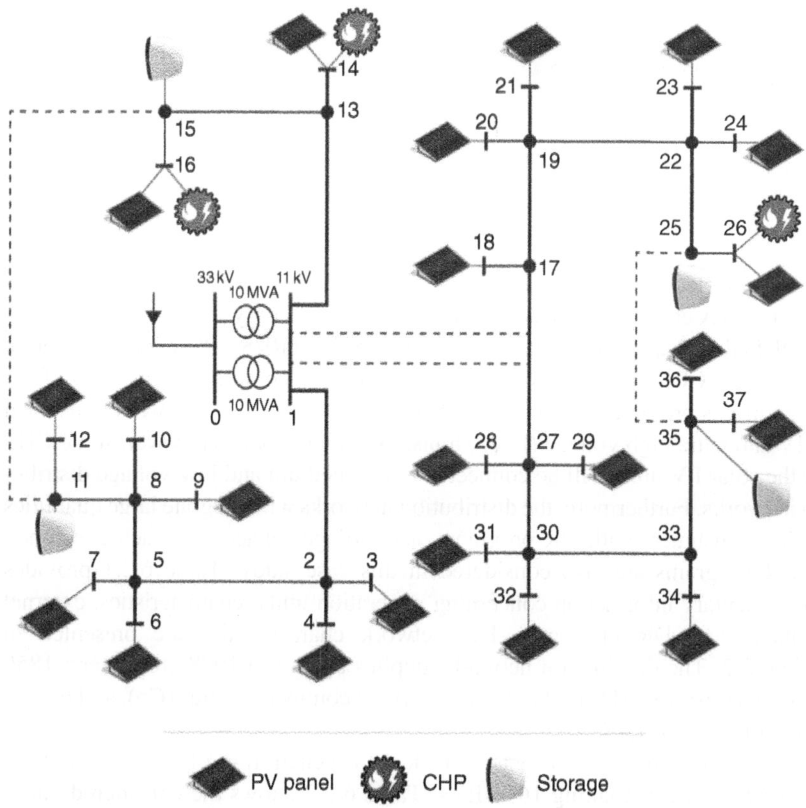

Figure 6.2.1 37-Bus distribution network. Source: adapted from [30].

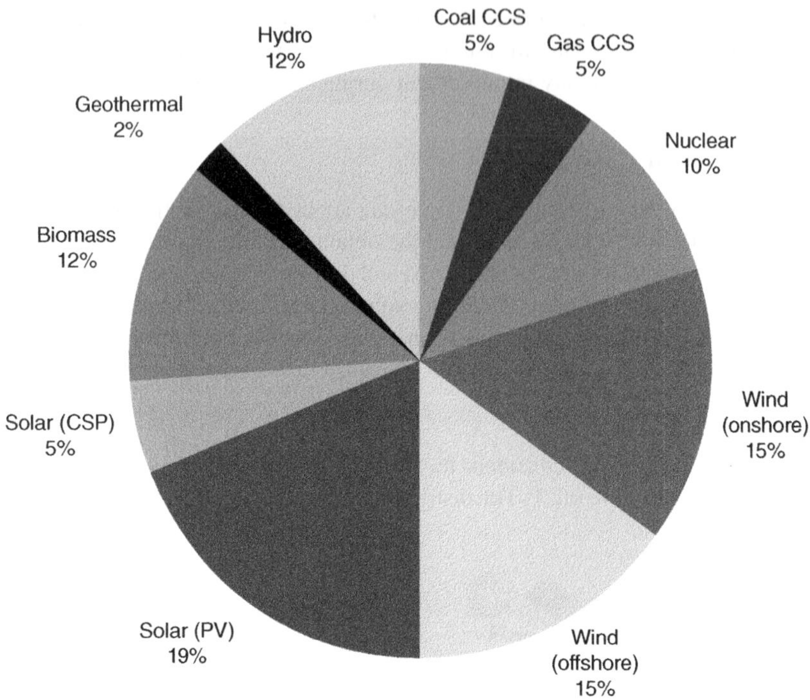

Figure 6.2.2 Energy mix in 2050 [29].

that is connected to the high voltage (HV) network through two power transformers with 10 MVA each one. It is used in this network an energy mix in 2050 proposed in [29]. In this chapter the scenario considering 80% of RES, 10% of carbon capture and storage (CCS), and 10% of nuclear is used. The complete distribution of the energy mix is presented in Figure 6.2.2. Most of the technologies will be installed in HV and extra high voltage (EHV) transmission networks. However, some CHP and the solar PV units will be connected to the medium and low-voltage distribution networks. Furthermore, the distribution networks will integrate large quantities of EVs, consumers with DR programs, and storage capacity. In each consumer, the DR programs are also considered in this case study. Table 6.2.1 provides complementary information concerning generation units' characteristics, external suppliers, and DR programs. The network characteristics are presented in Table 6.2.2. The distribution network supplies energy to 1908 consumers: 1850 domestic consumers (DM), 2 industries (In), 50 commerce stores (Co), and 6 service buildings (SB) [30].

Regarding the EVs, it was considered a penetration of 40% of the total number of vehicles making 1616 EVs. Table 6.2.3 shows the considered quantity of vehicles for each type of consumer. The EVs' characteristics were

TABLE 6.2.1 Energy Resources Characteristics and Costs

Technology		Number of Units	Total Installed Power	Total Installed Capacity	Mean Linear Cost (m.u./kWh)
Photovoltaic		22	7.74 (MWp)		0.0800
CHP		3	1.5 (MVA)		0.0500
External suppliers		2	10 (MVA)		0.0667
Demand response programs (curtailment)		22	0.45 (MW)		0.2000
Demand response programs (reduction)		22	0.65 (MW)		0.1500
Storage systems	Charge	4		1 (MWh)	0
	Discharge				0.0650
Electric vehicles	Charge	1616		25 (MWh)	0
	Discharge				0.0489

TABLE 6.2.2 Network Characteristics

Line	Bus	Bus	Distance (km)	R (p.u.)	X (p.u.)	Maximum Power Limit (MVA)
1	1	2	0.60	0.001 59	0.000 56	8
2	2	3	0.75	0.001 98	0.000 70	3.2
3	2	4	0.80	0.002 12	0.000 75	3.2
4	2	5	0.75	0.001 98	0.000 70	4.5
5	5	6	0.80	0.002 12	0.000 75	3.2
6	5	7	0.60	0.001 59	0.000 56	3.2
7	5	8	0.75	0.001 98	0.000 70	4.5
8	8	9	0.80	0.002 12	0.000 75	3.2
9	8	10	0.75	0.001 98	0.000 70	3.2
10	8	11	0.60	0.001 59	0.000 56	4.5
11	11	12	0.80	0.002 12	0.000 75	3.2
12	1	13	0.75	0.001 98	0.000 70	4.5
13	13	14	0.80	0.002 12	0.000 75	3.2
14	13	15	0.60	0.001 59	0.000 56	4.5
15	15	16	0.80	0.002 12	0.000 75	3.2
16	1	17	0.75	0.001 98	0.000 70	8
17	17	18	0.60	0.001 59	0.000 56	3.2
18	17	19	0.80	0.002 12	0.000 75	4.5
19	19	20	0.75	0.001 98	0.000 70	3.2

(Continued)

TABLE 6.2.2 (Continued)

Line	Bus	Bus	Distance (km)	*R* (p.u.)	*X* (p.u.)	Maximum Power Limit (MVA)
20	19	21	0.80	0.002 12	0.000 75	3.2
21	19	22	0.60	0.001 59	0.000 56	4.5
22	22	23	0.75	0.001 98	0.000 70	3.2
23	22	24	0.80	0.002 12	0.000 75	3.2
24	22	25	0.75	0.001 98	0.000 70	4.5
25	25	26	0.60	0.001 59	0.000 56	3.2
26	1	27	0.80	0.002 12	0.000 75	8
27	27	28	0.75	0.001 98	0.000 70	3.2
28	27	29	0.60	0.001 59	0.000 56	3.2
29	27	30	0.75	0.001 98	0.000 70	4.5
30	30	31	0.60	0.001 59	0.000 56	3.2
31	30	32	0.80	0.002 12	0.000 75	3.2
32	30	33	0.75	0.001 98	0.000 70	4.5
33	33	34	0.80	0.002 12	0.000 75	3.2
34	33	35	0.60	0.001 59	0.000 56	4.5
35	35	36	0.75	0.001 98	0.000 70	3.2
36	35	37	0.80	0.002 12	0.000 75	3.2

TABLE 6.2.3 Electric Vehicles by Consumer Type

Load	Domestic Consumer	Industry	Commerce	Services Building	Total
Number of consumers	1850	2	50	6	1908
Vehicles per consumer	2	10	4	20	
Number of vehicles	3700	20	200	120	4040
Number of EVs	1480	8	80	48	1616

determined based on the characteristics of some real models. The driving pattern was obtained using the EVeSSI simulation tool [31]. Table 6.2.4 shows the consumption characteristics for each consumption bus. Five consumption profiles, presented in Figure 6.2.3, are implemented in order to create a more realistic scenario.

The hourly energy price for the energy supplied by the external suppliers, shown in Figure 6.2.4, was obtained considering the average price in day-ahead Nord Pool market during the year 2013 week-days.

TABLE 6.2.4 Consumers Characteristics

Load	Bus	Consumer Demand (kW) Min.	Max.	Mean	Q/P	Consumers Type	Consumers Profile
1	3	324.2	866.8	551.8	0.3	210 DM	1
2	4	324.2	866.8	551.8	0.3	210 DM	2
3	6	324.2	866.8	551.8	0.3	210 DM	3
4	7	342.9	916.7	583.5	0.3	1 SB	1
5	9	342.9	916.7	583.5	0.3	1 SB	4
6	10	280.6	750.0	477.4	0.3	10 Co	5
7	12	280.6	750.0	477.4	0.3	10 Co	1
8	14	609.0	1627.9	1036.3	0.3	1 In	2
9	16	700.3	1872.1	1191.7	0.3	1 In	5
10	18	324.2	866.8	551.8	0.3	210 DM	4
11	20	324.2	866.8	551.8	0.3	210 DM	3
12	21	272.7	729.1	464.1	0.3	200 DM	1
13	23	342.9	916.7	583.5	0.3	1 SB	5
14	24	342.9	916.7	583.5	0.3	1 SB	4
15	26	280.6	750.0	477.4	0.3	10 Co	1
16	28	280.6	750.0	477.4	0.3	10 Co	2
17	29	272.7	729.1	464.1	0.3	200 DM	3
18	31	272.7	729.1	464.1	0.3	200 DM	1
19	32	272.7	729.1	464.1	0.3	200 DM	5
20	34	342.9	916.7	583.5	0.3	1 SB	3
21	36	342.9	916.7	583.5	0.3	1 SB	2
22	37	280.6	750.0	477.4	0.3	10 Co	1

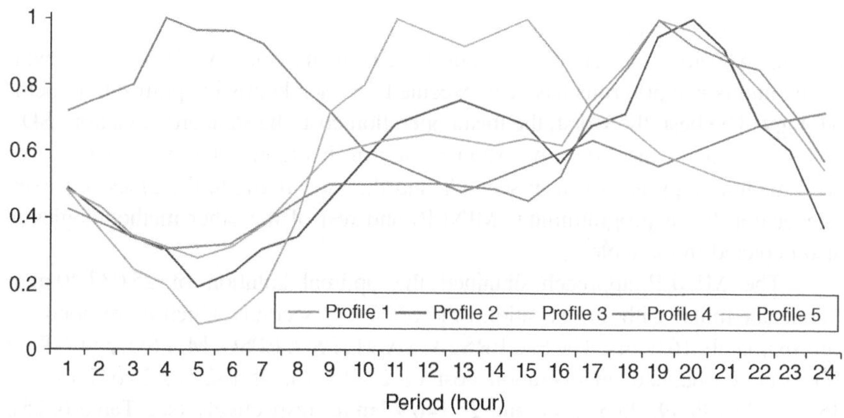

Figure 6.2.3 Consumers' profiles.

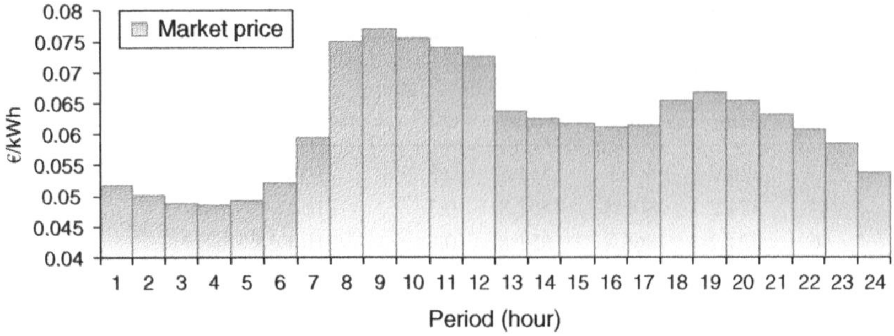

Figure 6.2.4 External suppliers' price based on Nord Pool day-ahead market.

TABLE 6.2.5 Consumers Characteristics

	Operation Cost (m.u.)				Mean Execution Time	
Methodology	Best	Worst	Mean	STD	Seconds	Ratio
MINLP	25 037.40	—	—	—	57 688 (16.02 h)	
SA	25 196.16	25 292.39	25 270.60	8.41	42.96	1343
ERS2A	25 142.53	25 255.46	25 189.29	24.02	47.70	1209
SADT	25 047.14	25 047.94	25 047.51	0.15	59.21	974
PSO	27 454.11	29 816.93	27 750.09	509.59	170.96	337
EPSO	25 789.49	30 022.41	26 624.00	777.95	363.58	159
MoPSO	25 574.53	26 668.63	25 722.02	190.00	225.34	256
2sPSO	25 146.77	25 146.87	25 146.81	0.05	28.00	2060

Methodologies Comparison

This section presents the comparison of the obtained results with each technique referred in Sections 6.2.3 and 6.2.4. Table 6.2.5 shows the objective function results of all methodologies tested in the 37-bus distribution network. A robustness test with a total number of 500 runs has been executed for each heuristic optimization methodology. The best, the worst, the mean operation cost, the standard deviation (SD), and the average execution time are presented in this table. The execution time of each methodology is shown in seconds and the ratio between the times of mixed integer non-linear programming (MINLP) and rest of the other methodologies is also depicted in the table.

The MINLP approach obtained the optimal solution of 25 037.40 m.u. (monetary units) with an execution time of 57 688 seconds which corresponds to approximately 16 hours. The SA, ERS2A, SADT, PSO, EPSO, MoPSO, and 2sPSO approaches obtained an operation cost of 25 196.16, 25 142.53, 25 047.14, 27 454.11, 25 789.49, 25 574.53, and 25 146.77 m.u., respectively (see Table 6.2.5). The SADT, ERS2A, and 2sPSO found solutions that are closer to the MINLP result, presenting a small difference of 0.04, 0.42, and 0.44%, respectively. These differences are below 0.5%. These results represent an outstanding performance for this three meta-heuristics-based approaches taking into account that the 0.5% improvement in the solution cost requires about 16 hours of computation time. The other methodologies obtained satisfactory solutions when compared with the MINLP result with a difference higher than 0.5%, achieving a cost of 27 454.11 m.u. (PSO), 25 789.49 m.u. (EPSO), 25 574.53 m.u. (MoPSO), and 25 196.16 m. u. (SA), respectively. The STD of SA, ERS2A, SADT, PSO, EPSO, MoPSO, and 2sPSO approaches was 8.41, 24.02, 0.15, 509.59, 777.95, 190.00 and 0.05 m.u., respectively (see Table 6.2.5). The SADT, ERS2A, and 2sPSO methodologies present a robust behavior with a low STD, mainly the SADT with a very small value (0.15 m.u.). The SA also presents a low STD value with 8.41 m.u.

Analyzing the execution time results, the 2sPSO, SA, ERS2A, and SADT present the best results with a time of 28.00, 42.96, 47.70, and 59.21 seconds, respectively. These four methodologies presented a time's ratio (see Table 6.2.5) of 2060, 1343, 1209, and 974, respectively. For instance, the 2sPSO has an average execution time of 28 seconds, approximately 2060 faster than the MINLP approach. The PSO, EPSO, and MoPSO present an average execution time of 171, 363, and 225 seconds, respectively. Briefly, it is possible to refer that the SADT, ERS2A, and 2sPSO are the ones that present the best results in terms of operation cost (with close solutions to the optimal one) and in terms of execution time (with a low time).

Regarding the energy resource scheduling during the 24 periods, Figure 6.2.5 presents the obtained results by the MINLP methodology. The line represents the consumers' demand. The MINLP approach selected the off-peak periods to charge the batteries of the EVs, because these periods have available resources with low operation cost. However, some EVs charge during the peak hours due to the high generation provided by PV panels. The EVs' discharge appears in the period 20 and 21 because in these periods the power demand is

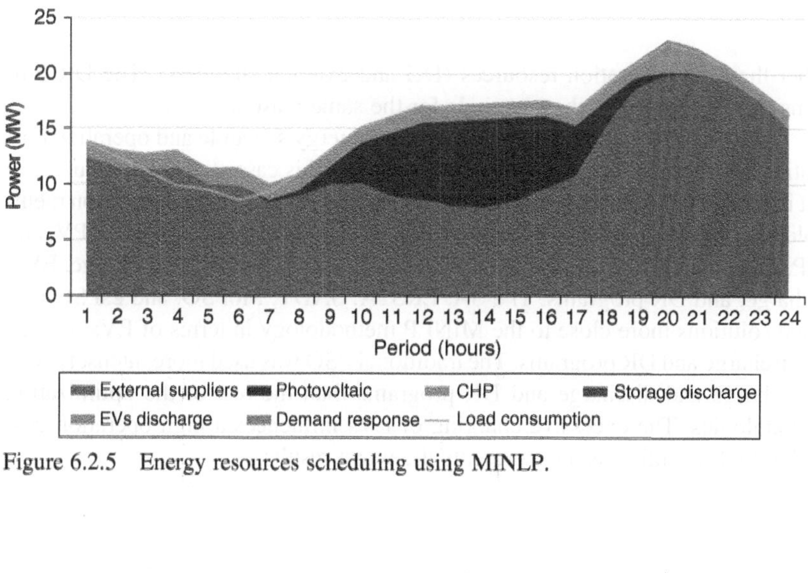

Figure 6.2.5 Energy resources scheduling using MINLP.

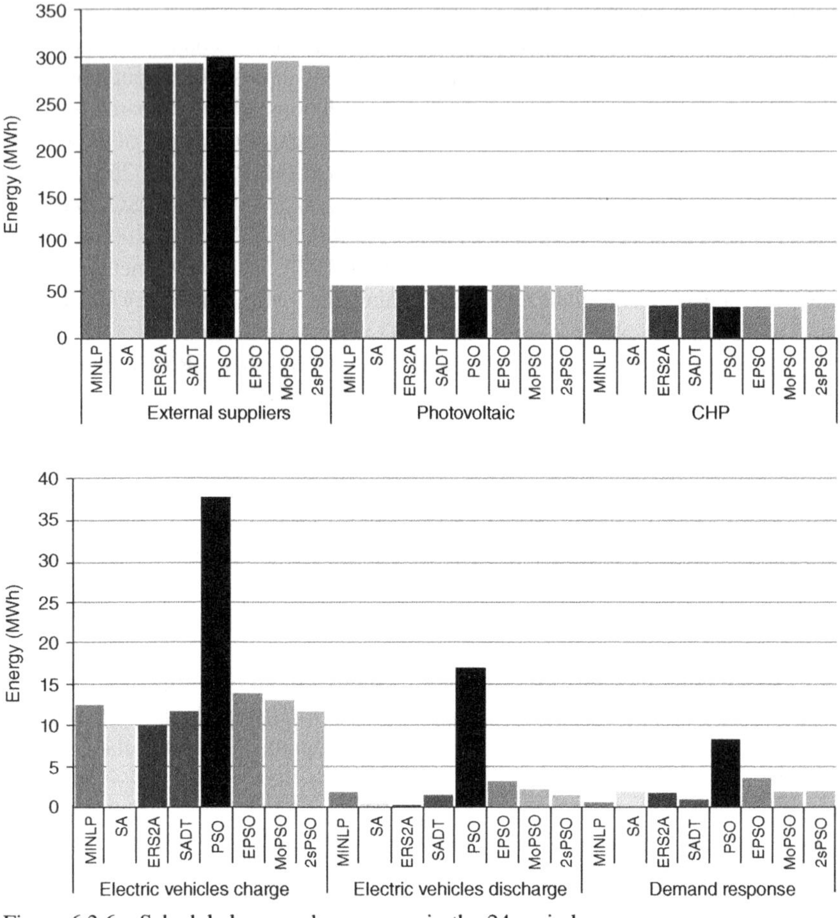

Figure 6.2.6 Scheduled energy by resource in the 24 periods.

higher than the generation resources (DG and external suppliers). The DR programs were also used in these periods for the same reason.

Figures 6.2.6 and 6.2.7 depict the global energy schedule and operation cost solution by resource for all methodologies used in this case study. The figures are split in two parts (a) and (b) to better understand the values. The optimization methodologies obtained similar scheduling solutions for the external suppliers, PV, and CHP. The main differences are in the scheduling results for the EVs' charge, EVs' discharge, and DR programs. The SA, ERS2A, SADT, MoPSO, and 2sPSO presented solutions more close to the MINLP methodology in terms of EVs' charge and discharge and DR programs. The traditional PSO was used more intensely with EVs' charge and discharge and DR programs than the rest of the optimization methodologies. The excessive charging of EVs influences a high operation cost for the PSO, for this reason it presents the worst results.

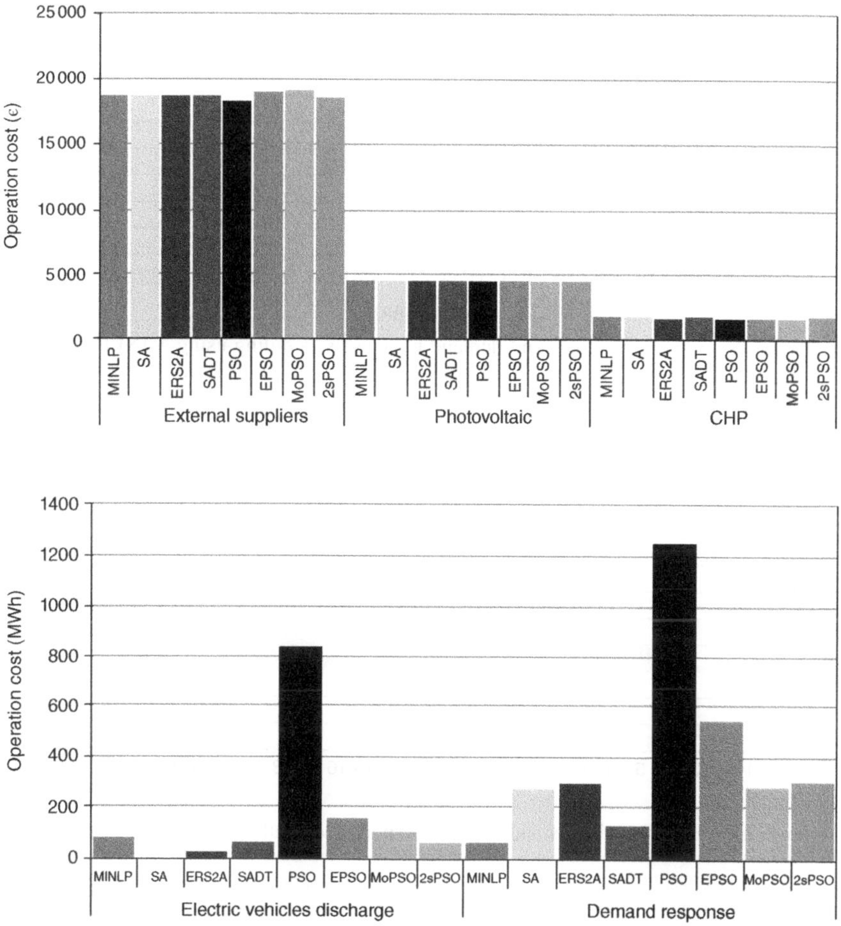

Figure 6.2.7 Operation costs in the 24 periods.

In order to evaluate the behavior of each methodology regarding the EVs' charge and discharge and DR scheduling, Figures 6.2.8–6.2.15 present the detailed scheduling results for these resources for each tested methodology. In these figures it is possible for MINLP, SADT, and 2sPSO to use more EVs' discharge than DR in periods 20 and 21, and the other methodologies use more DR, mainly the SA. The PSO methodologies are penalized due to the use of DR and EVs' discharge during all periods. The use of more expensive resources penalizes the operation cost and in the case of EVs energy, discharge implies the increased necessity of EVs' charge penalizing two times the operation cost. Even the 2sPSO methodology has worst results than the SADT due to the DR scheduling during the off-peak periods. This happens because PSO is a global heuristic and has more difficulty to refine the solutions more coherently than the local heuristic like the SA. On the other side, the SA has more difficulty to find a good solution when random

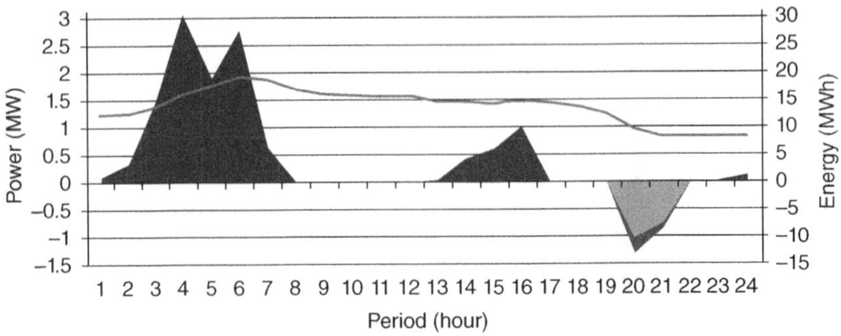

Figure 6.2.8 Electric vehicles and demand response scheduling using MINLP.

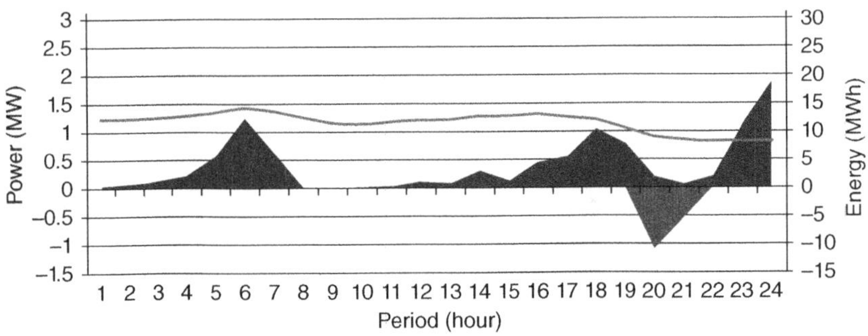

Figure 6.2.9 Electric vehicles and demand response scheduling using SA.

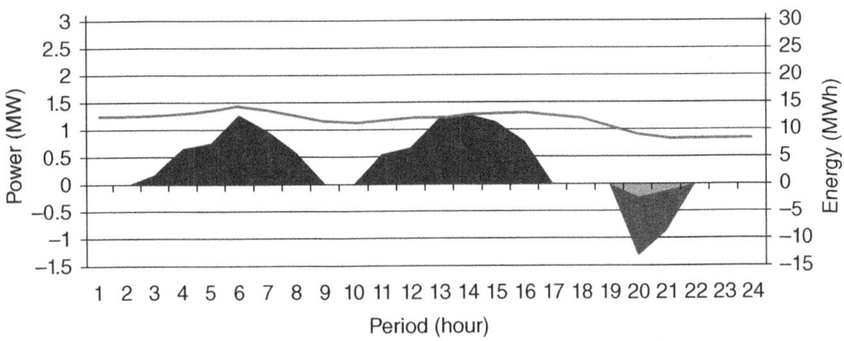

Figure 6.2.10 Electric vehicles and demand response scheduling using SADT.

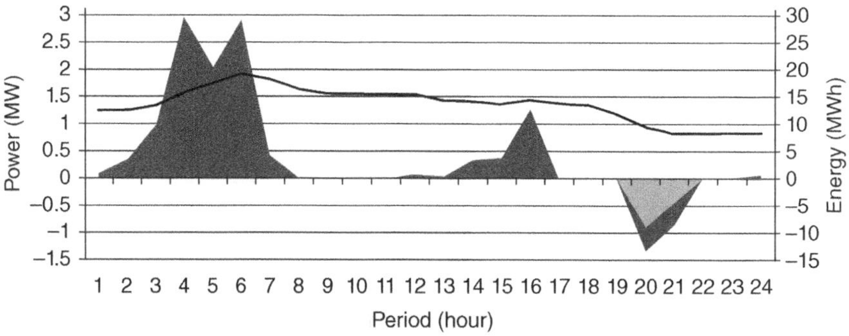

Figure 6.2.11 Electric vehicles and demand response scheduling using ERS2A.

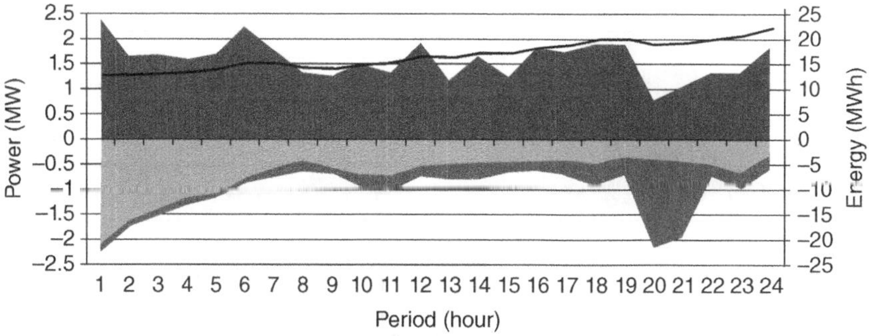

Figure 6.2.12 Electric vehicles and demand response scheduling using PSO.

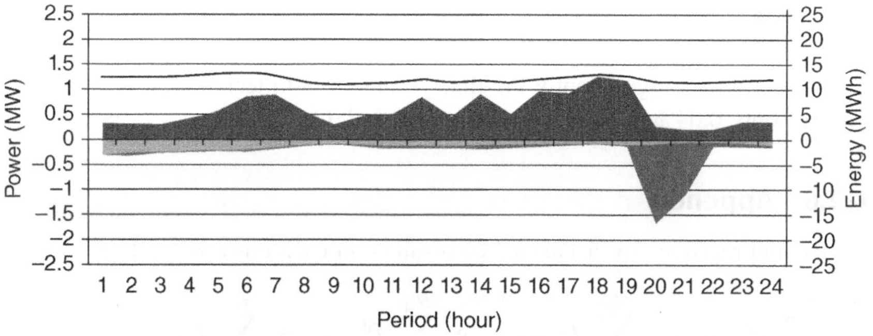

Figure 6.2.13 Electric vehicles and demand response scheduling using EPSO.

variables are used as initial solutions. In this sense, a careful and innovative heuristic is used in SA and ERS2A to provide good solutions. The PSO is, normally, less penalized by the initial solution and more effort is used to improve the internal processes of PSO like the mutation or the particles' velocities. However, in both heuristics techniques, PSO and SA, the use of relaxed deterministic results as initial

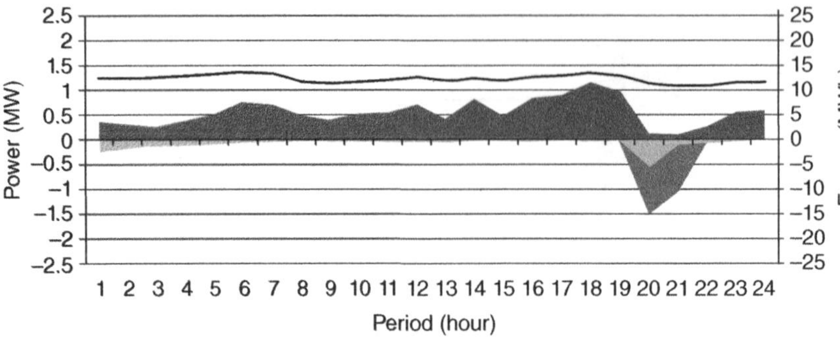

Figure 6.2.14 Electric vehicles and demand response scheduling using MoPSO.

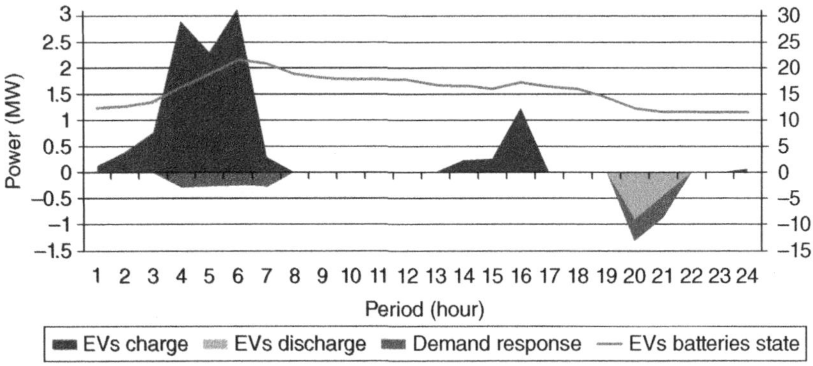

Figure 6.2.15 Electric vehicles and demand response scheduling using 2sPSO.

solution improves significantly the final results of both techniques, providing an error of only 0.04% in SADT and 0.44% in 2sPSO.

6.2.6 Appendix

Active and reactive power balance constraints, in each bus i, respectively:

$$\begin{aligned}&\sum_{DG=1}^{N_{DG}^{i}} P_{DG(DG,t)}^{i} - \sum_{DG=1}^{N_{DG}^{i}} P_{GCP(DG,t)}^{i} + \sum_{SP=1}^{N_{SP}^{i}} P_{SP(SP,t)}^{i} + \sum_{EV=1}^{N_{EV}^{i}} P_{Dch(EV,t)}^{i}\\&+ \sum_{ST=1}^{N_{ST}^{i}} P_{Dch(ST,t)}^{i} - \sum_{L=1}^{N_{L}^{i}} P_{Load(L,t)}^{i} - \sum_{EV=1}^{N_{EV}^{i}} P_{Ch(EV,t)}^{i} - \sum_{ST=1}^{N_{ST}^{i}} P_{Ch(ST,t)}^{i}\\&+ \sum_{L=1}^{N_{L}^{i}} P_{DR\_A(L,t)}^{i} + \sum_{L=1}^{N_{L}^{i}} P_{DR\_B(L,t)}^{i} + \sum_{L=1}^{N_{L}^{i}} P_{NSD(L,t)}^{i} = P_{i(t)}\\&\forall t \in \{1,\ldots,T\};\quad \forall i \in \{1,\ldots,N_{B}\}\end{aligned} \tag{6.2.4}$$

$$\sum_{DG=1}^{N_{DG}^{i}} Q_{DG(DG,t)}^{i} + \sum_{SP=1}^{N_{SP}^{i}} Q_{SP(SP,t)}^{i} - \sum_{L=1}^{N_{L}^{i}} Q_{Load(L,t)}^{i} + \sum_{L=1}^{N_{L}^{i}} Q_{DR\_A(L,t)}^{i} + \sum_{L=1}^{N_{L}^{i}} Q_{DR\_B(L,t)}^{i} + \sum_{L=1}^{N_{L}^{i}} Q_{NSD(L,t)}^{i} = Q_{i(t)}$$
$$\forall t \in \{1,\ldots,T\}; \quad \forall i \in \{1,\ldots,N_B\} \qquad (6.2.5)$$

The power flow formulation:

$$P_{i(t)} = G_{ii} \times V_{i(t)}^{2} + V_{i(t)} \times \sum_{j \in L^{i}} V_{j(t)}\left(G_{ij}\cos\left(\theta_{i(t)} - \theta_{j(t)}\right) + B_{ij}\sin\left(\theta_{i(t)} - \theta_{j(t)}\right)\right)$$
$$\forall t \in \{1,\ldots,T\}; \quad \forall i, \quad j \in \{1,\ldots,N_B\} \qquad (6.2.6)$$

$$Q_{i(t)} = V_{i(t)} \times \sum_{j \in L^{i}} V_{j(t)}\left(G_{ij}\sin\left(\theta_{i(t)} - \theta_{j(t)}\right) - B_{ij}\cos\left(\theta_{i(t)} - \theta_{j(t)}\right)\right) - B_{ii} \times V_{i(t)}^{2}$$
$$\forall t \in \{1,\ldots,T\}; \quad \forall i, \quad j \in \{1,\ldots,N_B\} \qquad (6.2.7)$$

Bus voltage magnitude and bus voltage angle limits:

$$V_{i}^{\min} \le V_{i(t)} \le V_{i}^{\max}; \qquad \forall t \in \{1,\ldots,T\} \qquad (6.2.8)$$

$$\theta_{i}^{\min} \le \theta_{i(t)} \le \theta_{i}^{\max}; \qquad \forall t \in \{1,\ldots,T\} \qquad (6.2.9)$$

Line thermal limits:

$$\left|\overline{U_{i(t)}} \times \left[\overline{y_{ij}} \times \left(\overline{U_{i(t)}} - \overline{U_{j(t)}}\right) + \overline{y_{sh\_i}} \times \overline{U_{i(t)}}\right]^{*}\right| \le S_{Lk}^{\max}$$
$$\left|\overline{U_{j(t)}} \times \left[\overline{y_{ij}} \times \left(\overline{U_{j(t)}} - \overline{U_{i(t)}}\right) + \overline{y_{sh\_j}} \times \overline{U_{j(t)}}\right]^{*}\right| \le S_{Lk}^{\max}$$
$$\forall t \in \{1,\ldots,T\}; \quad \forall i, \quad j \in \{1,\ldots,N_B\}; \quad i \ne j; \quad \forall k \in \{1,\ldots,N_K\} \qquad (6.2.10)$$

External suppliers limits:

$$P_{SP(SP,t)} \le P_{Max(SP,t)} \times X_{SP(SP,t)}; \quad \forall t \in \{1,\ldots,T\}; \quad \forall SP \in \{1,\ldots,N_{SP}\} \qquad (6.2.11)$$

$$Q_{SP(SP,t)} \le Q_{Max(SP,t)} \times X_{SP(SP,t)}; \quad \forall t \in \{1,\ldots,T\}; \quad \forall SP \in \{1,\ldots,N_{SP}\} \qquad (6.2.12)$$

Generation limits:

$$P_{Min(DG,t)} \times X_{DG(DG,t)} \le P_{DG(DG,t)} \le P_{Max(DG,t)} \times X_{DG(DG,t)}$$
$$\forall t \in \{1,\ldots,T\}; \quad \forall DG \in \{1,\ldots,N_{DG}\} \qquad (6.2.13a)$$

$$0 \le P_{DG(DG,t)} \le P_{DGForecast(DG,t)}$$
$$\forall t \in \{1,\ldots,T\}; \quad \forall DG \in \{1,\ldots,N_{DG}\} \qquad (6.2.13b)$$

$$P_{\mathrm{DG(DG},t)} + P_{\mathrm{GCP(DG},t)} = P_{\mathrm{DGForecast(DG},t)} \quad \forall t \in \{1, \ldots, T\}; \quad \forall \mathrm{DG} \in \{1, \ldots, N_{\mathrm{DG}}\} \tag{6.2.13c}$$

$$Q_{\mathrm{Min(DG},t)} \times X_{\mathrm{DG(DG},t)} \leq P_{\mathrm{DG(DG},t)} \leq Q_{\mathrm{Max(DG},t)} \times X_{\mathrm{DG(DG},t)} \quad \forall t \in \{1, \ldots, T\}; \quad \forall \mathrm{DG} \in \{1, \ldots, N_{\mathrm{DG}}\} \tag{6.2.14}$$

$$P_{\mathrm{DG(DG},t)} - P_{\mathrm{DGPrevious(DG},t)} \leq R_{\mathrm{SU(DG)}} \quad \forall t \in \{1, \ldots, T\}; \quad \forall \mathrm{DG} \in \{1, \ldots, N_{\mathrm{DG}}\} \tag{6.2.15}$$

$$P_{\mathrm{DGPrevious(DG},t)} - P_{\mathrm{DG(DG},t)} \leq R_{\mathrm{SD(DG)}} \quad \forall t \in \{1, \ldots, T\}; \quad \forall \mathrm{DG} \in \{1, \ldots, N_{\mathrm{DG}}\} \tag{6.2.16}$$

DR limits:

$$P_{\mathrm{DR\_A}(L,t)} \leq P_{\mathrm{Max}(L,t)}; \quad \forall t \in \{1, \ldots, T\}; \quad \forall L \in \{1, \ldots, N_L\} \tag{6.2.17}$$

$$P_{\mathrm{DR\_B}(L,t)} \leq P_{\mathrm{Max}(L,t)} \times X_{\mathrm{DR\_B}(L,t)}; \quad \forall t \in \{1, \ldots, T\}; \quad \forall L \in \{1, \ldots, N_L\} \tag{6.2.18}$$

EVs' limits:

$$E_{\mathrm{BatMin(EV},t)} \leq E_{\mathrm{Stored(EV},t)} \leq E_{\mathrm{BatMax(EV},t)} \tag{6.2.19}$$

$$P_{\mathrm{Dch(EV},t)} \leq P_{\mathrm{Max(EV},t)} \times X_{\mathrm{Dch(EV},t)} \times Z_{\mathrm{EV(EV},t)} \quad \forall t \in \{1, \ldots, T\}; \quad \forall \mathrm{EV} \in \{1, \ldots, N_{\mathrm{EV}}\}; X_{\mathrm{Dch(EV},t)} \text{ and } Z_{\mathrm{EV(EV},t)} \in \{0, 1\} \tag{6.2.20}$$

$$P_{\mathrm{Ch(EV},t)} \leq P_{\mathrm{Max(EV},t)} \times X_{\mathrm{Ch(EV},t)} \times Z_{\mathrm{EV(EV},t)} \quad \forall t \in \{1, \ldots, T\}; \quad \forall \mathrm{EV} \in \{1, \ldots, N_{\mathrm{EV}}\}; X_{\mathrm{Ch(EV},t)} \text{ and } Z_{\mathrm{EV(EV},t)} \in \{0, 1\} \tag{6.2.21}$$

$$X_{\mathrm{Ch(EV},t)} + X_{\mathrm{Dch(EV},t)} \leq 1 \quad \forall t \in \{1, \ldots, T\}; \quad \forall \mathrm{EV} \in \{1, \ldots, N_{\mathrm{EV}}\} \quad X_{\mathrm{Ch(EV},t)} \text{ and } X_{\mathrm{Dch(EV},t)} \in \{0, 1\} \tag{6.2.22}$$

$$E_{\mathrm{Stored}(V,t)} = E_{\mathrm{Stored}(V,t-1)} - E_{\mathrm{Trip}(V,t)} + \left(\eta_{c(V)} \times P_{\mathrm{Ch}(V,t)} - \frac{1}{\eta_{d(v)}} \times P_{\mathrm{Dch}(V,t)} \right) \times \Delta t \quad \forall t \in \{1, \ldots, T\}; \quad \forall V \in \{1, \ldots, N_V\} \quad t = 1 \rightarrow E_{\mathrm{Stored}(V,t-1)} = E_{\mathrm{Initial}(V)} \tag{6.2.23}$$

6.2.7 Conclusions

RES is the path toward a sustainable development, constraining GHG emissions. RES integration is already significant in some countries. The integration of RES in power systems and their operation requires important changes in the operation methods and business models. The main problems verified with RES are the variability and unpredictable nature of the primary natural energy sources. The power output is directly related to climate and weather conditions. Therefore, the operation is constrained by these natural conditions. The increasing number of RES units converts the traditional dispatch problem into a very large optimization problem for which several computational intelligence approaches have been proposed in the literature.

This section presented an ERS problem of large-scale nature solved with different kinds of algorithms from deterministic, hybrid, and meta-heuristics. The case study considered is a realistic 37 bus distribution network projected for a 2050 scenario. The network supplies energy to 1908 consumers and 1616 EVs (at 40% penetration). Several algorithms are compared including MINLP, SA, ERS2A, PSO, EPSO, MOPSO, and 2sPSO. The best quality solution is achieved by MINLP (25 037 m.u.) without variance but at a very expensive computational cost (16 hours) while meta-heuristics took close to five minutes in the worst-case algorithm. The best execution time is achieved by 2sPSO algorithm under 30 seconds, which is also the algorithm with the lowest mean cost (25 147 m.u.) in the meta-heuristic group. ERS2A is second-best within the meta-heuristic group with a mean execution time of 48 seconds and with 25 189 m.u. The best balance is achieved with SADT, a hybrid SA and mathematical technique that is able to deliver on average a cost of 25 048 m.u. with very little variance and under 60 seconds.

6.3 OPERATION AND CONTROL OF SMART GRID

Koichi Nara

Ibaraki University, Ibaraki, Japan

6.3.1 Introduction

Recently, many discussions have been proposed about such new type power delivery systems as Smart Grid (SmartGrids) [32–39], MicroGrid [40–43], Flexible, Reliable, and Intelligent ENergy Delivery System (FRIENDS) [44–46], Power park [47], Virtual power plant, and Demand Area Network [48]. The purposes of these new type power delivery systems are to connect a lot of DG systems and distributed energy storage systems (DESSs) into the existing power system, and to utilize them effectively for an efficient and reliable electric power supply. They also intend to maximize the capability of power supply through effectively using intelligent (smart) control and bidirectional communication technique.

Many problems must be solved to realize the above systems. The problems to be solved relating to develop them can be systematically classified into the following categories: namely, (i) systems configuration or systems design, (ii) systems operation and systems control, and (iii) systems management. The published papers in which intelligent systems are applied to these new type power delivery systems are not so many. Those published in IEEE PES major journals and conferences are shown below as reference numbers arranged according to the above (i)–(iii). In most of these papers, Tabu search (TS), PSO, multi-agent systems, fuzzy systems, and heuristic approaches are applied in planning, operation, and control of the new type power delivery system. However, many areas and problems are untouched by intelligent or heuristic approach (are solved by other methods) as shown in the beginning of Section 6.3.2, and more research through intelligent approaches may be expected in such areas and problems.

6.3.2 Problems for Systems Configuration or Systems Design

A specific system design technology for some of the above concepts has been proposed according to their characteristics. How to configure the specific system and how to realize the proposed functions are published in several papers [49–51]. By using the proposed technology, several demonstrative new type power delivery systems have been constructed and tested [43]. A heuristic technique and other intelligent methods are applied to the optimal DG allocation problems [52] and DESS allocation problems [53].

Among the above papers, in reference [50], a method to find the optimal network configuration of FRIENDS is proposed in which the cost of power interruptions is taking into consideration. The concept of FRIENDS is shown in Figure 6.3.1.

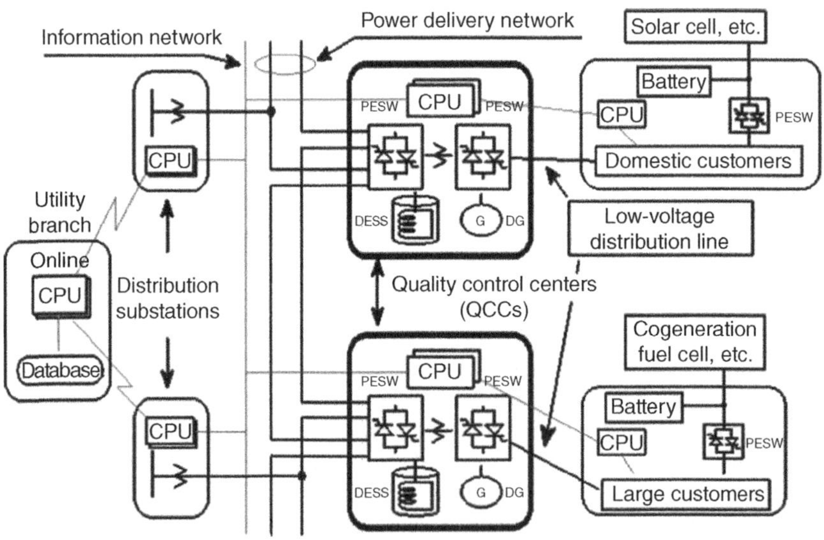

Figure 6.3.1 Concept of FRIENDS.

The problem is to determine HV power line configuration and DG's installations in power quality control center (QCC) so as to minimize the weighted sum of total facility installation cost (line installation cost and DG's installation cost) and cost of distribution loss under the constraints of expected interruption cost, power flow equation, etc. The problem can be mathematically formulated as follows under the following assumptions:

1. Location and capacity of each QCC are known.
2. Daily load pattern of QCC are known.
3. Available power delivery lines, available DGs, and their installation costs are known.
4. Objective function:

$$\text{Min.}\quad \alpha\left(\sum_{n=1}^{\text{ND}}(aX_n + bYN_n) + \sum_{m=1}^{\text{BR}} c_m YL_m\right) + \beta \sum_{t=1}^{T} \text{Oloss}^t \tag{6.3.1}$$

Constraints – (DG's maximum capacity)

$$X_n \in \left\{x_n^1, x_n^2, \ldots, x_n^i, \ldots, x_n^L\right\} \quad (n = 1, \ldots, \text{ND}) \tag{6.3.2}$$

(Expected power interruption cost)

$$\sum_{t=1}^{T}\sum_{r=1}^{\text{FLT}} \frac{1}{T} p_r \text{BLCost}^{rt} \leq \varepsilon \tag{6.3.3}$$

(Power supply to all the loads)

$$\exists R_{sn}^r \tag{6.3.4}$$

where

ND: total number of QCCs,
BR: total number of potential lines,
X_n: discrete variable for DG's capacity in nth QCC,
x_n^i: ith available capacity of DG for nth QCC (X_n must be selected from x_n^i),
YN_n: =1 if X_n is not 0; =0 otherwise,
YL_m: 0–1 variable for installation of line m (1: if line m is installed; 0: otherwise),
FLT: number of fault cases,
a: variable cost for capacity of DG installed at nth QCC,
b: fixed cost of newly installed DG at nth QCC,
C_m: installation cost of line m,
R_{sn}^r: power supply route from every node n to power source (substation) must exist in any fault case r,
Oloss^t: distribution loss at load pattern t,
T: number of load patterns under consideration,
ε: limit of expected amount of line overload,
α, β: weighting coefficients.

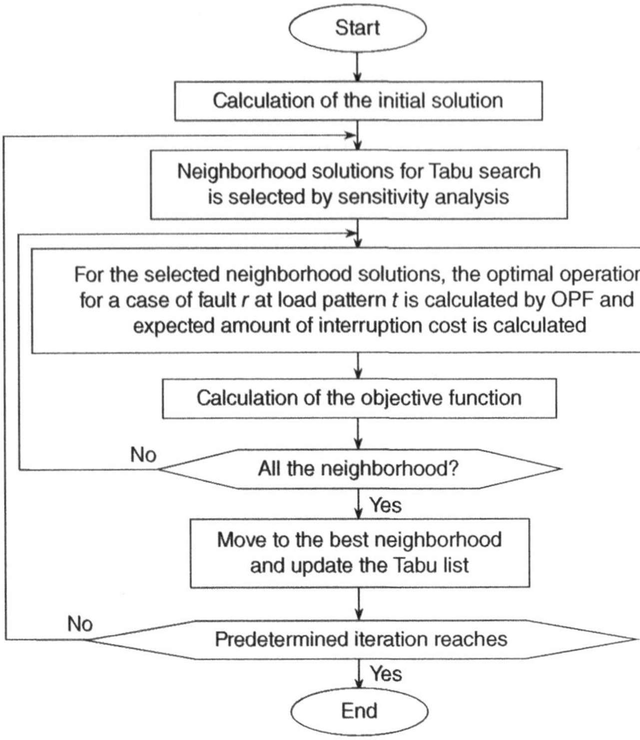

Figure 6.3.2 Solution algorithm of planning problem.

Here, BLCost^{rt} in Eq. (6.3.3) means the sum of power interruption costs of customers in fault case r at load pattern t. Since the problem is mathematically formulated as a complex mixed-integer programming problem, the TS method and the discrete optimal power flow (OPF) are employed to find the optimal network configuration.

The solution algorithm is shown in Figure 6.3.2.

The solution algorithm is applied to the real-scale distribution system and the optimal solution is shown in the paper.

6.3.3 Systems Operation and Systems Control

Several research results by using intelligent or heuristic techniques have been published for the following problems relating to systems operation and control of the new type power delivery systems. However, the following problems are so complex, and more research effort is required to realize the systems:

1. Power supply reliability enhancement by DGs and DESSs.
2. Trade-off of power fluctuation of generation by renewable energy resources (RES) and stable or controllable power generation by conventional ones [54].

3. Voltage control and evaluation of the value of reactive power [55, 56].
4. Evaluation and control of harmonics.
5. Complex distribution systems analysis with DGs.
6. Seamless connection and disconnection (islanded operation) of a part of the system from the main grid [57–61].
7. Congestion management of a new type system.
8. Autonomous control scheme of a new type system [62–66].
9. New protective relaying scheme [67–69].
10. Optimal operation technology of DGs and DESSs (including plug-in storage) [70–75].
11. Advanced metering and sensing.
12. Contingency analysis with DGs and DESSs.

Among these papers, reference [62] proposes a new voltage control method by autonomously controlling the power outputs of the distributed generators. Namely, the authors introduced a new idea named "Control Area of Distributed Generators" as shown in Figure 6.3.3 in the autonomous control algorithm.

The voltage deviation sensitivity factor is used to determine the "control area." The most sensitive generator to control the specific bus voltage is selected, and the resulting set of buses which is controlled by the same generator is defined as the "control area." The voltage deviation sensitivity factors for all the buses can be defined as the elements of impedance matrix.

In the method, first, each DG tries to eliminate voltage limit violations of the own control area simultaneously. Therefore, the control algorithm works as a multi-agent system. If all the violations are eliminated, then the algorithm quits. If DG cannot eliminate all the voltage violations in its own area, it requests to the neighboring DGs to control reactive power so as to eliminate the voltage violations of the area. The neighboring DGs accept or reject the "request" according to their status. The conditions to reply "reject" from the neighboring DG are as follows:

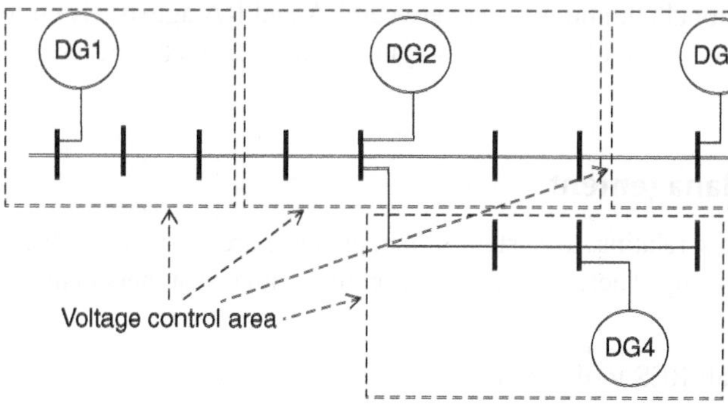

Figure 6.3.3 Voltage control area.

1. One's reactive power generation has already hit its limit.
2. Other DG requests the one reactive power of opposite direction.
3. One is controlling its reactive power to the opposite direction to eliminate the violations in its own area.
4. One has created voltage violation in its own area when one has controlled reactive power to the same direction by the same request, and currently is restoring the situation.

In reference [62], the simulation results of the method are compared with the former autonomous control algorithm by the authors, and the performance comparison of the methods is shown.

In reference [73], a coordination method of EVs charging in which a multi-agent system and a NN are used is proposed. Four types of agents are defined and used in the method: namely, the distribution system operator (DSO) agent, the regional aggregator (RA) agents, the local aggregator (LA) agent, and the EV agent.

The DSO agent is located at the primary substation, and responsible for the technical operation of the distribution network. The RA agent is located at the primary substation, and is responsible for managing the EV demand under satisfying distribution line capacity. The LA agents located at secondary substation are used as mediators between EV agents and RA agent. The EV agent represents the EV owners who determine EV connection periods, desired state of charges (SoCs), and vehicle to grid (V2G) capabilities.

The DSO agent produces EV charging of each hour of day under the constraints of LV cable loading and feeder voltages. The RA agent sends the available LV feeder capacity to each LV agent which chooses the schedule for EV battery charging. A network monitoring system provides the DSO agent with real-time measurements of the distribution grid so as to determine network flow, customer's load forecast, and emergency corrective actions.

The interaction diagram of agent operations during normal and emergency states are shown in Figures 6.3.4 and 6.3.5.

The authors claim that the following five aspects are not considered in the method. They are: (i) EV load forecasting, (ii) pricing framework and price signals to EV agent, (iii) additional communication between LA and RA agents, (iv) fair billing framework, and (v) fault tolerance investigation. More research is necessary for the above aspects.

6.3.4 System's Management

The following problems relating to system's management are normally solved politically and economically. Therefore, the collaboration with researchers in these areas is necessary.

1. Incentive to install RES to the grid.
2. Power management and DR [76, 77].

3. Who should pay investment cost?
4. Value of power wheeling or power backing.
5. Optimal operation of distributed generators [78–80].
6. Asset management and predictive maintenance.
7. Load modeling or prediction through smart metering.
8. Dispatcher training simulator of a new system [81].

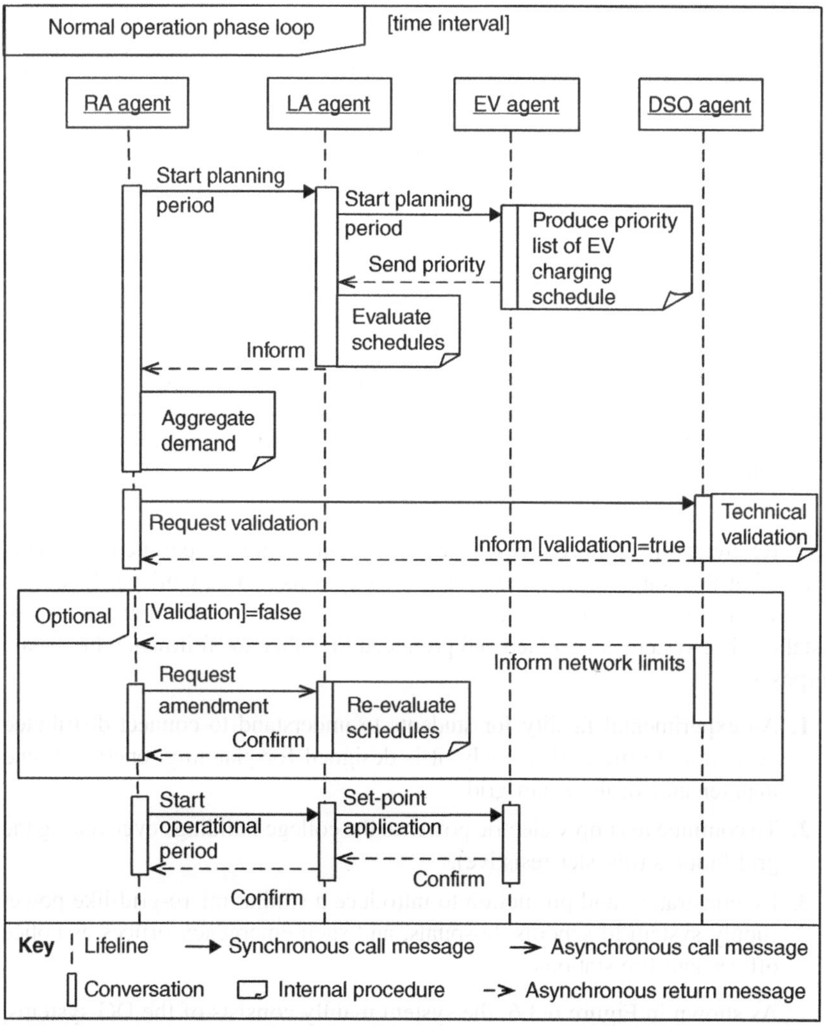

Figure 6.3.4 The interaction diagram of agent operations during normal states. Source: the figure is modified from Ref. [73].

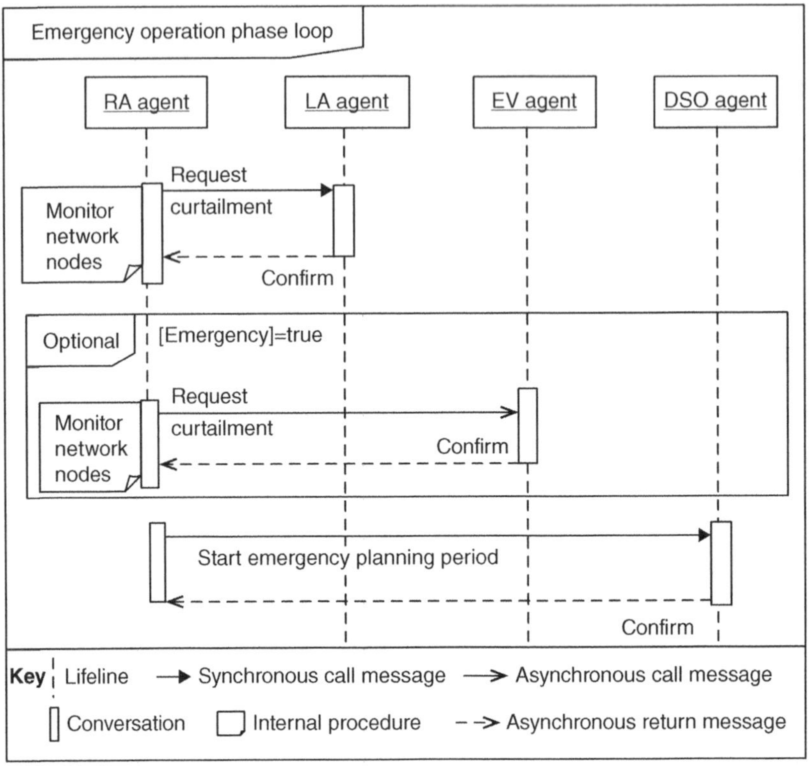

Figure 6.3.5 The interaction diagram of agent operations during emergency states. Source: the figure is modified from Ref. [73].

Reference [81] introduces a real-scale smart grid experimental system installed at the National Institute of Technology, Fukushima College, Japan. So as to develop new intelligent functions of smart grid, real-time simulator is also installed. The system is designed and produced to realize the following three main purposes:

1. An experimental facility for students to understand to connect distributed generators to the grid: namely, it is designed for planning, operation, and maintenance of the smart grid.
2. To continue to supply electric power to the college buildings even during the grid failures (disaster resistive).
3. Demonstration and promotion to introduce a similar micro-grid-like power supply system to schools, hospitals, and such emergency offices as police offices and fire stations.

As shown in Figure 6.3.6, the system mainly consists of the DG systems, uninterruptible power supply (UPS) system, a bidirectional inverter for batteries, and a supervisory control system. The distributed generators include a

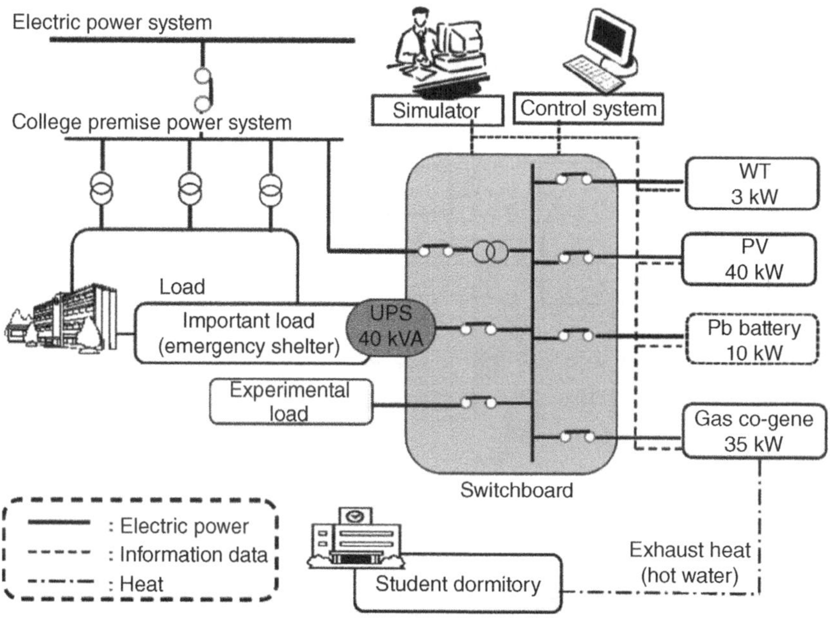

Figure 6.3.6 Structure of real-scale smart grid experimental system.

gas-engine cogeneration system, a wind power generation system, and a solar power generation system. In normal operation, the whole electricity from the distributed generators is supplied to the load in the college. Exhausted heat from the gas-engine cogeneration system is used to supply hot water to the student dormitory. When a blackout occurs in the utility power system, the system mode can automatically shift to the islanded operation mode to supply electricity to an important load of the main building through uninterrupted power supply system. The frequency and voltage of the islanded system are controlled through the constant voltage constant frequency (CVCF) inverter of the gas-engine cogeneration system. An experimental load connection panel (switchboard) is also facilitated in the system so as to examine such load control apparatus as smart meters, etc.

The newly installed distributed generators are connected to the existing power receiving system through newly installed 6.6 kV/200 V transformer. The system structure and a protective relaying system installed in the system are shown in Figure 6.3.7. A relationship of the tripping signals is shown by dotted line in the figure. A reverse power flow to the power system is not allowed because the power from the distributed generators is so small compared with the total load of the college. The reverse power flow protective relay and a under power detection relay are installed for this purpose. The protective relaying scheme is followed according to the Japanese grid code: namely, a fault in the utility system is detected through protective relaying system, and SW2 in Figure 6.3.7 is opened to disconnect distributed generators from the utility system.

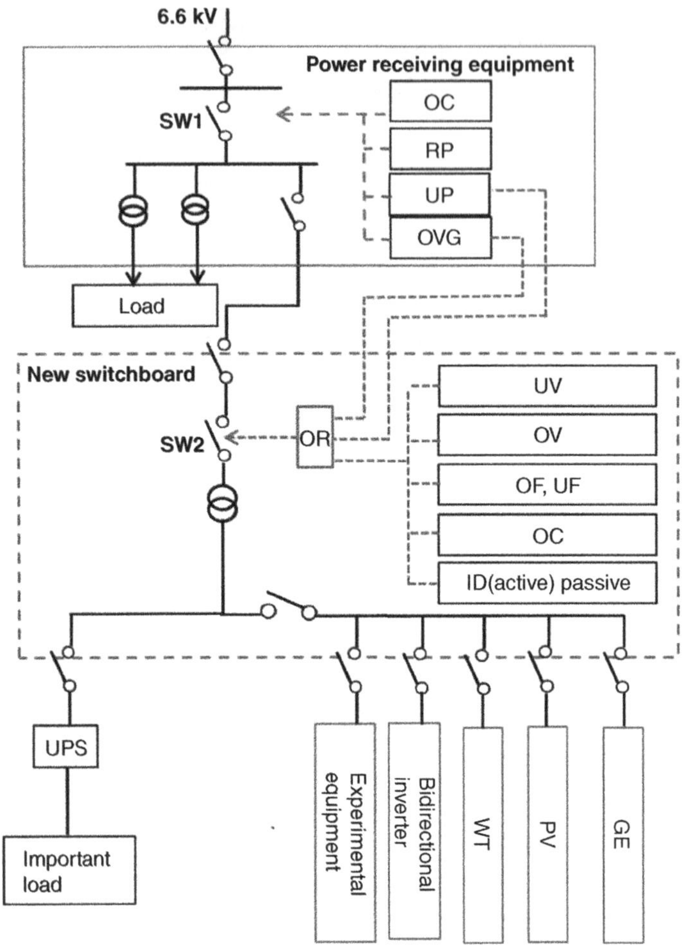

Figure 6.3.7 Protection scheme of the system.

The system has the following supervisory control and data acquisition (SCADA) functions: (i) supervision and control of facilities and systems, (ii) data logging, (iii) reporting, and (iv) daily operation scheduling. The system is also equipped with the simulator which can simulate the behavior of the real smart grid system. The simulator can be used for an experiment and training of the students, and for a research and development of new intelligent control functions of smart grid. It can simulate monitoring, control, and operational functions of the system by using a real-time or past data which can be served from the hard disk drive (HDD). The simulator also has data logging functions. Since simulation results can be memorized on the HDD, if necessary, detailed analysis of the simulation results through intelligent control can be available after the simulation.

Since the simulator has all the SCADA functions of the real system, it is designed to be able to back up all the SCADA functions when a human interface computer is inoperative.

The simulator software consists of the following simulation sub-models:

1. Gas-engine cogeneration model
2. PV power generator model
3. Wind-power generation model
4. UPS system model
5. Switchboard (Metal-clad) model
6. SCADA functions
7. Others

6.3.5 Conclusion

In this section, a concept of the smart community is not included although various definitions of the "smart grid" exist. Namely, intelligent systems applications to the systems configuration and design, operation and control, and management of the micro-grid are specifically illuminated. Some typical optimization methods are illustrated rather precisely in the area of design, operation, control, and management of the micro-grid.

Relatively small number of papers can be found in the area of intelligent systems application to the smart grid so far. Since the application areas of asset management, load modeling and prediction, etc. in the management area are the intelligent system's home ground, and only few papers are published in the management area of the smart grid so far, further advancement of the intelligent systems application to the area can be expected.

6.4 COMPLIANCE OF REACTIVE POWER REQUIREMENTS IN WIND POWER PLANTS

Aimilia-Myrsini Theologi[1], *Jose Rueda*[2], *Mario Ndreko*[3], *and István Erlich*[4]

[1]*Jedlix Smart Charging, Rotterdam, The Netherlands*
[2]*TU Delft, Delft, The Netherlands*
[3]*TenneT TSO GmbH, Bayreuth, Germany*
[4]*University of Duisburg-Essen, Duisburg, Germany*

6.4.1 Introduction

Nowadays, offshore wind is a competitive power source and an increasingly attractive investment with stable income returns and various benefits of the electric power generation. Europe is considered as the front-runner in this field, where during the year 2015 new offshore capacity of 3.02 GW was connected to the grid [82]. However, the high penetration of the wind power into the energy systems holds many technical/operational challenges, which require further analysis. Offshore wind power plants are required to provide reactive power support during both the steady state as well as during AC fault conditions.

The transmission system operators (TSO) of each country have defined Grid Code Requirements in order to ensure the safe, secure, and reliable operation of power systems. Traditionally, the optimal reactive power management sources in the synchronous transmission systems are designed for operation in an uncoordinated manner, i.e. meeting local targets as seen at the terminal bus of each device. Although the reactive power requirement at the point of common coupling (PCC) can be achieved without major drawbacks, the traditional approach mentioned above is quite conservative. The emerging approach involving coordinated management of reactive sources, however, allows the achievement of several operational objectives, such as minimum power losses and reduction of stress or disturbances for the controllable devices, i.e. transformers, simultaneously [83]. The existing technologies for data communication and acquisition render the coordinated planning feasible. Since, reactive power management appertains to the mixed-integer optimization problem with restricted computing budget, a new heuristic algorithm called mean–variance mapping optimization (MVMO) is used.

In this chapter, two different approaches for the optimal dispatch of reactive power sources are suggested. According to the first approach, the optimization is performed for every current operating point and results in minimum transmission losses. However, the cost of the onload tap changer (OLTC) is not considered. In order to solve this problem, a second approach is proposed, which includes the number of tap changes in the objective function. Besides, the optimization is performed over a predicted time period by incorporating wind speed forecasting, which is based on NNs and accounts for the active power per each wind turbine.

Simulations have been conducted for the AC-connected Dutch nearshore wind farm Borssele, located in North Sea. Several test cases have been investigated, in order to demonstrate the effectiveness of MVMO.

6.4.2 Problem Definition

Mathematically, optimal reactive power management refers to finding a set of optimal decision vectors over a planning horizon that minimizes an objective function under several predefined constraints [84]. The problem to be solved has two different formulations in the presented approach:

1. *Optimization for the current operating point* – The optimization task aims to the minimization of the total transmission losses in the system and the objective function is formulated as follows [85]:

$$\text{Minimize OF} = P_{L,t}, \quad t = 1, 2, \ldots, 24 \tag{6.4.1}$$

where, OF: objective function,
$P_{L,t}$: total real power losses for hour t.

2. *Optimization for a predicted time horizon* – The optimization is performed for a given scenario, which includes a set of future operating points on a 24-hour time horizon.

As presented in Figure 6.4.1, the wind speed prediction for 24-time steps ahead is received by the optimization algorithm as input and then, MVMO suggests the optimal OLTC tap settings of both offshore and onshore transformers together with the optimal reactive power reference for each wind turbine. The formulation of the objective function is multi-objective and is given by Eq. (6.4.2), although the problem is treated as single objective due to the use of the weight coefficients [86].

$$\text{Minimize } \text{OF} = \sum_{t=1}^{24} (w_1 \cdot P_{L,t} + w_2 \cdot \text{OLTC}_{\text{cost},t}) \tag{6.4.2}$$

where,

$P_{L,t}$: total real power losses of hour t,

$\text{OLTC}_{\text{cost},t}$: total operation cost of the OLTC for hour t, given by

$$\text{OLTC}_{\text{cost},t} = w_3 \cdot |\text{tap}_t - \text{tap}_{t-1}| \tag{6.4.3}$$

where,

w_1, w_2, w_3: weight coefficients corresponding to cost values,

tap_t, tap_{t-1}: tap position for hour t and $t-1$.

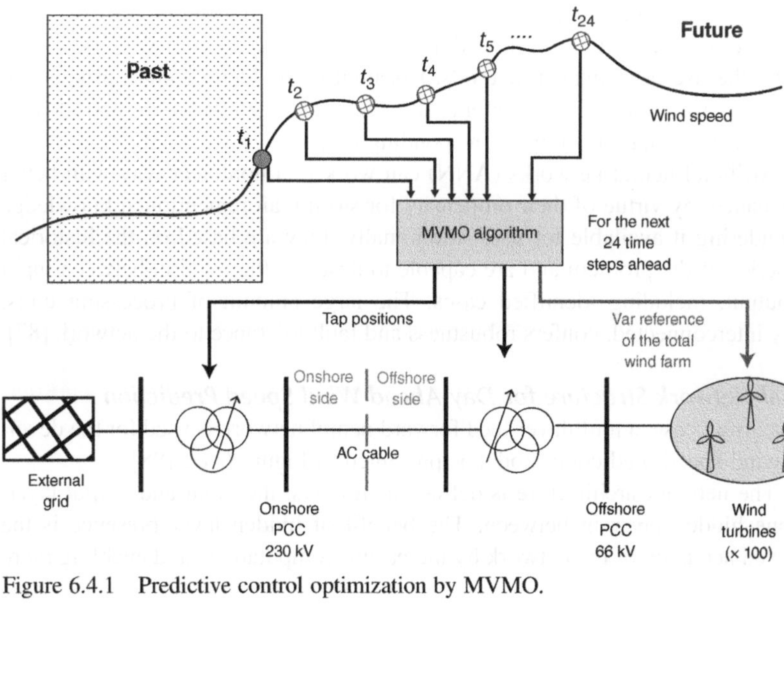

Figure 6.4.1 Predictive control optimization by MVMO.

Eventually, the generic formulation of the problem in both approaches is stated as follows:

$$\begin{aligned}&\text{Min.}\quad \text{objective function (OF)}\\&\text{Subject to}\quad v_{\min} \le v \le v_{\max}\end{aligned} \tag{6.4.4}$$

$$i \le i^{\text{lim}} \tag{6.4.5}$$

$$s \le s^{\text{lim}} \tag{6.4.6}$$

$$\text{and search space given by}\quad q_{\text{WTG}}^{\min} \le q_{\text{WTG}} \le q_{\text{WTG}}^{\max} \tag{6.4.7}$$

$$\text{tap}_{\text{Tr, min}} \le \text{tap}_{\text{Tr}} \le \text{tap}_{\text{Tr, max}} \tag{6.4.8}$$

where,

v: voltage magnitude of the buses,

i: current flow through the cables, line, and transformers,

s: power flow through the transmission line,

q_{WTG}: reactive power reference of wind turbine WTG,

tap_{Tr}: tap position of transformer Tr.

6.4.3 NN-Based Wind Speed Forecasting Method

As described in the preceding section, the input of MVMO proceeds from wind speed prediction algorithm. Thus, for the optimal operation of the wind farm, an accurate wind speed prediction tool is beneficial. The estimation of the wind energy requires configuration procedures, the implementation of which is facilitated by the use of statistical methods. Essentially, the identification problem of wind speed is comprised of the appraisal of the remarkable behavior and the correlation of this behavior to the corresponding output [87].

Artificial neural networks (ANN) can work in a nonlinear way with better performance, by virtue of their propensity for storing aforementioned knowledge and rendering it available for use. Additionally, they are based on the historical time series of the problem and are capable to describe the relation between input and output, including identified cases. The large amount of processing units, locally interconnected, confers robustness and fault tolerance to the network [87].

Neural Network Structure for Day-Ahead Wind Speed Prediction

In the current case, a multilayer feed forward neural network is used for the developed wind speed prediction model as presented in Figure 6.4.2 [88].

The network architecture is defined as follows: the input and output layer, and one hidden layer in between. The benefit of hidden layer presence is the improved accuracy of the network by increasing computations and enabling more

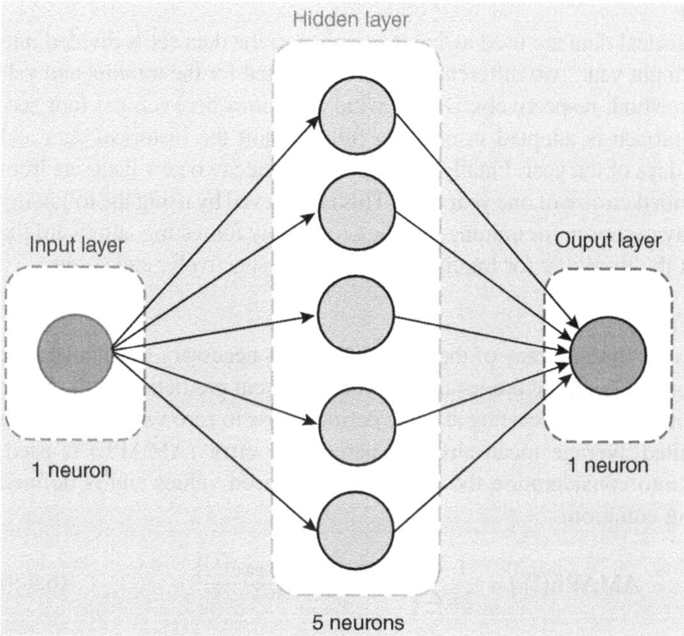

Figure 6.4.2 Multilayer perceptron.

complex operations. It is indicated that feed forward types of network permits only unidirectional data flow from the input to the output layer, without feedback connections. Each neuron can be identified with a simple logistic regression, whose input values are weighted with the appropriate weights, subsequent to their import. Ultimately, the output is resulting from a sigmoid transfer function, in which the summation of the weights and bias is used as input.

Implementation

The NN-based method was implemented in MATLAB by using the "Neural Network Toolbox" and the work flow of the NN design process is stated as follows [89]:

- Data collection
- Network creation
- Network configuration
- Weights and biases initialization
- Network training
- Network validation
- Network testing

Data Partition

One year of historical data are used as input sample and the data set is divided into days. From the input year, two different subsets are created for the training and validation of NN method, respectively. Due to wind variations between the four seasons, a new approach is adopted in order to fully exploit the historical data and consider all the days of the year. Finally, we choose to create two new data sets from the input year with data size of one year each. This is achieved by using the following division: first day is chosen for training and the second day for testing, afterward the second day and the third day for training and testing, respectively, and so on.

Evaluation Criteria

In order to evaluate the accuracy of the trained NN, it is necessary to calculate the forecasting error and compare the results between different prediction tools. Since in the case of wind speed forecasting at some periods close to zero values occur, the error metric called average mean absolute percentage error (AMAPE) is used. AMAPE takes into consideration the mean of wind speed values and is defined by the following equation:

$$\text{AMAPE}(\%) = \frac{1}{24}\sum_{t=1}^{24}\frac{\left|v_{m,\text{act}}(t) - v_{m,\text{pred}}(t)\right|}{v_{m,\text{act}}^{\text{AVERAGE}}} \tag{6.4.9}$$

where,

$v_{m,\text{act}}(t)$: actual wind speed for hour t,

$v_{m,\text{pred}}(t)$: predicted wind speed for hour t,

$v_{m,\text{act}}^{\text{AVERAGE}}$: average actual wind speed for 24-hour horizon, given by

$$v_{m,\text{act}}^{\text{AVERAGE}} = \frac{1}{24}\sum_{h=1}^{24} v_{m,\text{act}}(t) \tag{6.4.10}$$

6.4.4 Mean Variance Mapping Optimization Algorithm

MVMO belongs to the population-based optimization algorithms and can be applied in multi-objective mixed-integer nonlinear problems, such as optimal tuning of controllers, due to its conceptual simplicity, easy adaptability, and reduced human intervention. The enhanced performance of MVMO in comparison with other evolutionary algorithms appears in terms of convergence speed. This is mainly attributed to the so-called mapping function evolutionary operator, which adaptively switches the search priority between exploration to exploitation according to recorded statistics of the evolved best solutions so far in a continuously updated "solution archive" (i.e. adaptive memory) [90].

Flowchart

In Figure 6.4.3, the methodology of MVMO for the proposed approach is described. The procedure starts with the initialization of the parameter settings, such as the

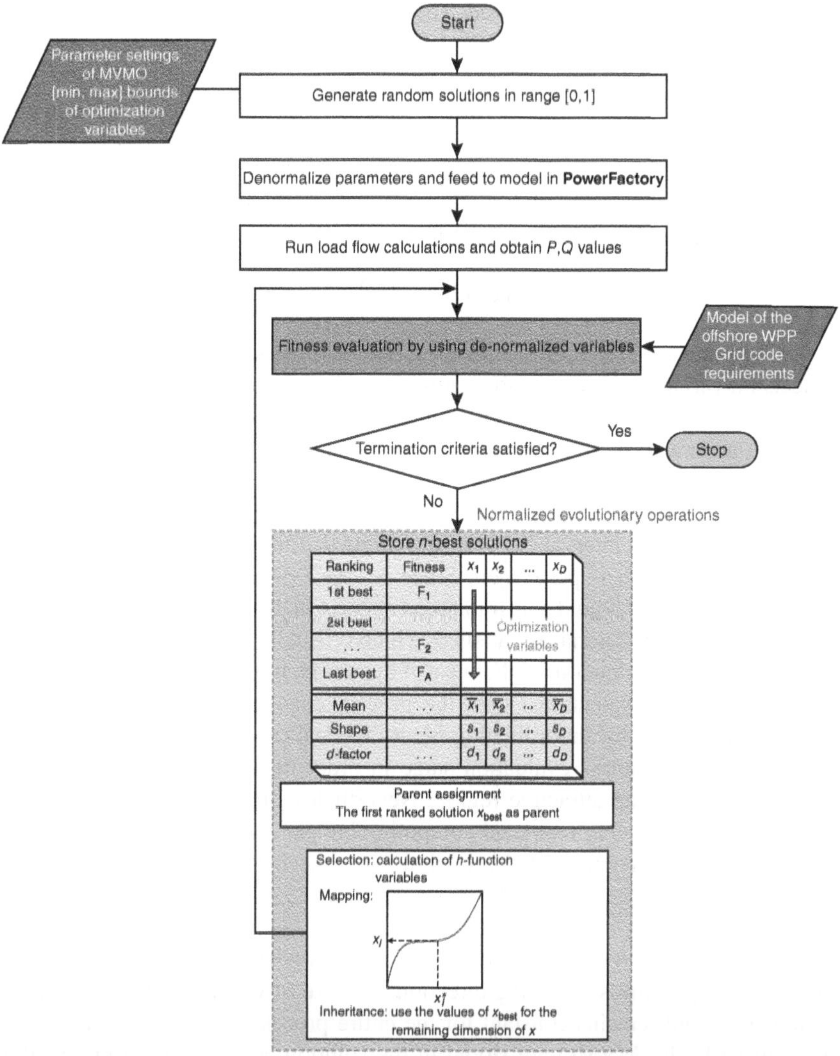

Figure 6.4.3 MVMO-based procedure for optimal reactive power management.

archive size, the selection method, and the maximum number of iterations. The searching space of all variables is confined in [0, 1] and therefore the real min/max have to be normalized to this interval. During every iteration step, the solution vector cannot violate the required boundaries and only a single offspring is generated. Thus, the characterization of MVMO as single-agent search algorithm is owed to the latter property. With respect to other heuristic techniques, MVMO uses a special mapping function described by mean and shape variables, which transforms a variable x_i^* with unity distribution to another variable x_i. Subsequently, during fitness evaluation, the archive is updated only if the new solution is better compared with the previously stored. The major advantage of MVMO is the minimization of the risk associated with premature convergence, which contributes to the confrontation of zero-variance [91].

MVMO-Based Procedure

Initialization In the first approach, where the optimization is performed for the current operating point, the initial candidate solution is randomly generated between the specified boundaries as follows:

$$x_i^{\text{init}} = x_i^{\min} + \text{rand}\left(x_i^{\max} - x_i^{\min}\right), \quad i = 1, 2, \ldots, D \tag{6.4.11}$$

The index $i = 1, 2, \ldots, D$ stands for problem dimension and D for the number of decision variables. However, the optimization is performed in a predictive manner, after the first hour of the day, the initial candidate solution for the subsequent hours are generated by the best solutions obtained from the previous hour.

Fitness Evaluation and Local Search The decision variables are de-normalized from the interval [0, 1] to the original [min, max] boundaries before the fitness evaluation is performed. Since MVMO performs within normalized range, no violation of bound constraints can occur. Finally, the search process stops after the termination criteria are satisfied, which is usually specified as a predefined number of fitness evaluations. Otherwise, if there is no improvement of fitness over successive fitness evaluations, then the process can be also terminated. Local search strategy, e.g. by subordinating other classical or heuristic algorithms, can be added into the fitness evaluation stage in order to intensify the search once MVMO has found an attractive region. Nevertheless, this option is not used in this work due to the very restricted computing budget and in order to exclusively ascertain the effectiveness of the evolutionary mechanism of MVMO in this pure form.

Solution Archive The solution archive serves as the knowledge base for guiding the algorithm's searching direction, the size of which is set in the initialization stage and remains constant for the entire process. The n best individuals obtained so far by MVMO are stored in the solution archive. The filling of the archive obeys a descending order of fitness over the iterations, as presented in Figure 6.4.4. Consequently, the overall best found so far is always the first ranked individual. Once the archive is full, an update is conducted only if the solution fitness evaluation revealed that the new solution is better than these already stored in the archive. Finally, since the fitness improves over the iterations, the stored solutions in the archive keep changing.

After every update of the archive for each optimization variable, the mean and shape variable are computed, respectively:

$$\bar{x}_i = \frac{1}{n}\sum_{j=1}^{n} x_i(j) \tag{6.4.12}$$

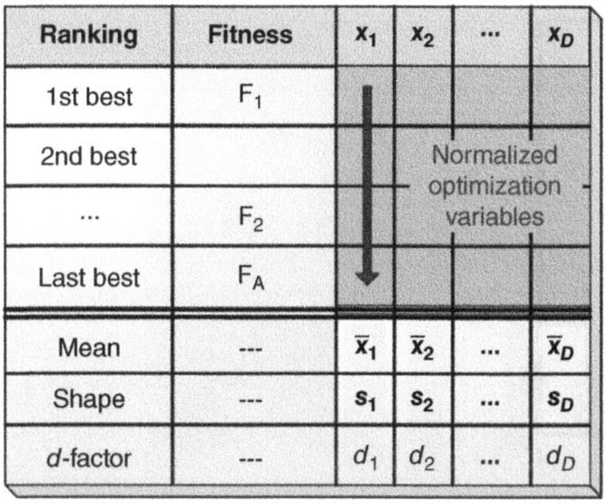

Figure 6.4.4 Solution archive.

$$s_i = -\ln(v_i) \cdot f_s \tag{6.4.13}$$

where the variance is calculated only for different variables in the archive as follows:

$$v_i = \frac{1}{n}\sum_{j=1}^{n}\left(x_i(j) - \overline{x}_i\right)^2 \tag{6.4.14}$$

At the beginning v_i is set to 1, since $\overline{x_i}$ corresponds with the initialized value of x_i. The shape variable computed s_i is one of the mapping function inputs with strong influence on its geometric characteristic shape. For this reason, the scaling factor f_s, which allows controlling the form of the mapping function and the search process, is involved in the calculation of s_i.

Offspring Generation MVMO uses a random sampling function for creating an offspring. In order to generate new solution, in every iteration the solution with the best fitness so far is used. The distribution of the new variable x_i does not correspond with any of the well-known distribution functions. Given a random number x_i^* from the interval [0, 1], the new value of each selected dimension x_i is described mathematically by

$$x_i = h_x + (1 - h_1 + h_0) \cdot x_i^* - h_0 \tag{6.4.15}$$

where h_x, h_1, and h_0 are the inputs of the mapping function based on different inputs given by

$$h_x = h\left(x = x_i^*\right) \tag{6.4.16}$$

$$h_1 = h(x = 1) \tag{6.4.17}$$

$$h_0 = h(x = 0) \tag{6.4.18}$$

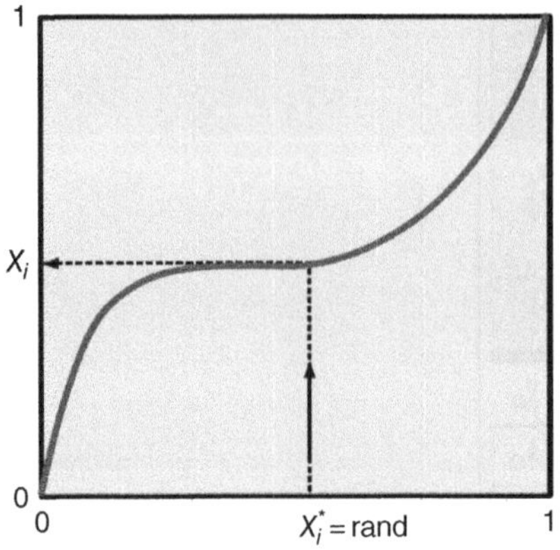

Figure 6.4.5 Variable mapping.

Both input and output of the mapping function are always between the range [0, 1]. The definition of the transformation mapping h-function is the following:

$$h(\overline{x_i}, s_1, s_2, x) = \overline{x_i} \cdot (1 - e^{-x \cdot s_{i1}}) + (1 - \overline{x_i}) \cdot e^{-(1-x) \cdot s_{i2}} \tag{6.4.19}$$

As illustratively shown in Figure 6.4.5, the h-function transforms the variable x_i^* varied randomly with unity distribution to another variable x_i, which is concentrated around the mean value calculated from the archive. The variation of $\overline{x_i}$ implies shifting of the curve between the original lower and upper boundaries of the search range, while the variation of $s_{i,1}$ and $s_{i,2}$ affects the bent shape of the curve, i.e. emphasized either exploration or exploitation.

When the accuracy need to be improved or more global search is required, the factor f_s should be increased ($f_s > 1$) and decreased ($f_s < 1$), respectively. Therefore, f_s can be used to change the shape of the function.

6.4.5 Case Studies

A software-based platform for automated calculations, e.g. forecasting, power flow calculations, and iterations via optimization algorithm, is built by creating special routines/scripts in MATLAB R2015b, Python 3.5, and DIgSILENT PowerFactory 2016.

Test System Description

The nearshore wind power plant used in this chapter is presented in Figure 6.4.6. The test system layout used is taken from a future wind power plant project to be connected in the Dutch transmission system, termed as the Borssele plant. The plant consists of two zones, I and II, connected via a 22 km HVAC cable. One hundred fully rated-converter wind turbines each rated at 6 MW are used for the simulations. The internal power transmission of each 300 MW zone of the wind farm is

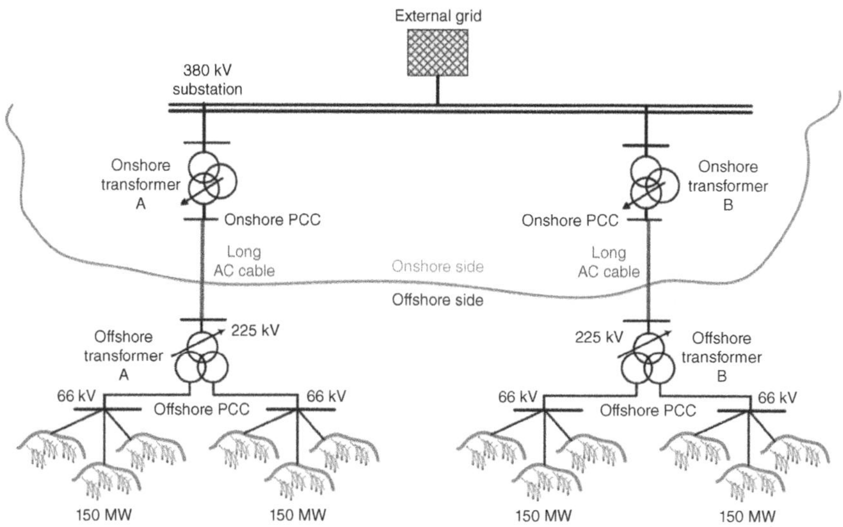

Figure 6.4.6 Borssele wind farm layout with AC cable.

realized by the wind turbine transformers each rated at 0.69/66 kV, multiple cables with different lengths, and four step-up onload tap-changing transformers of 6.7 MVA, at the offshore and onshore side, rated at 230/66/66 kV and 380/225/33 kV, respectively [92].

Parameters for MVMO

The parameters required for the MVMO algorithm are set to predefined values, which are summarized in Table 6.4.1. It is worth mentioning that these parameters were tuned by performing extensive analysis with respect to a single-parameter change within few optimization trials.

Study Cases

As described in the previous section, the grid was optimized for only transmission losses, and then for losses and number of tap changes together. The characteristics of the different cases under examination are summarized in Table 6.4.2.

TABLE 6.4.1 MVMO Parameters

Parameters	Values
Size of archive	4
Number of variables changed randomly	15
Maximum number of iterations	1000
Scaling factor	2

TABLE 6.4.2 Study Cases for the AC-connected Wind Farm

	Optimization For	Optimize Tap Positions For	Optimization Variables
Case 1	Current operating point	Offshore transformers (x2)	102
Case 2	A predicted time horizon		
Case 3	Current operating point	Onshore transformers (x2)	102
Case 4	A predicted time horizon		
Case 5	Current operating point	On- and off-shore transformers (x2)	104
Case 6	A predicted time horizon		

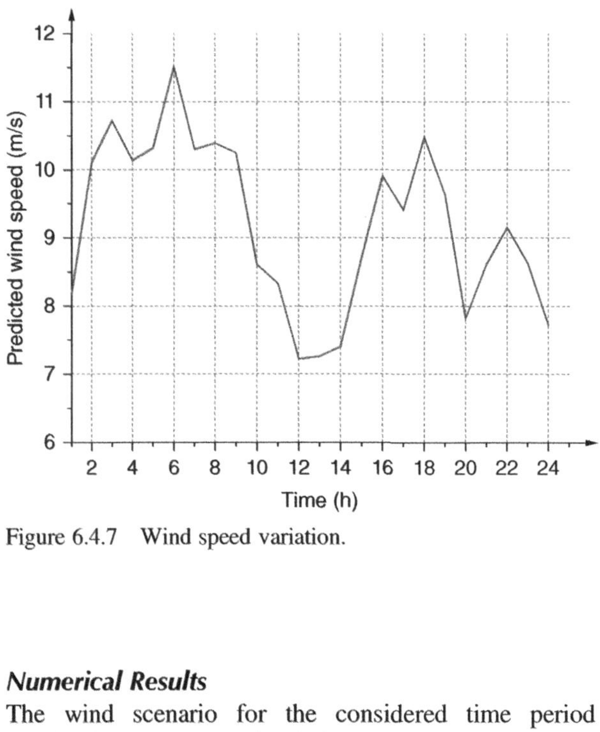

Figure 6.4.7 Wind speed variation.

Numerical Results

The wind scenario for the considered time period of 24 hours is the result of the NN-based wind speed forecasting method. The wind profile, shown in Figure 6.4.7, is used for the simulations in the case of the nearshore wind farm.

The problem in this study case has 102 (or 104) optimization variables, comprising 100 continuous variables associated with wind turbine reactive power setpoints and 2 (or 4) discrete variables associated with stepwise adjustable onload transformers' tap positions. The computing budget is restricted to 1000 problem evaluations (i.e. 1000 AC power flow calculations).

CASE 1

A real-time reactive power management is performed, where the formulation of the objective function is defined by Eq. (6.4.1) and the vector of the parameters to be optimized is the following:

$$x = \left[Q_{\mathrm{WTG},1}, \ldots, Q_{\mathrm{WTG},100}, \mathrm{tap}_{\mathrm{on,A}}, \mathrm{tap}_{\mathrm{on,B}}\right] \tag{6.4.20}$$

The reactive power reference for each wind turbine derived from the optimization are according to the Grid Code Requirements, as presented in Figure 6.4.8a [93], due to the normalized search space of MVMO, which ensures that bound constraints are never violated. This is an advantage with regard to other algorithms, since the algorithm does not require extra computing effort to repair solutions to lie within the [min, max] boundaries. Each set of points arranged in the same horizontal line refers to different values of wind speed.

As presented in Figure 6.4.8b, the smallest hourly reduction of the real power losses with regard to the initial losses calculated for zero reactive power reference at every wind turbine is around 15% and occurs for higher wind speed values between the range of 8.5–11.5 m/s (cf. Figure 6.4.7). By contrast, for lower wind speed values, e.g. between the range of 7–8.5 m/s, the hourly reduction percentage of the losses is much higher and reaches a maximum value of 48%.

While the reactive power levels of the MV- and LV-side of the transformers (66 kV buses) are maintained within the predefined envelope, as shown in Figure 6.4.8d, it is observed that the requirements for specific reactive power absorption/injection at the onshore PCC are not satisfied according to Figure 6.4.8e. This is due to the transformer tap positions, shown in Figure 6.4.8c as derived from the optimization algorithm. The total number of the tap movements within 24 hours is 38 and 85 for A and B onshore transformer, respectively. Consequently, the change of the tap positions exclusively at the onshore transformers is not feasible, since the deviation of the reactive power is more than the acceptable ±0.1 p.u. (=±30 MVAr) at the onshore PCC.

CASE 2

A predictive reactive power management scheme is applied. Here, the vector of the parameters to be optimized is defined as in Case 1. The formulation of the objective function is given by Eq. (6.4.2).

The reactive power reference per each wind turbine is derived from the optimization according to the Grid Code Requirements, since all the values are inside the curve presented in Figure 6.4.9a. Each set of points arranged in the same horizontal line refer to different values of wind speed.

As presented in Figure 6.4.9b, the difference between the cumulative initial and optimum cost is estimated around 38.16% for the 24-hour time horizon under investigation. The reactive power levels of the MV- and LV-side of the offshore

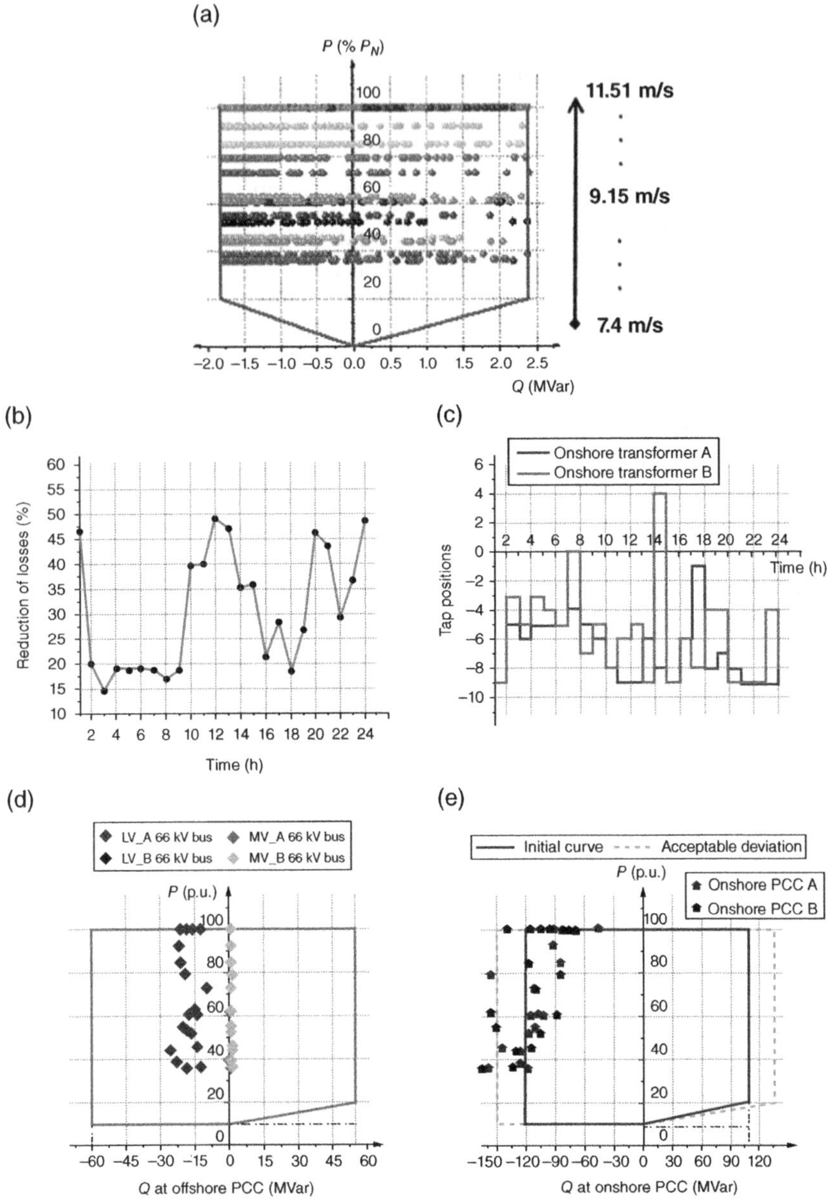

Figure 6.4.8 Case 1: (a) Hourly Q set-points of every wind turbine, (b) hourly reduction of wind farm active power losses, (c) OLTC tap positions – onshore transformers, (d) reactive power at the offshore PCC, (e) reactive power at the onshore PCC.

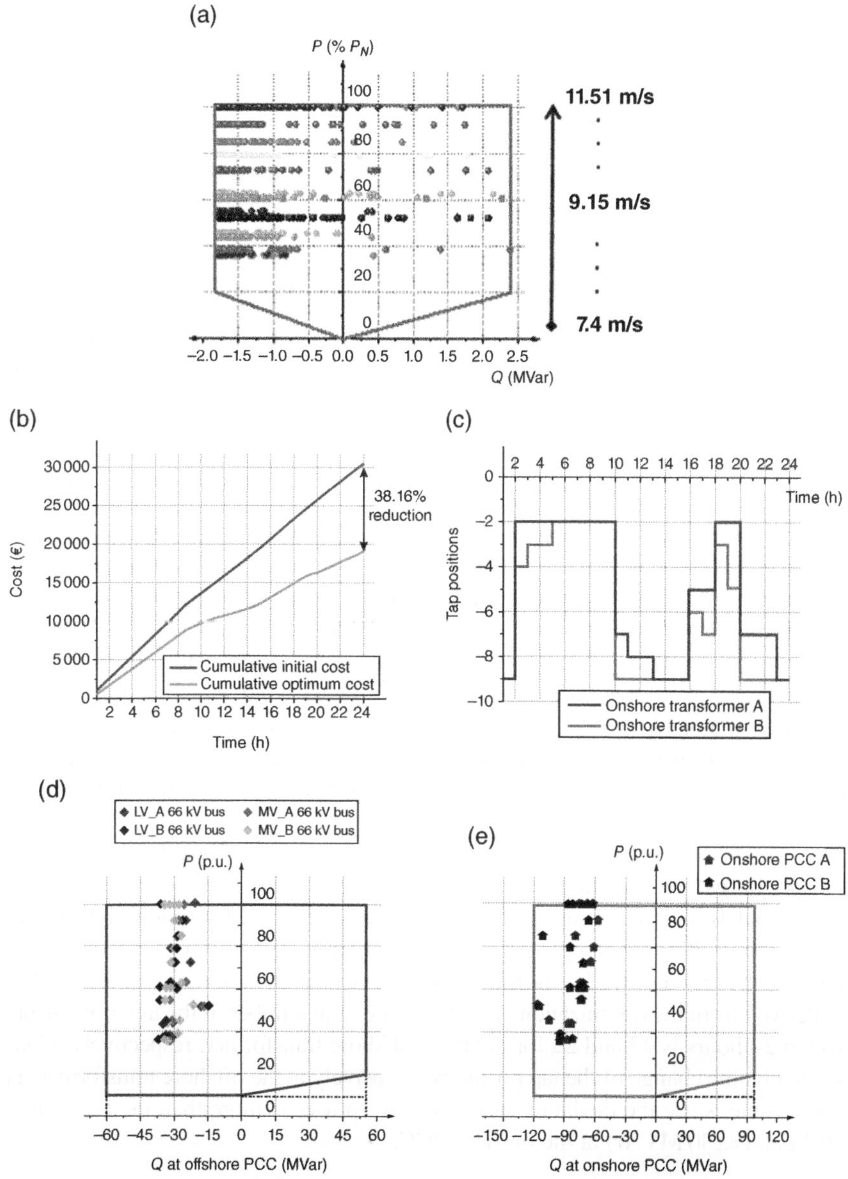

Figure 6.4.9 Case 2: (a) Hourly Q set-points of every wind turbine, (b) reduction of cumulative cost in the wind farm, (c) OLTC tap positions – onshore transformers, (d) reactive power at the offshore PCC, (e) reactive power at the onshore PCC.

transformers (66 kV buses) are maintained within the predefined envelope, as shown in Figure 6.4.9d. In addition, it is observed that the reactive power absorption/injection at the onshore PCC is within the required boundaries according to Figure 6.4.9e, in comparison with Case 1. Eventually, the main target of this approach, for minimizing the OLTC tap changes during daily operation is achieved. As displayed in Figure 6.4.9c, the number of the total tap movements within 24 hours was reduced to 28 for each of the A and B onshore transformers.

CASE 3

A real-time reactive power management is performed, where the formulation of the objective function is defined by Eq. (6.4.1) and the vector of the parameters to be optimized is the following:

$$x = \left[Q_{\text{WTG},1}, \ldots, Q_{\text{WTG},100}, \text{tap}_{\text{off,A}}, \text{tap}_{\text{off,B}}\right] \qquad (6.4.21)$$

The reactive power reference per each wind turbine is derived from the optimization according to the Grid Code Requirements, since all the values are inside the curve presented in Figure 6.4.10a. Each set of points arranged in the same horizontal line refer to different values of wind speed.

As presented in Figure 6.4.10b, the minimum hourly reduction of the real power losses is around 18.5% and occurs for higher wind speed values between the range of 8.5–11.5 m/s. The minimum point corresponds to an hourly reduction of 17.5%. For lower wind speed values between the range of 7–8.5 m/s, the hourly reduction percentage of the losses is much higher and reaches a maximum value of 32%.

While the reactive power levels of the MV- and LV-side of the transformers (66 kV buses) are maintained within the predefined envelope, as shown in Figure 6.4.10d, it is observed that the reactive power absorption/injection at the onshore PCC is completely out of the required boundaries according to Figure 6.4.10e. This owed to the transformers' tap positions, shown in Figure 6.4.10c as derived from the optimization algorithm. The total number of the tap movements within 24 hours is 43 and 25 for A and B offshore transformer, respectively. Consequently, the change of the tap positions exclusively at the offshore transformers is not feasible, since the deviation of the reactive power is more than the acceptable ±0.1 p.u. (=±30 MVAr) at the onshore PCC.

CASE 4

A predictive reactive power management scheme is applied. Here, the vector of the parameters to be optimized is defined as in Case 3. The formulation of the objective function is given by Eq. (6.4.2).

The reactive power reference per each wind turbine is derived from the optimization according to the Grid Code Requirements, since all the values are inside

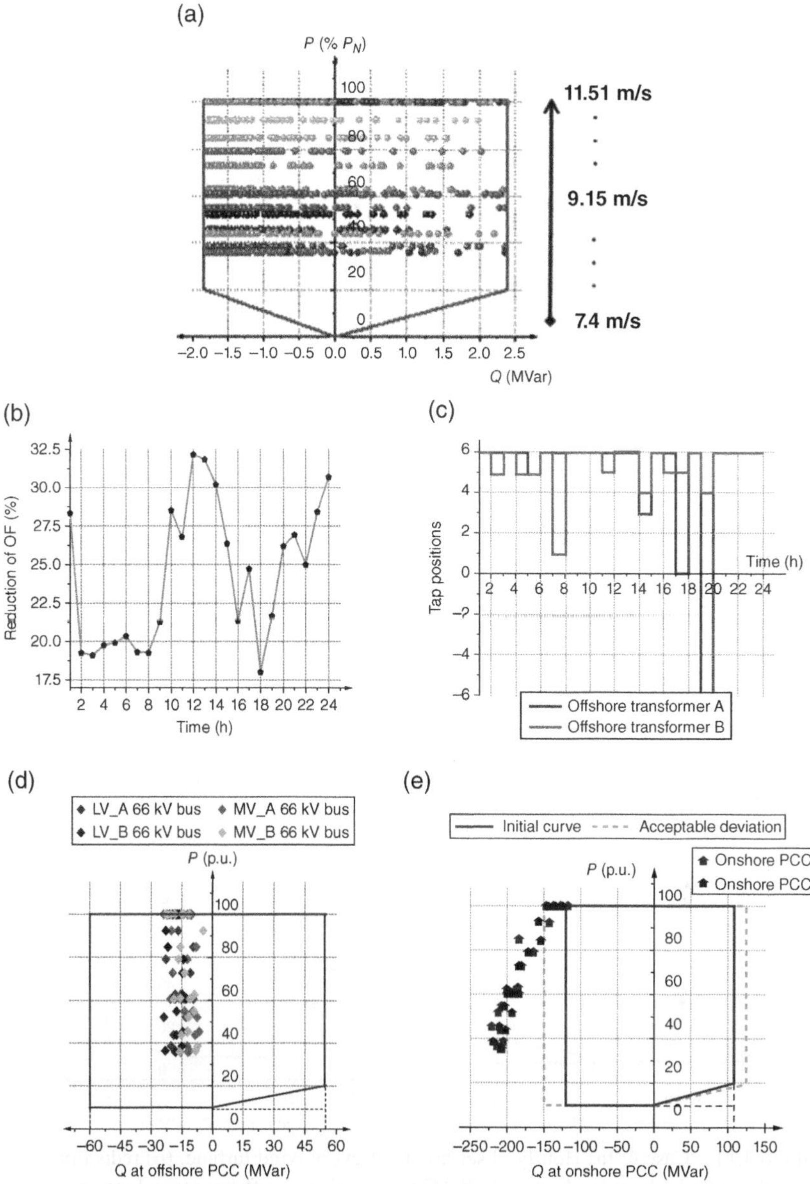

Figure 6.4.10 Case 3: (a) Hourly Q set-points of every wind turbine, (b) hourly reduction of wind farm active power losses, (c) OLTC tap positions – offshore transformers, (d) reactive power at the offshore PCC, (e) reactive power at the onshore PCC.

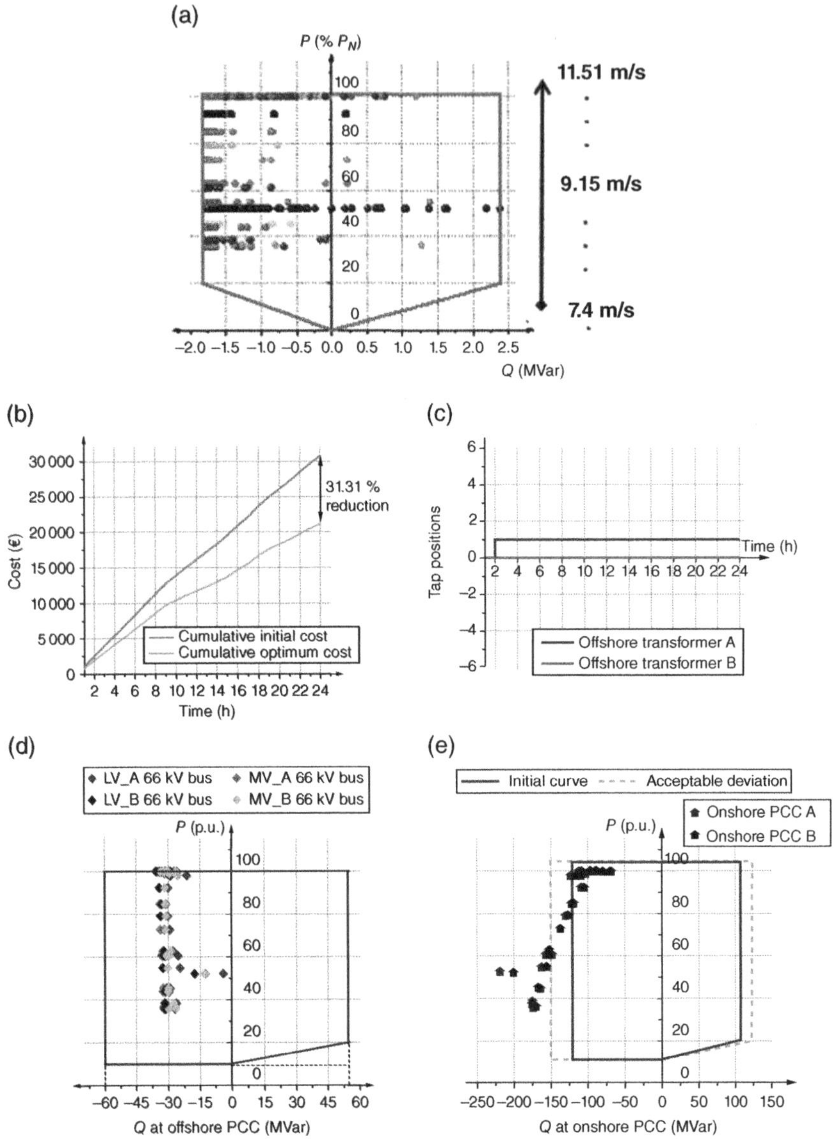

Figure 6.4.11 Case 4: (a) Hourly Q set-points of every wind turbine, (b) reduction of cumulative cost in the wind farm, (c) OLTC tap positions – offshore transformers, (d) reactive power at the offshore PCC, (e) reactive power at the onshore PCC.

the curve presented in Figure 6.4.11a. Each set of points arranged in the same horizontal line refer to different values of wind speed.

As presented in Figure 6.4.11b, the difference between the cumulative initial and optimum cost is estimated around 31.1% for the 24-hour time horizon under investigation. The reactive power levels of the MV- and LV-side of the offshore transformers (66 kV buses) are maintained within the predefined envelope, as shown in Figure 6.4.11d. In addition, it is observed that the reactive power absorption/injection at the onshore PCC is again completely outside the required boundaries according to Figure 6.4.11e, since the deviation is more than the acceptable tolerance (±0.1 p.u.). Although, the main target of this approach, for minimizing OLTC tap changes during daily operation is achieved, as displayed in Figure 6.4.11c, the big deviation of the reactive power proves that the presented predictive optimization by changing only offshore transformers' tap positions does not satisfy the Grid Code Requirements.

CASE 5

A real-time reactive power management is performed, where the formulation of the objective function is defined by Eq. (6.4.1) and the vector of the parameters to be optimized is the following:

$$x = \left[Q_{\mathrm{WTG},1}, \ldots, Q_{\mathrm{WTG},100}, \mathrm{tap}_{\mathrm{on,A}}, \mathrm{tap}_{\mathrm{on,B}}, \mathrm{tap}_{\mathrm{off,A}}, \mathrm{tap}_{\mathrm{off,B}}\right] \tag{6.4.22}$$

The reactive power reference per each wind turbine is derived from the optimization according to the Grid Code Requirements, since all the values are inside the curve presented in Figure 6.4.12a. Each set of points arranged in the same horizontal line refer to different values of wind speed.

As presented in Figure 6.4.12b, the hourly reduction of the real power losses is around 17.5% and occurs for higher wind speed values between the range of 8.5–11.5 m/s. The minimum point corresponds to an hourly reduction of 15%. On the other hand, when the wind speed is lower between the range of 7–8.5 m/s, the hourly reduction percentage of the losses is much higher and reaches the remarkable maximum value of 55%.

The reactive power levels of the MV- and LV-side of the offshore transformers (66 kV buses) are maintained within the predefined envelope, as shown in curve Figure 6.4.12e. In addition, it is observed that the reactive power absorption/injection at the onshore PCC is completely between the required boundaries according to Figure 6.4.12f, while the onshore and offshore transformer tap settings, as derived from the optimization algorithm, are shown in Figure 6.4.12c and d, respectively. The total number of the tap movements within 24 hours is 42 for each of the A and B onshore transformers. Concerning also the offshore transformers A and B, the total number of the tap movements is 43 and 26, respectively. The small deviation displayed in Figure 6.4.12f is acceptable, because the offshore wind turbines are expected to contribute to the fine regulation of reactive power at PCC with ±0.1 p.u. (=±30 MVAr) according to the Grid Code Requirements defined by TenneT TSO.

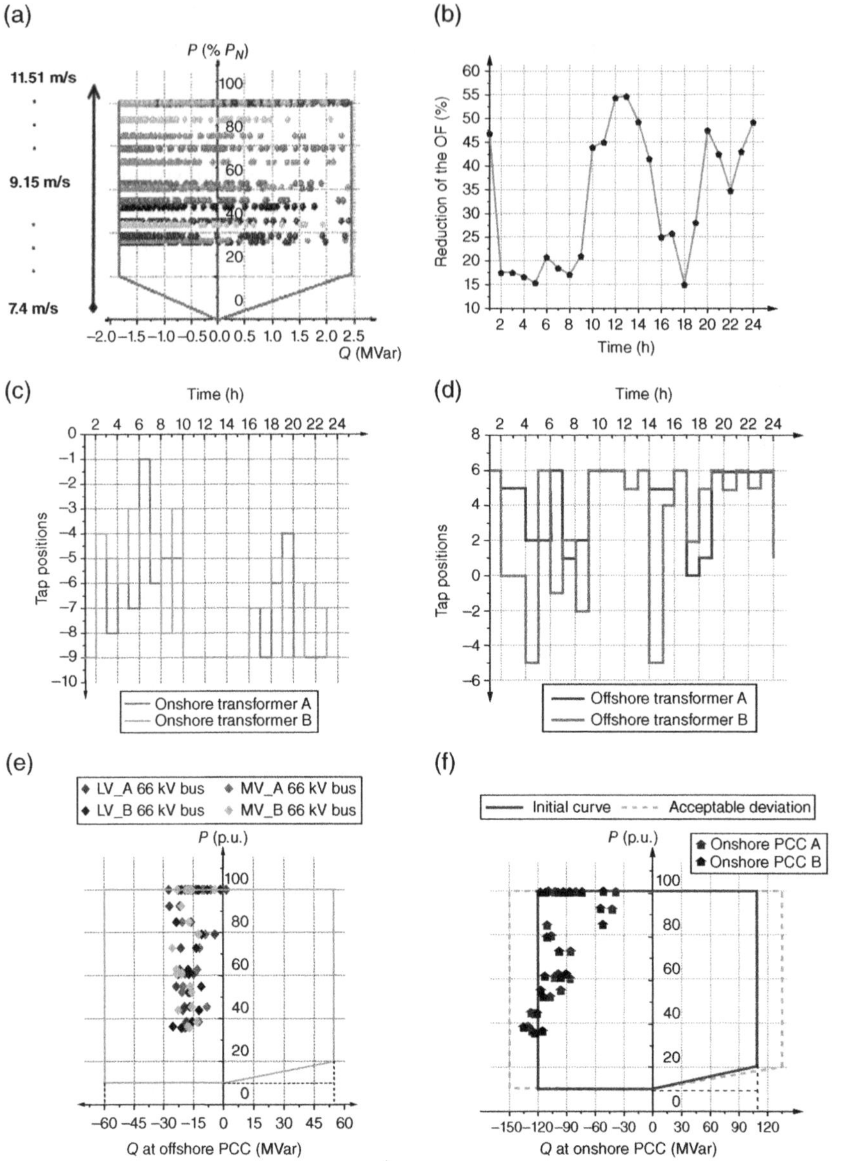

Figure 6.4.12 Case 5: (a) Hourly Q set-points of every wind turbine, (b) hourly reduction of wind farm active power losses, (c) OLTC tap positions – onshore transformers, (d) OLTC tap positions – offshore transformers, (e) reactive power at the offshore PCC, (f) reactive power at the onshore PCC.

CASE 6

A predictive reactive power management scheme is applied. Here, the vector of the parameters to be optimized is defined as in Case 5. The formulation of the objective function is given by Eq. (6.4.2).

The reactive power reference per each wind turbine is derived from the optimization according to the Grid Code Requirements, since all the values are inside the curve presented in Figure 6.4.13a. Each set of points arranged in the same horizontal line refer to different values of wind speed.

As presented in Figure 6.4.13b, the difference between the cumulative initial and optimum cost reaches a percentage of 41.12% for the 24-hour time horizon under investigation. The reactive power levels of the MV- and LV-side of the offshore transformers (66 kV buses) are maintained within the predefined envelope, as shown in Figure 6.4.13e. In addition, it is observed that the reactive power absorption/injection at the onshore PCC is also between the required boundaries according to Figure 6.4.13f, since the deviation is lower than the accepted ±0.1 p.u. Eventually, the main target of this approach, for minimizing the OLTC tap changes during daily operation, is achieved. In comparison with Case 5, as displayed in Figure 6.4.13c and d, the tap positions of the A and B transformers within 24 hours are constant, while the number of total tap movements was reduced to 2 and 0 for the onshore transformers, respectively.

6.4.6 Conclusions

The main goal of the approach presented in this chapter is to minimize the wind farm power losses, as well the variations of the transformers' tap positions, while the reactive power set-points of individual wind turbines in the plant are set utilizing an optimal reactive power management scheme. The scheme ensured grid code compliance with steady-state reactive power requirements, while at the same time minimizing the electrical losses in the wind power plant. The first step was the implementation of a NN-based wind speed forecasting method, which provides the inputs to the optimization algorithm. Then, two different reactive power optimization approaches are presented. Optimization for any given operating point results in minimum losses, but it does not allow the consideration of the OLTC costs. Optimization over a predicted 24-hour time horizon solves this problem by including in the objective function the number of tap changes. The Var reference of the entire wind farm and the tap positions of the transformers are provided by the optimization algorithm. The meta-heuristic optimization method called MVMO is used for all the introduced optimization tasks.

It was presented by means of numerical results on a real offshore wind power plant in the Netherlands that the application of MVMO in the optimal coordination of reactive power sources in the nearshore wind farm under investigation entails

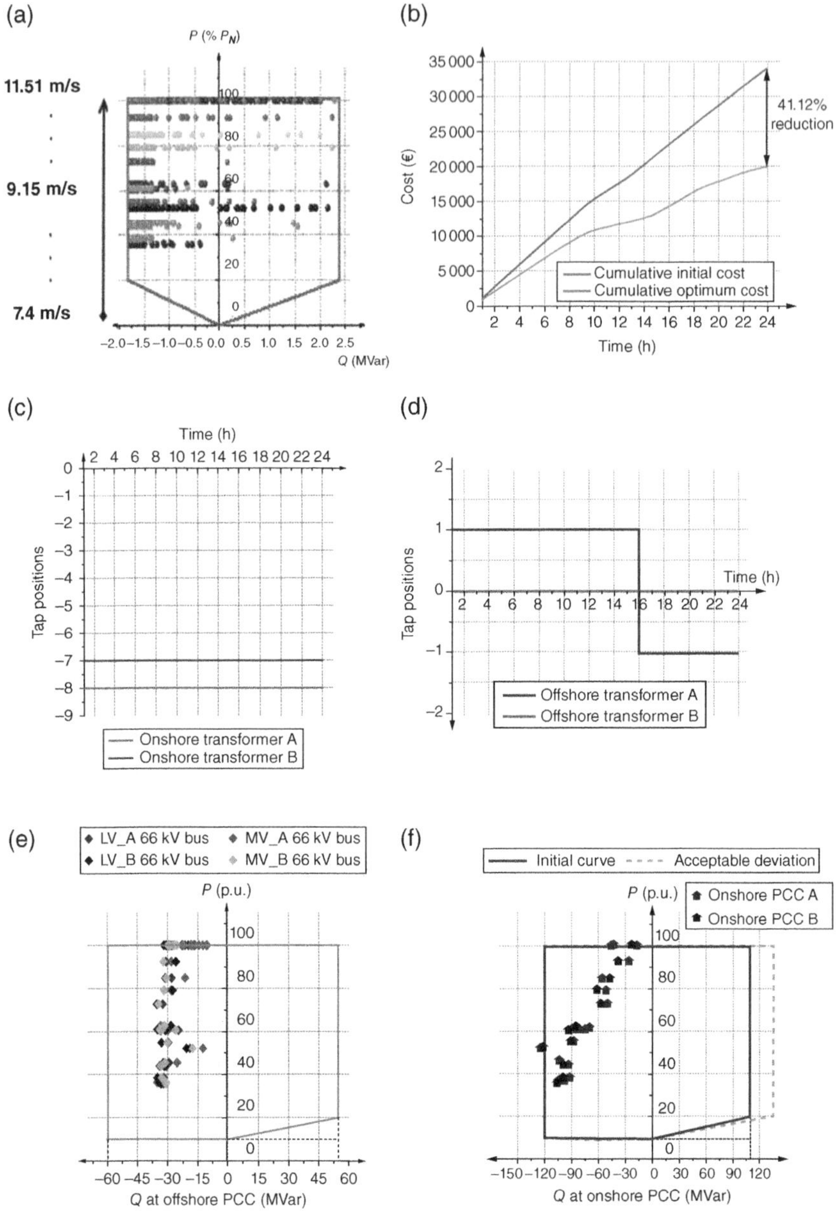

Figure 6.4.13 Case 6: (a) Hourly Q set-points of every wind turbine, (b) reduction of cumulative cost in the wind farm, (c) OLTC tap positions – onshore transformers, (d) OLTC tap positions – offshore transformers, (e) reactive power at the offshore PCC, (f) reactive power at the onshore PCC.

robustness and enhanced performance in terms of convergence speed. The fast convergence behavior and the quick evolution of the optimum solution with minimum risk of premature convergence are revealed, thanks to the well-designed balance between search diversification and intensification of MVMO. In addition, such an approach can lead to reduction of operational costs for large offshore wind plants. Furthermore, it ensures compliance for the plant without the need for additional hardware (i.e. STATCOM) as the compliance is ensured by means of optimal reactive power set-points per each wind turbine. Finally, the proposed approach for performing optimization over a predicted time horizon leads to significant reduction of total losses and minimum number of transformer's tap changes. As a matter of fact, the lifetime of the switchable devices (e.g. transformers) is reduced and consequently, the total operation cost of the wind farm is reducing.

6.5 PHOTOVOLTAIC CONTROLLER DESIGN

Sukumar Mishra[1] *and Deepak Pullaguram*[2]

[1]*Indian Institute of Technology Delhi, New Delhi, India*
[2]*National Institute of Technology, Warangal, Telangana, India*

6.5.1 Introduction

The annual increase in electrical load demand and carbon emission leads to exploring alternative (mostly renewable) energy resources such as PV, micro hydro (MH), wind, geothermal, biomass, ocean wave and tides, and other clean power generation technologies such as microturbines (MTs) and fuel cells (FCs) for generation of electric power.

Among all the renewable power, generations from PV sources has the advantage of being copious, eco-friendly, and free source of energy; requires low maintenance; and can be installed at any place easily without occupying much additional area as these can be placed over rooftop of buildings. Installation of 1 kW of rooftop system generally requires an area of 12 m^2 (130 ft^2) [94]. During a sunny day, over a shiny region of earth, the average solar energy density is nearly 4–7 kWh/m^2. The solar PV generates DC electric energy by converting solar irradiance energy falling on the panels. In the last two decades, PV technology has shown tremendous growth, with growth rate of 25–30% annually. The potential of solar energy source is already being recognized and everyone of interest is behind the extraction of this energy as maximum as possible by setting up large-capacity PV generation systems [95]. Besides these, it is very much vital to have maximum utilization of available solar energy due to the high installation cost of the PV panels [96–100]. Hence, typically PV systems are expected to operate at maximum power point (MPP).

6.5.2 Maximum Power Point Tracking in PV System

PV Array Modeling

A basic PV array can be modeled by single diode model in which photon-generated current source (I_{ph}) is connected in antiparallel to a reverse biased diode [101, 102] as shown in Figure 6.5.1.

The output current (I_{PV}) of PV cell will be

$$I_{PV} = I_{ph} - I_0\left(\exp\left(\frac{q}{N_s k T_c A}(V_{PV} + IR_s)\right) - 1\right) - \frac{V_{PV} + IR_s}{R_{sh}} \tag{6.5.1}$$

where

I_{ph}: Photon current

I_0: Reverse saturation current of a p–n junction diode

q: Charge of an electron

k: Boltzmann's constant

T_c: Operating temperature

A: Ideality factor

N_s: Number of cells connected in series

R_{sh}: Shunt resistance indicating the leakage current in the cell

R_s: Series resistance accounting the internal resistance of the cell

The photon current is a function of both insolation and temperature which is given by

$$I_{ph} = (I_{sc} + K_T(T_c - T_n))\frac{G}{G_n} \tag{6.5.2}$$

where

T_n: Nominal temperature (usually 278 K)

T_c: Operating temperature (K)

G: Irradiation on the device surface (W/m^2)

G_n: Nominal irradiance (usually 1000 W/m^2)

I_{sc}: Short-circuit current at nominal temperature and irradiation

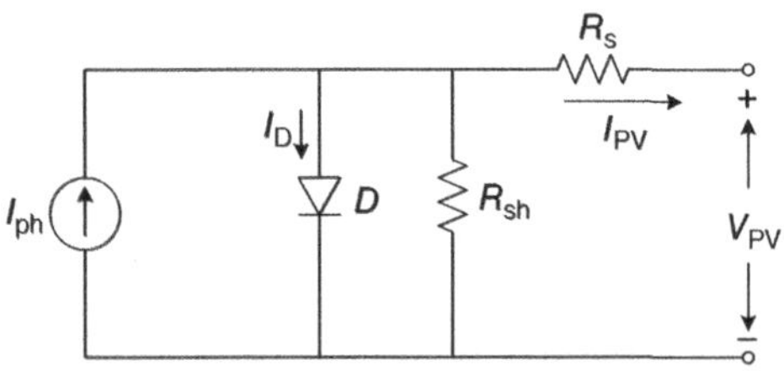

Figure 6.5.1 Single diode equivalent of PV module.

The temperature dependency on the diode saturation current I_0 may be expressed by

$$I_0 = I_{0,n}\left(\frac{T_n}{T}\right)^3 \exp\left(\frac{qE_g}{Ak}\left(\frac{1}{T_n} - \frac{1}{T_c}\right)\right) \tag{6.5.3}$$

where E_g is the band gap energy of the semiconductor ($E_g = 1.12$ eV for the polycrystalline Si at 25 °C) and $I_{0,n}$ is the nominal saturation current given by

$$I_{0,n} = \frac{I_{sc,n}}{\exp\left(V_{O.C,n}/AV_{t,n}\right) - 1} \tag{6.5.4}$$

with $V_{t,n}$ being the terminal voltage of all series connected cells in an array at T_n.

The current–voltage (*I-V*) and power–voltage (*P-V*) characteristics of a PV array with varying irradiance is shown in the Figure 6.5.2a and b, respectively, and with varying temperature is shown in Figure 6.5.3a and b, respectively. Furthermore, most of the time, the PV module is operated around the nearly constant voltage zone because the variation of the current during the variation of insolation is very high as compared to variation of voltage. Usually, there are fast dynamics of irradiance because of fast-moving clouds. If the current of the PV module is used as the operating point, a fast tracking MPP is required to follow a wide operating current range from 0 A to the short-circuit current, based on irradiance conditions. Nevertheless, the change in voltage with variation in irradiance is very less as observed in Figure 6.5.2a. Although the variation in voltage is high for the temperature variation as compared to current, the temperature variation is a very slow process which discards the requirement of the fast-acting MPP tracker [103].

A major challenge in PV generation system is to obtain a unique global MPP on its *P-V* curve using the nonlinear relationship between current–voltage (*I-V*). As the power generated form PV array mostly depends on temperature and irradiance which are time-varying quantities, so there is an absolute necessity of a real-time MPP tracking system. Over a period of time many real-time maximum power point tracking (MPPT) algorithms have been proposed and explained [104–106] which are:

1. Perturb and observe (P&O)
2. Incremental conductance
3. Fractional short-circuit current
4. Fractional open-circuit voltage
5. Fuzzy logic control
6. Ripple correlation approaches

Many of these methods are effective and tested to obtain optimal PV module operation by varying terminal voltage corresponding to the MPP value considering uniform solar insolation over the entire PV array [104–106]. All these algorithms have differences in accuracy, complexity, and speed.

Generally, multiple PV modules are connected in series and parallel combination over a wide area to obtain the required amount of voltage and power. Due

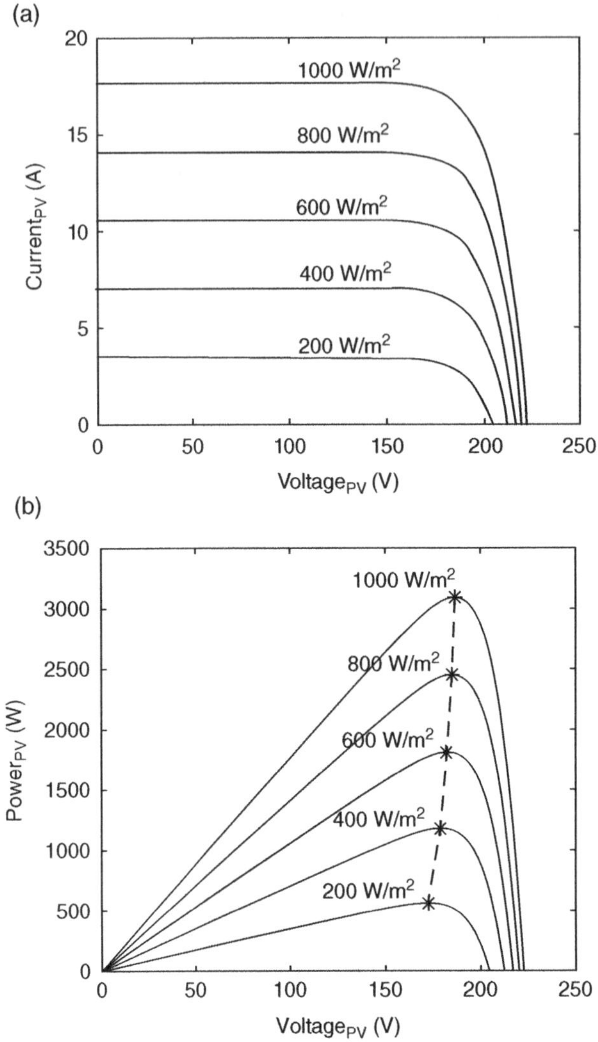

Figure 6.5.2 (a) *I-V* characteristics and (b) *P-V* characteristics with irradiance variation.

to the presence of multiple panels in the PV system, partially shaded condition will be an inevitable condition as some parts of the array may receive less irradiation due to movement of clouds, shadows of buildings, tree shadows, and other light-obstructing substances.

Based on the PV module configuration, bypass diode incorporated into the PV modules, and shading pattern, partial shading may have a significant impact on the output power. During partial shading, PV modules belonging to the same string experience different insolation resulting in complex *P-V* characteristics with multiple peaks as compared to the partial shading in parallel strings [107–109].

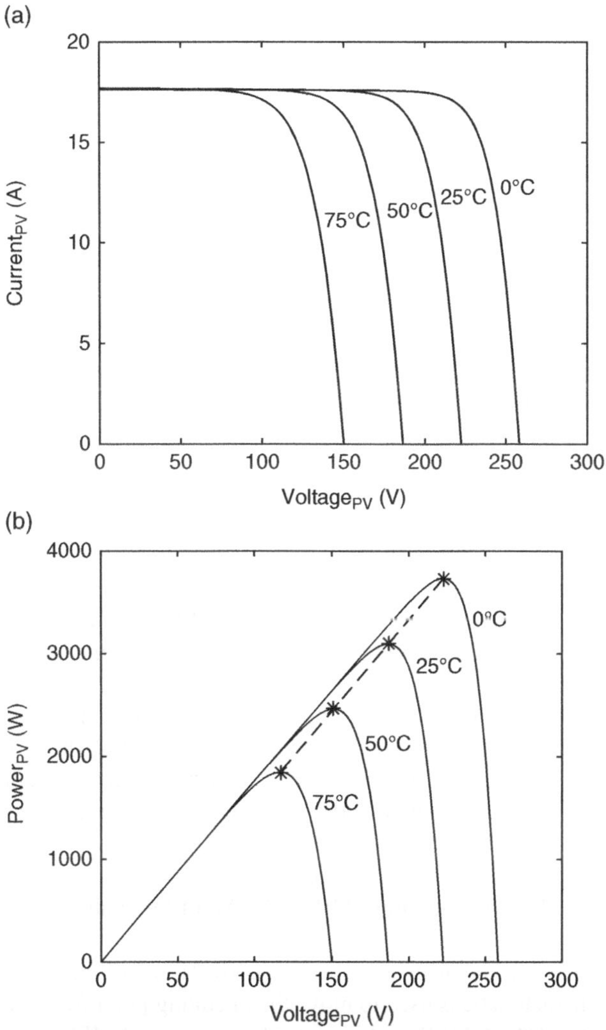

Figure 6.5.3 (a) *I-V* characteristics and (b) *P-V* characteristics with temperature variation.

The *I-V* and *P-V* characteristics of the partially shaded PV system is depicted in Figure 6.5.4a and b, respectively. This is mainly because, whenever there is an irradiance change, the change in current is very much high as compared to that of the voltage change as shown in Figure 6.5.2a. However, if there are no bypass diodes across PV modules, the current in the complete series connected string will experience single peak *P-V* curve with the minimum possible current flowing through them as there are no other paths available to flow the excess current generated by the module that get high irradiance. Bypass diodes across the low irradiance module provide the path for the excess current generated by the module with high

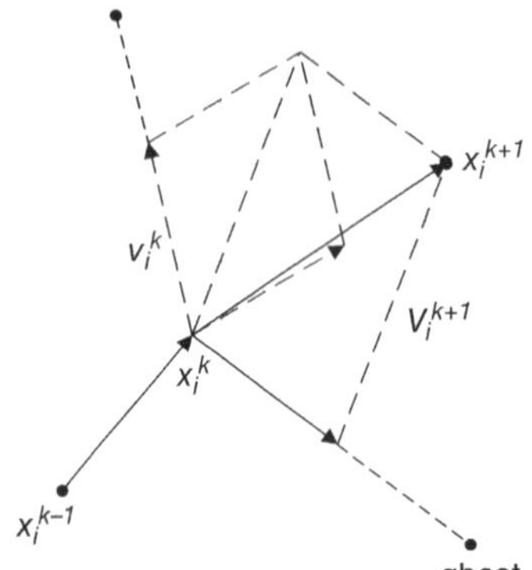

Figure 6.5.4 Movement of particles during optimization.

irradiance. This makes the string to have the multiple peaks in *P-V* characteristics at different voltage levels.

The conventional algorithms described above are designed considering the "hill-climbing" principle in which the operating point is moved in such a direction, that increases power output, till it reaches the MPP. This principle is well valid for the *P-V* characteristics having single peak, but during the partial shading condition these algorithms do not work effectively due to the presence of multiple peaks. Most of the time these algorithms may get stuck at the local MPP point in partial shaded condition. As in a solar park the partial shading condition is most common, there is a requirement of a MPPT algorithm that can track global maximum power point (GMPP) among the multiple peaks under partial shaded conditions.

Many algorithms are developed and proposed in the literature to obtain the GMPP point during partial shaded conditions [110–124]. All of these algorithms have their own advantages and drawbacks.

In [110, 111], GMPP is tracked using a double-stage MPPT algorithm. In the first stage, the load line characteristic is used to move the operating point from the current position to vicinity of the GMPP, and in second stage normal MPP algorithm is used to obtain the GMPP. But these methods fail to obtain GMPP if the GMPP point lies left side of the operating point. A Fibonacci sequence search-based approach is utilized in [102] and DIRECT search approach is used in [113] to obtain GMPP operation. In these two methods power measured at two different operating points are used to determine the movement of operating point. These techniques are equivalent to the P&O method with variable step size. But these approaches also fail to guarantee the GMPP operation under all conditions, especially when GMPP falls at the right side of the present operating point, with the present operating point being at local MPP.

In [114], the irradiance factor (determines the current value) and temperature or voltage factors (determines the voltage) are used to determine the GMPP. The voltage factors are calculated at least once and stored as a lookup for all MPP conditions before going to the GMPP tracking algorithm which makes it to be system dependent. To track the GMPP, a double-stage MPP technique was also proposed in [115, 116]. In these methods the first stage is used to obtain the local region in which the GMPP is present and then conventional MPP algorithm is used to achieve the GMPP. These methods fail to track the GMPP when there are fast varying shading conditions like fast cloud movements. Hence, in these methods, all local MPPs should be obtained and compared to get the GMPP which makes the technique slow. In [117], an extremum-seeking control technique for global MPPT is proposed. This control is developed using an approximate PV model that has similar characteristics of PV under partial shaded conditions. The search ranges are defined and a segmental search is used which improves the computational efficiency as compared to conventional sweep search. Although this method is fast, it is system dependent and possesses steady-state errors.

Along with the deterministic approaches, many artificial intelligence- based techniques [118–124] are developed by researchers to obtain the GMPP in the PV system. An ANN with polar information and fuzzy logic controller-based MPPT algorithm are proposed in [118]. A three-layer ANN is considered, and it is trained for various partial shading conditions against the voltage reference of the GMPP point which makes it system dependent. Moreover, to train the ANN operating voltage point, the GMPP point should be known. Further, the input to the ANN is the temperature and the insolation, but this information is often not available. Biological swarm chasing-based MPPT is proposed in [119] which is applied for the uniform-shaded PV system, which makes it to fail for the partially shaded condition.

An alternative approach for obtaining MPPT in the PV system is to employ meta-heuristic techniques. Due to the ability to obtain the optimal solution for complex nonlinear objective functions using meta-heuristics, it is very effective to deal with the MPPT problem for the partially shaded PV systems. PSO is one of the most recognized and reliable meta-heuristic technique due to its easy implementation, and fast computation capability and computational efficiency [120]. Since PSO is a random search optimization technique, it can be easily applicable to obtain the GMPP for any type of *P-V* curve irrespective of irradiance pattern of the solar panels.

Miyatake et al. [120] used the PSO algorithm to obtain GMPP. In this paper, a centralized MPPT control is realized for the PV systems in which PV panels are cascaded by multiple converters. However, it is economical to use only one central high-power single converter-based PV system. A MPPT algorithm based on adaptive perceptive particle swarm optimization (APPSO) is presented in [121]. In APPSO, for an n-dimensional optimization problem, $(n + 1)$ dimensional search space is considered. An extra dimension is used to represent the underlying particle performance at their respective positions in n-dimensional search space. A complete algorithm of an APPSO is provided in [122].

In [123] an improved PSO-based MPPT algorithm which uses PSO in conjunction with the direct duty cycle control has been proposed. In [124] a MoPSO-based MPPT algorithm is proposed to meet the practical consideration for partially shaded PV.

6.5.3 Particle Swarm Optimization

PSO is a stochastic swarm intelligence algorithm for optimization problem developed by Eberhart and Kennedy in 1995 [125], developed based on the inspiration of social behavior of bird flocks. PSO is a global optimization algorithm to obtain the optimal solution for a complex nonlinear surface problem with n-dimensional search space. Unlike GA, PSO has no evolution operators such as crossover and mutation but the potential solutions, called particles try to fly toward best particle by following the current optimum particles with a varying velocity and position. In this optimization technique, a set of cooperative particles are used, all these particles share position and fitness information obtained during their respective search. All the particles follow two rules to update themselves:

1. To move toward the global best particle with best performance
2. To move toward its respective best position.

Therefore, the position of a particle is influenced by its own best position pbest providing the local search and by the gbest which is the best solution found among all the particles in the entire population. The particle position x_i in the kth iteration is adjusted using Eq. (6.5.5),

$$x_i^{k+1} = x_i^k + v_i^{k+1} \tag{6.5.5}$$

where v_i is the velocity component with which the particle is moving. The velocity component is calculated by using

$$v_i^{k+1} = wv_i^k + c_1 r_1 \left(P_{\text{best}i} - x_i^k\right) + c_2 r_2 \left(G_{\text{best}} - x_i^k\right) \tag{6.5.6}$$

where c_1 and c_2 are the acceleration coefficients which provide the acceleration in the direction of gbest and pbest, w is the inertia weight, r_1, r_2 is an uniform random number between (0, 1), pbest$_i$ is the particle I respective best position, and gbest is the global best position among all particles in the entire population.

The typical movement of the particles during optimization process can be visualized by using Figure 6.5.4.

6.5.4 Application of Particle Swarm Optimization in MPPT

To illustrate the application of PSO optimization, in obtaining the MPPT, a PV system as depicted in Figure 6.5.5 is considered. The PV system consists of series connected PV string in conjunction with boost converter used to employ MPPT scheme based on the duty cycle obtained from the MPPT algorithm. As discussed in Section 6.5.2, the MPP of the PV system continuously varies based on the irradiance falling on PV modules and temperature of the modules.

Due to the distinctiveness of obtaining GMPP in partial shaded condition, the standard version of PSO algorithm is needed to be modified to meet the practical consideration of PV system under irradiance variation. The feasibility of PSO for MPPT problem is explained below.

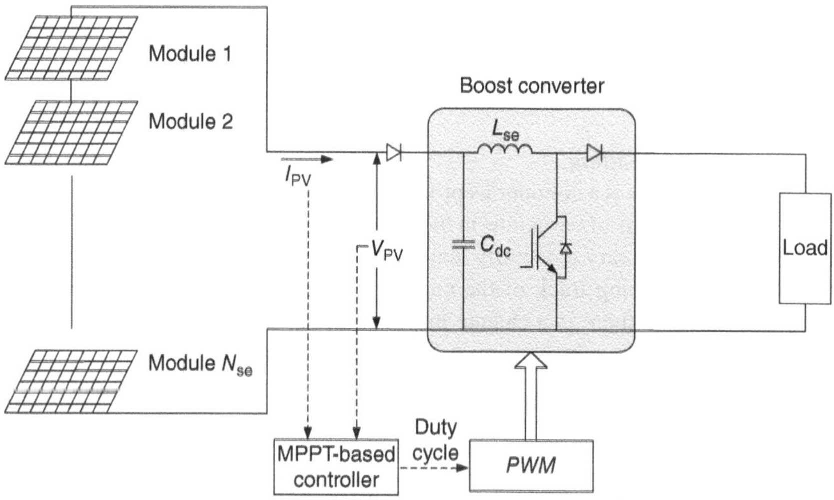

Figure 6.5.5 PV system with boost converter.

Parameter Selection

For any kind of optimization problem, the parameter selection is one of the major hurdle, the same goes with the MPPT problem. For the system considered here, the position of a particle is defined as the duty cycle value that is to be provided for the dc–dc converter, while the velocity shows the perturbation required to obtain the next duty cycle.

In general, the interval that the duty cycle always falls is within the range of (0, 1). However, this range can be restricted to much smaller $D_{min} > 0$, $D_{max} < 1$, in practice [113]. The limits of duty cycle D_{min} and D_{max} can be estimated by using the known boundary parameters like maximum and minimum values of the open-circuit voltage, of the short-circuit current, and of the equivalent resistive load.

In general, the MPP of the *P-V* curve occurs nearly at multiples of 80% of the module open voltage $V_{O.C,\ module}$ [114] and successive peaks are displaced by nearly 80% of $V_{O.C,\ module}$. Hence, instead of the initialization of the particles, it is better to adopt the fixed position at equal distance initialization that covers search space $[D_{min}, D_{max}]$.

Objective Function

The proper selection of objective function identification is very much important to have easy and simple search in optimization problem that enforces all constraints. As the MPP problem is referred to tracking of the MPP, the objective function will be the power flowing out of the PV module after the digital controller output provides the pulse width modulation (PWM) command according to the position of particle "*i*."

To avoid the measurement noise, the voltage measurements "V_{PV}" and current measurements "I_{PV}" are filtered using digital finite impulse response filters and then used to calculate the P_{PV}, which acts as fitness of particle "*i*" in PSO.

$$P_{\mathrm{PV}} = V_{\mathrm{PV}} I_{\mathrm{PV}} \tag{6.5.7}$$

The samples of voltage and current should be acquired only after the settling time of the power converter to have the accurate power measurement.

Identification of Shading

As the MPPT process is a continuous process, the algorithm should be able to track the MPPT for all kinds of variations in the irradiance. To track the MPPT for all the conditions, it is necessary to identify that there is a variation in irradiance. This can be achieved by keeping track of the change in power between two consecutive samples. Whenever there is a change in power observed over a threshold value, then it can be treated as the change in MPP [124].

$$\left|P_{\mathrm{PV}}^{k+1} - P_{\mathrm{PV}}^{k}\right| \geq \Delta P_{\mathrm{Threshold}} \tag{6.5.8}$$

6.5.5 Illustration of PSO Technique for MPPT During Different Irradiance Conditions

To illustrate the PSO technique, equidistant three duty cycles between $[D_{\mathrm{min}}\ D_{\mathrm{max}}]$ are considered. For this, from here on the particle and the duty both indicate the same meaning.

$$x_i^k = [d_1, d_2, d_3] \tag{6.5.9}$$

where x_i^k is the ith particle in the kth iteration.

MPPT for System with Uniform Irradiance on PV Modules

Consider the PV system having the uniform irradiance over the all PV panels. Then, for the first iteration the duty cycle considered itself acts as the pbest_i as there are no preceding particles available. It can be observed that among all other particles d_2 provides the maximum power from the PV array as depicted in Figure 6.5.6a; hence, it is to be considered as gbest. The new particles for the second iteration are obtained by using Eqs. (6.5.5) and (6.5.6) in which Eq. (6.5.6) is used to get the velocity of the particle which provides the information of magnitude and direction of the movement of the particle. But the resulting velocity is only updated due to the gbest term as the $(\mathrm{pbest}_i - x_i^k)$ component is zero [123]. Moreover, it can also be observed that the $(\mathrm{gbest} - x_i^k)$ is zero for the particle "d_2." This results in no variation in the duty cycle as that of previous iteration. In consequence, the particle "d_2" does not participate in exploring the solution. To avoid this situation, a small perturbation is made in the duty cycle for that particle, to have a change in power value, as shown in Figure 6.5.6b. From the same figure it can also be observed that the fitness values of all the particles are improved as compared to previous iteration. This is because there is only one peak point available in the P-V characteristics. Thus, all the duty cycles are moving toward that global peak and also the pbest_i is updated with the new best fitness values.

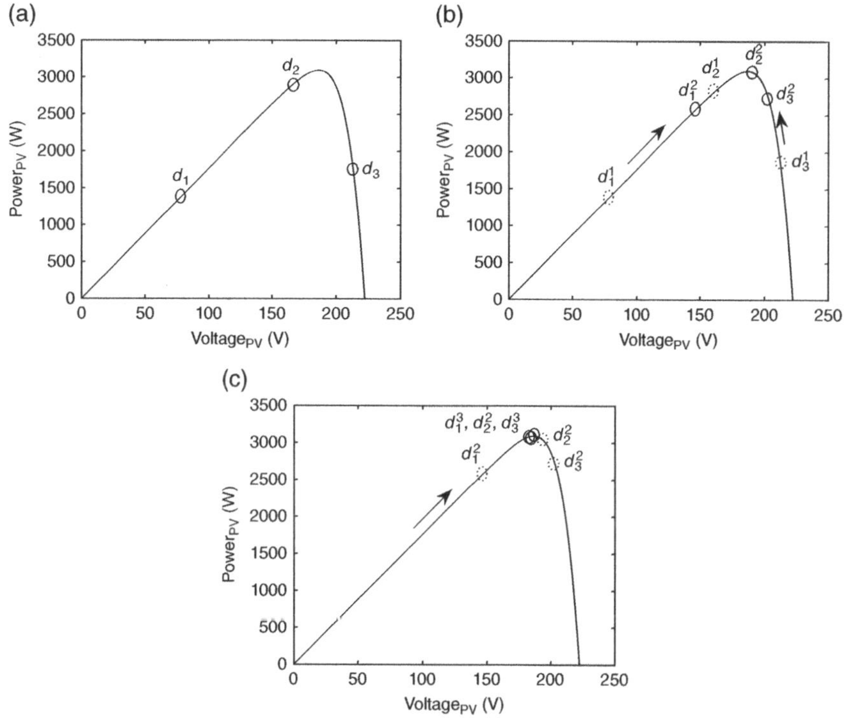

Figure 6.5.6 Duty cycle movement during uniform shading in (a) iteration 1, (b) iteration 2, (c) iteration 3.

Figure 6.5.6c shows the movement of the particles in the third iteration. As there was an improvement in the fitness in second iteration, the velocity direction of these duty cycles remains the same and subsequently helps them to move toward gbest in the same direction. After the end of the third iteration, all duty cycles (*di*, $i = 1, 2, 3$) may arrive closest to the MPP with the reduced value of velocity as the component ($\text{pbest}_i - x_i^k$) is zero and component ($\text{gbest} - x_i^k$) tending toward zero. In the subsequent iteration, as a result of reduced velocities, the value of the duty cycle almost remains as constant. Thus, the system operating point will remain constant at the MPP with reduced oscillations.

MPPT for System with Partial Shaded PV Modules (Nonuniform Irradiance)

In the previous case the PV curve exhibits only single maximum peak point, whereas during partial shaded condition there will be multiple peaks as explained in Section 6.5.2. Due to these multiple peaks most of the deterministic MPPT algorithms fail to obtain the global MPP. But for PSO MPPT, this scenario has got no difference as compared to the procedure explained for the uniform irradiance which can be observed from Figure 6.5.7a.

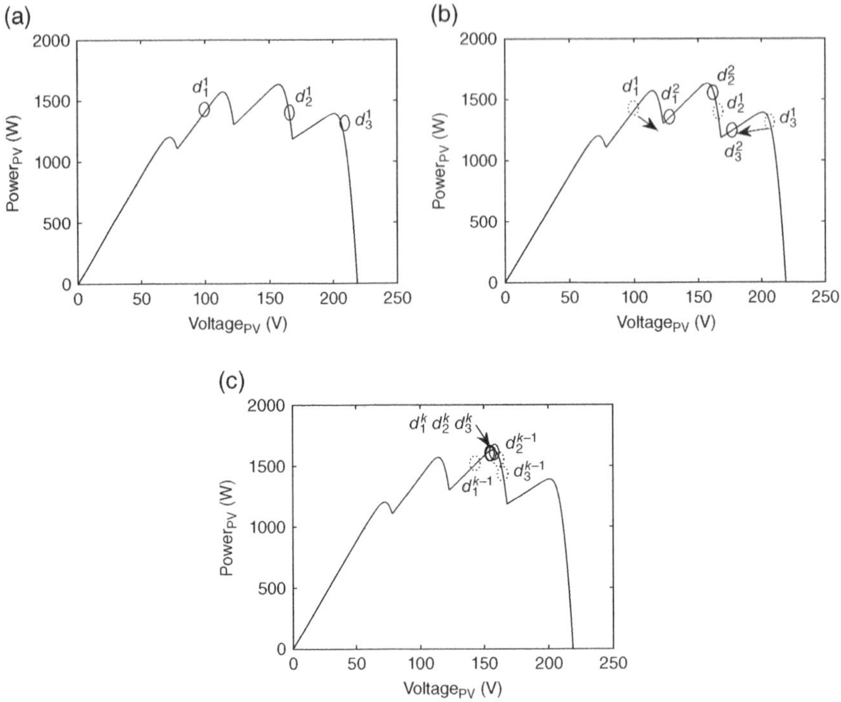

Figure 6.5.7 Duty cycle movement during particle shading in (a) iteration 1, (b) iteration 2, (c) iteration 3.

The first iteration and second iteration particle generation for the partial shading case will be similar to that of uniform irradiance. The fitness obtained for the duty cycles in the second iteration during partial shaded case may not be the best as that in the uniform case as shown in Figure 6.5.7b. This makes that the pbest_i to remain the same as that of the first iteration. The new velocity obtained in the second iteration now depends on ($\text{pbest}_i - x_i^k$) and ($\text{gbest} - x_i^k$) components, acceleration parameter c_1, c_2, and random parameter r_1, r_2. Based on these values the direction in which the particle should move is decided. After few iterations, the pbest_i of all the particles will slowly move toward the hill that has the global peak as in Figure 6.5.7c. Once this peak has reached, the problem will be as similar as that of third iteration in the uniform shaded case resulting in a constant global MPP with reduced oscillations. Figure 6.5.8 shows the flowchart.

6.5.6 Conclusion

This chapter provides an insight of the application random search-based MPPT algorithms for MPP operation of the PV system. PV systems are usually expected to operate at the MPP, so a good MPPT algorithm is needed to have the best

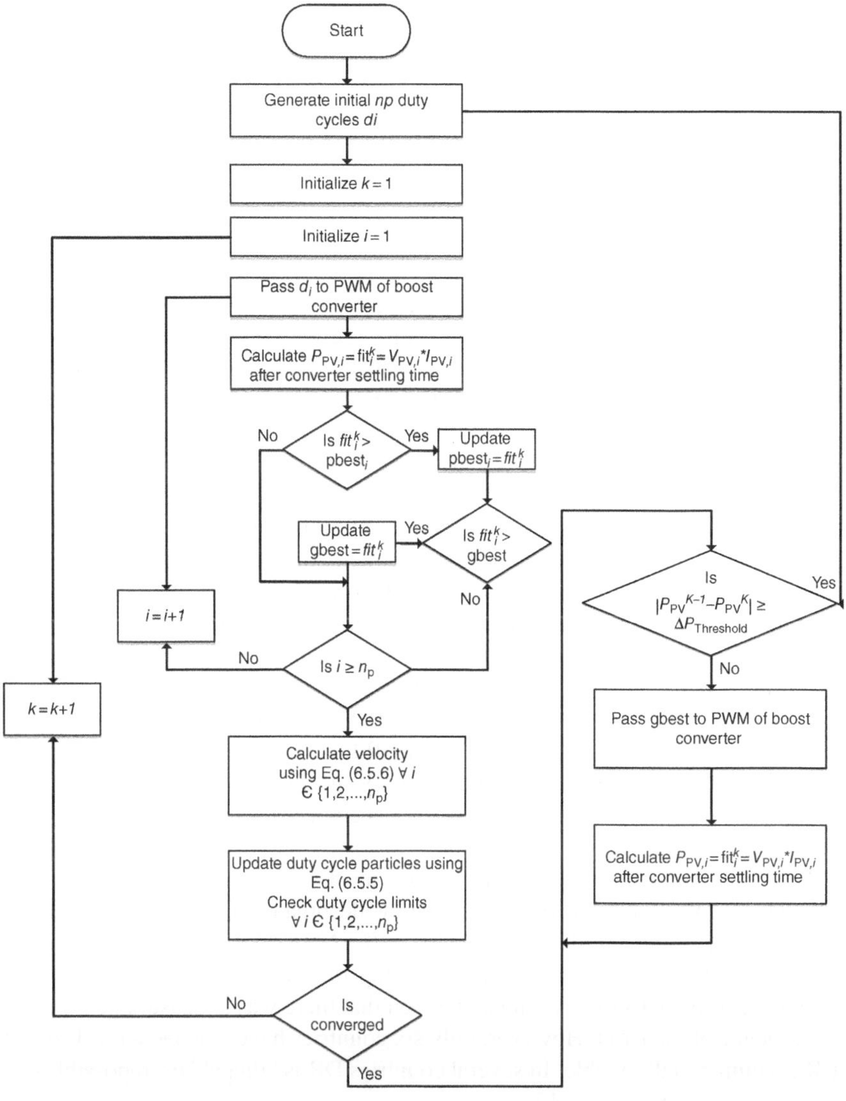

Figure 6.5.8 Flowchart to obtain optimal maximum power point for PV systems using PSO.

operation in all the conditions. As most of the PV systems seek the multiple peak characteristics, the normal conventional MPPT algorithms fail to obtain the MPP operation. Hence, the random search technique can be used to solve this nonlinear multi peak problem. The application of the PSO technique in solving this problem is explained.

6.6 DEMAND SIDE MANAGEMENT AND DEMAND RESPONSE

Zita A.Vale, Pedro Faria, and João Soares

Polytechnic of Porto, Porto, Portugal

6.6.1 Introduction

DR is a very important topic in the planning and operation of electricity markets and of power systems in general. It has been largely fostered in the past few years, in approaches and business models that consider opportunities beyond the traditional use of DR that was based on distinct electricity tariffs for different periods of the day. The changes in the power systems operation and in the electricity market (EM) organization have required the development of new DR programs in order to take full advantage of DR.

DR refers to the change of the consumption pattern by the end consumers in response to market price signals or incentives due to reliability or economic reasons [126, 127]. This concept has gained more relevance leading to the implementation of DR programs in electricity markets around the world. The most relevant example is the United States where several independent system operators (ISO) have put in place DR programs that achieved remarkable power levels. In January 2014, the transmission operator "PJM Interconnection," one of the largest US ISO, activated about 2000 MW of DR for several hours on 7 January, and over 2500 MW for several hours on 23 January and 28 January. DR resources provided 21 MW on five occasions between December 2013 and February 2014 to ISO New England (ISO-NE), as part of its 2013–2014 Winter Reliability Program [128]. Without much public information, the most developed countries in Asia are also tracking this path. DR has also gained widespread policy support in Europe, namely through the Third Energy Package, the Network Codes, and the Energy Efficiency Directive (EED) [129]. Analyzing the evolution of the state of DR in Europe from 2013 to 2014, it can be seen that there was a measurable progress in sequence of the EED. However, only six countries have reached a level so that DR is commercially viable. In several countries DR is "illegal" or impossible due to the current regulations [130].

In fact, consumers' active participation is one of the bases for the successful implementation of smart grids [131–133]. However, that participation is still being envisaged as part of the traditional customer–supplier relationship and not with suitable models that consider more flexible approaches. Regarding smart grids, huge investments have already been made, namely in smart metering. Currently, about 45 million electricity smart meters are operating in only three EU State Members (Finland, Italy, and Sweden). An additional investment of 45 billion Euros is estimated to install more 200 million smart meters in the EU till 2020 [134].

While a repeated pattern of regulatory barriers remains and several additional problems are identified, 10 indicative rules have been listed for the success of DR in Europe [135]. Some of them are related to legal and regulatory issues. Many others refer to the problem of resources' aggregation and the respective market structures and to the communication protocols [131, 136, 137].

The identification of the main obstacles to the effective integration of DR, and of the most prominent potential solutions to overcome them, leads to the urgent need for further developments in order to enable the necessary step forward in the field. In fact, it is needed to put together solutions that cover a set of essential and connected fields, such as DR, consumer aggregation, communications, and resource optimization. The aim is to take full advantage of the potential of consumers' demand flexibility, while ensuring their proper remuneration and increasing the efficiency of the whole energy and power system [138, 139].

The participation of a consumer in a DR program or event is always a contribution to the reduction on the need of electricity in a specific period, in a specific area. This area can be seen as the whole power system or be related with the special case of transmission congestion in a small area. Consumers can react to electricity prices or DR incentive-based events [126] in three ways. First, one has the possibility of "Foregoing" which corresponds to reducing electricity usage at times of high prices or DR program events, without using it later. It can cause temporary loss of amenity or comfort. Second, the "Shifting" corresponds to the rescheduling of the electricity usage from periods of high prices or DR programs events to other periods. The service or amenity is re-established in the subsequent or rescheduled period. This reschedule can have costs regarding overtime payment or productivity losses due to the required adjustments in the production process. Finally, "Onsite generation," in which consumers who have an onsite or backup emergency generator may respond by using it to satisfy some or all of their or others' consumption needs.

The use of any of the referred strategies causes inconvenience, discomfort, and/or loss of productivity in the buildings' occupants or in the laboring process. Such factors should be included in the costbenefit analysis that supports DR decisions, even if some of them are not directly accounted. One of the approaches to measure the value that the consumers put in electricity consumption, namely when facing electricity price changes, is the price elasticity of demand. The price elasticity rate is a measure used in the economics to evaluate a good or service DR to a change in its price, i.e. the percentage change in the demanded quantity in response to 1% change in price [140]. This is a normalized measure of the intensity of how the usage of electricity changes when its price changes by 1%. The price elasticity rate can be of two types: the own-price elasticity which measures how customers will change the consumption due to changes in electricity price, regardless of the period of variation; the substitution elasticity which is related to the change in the period, of the day or the week, of the consumption of electricity.

The implementation of electricity markets gives place to the acting of several players in the market trying to reach individual and global goals. The main players are the consumers and the producers since the objective is to supply the consumers' demand. Traditionally, this has often been achieved by vertically integrated companies supplying to consumers the energy provided by producers. In the scope of the current EMs, a diversity of players such as distribution network operator (DNO), TSO, market operator (MO), virtual power player (VPP), curtailment service provider (CSP), and retailers interact to accomplish individual and common goals. These players' interaction can regard physical electricity flows and/or financial electricity exchanges [141].

Considering the participation of the new DERs connected to the networks in EMs, aggregators are required. VPP can aggregate several small-scale energy resources, as DG, storage, EVs, and DR, managing these resources and making them able to participate in electricity markets [142, 143]. Focusing on aggregating consumers' DR participation, the CSP is relevant in order to make the small consumers able to participate in DR programs designed for large consumers. Small consumers, without the reduction capacity required by the DR program operator (usually an ISO) make a contract with a CSP, which aggregates several small consumers participate in the DR program.

DR gained increasing relevance as it proved its effectiveness in practice and its ability to generate benefits for the whole power and energy systems as well as for the involved players [132, 144, 145]. Successful DR programs are currently implemented in electricity markets. As an example, in 2013, in the US, businesses and homeowners earned over $2.2 billion in revenues from DR, more than the avoided investment in grid infrastructure and power plants. However, some legal issues are putting the DR business models currently used in the US at risk, interfering with the already existing DR implementations [137, 139].

DR is a high-value resource with low cost, when compared with the other available substituting resources such as the large power plants and some DG [126, 129]. It has already been proved in practice that DR is able to adequately prevent and/or solve emergency situations [142, 146, 147]. One of the most relevant events occurred in the summer 2013. Due to a heat wave in the Eastern United States, a new record for peak demand (33 955 MW) has been registered for New York ISO (NYISO) on 19 July. In PJM, one of the largest US ISOs, emergency and economic DR helped to address transmission constraints, high loads, and unplanned outages. On 18 July, 1638 MW of DR resources have been dispatched. The largest DR event (5949 MW) for emergency purposes in PJM occurred on 11 September. On 19 July, at ISO-NE 200 MW of DR have been dispatched, in a context of 27 359 MW of peak demand [128].

In spite of having already proved to be a very valuable and flexible resource with low cost when compared with alternative resources [137, 145, 148], DR use is still at very low levels around the world. Unlocking a significant part of the full DR potential across the diverse segments of power and energy systems will bring increased efficiency to power and energy systems. This is the faster and most significant energy efficiency increase in the sector that can be envisaged for the coming years. For instance, it is estimated that US households alone have a potential of 65 GW of peak reduction – equivalent to $3 billion worth of annual capacity [139].

The use of heuristic approaches has been gaining increasing relevance in the context of smart grids, namely for obtaining the optimal scheduling of DG and DR resources. It is especially relevant when the performed studies involve a large set of resources and/or when the execution time is relevant. The work related to this thesis considers the use of distinct approaches concerning PSO. The traditional PSO has been used in [149–151]. Mutated PSO was used in [149], providing the comparison with traditional PSO, and the quantum PSO (QPSO) has been used in [127].

The PSO-based methodology presented in [149] aims at supporting the minimization of the operation costs of a VPP that manages the resources in a distribution network and the network itself. Those resources include the DERs and the energy from external energy suppliers. DR resources were divided into three incentive programs. Network constraints were considered using an AC power flow. The PSO approach used Gaussian mutation of the strategic parameters and self-parameterization of the maximum and minimum particle velocities, according to the context.

In [150], the proposed methodology considers DR in terms of electricity price variation imposed by the DNO in the presence of a reduction need due to reasons such as lack of generation or high market prices. Real time pricing (RTP) is used in this price-responsive approach enabling to overcome the difficulties related to assuring the consumers' response to the DR event, and to overcome lack of DLC capability when the smart grid is at an initial implementation stage. In the case of the methodology proposed in [151], the energy resource scheduling is done taking into account the energy and reserve needs in a distribution network operated by a VPP. The QPSO approach presented in [127] has been developed aiming to supporting the decisions of a VPP in the DR program's definition, namely in what concerns the number of DR events and the duration of each single event. It considers the impact of the load shifting in the subsequent periods. The VPP operation costs are minimized considering DR, DG, and supplier resources, through the optimization done using QPSO. A large set of scenarios has been implemented, enhancing the advantages of using a heuristic approach. A similar methodology has been implemented using differential search algorithm (DSA), as presented in [152].

6.6.2 Methodology for Consumption Shifting and Generation Scheduling

The methodology proposed deals with the minimization of the operation costs of a VPP that operates a distribution network, as well as the resources available in the network area. The resources include consumers (providing DR) and also the generation from DG units and from a supplier able to deliver electricity in the network area. The nomenclature is as follows:

$C_{DG(g,t)}$	Cost of DG unit g in period t (m.u./kW)
$C_{DR(t,i)}$	Cost of DR from period t to period i (m.u./kW)
$C_{NSP(t)}$	Non-supplied power cost in period t (m.u./kW)
$C_{Sp(t)}$	Supplier power cost (m.u./kW)
g	Each DG unit

G	Total number of DG units
i	Each shifting period
I	Maximum distant periods for shifting
$L_{(t)}$	Initial consumption in period t (kW)
OC	Virtual power player operation costs (m.u.)
$P_{DG(g,t)}$	Power schedule in DG unit g, in period t (kW)
$P_{DR(i,t)}$	DR incoming in period i from t (kW)
$P_{DR(t,i)}$	Scheduled DR from period t to period i (kW)
$PMax_{DG(g,t)}$	Maximum available power in DG unit g in period t (kW)
$PMax_{DR(t,i)}$	Maximum available DR from period t to period i (kW)
$PMax^{i}_{DR(t,i)}$	Maximum allowed DR from any period t to a specific period i (kW)
$PMax_{NSP(t)}$	Maximum allowed non-supplied power in period t (kW)
$PMax_{Sp(t)}$	Maximum available power from the supplier in period t (kW)
$PMax^{t}_{DR(t,i)}$	Maximum allowed DR to any period i from a specific period t (kW)
$P_{NSP(t)}$	Non-supplied power in period t (kW)
$P_{Sp(t)}$	Supplier power schedule in period t (kW)
T	Total number of periods
t	Each elementary period

The objective function presented in Eq. (6.6.1) considers, for all the periods in the defined time horizon T, the DG, the supplier, and the shifting costs in each period t. The cost of non-supplied power is also considered, as the price that the VPP must pay to the consumers not supplied due to lack of capacity of other resources. It is a noncontracted consumption reduction, which should be avoided.

Minimize

$$\mathrm{OC} = \sum_{t=1}^{T} \left[\begin{array}{l} P_{Sp(t)} \times C_{Sp(t)} + \sum_{g=1}^{G} P_{DG(g,t)} \times C_{DG(g,t)} \\ \quad + \sum_{t-I \le i}^{i \le t+I} P_{DR(t,i)} \times C_{DR(t,i)} + P_{NSP(t)} \times C_{NSP(t)} \end{array} \right] \tag{6.6.1}$$

The first constraint of the model, presented in Eq. (6.6.2), is the balance equation which includes the supplier, the DG, the load demand, and the DR balance in each period t. The balance equation in this model considers both the consumption shifted from period t to period i, and the incoming consumption in period t, shifted from period i.

$$\sum_{g=1}^{G} P_{DG(g,t)} + P_{Sp(t)} = L_{(t)} - P_{NSP(t)} - \sum_{t-I \le i}^{i \le t+I} P_{DR(t,i)} + \sum_{t-I \le i}^{i \le t+I} P_{DR(i,t)}; \quad \forall 1 \le t \le T \tag{6.6.2}$$

The remaining constraints of the model include the maximum capacity of each resource. The DR maximum capacity limit is presented in Eq. (6.6.3). It considers that the consumption reduction performed in period t can be transferred for a period i before or after t, constrained by I, the maximum number of distant periods that can be used for load shifting. In this way, Eq. (6.6.3) must be verified in each consumption reduction period t and in each shifting period i.

$$P_{\mathrm{DR}(t,i)} \leq P\mathrm{Max}_{\mathrm{DR}(t,i)} \quad \forall 1 \leq t \leq T \quad \forall -I \leq i \leq I \tag{6.6.3}$$

The maximum capacity limit in each period t concerning DG and the supplier resources are presented, respectively, in Eqs. (6.6.4) and (6.6.5).

$$P_{\mathrm{DG}(g,t)} \leq P\mathrm{Max}_{\mathrm{DG}(g,t)} \quad \forall 1 \leq t \leq T \tag{6.6.4}$$

$$P_{\mathrm{Sp}(t)} \leq P\mathrm{Max}_{\mathrm{Sp}(t)} \quad \forall 1 \leq t \leq T \tag{6.6.5}$$

It is important to note that the consumption curtailment (consumption that is not postponed nor planned to be performed before the DR event period) is considered by the proposed model when i is equal to t.

For improved characterization of the consumer's behavior, Eqs. (6.6.6) and (6.6.7) have also been implemented. In Eq. (6.6.4), it is defined the maximum limit of consumption that can be shifted from period t to period i. The constraint presented in Eq. (6.6.6) considers a maximum limit for the total consumption that can be shifted from the specific period t to all the shifted periods i. In this way, it is possible to shift several consumption sets to other periods i, but the sum of all these shifted consumption sets must be lower than a determined limit, as imposed by Eq. (6.6.6).

$$\sum_{t-I \leq i}^{i \leq t+I} P_{\mathrm{DR}(t,i)} \leq P\mathrm{Max}_{\mathrm{DR}(t)}^{t}; \quad \forall 1 \leq t \leq T \tag{6.6.6}$$

$$\sum_{t-I \leq i}^{i \leq t+I} P_{\mathrm{DR}(i,t)} \leq P\mathrm{Max}_{\mathrm{DR}(t)}^{i}; \quad \forall 1 \leq t \leq T \tag{6.6.7}$$

Similarly, one can specify a maximum limit for the consumption that can be shifted to each period i, as modeled by Eq. (6.6.7). It is important to remember that, in a certain period, i match t. In this case, the consumption reduction in period t is also limited by Eq. (6.6.7).

6.6.3 Quantum PSO

Classical PSO-based methods rely on the convergence to the global best position (best solution achieved so far), not taking into account the position of other particles in the swarm. This feature is the major limitation of PSO classical physics behavior when there are particles distant from the global best position. The concept of lagged particles can be seen in Figure 6.6.1. It can be seen that the personal bests of some particles are distant from the global best position. In QPSO the lagged

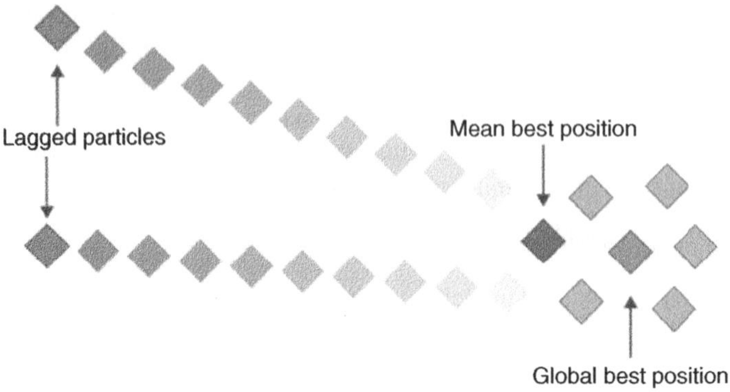

Figure 6.6.1 Lagged particles and wait among particles in QPSO.

particles are not abandoned by the swarm as in classical PSO. The lagged particles affect the mean best position, c.f. Eq. (6.6.9), and therefore the lagged particles are shifted toward the rest of the swarm. The particles' distribution affects the convergence rate; however, QPSO can provide stronger global search ability than traditional PSO [133].

Figure 6.6.2 illustrates how differently the lagging phenomena of QPSO and PSO particles are. In Figure 6.6.2, the big circle represents the global best position while the little circles represent the other particles, and the little circles with vertical lines represent the lagged particles. The arrows around the little circles represent the possible directions of the particles; the big arrowhead points to the direction in which the particle moves with high probability.

Equation (6.6.8) presents the local focus equation which is a random point located within the hyperplane constructed between $P_{i,n}$ and G_n in the search space. The local focus is then used in Eq. (6.6.9), which represents the movement equation. The contraction–expansion (CE) coefficient α is vital to the dynamical

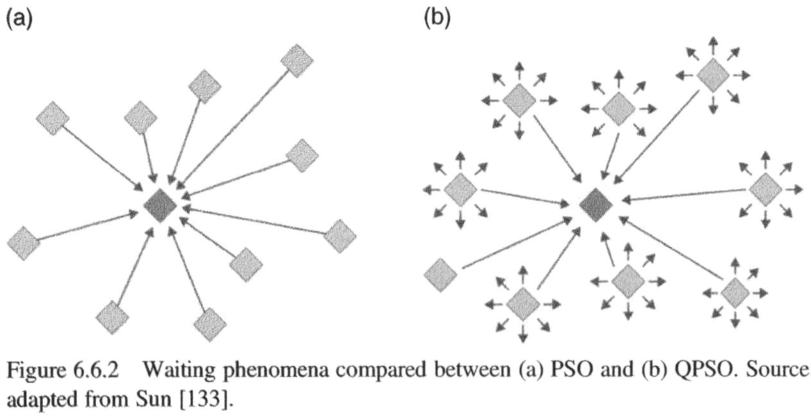

Figure 6.6.2 Waiting phenomena compared between (a) PSO and (b) QPSO. Source: adapted from Sun [133].

behavior of an individual particle and the convergence of the algorithm. The algorithm decides to add or subtract the second term of Eq. (6.6.9) using a probability of 50%.

$$p^j_{i,n+1} = \varphi^j_{i,n} P^j_{i,n} + \left(1 - \varphi^j_{i,n}\right) G^j_n \tag{6.6.8}$$

$$X^j_{i,n+1} = p^j_{i,n} \pm \alpha \left| X^j_{i,n} - C^j_n \right| \ln\left(\frac{1}{u^j_{i,n+1}}\right) \tag{6.6.9}$$

where

C^j_n: Mean best position which can be defined by the average of the personal best positions of all particles

G^j_n: The global best of particle at the nth iteration

$P^j_{i,n}$: The jth component of the personal best of particle i at the nth iteration

$p^j_{i,n}$: The jth component of the local focus $p_{i,n}$ of particle i at the nth iteration

$u^j_{i,n+1}$: Random uniformly distributed sequence [0,1]

$X^j_{i,n}$: Position of the particle ith at the nth iteration

$\varphi^j_{i,n}$: Random uniformly distributed sequence [0,1].

Finally, Figure 6.6.3 presents the flowchart of the QPSO algorithm. It follows the common approach of swarm intelligence methods, namely initialization of particles (usually randomly), evaluation of the initial positions of those particles, and identification of the best global position and personal bests before starting the optimization loop.

Also, before starting the optimization loop, all parameters of QPSO are initiated stage and verified. After the algorithm starts the optimization loop stage, the alpha parameter (α) will start to decay (alternatively the alpha parameter may remain equal during the search process and without decaying) to evolve from exploration to exploitation of the swarm, i.e. global search to local search. In the next step the particle movements (6.6.8) and (6.6.9) are applied. Then, those moved particles are re-evaluated. If they achieve best positions, the global best and personal bests of the particles are updated. Finally, depending on the stopping criteria, which usually is dynamic (the swarm search can stop if not evolving), the algorithm may end, otherwise it will continue until the stopping criteria is satisfied.

6.6.4 Numeric Example

The present section brings out the details on the implemented case study. A distribution network and the available resources are operated by a VPP, which makes use of the proposed methodology in order to determine the optimum

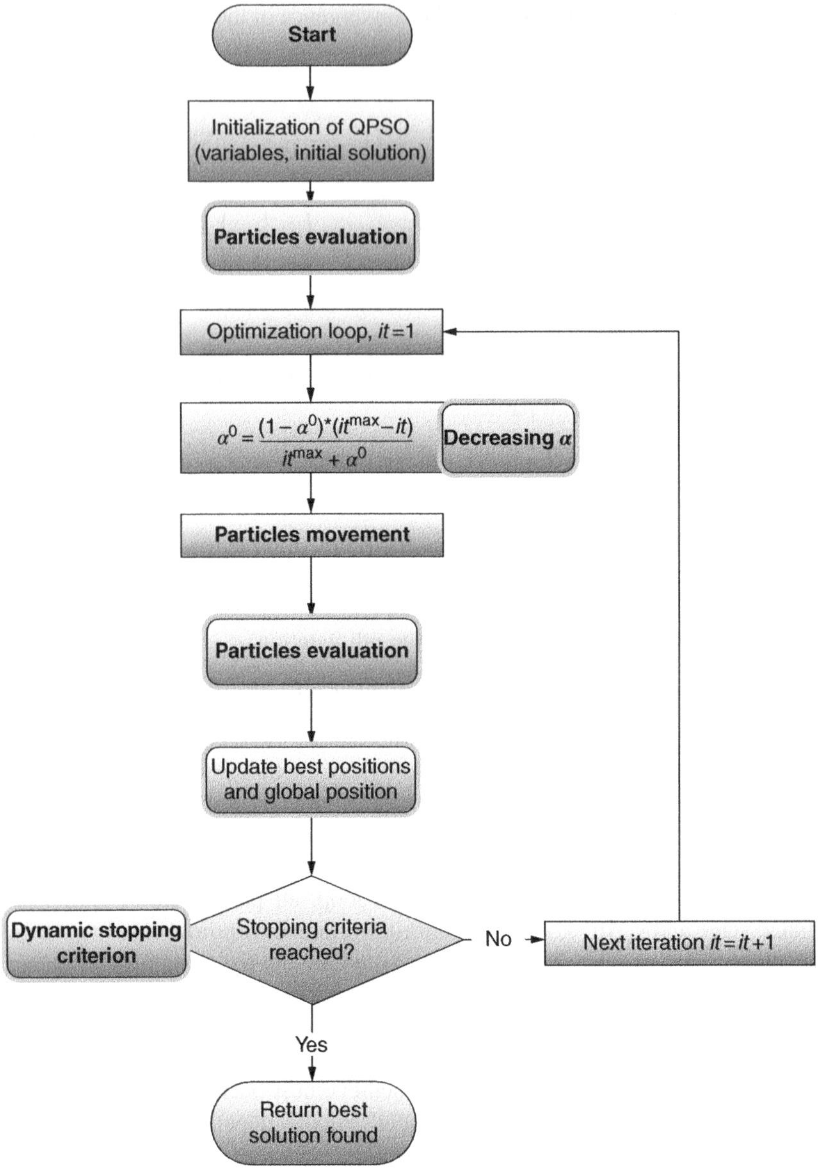

Figure 6.6.3 Flowchart of QPSO algorithm.

schedule of the resources. The scenario, which is based on the one presented in [153], is characterized by the information shown in Figure 6.6.4 for the resources' availability and demand. The information concerning the different parameters is presented in Table 6.6.1.

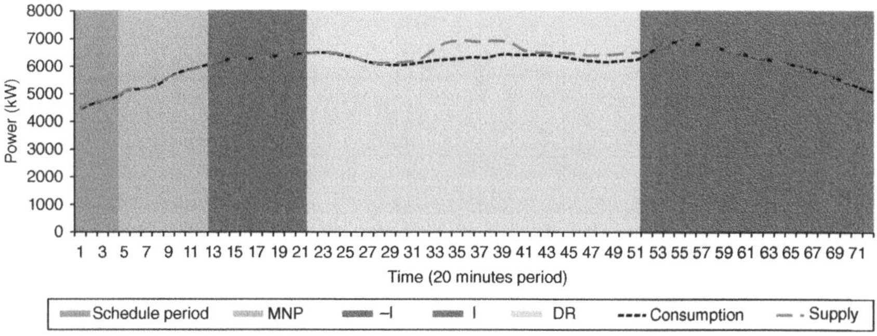

Figure 6.6.4 Generation and consumption diagram.

TABLE 6.6.1 Consumers and DR Programs

Parameter	Base
$C_{DG(g,t)}$	0.05; 0.06; 0.07; 0.08
$C_{DR(t,i)}$	0.06; 0.07; 0.08
$C_{NSP(t)}$	4
$C_{Sp(t)}$	0.04
$L_{(t)}$	According to Figure 6.6.4
$PMax_{DG(g,t)}$	500; 700; 800; 1000
$PMax_{DR(t,i)}$	20; 60
$PMax^{i}_{DR(t,i)}$	444.3
$PMax_{NSP(t)}$	8000
$PMax^{t}_{DR(t,i)}$	444.3

From Figure 6.6.4 one can see that the implemented case study concerns the period of one day composed of 72 periods of 20 minutes. It is considered that in the initial periods of the day, the VPP takes time for the resources' schedule optimization task. After that, the resources are informed about their expected schedule. The period between the end of the scheduling period and the beginning of DR use optimization, in which the resources are notified on the schedule results, is called the minimum notification period.

The DR event period set corresponds to periods of the consumption reduction scenarios. However, in periods before (–*I*) and after (*I*) this DR event period, additional consumption reduction can be scheduled due to the fact that some consumption is shifted in these periods, which are shown between periods 13 and 21 and also between 51 and 72 in Figure 6.6.4.

In the case of DG, the presented values concern four distinct DG units. In the case of *P*MaxDR(*t*,*i*), it has been assumed 60 kW for consumption shifting

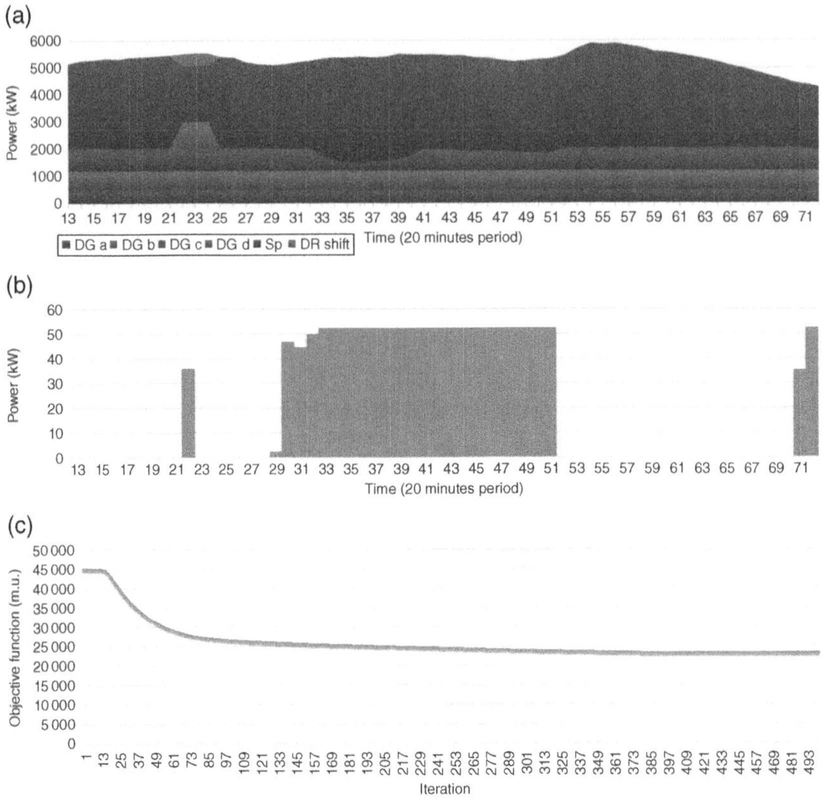

Figure 6.6.5 Scheduling results: (a) all resources; (b) shift DR income, and (c) QPSO convergence.

to periods before the DR event period, whereas 60 kW has been considered for the consumption shifting to periods after the DR event.

In what concerns the QPSO implementation, the modeling of the present case study includes 3960 variables. QPSO was implemented in MATLAB. The number of particles is 10, while the maximum number of iterations was set to 500. The alpha value is an important parameter in QPSO directly related with the method's convergence. This parameter varies from 1.0 to 0.5 during swarm iterations. The initial solution of the method is generated using an order of merit scheme, in which the resources with lower costs are scheduled first, including the DR shifting resources.

It is possible to see in Figure 6.6.5c the convergence performance of the QPSO algorithm. The convergence figure resulted from 1000 random runs. For each one of the 1000 runs, a maximum of 500 iterations has been defined. After the 1000 runs, the mean fitness was calculated. This resulted in just one mean value per iteration as can be seen in Figure 6.6.5c. One can see that, on average, the algorithm presents a good convergence with a quick rate between 30th and 200th iteration and a slower rate onward until it vanishes later after 450 iterations.

In what concerns the results obtained for the resources scheduling, Figure 6.6.5a presents the generation and supplier power schedule and Figure 6.6.5b presents the results for the consumption shifting in each one of the 60 periods according to Figure 6.6.4.

Focusing on the power supply resources, we have one supplier and four DG resources. The DG resources DGa and DGb are scheduled at constant power in the focused time set. One can see that in periods 22–24, DGd is activated and the consumption reduction reaches its maximum. DGc and the supplier assure the generation in the remaining periods at the needed levels.

In what concerns the DR results, it is possible to see in Figure 6.6.5b the consumption incoming in several periods due to the consumption shifting from periods 22 to 24.

6.6.5 Conclusions

DG includes a diversity of technologies that are largely based on renewable sources. DR can have also a distributed nature, contributing to the increase of the consumer's participation in the smartgrids concept.

In the present section, DR, namely load shifting was optimally scheduled together with DG, contributing to accommodate all the available renewable-based generation.

Due to the large amount of resources and to the complexity of the optimization problem, heuristic approaches are very relevant. In this section, QPSO was used, showing its effectiveness to deal with DG and DR scheduling.

6.7 EPSO-BASED SOLAR POWER FORECASTING

Hiroyuki Mori[1] *and Masato Takahashi*[2]

[1]*Meiji University, Nakano-city, Tokyo, Japan*
[2]*Fuji Electric Co., Hino-city, Tokyo, Japan*

6.7.1 Introduction

This section addresses an ANN-based method for dealing with the output prediction of solar PV systems. In recent years, global warming is one of the most important challenges in the world. As one of tools to suppress CO_2, renewable energy such as PV systems, wind power generation, etc., is widely spread in the world. For example, the government of Japan aimed at realizing a target that a large amount of PV systems of 53 GW should be introduced into the networks until 2030. The trend has been speeded up due to the accident of Fukushima in 2011. In fact, Japan has already had the PV capacity of 43 GW as of 31 December 2016. Also, there has been a recent trend that renewable energy is connected to smart grid in the world. To optimize the energy efficiency, renewable energy cooperates with energy management system (EMS) at buildings, houses, factories, communities, etc. From a standpoint of global warming, renewable energy plays a key role to provide clean energy with power systems. However, it is not easy to introduce renewable energy into power systems due to the intermittent generation characteristics. Renewable energy brings about a large amount of the uncertainties to power system operation and planning so that the accurate models are required for the forecasting methods. The conventional methods on generation output forecasting of PV system may be classified as follows:

a. Autoregressive moving average (ARMA) models [154–156]

b. Autoregressive integrated moving average (ARIMA)model [157]

c. Multilayer perceptron (MLP) [158–160]

d. Radial basis function network (RBFN) [161–163]

Methods (a) and (b) are based on classical time series analysis like the Box–Jenkins method [164]. They have a drawback that the parameters are not flexible to the sudden change of time series. In other words, they do not have good approximation for nonlinear time series or sudden changes. Method (c) is one of ANNs that have been widely spread in the engineering fields. Usually, it consists of three layers: input, hidden, and output. The input of each neuron is transformed into the output through the sigmoid function. It has better performance than Methods (a) and (b) due to the nonlinear approximation. Also, Method (d) is a kind of ANNs that makes use of linear combinations of radial basis functions (RBFs) that corresponds to the Gaussian ones [165, 166]. RBFN often consists of three layers like MLP. The nonlinear transformation is made between hidden and output layers. RBFN has better performance than MLP if the model is tuned up approximately. The tuning procedures of RBFN sometimes cause a problem that good results are not obtained due the poor parameter adjustment. To improve the performance of RBFN, generalized radial basis function network (GRBFN) was developed [167, 168]. The difference between GRBFN and RBFN is that GRBFN determines the center and the width of the RBF through updating them at the learning process like the weights between the neurons while RBFN employs them with the use of clustering results of learning data and does not use the learning process.

In this section, a GRBFN-based method is presented to handle the prediction of PV system generation output. The main difference between the proposed and the conventional GRBFN may be described as follows:

1. Deterministic annealing (DA) clustering [169–172] of global clustering is used to determine the initial solutions of the center and the width of the RBFs. It is important to estimate globally optimal initial solutions although *k*-means of local clustering is used in the conventional method.

2. The weight decay method (WDM) is used to avoid overfitting for learning data [173]. It is useful for realizing more robust learning for nonlinear complicated time series data of PV system generation output. As mentioned later, the cost function may be modified to suppress overfitting in optimizing the parameters in GRBFN.
3. EPSO [174, 175] of modern heuristics or meta-heuristics is applied to tune up the center and width of the RBFs as well as the weights between input and hidden layers. EPSO has function to evaluate better solution through the use of adaptive parameters with evolutional strategy in comparison with the conventional PSO. Specifically, EPSO minimizes the modified cost function to avoid the overfitting problem.

The method is successfully applied to real data of PV system generation output. The effectiveness of the proposed method is demonstrated in comparison with the conventional ANNs such as MLP, RBFN, and GRBFN.

6.7.2 General Radial Basis Function Network

This section describes GRBFN [167, 168] that is an extension of RBFN in a way that the center and the width of the RBF are evaluated by the learning process. Before describing GRBFN, RBFN is explained. It consists of three layers of input, hidden, and output as shown in Figure 6.7.1, where variables $x_1, x_2, \ldots, x_n$ are input variables, variable y is output, the weights between input and the hidden layers are unity, and variables$w_1, w_2, \ldots, w_n$ show those between the hidden and the output layers. The hidden layer plays a key role to transform input variables into the output through the nonlinear transform of the Gaussian function. The output of neuron i at the hidden layer may be written as

$$a_i = \exp\left(-\frac{\|x-u_i\|^2}{\sigma_i^2}\right) \tag{6.7.1}$$

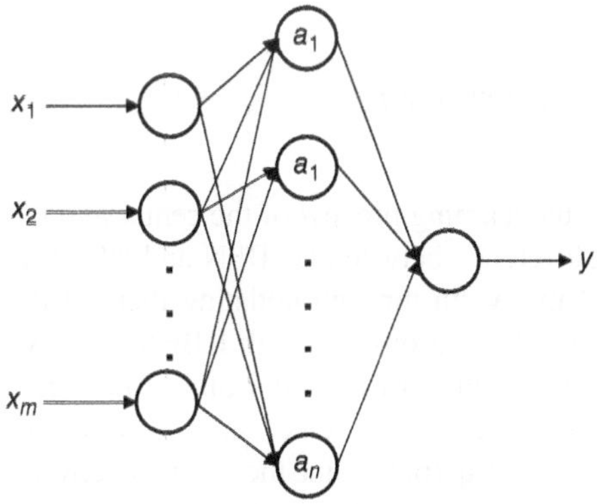

Figure 6.7.1 Structure of RBFN.

where,

a_i: output of neuron i at the hidden layer

x: input vector at the input layer

u_i: center of neuron i at the hidden layer

σ_i: width of Gaussian function at neuron i of the hidden layer

Also, output y at the output layer may be expressed as a sum of weighted output variables at the hidden layer.

$$y = \sum_{i=1}^{n} w_i a_i(x) \tag{6.7.2}$$

where,

y: output

w_i: weight of neuron i between hidden and output layers

n: number of hidden units

The clustering method technique is used to determine the center and the width of the Gaussian functions. For example, Moody et al. used k-means to evaluate the center [176]. However, it is well known that k-means gives a locally optimal solution. In other words, the solution of k-means is affected by the initial solution significantly. Thus, more advanced methods are required to determine the center.

Also, the width is evaluated by the results of clustering that make use of the distance between the center and the most distant point as the width. The weight between the hidden and the output layers may be calculated by minimizing the following cost function:

$$f = \frac{1}{2}\sum_{j=1}^{l} (y_j - t_j)^2 \tag{6.7.3}$$

where,

y_j: output j of RBFN obtained by input pattern j

t_j: target j of learning data

Next, GRBFN is explained. It has the learning process of the center and the width of the RBFs unlike RBFN. The difference between GRBFN and RBFN is that GRBFN determines the center and the width through updating them at the learning process like the weights between the neurons although RBFN employs the center and the width that correspond to clustering results of learning data and does not use the learning process. Specifically, the center and the width are determined by minimizing the cost function of Eq. (6.7.3) like the weights between the hidden and the output layers iteratively. As a result, it is expected that GRBFN gives better results than the conventional RBFN due to the additional learning process of the center and the width of the RBFs. It should be noted that the performance of GRBFN is affected by the evaluation of the clustering results, the weights between neurons at the hidden and the output layers, and the strategy for model overfitting.

6.7.3 *k*-Means

This section deals with *k*-means of clustering [177]. In this section, it is used to evaluate the mean and width of the RBF in RBFN. Clustering is a technique that classifies a set of given data into some clusters from a standpoint of data similarity. Now, let us consider clustering of data $\boldsymbol{x}_n = \{x_n\}$, $\{n = 1, \ldots, N\}$ into k clusters with center $\boldsymbol{\mu}_k = \{\mu_k\}$, $\{k = 1, \ldots, K\}$. Mathematically, clustering may be expressed as the optimization problem that minimizes the distance between data and the center at each attribute. The cost function may be written as

$$d = \sum_{n=1}^{N}\sum_{k=1}^{K} r_{nk}\left\|x_n - \mu_k\right\|^2 \tag{6.7.4}$$

where,

r_{nk}: attribute of cluster k such as

$$r_{nk} = \begin{cases} 1 & k = \arg\,\min_j \left\|x_n - \mu_j\right\|^2 \\ 0 & \text{otherwise} \end{cases} \tag{6.7.5}$$

Center $\boldsymbol{\mu}_k$ to minimize d may be obtained by differentiating (6.7.4) with respect to $\boldsymbol{\mu}_k$.

$$\boldsymbol{\mu}_k = \frac{\sum_n r_{nk} x_n}{\sum_n r_{nk}} \tag{6.7.6}$$

The algorithm of *k*-means may be summarized as follows:

Step 1: Set initial conditions.

Step 2: Prepare input data.

Step 3: Prepare the random initial solution of $\boldsymbol{\mu}_k$.

Step 4: Fix $\boldsymbol{\mu}_k$ and minimize J with respect to r_{nk}.

Step 5: Fix r_{nk} and minimize J with respect to $\boldsymbol{\mu}_k$.

Step 6: Repeat Steps 4–5 until the convergence criterion is satisfied.

6.7.4 Deterministic Annealing Clustering

This section describes DA clustering of global clustering method that is not influenced by the initial solution [170–172]. DA carries out the optimization process through changing the cost function at each temperature with the concept of the free energy in statistical mechanics so that a globally optimal solution is obtained. As temperature cools down, the cost function gradually approaches the original cost function. This process allows the users to evaluate a globally optimal solution. According to the principle of maximum entropy in statistical mechanics, DA clustering gives better clustering results. The cost function may be written as

$$d = \sum_{i=1}^{m} \sum_{j=1}^{k} P(x_i \in C_j) \|x_i - y_j\|^2 \tag{6.7.7}$$

where,

d: cost function

x_i: ith data

C_j: jth center

$P(x_i \in C_j)$: probability that data x_i belongs to cluster C_j.

It should be noted that the association state may be expressed as the association probability in DA clustering rather than the binary number in k-means. The algorithm of DA clustering may be summarized as follows:

Step 1: Set initial conditions (the inverse of the initial temperature $\beta_0 = 1/T_0$, where T_0 is the initial temperature and the upper bound of temperature β_{max}, and convergence criterion ε).

Step 2: Select the initial centers from a data set.

Step 3: Calculate the association probability by the following equation:

$$P\left(x_i \in C_j^{(t)}\right) = \frac{\exp\left(-\beta\|x_i - y_j\|^2\right)}{\sum_{j=1}^{k} \exp\left(-\beta\|x_i - y_j\|^2\right)} \tag{6.7.8}$$

where,

β: temperature parameter equal to $1/T$

t: iteration number

$C_j^{(t)}$: jth cluster at iteration t

Step 4: Calculate the new center by

$$y_j^{(t)} = \frac{\sum_{i=1}^{m} x_i P(x_i \in C_j)}{\sum_{i=1}^{m} P(x_i \in C_j)} \tag{6.7.9}$$

where,

$y_j^{(t)}$: center of jth cluster at iteration t

Step 5: Evaluate new association probability by (6.7.8).

Step 6: Calculate the cost function of (6.7.7) and go to Step 7 if the following equation holds:

$$\left|\frac{d^{(t+1)} - d^{(t)}}{d^{(t)}}\right| \leq \varepsilon \tag{6.7.10}$$

where,

$d^{(t)}$: cost function at iteration t

ε: convergence criterion

Otherwise, return to Step 4.

Step 7: Stop if $\beta \geq \beta_{\max}$. Otherwise, update β and return to Step 4.

6.7.5 Evolutionary Particle Swarm Optimization

In this section, EPSO is outlined [174, 175]. It is an extension of PSO [178] in a sense that the evolutionary strategy is applied to PSO to tune up the parameters with the evolutionary strategy. PSO stems from the analogy of behavior that flock of birds or a school of fish search for prey. In other words, PSO is one of multipoint search methods that make use of the group information of swarm intelligence, i.e. pbest_i and gbest to find out better solutions in the optimization process, where pbest_i implies the best solution of agent i and gbest indicates the best solution in a group of agents. Specifically, the moving rule of PSO may be written as

$$V_i^{t+1} = w_0 V_i^t + w_1 \cdot \text{rand} \cdot \left(\text{pbest}_i - S_i^t\right) + w_2 \cdot \text{rand} \cdot \left(\text{gbest} - S_i^t\right) \tag{6.7.11}$$

$$S_i^{t+1} = S_i^t + V_i^{t+1} \tag{6.7.12}$$

where,

V_i^{tt}: velocity of agent i at iteration t

S_i^t: placement of agent i at iteration t

$w_0 - w_2$: weights

pbest_i: best solution of agent i

gbest: best solution of swarm

rand: uniform random numbers of interval [0, 1]

It can be observed that each agent modifies the velocity through the distance to pbest_i and gbest.

However, PSO has a drawback that it easily gets stuck in local minimum due to inappropriate setting of the parameters to deteriorate the search process. To overcome the drawback, Miranda developed EPSO that combined PSO with the evolutionary strategy to escape from local minimum. EPSO has advantage that it evaluates better solutions through updating the parameters with the evolutionary strategy. After replicating agents from the original agents, the mutation is carried out for the copied agent. Information on better agents is preserved with the selection rule. The moving rule of EPSO may be written as

$$V_i^{t+1} = w_{i0}^* V_i^t + w_{i1}^* \left(\text{pbest}_i - S_i^t\right) + w_{i2}^* \left(\text{gbest}^* - S_i^t\right) \tag{6.7.13}$$

$$w_{ik}^* = w_{ik} + \tau N(0, 1) \tag{6.7.14}$$

$$\text{gbest}^* = \text{gbest} + w_{i3}^* \tag{6.7.15}$$

$$w_{i3}^* = w_{i3} + \tau' N(0, 1) \tag{6.7.16}$$

where,

V_i^t: velocity of agent i at iteration t

$w_{i0} - w_{i3}$: weights

$w_{i0}^* - w_{i3}^*$: weights with evolution process

gbest*: gbest with evolution process

S_i^t: placement of agent i at iteration t

τ, τ': learning rates of Gaussian noise for V_i and gbest

The algorithm of EPSO may be summarized as follows:

Step 1: Set the initial conditions.

Step 2: Copy the agents with the replication rate.

Step 3: Carry out the mutation for the weights of each agent.

Step 4: Move to the new solutions through the moving rule at each agent

Step 5: Evaluate each agent.

Step 6: Select the agent with the best cost function from the candidate of agents.

Step 7: Update pbest and gbest if the current pbest and gbest are better than the previous.

Step 8: Stop if the termination conditions are satisfied. Otherwise, return to Step 2.

6.7.6 Hybrid Intelligent Method

This section proposes a hybrid intelligent method for forecasting of PV system generation output. The simple RBFN does not necessarily give better results for complicated time series of PV system generation. Therefore, the prediction of PV system generation requires ingenious ideas to obtain better results. The proposed method consists of the following techniques:

1. GRBFN
2. DA clustering
3. The WDM
4. EPSO

To improve the prediction model accuracy of RBFN, this paper uses GRBFN that is an extension of RBFN in a way that the centers and the widths of the RBFs are updated by the learning process although those of RBFN are not updated. DA clustering of global clustering is used to estimate good initial solutions for the center and the width. As mentioned before, the clustering results of DA clustering are not affected by the initial conditions. The clustering results give the initial solutions of the center and the initial solutions of the width are evaluated by finding out the most distant point from the center in the cluster. The WDM is used to avoid overfitting for learning data. For complicated nonlinear complicated time series of PV system generation output, GRBFN is inclined to overfit learning data. To overcome the problem, more robust learning techniques are required to improve the accuracy of the prediction model. In this paper, the WDM is used to avoid the overfitting results [173]. The cost function may be rewritten as

$$E = \frac{1}{2}\sum_{k}(t_k - y_k)^2 + \lambda\frac{1}{2}\sum_{i} w_i^2 \tag{6.7.17}$$

where,

E: cost function

t_k: target signal

y_k: output of GRBFN

λ: penalty coefficient

It should be noted that the second term serves to avoid overfitting for learning data in the above equation. EPSO is employed to estimate better weights between the hidden and output layers though minimizing the cost function of Eq. (6.7.17) so that the proposed method easily escapes from a local minimum. The algorithm of the proposed method may be summarized as follows:

Step 1: Evaluate the center of the RBFs by DA clustering and the width through finding out the most distant point from the center in the obtained cluster.

Step 2: According to the procedure of GRBFN, calculate the weights between the hidden and the output layers, and the center and the width through the learning process.

Step 3: Regard the obtained weights, centers, and widths as the initial solutions of EPSO, and minimize the cost function of (6.7.17) with EPSO.

Step 4: Give unknown data to the proposed method.

6.7.7 Case Studies

Simulation Conditions

1. The proposed method was applied to real data of PV system generation output forecasting in Japan. The observed data were collected with the sampling

time of 1 (minutes) for five days from 2 May 2005 to 7 May 2006. As learning data, data for the first four days were used while the rest of data were assigned to test data. In other words, the number of learning and test data are 2700 and 900, respectively, where data from 8 p.m. to 5 a.m. were excluded because the generation was not available.

2. The input variables are given as follows:

   x_{1T}: PV system generation output at time T

   x_{2T}: temperature of PV system panel at time T

   x_{3T}: variance of x_{t1} for previous five minutes at time T

   x_{4T}: variance of x_{t2} for previous five minutes at time T

   x_{5T}: variance of x_{t1} for previous 10 minutes at time T

   x_{6T}: variance of x_{t4} for previous 10 minutes at time T

   x_{7T}: first-order difference between x_{1T} and x_{1T-1}

   x_{8T}: first-order difference between x_{2T} and x_{2T-1}

   x_{9T}: second-order difference between x_{7T} and x_{7T-1}

   x_{10T}: second-order difference between x_{8T} and x_{8T-1}

   Also, the output variable is PV system generation output at time $T + 30$ (x_{1T+30}).

3. To demonstrate the effectiveness, the proposed method was compared with other methods in terms of average and maximum errors and the SD of errors for test data. For convenience, the following methods are defined:

   Method A: MLP

   Method B: k-means-GRBFN (k-means and GRBFN; the conventional GRBFN)

   Method C: DA-GRBFN (hybrid method of DA clustering and GRBFN)

   Method D: DA-GRBFN-PSO (hybrid method of DA clustering, GRBFN, and PSO)

   Method E: DA-GRBFN-EPSO (hybrid method of DA clustering, GRBFN, and EPSO)

   Method F: DA-GRBFN-WDM (hybrid method of DA clustering, GRBFN, and WDM)

   Method G: DA-GRBFN-PSO-WD (hybrid method of DA clustering, GRBFN, WDM, and PSO)

   Method H: DA-GRBFN-EPSO-WD (hybrid method of DA clustering, GRBFN, WDM, and EPSO; proposed method)

   In the definition above, Method B means GRBFN that determines centers and width of the RBFs by k-means. Method E implies GRBFN that determines centers and width of the RBFs by DA clustering and optimizes the weights between the hidden and output layers with EPSO after evaluating them with the gradient method. Method H indicates GRBFN that avoids

TABLE 6.7.1 Parameters of DA-Clustering

Parameters				
No. of Clusters	β_0	β_{max}	$\Delta\beta$	Convergence Criterion ε
16	1	6000	0.1	0.001

TABLE 6.7.2 Parameters of ANNs

Methods	λ	No. of Hidden Units	No. of Learning Iterations
A		14	30 000
B		16	20 000
C		16	20 000
D		16	20 000
E		16	20 000
F	0.0001	16	20 000
G	0.0001	16	20 000
H	0.0001	16	20 000

TABLE 6.7.3 Parameters of PSO and EPSO

	Parameters									
Methods	w_{max}	w_{min}	w_1	w_2	w_3	r	τ	τ'	No. of Agents	No. of Iterations
PSO	1	0.1	1	1					80	800
EPSO	1	0.1	1	1	0.1	2	0.05	0.05	80	800

overfitting for learning data to improve the performance of Method E with WDM.

4. Tables 6.7.1 and 6.7.2 show the parameters of DA clustering and ANNs in the simulation, respectively. The parameters of PSO and EPSO are shown in Table 6.7.3.

Simulation Results

Table 6.7.4 shows the simulation results of the average and maximum errors, and the SD of errors for each method, where values in parenthesis mean the normalized data with that of MLP. Looking over Methods A, B, and C, it can be seen that Method C is better than others in terms of the average and maximum errors and the SD. It reduced 23.56 and 13.35% of the average errors for Methods A and B,

TABLE 6.7.4 Simulation Results of Each Method

Methods	Average Errors (p.u.) (Normalized)	Maximum Errors (p.u.) (Normalized)	SD of Errors (Normalized)
A MLP	0.0711 (1)	0.395 (1)	0.0577 (1)
B *k*-means-GRBFN	0.0628 (0.8821)	0.286 (0.723)	0.0545 (0.9445)
C DA-GRBFN	0.0544 (0.7644)	0.28 (0.7087)	0.0511 (0.8861)
D DA-GRBFN-PSO	0.0508(0.7134)	0.256 (0.6488)	0.0404 (0.6998)
E DA-GRBFN-EPSO	0.0485 (0.6813)	0.249 (0.6299)	0.0396 (0.6862)
F DA-GRBFN-WD	0.0497(0.6984)	0.276 (0.6990)	0.0503 (0.8716)
G DA-GRBFN-PSO-WD	0.0465(0.6539)	0.253 (0.6396)	0.0385 (0.6672)
H DA-GRBFN-EPSO-WD	0.0464 (0.6525)	0.228 (0.5775)	0.0384 (0.6649)

Note: Values in parenthesis mean the normalized data by that of MLP.

respectively. Regarding the maximum errors, Method C succeeded in cutting down 29.13 and 1.97% for Methods A and B, respectively. It reduced 11.39 and 6.18% of the SD for Methods A and B, respectively. Therefore, DA clustering is effective for determining the parameters of the RBFs. A comparison of Methods C, D, and E was made to demonstrate the effectiveness of EPSO. Method D reduced 6.6 and 10.88% of the average error for Methods C and D, respectively. Regarding the maximum errors, it succeeded in cutting down 8.46 and 11.13% for Methods C and D, respectively. It reduced 21.02 and 22.55% of the SD for Methods C and D, respectively. Method D did not give good results due the fixed parameter unlike EPSO. Thus, EPSO worked so well in the learning process due to the use of adaptive parameters.

Furthermore, to investigate the effectiveness of the WDM, a comparison between Methods C and F was made. Method F was better than Method C in terms of all the error indices. Finally, Method H of the proposed method was compared with Methods A and B. It can be observed that the proposed method reduced 34.75 and 26.03% of the average error for Methods A and B, respectively. Also, it decreased 42.24 and 20.12% of the maximum error for Methods A and B, and succeeded in cutting down 33.51 and 29.60% of the SD, respectively. As a result, the proposed GRBFN with DA clustering, the WDM, and EPSO provided much better results than the conventional MLP and GRBFN.

Figure 6.7.2 shows an example of comparison between the predicted value by the proposed method and observed data. It can be seen that the predicted values look a little bit conservative between 180 and 500 (minutes). On the other hand, the proposed method gave the overestimated value like spikes at 560 and 640 (minutes) because of large variance of PV system generation output during the time zone.

Figure 6.7.3a and b give the behavior of errors for MLP and the proposed method. It can be seen that most of the errors for the proposed method exist from −0.22 to 0.18 (p.u.) while those for MLP exist in wider area from −0.40 to 0.17 (p.u.).

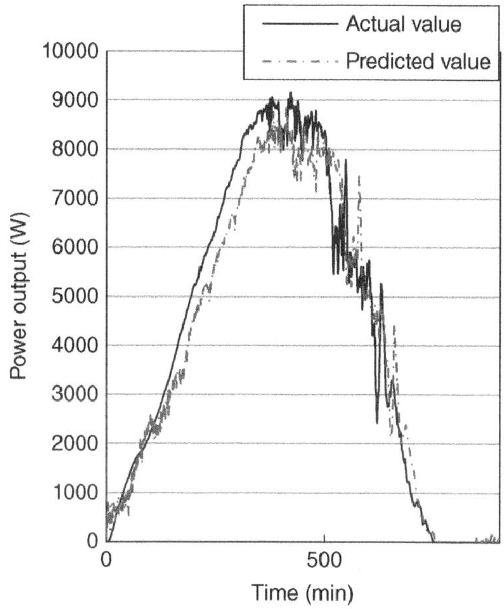

Figure 6.7.2 Comparison between actual value and predicted value by proposed method.

Therefore, the simulation results have shown that the proposed method gave better results than other methods. In particular, it is noteworthy that the proposed method significantly has succeeded in reducing errors in comparison with the conventional MLP, RBFN, and GRBFN.

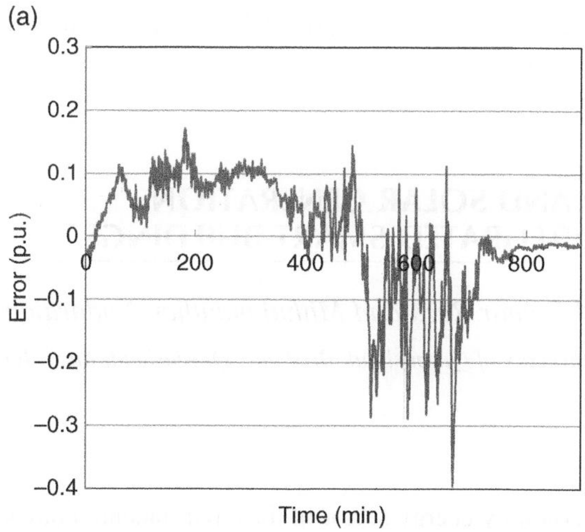

Figure 6.7.3 Behavior of errors of MLP and the proposed method, (a) case of MLP and (b) case of the proposed method.

(b)

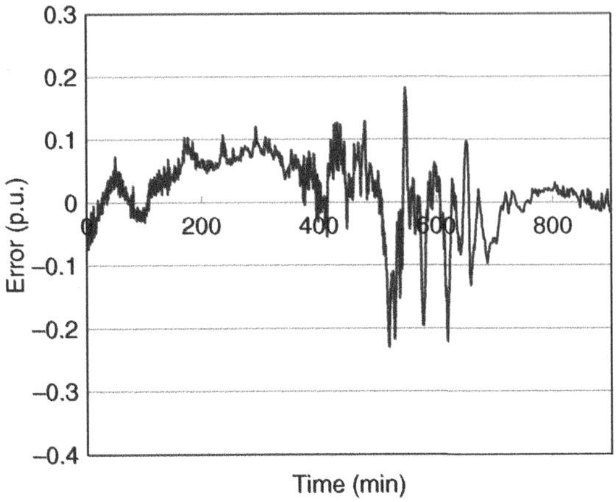

Figure 6.7.3 (Continued)

6.7.8 Conclusion

This paper proposed a new hybrid intelligent method for PV system generation output forecasting. The proposed method is based on GRBFN that consists of DA clustering for determination of the center and the width of the RBFs, the WDM for avoiding overfitting for learning data of complicated nonlinear time series, and EPSO for evaluating better weights between neurons. The proposed method was successfully applied to real data of PV system generation output. A comparison was made of the conventional ANNs such as MLP, RBFN, and GRBFN in terms of the average, maximum errors, and SD. The simulation results have indicated that the proposed method significantly reduced the errors in comparison with the conventional ANNs.

6.8 LOAD DEMAND AND SOLAR GENERATION FORECAST FOR PV INTEGRATED SMART BUILDINGS

Muhammad Qamar Raza and Mithulananthan Nadarajah

The University of Queensland, Brisbane, Queensland, Australia

6.8.1 Introduction

Driven by excessive CO_2 emission by energy sector in the environment, today's world is facing several challenges such as global warming and unpredictable swing in annual seasons [179–181]. In addition, drastically increasing energy demand

and fast depleting fossil fuel energy resources are also major concerns. It is expected that this will continue in future as well [182]. There is an urgent need to pay attention toward these issues and take more practical steps to mitigate the impacts on the environment. In order to make a contribution in this direction, International Energy Agency (IEA) raised the concerned on over Energy, Economy, and Environment, which are popularly referred to as "3E." Several developed and developing countries are putting efforts on the energy and environmental concerns to align with 3E's. While some countries set aggressive targets, some of them set their own target based on their national energy and economic interests. The EU has also set their own energy targets and expected to achieve the target by 2020 [181]. These targets are (i) 20% contribution of RES in EU energy consumption, (ii) reduction in energy consumption by 20% in comparison with projected energy demand, and (iii) 20% reduction in GHG emission as compared to the 1990 level. United States energy department also define their own goals for energy efficiency and conservation.

If one looks at the sources of energy consumptions, the buildings are one of the major energy consumers and contributors of CO_2 emission. Figure 6.8.1 highlights the building energy consumption and CO_2 emission in the environment in selected countries. According to the literature, buildings consume about 40% of total energy in the United States. In addition, buildings are responsible for

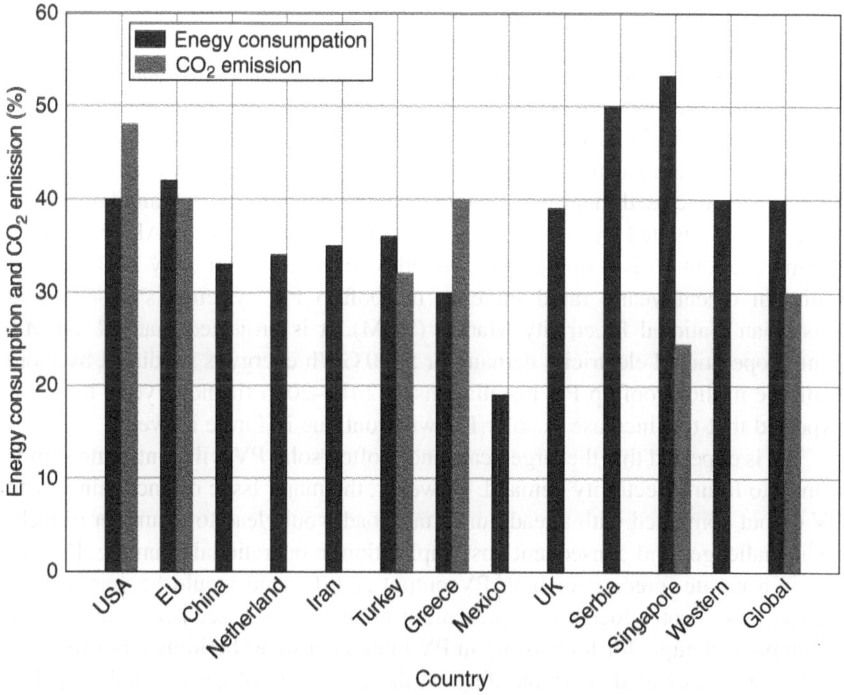

Figure 6.8.1 Global energy consumption and CO_2 emission scenario.

approximately 48% of total CO_2 emission in the environment and it is the highest emission due to buildings in the world. Similar kind of energy consumption and CO_2 emission trend is observed in the EU. The highest amount of energy (approximately 52%) is consumed in buildings in Singapore as compared to rest of selected countries. Globally, buildings are accountable for approximately 40% consumption and 30% CO_2 emission. Energy consumption in buildings can be reduced dramatically by implementing saving strategies and this may account for the entire transport sector's demand [183]. Moreover, a significant amount of energy can be saved and CO_2 emission can be reduced with installation of solar PV in buildings.

A remarkable growth in the PV system was observed during the last decade. Since 2000, the installed capacity of solar PV capacity is increased by 100 times and it is also expected to continue in the future as well [184]. An addition of 75 GW in global solar PV was observed in 2016 according to the REN21 report. According to the research report, an estimated more than 31 000 solar panels are added every hour [185]. Just a decade ago, yearly market addition was more than 50 times less in size of aggregated today's world solar addition.

China, Japan, Germany, United States, and Italy are leading countries in terms of solar PV capacity as shown in Section 6.8.3. In 2015, approximately 34.5 GW is added in China solar PV system and its total capacity reached to nearly 79 GW. Therefore, China solar PV generation capacity is the highest in the world with a global share of 46%. With new addition in China solar PV, it overtakes the long-time solar PV leader Germany in 2015. In addition, other top countries are United Kingdom, India, France, Australia, and Spain.

Australian utilities are also expecting a large penetration of solar PV into the system. In 2016, approximately 0.9 GW was added to the Australian energy market and ranked ninth globally for new installations. The total installation capacity of solar PV system in Australia reached up to 5.8 GW in 2015. Australian solar market has been enclosed mainly by residential solar PV with rooftop and large-scale solar plants, initiated by Australian Renewable Energy Agency (ARENA). Furthermore, about 16% of homes are covered with rooftop as of early 2016 in Australia. In recent years, rapid increase in rooftop PV systems is observed in Australian National Electricity Market (NEM). It is projected that 3% NEM's annual operational electricity demand or 5000 GWh energy is produced by more than one million rooftop PV installations in 2015–2016 financial year. It is also expected that this increase in solar PV will continue in future as well.

It is expected that the large-scale and rooftop solar PV will contribute significantly to future electricity demand. However, the major issue of uncertainty with PV output combined with already uncertain load would lead to a number of technical challenges and consequent cost implication in operational planning. Providing an accurate forecast for both PV output and demand would be very useful. Hence, this chapter focuses on presenting a case study based on an innovative ensemble technique for forecasting in PV integrated smart buildings. For the case study, a PV integrated smart buildings at the University of Queensland (UQ) has been chosen.

Rooftop PV Integrated Smart Buildings at UQ

The UQ, Australia installed the rooftop and large solar PV system at different sites such as St. Lucia campus, Gatton campus, Heron, and North Stradbroke island with installation capacity of 2.14 megawatt (MW), 3.275 MW, 54.6 kilowatt peak (kWp), and 40 kWp, respectively. The solar PV generation unit at UQ St. Lucia and Gatton are larger in terms of generation capacity. An overview of UQ solar rooftop PV (at St. Lucia) and fixed tilt array (at Gatton campus) network are shown in Figure 6.8.2.

The power generated from solar PV array is back wheeled into distribution network when the internal demands are not high. In addition, PV integrated smart buildings are also connected with utility electric supply. The power supply data are recorded using digital meters and stored on online UQ-Solar data management system. Global Climate Institute (GCI) and Advance Engineering Building (AEB) are PV integrated smart buildings built at UQ St Lucia campus. In this study, real-time load demand, PV output power, and meteorological data were used to forecast framework validation. The AEB and GCI rooftop PV integrated smart buildings are shown in Figures 6.8.3 and 6.8.4. The design objective of GCI and AEB building is to closely work with the natural environment by achieving the target of zero carbon emission and zero net energy building. The GCI building is equipped with modern sun shading, which tracks the sunlight and provides the natural ventilation for reduction in energy consumption.

The zero net energy objective is achieved by cross ventilation system, shading control from solar heat, optimal lighting, and solar PV integration. These PV arrays are connected with batteries and solar hot water. The 138 kW rooftop PV was installed at the GCI building and excess energy is stored in batteries. The electrical power flow diagram of the GCI building is shown in Figure 6.8.5. AEB at the UQ St. Lucia campus is designed in energy-efficient way by implementing the mix-mode air conditioning system. In this building, 95.75 kW rooftop PV system comprises of tilt-mounted 383 modules. The AEB building is connected to 11 kV UQ internal grid and rooftop PV.

The voltage is step down to 415 V for AEB distribution network. In addition, PV is connected to low voltage (LV) distribution network of AEB. The electrical power is absorbed by the electrical load and chilled water production system. The pure electrical load consists of the building lighting systems, lifts, fans, etc. In addition, chilled water production system that consists of a chiller station is also run using electrical power.

Uncertain PV Output Power and Volatile Load Demand

A large penetration of solar PV in current power network raises different technical and stability challenges for the power system directly or indirectly. Although the solar PV (rooftop, small and large units) installation helps to reduce the GHG emission, it creates several challenges for power system stability and safe operation. Furthermore, power generation from solar PV does not contribute in system inertia and also solar generation is not dispatchable [186]. As a result, it can create different power system stability issues. In addition, generation is not directly measured from

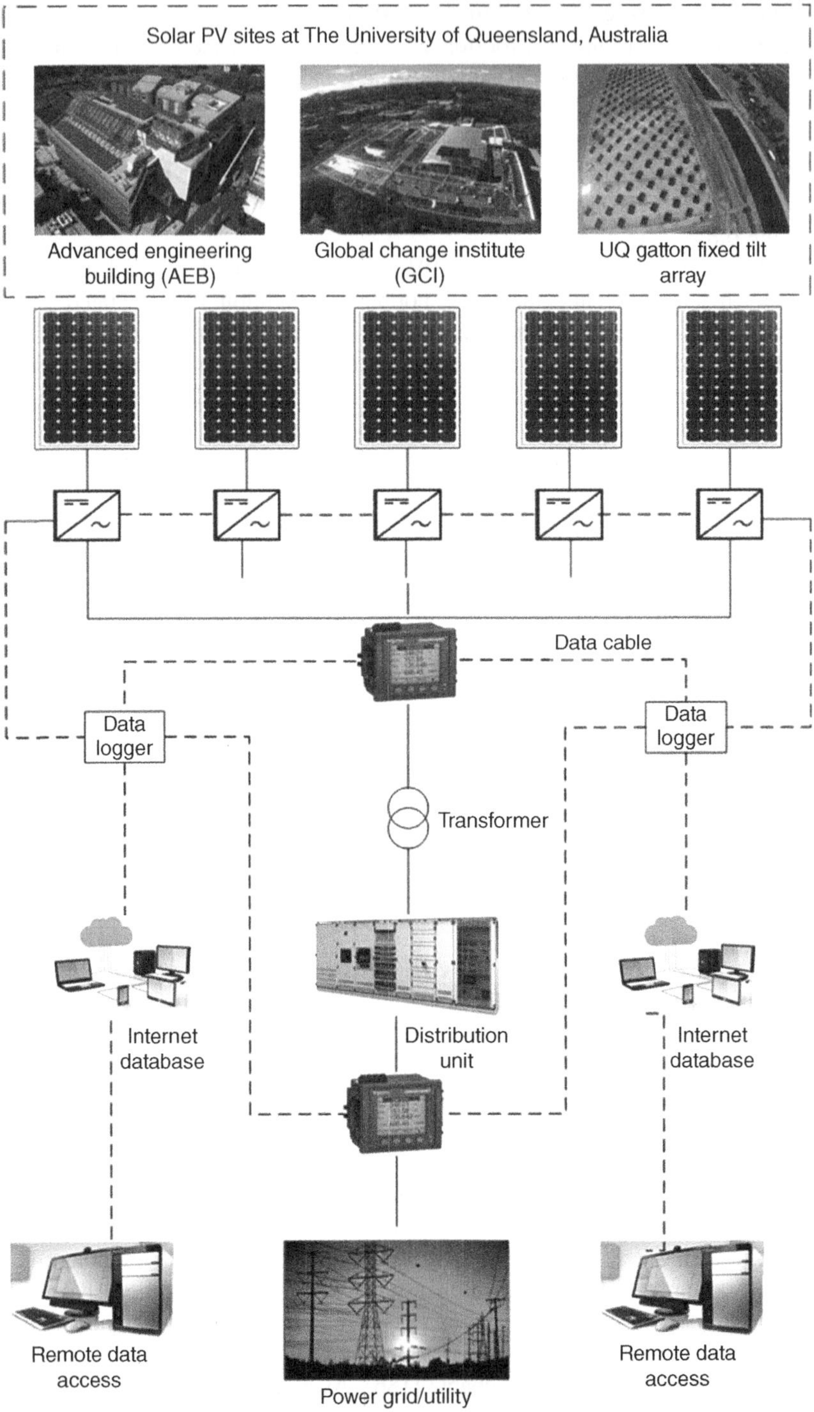

Figure 6.8.2 Overview of UQ solar PV network connected with power grid.

Figure 6.8.3 AEB PV integrated smart building at UQ St. Lucia Campus.

Figure 6.8.4 GCI PV integrated smart building at UQ St. Lucia Campus.

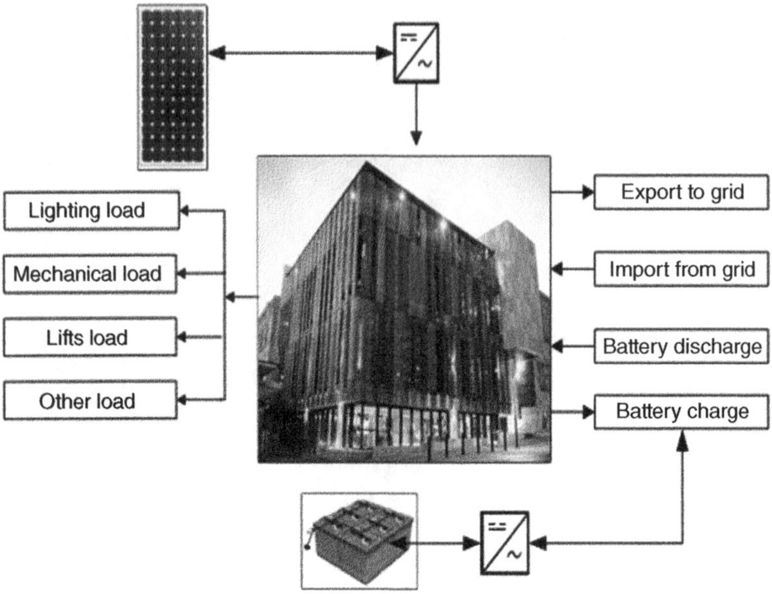

Figure 6.8.5 Electrical power flow diagram of GCI building.

solar PV forms and therefore is more uncertain form of generation than the other sources. Furthermore, in most of the cases only net energy is calculated for PV integrated smart building scenarios. This net metering practice only measures the overall net load demand of building, which is the difference of locally installed PV generation and building demand. Therefore, it is difficult to measure the PV generation in PV integrated buildings. In addition, the PV output is highly uncertain and variable due to solar irradiance and other meteorological variables such as wind speed, cloudy cover, humidity, and temperature. Furthermore, dust cover on solar panels in desert, irregular deterioration of panel performance, and solar technology efficiency also contribute to uncertainty of solar PV generation. Therefore, uncertain PV output power has a large impact on electricity market operation, planning, and management. Uncertain PV generation also leads to volatile load demand of PV integrated smart buildings from grid perspective. The grid load profile and peak demand curves will be more unpredictable due to indeterminate entry and exit of PV integrated smart buildings. As a result, it may create different issues for overall power system stability and operation. The load demand of the PV integrated smart buildings is volatile due to uncertain PV generation and also other factors such as meteorological variables, social events, random usage pattern, and occupant's comfort requirements. The main meteorological variables, which affect the load demand, are temperature, wind speed, humidity, dew point, and dry bulb. The unpredictable load demand of houses or buildings with rooftop PV (as in Australia) creates problems in power system stability, operation, management, and scheduling [187, 188]. Therefore, it is utmost important to forecast the load demand and solar output power generation of PV integrated smart buildings.

Potential Benefits of an Accurate Forecast

Forecasting the load demand of PV integrated smart buildings is important for modeling and characterization to implement the energy efficiency plans. CO_2 and energy consumption of those buildings can also be reduced by correctly characterizing their generation and consumptions of energy. The accurate load demand forecast will provide several benefits in terms of economics and environment. It is also beneficial for several energy efficiency and management applications, such as appliance scheduling, better thermal comfort management, improvement of building energy efficiency, infrastructure planning, comfort optimization, and improvement in load profile of building. As a result, an accurate load demand forecast of PV integrated smart buildings can also contribute to increase the energy security in comparison with conventional buildings. Furthermore, a building facility manger is able to do better energy planning with the help of accurate load demand forecast of buildings. The load demand of PV integrated smart buildings is volatile from grid perspective as PV output power is uncertain and variable. In addition, it is due to multiple factors affecting it such as solar irradiance, temperature, wind speed, humidity, cloud cover, and solar system efficiency. Therefore, it is also essential to forecast the PV output power for precise characterization of PV integrated smart buildings.

An accurate PV output forecast over the spectrum of forecast horizon is required for independent power producing (IPP) and managing companies or equivalent grid balancing authorities. The accurate PV forecast will help power distribution companies for better energy planning, scheduling, and management. In addition, accurate forecast will be beneficial in terms of smart integration of the PV generation with current grid with higher system reliability [189, 190]. Therefore, the importance of PV output forecast is further increased as it is required to achieve large penetration of solar power technology in a systematic manner. It will also contribute to minimize the reliance on fossil fuel resources. In addition, grid regulation, power scheduling, unit commitment, and EMS can be designed in effective manner with accurate PV forecast. The PV output forecast can be divided into different categories based on forecast horizon such as very short-term, short-term, medium-term, and long-term PV output forecast.

In large penetrated solar PV scenario, short-term PV output forecast can be utilized for load balancing, automatic generation control (AGC), better unit dispatching, and power plant operational management. ISOs and utilities are more interested in relatively longer forecast horizons for grid balancing, scheduling, and unit commitment. The distribution and transmission grids' operational planning and balancing require the accurate forecast over the spectrum of time horizon for efficient management. Accurate forecast also helps the grid to reduce the ancillary costs associated with weather dependency and quality of dispatch energy. It will increase PV penetration in existing power grid network and also help to reduce the overall CO_2 emission. From the grid perspective, reduction of the power system operational costs is a main factor to design accurate PV forecast models. PV output power forecast may also be part of smart grid (future generation power grids) EMS along with wind and load demand forecast.

TABLE 6.8.1 Classification of Prediction Techniques Based Forecast Time Horizon

Forecast Horizon	Time Range	Potential Applications
Very short term	Few seconds or 1–30 min ahead	• Electricity market regulation • Market clearing
Short term	30 min to 1 d or a week ahead	• Load management and balance • Load dispatch planning • Operational planning
Medium term	1 wk to 6 mo ahead	• Generator maintenance scheduling • Generation expansion planning
Long term	6 mo to 10 yr ahead	• Long-term generation planning • Energy policy implementation • Future energy programs

Classification of Prediction Techniques Based on Forecast Time Horizon

The forecast methods can be classified based on forecast horizon (prediction time span). In general, there are no defined criteria to classify the forecasting techniques based on forecast horizon in different groups. In [191], authors divide the load demand forecasting techniques in three categories named as short, medium, and long-term forecast. However, some of the researchers divide the forecast techniques into four groups named as very short, short, medium, and long-term forecast [192]. Broadly, load forecasting techniques can be divided in four groups as given in Table 6.8.1.

It is important to mention that classification of forecasting techniques based on time horizon may vary based on the forecast application. Therefore, forecast horizon could be varied for load demand and PV output power forecast.

Classification of Forecast Methods

Generally, forecast methods can be divided in different into clusters based on the working principles. These clusters are shown in Figure 6.8.6 and subclusters are given below [187, 191, 192]. The details of different prediction forecast methods can be found in [187].

Persistence and Naïve Method Persistence and Naïve methods are considered as benchmark forecast models. Researchers utilized these methods to compare the performance of any new models.

Physical Approaches The physical approaches include global forecast system, MM5, predictor, and high resolution limited area model (HIRLAM). Some of the physical approaches are used along numerical weather prediction (NWP)-based techniques for PV output power forecast.

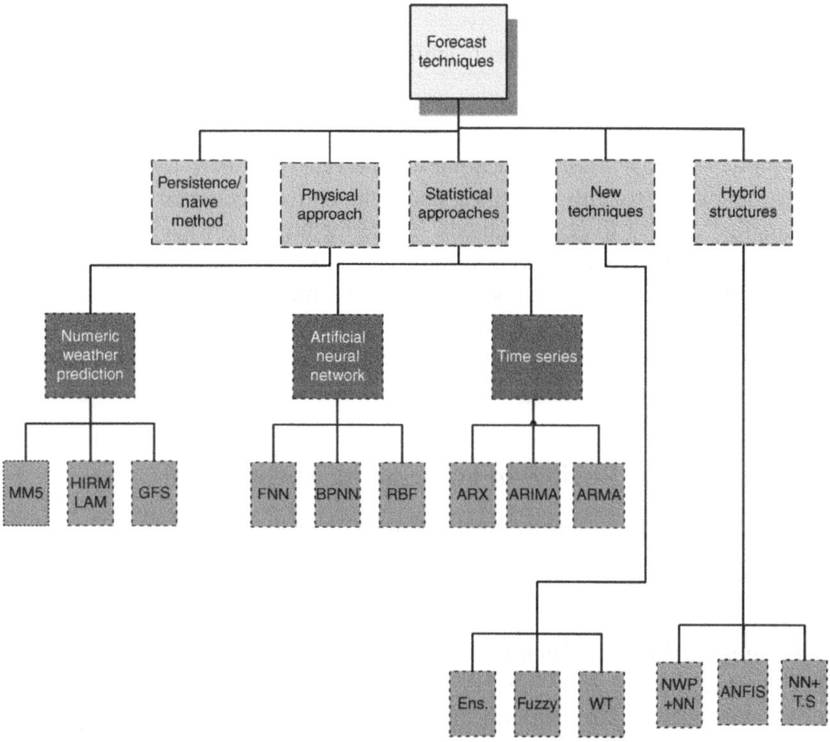

Figure 6.8.6 Classification of forecast techniques.

Statistical Approaches Statistical techniques can be further divided into two groups named as ANN and time series forecast techniques. The popular ANN-based forecast models are feed-forward networks (FNN), backpropagation neural network (BPNN), RBF, Kohonen self-organized maps (SOMs), and recurrent neural network (RNN). Time series-based forecast techniques are autoregressive (AR, ARMA, ARIMA, and moving average (MA).

Hybrid Forecast Methods Hybrid forecast model are designed by hybridizing the two or more forecasting techniques. These models are designed to take advantage of each predictor in hybrid architecture. Some of the hybrid forecast models are given below.

- ANN with GAs
- ANN with fuzzy and GA
- ANN with gradient-based learning techniques
- Adaptive neuro-fuzzy inference system (ANFIS)
- ANN with wavelet transform (WT) and time series
- ANN with expert system (ES) and regression technique

- NWP with NN
- ANN with support vector machine (SVM) and artificial immune system (AIS)

New Forecast Methods Some of the relatively new forecast methods which cannot be explicitly classified under the above-mentioned categories can be classified as new forecast methods. These methods are wavelet transformed models and ensemble-based models. In ensemble-based forecast models, multiple types of forecast predictors can be integrated to achieve the higher forecast accuracy [193, 194]. Majority of forecast models requires historical recorded and other exogenous variables as forecast model inputs [195]. The possible classification of forecast models is given in Figure 6.8.6.

6.8.2 Literature Review of Forecasting Techniques

The literature studies of load demand and solar generation of PV integrated smart buildings can be divided into two groups. These literature groups are named as load demand and solar PV output forecast techniques.

Load Demand Forecast Techniques

Load demand forecast of PV integrated smart buildings is important for building energy management and planning. In [196], authors proposed a long-term load demand forecast model for a residential area in South West China. In this research, the proposed forecast model based on ANN was compared with other models such as polynomial model, polynomial regression model, regression model, and grey model. The forecast results of residential area indicate that the proposed ANN-based forecast model achieves higher prediction results in different forecast studies. Case-based reasoning (CBR) forecast model is proposed to predict the load demand of office buildings located in Varennes, Quebec, Canada [197]. This research analyzes the performance of the forecast model over multiple forecast horizon such as 3-, 6-, and 24-hours ahead. The forecast results indicate that the prediction error is reduced with the decrease in forecast horizon. Therefore, the prediction model gives higher forecast accuracy for 3 hours ahead case study than the 6 and 24 hours ahead. Statistical-based ANN predication model is proposed for building demand forecast. This research study analyzes the impact of meteorological variations and occupant usage on load profile of building. This study concludes that these factors need to be considered as forecast model inputs because they contribute to the significant change in the building load profile.

Other research studies also report that the performance of ANN models is better than the comparative forecast models. In [198], authors proposed NN ensemble load forecast framework for 24 hours ahead prediction. The performance of the proposed ensemble model is compared with time series seasonal autoregressive moving average (SRIMA) model. The proposed NN ensemble model gives higher forecast accuracy than the benchmark models up to 50%. However, load demand in

PV integrated smart building is highly volatile as compared to without solar PV system. This is due to uncertain and variable PV output power. In addition, the load demand of PV integrated smart buildings is also indeterminate due to variable meteorological conditions, uncertain occupant's usage behavior, building comfort level, and structure.

By considering all the constraints of PV integrated smart buildings, it is a difficult task to accurately forecast the load demand for multiple forecast horizons. Therefore, some research studies propose hybrid forecast models with combination of two or more techniques for higher forecast accuracy. In [199], authors propose a SOM and SVM-based two-stage hybrid load forecast model. The proposed hybrid forecast model demonstrates better results than single SVM model for different time series data sets. An ANN integrated with chaotic particle swarm optimization (CPSO)-based forecast model was proposed in [200]. The results demonstrate that the proposed model produces better forecast accuracy compared with the NN model with Levenberg–Marquardt (LM) learning technique. Some of the major load demand forecasts were reviewed in [201]. However, there is still scope to improve the prediction accuracy, especially in PV integrated smart buildings.

Solar PV Output Power Forecast

Solar PV output power is variable and uncertain due to different factors affecting the PV output power such as solar irradiance, temperature, humidity, wind speed, and direction. Uncertain solar output power adds more uncertainty in load demand of PV integrated smart buildings environment. Therefore, buildings with solar PV may create serious power stability and energy management issues due to uncertain entry and exit of these types of loads. Therefore, it is vital to forecast the load demand of PV integrated smart buildings in order to capture the maximum solar energy and reduce the tariff.

Several methods have been proposed to accurately forecast the PV output power. These PV output power forecasting methods can be classified into different categories named as persistence method, physical techniques, statistical approaches, hybrid forecast methods, and some new forecast techniques. Majority of forecast techniques utilized the historical meteorological data and other exogenous variables as forecast model input. In [202, 203], time series prediction techniques were proposed to improve the forecast accuracy of PV output power. In [203], NWP-based forecast model was implemented with different learning techniques named as adaptive neuro-fuzzy, *k*-nearest neighbors (KNN), and ARIMA. In this research study, different case studies were designed to forecast the PV output power forecast model performance different from 1 to 39 hours ahead. In addition, the proposed hybrid technique is less as compared to comparative benchmark in most of the case studies.

Another research proposes a hybrid model to predict the global solar radiation using coral reefs optimization and extreme learning machine (CRO-ELM) algorithm with different meteorological data as model inputs [204]. However, the forecast accuracy of proposed techniques is not up to the mark level as compared to load demand forecast. There are several reasons for lower performance of hybrid models such as partial inability of a single model or weak compatibility between each other. NN-based PV output forecast model or combination of NN could be the better option to accurately forecast over the spectrum of forecast horizon. In [205], radial basis function neural network (RBFNN)-based solar PV output generation forecast model was designed for 24 hours ahead prediction. In this research, SOM technique was used to classify the different input variables such as air temperature, wind speed, humidity, and average of solar irradiance. However, the forecast accuracy of the proposed model was not consistent in different case studies. It was due to inability of the single model to capture the sharp variations.

In [206], authors designed a NN with backpropagation (BP) learning technique-based PV output power forecast model. Historical PV output and meteorological data with aerosol index (AI) were used to train the forecast model. However, NN with BP learning technique does not ensure the higher forecast performance due to incapability of the BPNN network. There is also possibility that the network may be trapped in local minima during the training process of NN and it leads to improper NN training. As a result, the forecast accuracy decreases significantly. The accuracy of the prediction model can be improved further by optimizing the NN learning and overcoming the shortcomings of stand-alone NN network. Some other researchers propose the forecast techniques, which are mainly focused to enhance the forecast accuracy [206–208]. However, there is still room to improve the forecast accuracy by exploring the more possible ways to accommodate the affecting variables and enhance the neural predictor's capability.

Motivation for Ensemble Forecast Framework

In previously reported research, the hybrid model tries to precisely forecast load demand by integrating two or more models. The forecast performance of hybrid models is affected due to dependence on each other. As a result, the overall forecast accuracy is affected due to bad performance of any of the models. The NN has demonstrated its capabilities as good alternative to a wide range of conventional time series techniques. The NN can adapt according to new situation and is robust in nature due to its learning process. However, a single NN cannot provide higher generalization due to performance limitations. Therefore, there is a need to design a multi predictor-based forecast model, in which each predictor does not affect the performance of each other. It is reported that the forecast performance of independent predictors, even with low-quality solutions in ensemble network, will increase the overall forecast accuracy [194, 209, 210]. The individual predictor produces different forecast outputs with the same input data due to different operation principles. The diverse output of an individual independent predictor provides an opportunity to enhance the overall forecast output by exploring other possible solutions. Multiple NNs organized in a systematic way called NN ensemble can provide better generalization capability as compared to single network. NN ensemble is a way to create multiple networks and train them for specific output. The output of the network is fused or combined using aggregation technique. Therefore, the higher forecast accuracy can be achieved by using NN ensembles and better training of network. The improved load demand and PV output power forecast model can play a vital role in planning and management of smart building in smart grid environment.

Summarizing the main points of the literature, there is a gap to investigate the performance of solar generation and demand forecast models for PV integrated smart buildings. Future generation smart buildings will be equipped with RES such as solar energy to meet the power requirements. In some meteorological conditions, the load demand of PV integrated smart buildings will be raised up to the peak level. In the worst-case scenario, smart buildings will need total supply from grid. It may affect the power system stability and power quality. As a result, the grid may face serious stability issues without prior knowledge of smart building electricity consumption and PV generation. Therefore, accurate load demand and PV output power forecast framework will assist to design better EMS and DR for this configuration.

The main contribution of this research chapter is as follows:

- Design and validate the feed forward neural ensemble network-based load demand and PV output power forecast framework.
- Three-level wavelet transformation of historical load demand and PV output power practical data of UQ AEB.
- Training of FNN predictors in ensemble network framework with PSO to improve the training for higher predication accuracy.
- Significant improvement in load demand and PV output power forecast accuracy in comparison with existing models.

6.8.3 Ensemble Forecast Methodology for Load Demand and PV Output Power

This chapter focuses on accurate load demand and solar power generation forecast of PV integrated smart buildings due to its enormous benefits. In this work, a novel ensemble forecast framework is proposed for both load demand and PV generation forecast. The proposed forecast framework is based on ensemble combination of FNN and trained with PSO. The output of each predictor in the ensemble network is aggregated with equal weight combination to best performing models. The WT technique is applied to handle the sharp spikes and fluctuations in historical PV output and load demand data. Correlated variables such as historical PV output power, load demand data, solar irradiance, temperature, humidity, and wind speed are applied as inputs for higher prediction accuracy. It is worth mentioning here that two different ensemble forecast frameworks are PV generation and load demand forecast. Historical load demand data are modeled as load demand forecast framework input, while, PV generation forecast framework utilized the historical PV output power data. The different components of the proposed load demand and PV output forecast framework is described below.

Wavelet Decomposition and Composition

As mention earlier, load demand data are volatile due to added uncertainty of solar generation and other meteorological factors. In historical PV output power, data also contain various oscillations, peaks, and different types of nonstationary data components. These are due to the unexpected changes in meteorological and other exogenous variables. Therefore, WT technique has potential to apply on historical load demand and PV output power data to reduce the effect of sharp oscillations on predictor's training. The WT technique can be separated into two groups named as discrete WT (DWT) and continuous WT (CWT). The historical load demand and PV output power data can be decomposed into a series of constitutive components using WT as shown in Figure 6.8.7. These transformed constitutive components depict more stable behavior with less variations as compared to without transformed data. This will increase the training capability of neural predictors. Ultimately, its leads to better load demand and PV output power forecast accuracy. WT technique can be divided into two basic functions named as mother wavelet signal and scaling function. The series of function can be derived as given in Eqs. (6.8.1) and (6.8.2) [211].

$$\varphi_{j,k}(t) = 2^{j/2} * \phi\left(2^{j/2} - k\right) \tag{6.8.1}$$

$$\psi_{j,k}(t) = 2^{j/2} * \psi\left(2^{j/2} - k\right) \tag{6.8.2}$$

In the above set of equations, the scaling and translating integer variables are j and k. The signal $s(t)$ can be expressed by using the scaling $\phi_{j,k}(t)$ and wavelet function $\psi_{j,k}(t)$ as

$$S(t) = \sum_{k} C_{j_0}(k) 2^{j_0/2} \phi(2^{j_0}t - k) + \sum_{k} \sum_{j=j_0}^{\infty} d_j(k) 2^{j/2} \psi(2^j t - k) \tag{6.8.3}$$

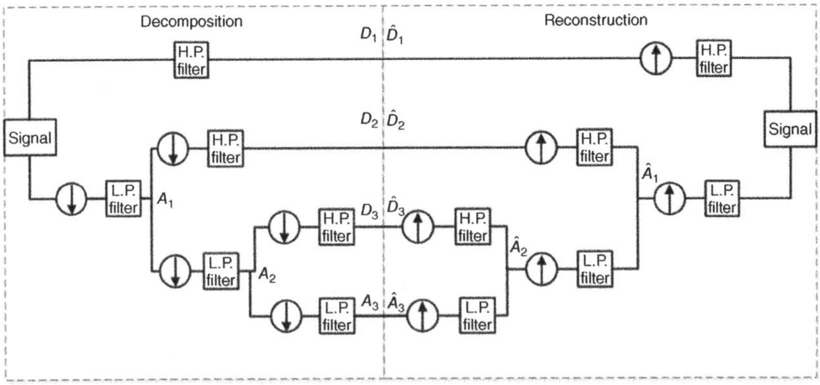

Figure 6.8.7 Wavelet decomposition and composition process of load and PV signal.

where d_j and C_{j_0} represent the detail and approximation coefficients, respectively. The pre-scaling coefficient is j_0 in the above signal Eq. (6.8.3). The first term of signal $s(t)$ at the predefined scale j_0 gives the low resolution. The second term at the predefined scale elaborates the higher resolution component.

Mallat's multiresolution analysis is technique to implement the WT using filters by a decomposition and reconstruction process [212]. In the decomposition process, the original is decomposed into different components by using low- and high-pass filters as shown in Figure 6.8.7. In [211], two-level WT decomposition was used for historical load data for short-term load demand forecast. However, another research study utilized the three-level WT decomposition for electricity price forecasting [213]. The forecast results demonstrate the significance of three-level WT decomposition in terms of improved prediction accuracy. Therefore, the three-level decomposition is applied to historical load demand and PV output power data in this study. However, WT is not applied to meteorological-related variables.

The detail or the high-frequency coefficients D_1, D_2, and D_3 are obtained with high pass (HP). The approximation signal (low-frequency components) A_3 can be obtained by down sampling with low-pass filters. The individual-decomposed historical load demand data signals (A_1, D_1, D_2, and D_3) are applied as forecast model inputs along with meteorological data. The output of individual forecast NN ensembles can be composed to generate the output. The reconstruction process on approximation A_3 and detail coefficients (D_1, D_2, andD_3) are applied to generate the forecast results. In this study, Daubechies-type function of order four is used as the mother wavelet as mentioned in [212].

Neural Network Ensemble Forecast Framework for Solar Generation and Load Demand Prediction

The proposed neural ensemble forecast model can be divided into five different processes named as input data preprocessing, WT decomposition, prediction, WT reconstruction, and output aggregation. The NN ensemble-based load demand and PV output power forecast framework are shown in Figure 6.8.8. The WT is applied to enhance the quality of training data and each individual predictor (FNN) is trained using PSO for improved training performance of predictors. The working of proposed PV output power and load demand forecast framework is described below.

1. Input Selection and Data Preprocessing: The real time recorded of PV integrated smart buildings at UQ St. Lucia campus data from January 2012 to December 2015 were collected to train and validate the proposed framework. The collected parameters are historical PV output power and load demand, temperature, solar irradiation, wind speed, and humidity. The historical PV output power, temperature, solar irradiation, wind speed, and humidity were applied as forecast model inputs of solar generation forecast model. Furthermore, historical load demand, temperature, wind speed, humidity, day of the week (DW), type (*D*) (working day or off day), and time of the day (*H*) were

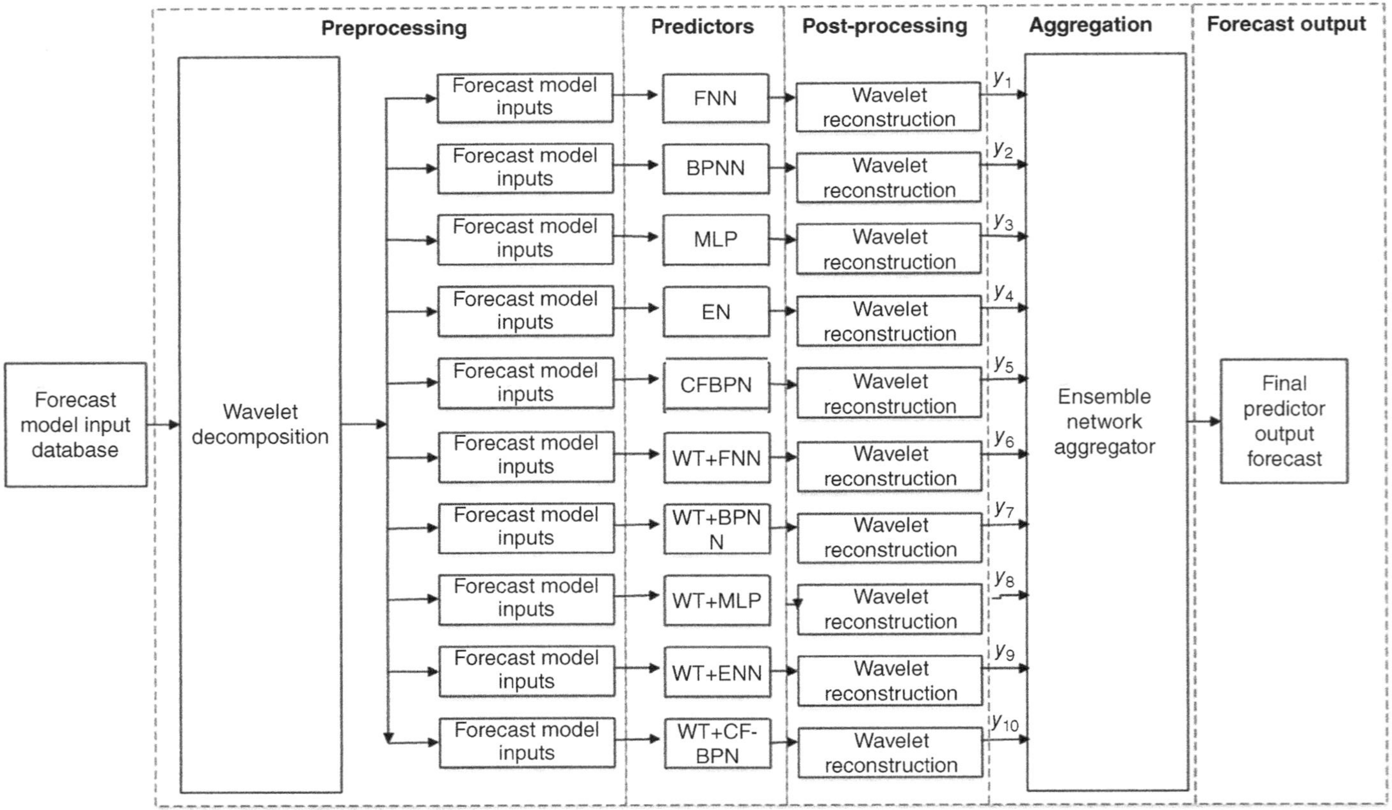

Figure 6.8.8 Proposed neural network ensemble-based forecast framework.

used as load demand forecast framework model inputs. This input was selected based on correlation analyses. Relatively highly correlated historical data, and meteorological and exogenous variables were selected as model inputs. These variables were applied as input to train the prediction model. In addition, selected data of load demand and PV output power of selected days were used to analyze the performance of both designed frameworks.

2. WT Decomposition: The historical data of PV output power data is fluctuating and uncertain. In addition, historical load demand data of PV integrated smart buildings are volatile due to multiple factors affecting it. However, the sharp changes and spikes in historical PV output power data were relatively higher as compared to the load demand data. These sharp variations affect the training of FNN predictors in the ensemble feedforward NN. Therefore, WT technique was applied to both historical PV output and load demand data which were applied as forecast framework inputs. Wavelet transformed historical load demand and PV output power data provide better training performance in comparison with untransformed data. The historical PV output data were transformed into detailed (D_1, D_2, and D_3) and approximate component (A_3) using WT using high- and low-pass filters, respectively.
3. Construction of NNE and Training: The proposed framework of neural network ensemble (NNE) network for load demand and PV output power forecast is shown in Figure 6.8.8. In the proposed framework of PV output forecast, 6 structures are initialized and each structure contains 20 individual predictors. However, 5 structures are initialized and each structure contains 15 individual predictors in load demand forecast framework. The optimal number of individual predictors in each structure and structures were identified by designing and implementing test case studies. However, further research can be carried to identify the type of the predictors, optimal number of predictors, and structures.

It is observed from different experiments that the overall framework load demand and PV output power forecast vary with change in hidden layers' neurons in neural predictor, number of structures, and individual predictors in it. Therefore, in order to get diverse output from each structure to explore more possible forecast solution, different number hidden layers are selected in each structure. It is reported that the forecast model with similar structure outperforms the comparative benchmark models [214–216]. The number of individual predictor in load demand and PV output power forecast structure is given in Eqs. (6.8.4) and (6.8.5), respectively.

$$NN\_Struc\_demand = \sum_{n=1}^{15} \{p_1 1, p_2, p_3, \ldots, p_5\} \tag{6.8.4}$$

$$NN\_Struc\_PV = \sum_{n=1}^{20} \{p_1 1, p_2, p_3, \ldots, p_6\} \tag{6.8.5}$$

It is important to mention that each structure contains the same number of hidden layers named as input, hidden, and output layer.

In order to achieve the diverse forecast output, FNN structures are trained with two different learning techniques of the network named as LM and PSO. In PV output power forecast framework, structures 1–3 were trained using LM algorithm. PSO was used to train structures 3–6. Structures 1–2 and 3–5 were trained with LM and PSO, respectively, in load demand forecast framework. The NN predictors in the ensemble network produce diverse forecast output. This diverse output enhances overall forecast accuracy by exploring multiple solutions.

4. WT Reconstruction: The output of each predictor is produced using the WT reconstruction process. Every individual predictor in each structure may give different forecast output due to different model training, forecast performance, and affecting variables on it. The detailed ($\widehat{D_1}$, $\widehat{D_2}$, and $\widehat{D_3}$) and approximate ($\widehat{A_3}$) components are combined by up sampling them using low- and high-pass filters. The wavelet decomposition and reconstruction process was the same in load demand and PV output forecast framework, except the input signal.
5. Output Aggregation: The output of each predictor in the ensemble network is combined using aggregation or fusion technique. In this chapter, equal weighting technique to best-performing predictors was applied to aggregate the output of each predictor in the ensemble network.

Equal-Weights Aggregation of Best Performing NN Models

The output of each predictor in both ensemble networks is combined using aggregation technique. In this research, equal weighted average technique to best performing models was applied to combine the forecast output of load demand and PV output power ensemble frameworks. Equal weighted average of best performing is relatively simplest way to aggregate the output of the ensemble network. Another research studies also demonstrate the potential benefits to apply on forecast applications [214, 217]. In the equal weight aggregation technique, the best performing predictors were selected. The number of best forecast predictors (nBest) are selected based on the forecast nMAPE from all models (nTotal) using validation data (Dvald). The output of (nBest) models is combined using average formula by giving the equal weight to each best performing models.

6.8.4 Numerical Results and Discussion

Load Demand and PV Output Power Forecast Case Studies

The proposed neural ensemble framework was applied for both forecast named load demand and generation output for PV integrated smart buildings at UQ St. Lucia campus named as AEB. The real time recoded load demand and PV output power data along with meteorological variables were used to train and validate the forecast framework. To assess the performance of proposed forecast framework, persistence, ARIMA, and wavelet transformed backpropagation neural network (WT + BPNN) models were applied to each individual predictor in the ensemble network. The historical two years (2012–2014) data of load demand and solar PV output along with meteorological were used. The historical load demand, wind speed (WS_t), humidity (HD_t), temperature (T_t), type (D), day (DW), and hour of the day (H) were applied as inputs of neural predictors in the proposed ensemble network. However, historical PV output power, humidity (HD_t), temperature (T_t), wind speed (WS_t), and solar irradiance were as inputs of solar generation forecast. The output of each independent predictor in the ensemble framework was combined using aggregation techniques as discussed earlier. In this study, normalized root-mean-square error (NRMSE) is calculated to evaluate the performance of predictors as given in Eq. (6.8.6) [218, 219].

$$\text{NRMSE}(\%) = \sqrt{\frac{1}{N} * \sum_{t=1}^{N} \left(\frac{\text{LD}_t^{\text{Actual}} - \text{LD}_t^{\text{Forecast}}}{\text{LD}_t^{\text{Peak}}} \right)^2} * 100 \tag{6.8.6}$$

where N is the number of load demand and PV output power data points for one day ahead forecast case study. The $\text{LD}_t^{\text{Actual}}$ is actual demand at time t, $\text{LD}_t^{\text{Forecast}}$ is forecasted demand, and $\text{LD}_t^{\text{Peak}}$ represents peak demand at hour t. In addition, normalized mean absolute error (NMAE) is calculated for each predictor and case study in order to assess the forecast performance of proposed and other models as given in Eq. (6.8.7).

$$\text{NMAE}(\%) = \frac{1}{N} * \sum_{t=1}^{N} \left| \frac{\text{LD}_t^{\text{Actual}} - \text{LD}_t^{\text{Forecast}}}{\text{LD}_t^{\text{Peak}}} \right| * 100 \tag{6.8.7}$$

In addition, R^2 is also calculated to find the accuracy of proposed model as given in Eq. (6.8.8).

$$R^2 = 1 - \left(\frac{\text{Var}\left(\hat{Z} - 1\right)}{\text{Var}(Z)} \right) \tag{6.8.8}$$

A Day-Ahead Load Demand Forecast

In this chapter, seasonal one day ahead forecast case studies were designed to assess the performance of the proposed forecast ensemble forecast model along with other comparative predictors. One day is selected from each season (winter, spring, summer, and autumn) in day-ahead seasonal forecast case study as considered in [220]. In order to evaluate the forecast model's performance, one day is selected from each season to forecast the load demand of AEB. For load demand case study, the selected days are December 13 (Summer), April 6 (Autumn), July 13 (winter), and October 24 (spring) of 2014. The nMAE and nRMSE are calculated for the ensemble network and benchmark forecast techniques to analyze the performance. A day-ahead forecast results of the proposed framework are compared with the persistence model, BPNN, ARIMA, and WT + NNE + PSO.

TABLE 6.8.2 Daily Forecast Error Comparison for AEB

	Summer		Autumn		Winter		Spring		Average
	E_1	E_2	E_1	E_2	E_1	E_2	E_1	E_2	*n*RMSE
Persistence	9.25	7.24	8.38	6.49	9.67	7.73	9.34	7.4	9.16
BPNN	10.43	8.65	8.93	7.14	9.81	7.76	9.35	7.6	9.63
ARIMA	7.85	6.16	7.86	5.81	8.64	7.08	7.32	5.66	7.91
WT + FNN + PSO	7.3	5.75	6.68	5.01	7.72	6.24	6.5	4.82	7.05
Proposed WT + NNE + PSO	5.27	3.98	4.89	3.6	5.04	3.85	4.62	3.39	4.95

E_1 = *n*RMSE, E_2 = *n*MAE

Table 6.8.2 demonstrates the seasonal daily forecast nRMSE and nMAE comparison of the proposed ensemble forecast framework with comparative model for AEB.

It can be observed that the persistence method produces higher nRMSE (9.67, 9.25, 8.38, and 9.34%) in comparison with the proposed ensemble forecast framework for selected seasonal days. In addition, BPNN predictors produce higher forecast nRMSE than the persistence method for all seasonal days as given in Table 6.8.2. ARIMA forecast models produce nRMSE of 7.85, 7.86, 8.64, 7.32% in comparison with WT + NNE + PSO, which gives nRMSE of 7.72, 7.3, 6.68, and 6.5% in the seasonal day-ahead forecast case study. The proposed ensemble forecast gives lower forecast nRMSE of 5.27, 4.89, 5.04, and 4.62% as compared to all other benchmark comparative model. It can also be observed that the forecast performance of the proposed and comparative model is inconsistent during different seasonal days.

The actual and prediction load demand plots of the proposed framework for summer and spring day are shown in Figures 6.8.9 and 6.8.10. The proposed forecast framework gives the R squared value of 0.9636 for spring in comparison with summer day (0.9512).

It is also observed that the forecast performance of the ensemble framework and other models vary during the seasonal daily load forecast. It is concluded from daily seasonal forecast that the forecast model performance is sensitive to seasonal variations. Therefore, inconsistent forecast results are obtained.

PV Output Power Forecast

PV output power is variable during different days and season of the year. It is mainly due to variation in solar irradiation and temperature at the solar panel surface. In addition, other meteorological factors such as wind speed, cloud cover, and humidity also affect it. Therefore, seasonal one day ahead forecast case study is designed to analyze the performance of the proposed framework. Seasonal days can be alienated into three different groups based on solar irradiations named as clear day (CD), partially cloudy day (PCD), and cloudy day (CLD) [221]. These

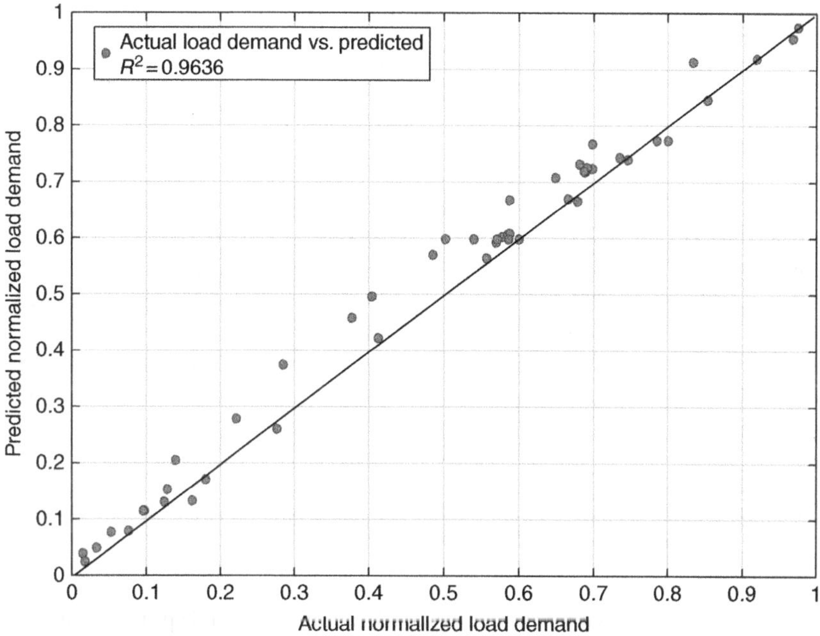

Figure 6.8.9 Actual and predicted load demand of proposed spring day.

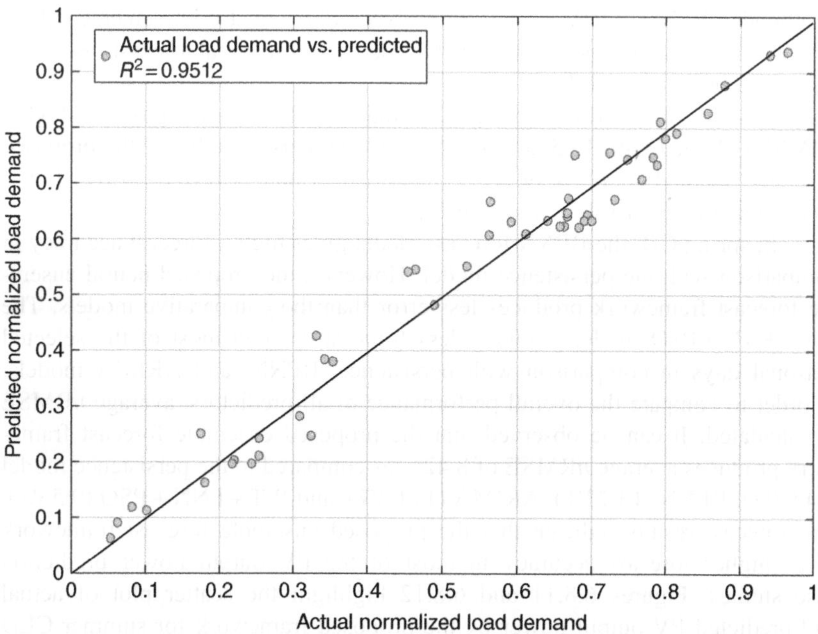

Figure 6.8.10 Actual and predicted output of proposed framework for summer day.

TABLE 6.8.3 Seasonal Day Ahead Forecast Error Comparison

Season	Day	Persistence	BPNN	ARIMA	WT + FNN + PSO	Proposed
Summer	CD	14.77	14.25	11.64	10.12	8.99
Summer	PCD	15.2	14.62	11.68	10.24	9.12
Summer	CLD	17.07	15.65	12.09	10.36	9.09
Autumn	CD	12.5	13.49	11.91	10.41	9.39
Autumn	PCD	13.77	13.35	12.32	10.93	9.73
Autumn	CLD	15.22	14.9	13.95	12.27	11.06
Winter	CD	13.06	13.5	10.05	8.28	7.23
Winter	PCD	12.03	10.96	8.7	7.14	6.01
Winter	CLD	13.99	14.54	11.14	9.81	8.88
Spring	CD	11.72	11.68	8.8	7.49	6.56
Spring	PCD	14.09	13.82	10.14	8.59	7.48
Spring	CLD	15.03	13.89	10.02	8.72	7.5
Avg. *n*RMSE		14.03	13.72	11.03	9.53	8.42

CD, clear day; PCD, partially cloudy day; CLD, cloudy day.

days are selected from winter (W), summer (S), autumn (A), and spring (S) of 2014 to analyze the performance. As reported in [221], the selected seasonal days may not be necessarily correlated to real "CD," "PCD," and "CLD" due to incapability of calculation technique, which is mainly based on solar irradiation. In this case study, three days from each season was collected randomly. In addition, persistence model, BPNN, ARIMA, and WT + FNN + PSO are also implemented along with the proposed forecast framework. One day ahead seasonal forecast case study results are provided in Table 6.8.3.

It can be observed from Table 6.8.3 that the BPNN model produces higher nRMSE (14.25, 13.49, 13.5, and 11.68%) in comparison to that of the proposed ensemble method (8.99, 9.39, 7.23, and 6.56%) and ARIMA (11.64, 11.91, 10.05, and 8.8%) for selected seasonal CD.

In spring CD, the BPNN forecast model gives higher forecast accuracy in comparison with the persistence model. However, the proposed neural ensemble forecast framework produces less error than the comparative models. The WT + FNN + PSO model produces less forecast error in most of the selected seasonal days in comparison with persistence, BPNN, and ARIMA models. In order to compare the overall performance of all predictors, average *n*RMSE is calculated. It can be observed that the proposed ensemble forecast framework produces average *n*RMSE of 8.42% as compared to the persistence model (14.03%), BPNN (13.72%), ARIMA (11.03%), and WT + FNN + PSO (9.53%). The forecast results indicate that the proposed ensemble forecast framework gives higher forecast accuracy in most of the PV output power prediction case studies. Figures 6.8.11 and 6.8.12 highlight the scatter plot of actual and predicted PV output power by the proposed framework for summer CLD and spring CD.

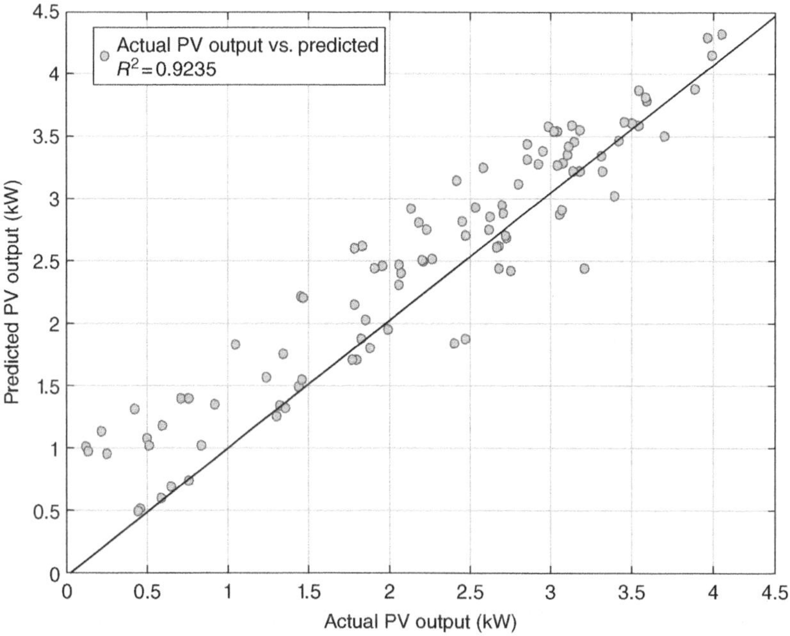

Figure 6.8.11 Actual and predicted output of proposed framework for summer CLD.

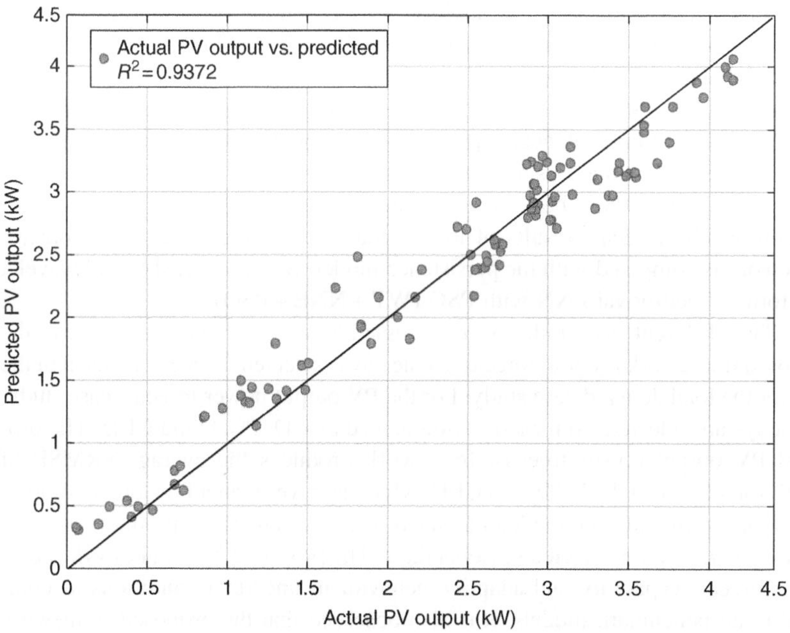

Figure 6.8.12 Actual and predicted output of proposed framework for spring CD.

It is observed from data analysis that the number of underestimated and overestimated PV output power produced during summer CLD day is more than the spring CD. This is due to uncertainty addition in CLD by meteorological conditions prevailing at the PV site. As a result, the performance of the forecast model is affected.

The proposed forecast framework gives the *R* squared value of 0.9372 for spring CD in comparison with summer CLD (0.9235). This also indicates that the proposed NNE gives better forecast results for spring CD than summer CLD. However, the forecast performance is variable in different days.

6.8.5 Conclusions

A significant role of RES such as solar PV is substantially important for sustainable built environment. Therefore, the large penetration of solar PV is observed in buildings with intelligent control called PV integrated smart buildings. The load demand of PV integrated smart buildings is volatile due to several factors affecting it such as uncertain PV generation, meteorological variables, unpredictable occupant's usage, and comfort requirements. Therefore, it is vital to accurately forecast the load demand and solar generation of PV integrated smart buildings. In this chapter, a novel NN ensemble load demand and PV output power forecast framework is implemented and validated.

In the proposed ensemble forecast framework, WT technique is applied to smoothing the historical load demand and PV output power forecast data. Wavelet transformed data historical load demand and PV output power data trained the forecast models in a better way. In addition, NN ensemble forecast framework is applied for both forecasts, which consist of multiple NN combined with intelligent. The performance of individual predictor in the ensemble framework is enhanced by training with PSO. Further, LM training techniques are also used to achieve diverse forecast output of neural predictors. The output of neural predictors in the ensemble is combined using equal weights aggregation technique to the best performing model. These best predictors are selected based on the forecast error. In addition, the forecast results of load demand and PV output power forecast framework is compared with the persistence model, BPNN, ARIMA, and wavelet transformed feedforward NN with PSO (WT + NNE + PSO).

Two different case studies were designed to analyze the performance of the proposed framework for both forecasts. One day is selected from each season of the year for the load demand case study. For the PV output power forecast case study, three days are selected from each season named as CD, PCD, and CLD. The proposed PV output power forecast framework produces the average nRMSE of 8.42% for seasonal CD, PCD, and CLD, which is lower than all the other comparative models. In addition, the load demand forecast framework produces average nRMSE of 4.95% for selected seasonal days. The proposed NN framework shows better forecast capability and adaptive behavior in uncertain conditions as compared to all benchmark models. It is also observed that the proposed framework gives higher prediction accuracy for load demand forecast in comparison with

PV output power. It is due to large uncertainity of PV output power due to mainly large variations in solar irradiation and temperature. It is observed that the load demand profile of (AEB) PV integrated smart building is relatively more stable than the PV output power profile. The PV output power is highly fluctuating mainly due to solar irradiation and temperature along with other meteorological variables. The proposed framework for load demand and PV output power forecast can be utilized for building energy management, planning, and efficient utilization. In future, both accurate forecasts can also be used for improving occupant's comfort and reduction in energy consumption of PV integrated smart buildings.

6.9 MULTI-OBJECTIVE PLANNING OF PUBLIC ELECTRIC VEHICLE CHARGING STATIONS

Ruifeng Shi[1] *and Kwang Y. Lee*[2]

[1]*North China Electric Power University, Beijing, China*
[2]*Baylor University, Waco, TX, USA*

6.9.1 Introduction

Transportation electrification has become a trend during the past few decades, due to the development of clean, efficient, and economical EV industry. Not only the developed countries like the United States, United Kingdom, Dutch, France, and Japan but also many developing countries, such as China and Iran, have shown great interest within the EV charging station planning research field [222–224]. The planning methods include operations research, game theory, meta-heuristic algorithms, graph theory, sensitivity analysis, simulation technique, and modern heuristic optimization techniques [225–231].

Some researchers investigated the EV charging station infrastructure in highways or in urban area [232], while more attention is focused on the techniques with the interaction between EV charging network and the power grid [233–236]. The EV charging pricing problem is also taken into consideration by many economists and EV charging station operation researchers [224, 237, 238]. Besides, the multistage EV charging station planning problem has also been investigated during the past few years [239]. Only very recently, the multi-objective EV charging station siting problem is addressed by some researchers [240], yet the objectives considered are more technically oriented than the human–system interaction interface. Therefore, what can well address the EV users' demand in an easy and convenient way should be considered during the EV charging station layout planning stage.

In this chapter, Section 6.9.2 presents a mathematical model for a three-objective EV charging station layout planning, Section 6.9.3 proposes an improved SPEA2 algorithm to solve the problem, and Section 6.9.4 shows a case study to demonstrate the effectiveness of the method proposed in this study.

6.9.2 Multi-Objective Electric Vehicle Charging Station Layout Planning Model

EV charging station layout planning is a combinatorial optimization problem, which should not only seek the optimal economical cost for either the construction or its later operation but also should enable the EV users to obtain an efficient and convenient service. In order to satisfy these different goals simultaneously, a multi-objective electric vehicle charging station layout planning model (MOEVCSLP) is proposed in this chapter to address this issue in a mathematical way.

Assumptions

In order to make it feasible to build the model, several assumptions are employed to reduce the complexity of the modeling process.

- Demands' distribution is taken into consideration for identifying the potential candidate locations of charging stations, which helps to satisfy the requirements of construction safety and environmental laws.
- The demand on charging in each candidate location is relative to the total number of EVs in this area.
- EV users are allocated to a given charging station to be considered as rational consumers.
- The distribution of demand to charging stations is in accordance with the proximity principle.

Optimization Objectives of the Mathematical Model

There are three objectives that are taken into consideration in this study: minimizing the economical cost (EC), maximizing the average utilization rate of charging poles (AURCP), and maximizing the users' charging convenience (UCC).

Economical Cost: $\mathbf{f_1}$ The aim of minimizing the EC objective is to find a proper economical balance between the satisfaction of charging requirements and its construction, maintenance, and operation, to obtain a global minimal cost of the resource with certain charging demand. The objective includes two parts, one is related with the investment of EV charging station construction and its annual operation cost, and the other is with the EV users' charging cost:

$$\min F = \min \left(F_1 + F_2\right) \tag{6.9.1}$$

1. Charging station construction investment and its annual operation cost:

The definition considers the fixed construction investment and the annual running cost F_1 simultaneously, where the construction investment takes the initial construction and installation costs into account. It includes the cost of purchasing chargers, transformers, and other auxiliary equipment, and it also includes the expense on the land acquisition and road construction, etc. The annual operation cost includes the daily operation and maintenance expense of an EV charging station, staff wages, etc.

The mathematical function to calculate the fixed construction investment is modeled as follows in this study:

$$F_1 = \sum_{j \in J} C_j \left[T(N_j) \frac{r_0 (1 + r_0)^{n_{\text{year}}}}{(1 + r_0)^{n_{\text{year}}} - 1} + Y(N_j) \right] \tag{6.9.2}$$

where J is the set of potential EV charging station candidate locations, C_j is the 0–1 decision variables to determine whether an EV charging station should be constructed at location j, N_j is the number of chargers at location j, $T(N_j)$ is the fixed investment at location j, $Y(N_j)$ is the annual maintenance cost at location j, and r_0 is the financial interest.

2. Users' charging cost: Users' annual charging cost can be estimated with two aspects.

The one is the expense on the round trip from users' home to the charging station, and the other is the expenses on the electricity. The following gives the estimation of the users' cost:

$$F_2 = 365 \left[\overline{\omega} \sum_{j \in J} \sum_{i \in I} X_{ij} n_i \lambda d_{ij} + \kappa \sum_{i \in I} n_i \right] \tag{6.9.3}$$

where I is the set of charging demands, $\overline{\omega}$ is the charging cost per unit distance on the way to the charging station, X_{ij} is a 0–1 decision variable to determine whether the demand at location i is served by station j, n_i is the number of daily EV charging demands in location i, λ is a nonlinear coefficient for different city road, λ is the distance between locations i and j, and κ is the unit electric price in a fast EV charging station.

Average Utilization Rate of Charging Poles: $\mathbf{f_2}$ In order to get a balance between maximizing efficiency of charging poles and minimizing the EV users' queuing time, an AURCP index is introduced to address the idleness of the charging poles.

Parameter η_j is defined to describe the spare time of a charging pole in hour j, as follows:

$$\eta_j = \frac{K \cdot \sum_{i \in I} X_{ij} n_i}{T} \tag{6.9.4}$$

where K is the ratio of served EVs in a busy time to the total served EVs of that day and T is the interval of a busy time.

Thus, we can estimate the AURCP with

$$\min \psi = \left| \frac{1}{M} \sum_{j \in J} \frac{\eta_j}{N_j \mu} - 1 \right| \tag{6.9.5}$$

where M is the number of total EV charging stations and μ is the average service rate of a charging pole per hour.

User Charging Convenience: $\boldsymbol{f_3}$ In order to estimate the performance of how an EV charging station satisfies its EV customers, a UCC index is defined as

$$\min \sigma = \frac{\sum_{j\in J}\sum_{i\in I} X_{ij} d_{ij} n_i}{\sum_{i\in I} n_i} \tag{6.9.6}$$

Constraints of the Mathematical Model

Decision Variable Constraints Users in the same charging demand location are required to recharge at the same station, as defined by

$$\sum_{i\in I, j\in J} X_{ij} = 1, X_{ij} \in \{0,1\} \tag{6.9.7}$$

Only when the candidate location is chosen as a station, can it serve the EV users, as shown by

$$X_{ij} \le C_j, \quad \forall i \in I, \quad j \in J, \quad C_j \in \{0,1\} \tag{6.9.8}$$

Reasonable Charging Poles' Amount The charging poles' amount in a charging station should be limited by the distribution power system, as shown by

$$N_{\min} \le N_j \le N_{\max}, \quad \forall j \in J \tag{6.9.9}$$

where $N_{\min}$ and $N_{\max}$ are, respectively, the minimum and maximum charging pole numbers for a given charging station.

Distance Between Stations In order to obtain a well-distributed EV charging station layout plan, a minimum distance between each pair of EV charging stations is introduced to guarantee the performance, as defined by

$$\lambda D_{jj'} \ge D_{\min}, \quad \forall j, \quad j' \in J; \quad j \ne j' \tag{6.9.10}$$

where $D_{jj'}$ is the distance between station j and j'; and $D_{\min}$ is the minimum distance requirement.

Distance Between Demand and Charging Station EV users prefer to get their EVs recharged within an acceptable/reasonable distance, as defined by

$$X_{ij} d_{ij} \le d_{\max}, \quad \forall i \in I, \quad j \in J \tag{6.9.11}$$

where $d_{\max}$ is the maximum acceptable travel distance in charging an EV.

6.9.3 An Improved SPEA2 for Solving EVCSLP Problem

Compared with traditional optimization methods, multi-objective evolutionary algorithms, such as non-dominated sorting genetic algorithm (NSGA) and strength Pareto evolutionary algorithm (SPEA), have been widely adopted in many industrial optimization applications [241]. We develop an improved SPEA2 in this study to solve the MOEVCSLP model.

Basic SPEA2 Algorithm Procedure

The basic SPEA2 algorithm can be carried out with the following steps:

Step 1: Create an initial population P_0, population size as N, a null external file A_0, and set generation count as $t = 0$.

Step 2: Calculate the individual fitness of population P_t and external file A_t.

Step 3: Define $A_{t+1} = \{x^i \mid x^i \in P_t \cup A_t\}$; if the archived Pareto solution's size of A_{t+1} is over the maximum limit $\overline{N}$, cut the size to $\overline{N}$; but if the size of A_{t+1} is less than $\overline{N}$, join the dominated solutions in the P_t and A_t to the A_{t+1} until the size of A_{t+1} is equal to $\overline{N}$.

Step 4: If $t > T$, output the external file A_{t+1} as the final Pareto set, stop the search process; otherwise, turn to Step 5.

Step 5: Select individuals from A_{t+1} into the mating pool with substituted binary championship rule.

Step 6: Implement crossover and mutation operator to the individuals of mating pool and population P_{t+1}, let $t = t + 1$, turn to Step 3.

An Improved SPEA2 Algorithm

The two kinds of decision variables in our MOEVCSLP model, C_j and X_{ij}, are both integers. A hybrid coding strategy is designed to combine the constraints into consideration. Corresponding crossover and mutation evolutionary operators are also proposed to guarantee that the derived solutions are feasible during the optimization process.

Chromosome Coding Strategy A two-layer chromosome encoding strategy is introduced in this chapter, shown as Figure 6.9.1.

Genetic Operators

1. Selection strategy: The proportional roulette wheel selection is selected for application in the study.
2. Crossover operator: In order to guarantee the legality and effectiveness of an offspring derived by crossover operator, a specific crossover strategy is designed in this chapter as shown in Figure 6.9.2.

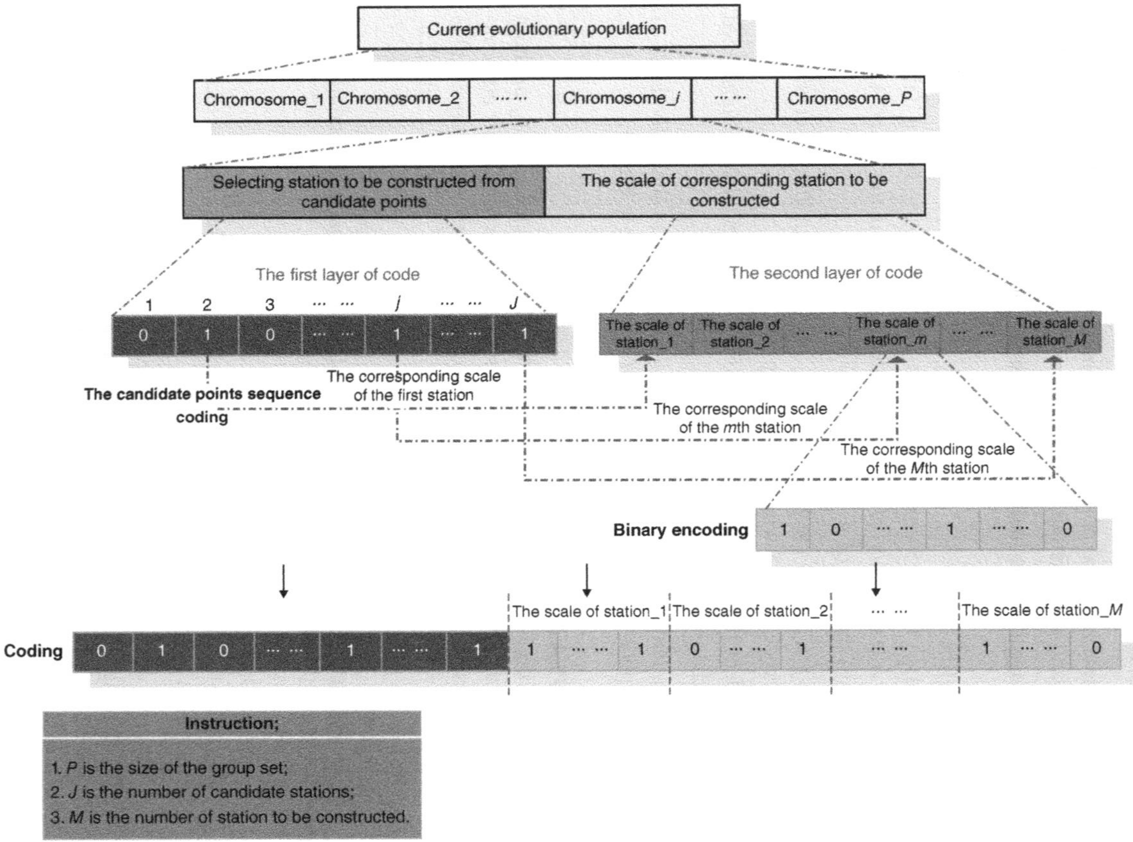

Figure 6.9.1 The two-layer chromosome coding structure.

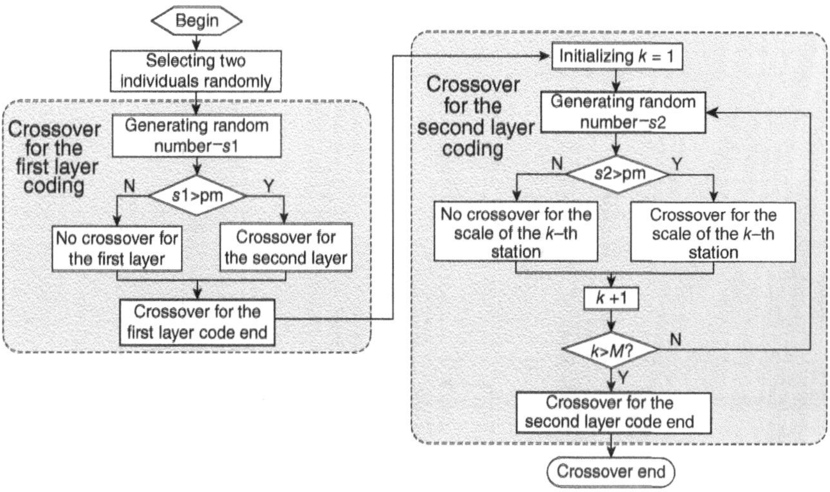

Figure 6.9.2 Flowchart of the crossover operator for improved SPEA2.

3. Mutation operator: In order to promote the mutation efficiency, a specifically designed mutation rule is employed to promote the mutation performance, as shown in Figure 6.9.3.

Fitness Evaluation [21] We can obtain the values of the three objective functions of an individual i in the population and the external file with Eqs. (6.9.1), (6.9.5), and (6.9.6), and calculate its strength value $S(i)$, which shows the amount of solutions that are dominated by individual i, by

$$S(i) = \left|\{j \mid x^j \in P_t + A_t, x^i \succ x^j\}\right| \tag{6.9.12}$$

The original fitness $R(i)$ of individual i is equal to the sum of strengths of all individuals dominated by it, as shown by [241]:

$$R(i) = \sum_{x^j \in P_t + A_t, x^j \succ x^i} S(j) \tag{6.9.13}$$

We also employ the k-proximity method [242] to calculate the density value $D(i)$ for individual i, as

$$D(i) = \frac{1}{\sigma_i^k + 2} \tag{6.9.14}$$

where σ_i^k is the distance in the objective space between an individual i and the kth adjacent individuals to i, $k = \sqrt{N + \overline{N}}$, where N is the population size, and $\overline{N}$ is maximum archived Pareto set limit.

Finally, we calculate the fitness $F(i)$with

$$F(i) = R(i) + D(i) \tag{6.9.15}$$

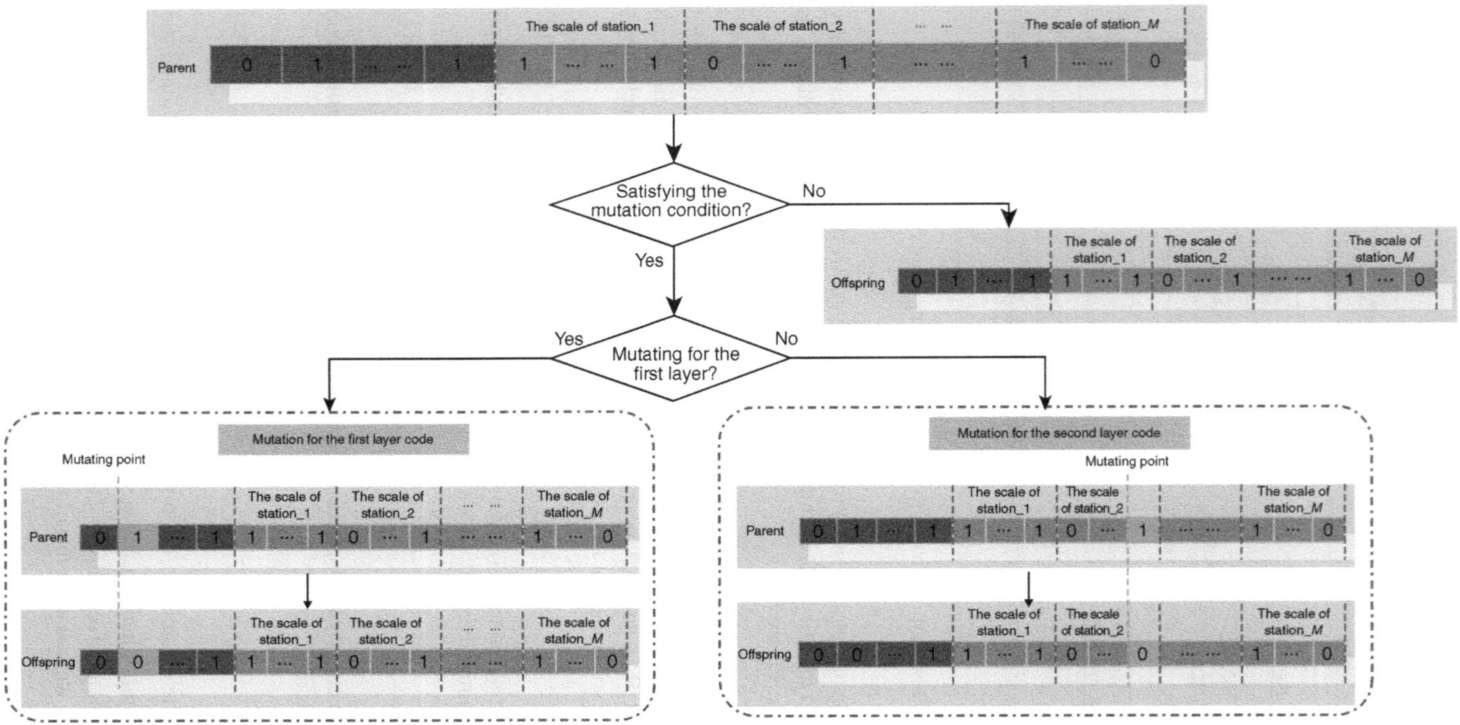

Figure 6.9.3 Flowchart of the mutation operator for improved SPEA2.

6.9.4 Case Study

Problem Description

In order to verify the effectiveness of the model and the improved SPEA2, a numerical case study is taken for demonstration. The EV urban area of the city in the study is 10.5 km^2. According to residential, commercial, and industrial area, we divide the zone into 40 subareas (40 independent EV charging demand points), and we choose 20 locations as EV charging station candidate points. The aim of this study is selecting 8 locations for constructing EV charging stations within the 20 candidates. Table 6.9.1 shows the demand coordinates and the EV amount in each point. Table 6.9.2 shows the coordinates of each EV charging station candidate. The distribution of the EV charging demand and the EV charging station points are illustrated in Figure 6.9.4.

Model Parameters

A binomial form is employed to express the fixed investment of charging station N_j as follows:

$$T(N_j) = W + qN_j + eN_j^2 \tag{6.9.16}$$

TABLE 6.9.1 The EV Charging Demand Coordinates and Their EV Amounts

Demand Index	X	Y	EV Amounts	Demand Index	X	Y	EV Amounts
1	11.70	8.60	37	21	11.00	3.80	44
2	5.70	5.30	48	22	5.50	8.50	51
3	1.70	12.00	37	23	12.00	11.30	42
4	2.10	1.50	42	24	5.70	1.30	37
5	2.40	6.70	49	25	6.90	8.90	43
6	12.20	9.30	54	26	6.78	9.70	38
7	12.00	2.00	38	27	7.60	5.60	35
8	6.30	3.30	46	28	10.90	6.40	46
9	6.60	11.80	54	29	2.40	8.00	44
10	7.70	2.10	50	30	10.50	6.90	40
11	12.60	9.40	35	31	1.00	10.30	38
12	11.80	4.30	55	32	7.50	1.00	41
13	10.70	7.10	55	33	2.20	4.80	50
14	12.60	2.30	52	34	1.80	10.80	40
15	3.20	8.00	49	35	6.70	7.20	47
16	3.60	7.80	47	36	10.30	8.10	52
17	6.30	8.80	49	37	12.20	2.50	53
18	4.50	9.40	49	38	4.40	3.90	39
19	3.70	4.70	42	39	6.90	6.70	40
20	2.80	8.90	39	40	10.80	1.10	49

TABLE 6.9.2 EV Charging Station Candidates' Location Coordinates

Candidate Index	X	Y	Candidate Index	X	Y
1	11.40	12.20	11	9.00	2.50
2	11.30	4.70	12	6.30	4.70
3	12.90	7.90	13	7.60	6.80
4	6.10	3.00	14	4.50	5.50
5	2.30	3.70	15	4.90	9.80
6	10.70	9.60	16	12.30	1.60
7	2.50	8.80	17	2.70	11.80
8	1.50	3.30	18	6.80	12.40
9	8.40	10.90	19	4.10	2.20
10	7.80	4.60	20	10.10	6.90

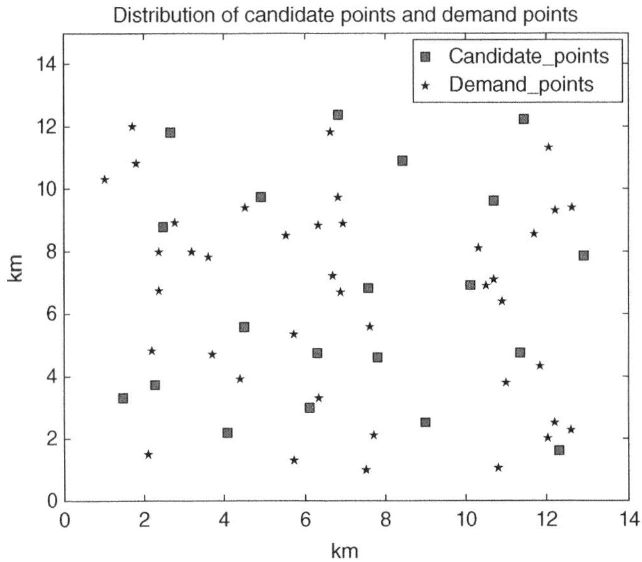

Figure 6.9.4 Distribution of EV charging demand points and potential EV charging station candidate points.

where W is the fixed investment, which is set as 16.7 million USD; q is the unit charger investment in a station, which is set as 0.83 million USD/set; and e is the equivalent investment coefficient about charger amount, including land acquisition and auxiliary facilities' costs, which is set as 0.33 million USD/set^2.

The value of annual running cost $Y(N_j)$ is set as 10% of the fixed investment; the depreciation period of a station is set as 20; the financial rate r_0 is set as 0.08; the charging cost per unit distance on the way to charging station ϖ is set as 1.33 USD/km (namely, 0.000 13 million RMB/km); nonlinear coefficient for city road λ is set as 1.2; average price of fast charging for an EV at the present stage κ is 0.67 USD/set (namely, 0.000 067 million USD/km); the ratio of number of vehicles served in busy time to the number of vehicles served all day K is set as 0.9; and the time length of busy time T is set as 16 hours. Besides, some more assumptions have been made to help the evaluation of the model, such as a general EV battery storage capacity can reach 80–90% in 15–20 minutes, so the average service time is set as 15 minutes, that is, the average service rate of charger μ is 4 set/h; the number of stations to be constructed, N_{CH}, is 8.

The minimum and maximum charging poles' number for charging station configuration, N_{min} and N_{max}, is set as 2 and 32, respectively; the minimum distance between stations, D_{min}, is set as 0.5 km; and the maximum travel distance for charging, d_{min}, is set as 2 km.

Results Analysis

Improved SPEA2 Algorithm Parameters Setting The improved SPEA2 algorithm that was introduced above is carried out with Python language, and the optimization parameter settings for the algorithm is: maximum optimization iteration is set as 150; the scale of internal and external population is, respectively, set as 200 and 20; the crossover probability is set as 0.6; and the mutation probability is set as 0.08.

Results After running over 20 independent times with improved SPEA2, we obtain the final Pareto solutions for the MOEVCSLP problem as shown in Figure 6.9.5, in which the Stars ("*") are the Pareto front solutions obtained at the 20th generations; while the Cross ("+") are the final frontier solutions.

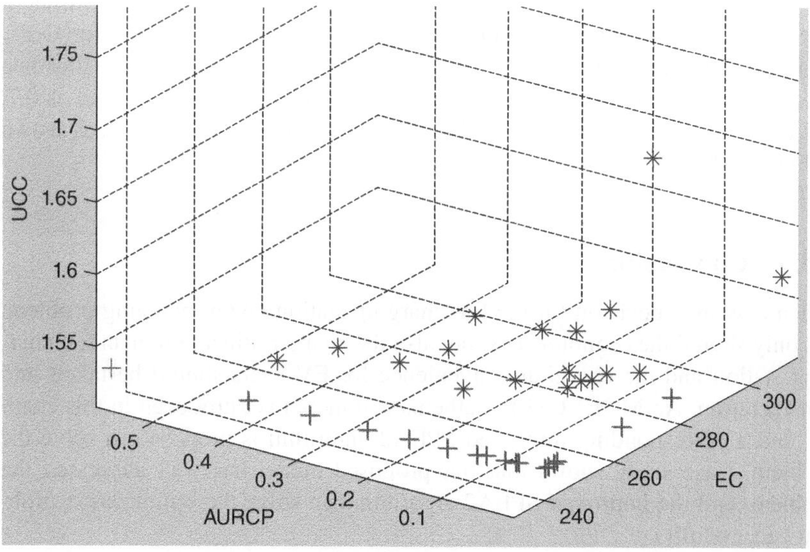

Figure 6.9.5 Pareto front of the tri-objective EV charging station layout planning.

TABLE 6.9.3 Typical Solutions in the Pareto Set

Item	Scheme 1	Scheme 2	Scheme 3	Scheme 4
[location, scale] (station #, charger #)	[4, 4] [5, 3] [6, 3] [7, 4] [13, 2] [15, 3] [16, 3] [20, 3]	[4, 3] [5, 3] [6, 3] [7, 4] [13, 3] [15, 3] [16, 3] [20, 3]	[4, 2] [5, 2] [6, 2] [7, 2] [13, 2] [15, 2] [16, 2] [20, 2]	[1, 3] [2, 3] [6, 2] [9, 5] [10, 2] [11, 11] [12, 3] [17, 3]
f_1: economy USD	240.67	240.53	235.86	338.40
f_2: average. Charging utility	0.000 39	0.002 10	0.578 52	0.000 009
f_3: charging convenience (km)		1.529 22	1.529 22	2.379 86

Since the maximum number of Pareto front solutions is set as 20, we choose four typical solutions from the final Pareto set to demonstrate the diversity (shown in Table 6.9.3), in which Scheme 1 is the optimization solution selected from the 2-objective optimization, and Schemes 2–4 are selected from the three-objective optimization results.

Table 6.9.3 shows that Scheme 1 obtains preferable result both on f_1 and f_2, but the objective f_3 is ignored. Scheme 2 is a compromise solution on the three objectives. Scheme 3 achieves very good performance in f_1, while the worst performance in f_2. Scheme 4 obtains the best performance in f_2, with poor performance in f_1 and f_3.

In real decision scenario, decision maker (DM) may compromise different objectives simultaneously; we simply employ the classic Technique for Order Preference by Similarity to Ideal Solution (TOPSIS) decision method to illustrate the decision process. The weight coefficients of objectives f_1, f_2, f_3 are set as 0.7, 0.2, 0.1, respectively. The final solution chosen by TOPSIS is Scheme 2, as shown in Figure 6.9.6.

6.9.5 Conclusion

When making a decision for the EV charging station layout planning problem, not only should the economic cost but also the charger utilization in an EV charging station and the charging convenience for EV users should be taken into consideration. An MOEVCSLP mathematical model is introduced in this chapter, and a corresponding improved SPEA2 algorithm is proposed to solve the problem. Case study shows that our proposed model has well addressed the problem, and the improved SPEA2 algorithm can solve the optimization problem successfully.

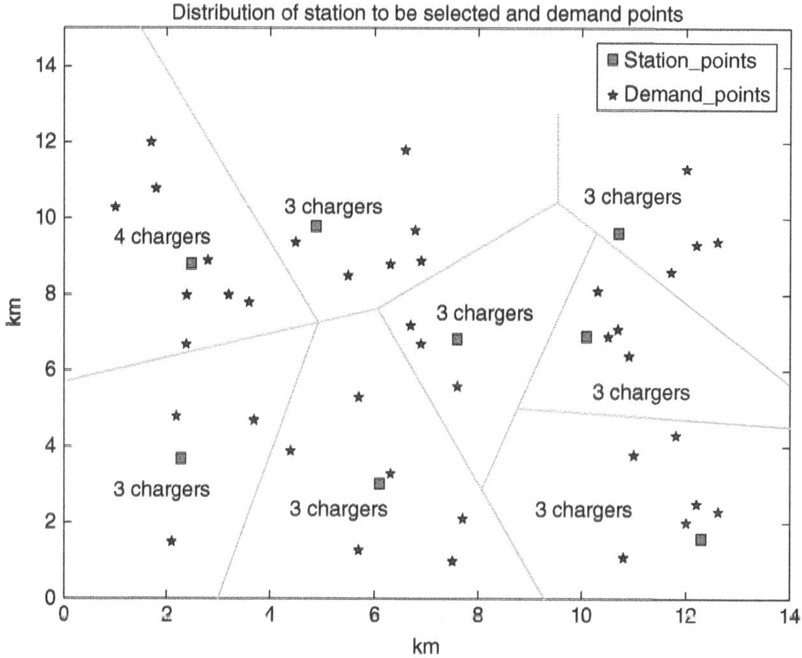

Figure 6.9.6 The final EV charging station layout planning scheme obtained by TOPSIS decision method.

6.10 DISPATCH MODELING INCORPORATING MANEUVER COMPONENTS, WIND POWER, AND ELECTRIC VEHICLES

Sergio Rivera[1] *and Andrés Romero*[2]

[1]*Universidad Nacional de Colombia, Bogotá, Colombia*
[2]*Universidad Nacional de San Juan, San Juan, Argentina*

6.10.1 Introduction

This chapter is intended to model an economic dispatch in a context in which maneuver components, power wind generators, and plug-in electric vehicles (PEVs) are connected to the power networks. Currently, wind energy has more widespread use in the power systems due to the need for alternatives to traditional energy generation [243, 244]. In the same way, a significant penetration of PEVs in the power networks implies an energy exchange with the grid, when they are plugged into a standard electric power outlet to charge/discharge [245, 246] their batteries [247]. The need for the proposed modeling approach is motivated due to the fact that the massive connection of both technologies will impact several aspects related to traditional power systems [243, 245, 248, 249]. For instance, the way in which the electric power infrastructure is operated and its expansion is planned, among others [247].

Such technologies bring several techno-economic challenges for power engineers that request to be faced and understood [247]. In this sense, research on wind generation and PEVs' penetration in the power networks becomes highly relevant. However, all the work that has been done needs to be organized to establish the state-of-the-art status in the topic, and to continue an adequate research into key aspects like the power grid operational optimization.

For this purpose, we present an adaptation of the traditional economic dispatch model in order to include the stochastic nature of wind generators and PEVs, due to uncertainty in wind speed and driven patterns, respectively. Additionally, the status of maneuver components is considered through decision variables affecting the generation cost optimization and constraints involved in the power system operation.

Heuristic optimization algorithms to solve these kinds of optimization problems are a convenient way to deal with challenges and complexities such as high dimensionality, nonlinearity, non-convexity, multi-variability, multimodality, combinatorial nature, and challenging constraints or when there is limited computation capacity [250–252].

Well-known formulations of optimal active–reactive power dispatch problems have been implemented using population-based heuristic optimization algorithms [253, 254]. Many of these implementations do not consider either the incorporation of wind-powered generators or the PEVs' penetration in the networks. Additionally, they disregard some important maneuver elements in the power system operation (like transformers with OLTCs and compensation devices) or contingency conditions.

The proposed approach uses the economic dispatch formulation presented in [243, 244], but considers further constraints related with the presence of transformers with stepwise OLTCs and shunt compensation devices in the network. Furthermore, the objective is to minimize the mathematical expectations of the generation costs (this expected value is deduced in detail) while satisfying constraints for non-contingency and contingency conditions ($N-1$). The constraints are described and classified according to how they are handled in the optimization algorithms. Additionally, the probability distributions of the charge/discharge behavior of PEVs are considered in the economic dispatch not only in its discharge condition as power generators (V2G) but also in its charge condition as negative power generators.

This chapter is organized as follows: Section 6.10.2 proposes an economic dispatch formulation considering conventional generators, wind power generators, and PEVs, as well as the presence of maneuver components (transformers with stepwise OLTCs and shunt compensation devices). Section 6.10.3 outlines a generic strategy to solve the proposed formulation and the heuristic algorithms used. Section 6.10.4 presents the test system and a comparative analysis of the results for each heuristic algorithm. In Section 6.10.5, we draw conclusions related to the formulation's usefulness in considering the elements outlined in this chapter.

6.10.2 Proposed Economic Dispatch Formulation

This section presents an optimization formulation (target function and constraints) of the economic dispatch problem in the context of wind power and PEVs' penetration. As previously mentioned, in this formulation, elements that are disregarded in previous studies [243, 244] are included, particularly: maneuver components as stepwise adjustable OLTCs and switchable shunt compensation devices. Additionally, mathematical expressions for the expected cost of wind power and PEVs' active power outputs are developed in deep detail, complementing the formulation presented in [243, 244].

Optimal Active–Reactive Power Dispatch Considering Wind Power and PEVs

In the formulation proposed, the objective is to minimize the expected value of the total generation cost (C_{total}):

$$\min \; E[C_{\text{total}}] \tag{6.10.1}$$

where $E[*]$ denotes a mathematical operator, given by the probability-weighted average of all possible values of a random variable [243]. The total generation cost is given by [243, 255]:

$$\begin{aligned} C_{\text{total}} = & \sum_{i=1}^{N_c} C_i(P_{s,i}) \\ & + \sum_{i=1}^{N_w} C_{w,i}(W_{s,i}) + \sum_{i=1}^{N_w} C_{w,u,i}(W_{s,i}, W_i) + \sum_{i=1}^{N_w} C_{w,o,i}(W_{s,i}, W_i) \\ & + \sum_{i=1}^{N_e} C_{e,i}(P_{e,s,i}) + \sum_{i=1}^{N_e} C_{e,u,i}(P_{e,s,i}, P_{e,i}) + \sum_{i=1}^{N_e} C_{e,o,i}(P_{e,s,i}, P_{e,i}) \end{aligned} \tag{6.10.2}$$

where N_c, N_w, N_e are the number of conventional generators, wind power generators, and number of buses with PEVs' facilities connected, respectively.

In this approach, the conventional generators have a quadratic cost function with $P_{s,i}$ as the scheduled power output of generator i, and this direct cost is given by:

$$C_i(P_{s,i}) = \alpha_i + \beta_i P_{s,i} + \gamma_i P_{s,i}^2 \tag{6.10.3}$$

where α_i, β_i, γ_i are the cost coefficients of generator i.

The cost of a wind power generator or a PEV bus is divided into three components. The direct cost ($C_{w,i}(W_{s,i})$ or $C_{e,i}(P_{e,s,i})$) paid by the system operator to the owner of the wind generator or PEV bus is the first component accounted for [244].

Factors for underestimation and overestimation of power availability are included in the formulation as the second and third components mentioned in the previous paragraph, due to the uncertainty of the wind power and the PEVs' power available at any given time [243, 244]. In order to analyze this uncertainty, the following reasoning is used [244, 256]: if the actual available power (W_i, $P_{e,i}$) is more than what was assumed (power scheduled: $W_{s,i}$, $P_{e,s,i}$), that power will be misused (in some cases can be used to package in energy storage systems), and the system operator must pay a cost to the renewable power producer for the unused of available renewable capacity [244, 256]. That can be formulated with an underestimation factor ($c_{w,u,i}$ or $c_{e,u,i}$). On the other hand, if a certain injection of power is scheduled and that power is not available, power must be purchased from an alternate source or loads must be shed (reserve requirement) [244]. That can be formulated with an overestimation factor ($c_{w,o,i}$ or $c_{e,o,i}$).

In this approach, the penalty cost for not using all the available power (underestimation case) is considered linearly related to the difference between the actual power used and the scheduled power [244, 256]. On the other hand, the penalty cost for reserve requirement (overestimation case) is considered linearly related to the difference between the available power and the actual power used [244].

Considering the previous analysis, the three components for wind generators and PEVs are given by ([243, 255]):

$$\begin{aligned} C_{w,i}(W_{s,i}) &= c_{w,i} \cdot W_{s,i} \\ C_{w,u,i}(W_{s,i}, W_i) &= c_{w,u,i}(W_i - W_{s,i}) \\ C_{w,o,i}(W_{s,i}, W_i) &= c_{w,o,i}(W_{s,i} - W_i) \end{aligned} \tag{6.10.4}$$

$$\begin{aligned} C_{e,i}(P_{e,s,i}) &= c_{e,i} \cdot P_{e,s,i} \\ C_{e,u,i}(P_{e,s,i}, P_{e,i}) &= c_{e,u,i}(P_{e,i} - P_{e,s,i}) \\ C_{e,o,i}(P_{e,s,i}, P_{e,i}) &= c_{e,o,i}(P_{e,s,i} - P_{e,i}) \end{aligned} \tag{6.10.5}$$

where $C_{w,i}(*)$, $C_{w,u,i}(*)$, and $C_{w,o,i}(*)$ are the direct cost, underestimated penalty cost, and overestimated penalty cost for wind generator i, respectively; and $C_{e,i}(*)$, $C_{e,u,i}(*)$, and $C_{e,o,i}(*)$ for PEVs at bus i. The direct, underestimated penalty, and overestimated penalty cost coefficients for wind generator i are $c_{w,i}$, $c_{w,u,i}$, and $c_{w,o,i}$, respectively, and $c_{e,i}$, $c_{e,u,i}$, and $c_{e,o,i}$ for PEVs at bus i. The PEVs' buses can be in two conditions: the discharge condition is a positive injection of power to the system known as V2G [257, 258], and the charge condition is a negative power injection or load to the system.

In order to solve the proposed formulation, several heuristic optimization algorithms can be used; in this way the target function is subject to the following kind of constraints depending on how they are handled in the optimization algorithm:

Equality Constraints Considered with a Power Flow Inside of the Optimization Algorithm These constraints are given by the active power flow balance Eq. (6.10.6) and the reactive power flow balance Eq. (6.10.7) at all buses:

$$P_{gi} - P_{di} - v_i \sum_{j \in N_i} v_j \left(g_{ij} \cos\theta_{ij} + B_{ij} \sin\theta_{ij} \right) = 0 \tag{6.10.6}$$

$$Q_{gi} - Q_{di} - v_i \sum_{j \in N_i} v_j \left(g_{ij} \sin \theta_{ij} + B_{ij} \cos \theta_{ij} \right) = 0 \qquad (6.10.7)$$

where $i = 1,\ldots,N_i$ (total number of buses); g_{ij} and B_{ij} are the conductance and susceptance of branch between buses i and j, respectively; v_i and v_j are the voltage magnitudes at buses i and j, respectively; θ_{ij} is the load angle difference between buses i and j. The active power generated and demanded in the bus i are P_{gi} and P_{di}, respectively. The reactive power generated and demanded in the bus i are Q_{gi} and Q_{di}, respectively. P_{gi} is given by (6.10.8), in case there are conventional or wind generators, or PEVs in the bus i. If there are not any generator or any PEVs, $P_{gi} = 0$.

$$P_{gi} = P_{s,i} + W_{s,i} + P_{e,s,i} \qquad (6.10.8)$$

These constraints are considered by solving a power flow inside of the optimization algorithm. Additionally, the formulation must select a slack bus, PV buses, and PQ buses.

Inequality Constraints Related with the Decision Variables The decision variables are divided into continuous and discrete variables. The first type of variables are related to generators and PEVs' active power outputs ($P_{s,i}$, $W_{s,i}$, $P_{e,s,i}$) and generators bus voltage set-points (v_{gi}). The discrete variables are related to stepwise adjustable onload transformers' tap positions, which are translated to a transformation ratio (T_k); and binary variables, related to switchable shunt compensation devices (Shunt_j).

These constraints are given by:

$$P_{s,i}^{\min} \le P_{s,i} \le P_{s,i}^{\max}, \quad i \in N_c - 1 \text{ (excluding the slack)} \qquad (6.10.9)$$

$$W_{s,i}^{\min} \le W_{s,i} \le W_{s,i}^{\max}, \quad i \in N_w \qquad (6.10.10)$$

$$P_{e,s,i}^{\min} \le P_{e,s,i} \le P_{e,s,i}^{\max}, \quad i \in N_e \qquad (6.10.11)$$

$$v_{gi}^{\min} \le v_{gi} \le v_{gi}^{\max}, \quad i \in N_g \qquad (6.10.12)$$

$$T_k^{\min} \le T_k \le T_k^{\max}, \quad k \in N_T \qquad (6.10.13)$$

$$\text{Shunt}_h = \begin{cases} 0 \\ \text{or} \\ \text{defaultShunt}_h \end{cases}, \quad h \in N_{CD} \qquad (6.10.14)$$

where N_g is the number of buses with conventional or wind generators, or PEVs ($N_g = N_c + N_w + N_e$); N_T is the number of branches with transformers with online tap changers. N_{CD} is the number of buses with compensation devices and defaultShunt_h is the reactive power injected at nominal tension by the compensation device in bus h. These constraints are considered in the optimization algorithm corresponding to the possible solution data, as described in Section 6.10.3.

Inequality Constraints Related with the Fitness Function Inside of the Optimization Algorithm These constraints are given by the active power generation limit at the slack bus (6.10.15), the reactive power generation limit at each generator bus (6.10.16), the voltage magnitude limit at each load bus (6.10.17), and the power flow limit constraint of each transmission line (6.10.18):

$$P_{\text{slack}} \le P_{\text{slack}}^{\max} \tag{6.10.15}$$

$$Q_{gi}^{\min} \le Q_{gi} \le Q_{gi}^{\max}, \quad i \in N_{\text{g}} \tag{6.10.16}$$

$$v_{\text{load}i}^{\min} \le v_{\text{load}i} \le v_{\text{load}i}^{\max}, \quad i \in N_{\text{B}} \tag{6.10.17}$$

$$s_{\text{l}} \le s_{\text{l}}^{\max}, \quad l \in N_{\text{l}} \tag{6.10.18}$$

where N_{B} represents the number of load buses, that is to say, buses where $P_{\text{d}i}$ is different than zero, and N_{l} represents the number of branches in the network.

These kinds of constraints are handled with a penalty factor in the fitness function and they are considered for non-contingency and contingency conditions ($N-1$), as described in Section 6.10.3.

Expected Cost of Wind Power

From previous research [243, 244], it is widely identified that the wind speed profile at a given location most closely follows a Weibull distribution over time. In this case the wind speed Weibull distribution, $f_{\text{v}}(v)$, is given by:

$$f_{\text{v}}(v) = \left(\frac{k}{c}\right)\left(\frac{v}{c}\right)^{k-1} e^{-\left(\frac{v}{c}\right)^k} \tag{6.10.19}$$

where v is the wind speed, c and k are the scale factor (units of wind speed) and shape factor (dimensionless) at a given location, respectively. Through different methods c and k can be estimated [259, 260]. For instance, in [261], the shape parameter varied from 1 to 3 and the scale parameter from 5 to 20 m/s. In this paper, the Weibull distribution with a shape factor of 2 and scale factor of $c = \sqrt{2}\sigma$ (where σ is the Rayleigh scale parameter), also known as the Rayleigh distribution given by (6.10.20), is used.

$$f_{\text{v}}(v) = \left(\frac{v}{\sigma^2}\right) e^{-\left(v/\sqrt{2}\sigma\right)^2} \tag{6.10.20}$$

The Rayleigh scale parameter can be calculated using the mean of a Weibull function, μ, given by:

$$\mu = c\Gamma\left(1 + k^{-1}\right) \tag{6.10.21}$$

where the gamma function is $\Gamma(z) = \int_0^\infty t^{z-1}e^{-t}\text{d}t$. Replacing $k = 2$ and $c = \sqrt{2}\sigma$ in (6.10.21), the mean, μ, is given by:

$$\mu = \sqrt{\frac{\pi}{2}}\sigma \tag{6.10.22}$$

Considering the forecasted wind speed (V_f) as the mean, the Rayleigh scale parameter is given by:

$$\sigma = \frac{V_f}{\sqrt{\pi/2}} \tag{6.10.23}$$

The power output of a wind turbine (W) with a given wind speed input (v) is known as wind turbine power curve, it can be stated as [244]:

$$W(v) = \begin{cases} 0, & \text{for } v < v_i \text{ or } v > v_o \\ a\,v + b, & \text{for } v_i < v < v_r \\ w_r, & \text{for } v_r < v < v_o \end{cases} \tag{6.10.24}$$

$$\text{with,} \quad a = \frac{w_r}{(v_r - v_i)}, \quad b = \frac{-w_r v_i}{(v_r - v_i)}$$

where there is no power output up to cut-in wind speed (v_i) neither with wind speeds greater than the cut-out wind speed (v_o). There is a linear power output relationship between cut-in and rated wind speed (v_r), and a constant rated power output (w_r) between v_r and v_o.

In order to develop the proposed formulation, it is needed to convert the wind speed distribution, $f_v(v)$, to a wind power distribution $f_W(W)$. The probabilities of the wind speed being smaller than v_i or larger than v_o are given by:

$$\text{Prob}(v \le v_i) = \int_0^{v_i} f_v(v)\,dv = 1 - e^{-\left(v_i^2/2\sigma^2\right)} \tag{6.10.25}$$

$$\text{Prob}(v_o \le v) = \int_{v_o}^{\infty} f_v(v)\,dv = e^{-\left(v_o^2/2\sigma^2\right)} \tag{6.10.26}$$

The probability of the wind power output being zero coincides with the sum of probabilities presented in Eqs. (6.10.25) and (6.10.26), in this way:

$$f_W(W = 0) = \left(1 - e^{-\left(v_i^2/2\sigma^2\right)} + e^{-\left(v_o^2/2\sigma^2\right)}\right)\delta(W) \tag{6.10.27}$$

where $\delta(\cdot)$ is the Dirac delta function that is introduced because the wind power distribution is discrete.

The probability of the wind speeds being greater than v_r but lower than v_o is:

$$\text{Prob}(v_r \le v \le v_o) = \int_{v_r}^{v_o} f_v(v)\,dv = e^{-\left(v_r^2/2\sigma^2\right)} - e^{-\left(v_o^2/2\sigma^2\right)} \tag{6.10.28}$$

Accordingly, the probability of the wind power generation being equal to w_r is given by:

$$f_W(W = w_r) = e^{-\left(v_r^2/2\sigma^2\right)} - e^{-\left(v_o^2/2\sigma^2\right)}\ \delta(W - w_r) \tag{6.10.29}$$

For $0 < W < w_r$, it is possible to use a variable change by means of the transformation property of probability density functions. Considering the transformation from variable v to variable W, where the relationship $W(v)$ is known, it is possible to get the respective wind power distribution given by:

$$f_W(W) = f_v\left(W^{-1}(W)\right)\frac{dW^{-1}(W)}{dW}, \quad 0 < W < w_r \tag{6.10.30}$$

where $W^{-1}(W) = v = (W - b)/a$, and $dW^{-1}(W)/dW = 1/a$, consequently:

$$f_W(0 < W < w_r) = \left(\frac{v}{\sigma^2}\right)e^{-\left(v/\sqrt{2}\sigma\right)^2}\frac{1}{a} = \frac{(W-b)}{a^2\sigma^2}e^{-\left(W-b/\sqrt{2}a\sigma\right)^2} \tag{6.10.31}$$

In this way, the wind power distribution, $f_W(W)$, is given by (6.10.27), (6.10.29), and (6.10.31).

Based on the wind power distribution, the expected value of the underestimated penalty cost of the wind power can be derived as

$$\begin{aligned}
E[C_{w,u,i}(W_{s,i}, W_i)] &= \int_{W_{s,i}}^{w_r} c_{w,u,i}(W_i - W_{s,i}) \cdot f_W(W_i)dW_i \\
&= \int_{W_{s,i}}^{w_r^-} c_{w,u,i}(W_i - W_{s,i}) \cdot f_W(W_i)dW_i \\
&\quad + \int_{w_r^-}^{w_r} c_{w,u,i}(W_i - W_{s,i}) \cdot f_W(w_r)dW_i \\
&= \int_{W_{s,i}}^{w_r^-} c_{w,u,i}(W_i - W_{s,i}) \cdot \frac{(W_i - b)}{a^2\sigma^2}e^{-\left((W_i-b)/\sqrt{2}a\sigma\right)^2}dW_i \\
&\quad + \int_{w_r^-}^{w_r} c_{w,u,i}(W_i - W_{s,i}) \cdot \left(e^{-\left(v_r^2/2\sigma^2\right)} - e^{-\left(v_o^2/2\sigma^2\right)}\right)\delta(W_i - w_r)dW_i \\
&= \frac{c_{w,u,i}}{2}\left(\sqrt{2\pi}a\sigma\left(\operatorname{erf}\left(\frac{w_r - b}{\sqrt{2}a\sigma}\right) - \operatorname{erf}\left(\frac{W_{s,i} - b}{\sqrt{2}a\sigma}\right)\right)\right. \\
&\quad \left. + 2(W_{s,i} - w_r) \cdot e^{-(w_r-b)^2/2a^2\sigma^2}\right) \\
&\quad + c_{w,u,i}(w_r - W_{s,i})\left(e^{-v_r^2/2\sigma^2} - e^{-v_o^2/2\sigma^2}\right)
\end{aligned} \tag{6.10.32}$$

where w_r^- represents a number that infinitely approaches w_r from the left side, erf(∗) is the Gauss error function given by (6.10.33); the erf(∗) function can be easily calculated with numerical integration and can be modeled in MATLAB using the built-in erf function.

$$\operatorname{erf}(x) = \frac{2}{\sqrt{\pi}}\int_0^x e^{-t^2}dt \tag{6.10.33}$$

The deduction presented in (6.10.32) uses a property of the delta function given by (6.10.34), and uses the defined integrals presented in (6.10.35) and (6.10.36). Additionally, Eq. (6.10.35) was used to deduce Eqs. (6.10.25), (6.10.26), and (6.10.28).

$$\int_a^b f(x)\delta(x-x_0)\mathrm{d}x = \begin{cases} f(x_0) & \text{if} \quad a < x_o < b \\ 0 & \text{if} \quad x_0 < a \text{ or } x_0 > b \end{cases} \tag{6.10.34}$$

$$\int_a^b t.e^{-t^2}\mathrm{d}t = \frac{1}{2}\left(\mathrm{e}^{-a^2} - \mathrm{e}^{-b^2}\right) \tag{6.10.35}$$

$$\int_a^b t^2 \cdot e^{-t^2}\mathrm{d}t = \frac{1}{4}\left(\sqrt{\pi}(\mathrm{erf}(b) - \mathrm{erf}(a)) + 2ae^{-a^2} - 2be^{-b^2}\right) \tag{6.10.36}$$

In the same way, the expected value of the overestimated penalty cost can be derived as

$$\begin{aligned} E[C_{\mathrm{w,o},i}(W_{\mathrm{s},i}, W_i)] &= \int_0^{W_{\mathrm{s},i}} c_{\mathrm{w,o},i}(W_{\mathrm{s},i} - W_i) \cdot f_{\mathrm{W}}(W_i)\mathrm{d}W_i \\ &= \int_0^{0^+} c_{\mathrm{w,o},i}(W_{\mathrm{s},i} - W_i) \cdot f_{\mathrm{W}}(0)\mathrm{d}W_i \\ &\quad + \int_{0^+}^{W_{\mathrm{s},i}} c_{\mathrm{w,o},i}(W_{\mathrm{s},i} - W_i) \cdot f_{\mathrm{W}}(W_i)\mathrm{d}W_i \\ &= \int_0^{0^+} c_{\mathrm{w,o},i}(W_{\mathrm{s},i} - W_i) \cdot \left(1 - e^{-\left(v_i^2/2\sigma^2\right)} + e^{-\left(v_o^2/2\sigma^2\right)}\right)\delta(W_i)\ \mathrm{d}W_i \\ &\quad + \int_{0^+}^{W_{\mathrm{s},i}} c_{\mathrm{w,o},i}(W_{\mathrm{s},i} - W_{\mathrm{i}}) \cdot \frac{(W_i - b)}{a^2\sigma^2} e^{-\left((W_i - b)/\sqrt{2}a\sigma\right)^2}\mathrm{d}W_i \\ &= c_{\mathrm{w,o},i} W_{\mathrm{s},i}\left(1 - e^{-\left(v_i^2/2\sigma^2\right)} + e^{-\left(v_o^2/2\sigma^2\right)} + e^{-\left(b^2/2a^2\sigma^2\right)}\right) \\ &\quad - \frac{c_{\mathrm{w,o},i}\sqrt{2\pi}a\sigma}{2}\left(\mathrm{erf}\left(\frac{W_{\mathrm{s},i} - b}{\sqrt{2}a\sigma}\right) - \mathrm{erf}\left(\frac{-b}{\sqrt{2}a\sigma}\right)\right) \end{aligned} \tag{6.10.37}$$

where 0^+ represents a number that infinitely approaches 0 from the right side.

Expected Cost of PEVs' Active Power Outputs

The PEVs' active power outputs used in this paper follow a normal distribution (6.10.38), according to a simulation-based approach to study the discharge/charge behavior of PEVs presented in [243, 262, 263].

$$f_{P_e}(P_e) = \frac{1}{\sqrt{2\pi\phi^2}} e^{-(P_e-\mu)^2/2\phi^2} \tag{6.10.38}$$

where P_e represents the available PEVs' power, μ and ϕ are the mean and SD of the PEVs' power normal distribution.

The normal distribution in (6.10.38) is a result of the convolution of the aggregated V2G power capacity ($P_{e,V2G}$) normal distribution and the aggregated load level ($P_{e,l}$) normal distribution, at buses with PEVs. This result is based on knowing that the probability of the difference ($P_e = P_{e,V2G} - P_{e,l}$) of two or more independent random variables is the convolution of their individual distributions [264]. Based on this PEVs' power distribution, the expected value of the underestimated penalty cost can be derived as

$$\begin{aligned} E[C_{e,u,i}(P_{e,i}, P_{e,s,i})] &= \int_{P_{e,s,i}}^{+\infty} c_{e,u,i}(P_{e,i} - P_{e,s,i}) \cdot f_P(P_{e,i}) dP_{e,i} \\ &= \frac{c_{e,u,i}}{2}(\mu - P_{e,s,i})\left(\text{erf}\left(\frac{\mu - P_{e,s,i}}{\sqrt{2}\phi}\right) + 1\right) \\ &+ \frac{c_{e,u,i} \cdot \phi}{\sqrt{2\pi}} e^{-((\mu - P_{e,s,i})^2/2\phi^2)} \end{aligned} \tag{6.10.39}$$

where, in addition to the defined integrals presented in (6.10.35) and (6.10.36) the following is used:

$$\int_a^b e^{-t^2} dt = \frac{\sqrt{\pi}}{2}(\text{erf}(b) - \text{erf}(a)) \tag{6.10.40}$$

In the same way, the expected value of the overestimated penalty cost can be derived as

$$\begin{aligned} E[C_{e,o,i}(P_{e,i}, P_{e,s,i})] &= \int_0^{P_{e,s,i}} c_{e,o,i}(P_{e,s,i} - P_{e,i}) \cdot f_P(P_{e,i}) dP_{e,i} \\ &= \frac{c_{e,o,i} \cdot \phi}{\sqrt{2\pi}} \left(e^{-((\mu - P_{e,s,i})^2/2\phi^2)} - e^{-(\mu^2/2\phi^2)}\right) \\ &+ \frac{c_{e,o,i}}{2}(P_{e,s,i} - \mu)\left(\text{erf}\left(\frac{\mu}{\sqrt{2}\phi}\right) - \text{erf}\left(\frac{\mu - P_{e,s,i}}{\sqrt{2}\phi}\right)\right) \end{aligned} \tag{6.10.41}$$

6.10.3 Population-Based Optimization Algorithms

Population-based optimization algorithms are a kind of heuristic techniques based on rules that can follow different types of procedures like natural processes. These techniques employ a population of solutions and a search strategy in order to obtain a better value of the objective function. In addition, this process modifies the population of the solutions in each iteration. The flowchart for the heuristic algorithms used in this paper is shown in Figure 6.10.1.

Each algorithm to be used has its own *heuristic search parameters*. The initial population (*initialization*) and the updated population of the *heuristic search strategy* follow the inequality constraints (6.10.9)–(6.10.14), related to the decision variables: active power outputs of generators and PEVs, bus voltage set-points of generators and PEVs, tap positions of stepwise adjustable onload transformers, and condition of switchable shunt compensation devices.

For each solution in each population a *power flow*, an *objective evaluation*, and a *fitness function value* are calculated. The power flow calculation considers constraints (6.10.6) and (6.10.7), where the formulation must select a slack bus, PV buses, and PQ buses. The objective evaluation corresponds to the expected value of total generation cost given by (6.10.1). The fitness function value is the function to be optimized in the algorithm, and is given by the objective function and a penalty function (PF), obtained by (6.10.41), that considers constraints (6.10.15)–(6.10.18) through a penalty factor (ρ).

$$\mathrm{PF} = \rho \left\{ \begin{aligned} & \left(P_{\text{slack}}^{\max} - P_{\text{slack}}\right)^2 + \sum_{i=1}^{N_g} \left[\left(Q_{gi} - Q_{gi}^{\min}\right)^2 + \left(Q_{gi}^{\max} - Q_{gi}\right)^2 \right] \\ & + \sum_{i=1}^{N_B} \left[\left(v_{\text{load}i} - v_{\text{load}i}^{\min}\right)^2 + \left(v_{\text{load}i}^{\max} - v_{\text{load}i}\right)^2 \right] + \sum_{l=1}^{N_l} \left(s_l^{\max} - s_l\right)^2 \end{aligned} \right\} \tag{6.10.42}$$

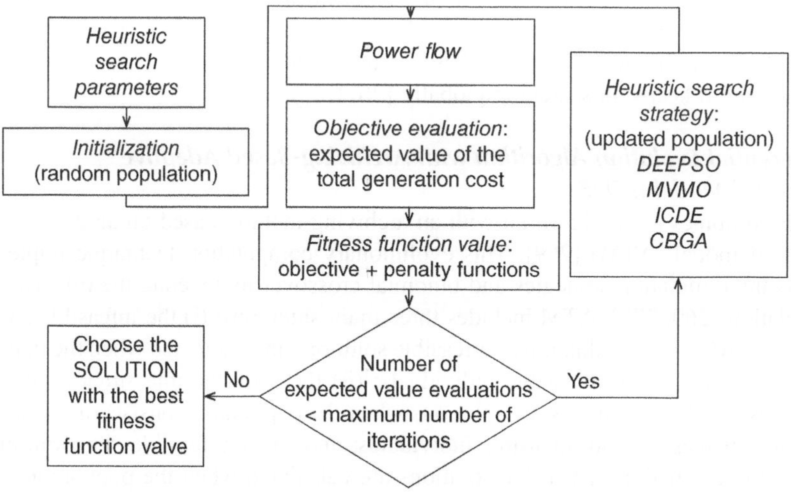

Figure 6.10.1 Heuristic algorithms flowchart.

The PF is calculated for non-contingency and contingency ($N-1$) conditions. After a maximum number of iterations, the solution with the best fitness function value will be selected. Four population-based optimization algorithms are used in this paper. In the following items the *heuristic search strategy* basis of each one is summarized, using the information presented in [90, 265–273].

Differential Evolutionary Particle Swarm Optimization (DEEPSO)

DEEPSO is a variation of EPSO heuristic technique, considering differential evolution (DE) [265]. In turn, EPSO is a mixed method that combines a PSO algorithm with an evolutionary programming (EP) approach by means of a self-adaptive recombination operator [266]. EP algorithms simulate the evolution of the solution in the populations based on procedures such as selection, recombination, and mutation. Additionally, the EP targets the subsistence of solutions with better performances in each generation. Alternatively, PSO is based on the social behavior of swarms, in which each particle (population of solutions) moves in the search space according to three different considerations: inertia, cooperation, and memory [266]. The hybrid DEEPSO keeps the self-adaptive procedures of EPSO but implements the concept of rough gradient process from DE algorithms [265].

Mean–Variance Mapping Optimization

MVMO constitutes an emerging heuristic population-based optimization algorithm [267]. This technique has an evolutionary mechanism implementing a population of a single parent–offspring pair solution along with a normalized range of the search space for all the decision variables [267, 268]. Besides, MVMO is characterized by an archive of n-best solutions determined by a mapping function. The recording in the function is defined by the mean and variance of the optimization variables [90].

The shape and location of the function (mapping curve) are adjusted according to the searching process. The particularity of the algorithm lies in the strategic transformation used for mutating the offspring based on mean–variance of the n-best dynamic population. Regardless of the orientation to the best solution, the algorithm keeps on searching globally [267].

Differential Evolution Algorithm with Archiving-Based Adaptive Trade-off Model (ICDE)

ICDE combines a DE algorithm with an archiving method based on an adaptive trade-off model (ATM) [269]. This evolutionary computation technique implements three mutation strategies and binomial crossover to generate the offspring population [269, 270]. ATM includes three main situations: (i) the infeasible situation, in which the evaluation of infeasible solutions are calculated when the population contains only infeasible individuals; (ii) the semi-feasible situation that balances feasible and infeasible solutions when the population consists of a combination of feasible and infeasible individuals; and (iii) the feasible situation, in which the evaluation of feasible solutions are calculated when the population is composed of feasible individuals only [271]. ICDE has the ability to maintain a worthy balance between the diversity and the convergence of the population during the evolution [270, 271].

Modified Chu–Beasley's Genetic Algorithm (CBGA)

The GA proposed by Chu and Beasley was first used to solve the generalized assignment problem [272]. Some of its characteristics are [272, 273]: a random initial population; objective function and unfeasibilities are considered separately; selection by tournament; in each iteration, only one descendant is generated using selection, recombination, mutation, and local improvement; the improved descendant replace one element in the population taking into consideration unfeasibility, quality, and diversity criteria. The Chu–Beasley algorithm was modified in the following ways [272, 273]: the initial population is generated randomly, but the individuals are improved by a local search; and the mutation operator is not used.

6.10.4 Test System and Results Analysis

The optimization model was solved with the above-mentioned four population-based algorithms, outlined in the previous section, using a maximum number of iterations of 100 000. In this section, the test system used and a comparative assessment of these heuristic optimization algorithms are presented. The proposed economic dispatch model is tested with a slightly modified version (Table 6.10.1) of IEEE 118-bus test system [274].

In this system, there are 130 decision variables, 77 are continuous variables related to 47 active power outputs of conventional generators (excluding the slack), which have the power maximum and minimum limits ($P_{s,i}^{\min}$ and $P_{s,i}^{\max}$) presented in [274]; two active power outputs of wind power generators and four active power outputs of PEVs; and 48 + 2 generators and four PEVs' bus voltage set-points (which have the following maximum and minimum limits: $v_{gi}^{\min} = 0.95$ p.u. and $v_{gi}^{\max} = 1.05$ p.u., respectively). Nine are discrete variables associated with stepwise adjustable onload transformers' tap positions, with minimum and maximum limits $T_k^{\min} = 0.9$ and $T_k^{\max} = 1.1$, with steps of 0.01 in the transformation ratio.

TABLE 6.10.1 Composition of IEEE 118-Bus Test System

Item	Modified IEEE 118-Bus Test System
Conventional generators	48
Wind Power generators	2
Buses with PEVs' facilities connected	4
Loads	99
Lines/cables	177
Transformers with stepwise onload tap changers	9
Shunt compensation devices (binary on/off)	14

14 variables are binary variables associated with switchable shunt compensation devices.

It is assumed that there are two wind generators installed at buses 24 and 26, and four buses with facilities for PEVs' connection at buses 59, 80, 90, and 116. The parameters of the wind speed Rayleigh distribution and the wind turbine power curve, during the time period of analysis (peak day time 11 : 00–12 : 00), are given in Table 6.10.2. Thus, the power minimum and maximum limits are $W^{\min}_{s,22} = W^{\min}_{s,24} = 0$ MW and $W^{\max}_{s,22} = W^{\max}_{s,24} = 150$ MW, respectively.

The aggregated V2G power capacity at buses 59, 80, 90, and 116, during the peak day time, is considered normally distributed with a mean of 19.58 MW and a SD of 0.58 MW. In these buses, the aggregated load of PEVs follows a normal distribution with a mean of 21.85 MW and a SD of 1.38 MW. Thus, the mean (μ) and SD (ϕ) of the distributions of PEVs' active power outputs are −2.2 MW and 1.93 MW, respectively. The power minimum and maximum limits are $P^{\min}_{e,s,59} = P^{\min}_{e,s,80} = P^{\min}_{e,s,90} = P^{\min}_{e,s,116} = -8$ MW and $P^{\max}_{e,s,59} = P^{\max}_{e,s,80} = P^{\max}_{e,s,90} = P^{\max}_{e,s,116} =$ 3.6 MW, respectively.

Table 6.10.3 shows the direct, underestimated penalty, and overestimated penalty cost coefficients of wind generators and PEVs during the time period of analysis. Additionally, the considered contingencies ($N-1$ conditions) are outages of branches 8 and 50. Table 6.10.4 shows the expected value of the total generation cost for 16 runs in each population-based algorithm. The expected cost

TABLE 6.10.2 Wind Turbine Power Curve and Wind Speed Rayleigh Distribution Parameters During the Time Period of Analysis

Symbol	Parameter	Value
V_f	Forecasted wind speed (m/s)	20
σ	Rayleigh scale parameter (m/s)	15.9577
v_i	Cut-in wind speed (m/s)	5
v_r	Rated wind speed (m/s)	15
v_o	Cut-out wind speed (m/s)	45
w_r	Rated wind power	150
a	Linear coefficient (MW/m/s)	15
b	Linear coefficient (MW)	−75

TABLE 6.10.3 Direct, Underestimated Penalty, and Overestimated Penalty Cost Coefficients of Wind Generators and PEVs During the Time Period of Analysis

Power Injection Type	Direct Cost Coefficient ($/MW)	Underestimated Penalty Cost Coefficient ($/MW)	Overestimated Penalty Cost Coefficient ($/MW)
Wind generator	10	30	70
PEV	65	30	70

TABLE 6.10.4 Expected Value of the Total Generation Cost for 16 Runs in Each Population-Based Algorithm During the Time Period of Analysis

	$E[C_{total}]$ ($) run							
Algorithm	1	2	3	4	5	6	7	8
DEEPSO	118 707	118 266	118 071	118 845	118 897	119 109	118 604	118 604
MVMO	117 263	117 263	117 243	117 263	117 263	117 253	117 263	117 242
ICDE	123 712	123 353	118 565	125 797	123 212	120 829	120 163	121 396
CBGA	120 605	117 478	118 642	119 191	117 890	117 407	118 469	117 636
Algorithm	9	10	11	12	13	14	15	16
DEEPSO	118 576	118 305	117 957	118 305	118 528	118 336	118 670	118 009
MVMO	117 240	117 242	117 253	117 242	117 242	117 259	117 242	117 242
ICDE	132 422	120 829	120 829	121 447	120 829	120 641	120 899	120 703
CBGA	119 588	118 406	117 385	118 389	120 584	119 304	118 388	117 745

for each run is different, and not always similar, that means that sometimes the algorithms reach a near-to-global-optimum and sometimes a local optimum. The decision variables related with the wind generator schedules and PEVs' schedules are very similar in the cells with the same color.

For instance, in the DEEPSO algorithm case, there are two colors: in the green cells the decision variables range for wind schedules, in the 16 runs, are $W_{s,22}$ = 149.986 to 150 MW, $W_{s,24}$ = 149.991 to 150 MW; and for the PEVs' schedules are $P_{e,s,59}$ = −0.436 to −0.362 MW, $P_{e,s,80}$ = −0.553 to −0.420 MW, $P_{e,s,90}$ = −0.479 to −0.404 MW, and $P_{e,s,116}$ = −0.460 to −0.412 MW. In the yellow cells the decision variables range for wind schedules, in the 16 runs, are $W_{s,22}$ = 149.988 to 150 MW, $W_{s,24}$ = 149.989 to 150 MW; and for the PEVs' schedules are $P_{e,s,59}$ = −0.487 to −0.420 MW, $P_{e,s,80}$ = −0.501 to −0.3 75 MW, $P_{e,s,90}$ = −7.983 to −7.934 MW, and $P_{e,s,116}$ = −0.498 to −0.198 MW.

In the MVMO algorithm case, there is only one color since the wind generator schedules and PEVs' schedules are very similar, here the ranges, in the 16 runs, are: $W_{s,22}$ = 150 to 150 MW, $W_{s,24}$ = 150 to 150 MW, $P_{e,s,59}$ = −0.438 to −0.438 MW, $P_{e,s,80}$ = −0.445 to −0.433 MW, $P_{e,s,90}$ = −0.459 to −0.450 MW, and $P_{e,s,116}$ = −0.451 to −0.446 MW. In the ICDE algorithm case, there are four colors and the expected costs are bigger than the best values reached with MVMO and DEEPSO algorithms. In the CBGA case, there are three colors, and only two expected costs (run 11 and run 16) reach a similar value compared with the best values of MVMO and DEEPSO algorithms. The MVMO algorithm reaches the best expected costs (near-to-global-optimum) and always give similar results in the decision variables, but this algorithm has the biggest running time as shown in Table 6.10.5.

In order to compare the advantages in the power system operation, when the presence of transformers with stepwise OLTCs and shunt compensation devices in

TABLE 6.10.5 Average Running Time for Each Population-Based Algorithm

DEEPSO	MVMO	ICDE	CBGA
878.47 s	6578.95 s	1168.57 s	2852.58 s

the network is considered, the economic dispatch model is solved with MVMO algorithm, but fixing the decision variables related to these elements to the values presented in the original systems [274]. In this case the average expected cost is \$119 163. There is an improvement of 1.63% in the expected cost, comparing this value (\$119 163) with the average expected cost for the MVMO algorithm (\$117 250.9) considering tap changers and compensation devices.

Finally, the aggregated V2G power capacity and the aggregated load of PEVs at buses 59, 80, 90, and 116 are changed in order to obtain different minimum and maximum limits. Thus, the power limits used are $P^{\min}_{e,s,59} = P^{\min}_{e,s,80} = -8$ MW, $P^{\min}_{e,s,90} = -3$ MW, $P^{\min}_{e,s,116} = 7$ MW and $P^{\max}_{e,s,59} = P^{\max}_{e,s,80} = 3.6$ MW, $P^{\max}_{e,s,90} =$ 15 MW, $P^{\max}_{e,s,116} = 13$ MW. In this case, the average expected cost, during the time period of analysis, for MVMO algorithm is \$120 116. The decision variables range in the 16 runs for the PEVs' schedules are: $P_{e,s,59} = -5.928$ to -5.841 MW, $P_{e,s,80} = -2.342$ to -2.339 MW, $P_{e,s,90} = -0.555$ to -0.542 MW, and $P_{e,s,116} = 10.460$ to 10.483 MW.

6.10.5 Conclusion

New elements, like wind generators and PEVs, in the power networks have introduced several challenges to the optimal power system operation. One important part of this operation is the economic dispatch. In this way, the proposed optimal active–reactive power dispatch formulation is able to consider new elements to be incorporated or already incorporated (like the dynamic or maneuver components) into the current power networks. The formulation considers the uncertainties of both wind power generators and PEVs in order to model the stochastic nature of these elements.

The optimization objective is able to minimize the mathematical expectations of the generation costs while satisfying constraints for non-contingency and contingency conditions ($N-1$). The constraints are able to include the presence of transformers with stepwise OLTCs and shunt compensation devices in the network. These components are dynamic elements important for the network operation.

Since four population-based optimization algorithms were used, several runs were calculated. The expected cost for each run is different, and not always similar, which means that sometimes the algorithms reach a near-to-global-optimum and sometimes a local optimum. These algorithms are convenient for solving the proposed formulation, allowing testing of several combinations of the decision variables through the heuristic search strategy of each algorithm. The MVMO algorithm gave the best results always reaching a near-to-global-optimum, but its average running time is the biggest.

Acknowledgement
The authors would like to thank UNIVERSIDAD NACIONAL DE COLOMBIA and INTITUTO DE ENERGIA ELECTRICA, UNIVERSIDAD NACIONAL DE SAN JUAN (where they are affiliated with) and the Cyted Network: RED IBEROAMERICANA PARA EL DESARROLLO Y LA INTEGRACION DE PEQUEÑOS GENERADORES EOLICOS (MICRO-EOLO) for the continued support during the development of this work.

6.11 CONCLUSIONS

Focusing on renewable energy, and smart grids, the present chapter discussed operation and control of different sources, demand side, and storage means including DR and EVs. MHO techniques have been applied and compared also in the scope of generation and consumption forecasting.

Some MHO techniques have been used for different topics, along the different sections, showing that in specific applications, and in some cases depending on the specific scenario, different methods can fit better the objective whether it is the computation time or the need of a very good solution. Solutions of MHO have in some cases been compared with the deterministic solution implemented for linear and nonlinear problems, in some cases with integer variables.

In order to adequately cope with the requirements of the specific applications discussed in the present chapter, techniques like PSO in a close to original version but also in innovative variants as EPSO and QPSO have been used. Hybrid approaches have also been developed with good results.

REFERENCES

1. Bãnos, R., Manzano-Agugliaro, F., Montoya, F. et al. (2011). Optimization methods applied to renewable and sustainable energy: a review. *Renewable & Sustainable Energy Reviews* 15: 1753–1766.
2. Poullikkas, A., Kourtis, G., and Hadjipaschalis, I. (2011). A hybrid model for the optimum integration of renewable technologies in power generation systems. *Energy Policy* 39: 926–935.
3. Chakraborty, S., Senjyu, T., Saber, A. et al. (2009). Optimal thermal unit commitment integrated with renewable energy sources using advanced particle swarm optimization. *IEEJ Transactions on Electrical and Electronic Engineering* 4: 609–617.
4. Kuo, C. (2010). Wind energy dispatch considering environ-mental and economic factors. *Renewable Energy* 35: 2217–2227.

5. Vale, Z., Morais, H., and Khodr, H. (2011). Renewable-based generation integration in electricity markets with virtual power producers. In: *Handbook of Renewable Energy Technology* (eds. A.F. Zobaa and R.C. Bansal), 637–670. World Scientific.
6. Saber, A. and Venayagamoorthy, G. (2011). Plug-in vehicles and renewable energy sources for cost and emission reductions. *IEEE Transactions on Industrial Electronics* 58: 1229–1238.
7. Saber, A. and Venayagamoorthy, G. (2010). Efficient utilization of renewable energy sources by gridable vehicles in cyber-physical energy systems. *IEEE Systems Journal* 4: 285–294.
8. Sousa, T., Morais, H., Vale, Z. et al. (2012). Intelligent energy resource management considering vehicle-to-grid: a simulated annealing approach. *IEEE Transactions on Smart Grid* 3: 535–542.
9. Soares, J., Sousa, T., Morais, H. et al. (2013). Application-specific modified particle swarm optimization for energy resource scheduling considering vehicle-to-grid. *Applied Soft Computing* 13: 4264–4280.
10. Soares, J., Morais, H., Sousa, T. et al. (2013). Day-ahead resource scheduling including demand response for electric vehicles. *IEEE Transactions on Smart Grid* 4 (1): 596–605.
11. Niknam, T. and Firouzi, B. (2009). A practical algorithm for distribution state estimation including renewable energy sources. *Renewable Energy* 34: 2309–2316.
12. AlRashidi, M. and El-Hawary, M. (2009). A survey of particle swarm optimization applications in electric power systems. *IEEE Transactions on Evolutionary Computation* 13: 913–918.
13. Kennedy, J. and Eberhart, R. (1995). Particle swarm optimization. *IEEE International Conference on Neural Networks Proceedings* 1–6: 1942–1948.
14. Eberhart, R. and Yuhui, S. (2001). Particle swarm optimization: developments, applications and resources. *Proceedings of the 2001 Congress on in Evolutionary Computation* 1: 81–86.
15. Bergh, F. (2002). An analysis of particle swarm optimizers. PhD dissertation. University of Pretoria, Pretoria, South Africa.
16. del Valle, Y., Venayagamoorthy, G., Mohagheghi, S. et al. (2008). Particle swarm optimization: basic concepts, variants and applications in power systems. *IEEE Transactions on Evolutionary Computation* 12: 171–195.
17. Miranda, V. and Fonseca, N. (2002). EPSO – evolutionary particle swarm optimization, a new algorithm with applications in power systems. In: *IEEE/PES Transmission and Distribution Conference and Exhibition 2002: Asia Pacific, vols. 1–3, Conference Proceedings*, 745–750. IEEE.
18. Fan, H. and Shi, Y. (2001). Study on Vmax of particle swarm optimization. In: *Proceedings of the Workshop on Particle Swarm Optimization* (April 2001). Indianapolis, IN: Purdue School of Engineering and Technology.
19. Soares, J., Silva, M., Sousa, T. et al. (2012). Distributed energy resource short-term scheduling using signaled particle swarm optimization. *Energy* 42: 466–476.
20. Kirkpatrick, S., Gelatt, C., and Vecchi, M. (1983). Optimization by simulated annealing. *Science* 220: 671–680.
21. Laarhoven, P. and Aarts, E. (1987). *Simulated Annealing: Theory and Applications.* Dordrecht, Lancaster: Reidel.
22. Sousa, T., Pinto, T., Morais, H., and Vale, Z. (2012). Intelligent electric vehicle heuristic for energy resource management using simulated annealing. *3rd IEEE PES Innovative Smart Grid Technologies Europe (2012 ISGT EU)*, Berlin, Germany (14–17 October 2012).

23. Trovão, J., Pereirinha, P., Jorge, H., and Antunes, C. (2013). A multi-level energy management system for multi-source electric vehicles – an integrated rule-based meta-heuristic approach. *Applied Energy* 105: 304–318.
24. Kokate, P., Das, M., and Loh, R. (1992). Rule-based control of slowly varying systems using compensator segmentation determined by simulated annealing. *IEEE Transactions on Industrial Electronics* 39: 359–367.
25. Zäpfel, G., Braune, R., and Bögl, M. (2010). *Metaheuristic Search Concepts: A Tutorial with Applications to Production and Logistics*. Berlin: Springer.
26. Thukaram, D., Banda, H., and Jerome, J. (1999). A robust three phase power flow algorithm for radial distribution systems. *Electric Power Systems Research* 50: 227–236.
27. Michalewicz, Z. and Schoenauer, M. (1996). Evolutionary algorithms for constrained parameter optimization problems. *Evolutionary Computation* 4: 1–32.
28. Uemura, K., Nakashima, N., Nagata, Y., and Ono, I. (2013). A new real-coded genetic algorithm for implicit constrained black-box function optimization. *IEEE Congress on Evolutionary Computation (CEC)*: 2887–2894.
29. ECF (2010). Roadmap 2050 – a practical guide to a prosperous, low-carbon Europe.
30. Allan, R., Billinton, R., Sjarief, I. et al. (1991). A reliability test system for educational purposes-basic distribution system data and results. *IEEE Transactions on Power Systems* 6 (2): 813–820.
31. Soares, J., Canizes, B., Lobo, C. et al. (2012). Electric vehicle scenario simulator tool for smart grid operators. *Energies* 5 (12): 1881–1899.
32. Amin, S.M. and Wollenberg, B. (2005). Toward a smart grid. *IEEE Power & Energy* 3 (5): 34–41.
33. Vojdani, A. (2008). Smart integration. *IEEE Power & Energy* 6 (6): 71–79.
34. Garrity, T.F. (2008). Getting smart. *IEEE Power & Energy* 6 (2): 38–45.
35. Ipakchi, A. and Albuyeh, F. (2009). Grid of the future. *IEEE Power & Energy* 7 (2): 52–62.
36. Fan, J. and Borlase, S. (2009). The evolution of distribution. *IEEE Power & Energy* 7 (2): 63–68.
37. Farhangi, H. (2010). The path of the smart grid. *IEEE Power and Energy* 8 (1): 18–28.
38. Collier, S.E. (2010). Ten steps to smart grid. *IEEE Industry Applications Magazine* 16: 62–68.
39. European Commission (2006). *EUR-22040 – European Technology Platform SmartGrids*. European Commission.
40. Marnay, C., Asano, H., Papathanassiou, S., and Strbac, G. (2008). Policymaking for microgrid. *IEEE Power & Energy* 6 (3): 66–77.
41. Venkataramanan, G. and Marnay, C. (2008). A large role for microgrid. *IEEE Power & Energy* 6 (3): 78–82.
42. Marnay, C. and Venkataramanan, G. (2006). Microgrid in the evolving electricity generation and delivery interface. *Proceedings of 2006 IEEE PES GM*, Montreal, Canada (June).
43. Kroposki, B., Lasseter, R., Ise, T. et al. (2008). Making microgrid work. *IEEE Power & Energy* 6 (3): 40–53.
44. Nara, K. and Hasegawa, J. (1995). An advanced flexible and reliable distribution system. *Proceedings of EMPD'95*, Singapore (November).
45. Nara, K. and Hasegawa, J. (1997). A new flexible, reliable and intelligent electrical energy delivery system. *Electrical Engineering in Japan* 121 (1): 26–34.
46. Nara, K., Hasegawa, J., Oyama, T. et al. (2000). FRIENDS – Forwarding to Future Power Delivery System. *Proceedings of International Conference on Harmonics and Quality of Power'2000*.

47. Hingorani, N. (1998). Overview of custom power applications. *IEEE PES SM'98*, San Diego.
48. Ichikawa, T. and Rehtanz, C. (2002). Recent trends in distributed generation – technology, grid integration, system operation. *14-th Power System Computation Conference*.
49. Nara, K., Satoh, T., and Mishima, Y. (2005). Optimal network configuration of RIENDS taking power supply reliability into account. *IEEJ Transactions on Power and Energy* 125 (1): 45–52.
50. Mishima, Y., Matsuura, Y., Satoh, T., and Nara, K. (2003). Optimal network configuration of FRIENDS by Tabu search taking cost of power interruption into consideration. *Proceedings of ISAP 2003*.
51. Mitra, J. (2006). Reliability stipulated microgrid architecture using particle swarm optimization. *Proceedings of 9th International Conference on Probabilistic Method Applied to Power System*, Stockholm.
52. Armas, J.M. and Suryanarayanan, S. (2009). A heuristic technique for scheduling a customer-driven residential distributed energy resource installation. *Proceedings of ISAP09*.
53. Calpinelli, G., Celli, G., Mocci, S. et al. (2013). Optimal integration of distributed energy storage device in smart grid. *IEEE Transactions on Smart Grid* 4 (2): 985–995.
54. Bevrani, H., Habibi, F., Babahajyani, P. et al. (2012). Intelligent frequency control in an AC microgrid: online PSO-based fuzzy tuning approach. *IEEE Transactions on Smart Grid* 3 (4): 1935–1944.
55. Ma, H., Chan, K.W., Liu, M. et al. (2013). An intelligent control scheme to support voltage of smart power systems. *IEEE Transactions on Industrial Informatics* 9 (3): 1405–1414.
56. Aquino-Lugo, A.A., Klump, R., and Overbye, T.J. (2011). A control framework for the smart grid for voltage support using agent-based technologies. *IEEE Transactions on Smart Grid* 2 (1): 173–180.
57. Katiraei, F., Iravani, M.R., and Lehn, P.W. (2005). Micro-grid autonomous operation during and subsequent to islanding process. *IEEE Transactions on Power Delivery* 20 (1): 248–257.
58. Pipattanasomporn, M., Feroze, H., and Rahman, S. (2009). Multi-agent systems in a distributed smart grid: design and implementation. *Proceedings of IEEE PSCE09* (March).
59. Datta, M., Senjyu, T., Yona, A., and Funabashi, T. (2009). A fuzzy control based coordinated method for isolated power utility connected clustered photovoltaic systems to provide frequency control. *Proceedings of ISAP 09*, Curitiba, Brazil (November).
60. Sotkiewicz, P.M. and Vignolo, J.M. (2006). Nodal pricing for distribution networks: efficient pricing for efficiency enhancing DG. *IEEE Transactions on Power Systems* 21 (2): 1013–1014.
61. Nquyen, C.P. and Flueck, A.J. (2012). Agent based restoration with distributed energy storage support in smart grid. *IEEE Transactions on Smart Grid* 3 (2): 1029–1038.
62. Nara, K., Ishizu, S., Funada, T., et al. (2006). Voltage control of distribution system by autonomous control of distributed generators. *Proceedings of ICEE* (July).
63. Pogaku, N., Prodanovic, M., and Green, T.C. (2007). Modeling, analysis and testing of autonomous operation of an inverter-based microgrid. *IEEE Transactions on Power Electronics* 22 (2): 613–625.
64. Colson, C.M. and Nehir, M.H. (2013). Comprehensive real-time microgrid power management and control with distributed agent. *IEEE Transactions on Smart Grid* 4 (1): 617–627.

65. Kahrobaee, S., Rajabzadeh, R.A., Soh, L.-K., and Asgarpoor, S. (2013). A multiagent modeling and investigation of smart homes with power generation, storage, and trading feature. *IEEE Transactions on Smart Grid* 4 (2): 659–668.
66. Dou, C.-X. and Liu, B. (2013). Multi-agent based hierarchical Hyblid control for smart grid. *IEEE Transactions on Smart Grid* 4 (2): 771–778.
67. Zeineldin, H.H., El-Saadany, E., and Salama, M.M.A. (2006). Protective relay coordination for micro-grid operation using particle swarm optimization. *Proceedings of Conference on Power Engineering 2006, Large Engineering Systems*, pp. 152–157.
68. Tong, X., Wang, X., Wang, R., and Huang, F. (2013). The study of a regional decentralized peer-to-peer negotiation-based wide-area backup protection multi-agent system. *IEEE Transactions on Smart Grid* 4 (2): 1197–2006.
69. Ross, K.J., Hopkinson, K.M., and Pachter, M. (2013). Using a distributed agent-based communication enabled special protection system to enhance smart grid security. *IEEE Transactions on Smart Grid* 4 (2): 1216–1224.
70. Barton, J.P. and Infield, D.G. (2004). Energy storage and its use with intermittent renewable energy. *IEEE Transactions on Energy Conversion* 19 (2): 441–448.
71. Venayagamoorthy, G.K. (2009). Potentials and promises of computational intelligence for smart grids. *Proceedings of IEEE PES General Meeting*, Calgary, Alberta (July).
72. Saber, A. and Venayagamoothy, G.K. (2009). Optimization of vehicle-to-grid scheduling in constrained parking lots. *Proceedings of IEEE PES General Meeting*, pp. 26–30.
73. Papadopoulos, P., Jenkins, N., Cipcigan, L., and Unda, I.G. (2013). Coordination of the charging of electric vehicles using a multi-agent system. *IEEE Transactions on Smart Grid* 4 (4): 1802–1809.
74. Teng, J., Luan, S.-W., and Huang, Y.-q. (2013). Optimal charging/discharging scheduling of battery storage systems for distribution systems interconnected with sizeable PV generation systems. *IEEE Transactions on Power Systems* 28 (2): 1425–1433.
75. Yang, S., Tan, S., and Xu, J. (2013). Consensus based approach for economic dispatch problem in a smart grid. *IEEE Transactions on Power Systems* 28 (4): 4416–4426.
76. Logenthiran, T., Srinivasan, D., and Aung, H.N. (2012). Multi-agent system for real-time operation of a microgrid in real-time digital simulator. *IEEE Transactions on Smart Grid* 3 (2): 925–933.
77. Wang, L., Wang, Z., and Yang, R. (2012). Intelligent multi-agent control system for energy and comfort management in smart grid and sustainable buildings. *IEEE Transactions on Smart Grid* 3 (2): 605–617.
78. Dimeas, A.L. (2007). Agent based control of Virtual Power Plant. *Proceedings of ISAP07*, Kaohsiung, Taiwan (November).
79. Warmer, C.J. (2007). A field test using agent for coordination of residential micro-chp. *Proceedings of ISAP07*, Kaohsiung, Taiwan (November).
80. Ramachandran, B., Srivastava, S.K., Edrington, C.S., and Cartes, D.A. (2011). An intelligent auction scheme for smart grid market using a hybrid immune algorithm. *IEEE Transactions on Industrial Electronics* 58 (10): 4603–4612.
81. Higuchi, N., Nara, K., Yamamoto, T. et.al. (2014). Functions and operational experiences of real scale smart grid experimental system. *Proceedings of 2014 IEEE Smart Grid Technologies Conference – Asia (ISGA-Asia)*, Kuala Lumpur.
82. European Wind Energy Association (2016). The European offshore wind industry – key trends and statistics 2015. http://www.ewea.org/statistics/offshore (accessed 3 October 2019).
83. Pham, H.V., Rueda, J.L., and Erlich, I. (2014). Online optimal control of reactive sources in wind power plants. *IEEE Transactions on Sustainable Energy* 5: 608–616.

84. Torres, J.L.R., Theologi, A.M., Ndreko, M. et al. (2018). Metaheuristic approach for online optimal reactive power management in near-shore wind power plants. In: *2017 IEEE PES Innovative Smart Grid Technologies Conference – Latin America (ISGT Latin America), Quito*, 1–6. IEEE https://doi.org/10.1109/ISGT-LA.2017.8126706.
85. Erlich, I., Nakawiro, W., and Martinez, M. (2011). Optimal dispatch of reactive sources in wind farms. In: *2011 IEEE Power and Energy Society General Meeting*, 1–7. IEEE.
86. Pappala, V.S., Nakawiro, W., and Erlich, I. (2010). Predictive optimal control of wind farm reactive sources. In: *IEEE PES T&D 2010*, 1–7. IEEE.
87. Wu, Y.-K. and Hong, J.-S. (2007). A literature review of wind forecasting technology in the world. In: *2007 IEEE Lausanne Power Technology*, 504–509. IEEE.
88. Catalão, J.P.S., Pousinho, M.I., and Mendes, V.M.F. (2009). An artificial neural network approach for short-term wind power forecasting in Portugal. *Engineering Intelligent Systems* 1: 5–11.
89. Demuth, H., Beale, M., and Hagan, M. (2000). *Neural Network Toolbox™ 6 User's Guide*. https://www.mit.bme.hu/system/files/oktatas/targyak/9109/MATLAB_-_Neural_Network_Toolbox_6_-_Users_Guide_2009.pdf (accessed 3 October 2019).
90. Erlich, I., Venayagamoorthy, G.K., and Worawat, N. (2010). A mean-variance optimization algorithm. In: *IEEE Congress on Evolutionary Computation, Barcelona*, 1–6. IEEE https://doi.org/10.1109/CEC.2010.5586027.
91. Rueda, J.L. and Erlich, I. (2015). MVMO for bound constrained single-objective computationally expensive numerical optimization. In: *2015 IEEE Congress on Evolutionary Computation (CEC), Sendai*, 1011–1017. IEEE https://doi.org/10.1109/CEC.2015.7257000.
92. Netherlands Enterprise Agency (2014). Borssele wind farm zone, wind energy.
93. Tennet TSO GmbH. (2012). Requirements for offshore grid connections in the grid of TenneT TSO GmbH, p. 14.
94. Solar Energy Corporation of India Limited. (2018). Frequently asked questions. http://seci.gov.in/upload/uploadfiles/files/FAQ.pdf (accessed 3 October 2019).
95. Bouzguenda, M. and Rahman, S. (1993). Value analysis of intermittent generation sources from the system operations perspective. *IEEE Transactions on Energy Conversion* 3 (2): 484–490.
96. Chiu, H.J., Lo, Y.K., Yao, C.J. et al. (2011). A modular self-controlled photovoltaic charger with interintegrated circuit (I2C) interface. *IEEE Transactions on Energy Conversion* 26 (1): 281–289.
97. Farivar, G. and Asaei, B. (2011). A new approach for solar module temperature estimation using the simple diode model. *IEEE Transactions on Energy Conversion* 26 (4): 1118–1126.
98. Thounthong, P. (2011). Model based-energy control of a solar power plant with a supercapacitor for grid-independent applications. *IEEE Transactions on Energy Conversion* 26 (4): 1210–1218.
99. Maki, A. and Valkealahti, S. (2012). Power losses in long string and parallel-connected short strings of series-connected silicon-based photovoltaic modules due to partial shading conditions. *IEEE Transactions on Energy Conversion* 27 (1): 173–183.
100. Lopez-Lapena, O., Penella, M.T., and Gasulla, M. (2012). A closed-loop maximum power point tracker for subwatt photovoltaic panels. *IEEE Transactions on Industrial Electronics* 59 (3): 1588–1596.
101. Chan, D.S.H. and Phang, J.C.H. (1987). Analytical methods for the extraction of solar-cell single- and double-diode model parameters from I-V characteristics. *IEEE Transactions on Electron Devices* 34 (2): 286–293.

102. Villalva, M.G., Gazoli, J.R., and Filho, E.R. (2009). Comprehensive approach to modeling and simulation of photovoltaic arrays. *IEEE Transactions on Power Electronics* 24 (5): 1198–1208.
103. Xiao, W., Dunford, W.G., Palmer, P.R., and Capel, A. (2007). Regulation of photovoltaic voltage. *IEEE Transactions on Industrial Electronics* 54 (3): 1365–1374.
104. Hohm, D.P. and Ropp, M.E. (2003). Comparative study of maximum power point tracking algorithms. *Proceedings of IEEE Photovoltaic's: Research and Applications* 11 (1): 47–62.
105. Salas, V., Olías, E., Barrado, A., and Lázaro, A. (2006). Review of the maximum power point tracking algorithms for stand-alone photovoltaic systems. *Solar Energy Materials & Solar Cells* 90 (11): 1555–1578.
106. Esram, T. and Chapman, P.L. (2007). Comparison of photovoltaic array maximum power point tracking techniques. *IEEE Transactions on Energy Conversion* 22 (2): 439–449.
107. Patel, H. and Agarwal, V. (2008). MATLAB-based modeling to study the effects of partial shading on PV array characteristics. *IEEE Transactions on Energy Conversion* 23 (1): 302–310.
108. Wang, Y.J. and Hsu, P.C. (2010). Analytical modelling of partial shading and different orientation of photovoltaic modules. *IET Renewable Power Generation* 4 (3): 272–282.
109. Paraskevadaki, E.V. and Papathanassiou, S.A. (2011). Evaluation of MPP voltage and power of mc-Si PV modules in partial shading conditions. *IEEE Transactions on Energy Conversion* 26 (3): 923–932.
110. Kobayashi, K., Takano, I., and Sawada, Y. (2006). A study of a two stage maximum power point tracking control of a photovoltaic system under partially shaded insolation conditions. *Solar Energy Materials & Solar Cells* 90 (18–19): 2975–2988.
111. Ji, Y.H., Jung, D.Y., Kim, J.G. et al. (2011). A real maximum power point tracking method for mismatching compensation in PV array under partially shaded conditions. *IEEE Transactions on Power Electronics* 26 (4): 1001–1009.
112. Ahmed, N.A. and Miyatake, M. (2008). A novel maximum power point tracking for photovoltaic applications under partially shaded insolation conditions. *Electric Power Systems Research* 78 (5): 777–784.
113. Nguyen, T.L. and Low, K.S. (2010). A global maximum power point tracking scheme employing DIRECT search algorithm for photovoltaic systems. *IEEE Transactions on Industrial Electronics* 57 (10): 3456–3467.
114. Carannante, G., Fraddanno, C., Pagano, M., and Piegari, L. (2009). Experimental performance of MPPT algorithm for photovoltaic sources subject to inhomogeneous insolation. *IEEE Transactions on Industrial Electronics* 56 (11): 4374–4380.
115. Patel, H. and Agarwal, V. (2008). Maximum power point tracking scheme for PV systems operating under partially shaded conditions. *IEEE Transactions on Industrial Electronics* 55 (4): 1689–1698.
116. Renaudineau, H., Houari, A., Martin, J.P. et al. (2011). A new approach in tracking maximum power under partially shaded conditions with consideration of converter losses. *Solar Energy* 85 (11): 2580–2588.
117. Lei, P., Li, Y., and Seem, J.E. (2011). Sequential ESC-based global MPPT control for photovoltaic array with variable shading. *IEEE Transactions on Sustainable Energy* 2 (3): 348–358.
118. Karatepe, E. and Hiyama, T. (2009). Artificial neural network-polar coordinated fuzzy controller based maximum power point tracking control under partially shaded conditions. *IET Renewable Power Generation* 3 (2): 239–253.

119. Chen, L.R., Tsai, C.H., Lin, Y.L., and Lai, Y.S. (2010). A biological swarm chasing algorithm for tracking the PV maximum power point. *IEEE Transactions on Energy Conversion* 25 (2): 484–493.
120. Miyatake, M., Veerachary, M., Toriumi, F. et al. (2011). Maximum power point tracking of multiple photovoltaic arrays: a PSO approach. *IEEE Transactions on Aerospace and Electronic Systems* 47 (1): 367–380.
121. Chowdhury, S.R. and Saha, H. (2010). Maximum power point tracking of partially shaded solar photovoltaic arrays. *Solar Energy Materials & Solar Cells* 94 (9): 1441–1447.
122. Chowdhury, S.R., Chakrabarti, D., and Saha, H. (2009). Medical diagnosis using adaptive perceptive particle swarm optimization and its hardware realization using field programmable gate array. *Journal of Medical Systems* 33 (6): 447–465.
123. Ishaque, K., Salam, Z., Amjad, M., and Mekhilef, S. (2012). An improved particle swarm optimization (PSO)-based MPPT for PV with reduced steady-state oscillation. *IEEE Transactions on Power Electronics* 27 (8): 3627–3638.
124. Liu, Y.-H., Huang, S.-C., Huang, J.-W., and Liang, W.-C. (2012). A particle swarm optimization-based maximum power point tracking algorithm for PV systems operating under partially shaded conditions. *IEEE Transactions on Energy Conversion* 27 (4): 1027–1035.
125. Eberhart, R. and Kennedy, J. (1995). A new optimizer using particle swarm theory. *Proceedings of the Sixth International Symposium on Micro Machine and Human Science, 1995. MHS '95* (4–6 October 1995), pp. 39–43.
126. Department of Energy (2006). Benefits of demand response in electricity markets and recommendations for achieving them (February 2006). http://energy.gov/sites/prod/files/oeprod/DocumentsandMedia/DOE_Benefits_of_Demand_Response_in_Electricity_Markets_and_Recommendations_for_Achieving_Them_Report_to_Congress.pdf (accessed May 2015).
127. Faria, P., Soares, J., and Vale, Z. (2014). Quantum-based particle swarm optimization application to studies of aggregated consumption shifting and generation scheduling in smart grids. CIASG 2014 – Computational Intelligence Applications in Smart Grid (CIASG) at the IEEE SSCI 2014 (IEEE Symposium Series on Computational Intelligence), Orlando, FL, USA (9–12 December 2014).
128. FERC (2013). Assessment of demand response and advanced metering. https://www.ferc.gov/legal/staff-reports/2013/oct-demand-response.pdf (accessed April 2014).
129. European Commission (2013). Incorporing demand side flexibility, in particular demand response, in electricity markets. https://ec.europa.eu/energy/sites/ener/files/documents/com_2013_public_intervention_swd07_en.pdf (accessed 25 September 2019).
130. Saber, A. and Venayagamoorthy, G. (2012). Resource scheduling under uncertainty in a smart grid with renewables and plug-in vehicles. *IEEE Systems Journal* 6: 103–109.
131. Borlase, S. (ed.) (2013). *Smart Grids – Infrastructure, Technology, and Solutions*. CRC Press.
132. Ghazvini, M., Soares, J., Horta, N. et al. (2015). A multi-objective model for scheduling of short-term incentive-based demand response programs offered by electricity retailers. *Applied Energy* 151: 102–118.
133. Sun, J., Lai, C.-H., and Wu, X.-J. (2012). *Particle Swarm Optimisation: Classical and Quantum Perspectives*, Numerical Analysis and Scientific Computing Series. Chapman & Hall/CRC.

134. European Commission (2014). Benchmarking smart metering deployment in the EU-27 with a focus on electricity. https://ses.jrc.ec.europa.eu/publications/reports/benchmarking-smart-metering-deployment-eu-27-focus-electricity (accessed 25 September 2019).
135. SEDC (2014). Mapping demand response in Europe Today – tracking compliance with Article 15.8 of the energy efficiency directive (April 2014). http://www.smarten.eu/wp-content/uploads/2014/04/SEDC-Mapping_DR_In_Europe-2014-04112.pdf (accessed 25 September 2019).
136. NAVIGANT (2015). Demand response spending is expected to exceed $ 1.3 billion a year by 2024 (May 2015). https://www.peakload.org/demand-response-spending-is-expected-to-exceed-1-3-billion-a-year-by-2024 (accessed 25 September 2019).
137. PNNL (2014). Reference guide for a transaction-based building controls framework (April 2014). http://energy.gov/sites/prod/files/2014/06/f16/PNNL-23302_draft.pdf (accessed May 2015).
138. Kilkki, O., Alahaivala, A., and Seilonen, I. (2015). Optimized control of price-based demand response with electric storage space heating. *IEEE Transactions on Industrial Informatics* 11 (1): 281–288.
139. OPOWER (2014). Transform every customer into a demand response resource (October 2014). http://www.so-ups.ru/fileadmin/files/company/markets/dr/docs/dr_opower_white_pap.pdf (accessed June 2015).
140. Arnold, R. (2008). *Economics*, 8e. USA: Thomson.
141. Nguyen, D., Negnevitsky, M., and Groot, M. (2011). A new and separated market for demand response in restructured power systems. *Power Systems Conference and Exposition (PSCE)* (20–23 March 2011).
142. European Commission (2010). European SmartGrids Technology Platform – vision and strategy for Europe's electricity networks of the future. European Communities (April 2010).
143. Morais, H., Faria, P., and Vale, Z. (2014). Demand response design and use based on network locational marginal prices. *International Journal of Electrical Power & Energy Systems* 61: 180–191.
144. Fernandes, F., Morais, H., Vale, Z., and Ramos, C. (2014). Dynamic load management in a smart home to participate in demand response events. *Energy and Buildings* 82: 592–606.
145. Karangelos, E. and Bouffard, F. (2012). Towards full integration of demand-side resources in joint forward energy/reserve electricity markets. *IEEE Transactions on Power Systems* 27 (1): 280–289.
146. Woo, C. and Greening, L. (2010). Guest editors' introduction special issue on demand response resources: the US and international experience. *Energy* 35 (4): 1515–1517.
147. Faria, P. and Vale, Z. (2011). Demand response in electrical energy supply: an optimal real time pricing approach. *Energy* 36: 5374–5384.
148. Faria, P. and Vale, Z. (2014). Remuneration structure definition for distributed generation units and demand response participants aggregation. *2014 IEEE Power and Energy Society General Meeting*, National Harbor, USA (27–31 July 2014).
149. Faria, P., Soares, J.P., Vale, Z. et al. (2013). Modified particle swarm optimization applied to integrated demand response and DG resources scheduling. *IEEE Transactions on Smart Grid* 4 (1): 606–616.
150. Faria, P., Vale, Z., Soares, J., and Ferreira, J. (2013). Demand response management in power systems using a particle swarm optimization approach. *IEEE Intelligent Systems* 28 (4): 43–51.

151. Faria, P., Soares, T., Pinto, T., et al. (2013). Dispatch of distributed energy resources to provide energy and reserve in smart grids using a particle swarm optimization approach. *CIASG 2013 – Computational Intelligence Applications in Smart Grid (CIASG) at the IEEE SSCI 2013 (IEEE Symposium Series on Computational Intelligence)*, Singapore (15–19 April 2013).
152. Faria, P., Soares, J., and Vale, Z. (2015). Definition of the demand response events duration using differential search algorithm for aggregated consumption shifting and generation scheduling. *Intelligent System Applications to Power Systems, ISAP 2015*, Porto, Portugal (11–16 September 2015).
153. Faria, P. and Vale, Z. (2013). Demand response programs definition using demand price elasticity to define consumers aggregation for an improved remuneration structure. *4th IEEE PES ISGT Europe* (October 2013), pp. 7–10.
154. Moreno-Munoz, A., de la Rosa, J., Posadillo, R. and Pallares, V. (2008). Short term forecasting of solar radiation. *Proceedings of IEEE ISIE 2008*, Cordoba, Argentina (July 2008), pp. 1537–1541.
155. Hassanzadeh, M., Etezadi-Amoli, M. and Fadali, M.S. (2010). Practical approach for sub-hourly and hourly prediction of PV power output. *Proceedings of NAPS 2010*, Arlington, TX, USA (September 2010), pp. 1–5.
156. Goh, T.N. and Tan, K.J. (1977). Stochastic modeling and forecasting of solar radiation data. *Solar Energy* 19: 755–757.
157. Reikard, G. (2009). Predicting solar radiation at high resolutions: a comparison of time series forecasts. *Solar Energy* 83: 342–349.
158. Mekki, H., Mellit, A., Salhi, H. and Khaled, B. (2007). Modeling and simulation of photovoltaic panel based on artificial neural networks and VHDL-language. *Proceedings of IEEE ICECS 2007*, Marrakech, Morocco (July 2007), pp. 58–61.
159. Ghanbarzadeh, A., Noghrehabadi, A.R., Assareh, E. and Behrang, M.A. (2009). Solar radiation forecasting based on meteorological data using artificial neural network. *Proceedings of IEEE INDIN 2009*, Cardiff, MEC (June 2009), pp. 227–231.
160. Yu, T.-C. and Chang, H.-T. (2011). The forecast of the electrical energy generated by photovoltaic systems using neural network method. *Proceedings of IEEE ICEICE 2011*, Wuhan, China (April 2011), pp. 2758–2761.
161. Mellit, A., Menghanem, M., and Bendekhis, M. (2005). Artificial neural network model for prediction solar radiation data: application for sizing stand-alone photovoltaic power system. *Proceedings of IEEE PES GM 2005*, Hilton, CA, USA 1: 40–44.
162. Yona, A., Senju, T., Saber, A.T. et al. (2007). Application of neural network to one-day-ahead 24 hours generating power forecasting for photovoltaic system. *Proceedings of IEEE ISAP 2007*, Kaohsiung, Taiwan (November 2007), pp. 1–6.
163. Yona, A., Senju, T., Saber, A.T. et al. (2008). Application of neural network to 24-hour-ahead generating power forecasting for PV system. *Proceedings of IEEE PES GM 2008*, Pittsburgh, PA, USA (July 2008), pp. 1–6.
164. Box, G.E.P. and Jenkins, G.M. (1970). *Time Series Analysis; Forecasting and Control*. San Francisco, CA: Holden Day.
165. Powell, M.J.D. (1987). Radial basis functions for multivariable interpolation: a review. In: *Algorithm for Approximation of Functions and Data* (eds. J.C. Mason and M.G. Cox), 143–167. Oxford, UK: Oxford University Press.
166. Bishop, C.M. (1995). *Neural Networks for Pattern Reorganization*. Oxford University Press.
167. Poggio, T. and Girosi, F. (1990). Networks for approximation and learning. *Proceedings of the IEEE* 78 (9): 1481–1497.

168. Miyazaki, A. and Yamada, T. (1993). A systematic synthesis procedure for feedforward neural networks by using the GRBF (generalized radial basis function) network technique. *Proceedings of IEEE IJCNN93*, Nagoya, Japan (October 1993), pp. 363–366.
169. Bishop, C.M. (2006). *Pattern Recognition and Machine Learning*. New York, USA: Springer.
170. Rose, K., Grurewits, E., and Fox, G.C. (1990). A deterministic annealing approach to clustering. *Pattern Recognition Letters* 11 (9): 589–594.
171. Mori, H. and Yuihara, A. (2001). Deterministic annealing clustering for ANN-based short-term load forecasting. *IEEE Transactions on Power Systems* 16 (3): 545–551.
172. Mori, H., Aoyama, H., Yamanaka, T., and Urano, S. (2002). A fault detection technique with preconditioned ANN in power systems. *Proceedings of IEEE PES T&D 2002*, Yokohama, Japan 2: 758–763.
173. Hinton, G.E. (1987). Learning translation invariant recognition in a massively parallel network. *Proceedings PARLE Conference on Parallel Architectures and Languages Europe*, Berlin, pp. 1–13.
174. Miranda, V. and Fonseca, N. (2002). EPSO-best-of-two-worlds meta-heuristic applied to power system problems. *Proceedings of IEEE CEC 2002*, Honolulu, HI, USA (March 2002), pp. 1080–1085.
175. Miranda, V. and Fonseca, N. (2002). EPSO-evolutionary particle swarm optimization, a new algorithm with applications in power systems. *Proceedings of IEEE PES T&D 2002*, Yokohama, Japan 2: 745–750.
176. Moody, J. and Darken, C.J. (1989). Fast learning in networks of locally-tuned processing units. *Neural Computation* 1: 281–289.
177. MacQueen, J.B. (1967). Some methods for classification and analysis of multivariate observations. *Proceedings of 5th Berkeley Symposium on Mathematical Statistics and Probability* 1: 281–297.
178. Kennedy, J. and Eberhart, R. (1995). Particle swarm optimization. *Proceedings of IEEE International Conference on Neural Networks*, Perth, Australia IV: 1942–1948.
179. Sharma, A., Saxena, A., Sethi, M., and Shree, V. (2011). Life cycle assessment of buildings: a review. *Renewable and Sustainable Energy Reviews* 15 (1): 871–875.
180. AEMO (2016). National Electricity Forecasting Report.
181. Houghton, J., Ding, Y., Griggs, D.J. et al. (2001). *The scientific basis. IPCC Intergovernmental Panel on Climate Change. Contribution of Working Group 1 to the IPCC, Third Assessment*. Cambridge, UK: Cambridge University Press.
182. Solomon, S., Qin, D., Manning, M. et al. (2007). *Contribution of Working Group I to the Fourth Assessment Report of the Intergovernmental Panel on Climate Change*. Cambridge, UK and New York, NY, USA: Cambridge University Press.
183. Cabeza, L.F., Rincón, L., Vilariño, V. et al. (2014). Life cycle assessment (LCA) and life cycle energy analysis (LCEA) of buildings and the building sector: a review. *Renewable and Sustainable Energy Reviews* 29: 394–416.
184. Rana, M., Koprinska, I., and Agelidis, V.G. (2016). Univariate and multivariate methods for very short-term solar photovoltaic power forecasting. *Energy Conversion and Management* 121: 380–390.
185. REN21 (2017). Global Status Report, REN21 Secretariat, Paris, France. Technical Report.
186. Victor, M.J. and Ruelle, D. (2016). Rooftop PV Model Technical Report. The University of Melbourne.

187. Raza, M.Q., Nadarajah, M., and Ekanayake, C. (2016). On recent advances in PV output power forecast. *Solar Energy* 136: 125–144.
188. Dixon, R.K., McGowan, E., Onysko, G., and Scheer, R.M. (2010). US energy conservation and efficiency policies: challenges and opportunities. *Energy Policy* 38 (11): 6398–6408.
189. Helman, U., Loutan, C., Rosenblum, G. et al. (2010). Integration of renewable resources: operational requirements and generation fleet capability at 20% rps. California Independent System Operator.
190. Rodriguez, G.D. (2010). A utility perspective of the role of energy storage in the smart grid. *IEEE PES General Meeting*, pp. 1–2.
191. Raza, M.Q. and Khosravi, A. (2015). A review on artificial intelligence based load demand forecasting techniques for smart grid and buildings. *Renewable and Sustainable Energy Reviews* 50: 1352–1372.
192. Soman, S.S., Zareipour, H., Malik, O., and Mandal, P. (2010). A review of wind power and wind speed forecasting methods with different time horizons. *North American Power Symposium (NAPS)*, pp. 1–8.
193. Li, S., Goel, L., and Wang, P. (2016). An ensemble approach for short-term load forecasting by extreme learning machine. *Applied Energy* 170: 22–29.
194. Siwek, K. and Osowski, S. (2012). Improving the accuracy of prediction of PM10 pollution by the wavelet transformation and an ensemble of neural predictors. *Engineering Applications of Artificial Intelligence* 25 (6): 1246–1258.
195. Pelland, S., Galanis, G., and Kallos, G. (2013). Solar and photovoltaic forecasting through post-processing of the global environmental multiscale numerical weather prediction model. *Progress in Photovoltaics: Research and Applications* 21 (3): 284–296.
196. Farzana, S., Liu, M., Baldwin, A., and Hossain, M.U. (2014). Multi-model prediction and simulation of residential building energy in urban areas of Chongqing, South West China. *Energy and Buildings* 81: 161–169.
197. Monfet, D., Corsi, M., Choinière, D., and Arkhipova, E. (2014). Development of an energy prediction tool for commercial buildings using case-based reasoning. *Energy and Buildings* 81: 152–160.
198. Jetcheva, J.G., Majidpour, M., and Chen, W.-P. (2014). Neural network model ensembles for building-level electricity load forecasts. *Energy and Buildings* 84: 214–223.
199. Fan, S. and Chen, L. (2006). Short-term load forecasting based on an adaptive hybrid method. *IEEE Transactions on Power Systems* 21 (1): 392–401.
200. Lu, W.-Z. and Xue, Y. (2014). Prediction of particulate matter at street level using artificial neural networks coupling with chaotic particle swarm optimization algorithm. *Building and Environment* 78: 111–117.
201. Raza, M.Q. and Baharudin, Z. (2012). A review on short term load forecasting using hybrid neural network techniques. *2012 IEEE International Conference on Power and Energy (PECon)*, pp. 846–851.
202. Prema, V. and Rao, K.U. (2015). Development of statistical time series models for solar power prediction. *Renewable Energy* 83: 100–109.
203. Fernandez-Jimenez, L.A., Muñoz-Jimenez, A., Falces, A. et al. (2012). Short-term power forecasting system for photovoltaic plants. *Renewable Energy* 44: 311–317.
204. Salcedo-Sanz, S., Casanova-Mateo, C., Pastor-Sánchez, A., and Sánchez-Girón, M. (2014). Daily global solar radiation prediction based on a hybrid coral reefs optimization – extreme learning machine approach. *Solar Energy* 105: 91–98.

205. Chen, C., Duan, S., Cai, T., and Liu, B. (2011). Online 24-h solar power forecasting based on weather type classification using artificial neural network. *Solar Energy* 85 (11): 2856–2870.
206. Liu, J., Fang, W., Zhang, X., and Yang, C. (2015). An improved photovoltaic power forecasting model with the assistance of aerosol index data. *IEEE Transactions on Sustainable Energy* 6 (2): 434–442.
207. Rana, M., Koprinska, I., and Agelidis, V.G. (2015). 2D-interval forecasts for solar power production. *Solar Energy* 122: 191–203.
208. Ghayekhloo, M., Ghofrani, M., Menhaj, M.B., and Azimi, R. (2015). A novel clustering approach for short-term solar radiation forecasting. *Solar Energy* 122: 1371–1383.
209. Raza, M.Q., Nadarajah, M., and Ekanayake, C. (2016). An improved neural ensemble framework for accurate PV output power forecast. *2016 Australasian Universities Power Engineering Conference (AUPEC)*, pp. 1–6.
210. Raza, M.Q., Nadarajah, M., and Ekanayake, C. (2016). An improved WT and NN ensemble demand forecast model for PV integrated smart buildings. *2016 IEEE Innovative Smart Grid Technologies – Asia (ISGT-Asia)*, pp. 781–786.
211. Li, S., Wang, P., and Goel, L. (2015). Short-term load forecasting by wavelet transform and evolutionary extreme learning machine. *Electric Power Systems Research* 122: 96–103.
212. Catalao, J., Pousinho, H., and Mendes, V. (2011). Hybrid wavelet-PSO-ANFIS approach for short-term electricity prices forecasting. *IEEE Transactions on Power Systems* 26 (1): 137–144.
213. Mandal, P., Haque, A.U., Madhira, S.T., and Al-Hakeem, D.I. (2013). Applying wavelets to predict solar PV output power using generalized regression neural network. In: *North American Power Symposium (NAPS)*, 1–5. IEEE.
214. Hassan, S., Khosravi, A., and Jaafar, J. (2015). Examining performance of aggregation algorithms for neural network-based electricity demand forecasting. *International Journal of Electrical Power & Energy Systems* 64: 1098–1105.
215. Raza, M.Q., Nadarajah, M., and Ekanayake, C. (2016). An improved WT and NN ensemble demand forecast model for PV integrated smart buildings. In: *Innovative Smart Grid Technologies-Asia (ISGT-Asia), 2016 IEEE*, 781–786. IEEE.
216. Hassan, S., Khosravi, A., Jaafar, J., and Belhaouari, S.B. (2012). Load forecasting accuracy through combination of trimmed forecasts. In: *Neural Information Processing: 19th International Conference, ICONIP 2012, Doha, Qatar, November 12–15, 2012, Proceedings, Part I* (eds. T. Huang, Z. Zeng, C. Li and C.S. Leung), 152–159. Berlin, Heidelberg: Springer.
217. Muhammad Qamar Raza, M.N. and Ekanayake, C. (2017). A multivariate ensemble framework for short term solar photovoltaic output power forecast. Presented at the 2017 IEEE Power & Energy Society General Meeting, Chicago, USA.
218. Haque, A.U., Nehrir, M.H., and Mandal, P. (2014). A hybrid intelligent model for deterministic and quantile regression approach for probabilistic wind power forecasting. *IEEE Transactions on Power Systems* 29 (4): 1663–1672.
219. Raza, M.Q., Nadarajah, M., and Ekanayake, C. (2017). Demand forecast of PV integrated bioclimatic buildings using ensemble framework. *Applied Energy* 208: 1626–1638.
220. Raza, M.Q., Mithulananthan, N., Li, J. et al. (2019). An ensemble framework for day-ahead forecast of PV output power in smart grids. *IEEE Transactions on Industrial Informatics* 15 (8): 4624–4634.

221. Haque, A.U., Nehrir, M.H., and Mandal, P. (2013). Solar PV power generation forecast using a hybrid intelligent approach. In: *Power and Energy Society General Meeting (PES), 2013 IEEE*, 1–5. IEEE.
222. Hatton, C.E., Beella, S.K., Brezet, J.C., and Wijnia, Y.C. (2009). Charging station for urban settings: the design of a product platform for electric vehicle infrastructure in Dutch cities. *World Electric Vehicle Journal* (3): 1–13.
223. Benysek, G. and Jarnut, M. (2012). Electric vehicle charging infrastructure in Poland. *Renewable and Sustainable Energy Reviews* 16 (1): 320–328.
224. Li, Z. and Ouyang, M.G. (2011). The pricing of charging for electric vehicles in China-dilemma and solution. *Energy* 36: 5765–5778.
225. Zhou, H. and Li, H. (2011). Optimization model of electric vehicle charging station sitting based on game theory. *Science Technology and Industry* 11 (2): 51–54.
226. Feng, L., Ge, S., Liu, H. et al. (2012). The planning of charging stations on the urban trunk road. *Proceedings of 2012 IEEE International Conference on Innovative Smart Grid Technologies – Asia (ISGT Asia)*, Tianjin, pp. 1–4.
227. Jia, L., Hu, Z., Song, Y., and Luo, Z. (2012). Optimal siting and sizing of electric vehicle charging stations. *Proceedings of 2012 IEEE International Conference on Electric Vehicle Conference (IEVC-2012)*, Greenville, SC, pp. 1–6.
228. Mehar, S. and Senouci, S.M. (2013). An optimization location scheme for electric charging stations. *Proceedings of 2013 International Conference on Smart Communications in Network Technologies (SaCoNeT)*, Paris 6: 17–19.
229. Cui, S., Ai, X., Dong, R. et al. (2014). Electric vehicle charging station planning based on sensitivity analysis. *Proceedings of 2014 IEEE Conference and Expo on Transportation Electrification Asia-Pacific (ITEC Asia-Pacific)*, Beijing, pp. 1–5.
230. Shi, R., Zheng, S., Zhang, C., and Zhang, L. (2014). Study on EV charging station location planning based on the load balance principle with agent-based AnyLogic simulation. *Proceedings of the 26th Chinese Control and Decision Conference (2014 CCDC)*, Changsha, pp. 1515–1519.
231. Lam, A.Y.S., Leung, Y.W., and Chu, X. (2014). Electric vehicle charging station placement: formulation, complexity, and solutions. *IEEE Transactions on Smart Grid* 5 (6): 2846–2856.
232. Kim, S., Jung, H., Hwang, J. et al. (2013). A study on the construction of EV charging infrastructures in highway rest area. *Proceedings of 2013 IEEE Fourth International Conference on Power Engineering, Energy and Electrical Drives (POWERENG)*, Istanbul, pp. 396–400.
233. Crosier, R. and Wang, S. (2013). Modeling the closed-loop interaction between the grid and a multifunctional electric vehicle charging station integrated with an active power filter. *Proceedings of 2013 IEEE International Conference on Electric Vehicle Conference (IEVC-2013)*, Santa Clara, CA, pp. 1–7.
234. Xu, H., Miao, S., Zhang, C., and Shi, D. (2013). Optimal placement of charging infrastructures for large-scale integration of pure electric vehicles into grid. *International Journal of Electrical Power & Energy Systems* 53: 159–165.
235. Zhang, L., Brown, T., and Samuelsen, S. (2013). Evaluation of charging infrastructure requirements and operating costs for plug-in electric vehicles. *Journal of Power Sources* 240: 515–524.
236. Leeprechanon, N., Phonrattanasak, P., and Sharma, M.K. (2016). Optimal planning of public fast charging station on residential power distribution system. *Proceedings of 2016 IEEE Conference and Expo on Transportation Electrification, Asia-Pacific (ITEC Asia-Pacific)*, Busan, pp. 519–524.

237. Schroeder, A. and Traber, T. (2012). The economics of fast charging infrastructure for electric vehicles. *Energy Policy* 43: 136–144.
238. Sheppard, C.J.R., Harris, A., and Gopal, A.R. (2016). Cost-effective siting of electric vehicle charging infrastructure with agent-based modeling. *IEEE Transactions on Transportation Electrification* 2 (2): 174–189.
239. Luo, C., Huang, Y.F., and Gupta, V. (2015). A consumer behavior based approach to multi-stage EV charging station placement. *Proceedings of 2015 IEEE 81st Vehicular Technology Conference (VTC Spring)*, Glasgow, pp. 1–6.
240. Yao, W., Zhao, J., Wen, F. et al. (2014). A multi-objective collaborative planning strategy for integrated power distribution and electric vehicle charging systems. *IEEE Transactions on Power Systems* 7: 1811–1821.
241. Lee, K.Y. and El-Sharkawi, M.A. (eds.) (2008). *Modern Heuristic Optimization Techniques with Applications to Power Systems*. New York: Wiley-IEEE Press.
242. Zitzler, E.L., Laumanns, M., and Thiele, L. (2001). SPEA2: improving the strength pareto evolutionary algorithm. Technical report TIK-Report 103, Swiss Federal Institute of Technology.
243. Zhao, J.H., Wen, F., Dong, Z.Y. et al. (2012). Optimal dispatch of electric vehicles and wind power using enhanced particle swarm optimization. *IEEE Transactions on Industrial Informatics* 8: 889–899.
244. Hetzer, J., Yu, D.C., and Bhattarai, K. (2008). An economic dispatch model incorporating wind power. *IEEE Transactions on Energy Conversion* 23: 603–611.
245. Weihao, H., Chi, S., Zhe, C., and Bak-Jensen, B. (2013). Optimal operation of plug-in electric vehicles in power systems with high wind power penetrations. *IEEE Transactions on Sustainable Energy* 4: 577–585.
246. Pillai, J.R. and Bak-Jensen, B. (2011). Integration of vehicle-to-grid in the western Danish power system. *IEEE Transactions on Sustainable Energy* 2: 12–19.
247. Lopez, S., Caicedo, J., Mamani, M. et al. (2014). Literature review: potential impacts of plug-in electric vehicles on electric power systems. *2014 IEEE PES Transmission & Distribution Conference and Exposition – Latin America (PES T&D-LA)*, pp. 1–6.
248. Yu, H.J., Gu, W., Zhang, N., and Lin, D.Q. (2012). Economic dispatch considering integration of wind power generation and mixed-mode electric vehicles. *2012 IEEE Power and Energy Society General Meeting*, pp. 1–7.
249. Khatamianfar, A., Khalid, M., Savkin, A.V., and Agelidis, V.G. (2013). Improving wind farm dispatch in the Australian electricity market with battery energy storage using model predictive control. *IEEE Transactions on Sustainable Energy* 4: 745–755.
250. Seyedmahmoudian, M., Rahmani, R., Mekhilef, S. et al. (2015). Simulation and hardware implementation of new maximum power point tracking technique for partially shaded PV system using hybrid DEPSO method. *IEEE Transactions on Sustainable Energy* 6: 850–862.
251. Binetti, G., Davoudi, A., Naso, D. et al. (2014). A distributed auction-based algorithm for the nonconvex economic dispatch problem. *IEEE Transactions on Industrial Informatics* 10: 1124–1132.
252. Sedaghati, R. and Namdari, F. (2015). An intelligent approach based on meta-heuristic algorithm for non-convex economic dispatch. *Journal of Operation and Automation in Power Engineering* 3: 47–55.
253. Saraswat, A. and Saini, A. (2011). Optimal reactive power dispatch by an improved real coded genetic algorithm with PCA mutation. *IET Conference Publications*, pp. 310–315.

254. Huang, C.M., Chen, S.J., Huang, Y.C., and Yang, H.T. (2012). Comparative study of evolutionary computation methods for active–reactive power dispatch. *IET Generation Transmission and Distribution* 6: 636–645.
255. Ma, X.Y. and Wei, W. (2014). Research on power system stochastic economic dispatch considering interruptible load and wind power. *Advanced Materials Research* 986–987: 493–497.
256. Jin, J., Zhou, D., Zhou, P., and Miao, Z. (2014). Environmental/economic power dispatch with wind power. *Renewable Energy* 71: 234–242.
257. Thirugnanam, K., Ezhil Reena, T.P., Joy, M.S., and Kumar, P. (2014). Mathematical modeling of li-ion battery using genetic algorithm approach for V2G applications. *IEEE Transactions on Energy Conversion* 29: 332–343.
258. Zhou, C., Qian, K., Allan, M., and Zhou, W. (2011). Modeling of the cost of EV battery wear due to V2G application in power systems. *IEEE Transactions on Energy Conversion* 26: 1041–1050.
259. Chang, T.P. (2011). Performance comparison of six numerical methods in estimating Weibull parameters for wind energy application. *Applied Energy* 88: 272–282.
260. Arslan, T., Bulut, Y.M., and Altın Yavuz, A. (2014). Comparative study of numerical methods for determining Weibull parameters for wind energy potential. *Renewable and Sustainable Energy Reviews* 40: 820–825.
261. Safari, B. and Gasore, J. (2010). A statistical investigation of wind characteristics and wind energy potential based on the Weibull and Rayleigh models in Rwanda. *Renewable Energy* 35: 2874–2880.
262. Kadurek, P., Ioakimidis, C., and Ferrao, P. (2009). Electric vehicles and their impact to the electric grid in isolated systems. *International Conference on Power Engineering, Energy and Electrical Drives*, pp. 49–54.
263. Fluhr, J., Ahlert, K.-H., and Weinhardt, C. (2010). A stochastic model for simulating the availability of electric vehicles for services to the power grid. *43rd Hawaii International Conference on System Sciences*, pp. 1–10.
264. Otiniano, C.E.G., Sousa, T.R., and Rathie, P.N. (2013). Stable random variables: convolution and reliability. *Journal of Computational and Applied Mathematics* 242: 1–11.
265. Alves, R. (2013. Stochastic location of FACTS devices in electric power transmission networks. MS thesis, Department of Electrical Engineering, Universidade do Porto, Portugal.
266. Miranda, V. and Alves, R. (2013). Differential evolutionary particle swarm optimization (DEEPSO): a successful hybrid. *1st BRICS Countries Congress on Computational Intelligence*, Recife, pp. 368–374.
267. Erlich, I., Rueda, J.L., Wildenhues, S., and Shewarega, F. (2014). Solving the IEEE-CEC 2014 expensive optimization test problems by using single-particle MVMO. *IEEE Congress on Evolutionary Computation*, pp. 1084–1091.
268. Nakawiro, W., Erlich, I., and Rueda, J.L. (2011). A novel optimization algorithm for optimal reactive power dispatch: a comparative study. *4th International Conference on Electric Utility Deregulation and Restructuring and Power Technologies*, pp. 1555–1561.
269. Jia, G., Wang, Y., Cai, Z., and Jin, Y. (2013). An improved ($\mu + \lambda$)-constrained differential evolution for constrained optimization. *Information Sciences* 222: 302–322.
270. Wang, Y. and Runarsson, T.P. (2011). Evolutionary algorithms for complex continuous optimization problems. PhD thesis, School of Information Science and Engineering, Central South University, South Corea.

271. Wang, Y. and Cai, Z. (2011). Constrained evolutionary optimization by means of ($\mu\mu + \lambda\lambda$) – differential evolution and improved adaptive trade-off model. *Evolutionary Computation* 19: 249–285.
272. Romero, R., Rider, M.J., and Silva, I.d.J. (2007). A metaheuristic to solve the transmission expansion planning. *IEEE Transactions on Power Systems* 22: 2289–2291.
273. Correa, C.A., Bolaños, R.A., and Escobar, A.H. (2014). Multi-objective transmission expansion planning considering multiple generation scenarios. *International Journal of Electrical Power & Energy Systems* 62: 398–409.
274. Birchfield, A.B., Xu, T., Gegner, K.M. et al. (2017). Grid structural characteristics as validation criteria for synthetic networks. *IEEE Transactions on Power Systems* 32 (4): 3258–3265.

CHAPTER 7

ELECTRICITY MARKETS

7.1 INTRODUCTION

Zita A. Vale and Tiago Pinto

Polytechnic of Porto, Porto, Portugal

Societies are highly dependent on electricity use to ensure safe, reliable, and comfortable living. The increase of electricity demand is expected to continue in the future and it is considered a crucial requirement for economic development. Concerns about the impact of electricity use in the environment and about the eventual fuel-based primary source shortage are presently taken as very serious at scientific, economic, and political levels [1]. These concerns have led to intensive research and to new energy policies envisaging the increased use of renewable energy sources for electricity production and increased energy use efficiency.

The European Union (EU) has assumed a pioneer and leading role in energy matters, namely in what concerns the increase of renewable energy sources. EU as a whole has committed to reach its 20% renewable energy target for 2020. This target considers Member States' different starting points and potential for increasing renewables production, which range from 10% in Malta to 49% in Sweden [2]. Moreover, on 23 October 2014, EU leaders agreed on setting a revised target for increasing the share of renewable-based energy to at least 27% of the EU's energy consumption by 2030 [3]. The EU presents even more ambitious targets for 2050, with the commitment to reduce emissions to 80–95% below 1990 levels [4]. As a consequence of these policies and of the subsequent incentives that have been put in place, huge investments have been made in renewable-based electricity generation plants and equipment. However, increased renewable-based generation capacity does not directly ensure a corresponding increase in renewable-based energy use as several constraints limit not only the production but also its use. Wind- and solar-based generation are dependent on natural sources and are not

Applications of Modern Heuristic Optimization Methods in Power and Energy Systems,
First Edition. Edited by Kwang Y. Lee and Zita A. Vale.
© 2020 by The Institute of Electrical and Electronics Engineers, Inc.
Published 2020 by John Wiley & Sons, Inc.

as dispatchable as fuel-based thermal plants. Power systems face new challenges to deal with the integration of intermittent renewable sources [5]. Significant research work has already been done around these issues, producing valuable results but also evidencing the limitations of the approaches that are being used.

The electricity sector restructuring [6] aimed at obtaining public benefits, increasing the efficiency of the sector by providing consumers with reliable high-quality service at fair costs. This should be achieved by introducing a competitive market-based approach to replace the centralized, monopolistic, and/or state-owned paradigm that traditionally ruled the sector. Although the transition of some segments of the market toward a competitive market is being successful, some remain as a regulated monopoly (namely transmission and distribution). The changes were particularly difficult for a sector in which old and new big players want to take the lead, technical and economic factors are more closely interrelated than in most sectors, and for which storage capacity is still at a very low level due to the high-cost solutions that are currently available. Prices that, in many cases, do not reflect the costs and the lack of experience in a field for which the sector particularities make prices' behavior significantly different from already existing markets transacting other commodities and products marked the reform departing point and its subsequent evolution [7].

This restructuring process has been made of successes and failures, some of the latter with serious consequences, such as the so-called California's electricity crisis of 2000–2001, the 14 August 2003 blackout in the United States and the 4 November 2006 quasi-blackout affecting nine European countries and some African nations as well [8]. Such experiences are leading to successive model and rule changes. Ultimately, wholesale electricity markets are finally proving to be able to accomplish their goals even if only partially, and the reforms made at local and/or national level are being extended giving place to larger markets at the regional/international level, rapidly evolving to a coupling trend at the continental level. A relevant example is the case of the EU, whose policy aims at establishing the internal electricity market in Europe [9]. Significant steps have been undertaken in this direction with the day-ahead wholesale electricity markets of 19 EU countries presently price coupled allowing the simultaneous calculation of electricity prices and cross-border flows across a region accounting for 85% of European power consumption [10]. That achievement has been enabled by the Multi-Regional Coupling (MRC), a pan-European initiative dedicated to the integration of power spot markets in Europe. This is a major achievement, in a market type (day-ahead) that usually accommodates most of the energy transaction in Europe. Balancing markets and intraday markets are the next focus, enabling the trading of electricity in a time horizon nearer to real time, which is essential in order to cope with the increasing penetration of (variable) renewable energy sources and demand flexibility trading.

In such a dynamic, complex, and competitive environment as the power and energy sector, simulation and decision support tools are of crucial importance. Market players and regulators are very interested in foreseeing market behavior: regulators to test rules before they are implemented and to detect market inefficiencies; market players to understand market's behavior and act in order to maximize their results from market participation. The need for understanding those mechanisms and how the involved players' interaction affects the outcomes of the markets contributed to the growth of usage of simulation tools. Moreover, the opportunity to reach improved market results brings out the quest for adaptive tools, able to provide effective support to market negotiating players. Such tools provide the means for an actual improvement in players' market results. By using intelligent tools, capable of adapting to different market circumstances and negotiating contexts, players can change their behavior in a real market environment, and therefore pursuit the achievement of the best possible outcomes. These strategic approaches and how they apply to a market-based environment are explored in the following sections of this chapter. In specific, the remainder of the chapter is organized as follows. Section 7.2 addresses the topic of bidding strategies, which is essential for market players to enhance their market participation outcomes. Section 7.3 focuses on market analysis and clearing, thus providing the insight on how the market price is calculated and what affects the definition of prices and accepted and refused bids; this is accomplished through electricity market simulation. Section 7.4 addresses the topic of electricity market forecasting, which is a crucial aspect in current and future energy markets due to the participation of uncertain sources in both generation and consumption. Finally, Section 7.5 covers two significant topics in the electricity market domain: electric vehicles (EV) and ancillary services, which are some of the most promising assets in future markets, due to their capability of providing dynamic services to the grid.

7.2 BIDDING STRATEGIES

Tiago Pinto and Zita A. Vale
Polytechnic of Porto, Porto, Portugal

7.2.1 Introduction

Competitive electricity markets have produced relevant changes in the power industry, which has resulted in complex environments whose internal dynamics result from the behavior of a large number of interacting players of diverse nature. Each player has its own goals and should use adequate strategies in order to pursuit those goals, its strategic behavior being determinant for its success.

Players' strategies are of crucial importance for the operation of electricity markets and are determined by market models, including auction mechanisms, and by the overall context of supply and demand. Deciding on a player best strategy is a complex problem because decisions must be taken in face of incomplete information. This incompleteness results from the fact that market conditions are determined by the combined behavior of all the involved players. A player behavior exhibits changes in response to new information and knowledge; this may refer to its self-knowledge, to knowledge coming from the exterior and from the dynamic complex interactions of the heterogeneous individual entities. Each agent has only partial knowledge of other agents and makes its own decisions based on its partial knowledge of the system.

There has been considerable research on the topic of optimal bidding strategies, with most of the works addressing the problem from the producers' point of view. In 2000, David and Wen published a survey on this topic [11]. The first works in this area address the problem using operational research techniques and game theory. In order to cope with the complexity of the problem, several simplifications are usually assumed, such as neglecting demand variations and active demand participation, assuming that the final market clearing price is not dependent from the players' bids, considering single negotiation periods, limited time horizons, and a very limited number of players. Most works implicitly assume that each player has complete information about the other players so that the proposed methodologies lead to a deterministic optimal solution.

In recent years, artificial intelligence-based methods, including heuristic optimization, have been used to address the problem of electricity market players' strategic behavior. These approaches allow more realistic modeling of the problem considering a larger number of players and larger time horizons as well as the need of players' strategies evolution over time.

In 1999, Richter et al. [12] reported a work in which trading players used genetic algorithms (GA) to evolve appropriate bidding strategies for the current market conditions.

Multi-Agent Simulator of Competitive Electricity Markets (MASCEM) [13], a multi-agent based electricity market simulator, uses a combination of intelligent techniques to address agents' strategic behavior. All the relevant business opportunities, including bilateral contracts, energy day-ahead market, settlements and derivatives markets, ancillary services market, and derivative markets to hedge the risk, are considered [14, 15]. MASCEM combines methodologies for strategic bidding with dynamic behavior strategies able to take advantage from the knowledge concerning past experience and other players [16]. Using data-mining techniques, these data are used for knowledge discovery and machine learning, improving player models, and generating sophisticated bidding strategies. A hybrid approach is used together with scenario analysis to significantly increase the profits of the players. GA and swarm intelligence (SI) are used to help determining the most suitable bid set, depending on the players' goals and his aversion to risk, undertaking a detailed analysis in the neighborhood of the scenario selected by the game theory algorithm.

Fuzzy adaptive particle swarm optimization (FAPSO) is used in [17] to determine an optimal bidding strategy of a thermal generator for each trading period in a day-ahead market. Fuzzy evaluation is used to dynamically adjust inertia weight of the particle swarm optimization (PSO).

Two PSO algorithms are used in [18] to determine bid prices and quantities in a competitive power market. Opponents' bidding behavior is modeled with probabilistic estimation. Case studies allow demonstrating that, for nonlinear cost functions, PSO solutions provide higher expected profits than marginal cost-based bidding. This allows following the frequently changing conditions in the successive trading sessions of a real electricity market.

7.2.2 Context Analysis

The highly dynamic nature of electricity markets obligates participating players to act in a constantly changing environment, where actions and decisions must be dependent on the particular characteristics of each distinct market type, as well as on the circumstances that are found at each moment. For this reason, the sense of context awareness becomes essential for market-negotiating players to be able to suit their decisions appropriately.

While context is essential for processing information, it is roughly totally absent from the modern information technology infrastructure. In [19], context awareness capabilities are applied to computer systems, and [20] approaches the concept of context in the scope of multi-agent simulation.

Context awareness in the scope of electricity markets must consider the main variables or characteristics that define an electricity market negotiation period or day. The characterization of the negotiation environment should integrate these defining variables in a way that, together, they become able to distinguish different market negotiation contexts.

The main factor to consider is the electricity market type itself, i.e. if it is auction based, or if the power is transacted via bilateral negotiation; and the particularities of each different market (the types of offers, e.g. block offers [21], flexible offers [22, 23], complex conditions [24], among others).

The forecasting of electricity market prices is a widely studied problem. In [25–27], artificial neural networks (ANN) are used to achieve very good forecasting results, while in [28], support vector machines (SVM) are shown to be able to achieve solid forecasting results, with reduced execution times. Adequate electricity market prices' prediction is essential to achieve a solid awareness regarding the prices that are expected to be practiced in each market, at each time.

The expected market prices are highly influenced by the variation in power consumption and generation that are verified at each moment, making the prediction of the amount of negotiated power a particularly important issue. This is especially evident with the growing integration of generation based on renewable, intermittent sources, such as wind and solar-based generation. The uncertainty associated with the amount of produced generation at each moment brings an additional source of uncertainty to the outcomes of market sessions. Studies show that the wind intensity variation has a notorious influence on the electricity market prices [29, 30]. From these studies, one can conclude that the shortage of wind generation power can lead to prices' increase, while the excess of wind-generated power can lead to a high prices' decrease, in some cases even reaching the value of 0 €/MWh. For this reason, the wind intensity forecast plays an important role as well, with several works being developed to address this problem [31–33].

While the wind speed intermittence is noticed throughout all hours of all days, the solar intensity presents a somewhat stable pattern, mainly referring to the full absence during the night; and the growing tendency as the sun arises, and decreasing propensity as the sun sets. Also, this is usually a very seasonal type of production, presenting a great difference from the sunny summer days to the cloudy winter times. The seasonality and the patterns of variance throughout the course of a day are also evident in what concerns the power consumption. Hence this becomes another important aspect that must be considered for an adequate contextualization of the power negotiation process. Classifying the types of days, considering not only the season but also the week day is essential. An obvious example is the difference between the consumption patterns between business days and weekends; however, special cases, such as holidays and special event days (where the global consumption is expected to be different from the usual cases), should also be considered.

7.2.3 Strategic Bidding

The diverse negotiation contexts that players face when participating in different markets obligate electricity market players to adapt their behavior, in order to take full advantage of each distinct environment. As stressed before, many different approaches have been and are constantly being developed, devoted to the problem of strategic bidding in electricity markets [11–18, 34]. However, the very distinct nature of the different market negotiation contexts makes it very difficult for the strategic approaches to be able to perform at their best in all cases, as claimed by the authors in [16, 34]. This enforces the need of using several distinct approaches, from different natures, so that the strategic behavior does not get trapped in an environment for which it is not properly prepared. The use of different bidding strategies in a cooperative way diminishes the chances of single strategies' inadequacy.

In [6, 24], authors propose a multi-agent system to provide decision support capabilities to electricity market negotiating players in what concerns their strategic bidding. Adaptive Learning Strategic Bidding System (ALBidS) uses reinforcement learning algorithms to choose the most suitable strategic approach from a set of different alternative strategic bidding algorithms. The choice is done by analyzing the performance of each approach when used in different contexts of negotiation.

The integrated bidding strategies include some simpler approaches, which require very low execution times, such as averages and regressions of the historic log of electricity market prices. These strategies complement more complex ones, based on very different natures. A dynamic ANN [35], which is retrained in each iteration in order to always include the most recent data, is included, as well as SVM-based approach [36]. Strategies used by different electricity market simulators are also integrated. An important example is Agent-based Modeling of Electricity Systems (AMES) [37, 38], whose market agents use a bidding strategy based on the application of the Roth-Erev reinforcement learning algorithm [39], integrated with a simulated annealing (SA) [40] approach to accelerate the convergence process. An adaptation of this strategy has also been made, using the SA meta-heuristic on top of the Q-Learning reinforcement learning algorithm [41].

Other incorporated bidding strategies are the Adapted Derivative-Following strategy, proposed by Greenwald et al. [42]; an application of Game Theory for scenario analysis [16, 43]; and a strategic approach based on the principles of the Information Theory [44], in which the electricity market forecasting error is analyzed and minimized.

Some adaptations of theories from very distinct areas can also be found. An application of the Determinism Theory [45], from sociology and physics, is implemented, where several meta-heuristics are used to optimize a player's final bid to the market, taking into account the expected state of the "world" at each time. The considered meta-heuristics are PSO, SA, GA, and Tabu search. ALBidS also uses a strategy that adapts the main principles of forecasting from a company's scope: the internal economic data analysis, and the external, or sectorial, data analysis [46].

The performance of all bidding strategies provides an important opportunity to support the development of approaches based on the concept of metalearner [47, 48]. These metalearners use the outputs of the other strategic approaches to learn about the process of learning itself, and create their own output. The metalearning process can be based on the application of techniques such as ANN, or by adapting other decision methods, such as the Six Thinking Hats conflict resolution method for meetings [49, 50].

7.3 MARKET ANALYSIS AND CLEARING

Zita A. Vale and Tiago Pinto

Polytechnic of Porto, Porto, Portugal

7.3.1 Introduction

Traditionally, the power sector was based on regulated vertically operated utilities. The present situation is radically different because, similarly to other sectors, the power industry has been deregulated. Electricity markets have undergone a liberalization process; however, electricity markets are a special case of a commodity market, due to the difficulty on storing electrical energy and to the need of a constant balance [51]. This process brings a large set of new challenges and makes the coordination between technical and economic issues much more complex than before. The diversity of players acting in this context require new computational tools, with both electricity market and power system operation having to consider the physical constraints of power systems, market operation rules, and financial issues.

Electricity markets [52] are dynamically changing environments with very particular characteristics. The players that act in these environments must have a good understanding of how they work, and how they are affected by the interaction among the players involved in their operation [34]. Several electricity market simulators have been developed for this purpose [13, 37, 53]. Most of the developed simulators use meta-heuristic optimization to support some of their features. For instance, MASCEM [34] uses GA, PSO, SA, and Tabu search for supporting players' strategic bidding and supporting decisions concerning contracts. SI-based strategic bidding in competitive electricity markets is also proposed in [54].

An integrated ant colony optimization approach is used in [55] to compare electricity markets' clearing strategies. A method for ancillary market clearing based on GA is presented in [56]. Methods for supporting players' portfolio decisions using PSO and GA have been proposed, respectively, in [14, 15, 57].

In the new context of electricity markets, consumers should assume a much more active participation. Demand response (DR) is not a new concept but it is gaining a growing focus of attention and requiring adequate business models to be improved helping in ensuring the reliable power system functioning and keeping electricity prices under control. In [58], PSO is used for the optimized management of DR contracts.

A bibliographic survey concerning the transmission of electric energy is presented in [59], considering the new research topics that have arisen due to the reforms undertaken in electricity markets. Several applications of meta-heuristic techniques, such as differential evolution, GA, PSO, Fuzzy, and Tabu search, are listed and briefly described. Relevant topics in this area include transmission expansion planning [60] and congestion management [61] for which a diversity of meta-heuristic techniques have been proposed.

7.3.2 Electricity Market Simulators

The huge diversity of developed models, devoted to overcoming the vast number of problems that result from such a dynamic and complex reality such as the electricity markets, reaches its highest peak of utility when combined appropriately. For this reason, several modeling and simulation tools have been developed in the last few years, aiming at an advantageous integration of different models that can fruitfully come in aid of professionals that are involved in the electricity market sector [13, 34, 37, 53, 62–65].

Complete modeling tools are becoming essential assets for market players, not only regarding the decision support capabilities for market participation that result from the combination of a realistic representation of the electricity market environment, with adequate artificial intelligence techniques, particularly data-mining, machine learning, and decision support mechanisms but also in what concerns the experimentation, testing, and validation of the impact of alternative regulatory models, which cannot be performed directly in reality because of the enormous risks that affect the entire population. By using such complete simulation tools, operators and regulators are able to refine their underdevelopment policies in a coherent framework before applying them in real cases.

Multi-agent technology is being increasingly used to represent, model, and simulate complex and dynamic environments [13, 34, 37, 53, 65]. The possibility of representing different entities as independent software agents with their own particular behavior and objectives, and the opportunity for easily enlarging the represented models, are some of the main reasons why multi-agent technology is widely chosen as the best option for developing complex simulation tools for constantly evolving environments such as the electricity markets.

The data that can be gathered from real electricity markets' operation together with the information that is generated during simulations can be used for knowledge discovery and machine learning, using data-mining techniques. Adequate data analysis from all available sources is able to provide electricity markets' players with simulation tools able to overcome the little experience they have in electricity markets' operation, granting these players the possibility of acting with the guidance of tools that are adaptive to players' characteristics and goals. Some of the existent electricity markets' simulators are endowed with machine learning capabilities [34, 37, 53], making them much more than simple static simulators of a highly dynamic scene.

The advantage of the use of multi-agent technology with adequate simulation abilities to model and represent electricity markets has been confirmed by a number of multi-agent electricity market simulators all around the world [13, 34, 37, 53, 65]. The electricity market simulators that have been or that are still being continuously developed present very different characteristics and objectives. Some are more directed to the study of the electricity market mechanisms themselves, such as the different market clearing tools that are used worldwide; some have their peak of performance in the study of the physical implications, from a network stability standpoint; and yet other are more concerned with the complex interactions between the involved players and the comparison of different scenarios for future electricity market evolution.

Simulator for Electric Power Industry Agents (SEPIA) [62] is developed under a Plug and Play architecture, and offers the possibility of performing simulations using several processing units to allow distributing simulations through several machines inside a network. Simulations in SEPIA can be oriented and followed by specific mechanisms, which allow studying the interactions and behaviors of the involved players.

Electricity Market Complex Adaptive System (EMCAS) [53] is a multi-agent-based simulator that represents the distinct electricity market participating entities via software agents. The types of agents that can be modeled are vast, including generators, consumers, distribution and transmission companies, independent system operators, and regulators. The represented players' strategies are based on adaptive machine learning algorithms. Probably the most important and distinctive characteristic of EMCAS when compared to other electricity market simulators is the capability of performing simulations for very different time horizons, being able to perform simulations that reach several decades.

Power Web [63] is available as a Web platform, enabling users to interact from distinct geographical points. Simulations with Power Web are controlled by a centralized agent, which acts as an independent system operator that guarantees the reliability of the system. This simulator offers the possibility of simulating a large number of scenarios with different rules and constraints. There is also the possibility of participating in an open market, in which users can compete with players controlled by other users or by virtual entities.

Short–Medium run Electricity Market Simulator (SREMS) [64] simulates the electricity market in the short–medium term (ranging from one month to several months). This is an especially suitable simulator for the Italian electricity market. SREMS includes a scenario analysis algorithm based on the application of game theory with the objective of calculating negotiating players' optimal hourly bids. Simulations in SREMS are sensitive to independent bidding zones, including a physical validation of the interzonal transit constraints.

AMES [37] is an agent-based electricity market simulator, directed to the study of the market design of the US Federal Energy Regulatory Commission (FERC). The represented electricity market includes an independent system operator agent, generation, and load entities. The software agents can be distributed through the transmission grid in order to test the market efficiency. Generators in AMES are equipped with stochastic reinforcement learning capabilities as support for their bidding decisions in the market.

Genoa Artificial Power Exchange (GAPEX) [65] is able to reproduce the market clearing procedures of the most important European power-exchange markets. This agent-based framework is highly directed to the study of the power exchanges between electricity market negotiating entities.

MASCEM [13, 16, 34, 35] has been under constant development since it was first introduced to the scientific community, in 2003 [13]. MASCEM's multi-agent model includes several entities besides the commonly represented ones (e.g. generation and consumption companies, independent system operators, and regulators), such as Virtual Power Players (VPP) [35] and traders. MASCEM is able to recreate a high diversity of market clearing models, based on the mechanisms used in different countries all around the world. Negotiating players in MASCEM use ALBidS [16, 34] as a decision support tool to achieve the best possible results out of the market environment, using adaptive learning algorithms, game theory for scenario analysis, and data-mining approaches. The use of ontologies to support players' interactions allows MASCEM to be easily connected to other simulation platforms and enlarge the scope of action of the involved players. A reference example is the connection of MASCEM to Multi-Agent Smart Grid Platform (MASGriP) [66], which allows the participation of players in a Smart Grid [67] environment, and parallel participation in electricity markets. The data that MASCEM uses to create simulation scenarios are gathered from real electricity market operators' websites, by using an automatic data extraction tool [68], enabling MASCEM simulations to be constantly up-to-date with the reality.

The functionalities that the diversity of electricity market simulators provide are fairly complementary, enabling players that are involved in the electricity market sector to be supported by adequate simulation tools. However, there is still much work to be done, namely in what concerns the decision support capabilities for electricity market negotiations, the participation of large-scale distributed generation units in electricity markets, consumers' active participation, among many other problems that should be adequately addressed.

7.3.3 Didactic Example

The simulation capabilities of electricity market simulators provide the chance for a comprehensible analysis of the electricity market environment, including the market clearing mechanisms, and interactions between market participating players. However, a minimal level of knowledge about the electricity market sector is required for a coherent representation of the simulation scenario. Thus, electricity market simulators often integrate rather intuitive user interfaces, so that the least experienced users can still take as most advantage as possible from the simulators. Figure 7.3.1 presents the main user interface of the MASCEM simulator.

The main user interface of MASCEM allows specifying the most basic definitions to execute an electricity market simulation. A previously saved scenario can

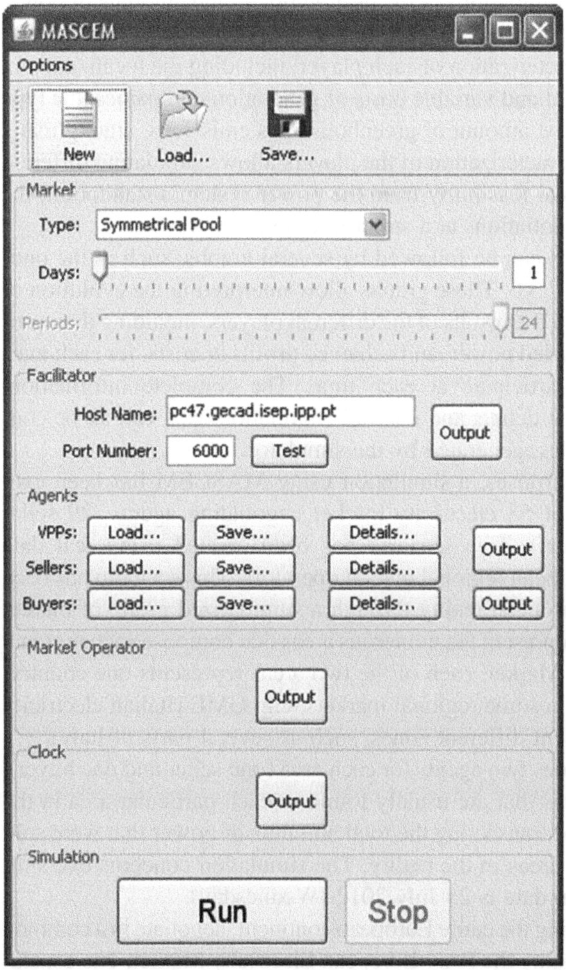

Figure 7.3.1 Main user interface of the MASCEM electricity market simulator.

be loaded, so that it can be repeated or slightly changed in order to study the impact of such changes. The electricity market type(s) desired for the current simulation can be chosen, including different market clearing mechanisms from different countries, or combinations of market types (e.g. day-ahead spot market with complex conditions, balancing market with six negotiating sessions, bilateral contracts, forwards market). The time horizon of the simulation can be also defined, being characterized by the number of simulation days, the number of daily trading periods, and also by the number of subdivisions per period (advanced option, which defines the number of proposals and counterproposals that can occur during a single negotiation period; this is especially important for direct negotiation processes, such as bilateral contracts, some types of forwards markets, and negotiations between VPPs and their members and/or potential members).

The number and characteristics of the players can also be defined. The essential characteristics to run a simulation are the desired prices and amounts of power to negotiate in the market. However, for a complete simulation, MASCEM allows defining the complete characterization of each player, including the localization in the power network, the fixed and variable costs of generation companies, the type of generation technology, the amount of greenhouse gas emissions, among many others [35]. A complete characterization of the players allows simulating different aspects, such as the technical feasibility from the power system's standpoint, the VPP operation, and the negotiations at a smart grid level.

The simulation results can be followed by several graphs, such as the ones presented in Figures 7.3.2–7.3.6. These graphs allow interpreting the evolution of the simulation by comparing the results of the different players, including their proposed prices, amounts of traded power, and achieved profits or costs, for each market type in which they participate at each time. The complete information, including network operation details and internal negotiation data, can all be consulted in specific output files, generated by the simulator.

For demonstration purposes, a simulation using MASCEM has been executed, comprising a total of 53 electricity market negotiating agents (29 seller agents and 24 buyer agents). This scenario has been created using real data extracted from several European regional market operators. This scenario has been created to represent the European reality through a summarized group of market negotiation agents, which represent the numerous areas that compose each regional market (e.g. in the Iberian Market, each of the two areas represents one country: Portugal and Spain; while in some regional markets, e.g. GME [Italian electricity market], these areas represent different zones, such as several parts of Italy).

The simulation includes two agents for each area (one seller and one buyer), practicing the average prices that are usually found in each particular area in the particular chosen cases, and transacting the total amounts of power that were sold or bought in each of these areas in the reality. The simulation concerns one simulation day, and the chosen date is 25 July 2012 (Wednesday).

All players, representing the entire European continent, negotiate in a common market environment, simulating the PAN-European Electricity Market. The considered market mechanisms is the MIBEL wholesale market [24], including the day-ahead spot market with the possibility of presenting complex conditions, the

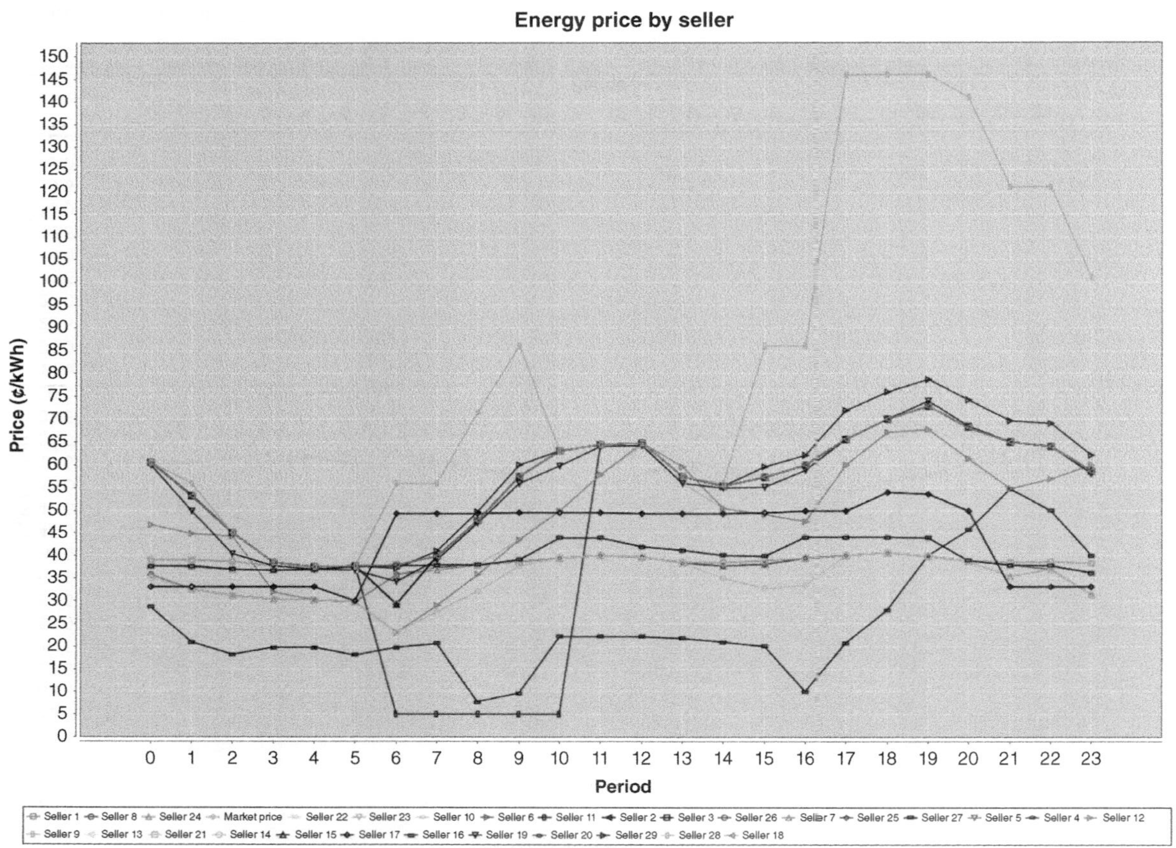

Figure 7.3.2 MASCEM output for sellers' bid information in the day-ahead spot market.

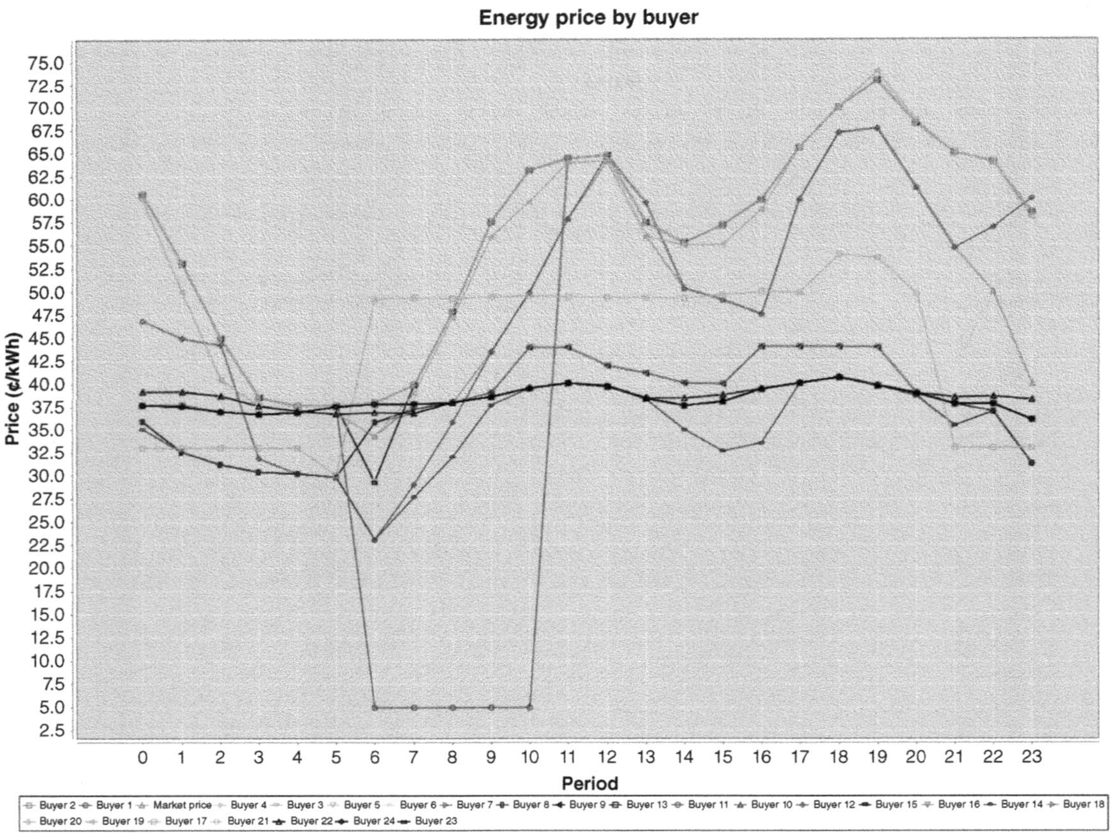

Figure 7.3.3 MASCEM output for buyers' bid information in the day-ahead spot market.

balancing market, where players can buy or sell the power amounts that they could not fulfill in the day-ahead spot market, and finally, the negotiation of bilateral contracts.

Figure 7.3.2 presents the MASCEM output graph that shows the comparison between the offers of all seller agents throughout the 24 hourly periods of the considered day in the day-ahead spot market. Figure 7.3.3 presents the same information regarding the buyer agents.

From the graphs presented in Figures 7.3.2 and 7.3.3, the user is able to analyze the bids that were submitted from the agents to the day-ahead spot market, and compare these values to the electricity market price throughout all negotiation periods. Figure 7.3.4 presents the individual output chart of agent Seller 22 in the day-ahead spot market. Seller 22 has included the charge gradient complex condition to its bid. The charge gradient complex condition defines the maximum value of variation between the sold amounts of power of a generation unit from one negotiation period to the following. Seller 22 defines the maximum variation value of 1000 MWh.

From Figure 7.3.4 it is visible that, in order to respect the charge gradient complex condition, the variation between the sold power of Seller 22 from each period to the following is never above 1000 MWh. Even in some periods where the agent's bid price is located below the market price (particularly from period 9 to period 12), which should mean that the player could sell the entire amount of power, Seller 22 sells only the maximum possible amount of power that respects the complex condition that was submitted. Since Seller 22 could not sell any power in period 6 because of its bid price being above the defined electricity market price, from that period onward the sold amount of power had to be incremented by a maximum value of 1000 MWh, as predefined.

Figure 7.3.5 presents the MASCEM output chart of Seller 7 in the day-ahead spot market. Seller 7's bid includes a very high minimum income complex condition. The minimum income complex condition defines the minimum value of revenue that the player accepts to achieve from its participation in the market. If in the total of all periods of the day the income amount does not reach the defined value, the player does not participate, and is excluded from the market negotiation of that day.

From Figure 7.3.5 it is visible that, in spite of Seller 7's bid being below the electricity market price in practically all periods of the considered day, this player does not sell any power throughout all day. This occurred because the amount of incomes that the player would achieve by selling all of its available power in all periods of the day would not reach the minimum value defined in its minimum income complex condition, hence this player was excluded from the market negotiations in the day-ahead spot market. Despite not having sold any power in the day-ahead spot market, Seller 7 is still able to participate in the balancing market to try selling its available power, or reach favorable agreements through bilateral contracts. Figure 7.3.6 presents the MASCEM output chart of Seller 7 when participating in the balancing market.

From Figure 7.3.6 one can see that Seller 7 was able to sell all of its available power in the balancing market, with the exception of one period (period 10).

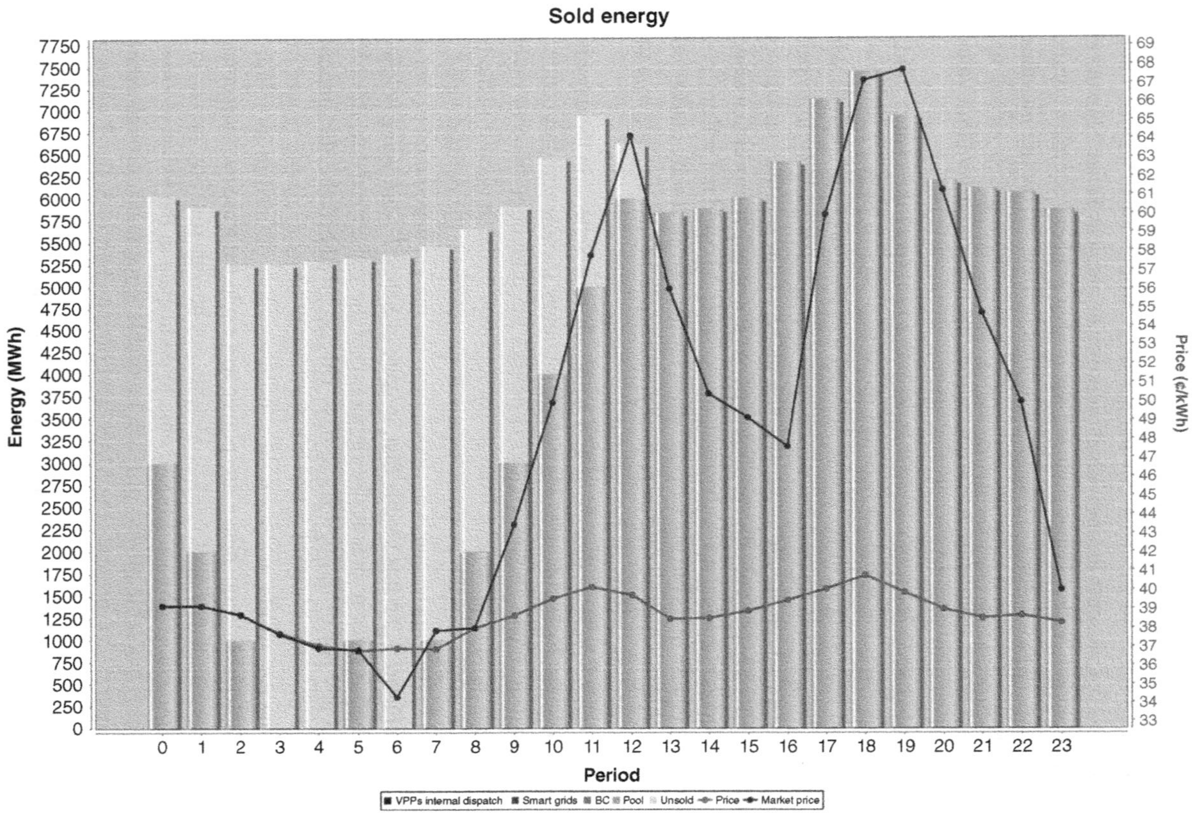

Figure 7.3.4 MASCEM output chart of Seller 22 for the day-ahead spot market.

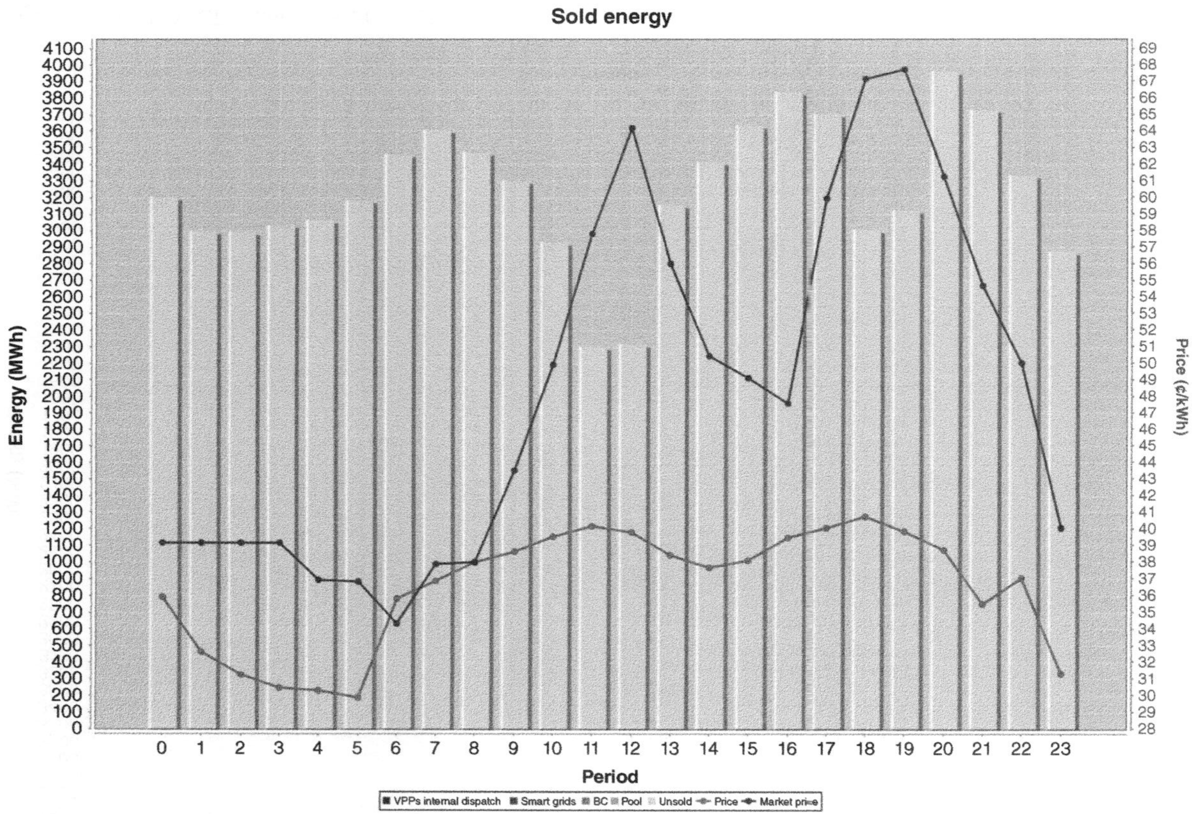

Figure 7.3.5 MASCEM output chart of Seller 7 for the day-ahead spot market.

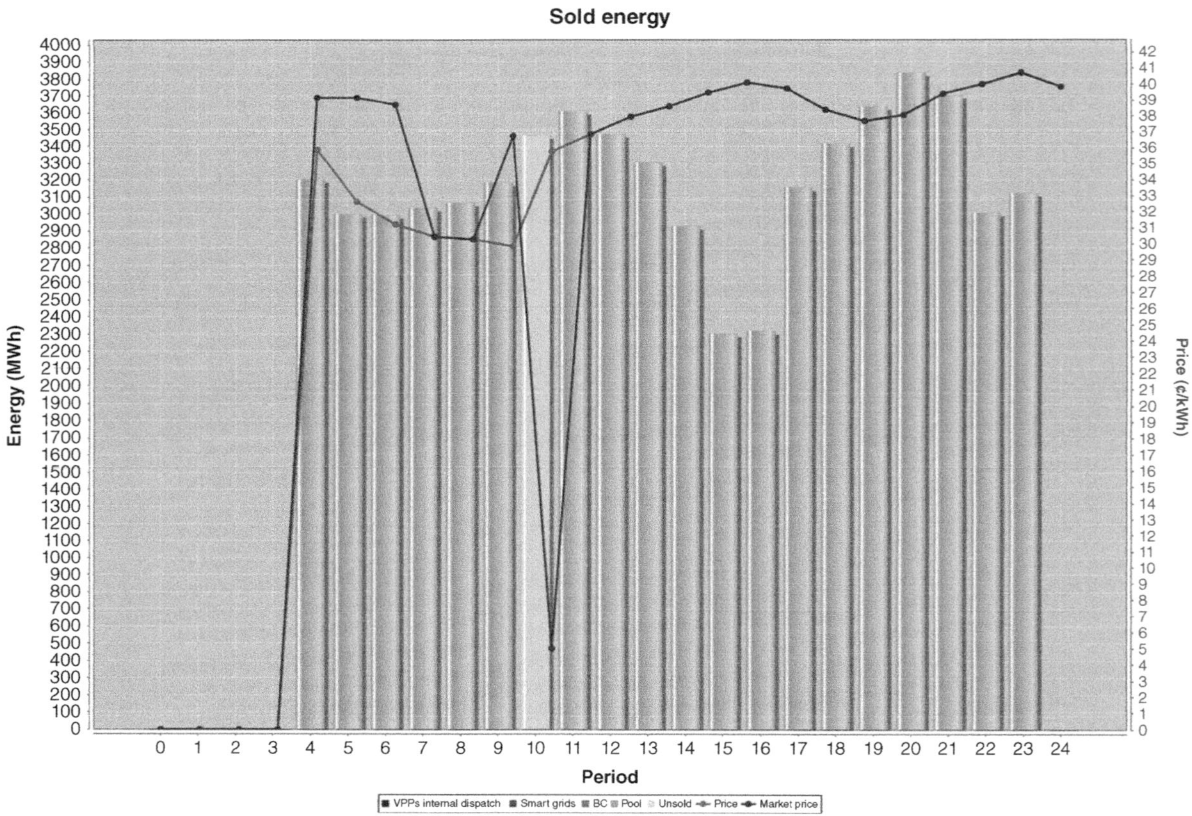

Figure 7.3.6 MASCEM output chart of Seller 7 for the balancing market.

However, the achieved market price, at which the power was sold, is below the values that were attained in the day-ahead spot market. This means that the complex condition submitted by Seller 7 in the day-ahead spot market caused a loss of incomes for this player. As for the periods in which the available power is null (the first four, and the last one), Seller 7 was able to reach good deals in bilateral contracts, and therefore did not need to participate in the balancing market to sell power in these periods.

Players' participation in multiple electricity market types emphasizes the need for decision support. Deciding whether to and to participate in each market type is critical for maximizing electricity market players' performance. This is where electricity market simulators can play a vital role, by realistically representing the electricity market environment, and analyzing all the available information in order to suggest potentially advantageous actions for market participants to take, in order to optimize their outcomes.

7.4 ELECTRICITY MARKET FORECASTING

Tiago Pinto and Zita A. Vale

Polytechnic of Porto, Porto, Portugal

7.4.1 Introduction

Electricity price forecasting in a competitive electricity market is essential to support market players in their decisions, enabling adequate risk management. Electricity price forecasting is rather difficult as electricity prices are dependent from a wide set of factors and evidence spikes unusually high even when compared with other commodities markets, mainly due to the electric energy non-storability in large quantities [51].

A variety of techniques is used in price forecasting. Due to their capacity to perform almost every complex function, ANNs have been extensively used in price forecast, e.g. in [69] that uses a combination of neural networks and fuzzy logic. A method to forecast day-ahead electricity prices based on self-organizing map (SOM) neural network and SVM models is proposed in [70]. SOM is used to cluster the data automatically according to their similarity and SVM models for regression are built on the categories clustered by SOM separately. Parameters of the SVM models are chosen by PSO algorithm. Results of the application of the proposed method to the Pennsylvania–New Jersey–Maryland (PJM) market are presented.

A hybrid method based on wavelet transform, auto-regressive integrated moving average (ARIMA) models, and radial basis function neural networks (RBFN) is proposed in [71] to forecast day-ahead price forecasting. This method uses PSO to optimize the network structure which makes the RBFN be adapted to the specified training set, reducing computation complexity and avoiding overfitting. Results of the application of this method to the electricity market of mainland Spain are presented and compared with some of the most recent price forecast methods. The wholesale price of the Ontario electricity market has been forecasted for each hour of the next day for a test period of three years by using a SVM [28].

Although a high variety of techniques can be applied to the electricity market price forecast problem, as claimed by the authors in [72], there is no consensual approach as being clearly the most effective in this scope. However, ANNs and SVM are the most commonly used techniques, showing the capability of providing very good results.

7.4.2 Artificial Neural Networks for Electricity Market Price Forecasting

ANNs are inspired on the human brain, consisting in a huge number of neurons with high interconnectivity. A NN is constituted by a series of nodes, or neurons, organized in different levels, and interconnected by numeric weights. NNs are similar to the human brain in two essential aspects: the NN knowledge being acquired from the surrounding environment through a learning process; and the network's nodes being interconnected by weights (synaptic weights), which are used to store the knowledge [73].

Each neuron executes a simple operation, the weighted sum of its input connections, which originates the signal that is sent to the other neurons. The network learns by adjusting the connection weights, in order to produce the desired output – the output layer values. The basic concept consists in providing the network with a large number of correct examples, so that through the comparison of the NN's output with the correct response, it can slowly change the connection weights until it is able to generate outputs that are coincident with the correct values. This way, the NN is able to extract basic rules from real data, differing from the programmed computation, where a set of rigid and predefined set of rules and algorithms is necessary [26].

An NN usually presents a high accuracy in classification and forecasting, even for complex problems. Also, NNs are advantageous when dealing with redundant attributes, since the weights assigned to these attributes are usually very small. Concerning the generated outputs, NNs present the advantage of being able to originate results based on discrete values, real values, or vectors of values [74]. Therefore, NNs are often applied to problems characterized by a high complexity and great variation in the problems' data [75], such as the electricity market prices forecast or in the load harmonics prediction [76].

Using a NN for a specific problem requires a solid analysis on the available data, so that it can be used appropriately. The definition of the NN topology is essential, considering the correlation between the available data.

In [35] a feedforward NN is proposed, with the objective of forecasting electricity market prices. This NN receives as inputs the market prices and total amount of negotiated energy in the market, referring to: the previous day to the one being forecasted, one week before, two weeks before, and three weeks before. The NN considers four nodes in the intermediate layer, and a single output – the forecasted electricity market price.

Another critical consideration is the amount of data that is used for the training of the NN. The same authors, in [77], define a *TrainingLimit* variable, which defines if the NN will use more or less values for training, meaning a faster but yet worse forecast, or a slower but more effective response. Considering an exaggerated amount of data may lead to overtraining, making the NN memorize the examples instead of learning the relationship between the data. Also, it may lead to the consideration of inadequate data, from a long time before, which possibly has no added value to the learning of the most recent tendencies of the data.

As the amount of data used for training increases, so does the execution time of the NN. In order to smooth this time's increase, parallel programming is proposed [77], dividing the required data into various matrices, and creating a different thread to fill each of these smaller matrices.

7.4.3 Support Vector Machines for Electricity Market Price Forecasting

In 1936, R. A. Fisher [78] created the first algorithm for pattern recognition. Subsequently, Vapnik and Lerner proposed the Generalized Portrait algorithm, in 1963 [79]. The generalization of the Generalized Portrait nonlinear algorithm led to the implementation of the first SVM. This was the first running kernel of SVM, able to solve only classification and linear problems.

The application of SVM using kernels got popular for several reasons, such as the concentration on convex problems, and allowing many linear algebra techniques to be used in a nonlinear way. Moreover, SVM have showed robustness in many application domains [21, 32], with the particularity of spending fewer resources and roughly half the time of NN.

The information to use when implementing a SVM should follow the following format: $(y_1, x_1), \ldots, (y_1, x_1)$, $x \in R^n$, $y \in R$, where each example x_i is a space vector example; y_i has a corresponding value to x_i; n is the size of training data [36]. In classification problems, y_i assumes finite values; in binary classifications, $y_i \in \{-1, 1\}$; and in regression, y_i is a real number ($y_i \in R$).

The implementation of SVM requires considering some important aspects, such as the *Feature Space*, which represents the method that can be used to construct a mapping into a high-dimensional feature space by the use of replicating *kernels*. The use of *Loss Functions* is also essential to a SVM implementation, since these functions allow quantifying the deviation between the forecasted values and the actual values. Finally, and arguably the most important, is the definition of the *Kernel Functions*. The main task of a *Kernel Function* is to find patterns and study the type of associations for general types of data. Some examples of *kernels* used in SVM are [36]: polynomial, Gaussian radial basis function, exponential radial basis function, multilayer perceptron, splines, and B splines.

The most widely applicable kernels for regression problems (as in the specific problem of the forecasting of electricity market prices) are the radial basis function (RBF) and the exponential radial basis function (eRBF). The definition of both these *kernels* requires the specification of the following parameters:

- *TrainingLimit* – number of days used for the training process;
- *e_Val* – value of ε-insensitive, or the degree of exigency of the forecast;
- C – limit of the function;
- $p1$ – angle (σ);
- $p2$ – offset.

An adequate combination of these parameters is essential for achieving the desired results. The most suitable combination is highly dependable on the characteristics and particularities of each distinct problem; therefore, an exhaustive sensibility analysis must be performed for each application, in order to achieve conclusions on the best combinations of parameters that should be used.

7.4.4 Illustrative Results

This illustrative example concerns the performance of SVM and NN when applied to the forecasting of electricity market prices of the Iberian electricity market – MIBEL [24]. The performance of the two techniques is evaluated under two aspects: (i) the quality of the forecasts, by comparing the forecasting errors, using the mean absolute percentage error (MAPE); and (ii) the execution time of the two approaches.

All the tests are referent to 61 consecutive days, starting on 1 September 2009, considering the day-ahead spot market, and the results are analyzed independently for each hourly period.

The topology and characterization of the considered NN is as described in Section 7.4.2. Regarding the SVM, two different kernels are considered: RBF and eRBF. After exhaustive sensibility analysis, the parameterizations that achieved the best results are

- kernel RBF: *TrainingLimit* = 20, $p1(\sigma) = 6$, ε-insensitive = 0, $C = \infty$, $p2 = 0$;
- kernel eRBF: *TrainingLimit* = 20, $p1(\sigma) = 18$, ε-insensitive = 0, $C = \infty$, $p2 = 0$.

Table 7.4.1 presents the average MAPE values of forecast of the SVM approach using each of two kernels (RBF and eRBF) and the NN, for the 61 considered days.

From Table 7.4.1 it is visible that the error values are always located below 1%. The forecast results are very similar when using the NN and the SVM with both kernels. In the total of all error values for the four considered periods of 61 days, the SVM approach using the RBF kernel achieved a higher error than the ANN, while the SVM approach using the eRBF kernel was able to achieve lower error values than the ANN. However, the error values are so similar.

TABLE 7.4.1 MAPE Forecast Error Values (%)

Period	SVM (RBF)	SVM (eRBF)	ANN
1	0.297 578	0.262 867	0.326 704
6	0.044 731	0.107 784	0.027 03
12	0.788 205	0.385 893	0.579 867
18	0.033 467	0.099 395	0.003 859
Total	1.163 981	0.855 939	0.937 46

TABLE 7.4.2 Average Execution Times of the SVM Approach

	Execution Times (in ms)		
TrainingLimit (days)	Minimum	Average	Maximum
5	4 776	5 022	9 501
10	4 730	4 984	5 830
15	4 721	5 107	7 437
20	4 789	5 092	10 136
35	4 793	4 942	23 595

TABLE 7.4.3 Average Execution Times of the ANN Approach

	Execution Time (ms)	
TrainingLimit (days)	Without Parallelism	With Parallelism
60	15 000	11 000
120	18 000	13 000
200	22 000	14 000
365	30 000	17 000
730	49 000	20 000

Table 7.4.2 presents the average execution times of the SVM approach with different amounts of training data, after 1000 run trials.

Table 7.4.2 shows that the average execution time of the SVM approach is about five seconds, regardless of the used kernel. From Table 7.4.2 it is also visible that the increase in considered days for training does not represent a considerable degradation in execution time. Table 7.4.3 presents the average execution times of the NN approach, for different amounts of training data.

Table 7.4.3 shows the average execution time of the NN. It is visible that, even with parallel programming for a faster access to data, the minimum average value is of 11 seconds (more than twice the average execution time of the SVM). Note that the amount of data that the ANN requires for its training is enormous when comparing to the SVM (NN: 60–730 days to achieve acceptable results; SVM: 5–35 days).

Comparing the performance of the two approaches in what concerns the quality of results, the conclusion is that the achieved results (error of prediction) are very similar. In what regards the execution times, the SVM approach is visibly faster to achieve results.

Finally, regarding the setup phase, both approaches require solid data analysis and experimentation test trials, in order to define the most appropriate topology (in the case of the NN), and the best combination of parameters (for the SVM), and in both cases, the most advantageous amount of historic data for the training process.

7.5 SIMULTANEOUS BIDDING OF V2G IN ANCILLARY SERVICE MARKETS USING FUZZY OPTIMIZATION

Ali T. Al-Awami

King Fahd University of Petroleum & Minerals, Dhahran, Saudi Arabia

7.5.1 Introduction

Recent years have witnessed a dramatic increase in the usage of EV and plug-in hybrid electric vehicles (PHEV). This has been a result of the desire to reduce air pollution, which affects the global climate severely. Also, achieving energy independence is another important reason for EV adoption, especially given that the usage of PHEV can reduce gasoline consumption by an estimated 6.5 million barrels/day [80]. These and the fact that EVs can be charged using power produced by renewable energy sources have encouraged several countries, such as the United States [81] and China [82], to promote EVs.

However, EV adoption is not free of challenges. For power system engineers, conventional means of transportation are never a concern since they are not part of the power system. In contrast, EVs are concerning since they may have potential negative impact on the electric power grid if the charging process is not regulated [83].

In order to turn this challenge into an opportunity, the vehicle-to-grid (V2G) concept has been proposed [84–87]. V2G turns the EV from just an ordinary electric load to a tool that can support the electric grid by controlling its charging pattern. It also enables discharging energy from the EV back to grid under certain constraints, which can be very helpful for the grid operator. In this regard, the EV owners must be compensated financially for their participation.

Using V2G, the EVs can participate in several electricity markets, such as the markets for energy, regulation, and reserves. However, a single EV does not have sufficient capacity to participate in an electricity market. Therefore, the participation of EV owner should be through an aggregator (EVA) who acts as an intermediary between a considerable number of EV owners and the electricity market [88]. In its day-to-day operation, the EVA needs to optimize its bidding strategy in the electricity markets in a way that maximizes its profits [89–91] and ensures the EV owners' satisfaction.

To provide V2G, the charging rate of each EV can be adjusted. This results in unidirectional V2G provision. Furthermore, the EVA may choose to drive the charging rate below zero, i.e. discharging the EV battery to the grid, providing bidirectional V2G services. Bidirectional V2G can provide a wider range of services to the grid. However, due to the difficulties associated with its implementation (e.g. battery degradation and the need for charger retrofitting), it is not expected to be implemented in the near future.

A considerable amount of work in unidirectional V2G has been reported. In [90, 91], several optimized charging algorithms have been proposed from an EVA perspective participating in day-ahead ancillary service markets, namely regulation and responsive reserve markets. These are, however, deterministic, i.e. no uncertainties are considered for the day-ahead aggregated bidding. In reality, an EVA should consider the uncertainties of the electricity markets, such as those associated with ancillary service prices and the deployment signals.

In order to handle uncertainties related to power system operations, different optimization approaches have been suggested. In [92], stochastic programming is used for formulating a coordinated bidding of thermal and wind power for an energy producer participating in the energy market. This has been extended in [93] for a load-serving entity that owns conventional and stochastic generating units and serves a load that has significant EV penetration. Only the energy market is considered in this work, though. In [94], robust optimization is suggested to maximize the benefits of participating in electricity markets for an EV battery swapping facility.

Fuzzy optimization (FO) has been used successfully to handle the uncertainties in regulated power system problems, such as unit commitment and economic dispatch [95, 96]. One major advantage of FO is that it efficiently incorporates uncertainties while maintaining the tractability of the problem size. This is because in FO, there is no need to represent each stochastic parameter by a large number of scenarios. Also, FO is particularly useful when historical data related to the uncertain parameters is limited or simply not existing.

In this section, the uncertainties associated with bidding V2G into the markets are modeled using FO [97]. This leads to the formulation of a fuzzy linear program for an EVA that coordinates the provision of ancillary services, namely regulation and spinning reserves, to electricity markets using unidirectional V2G. The different electricity market uncertainties, such as the ancillary service prices and the ancillary service deployment signals, are considered.

7.5.2 Fuzzy Optimization

FO is based on the application of fuzzy set theory on optimization. Fuzzy sets were first introduced in 1965 by Zadeh [98]. It is meant to facilitate the modeling of uncertainties that arise due to vagueness, imprecision, forecast errors, and so on.

Mathematically speaking, an element x can either belong or not belong to a crisp set X based on the definition of the set X. For example, let X be the set of all real numbers that are less than 6. The element $x_1 = 5$ is said to belong to X, while $x_2 = 8$ and $x_3 = 11$ do not belong to X. Now, define X to be the set of all real numbers that are less than 6. Also, define $\mu(x)$ to be the degree of closeness of x to 6, where $\mu(x)$ is defined as

$$\mu(x) = \begin{cases} 0 & x > 9 \\ \frac{9-x}{3} & 6 \le x \le 9 \\ 1 & x < 6 \end{cases} \tag{7.5.1}$$

The element $x_1 = 5$ belongs to the set X with a degree of 1. The element $x_2 = 8$ also belongs to X, but with a degree of 1/3 only, while $x_3 = 11$ belongs to X with a degree of 0. According to this definition, the set X is called a fuzzy set. The function $\mu(x)$ is called the satisfaction (or membership) function of X, see Figure 7.5.1. This simple example illustrates the fact that while a crisp set is defined by a crisp boundary, a fuzzy set is characterized by a fuzzy boundary.

This proves to be useful in optimization problems whenever there exist parameters that are not known with certainty, or whenever constraints can be considered "soft."

FO transforms the objectives and constraints into satisfaction functions of fuzzy sets. Optimality is achieved by maximizing the intersection of these satisfaction functions of the problem [99].

Consider the following optimization problem:

$$\begin{aligned} &\text{Maximize} \quad f(x) \\ &\text{subject to} \quad g_j(x) \le 0 \qquad j = 1, \ldots, M \end{aligned} \tag{7.5.2}$$

x is a decision variable vector. If the objective function $f(x)$ and all the constraints are crisp, this represents a crisp optimization problem. However, if the objective function and/or some of the constraints are functions of fuzzy parameters, this represents a FO problem. In general, the FO problem can be written as

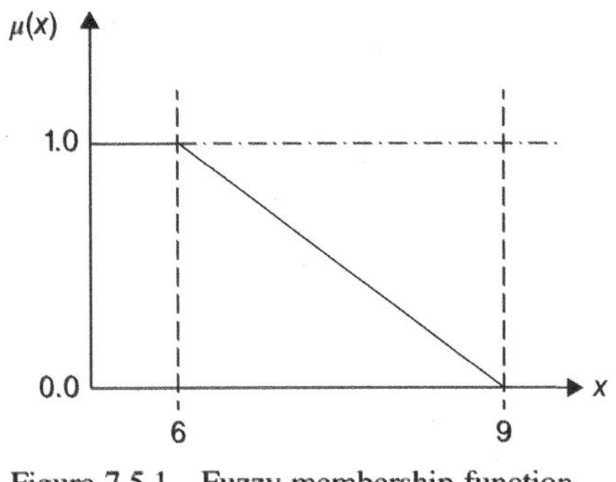

Figure 7.5.1 Fuzzy membership function.

$$\begin{aligned} &\text{Maximize} \quad f(x) \\ &\text{subject to} \quad g_j(x) \le 0 \quad j = 1, \ldots, m \\ &\qquad\qquad\quad\; g_j(x) \le 0 \quad j = m+1, \ldots, M \end{aligned} \tag{7.5.3}$$

where constraints $1, \ldots, m$ are fuzzy constraints, while constraints $m + 1, \ldots, M$ are crisp constraints.

Let the objective be associated with a fuzzy set $F = \{(u, \mu_F(u)) \mid u \in U\}$, where u is the value of the objective function and U is the objective space. $\mu_F(u)$ is the membership function that defines the degree of closeness of the objective to the optimal value. Similarly, let each fuzzy constraint be associated with a fuzzy set $G_j = \{(u_j, \mu_{Gj}(u_j)) \mid u_j \in U_j\}$. The subscript j refers to the jth constraint, u_j is the value the jth constraint assumes, and U_j is jth constraint space. $\mu_{Gj}(u_j)$ is the membership function that defines the degree of satisfaction of the jth constraint. The FO problem can be expressed as [99]:

$$\begin{aligned} &\text{Maximize} \quad \lambda = \min\left(\mu_F, \mu_{G1}, \ldots, \mu_{Gm}\right) \\ &\text{subject to} \quad g_j(x) \le 0 \qquad j = m+1, \ldots, M \end{aligned} \tag{7.5.4}$$

The min function determines the minimum of the satisfaction values. All the membership functions are defined in the range of [0, 1]. As λ is maximized, individual fuzzy satisfaction parameters relating to objectives and constraints are consequently optimized.

7.5.3 FO-based Simultaneous Bidding of Ancillary Services Using V2G

Consider an electric vehicle aggregator (EVA) who would like to charge a group of participating EVs in a timely manner. The EVA needs to buy energy from the day-ahead energy market. Also, the EVA can provide regulation and reserve services in their corresponding markets. It provides these Ancillary Services(AS) by altering the EV charging rates above or below the scheduled charging rates acquired from the energy market. For the ith EV, the scheduled charging rate is called the preferred operating point (POP_i). Once POP_i is decided, the EV can provide regulation-down by charging at a higher charging rate than POP_i. It can provide regulation-up and reserve by charging at a lower charging rate than POP_i. The charging rate is bound from above by the charger's power rating, MP_i, and is bound from below by zero, assuming unidirectional V2G is offered. Thus, the regulation-down capacity, $MxAP_i$, is bound from above by $MP_i - POP_i$, while regulation-up and reserve capacities, $MnAP_i$ and $RsRP_i$, are bound from below by POP_i. The EVA aims to maximize its profits by optimizing POP_i, $MxAP_i$, $MnAP_i$, and $RsRP_i$, while meeting the constraints imposed by the markets, EVs, and their owners' preferences. Given that the bids need to be sent to the markets a day ahead, the EVA is exposed to several uncertainties associated with energy and ancillary service prices, EV availability, and the AS real-time deployment signals.

Objective Function

The EVA aims to maximize its profits obtained from participating in the energy and AS markets while satisfying the EV owners in the best possible way and meeting all technical constraints associated with the EV. The EVA revenue streams are from:

1. Selling energy to the EVs at a predefined tariff, β. To provide an incentive for EV owners to participate in aggregation, β should be set at a lower value than the average retail price of energy.
2. Selling regulation and reserve capacities to their perspective markets, each at a market price to be defined once the market is cleared by the ISO.

Therefore, the EVA income is:

$$\begin{aligned} \mathrm{In} &= \sum_t (\mathrm{PU}_t \cdot \mathrm{RU}_t + \mathrm{PD}_t \cdot \mathrm{RD}_t + \mathrm{PR}_t \cdot \mathrm{RR}_t) \cdot \mathrm{EVPer}_t \\ &\quad + \beta \sum_i \sum_t \mathrm{Av}_{it} \cdot E[\mathrm{FP}_{it}] \cdot \mathrm{EVPer}_t \end{aligned} \tag{7.5.5}$$

PU, PD, and PR are the market prices for regulation-up, regulation-down, and reserve capacities, respectively. RU, RD, and RR are the regulation and reserve capacities. The EVA cost is:

$$C = \sum_i \sum_t \mathrm{Av}_{it} \cdot E[\mathrm{FP}_{it}] \cdot P_t \cdot \mathrm{EVPer}_t \tag{7.5.6}$$

P is the market price for energy, $E[\mathrm{FP}]$ is the expected power draw, and $\mathrm{EVPer} \leq 1$ is a factor used to account for unexpected departure of some EVs.

The fuzzy set for the EVA income is defined as:

$$\widetilde{\mathrm{In}} = \left\{[\mathrm{In}, \mu_{\mathrm{In}}], \underline{\mathrm{In}} \leq \mathrm{In} \leq \overline{\mathrm{In}}\right\} \tag{7.5.7}$$

This is based on the assumption that there is a minimum value of income, $\underline{\mathrm{In}}$, below which the EVA is not willing to participate, i.e. the membership function is zero if the income is below $\underline{\mathrm{In}}$. There is also an upper value of the income $\overline{\mathrm{In}}$ at which the EVA is completely satisfied. These limits must be decided upon by the EVA before the FO is carried out. A commonly used approach to set these limits is by solving the deterministic optimization first, and then define these limits as a certain percentage of the deterministic-case income, i.e. where the uncertain parameters are represented by their forecasted values only.

Therefore, the income's membership function, μ_{In} in (7.5.7), is defined as:

$$\mu_{\mathrm{In}} = \begin{cases} 0 & \text{if } \mathrm{In} \leq \underline{\mathrm{In}} \\ \dfrac{\mathrm{In} - \underline{\mathrm{In}}}{\overline{\mathrm{In}} - \underline{\mathrm{In}}} & \text{if } \underline{\mathrm{In}} \leq \mathrm{In} \leq \overline{\mathrm{In}} \\ 1 & \text{if } \mathrm{In} \geq \overline{\mathrm{In}} \end{cases} \tag{7.5.8}$$

This function is graphically presented in Figure 7.5.2.

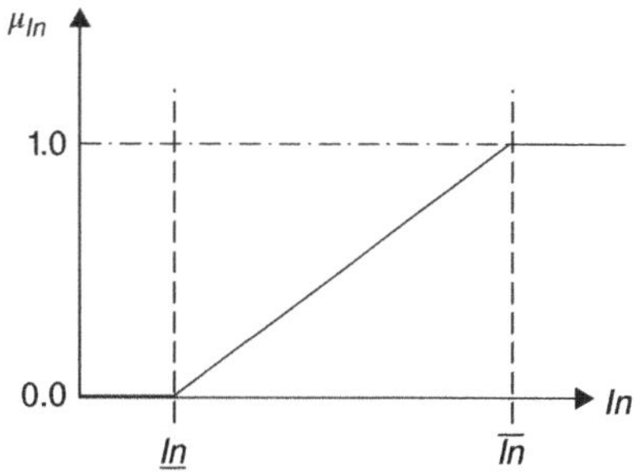

Figure 7.5.2 Fuzzy model of EVA income.

Fuzzy Model – Ancillary Service Prices and Deployments

ARIMA modeling is used to forecast the hourly AS prices for 24 hours on the day ahead. Based on historical data of the actual and forecasted prices, the MAPE are obtained. These are used as upper and lower limits for each uncertain price. The fuzzy model for the regulation-up prices is represented as:

$$\widetilde{\mathrm{PU}} = \left\{[\mathrm{PU}, \mu_{\mathrm{PU}}], \underline{\mathrm{PU}} \le \mathrm{PU} \le \overline{\mathrm{PU}}\right\} \tag{7.5.9}$$

The membership function for the regulation-up price is given in (7.5.10) and the graphical representation is similar to that in Figure 7.5.2. Note that these membership functions are all non-decreasing to indicate the desire of the EVA for higher prices.

$$\mu_{\mathrm{PU}} = \begin{cases} 0 & \text{if } \mathrm{PU} \le \underline{\mathrm{PU}} \\ \dfrac{\mathrm{PU} - \underline{\mathrm{PU}}}{\overline{\mathrm{PU}} - \underline{\mathrm{PU}}} & \text{if } \underline{\mathrm{PU}} \le \mathrm{PU} \le \overline{\mathrm{PU}} \\ 1 & \text{if } \mathrm{PU} \ge \overline{\mathrm{PU}} \end{cases} \tag{7.5.10}$$

Similarly, a fuzzy model is used for the regulation-down and responsive reserve prices.

The expected values of ancillary service deployments are calculated using the historical deployment signals from ERCOT [100]. The hourly actual averages are calculated, and the deviations from the forecasted values, which are forecasted using ARIMA, are obtained so that the membership functions of EU, ED, and ER can be defined. In the fuzzy model for the expected regulation-up deployments, EU is shown in (7.5.11) and its membership function in (7.5.12). The graphical representation of the expected regulation-up deployments membership function is shown in Figure 7.5.3. Note that the membership function of the expected AS deployments are assumed to be non-increasing since the EVA desires these signals to be low. Low deployment levels imply higher availability for the EVs to provide AS for the remaining charging period.

$$\widetilde{\mathrm{EU}} = \left\{[\mathrm{EU}, \mu_{\mathrm{EU}}], \underline{\mathrm{EU}} \le \mathrm{EU} \le \overline{\mathrm{EU}}\right\} \tag{7.5.11}$$

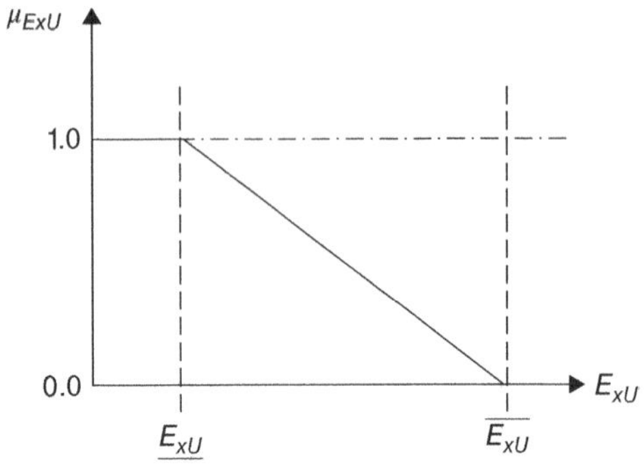

Figure 7.5.3 Fuzzy model of expected regulation-up deployments.

$$\mu_{\mathrm{EU}} = \begin{cases} 0 & \text{if } \mathrm{EU} \le \underline{\mathrm{EU}} \\ \dfrac{\overline{\mathrm{EU}} - \mathrm{EU}}{\overline{\mathrm{EU}} - \underline{\mathrm{EU}}} & \text{if } \underline{\mathrm{EU}} \le \mathrm{EU} \le \overline{\mathrm{EU}} \\ 1 & \text{if } \mathrm{EU} \ge \overline{\mathrm{EU}} \end{cases} \tag{7.5.12}$$

The fuzzy models for the expected regulation-down and reserve deployments are developed in a similar manner.

FO-Based Optimization The complete optimal FO-based formulation is stated below:

$$\textit{Maximize } \lambda \tag{7.5.13}$$

Subject to:
Relations (7.5.5) and (7.5.6)

$$\left(\overline{\mathrm{In}} - \underline{\mathrm{In}}\right)\lambda + \underline{\mathrm{In}} \le \mathrm{In} \tag{7.5.14}$$

$$\left(\overline{\mathrm{PU}} - \underline{\mathrm{PU}}\right)\lambda + \underline{\mathrm{PU}} \le \mathrm{PU} \tag{7.5.15}$$

$$\left(\overline{\mathrm{PD}} - \underline{\mathrm{PD}}\right)\lambda + \underline{\mathrm{PD}} \le \mathrm{PD} \tag{7.5.16}$$

$$\left(\overline{\mathrm{PR}} - \underline{\mathrm{PR}}\right)\lambda + \underline{\mathrm{PR}} \le \mathrm{PR} \tag{7.5.17}$$

$$\left(\overline{\mathrm{EU}} - \underline{\mathrm{EU}}\right)\lambda + \mathrm{EU} \le \overline{\mathrm{EU}} \tag{7.5.18}$$

$$\left(\overline{\mathrm{ED}} - \underline{\mathrm{ED}}\right)\lambda + \mathrm{ED} \le \overline{\mathrm{ED}} \tag{7.5.19}$$

$$\left(\overline{\mathrm{ER}} - \underline{\mathrm{ER}}\right)\lambda + \mathrm{ER} \le \overline{\mathrm{ER}} \tag{7.5.20}$$

$$\sum_{t}^{\mathrm{Trip}_i} \mathrm{Av}_{it} \cdot E[\mathrm{FP}_{it}] \cdot \mathrm{Comp}_{it} \cdot \mathrm{Ef}_{it} + \mathrm{SOCI}_i \le \mathrm{MC}_i \tag{7.5.21}$$

$$0.99\text{MC}_i \leq \sum_t^T \text{Av}_{it} \cdot \text{E}[\text{FP}_{it}] \cdot \text{Comp}_{it} \cdot \text{Ef}_{it} + \text{SOCI}_i - \text{Trip}_i \leq \text{MC}_i \tag{7.5.22}$$

$$(\text{MxAP}_{it} + \text{POP}_{it}) \cdot \text{Comp}_{it} \leq \text{Av}_{it} \cdot \text{MP}_i \tag{7.5.23}$$

$$\text{RsRP}_{it} \leq \text{POP}_{it} + \text{Av}_{it} \cdot \text{MP}_i - \text{MnAP}_{it} \tag{7.5.24}$$

$$\text{MnAP}_{it} \leq \text{POP}_{it} + \text{Av}_{it} \cdot \text{MP}_i \tag{7.5.25}$$

$$\text{SOCf}_i \geq \text{SOCfd}_i \tag{7.5.26}$$

$$\text{MxAP}_{it}, \text{MnAP}_{it}, \text{RsRP}_{it}, \text{POP}_{it} \geq 0 \tag{7.5.27}$$

$$E[\text{FP}_{it}] = \text{MxAP}_{it} \cdot \text{ED} + \text{POP}_{it} - \text{MnAP}_{it} \cdot \text{EU} - \text{RsRP}_{it} \cdot \text{ER} \tag{7.5.28}$$

$$\text{Comp}_{it} = 1 + \frac{\text{Dep}_{it}}{1 - \text{Dep}_{it}} \tag{7.5.29}$$

$$\text{EVPer}_t = \begin{cases} 1 - \dfrac{1}{\text{NEV}} \displaystyle\sum_{t=1}^{\text{Trip}_i} \sum_{i}^{\text{NEV}} \text{Dep}_{it} & \text{if } t < T\text{trip}_i \\ 1 - \dfrac{1}{\text{NEV}} \displaystyle\sum_{t=\text{Trip}_i}^{T} \sum_{i}^{\text{NEV}} \text{Dep}_{it} & \text{if } t \geq T\text{trip}_i \end{cases} \tag{7.5.30}$$

$$\text{RU}_t = \sum_i \text{MnAP}_{it} \tag{7.5.31}$$

$$\text{RD}_t = \sum_i \text{MxAP}_{it} \tag{7.5.32}$$

$$\text{RR}_t = \sum_i \text{RsRP}_{it} \tag{7.5.33}$$

The optimization formulation is constrained by the battery capacities as incorporated in (7.5.21)–(7.5.23). Relations (7.5.24)–(7.5.25) are due to the rate limitations. As this formulation is for the whole day, the EV availability and trip times have to be considered. If the EV is not available, then Av = 0 and that particular EV will not participate in providing services for that period. If the EV is available in the charging station, then Av = 1 and that particular EV will be available for bidding in the market. The compensation factor, Comp, which is less than one, is used to compensate for unexpected departure for a small percentage of EVs.

Relation (7.5.26) ensures that the EVs receive at least their desired SOC, SOCfd, by the end of the charging period. Since unidirectional V2G is assumed, relation (7.5.27) enforces non-negativity of POP and all AS capacities. Equation (7.5.28) defines how the expected energy received by each EV is calculated as a function of the bidding parameters. The equations to incorporate the unexpected EV departure and the compensation factors for unexpected departures are given by (7.5.29) and (7.5.30).

7.5.4 Case Study

The simulations are performed on the ERCOT market for the period 21 July–20 October 2010, with a group of 10 000 EVs used by commuters. Each simulation starts at 6 a.m. in the morning and ends at 6 a.m. next day. AS deployments have a five-minute resolution because of the available data.

The market prices and deployments are forecasted using ARIMA models. Using historical data of both actual and forecasted parameter values, the MAPE between the forecasted and actual values are calculated in order to incorporate these forecasting inaccuracies into the fuzzy formulation. The MAPEs during the three-month period are shown in the second column of Table 7.5.1. Using these MAPEs and the ARIMA-forecasted market parameters for the implementation day, the minimum and maximum limits of each fuzzy variable are obtained. As an example, columns 3–5 of Table 7.5.1 show the forecasted, maximum, and minimum limits for Hour 17 of 1 August 2010.

The maximum and minimum limits for *In* are determined as a predefined percentage of the deterministic profits. Unless stated otherwise, these limits are set at ±10% of the deterministic profits.

Among the EV group used, it is assumed that 50% are Nissan Leaf, 20% are Mitsubishi i-MiEV, and 30% are Tesla Model S. Each EV has a charging efficiency of 90%. The final desired SOC, SOCfd, is set to 99% for each EV for each day. In reality, the EV owners will have the freedom to change this preference at will.

Driving profiles are created using the 2009 National Household Travel Survey data [101]. More details on how the EV driving profile data are generated can be found in [90]. Figure 7.5.4 shows the average percentage of EVs available for providing V2G services in a 24 hour period. In this study, a hundred different EV driving profiles are considered.

TABLE 7.5.1 Fuzzy Parameters for the Forecasted Quantities

		Data for Hour 17 of 1 August 2010		
Electricity Market Parameters	MAPE (%)	Forecast	Max	Min
Regulation-up prices	8.3	32.45	35.1	29.7
Regulation-down prices	9.6	9.01	9.87	8.14
Responsive reserve prices	6.7	21.3	22.7	19.9
Regulation-up deployments	28.5	0.32	0.41	0.23
Regulation-down deployments	31.3	0.239	0.31	0.16
Responsive reserve deployments	24.8	0.001	0.001 2	0.000 75

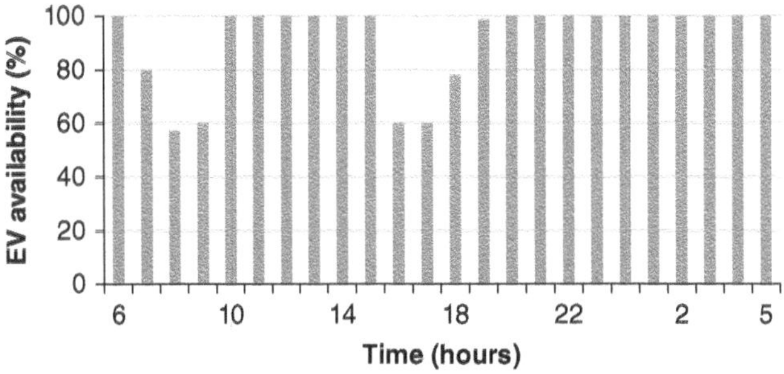

Figure 7.5.4 Average percentage of EV availability.

Two types of bidding algorithms are compared: the deterministic algorithm based on the one presented in [90] and the FO algorithm presented in [97]. The main difference between the deterministic algorithm and the FO algorithm is the incorporation of electricity market uncertainties in the model. For each algorithm, the expected day-ahead aggregator profits are obtained by evaluating the corresponding objective function. The expected profits are calculated using the forecasted market values. To further assess the effectiveness of the FO formulation, the actual aggregator profits on the bidding day are calculated for both the deterministic and FO algorithms by simulating the actual (realized) driving, charging, and ancillary deployments for the EVs with the ancillary service algorithm mentioned in [90]. That is, the actual aggregator profits are calculated using the actual (realized) market parameters, such as energy price, ancillary service prices, and regulation deployments.

The energy cost for the EV owner is assigned to be fixed at $\beta = \$0.05/\text{kWh}$, which is more than 50% lower than the average energy price paid by the residential customers in the ERCOT market in the year 2010 [102]. This gives the EV owners incentives to participate in the aggregation.

7.5.5 Results and Discussions

The deterministic and the FO algorithms are run each day from 6 a.m. to 6 a.m. next day for the whole test period. The average ancillary service prices during the test period are shown in Figure 7.5.5. The POPs for an average day during the test period using the deterministic and FO algorithms are shown in Figure 7.5.6. The POP is kept low in the early few hours, while it is set relatively high in the middle of the day to capture the higher mid-day regulation-up prices. The highest charging rates are scheduled just before the end of the simulation day in order to allow the EVs to perform AS for as many hours as possible.

The AS capacities provided by each algorithm are shown in Figures 7.5.7–7.5.9. As expected, the regulation-up and -down capacities follow their corresponding price signals. Regulation-down capacities tend to be higher than the regulation-up capacities. This is a result of the tendency to bid relatively low to just moderate POP levels for most of the early to mid-day hours. The reserve

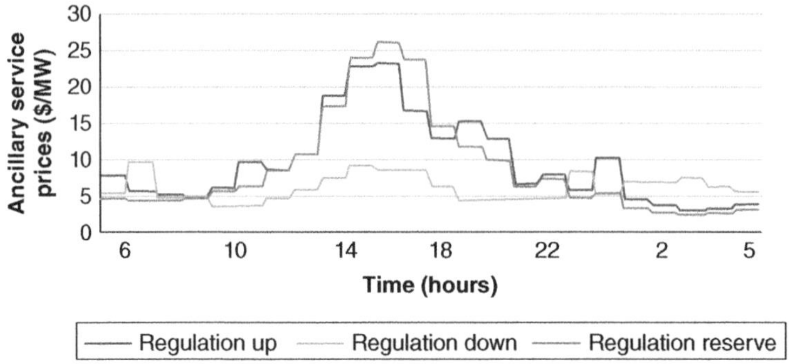

Figure 7.5.5 Average ancillary service prices for the simulation period with data from the ERCOT data archives [100].

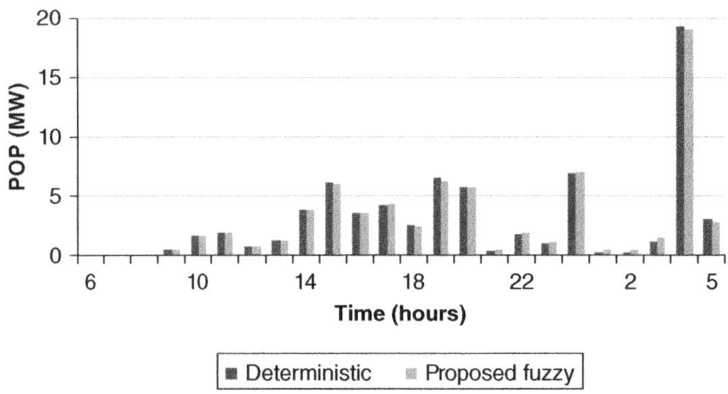

Figure 7.5.6 Average POP.

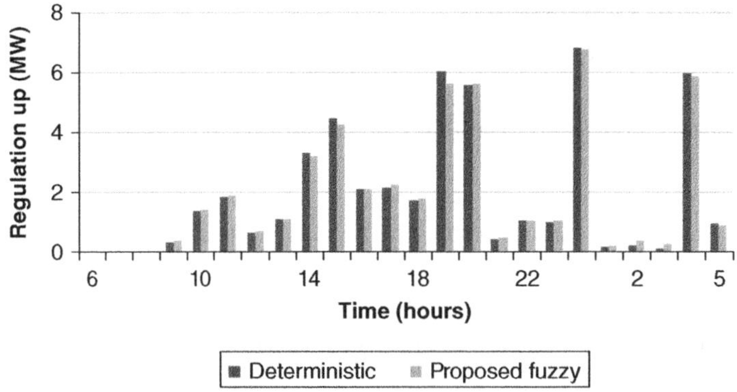

Figure 7.5.7 Average regulation-up.

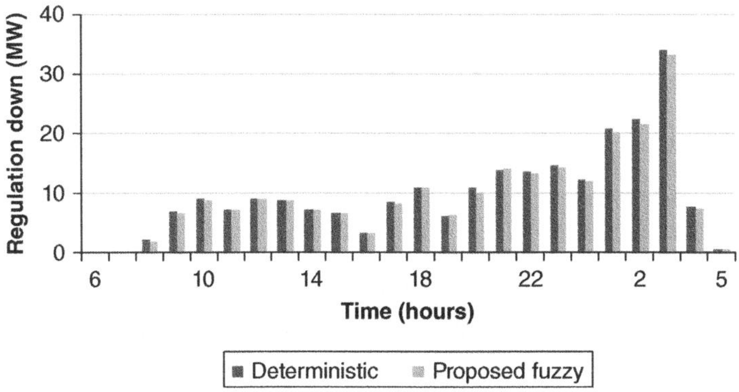

Figure 7.5.8 Average regulation-down.

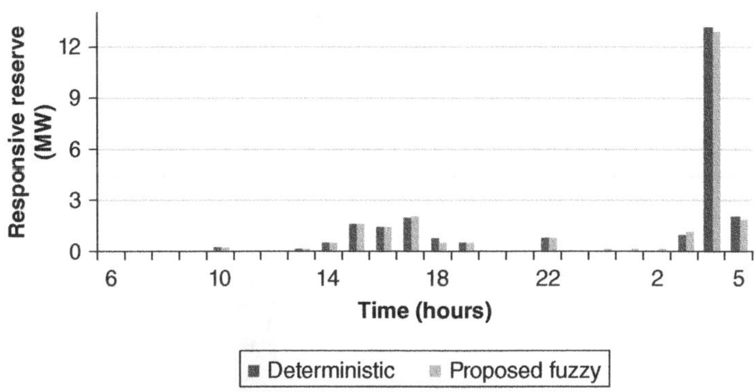

Figure 7.5.9 Average responsive reserve.

capacity is bid in the afternoon due to the high reserve price. However, this capacity is low because of the limited POP schedule and the nonzero regulation-up capacity bids during this period. At 4 : 00, a high reserve capacity is bid because of the high POP scheduled at this hour.

Figure 7.5.10 shows a comparison of EVA's total expected and actual profits over the simulation period for the deterministic and FO algorithms. The expected EVA profits for the deterministic case are about 3% higher than those of the expected FO profits. However, the actual (realized) profits obtained by the FO algorithm are 5.8% higher than those of the deterministic actual profits. This indicates that ignoring the uncertainties associated with this optimization may result in misleading expected profits that cannot be realized.

Table 7.5.2 shows the sensitivity on the profits of changing the fuzzy income limits, i.e. $\overline{\text{In}}$ and $\underline{\text{In}}$, which are prespecified before the optimization is carried out. These results indicate that setting these limits to ±10% of the nominal profits gives rise to the best outcomes. Note that the nominal profits are the profits obtained by running the deterministic algorithm.

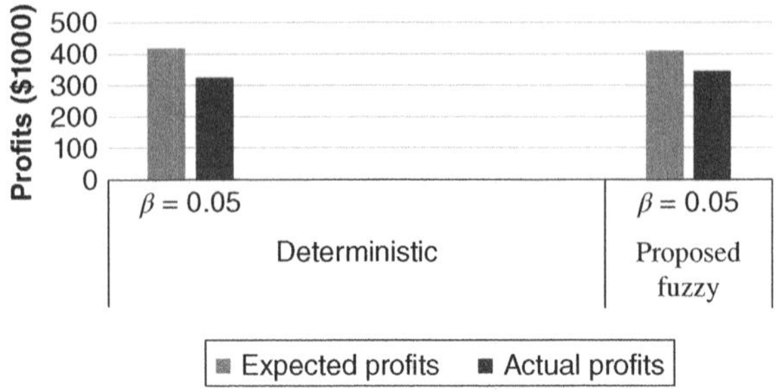

Figure 7.5.10 Total expected and actual EVA profits for different energy cost.

TABLE 7.5.2 Sensitivity of Profits to Fuzzy Income Limits

		Fuzzy Optimization			
	Deterministic Optimization	±5%	±10%	±15%	±20%
Expected profits	420.99	406.0	408.4	400.1	395.8
Realized profits	324.29	342.6	343.2	340.7	339.0

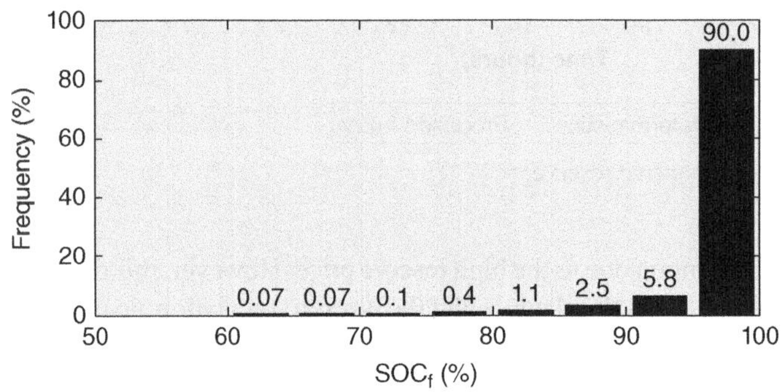

Figure 7.5.11 Final SOC statistics.

It is important that the EVs get charged to the desired final SOC, SOCfd, by the end of the charging period. As shown in Figure 7.5.11, SOCfd is less than 90% of the battery capacity for only 4% of the time, and less than 80% for only 0.7% of the time. The average SOCfd is 97.96% for the deterministic approach and as high as 98.50% for the FO approach. The standard deviation of SOCfd for each approach is 2.35 and 2.03%, respectively.

These results, though very encouraging, can be improved even further by assigning higher charging priority to the EVs with low SOC. This would not affect

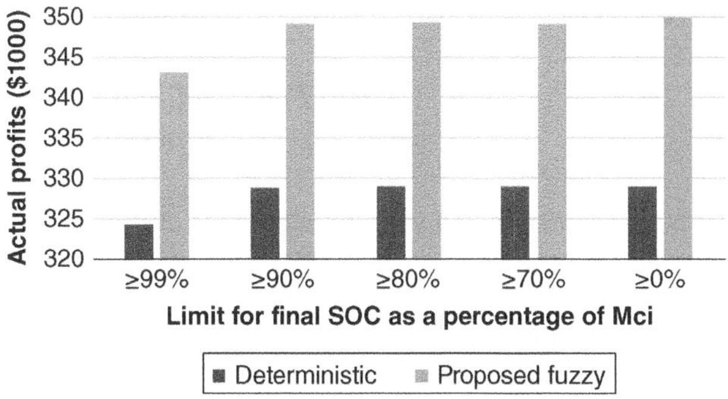

Figure 7.5.12 Total actual profits for different levels of final desired SOC.

the market bidding process or the profits, but it would help eliminate the small percentage of occurrences at which SOCfd does not reach 99% on the actual day.

Finally, the sensitivity of the actual profits to the level of final desired SOC is carried out. For simplicity, it is assumed that all EV owners request a uniform SOCfd in each case. The results presented in Figure 7.5.12 illustrate that as SOCfd decrease, the actual EVA profits increase. This is expected as lower SOCfd implies less constrained optimization, which is guaranteed to give rise to profits that are at least as high as the more constrained optimization. The first two columns to the right belong to the upper bound of the actual profits, at which no SOCfd limit is imposed.

7.5.6 Conclusion

In this work, the benefits of coordinated bidding of ancillary services for unidirectional V2G considering market uncertainties are investigated. The uncertainties associated with market prices and AS deployment signals are modeled using FO. The different parameters related to EVs are also considered, such as the EV availability, trip durations, and duration and time of trips. Simulation results and sensitivity analyses show that the FO formulation results in higher EVA's profits and lower deviation between the expected and actual profits. Statistical analysis of the final SOC due to the application of the fuzzy algorithm show satisfactory results. In fact, compared to deterministic optimization, the FO slightly improves the final SOC of the EVs.

ACKNOWLEDGEMENT

This work is supported by the Deanship of Scientific Research of King Fahd University of Petroleum & Minerals, under project no. RG1318-1 & -2. The author would like to acknowledge the contribution of E. Sortomme and M. A. H. Ansari.

7.6 CONCLUSIONS

Tiago Pinto and Zita A. Vale
Polytechnic of Porto, Porto, Portugal

Electricity markets have been characterized by multiple changes during the last decades. The deregulation and decentralization of market operation has caused a major restructuring of market operation at a worldwide level. On the other, the increasing penetration of generation based on renewable energy sources has brought by new challenges for power and energy system. This, in turn, has led to need of rethinking the current market design, in order to make the system able to deal with the new paradigm. Additionally, new players, resources, and approaches have been emerging with the main goal of enabling dealing with the variability of renewable generation. Diverse types of aggregators, EV with vehicle to grid capabilities, and DR are now central pieces in a market environment in constant mutation. This chapter has presented and discussed some of the most relevant issues in current and future electricity markets. The need for simulation and decision support solutions is explored, as a common direction to aid the decisions of intervenient entities, by helping them understand the market mechanisms and rules. An overview on market clearing and analysis through simulation is also presented. Bidding strategies are addressed as potential opportunities for players to improve their outcomes from the participation in the different market models; and the particular case of EV market participation is explored, including their actions ancillary services. Alternative forecasting approaches are also presented, and means to deal with the uncertainty associated with generation, consumption, and electricity market prices.

The overview of electricity markets' evolution addressed in this chapter enables identifying new paths for future research related to electricity market models, simulation, and decision support. Among the main identified gaps is the need for co-simulation between simulators directed to different specific problems, such as smart grids, buildings' energy management, or market negotiation; the need for improved probabilistic forecasting approaches; and the necessity for improved bidding and negotiation strategies for distinct players, and involving multiple market opportunities.

REFERENCES

1. Lund, H. (2014). *Renewable Energy Systems, Renewable Energy Systems – A Smart Energy Systems Approach to the Choice and Modeling of 100% Renewable Solutions*, 2e. Academic Press.
2. European Commission (2009). The 2020 climate and energy package. http://ec.europa.eu/clima/policies/package/index_en.htm (accessed October 2016).

3. European Commission (2014). 2030 framework for climate and energy policies. http://ec.europa.eu/clima/policies/2030/index_en.htm (accessed October 2016).
4. European Commission (2011). A roadmap for moving to a competitive low carbon economy in 2050. http://eur-lex.europa.eu/legal-content/EN/ALL/?uri=CELEX:52011DC0112R(01) (accessed October 2016).
5. Sheng, W., Liu, K.Y., Liu, Y. et al. (2016). Reactive power coordinated optimisation method with renewable distributed generation based on improved harmony search. IET Generation Transmission & Distribution 10 (13): 3152–3162.
6. Lin, J., Yu, W., and Yang, X. (2016). Towards multistep electricity prices in smart grid electricity markets. IEEE Transactions on Parallel and Distributed Systems 27 (1): 286–302.
7. Biggar, D.R. and Hesamzadeh, M.R. (eds.) (2014). *The Economics of Electricity Markets*, 1e. Wiley.
8. Sioshansi, F.P. (2013). *Evolution of Global Electricity Markets – New Paradigms, New Challenges, New Approaches*. Academic Press.
9. European Commission (2009). Directive 2009/28/EC of the European Parliament and of the Council (April 2009).
10. APX (2015). Italian borders successfully coupled. http://www.apxgroup.com/wp-content/uploads/2015-02-24_MRC_PCR_IBWT_Communication_SuccessfulGoLive.pdf (accessed October 2016).
11. David, A.K. and Wen, F. (2000). Strategic bidding in competitive electricity markets: a literature survey. Proceedings of IEEE Power Engineering Society Summer Meeting 4: 2168–2173.
12. Richter, C.W., Sheble, G.B., and Ashlock, D. (1999). Comprehensive bidding strategies with genetic programming/finite state automata. IEEE Transactions on Power Systems 14 (4): 1207–1212.
13. Praça, I., Ramos, C., Vale, Z.A., and Cordeiro, M. (2003). MASCEM: a multi-agent system that simulates competitive electricity markets (Special Issue on Agents and Markets). IEEE Intelligent Systems 18 (6): 54–60.
14. Azevedo, F., Vale, Z.A., and Moura Oliveira, P.B. (2007). A decision-support system based on particle swarm optimization for multi period hedging in electricity markets. IEEE Transactions on Power Systems 22 (3): 995–1003.
15. Azevedo, F., Vale, Z.A., Moura Oliveira, P.B., and Khodr, H. (2010). A long-term risk management tools for electricity markets using swarm intelligence. Electric Power Systems Research 80 (4): 380–389.
16. Pinto, T., Praça, I., Vale, Z. et al. (2013). Strategic bidding in electricity markets: an agent-based simulator with game theory for scenario analysis. Integrated Computer-Aided Engineering 20 (4): 335–346.
17. Bajpai, P. and Singh, S.N. (2007). Fuzzy adaptive particle swarm optimization for bidding strategy in uniform price spot market power systems. IEEE Transactions on Power Systems 22 (4): 2152–2160.
18. Yucekayaa, A.D., Valenzuelaa, J., and Dozierb, G. (2009). Strategic bidding in electricity market using particle swarm optimization. Electric Power Systems Research 79: 335–345.
19. Sama, M., Elbaum, S., Raimondi, F. et al. (2010). Context-aware adaptive applications: fault patterns and their automated identification. IEEE Transactions on Software Engineering 36 (5): 644–661.
20. Padovitz, A., Loke, S.W., and Zaslavsky, A. (2008). Multiple-agent perspectives in reasoning about situations for context-aware pervasive computing systems. IEEE

Transactions on Systems, Man and Cybernetics, Part A: Systems and Humans 38 (4): 729–742.
21. Ren, Y., Suganthan, P.N., and Srikanth, N. (2016). A novel empirical mode decomposition with support vector regression for wind speed forecasting. IEEE Transactions on Neural Networks and Learning Systems 27 (8): 1793–1798.
22. EPEX – European Power Exchange. www.epexspot.com (accessed April 2013).
23. Nord Pool. www.nordpoolspot.com (accessed April 2013).
24. MIBEL – Iberian Electricity Market Operator. www.omie.es (accessed April 2013).
25. Amjady, N., Daraeepour, A., and Keynia, F. (2010). Day-ahead electricity price forecasting by modified relief algorithm and hybrid neural network. IET Generation Transmission and Distribution.
26. Anbazhagan, S. and Kumarappan, N. (2013). Day-ahead deregulated electricity market price forecasting using recurrent neural network. IEEE Systems Journal 7 (4): 866, 872.
27. Wilamowski, B.M. and Yu, H. (2010). Neural network learning without backpropagation. IEEE Transactions on Neural Networks 21 (11): 1793–1803.
28. Aggarwal, S.K., Saini, L.M., and Kumar, A. (2009). Day-ahead price forecasting in Ontario electricity market using variable-segmented support vector machine-based model. Electric Power Components & Systems 37: 495–516.
29. Faria, P. and Vale, Z. (2013). Virtual power player using demand response to deal with unexpected low wind power generation. *EWEA Annual Event 2013 (EWEA 2013)*, Vienna, Austria (4–7 February 2013).
30. Faria, P., Vale, Z., and Morais, H. (2013). Maximizing the social welfare of virtual power players operation in case of excessive wind power. *EWEA Annual Event 2013 (EWEA 2013)*, Vienna, Austria (4–7 February 2013).
31. Pinto, T., Soares, J., Ramos, S., and Vale, Z. (2012). Artificial intelligence to predict short-term wind speed. *11th International Workshop on Large-Scale Integration of Wind Power into Power Systems as well as on Transmission Networks for Offshore Wind Power Plants (WIW12)*, Lisbon, Portugal (13–15 November 2012).
32. Pinto, T., Sousa, T.M., Praça, I. et al. (2016). Support vector machines for decision support in electricity markets' strategic bidding. Neurocomputing 172: 438–445.
33. Ramos, S., Soares, J., Pinto, T., and Vale, Z. (2013). Short-term wind forecasting to support virtual power player operation. *EWEA Annual Event 2013 (EWEA 2013)*, Vienna, Austria (4–7 February 2013).
34. Vale, Z., Pinto, T., Praca, I., and Morais, H. (2011). MASCEM: electricity markets simulation with strategic agents. IEEE Intelligent Systems 26 (2): 9–17.
35. Pinto, T., Morais, H., Oliveira, P. et al. (2011). A new approach for multi-agent coalition formation and management in the scope of electricity markets. Energy 36 (8): 5004–5015.
36. Gunn, S. (1998). *Support Vector Machines for Classification and Regression*. Technical Report. Image Speech and Intelligent Systems Research Group, University of Southampton.
37. Li, H. and Tesfatsion, L. (2009). Development of open source software for power market research: the AMES test bed. Journal of Energy Markets 2 (2): 111–128.
38. Sun, J. and Tesfatsion, L. (2007). Dynamic testing of wholesale power market designs: an open-source agent-based framework. Computational Economics 30 (3): 291–327.
39. Roth, A. and Erev, I. (1995). Learning in extensive form games: experimental data and simple dynamic models in the intermediate term. Games and Economic Behavior 8: 164–212.

40. Pinto, T., Sousa, T.M., Vale, Z., et al. (2012). Intelligent decision making in electricity markets: simulated annealing Q-learning. *IEEE Power and Energy Society General Meeting 2012*, San Diego, CA, USA (22–26 July 2012).
41. Bandyopadhyay, S. (2011). Multiobjective simulated annealing for fuzzy clustering with stability and validity. IEEE Transactions on Systems, Man, and Cybernetics, Part C: Applications and Reviews 41 (5): 682–691.
42. Greenwald, A. and Kephart, J.O. (1999). Shopbots and pricebots. *Proceedings of the Sixteenth International Joint Conference on Artificial Intelligence – IJCAI*, Stockholm (1999).
43. Rabinovich, Z., Naroditskiy, V., Gerding, E.H. et al. (2013). Computing pure Bayesian-Nash equilibria in games with finite actions and continuous types. Artificial Intelligence 195: 106–139.
44. Principe, J. (2010). *Information Theoretic Learning*, Information Science and Statistics Series. Springer.
45. Berofsky, B. (1971). *Determinism.* Princeton University Press.
46. Porter, M. (1979). How competitive forces shape strategy. Harvard Business Review.
47. Biggs, J.B. (1985). The role of meta-learning in study process. British Journal of Educational Psychology 55: 185–212.
48. Pinto, T., Sousa, T.M., Vale, Z. et al. (2012). Metalearning in ALBidS: a strategic bidding system for electricity markets. In: *Highlights on Practical Applications of Agents and Multi-Agent Systems – Advances in Intelligent and Soft Computing*, vol. 156 (eds. J. Pérez, M. Sánchez, P. Mathieu, et al.), 247–256. Berlin, Heidelberg: Springer.
49. de Bono, E. (1985). *Six Thinking Hats.* England: Penguin Books.
50. Barreto, J., Praça, I., Pinto, T., et al. (2013). Adapting meeting tools to agent decision. *IEEE Symposium on Evolving and Adaptive Systems 2013 (EIAS) at the IEEE SSCI 2013 (IEEE Symposium Series on Computational Intelligence)*, Singapore (15–19 April 2013).
51. Vale, Z.A. (2009). Intelligent power system. In: *Encyclopedia of Computer Science and Engineering*", 5-Volume set, vol. 3 (ed. B.W. Wah), 1604–1613. Hoboken, NJ: Wiley.
52. Shahidehpour, M., Yamin, H., and Li, Z. (2002). *Market Operations in Electric Power Systems: Forecasting, Scheduling, and Risk Management.* Wiley-IEEE Press.
53. Koritarov, V. (2004). Real-world market representation with agents: modeling the electricity market as a complex adaptive system with an agent-based approach. IEEE Power and Energy Magazine 2 (4): 39–46.
54. Bajpai, P., Punna, S.K., and Singh, S.N. (2008). Swarm intelligence-based strategic bidding in competitive electricity markets. IET Generation Transmission and Distribution 2: 175–184.
55. Azadeh, A., Skandari, M.R., and Maleki-Shoja, B. (2010). An integrated ant colony optimization approach to compare strategies of clearing market in electricity markets: agent-based simulation. Energy Policy 38: 6307–6319.
56. Vale, Z., Venayagamoorthy, G.K., Ferreira, J., and Morais, H. (2011). Computational intelligence applications for future power systems. In: *Computational Intelligence for Engineering Systems*, vol. 46 (eds. A. Madureira, J. Ferreira and Z. Vale), 176–193. Netherlands: Springer.
57. Feng, D., Yan, Z., Ostergaard, J. et al. (2010). Simulation embedded artificial intelligence search method for supplier trading portfolio decision. IET Generation Transmission and Distribution 4: 221–230.

58. Faria, P., Vale, Z., Soares, J., and Ferreira, J. (2011). Demand response management in power systems using a particle swarm optimization approach. IEEE Intelligent Systems.
59. Molina, J.D. and Rudnick, H. (2010). Transmission of electric energy: a bibliographic review. IEEE Latin America Transactions 8: 245–258.
60. Maghouli, P., Hosseini, S.H., Buygi, M.O., and Shahidehpour, M. (2009). A multi-objective framework for transmission expansion planning in deregulated environments. IEEE Transactions on Power Systems 24: 11.
61. Granelli, G., Montagna, M., Zanellini, F. et al. (2006). Optimal network reconfiguration for congestion management by deterministic and genetic algorithms. Electric Power Systems Research 76: 549–556.
62. Harp, S., Brignone, S., Wollenberg, B.F. et al. (2000). Sepia: a simulator for electric power industry agents. IEEE Control Systems Magazine 20 (4): 53–69.
63. Zimmermann, R. and Thomas, R.J. (2004). PowerWeb: a tool for evaluating economic and reliability impacts of electric power market designs. IEEE PES Power Systems Conference and Exposition 3: 1562–1567.
64. Migliavacca, G. (2007). *SREMS-Electricity Market Simulator Based Game Theory and Incorporating Network Constraints.* Lausanne, Swiss: IEEE Power Tech.
65. Cincotti, S. and Gallo, G. (2013). Genoa artificial power-exchange. In: *Agents Artificial Intelligence*, Communications in Computer and Information Science, vol. 358 (eds. J. Filipe and A. Fred). Berlin, Heidelberg: Springer.
66. Oliveira, P., Pinto, T., Morais, H. and Vale, Z. (2012). MASGriP – a multi-agent smart grid simulation platform. *IEEE Power and Energy Society General Meeting 2012*, San Diego CA, USA (22–26 July 2012).
67. Blumsack, S. and Fernandez, A. (2012). Ready or not, here comes the smart grid! Energy 37 (1): 61–68.
68. Praça, I., Sousa, T.M., Freitas, A., et al. (2012). Adaptive tool for automatic data collection of real electricity markets. *23rd International Workshop on Database and Expert Systems Applications (DEXA 2012)*, Vienna, Austria (3–6 September 2012).
69. Rodriguez, C. and Anders, G. (2004). Energy price forecasting in the Ontario competitive power system market. IEEE Transactions on Power Systems 19 (1).
70. Wu, D.D., Niu, D.X., and Liu, D. (2010). A soft computing system for day-ahead electricity price forecasting. Applied Soft Computing 10: 868–875.
71. Moghaddam, M.P., Shafie-khah, M., and Sheikh-El-Eslami, M.K. (2011). Price forecasting of day-ahead electricity markets using a hybrid forecast method. Energy Conversion and Management 52: 2165–2169.
72. Pinto, T., Vale, Z., Sousa, T.M. et al. (2014). Adaptive learning in agents behaviour: a framework for electricity markets simulation. Integrated Computer-Aided Engineering 21 (4): 399–415.
73. Cui, Y., Ahmad, S., and Hawkins, J. (2016). Continuous online sequence learning with an unsupervised neural network model. Neural Computation 28 (11): 2474–2504.
74. Bai, F., Wang, X., Liu, Y. et al. (2016). Measurement-based frequency dynamic response estimation using geometric template matching and recurrent artificial neural network. CSEE Journal of Power and Energy Systems 2 (3): 10–18.
75. Babu, G.S. and Suresh, S. (2013). Sequential projection-based metacognitive learning in a radial basis function network for classification problems. IEEE Transactions on Neural Networks and Learning Systems 24 (2): 194, 206.
76. Mazumdar, J., Harley, R.G., Lambert, F.C., and Venayagamoorthy, G.K. (2007). Neural network based method for predicting nonlinear load harmonics. IEEE Transactions on Power Electronics 22 (3): 1036–1045.

77. Pinto, T., Sousa, T.M., and Vale, Z. (2012). Dynamic artificial neural network for electricity market prices forecast. *IEEE 16th International Conference on Intelligent Engineering Systems (INES 2012)*, Costa de Caparica, Portugal (13–15 June 2012).
78. Fisher, R.A. (1936). The use of multiple measurements in taxonomic problems. Annals of Eugenics 7: 179–188.
79. Vapnik, V. and Lerner, A. (1963). Pattern recognition using generalized portrait method. Automation and Remote Control 24: 774–780.
80. Hadley, S.W. and Tsvetkova, A.A. (2009). Potential impacts of plug-in hybrid electric vehicles on regional power generation. The Electricity Journal 22 (10): 56–68.
81. Dickerman, L. and Harrison, J. (2010). A new car, a new grid. IEEE Power and Energy Magazine 8 (2): 55–61.
82. Cao, Y., Tang, S., Li, C. et al. (2012). An optimized EV charging model considering TOU price and SOC curve. IEEE Transactions on Smart Grid 3 (1): 388–393.
83. Stoeckl, G., Witzmann, R., and Eckstein, J. (2011). Analyzing the capacity of low voltage grids for electric vehicles. In: *2011 IEEE Electrical Power and Energy Conference (EPEC)*, 415–420. IEEE.
84. Kempton, W. and Letendre, S.E. (1997). Electric vehicles as a new power source for electric utilities. Transportation Research Part D: Transport and Environment 2 (3): 157–175.
85. Kempton, W. and Tomić, J. (2005). Vehicle-to-grid power fundamentals: calculating capacity and net revenue. Journal of Power Sources 144 (1): 268–279.
86. Kempton, W. and Tomić, J. (2005). Vehicle-to-grid power implementation: from stabilizing the grid to supporting large-scale renewable energy. Journal of Power Sources 144 (1): 280–294.
87. Tomić, J. and Kempton, W. (2007). Using fleets of electric-drive vehicles for grid support. Journal of Power Sources 168 (2): 459–468.
88. Escudero-Garzas, J.J., Garcia-Armada, A., and Seco-Granados, G. (2012). Fair design of plug-in electric vehicles aggregator for V2G regulation. IEEE Transactions on Vehicular Technology 61 (8): 3406–3419.
89. He, Y., Venkatesh, B., and Guan, L. (2012). Optimal scheduling for charging and discharging of electric vehicles. IEEE Transactions on Smart Grid 3 (3): 1095–1105.
90. Sortomme, E. and El-Sharkawi, M.A. (2011). Optimal charging strategies for unidirectional vehicle-to-grid. IEEE Transactions on Smart Grid 2 (1): 131–138.
91. Sortomme, E. and El-Sharkawi, M.A. (2012). Optimal combined bidding of vehicle-to-grid ancillary services. IEEE Transactions on Smart Grid 3 (1): 70–79.
92. Al-Awami, A.T. and El-Sharkawi, M.A. (2011). Coordinated trading of wind and thermal energy. IEEE Transactions on Sustainable Energy 2 (3): 277–287.
93. Al-Awami, A.T. and Sortomme, E. (2012). Coordinating vehicle-to-grid services with energy trading. IEEE Transactions on Smart Grid – Special Issue on Transportation Electrification and Vehicle-to-Grid Applications 3 (1): 453–462.
94. Sarker, M.R., Pandzic, H., and Ortega-Vazquez, M.A. (2015). Optimal operation and services scheduling for an electric vehicle battery swapping station. IEEE Transactions on Power Systems 30 (2): 901–910.
95. Venkatesh, B., Yu, P., Gooi, H.B., and Choling, D. (2008). Fuzzy MILP unit commitment incorporating wind generators. IEEE Transactions on Power Systems 23 (4): 1738–1746.
96. Venkatesh, B., Jamtsho, T., and Gooi, H.B. (2007). Unit commitment-a fuzzy mixed integer linear programming solution. IET Generation, Transmission & Distribution 1: 836–846.

97. Ansari, M., Al-Awami, A.T., Sortomme, E., and Abido, M.A. (2015). Coordinated bidding of ancillary services for vehicle-to-grid using fuzzy optimization. IEEE Transactions on Smart Grid 6 (1): 261–270.
98. Zadeh, L. (1965). Fuzzy sets. Information and Control 8: 338–353.
99. Zimmermann, H.J. (1986). Fuzzy set theory and mathematical programming. In: *Fuzzy Sets Theory and Applications* (eds. A. Jones, A. Kaufmann and H.J. Zimmermann), 99–114. Tokyo: Reidel.
100. ERCOT. Electric Reliability Council of Texas, Market information. http://www.ercot.com/mktinfo (accessed 15 February 2013).
101. F. H. Administration. National Household Travel Survey. http://www.nhts.ornl.gov (accessed 15 February 2013).
102. Texas Coalition for Affordable Power (TCAP) (2012). *Deregulation Electricity Texas* (December 2012). Texas Coalition for Affordable Power.

INDEX

Applications of Modern Heuristic Optimization Methods in Power and Energy Systems, First Edition. Edited by Kwang Y. Lee and Zita A. Vale.
© 2020 by The Institute of Electrical and Electronics Engineers, Inc.
Published 2020 by John Wiley & Sons, Inc.

IEEE Press Series on Power Engineering

Series Editor: M. E. El-Hawary, Dalhousie University, Halifax, Nova Scotia, Canada

The mission of IEEE Press Series on Power Engineering is to publish leading- edge books that cover the broad spectrum of current and forward-looking technologies in this fast-moving area. The series attracts highly acclaimed authors from industry/academia to provide accessible coverage of current and emerging topics in power engineering and allied fields. Our target audience includes the power engineering professional who is interested in enhancing their knowledge and perspective in their areas of interest.

1. *Electric Power Systems: Design and Analysis, Revised Printing*
Mohamed E. El-Hawary

2. *Power System Stability*
Edward W. Kimbark

3. *Analysis of Faulted Power Systems*
Paul M. Anderson

4. *Inspection of Large Synchronous Machines: Checklists, Failure Identification, and Troubleshooting*
Isidor Kerszenbaum

5. *Electric Power Applications of Fuzzy Systems*
Mohamed E. El-Hawary

6. *Power System Protection*
Paul M. Anderson

7. *Subsynchronous Resonance in Power Systems*
Paul M. Anderson, B.L. Agrawal, and J.E. Van Ness

8. *Understanding Power Quality Problems: Voltage Sags and Interruptions*
Math H. Bollen

9. *Analysis of Electric Machinery*
Paul C. Krause, Oleg Wasynczuk, and S.D. Sudhoff

10. *Power System Control and Stability, Revised Printing*
Paul M. Anderson and A.A. Fouad

Applications of Modern Heuristic Optimization Methods in Power and Energy Systems,
First Edition. Edited by Kwang Y. Lee and Zita A. Vale.
© 2020 by The Institute of Electrical and Electronics Engineers, Inc.
Published 2020 by John Wiley & Sons, Inc.

11. *Principles of Electric Machines with Power Electronic Applications, Second Edition*
Mohamed E. El-Hawary

12. *Pulse Width Modulation for Power Converters: Principles and Practice*
D. Grahame Holmes and Thomas A. Lipo

13. *Analysis of Electric Machinery and Drive Systems, Second Edition*
Paul C. Krause, Oleg Wasynczuk, and Scott D. Sudhoff

14. *Risk Assessment for Power Systems: Models, Methods, and Applications*
Wenyuan Li

15. *Optimization Principles: Practical Applications to the Operation of Markets of the Electric Power Industry*
Narayan S. Rau

16. *Electric Economics: Regulation and Deregulation*
Geoffrey Rothwell and Tomas Gomez

17. *Electric Power Systems: Analysis and Control*
Fabio Saccomanno

18. *Electrical Insulation for Rotating Machines: Design, Evaluation, Aging, Testing, and Repair, Second Edition*
Greg C. Stone, Ian Culbert, Edward A. Boulter, and Hussein Dhirani

19. *Signal Processing of Power Quality Disturbances*
Math H. J. Bollen and Irene Y. H. Gu

20. *Instantaneous Power Theory and Applications to Power Conditioning*
Hirofumi Akagi, Edson H. Watanabe, and Mauricio Aredes

21. *Maintaining Mission Critical Systems in a 24/7 Environment*
Peter M. Curtis

22. *Elements of Tidal-Electric Engineering*
Robert H. Clark

23. *Handbook of Large Turbo-Generator Operation and Maintenance, Second Edition*
Geoff Klempner and Isidor Kerszenbaum

24. *Introduction to Electrical Power Systems*
Mohamed E. El-Hawary

25. *Modeling and Control of Fuel Cells: Distributed Generation Applications*
M. Hashem Nehrir and Caisheng Wang

26. *Power Distribution System Reliability: Practical Methods and Applications*
Ali A. Chowdhury and Don O. Koval

27. *Introduction to FACTS Controllers: Theory, Modeling, and Applications*
Kalyan K. Sen and Mey Ling Sen

28. *Economic Market Design and Planning for Electric Power Systems*
James Momoh and Lamine Mili

29. *Operation and Control of Electric Energy Processing Systems*
James Momoh and Lamine Mili

30. *Restructured Electric Power Systems: Analysis of Electricity Markets with Equilibrium Models*
Xiao-Ping Zhang

31. *An Introduction to Wavelet Modulated Inverters*
S. A. Saleh and M. Azizur Rahman

32. *Control of Electric Machine Drive Systems*
Seung-Ki Sul

33. *Probabilistic Transmission System Planning*
Wenyuan Li

34. *Electricity Power Generation: The Changing Dimensions*
Digambar M. Tagare

35. *Electric Distribution Systems*
Abdelhay A. Sallam and Om P. Malik

36. *Practical Lighting Design with LEDs*
Ron Lenk and Carol Lenk

37. *High Voltage and Electrical Insulation Engineering*
Ravindra Arora and Wolfgang Mosch

38. *Maintaining Mission Critical Systems in a 24/7 Environment, Second Edition*
Peter M. Curtis

39. *Power Conversion and Control of Wind Energy Systems*
Bin Wu, Yongqiang Lang, Navid Zargari, and Samir Kouro

40. *Integration of Distributed Generation in the Power System*
Math H. J. Bollen and Fainan Hassan

41. *Doubly Fed Induction Machine: Modeling and Control for Wind Energy Generation*
Gonzalo Abad, Jesús López, Miguel Rodrigues, Luis Marroyo, and Grzegorz Iwanski

42. *High Voltage Protection for Telecommunications*
Steven W. Blume

43. *Smart Grid: Fundamentals of Design and Analysis*
James Momoh

44. *Electromechanical Motion Devices, Second Edition*
Paul Krause, Oleg Wasynczuk, and Steven Pekarek

45. *Electrical Energy Conversion and Transport: An Interactive Computer-Based Approach, Second Edition*
George G. Karady and Keith E. Holbert

46. *ARC Flash Hazard and Analysis and Mitigation*
J. C. Das

47. *Handbook of Electrical Power System Dynamics: Modeling, Stability, and Control*
Mircea Eremia and Mohammad Shahidehpour

48. *Analysis of Electric Machinery and Drive Systems, Third Edition*
Paul Krause, Oleg Wasynczuk, Scott Sudhoff, and Steven Pekarek

49. *Extruded Cables for High-Voltage Direct-Current Transmission: Advances in Research and Development*
Giovanni Mazzanti and Massimo Marzinotto

50. *Power Magnetic Devices: A Multi-Objective Design Approach*
S. D. Sudhoff

51. *Risk Assessment of Power Systems: Models, Methods, and Applications, Second Edition*
Wenyuan Li

52. *Practical Power System Operation*
Ebrahim Vaahedi

53. *The Selection Process of Biomass Materials for the Production of Bio-Fuels and Co-Firing*
Najib Altawell

54. *Electrical Insulation for Rotating Machines: Design, Evaluation, Aging, Testing, and Repair, Second Edition*
Greg C. Stone, Ian Culbert, Edward A. Boulter, and Hussein Dhirani

55. *Principles of Electrical Safety*
Peter E. Sutherland

56. *Advanced Power Electronics Converters: PWM Converters Processing AC Voltages*
Euzeli Cipriano dos Santos Jr. and Edison Roberto Cabral da Silva

57. *Optimization of Power System Operation, Second Edition*
Jizhong Zhu

58. *Power System Harmonics and Passive Filter Designs*
J. C. Das

59. *Digital Control of High-Frequency Switched-Mode Power Converters*
Luca Corradini, Dragan Maksimovic, Paolo Mattavelli, and Regan Zane

60. *Industrial Power Distribution, Second Edition*
Ralph E. Fehr, III

61. *HVDC Grids: For Offshore and Supergrid of the Future*
Dirk Van Hertem, Oriol Gomis-Bellmunt, and Jun Liang

62. *Advanced Solutions in Power Systems: HVDC, FACTS, and Artificial Intelligence*
Mircea Eremia, Chen-Ching Liu, and Abdel-Aty Edris

63. *Operation and Maintenance of Large Turbo-Generators*
Geoff Klempner and Isidor Kerszenbaum

64. *Electrical Energy Conversion and Transport: An Interactive Computer-Based Approach*
George G. Karady and Keith E. Holbert

65. *Modeling and High-Performance Control of Electric Machines*
John Chiasson

66. *Rating of Electric Power Cables in Unfavorable Thermal Environment*
George J. Anders

67. *Electric Power System Basics for the Nonelectrical Professional*
Steven W. Blume

68. *Modern Heuristic Optimization Techniques: Theory and Applications to Power Systems*
Kwang Y. Lee and Mohamed A. El-Sharkawi

69. *Real-Time Stability Assessment in Modern Power System Control Centers*
Savu C. Savulescu

70. *Optimization of Power System Operation*
Jizhong Zhu

71. *Insulators for Icing and Polluted Environments*
Masoud Farzaneh and William A. Chisholm

72. *PID and Predictive Control of Electric Devices and Power Converters Using MATLAB®/Simulink®*
Liuping Wang, Shan Chai, Dae Yoo, Lu Gan, and Ki Ng

73. *Power Grid Operation in a Market Environment: Economic Efficiency and Risk Mitigation*
Hong Chen

74. *Electric Power System Basics for the Nonelectrical Professional, Second Edition*
Steven W. Blume

75. *Energy Production Systems Engineering*
Thomas Howard Blair

76. *Model Predictive Control of Wind Energy Conversion Systems*
Venkata Yaramasu and Bin Wu

77. *Understanding Symmetrical Components for Power System Modeling*
J. C. Das

78. *High-Power Converters and AC Drives, Second Edition*
Bin Wu and Mehdi Narimani

79. *Current Signature Analysis for Condition Monitoring of Cage Induction Motors: Industrial Application and Case Histories*
William T. Thomson and Ian Culburt

80. *Introduction to Electric Power and Drive Systems*
Paul Krause, Oleg Wasynczuk, Timothy O'Connell, and Maher Hasan

81. *Instantaneous Power Theory and Applications to Power Conditioning, Second Edition*
Hirofumi Akagi, Edson Hirokazu Watanabe, and Mauricio Aredes

82. *Practical Lighting Design with LEDs, Second Edition*
Ron Lenk and Carol Lenk

83. *Introduction to AC Machine Design*
Thomas A. Lipo

84. *Advances in Electric Power and Energy Systems: Load and Price Forecasting*
Mohamed E. El-Hawary

85. *Electricity Markets: Theories and Applications*
Jeremy Lin and Fernando H. Magnago

86. *Multiphysics Simulation by Design for Electrical Machines, Power Electronics, and Drives*
Marius Rosu, Ping Zhou, Dingsheng Lin, Dan Ionel, Mircea Popescu, Frede Blaabjerg, Vandana Rallabandi, and David Staton

87. *Modular Multilevel Converters: Analysis, Control, and Applications*
Sixing Du, Apparao Dekka, Bin Wu, and Navid Zargari

88. *Electrical Railway Transportation Systems*
Morris Brenna, Federica Foiadelli, and Dario Zaninelli

89. *Energy Processing and Smart Grid*
James A. Momoh

90. *Handbook of Large Turbo-Generator Operation and Maintenance, Third Edition*
Geoff Klempner and Isidor Kerszenbaum

91. *Advanced Control of Doubly Fed Induction Generator for Wind Power Systems*
Dehong Xu, Frede Blaabjerg, Wenjie Chen, and Nan Zhu

92. *Electric Distribution Systems, Second Edition*
Abdelhay A. Sallam and Om P. Malik

93. *Power Electronics in Renewable Energy Systems and Smart Grid: Technology and Applications*
Bimal K. Bose

94. *Power System and Control and Stability, Third Edition*
Vijay Vittal, James D. McCalley, Paul M. Anderson, and A. A. Fouad

95. *Electromechanical Motion Devices: Rotating Magnetic Field-Based Analysis and Online Animations, Third Edition*
Paul Krause, Oleg Wasynczuk, Steven D. Pekarek, and Timothy O'Connell

96. *Applications of Modern Heuristic Optimization Methods in Power and Energy Systems*
Kwang Y. Lee and Zita A. Vale

Printed and bound by CPI Group (UK) Ltd, Croydon, CR0 4YY

17/09/2026

14964318-0008